THEORY OF

THEORY
OF
STRUCTURES

[A Text book for the students of U.P.S.C. (Engg. Services)
B.Sc., Engg.; Section 'B' of A.M.I.E.(I); and
Diploma Courses.]

in
M.K.S. UNITS

By
R. S. KHURMI

S. CHAND
AN ISO 9001 : 2000 COMPANY

S. CHAND & COMPANY LTD.
RAM NAGAR, NEW DELHI-110 055

S. CHAND & COMPANY LTD.
(An ISO 9001 : 2000 Company)

Head Office : 7361, RAM NAGAR, NEW DELHI - 110 055
Phones : 23672080-81-82, 9899107446, 9911310888;
Fax : 91-11-23677446
Shop at: **schandgroup.com**   E-mail: **schand@vsnl.com**

Branches :

- 1st Floor, Heritage, Near Gujarat Vidhyapeeth, Ashram Road,
  Ahmedabad-380014. Ph. 27541965, 27542369. ahmedabad@schandgroup.com
- No. 6, Ahuja Chambers, 1st Cross, Kumara Krupa Road,
  Bangalore-560001. Ph : 22268048, 22354008, bangalore@schandgroup.com
- 238-A M.P. Nagar, Zone 1, **Bhopal** - 462 011. Ph : 4274723, bhopal@schandgroup.com
- 152, Anna Salai, **Chennai**-600 002. Ph : 28460026, chennai@schandgroup.com
- S.C.O. 2419-20, First Floor, Sector- 22-C (Near Aroma Hotel), **Chandigarh**-160022,
  Ph-2725443, 2725446, chandigarh@schandgroup.com
- 1st Floor, Bhartia Tower, Badambadi, **Cuttack**-753 009, Ph-2332580; 2332581,
  cuttack@schandgroup.com
- 1st Floor, 52-A, Rajpur Road, **Dehradun**-248 001. Ph : 2740889, 2740861,
  dehradun@schandgroup.com
- Pan Bazar, **Guwahati**-781001. Ph : 2738811, guwahati@schandgroup.com
- Sultan Bazar, **Hyderabad**-500 195. Ph : 24651135, 24744815, hyderabad@schandgroup.com
- Mai Hiran Gate, **Jalandhar** - 144008 . Ph. 2401630, 5000630, jalandhar@schandgroup.com
- A-14 Janta Store Shopping Complex, University Marg, Bapu Nagar, **Jaipur** - 302 015,
  Phone : 2719126, jaipur@schandgroup.com
- 613-7, M.G. Road, Ernakulam, **Kochi**-682 035. Ph : 2378207, cochin@schandgroup.com
- 285/J, Bipin Bihari Ganguli Street, **Kolkata**-700 012. Ph : 22367459, 22373914,
  kolkata@schandgroup.com
- Mahabeer Market, 25 Gwynne Road, Aminabad, **Lucknow**-226 018. Ph : 2626801, 2284815,
  lucknow@schandgroup.com
- Blackie House, 103/5, Walchand Hirachand Marg, Opp. G.P.O., **Mumbai**-400 001.
  Ph : 22690881, 22610885, mumbai@schandgroup.com
- Karnal Bag, Model Mill Chowk, Umrer Road, **Nagpur**-440 032 Ph : 2723901, 2777666, 2720523
  nagpur@schandgroup.com
- 104, Citicentre Ashok, Govind Mitra Road, **Patna**-800 004. Ph : 2300489, 2302100,
  patna@schandgroup.com
- 291/1, Ganesh Gayatri Complex, 1st Floor, Somwarpeth, Near Jain Mandir, **Pune**-411011.
  Ph : 64017298, pune@schandgroup.com
- Flat No. 104, Sri Draupadi Smriti Apartment, East of Jaipal Singh Stadium, Neel Ratan Street,
  Upper Bazar, **Ranchi**-834001. Ph:2208761, ranchi@schandgroup.com
- Kailash Residency, Plot No. 4B, Bottle House Road, Shankar Nagar, **Raipur**. Ph. 09981200834
  raipur@schandgroup.com

First Edition 1961
Subsequent Editions and Reprints 1963, 65, 70, 72, 74, 76, 77, 83, 86, 89, 92, 94, 96, 99, 2002, 2004, 2005, 2007
Reprint 2008

ISBN : 81-219-0520-6
Code : 10 030

PRINTED IN INDIA
By Rajendra Ravindra Printers (Pvt.) Ltd., Ram Nagar, New Delhi-110 055 and published by S. Chand & Company Ltd., 7361, Ram Nagar, New Delhi-110 055

Preface to the Tenth Edition

I feel elevated in presenting the New edition of this standard treatise. The favourable reception, which the previous editions and reprints of this book have enjoyed, is a matter of great satisfaction for me.

I wish to express my sincere thanks to numerous professors and students for their valuable suggestions and recommending the book to their students and friends. I hope, they will continue to patronise this standard treatise in the future also.

Any errors, omissions and suggestions for the improvement of this volume, brought to my notice, will be thankfully acknowledged and incorporated in the next edition.

R. S. KHURMI

PREFACE TO FIRST EDITION

I take an opportunity to present *Theory of Structures* to the students of A.M.I.E. (I) section 'B', Degree and Diploma of all the Indian and foreign Universities. The object of this book is to present, a companion volume, to my popular book Strength of Materials in a most concise, compact to the point and lucid manner.

While writing the book, I have constantly kept in mind the examination requirements of A.M.I.E. (I) section 'B' Civil Engineering students, who find lot of difficulties for preparing themselves adequately for 'Theory of Structures', as it contains the solutions of their examination papers up to Nov. 1971. I have also kept in mind the requirements of my other numerous students, preparing for their Degree and Diploma examination, at home and abroad. To make the book really useful at levels, it has been written in an easy style, as well as with a simple and systematic mathematical approach. The subject matter has been divided into two parts—part I dealing with statically determinate structures and part II dealing with statically indeterminate structures. The subject matter has been amply illustrated by incorporating a good number of solved, unsolved and well-graded examples of almost every variety. These examples have been taken from the recent examination papers of Indian and foreign universities, as well as professional examining bodies, in order to make the students familiar, with the types of questions usually set in their examinations. At the end of each topic, a few exercises have been added, for the students to solve them independently. Answers have been provided to these problems ; but it is too much to hope that these are entirely free from errors. At the end of each chapter '*Highlights*' is added, for quick revision of the text before the examination. In short, it is hoped that this volume will also earn the appreciation of thousands of students and teachers, like my previous publications, at home and abroad.

Although every care has been taken to check mistakes and misprints, yet it is difficult to claim perfection. Any errors, omissions and suggestions, for the improvement of this volume brought to my notice, will be thankfully acknowledged and incorporated in the next edition.

R. S. KHURMI

Acknowledgements

The author is thankful to the following Indian and foreign universities as well as examining bodies, whose examination papers have been included in the text of the subject by way of illustration. Moreover, their syllabi have also been kept in view while writing this treatise.

1. Agra University	2. Allahabad University
3. Andhra University	4. Annamalai University
5. Banaras University	6. Bangalore University
7. Baroda University	8. Bhopal University
9. Bihar University	10. Bombay University
11. Calcutta University	12. Delhi University
13. Gauhati University	14. Gorakhpur University
15. Gujarat University	16. Indore University
17. Jabalpur University	18. Jadavpur University
19. Jiwaji University	20. Jodhpur University
21. Kanpur University	22. Karnatak University
23. Kerala University	24. Madras University
25. Madurai University	26. Mysore University
27. Nagpur University	28. Osmania University
29. Patna University	30. Pune University
31. Punjab University	32. Punjabi University
33. Rajasthan University	34. Ranchi University
35. Roorkee University	36. Uambridge University
37. London University	38. Oxford University
39. Institution of Engineers (I)	40. U.P.S.C. (Engg. Services)

SYLLABUS OF THEORY OF STRUCTURES
For
A.M.I.E. (India) Section 'B' Effective from May 1974.

Slopes and deflections in simply supported beams, double integration method and moment area method. Theorem of 3 moments, fixed and continuous beams. Eccentric loads on short columns. Long columns secant and empirical formulae. Columns subjected to lateral loads. Basic elastic theorems. Deflection of framed structures. Redundant framed structures. Moving loads on simply supported beams. Influence lines for bending moment and shear force in statically determinate beams and for forces in members of framed structures. Moment distribution and slope deflection methods. Arches, 3-hinged, 2-hinged. Suspension bridges with stiffening girders.

CONTENTS

CONTENTS

PART I

STATICALLY DETERMINATE STRUCTURES

INTRODUCTION

1. *Definition.* 2. *Fundamental units.* 3. *System of units.*
4. *C.G.S. units.* 5. *F.P.S. units.* 6. *M.K.S. units.* 7. *Metre.*
8. *Kilogram.* 9. *Second.* 10. *Useful data.* 11. *Algebra.* 12. *Trigonometry.* 13. *Differential Calculus.* 14. *Integral Calculus.*
15. *Applied Mechanics.* 16. *Strength of Materials.* 17. *Types of Structures.* 18. *Statically determinate structures.* 19. *Statically indeterminate structures.* 20. *Internally indeterminate structures.*
21. *Externally indeterminate structures.*

1·1. Definition

In day-to-day work, an engineer comes across various types of structures *i.e.*, beams, girders, arches etc. While designing a structure, an engineer has to study the effect of various types of forces acting on it. Sometimes, the effect of forces is determined by the application of the various laws of statics. But, sometimes, the effect of forces is determined by mathematical analysis, after making some suitable assumptions. A detailed study of structures, including their design for the safe working conditions is known as *Theory of structures.* This subject is closely related to the subject of strength of materials, and any boundary between the two, is just an arbitrary one. As a matter of fact, the study of Theory of structures is very essential for an engineer, to enable him in designing his all types of simple as well as complex structures.

1·2. Fundamental units

The measurement of physical quantities is one of the most important operations in engineering. Every quantity is measured in terms of some arbitrary, but internationally accepted units, called *fundamental units.* All physical quantities, met with, in Theory of structures, are expressed in the following three fundamental quantities *i.e.*

(1) length, (2) mass, and (3) time.

1·3. System of units

There are three systems of units, which are commonly used and internationally recognised. These are known as :

(1) C.G.S. units, (2) F.P.S. units, and (3) M.K.S. units.

1·4. C.G.S. units

In this system, the fundamental units of length, mass and time are *centimetre*, *gram* and *second* respectively.

1·5. F.P.S. units

In this system, the fundamental units of length, mass and time are *foot*, *pound* and *second* respectively.

1·6. M.K.S. units

Strictly speaking, the fundamental units of this system are not separate ones, but are the multiples of C.G.S. units. In this system the units of length, mass and time are *metre*, *kilogram* and *second* respectively. The international metre, kilogram and second are defined here.

1·7. Metre

The international metre may be defined as the shortest distance (at 0°C) between two parallel lines engraved upon the polished surface of a platinum-iridium bar, kept at the International Bureau of Weights and Measures at Sevres, near Paris.

1·8. Kilogram

The international kilogram may be defined as the mass of a platinum-iridium bar, which is also kept at the International Bureau of Weights and Measures at Sevres, near Paris.

1·9. Second

The fundamental unit of time for all the three systems of units is second, which is $1/24 \times 60 \times 60 = 1,86,400$th of the mean solar day. A solar day may be defined as the interval between the instants at which the sun crosses a meridian on two consecutive days. This value varies throughout the year. The average of all the solar days, of one year, is called the mean solar day.

1·10. Useful data

The following data summarises the previous memory and formulae, the knowledge of which is very essential at this stage.

1·11. Algebra

(1) $\qquad a^\circ = 1, \ x^\circ = 1$

(*i.e., any thing raised to the power zero is one.*)

(2) $\qquad x^m \times x^n = x^{m+n}$

(*i.e., If bases are same, in multiplication, the powers are added.*)

(3) $\qquad \dfrac{x^m}{x^n} = x^{m-n}$

(*i.e. If bases are same, in division the powers are subtracted.*)

(4) If $ax^2 + bx + c = 0$

then

$$x = \frac{-b + \sqrt{b^2 - 4ac}}{2a}.$$

where

a is the coefficient of x^2,

b is the coefficient of x, and

c is the constant term.

1·12. Trigonometry

In a right-angled triangle ABC.

(1) $\quad \dfrac{b}{c} = \sin \theta$

(2) $\quad \dfrac{a}{c} = \cos \theta$

(3) $\quad \dfrac{b}{a} = \dfrac{\sin \theta}{\cos \theta} = \tan \theta$

(4) $\quad \dfrac{c}{b} = \dfrac{1}{\sin \theta} = \mathrm{cosec}\ \theta$

(5) $\quad \dfrac{c}{a} = \dfrac{1}{\cos \theta} = \sec \theta$

(6) $\quad \dfrac{a}{b} = \dfrac{\cos \theta}{\sin \theta} = \cot \theta$

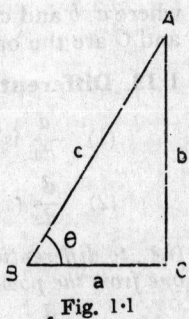

Fig. 1·1

(7) Following table shows the values of some trigonometrical functions, for some typical angles :

Angle	0°	30°	45°	60	90°
$\sin \theta$	0	$\dfrac{1}{2}$	$\dfrac{1}{\sqrt{2}}$	$\dfrac{\sqrt{3}}{2}$	1
$\cos \theta$	1	$\dfrac{\sqrt{3}}{2}$	$\dfrac{1}{\sqrt{2}}$	$\dfrac{1}{2}$	0
$\tan \theta$	0	$\dfrac{1}{\sqrt{3}}$	1	$\sqrt{3}$	∞

or in other words, for sin write

0°	30°	45°	60°	90°
$\dfrac{\sqrt{0}}{2}$	$\dfrac{\sqrt{1}}{2}$	$\dfrac{\sqrt{2}}{2}$	$\dfrac{\sqrt{3}}{2}$	$\dfrac{\sqrt{4}}{2}$

i.e. $\quad$ 0 $\qquad \dfrac{1}{2} \qquad \dfrac{1}{\sqrt{2}} \qquad \dfrac{\sqrt{3}}{2} \qquad$ 1

Now, for cos write the values in reverse order, and for tan divide the values of sin by cos for the respective angles.

(8) In the first quadrant (*i.e.* 0° to 90°) all the trigonometrical ratios are positive.

(9) In the second quadrant (*i.e.* 90° to 180°) only sin θ or cosec θ are positive.

(10) In the third quadrant (*i.e.* 180° to 270°) only tan θ and cot θ are positive.

(11) In the fourth quadrant (*i.e.* 270° to 360°) only cos θ and sec θ are positive.

(12) In any triangle ABC,

$$\frac{a}{\sin A}=\frac{b}{\sin B}=\frac{c}{\sin C}$$

where a, b and c are the lengths of three sides of the triangle. A, B and C are the opposite angles to the sides a, b and c.

1·13. Differential calculus

(1) $\frac{d}{dx}$ is the sign of differentiation.

(2) $\frac{d}{dx}(x)^n=nx^{n-1}$; $\frac{d}{dx}(x)^8=8x^7$

(*i.e., to differentiate any power of x, write power before x, and subtract one from the power.*)

(3) $\frac{d}{dx}(ax+b)^n=n(ax+b)^{n-1}\times a$

(*i.e., to differentiate any bracket with power, write power before the bracket and subtract one from the power. Also multiply by the differential coefficient of whatever is within the bracket.*)

(4) $\frac{d}{dx}(C)=0$ $\frac{d}{dx}(7)=0$

(*i.e., differential coefficient of a constant is zero.*)

(5) $\frac{d}{dx}(\sin x)=\cos x$

$\frac{d}{dx}(\cos x)=-\sin \theta$

$\frac{d}{dx}(\tan x)=\sec^2 x$

$\frac{d}{dx}(\cot x)=-\text{cosec}^2 x$

$\frac{d}{dx}(\sec x)=\sec x \tan x$

$\frac{d}{dx}(\text{cosec} x)=-\text{cosec} x \cot x$

(5) If the differential coefficient of a function is zero, the function is either maximum or minimum, *conversely*, if the maximum or minimum value of a function is required, then differentiate the function and equate it to zero.

1·14. Integral calculus

(1) $\int dx$ is the sign of integration.

(2) $\int x^n dx = \frac{x^{n+1}}{n}$; $\int x^8 dx \quad \frac{x^9}{9}$

(*i.e., to integrate any power of x, add one to the power and divide by the new power.*)

(3) $\int 7dx = 7x$; $\int Cdx = Cx$

(*i.e., to integrate any constant, multiply the constant by x.*)

(4) $\int (ax+b)^n dx = \frac{(ax+b)^{n+1}}{(n+1)\times a}$

(*i.e., to integrate any bracket with power, add one to the power and divide by the new power. Also divide by the coefficient of x with the bracket.*)

(5) $\int_0^l x^3 dx = \left[\frac{x^4}{4}\right]_0^l = \frac{l^4}{4} - \frac{0}{4} = \frac{l^4}{4}$

(*i.e., to integrate any power of x between the two limits, integrate the function and substitute the upper limit for x. From this, subtract the value by substituting the lower limit for x.*)

1·15. Applied Mechanics

(1) *Force.* It is an important factor in the field of engineering science, which may be defined as an agent, which produces or tends to produce, destroys or tends to destroy motion.

(2) *Resultant force.* If a number of forces P, Q, R...etc. are acting simultaneously on a particle, then a single force which will produce the same effects as that of all the given forces, is known as *resultant force.* The given forces P, Q, R......are called component forces.

(3) *Moment of a force.* It is the turning effect produced by a force on a body on which it acts. The moment of a force is equal to the product of the force and the perpendicular distance between the line of action of the force and the point about which the moment is required.

(4) *Lami's theorem.* It states "*If three coplaner forces acting on a point be in equilibrium, then each force is proportional to the sine of the angle between the other two.*"

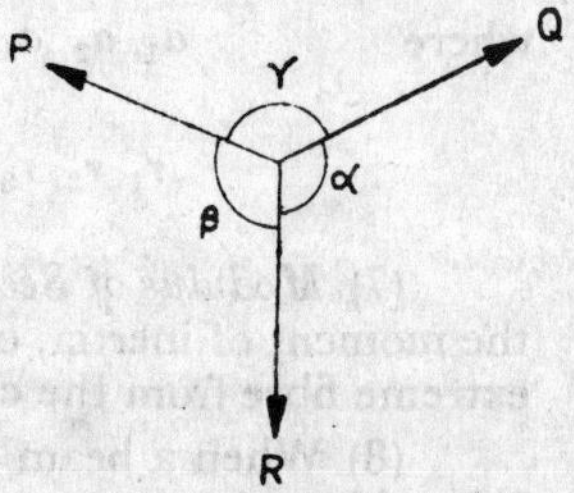

i.e. $\frac{P}{\sin \alpha} = \frac{Q}{\sin \beta} = \frac{R}{\sin \gamma}$

where P, Q and R are three forces and α, β and γ are the angles as shown in Fig. 1·2.

Fig. 1·2

(5) *Conditions of equilibrium of any system of force.* Following are the three conditions of equilibrium of any system of forces :

(*i*) the algebraic sum of the horizontal components of all the forces should be zero.

i.e. $\Sigma H = 0.$

(*ii*) the algebraic sum of the vertical components of all the forces should be zero.

   i.e. $\Sigma V = 0$

(*iii*) the algebraic sum of the moments of all the forces about any point should be zero.

   i.e. $\Sigma M = 0$

1·16. Strength of Materials

(1) *Stress.* It is the resistance to deformation per unit area, due to the action of external force. Mathematically stress,

$$p = \frac{P}{A}$$

where P = Load or force acting on the body, and

   A = Cross-sectional area of the body.

(2) *Strain.* It is the deformation per unit length.

(3) *Elasticity.* It is the property, by virtue of which a deformed body springs back to its original position, after removing the external forces

(4) *Deformation of a body.* When a body is subjected to a tensile or compressive force, it undergoes some deformation. Mathematically,

$$\delta l = \frac{Pl}{AE}$$

where δl = Deformation (*i.e.*, extension or contraction) of the body,

   P = Force acting on the body,

   l = Length of the body,

   A = Cross-sectional area of the body, and

   E = Modulus of elasticity of the material.

(5) *Centre of gravity.* It is a point, through which the whole weight of the body acts, irrespect of the position of the body.

(6) The moment of inertia of a plane area about a fixed line,

$$I = \Sigma \, ar^2$$

where $a_1, a_2, a_3 \ldots$ = Areas into which the whole plane has been divided.

   $r_1, r_2, r_3 \ldots$ = Distance of the centres of gravity of the areas $a_1, a_2, a_3 \ldots$ from the fixed line.

(7) *Modulus of Section.* It is the quantity obtained by dividing the moment of inertia, of a body about its c.g., by the distance of the extreme fibre from the centroidal axis. It is denoted by Z.

(8) When a beam or a cantilever is subjected to some loading, it is subjected to a moment. Then

$$M = f.Z$$

   M = Moment at a section,

   f = Stress at the outer layer of the section, and

   Z = Modulus of the section.

1 17. Types of structures

A structure may be defined as an assembly of members, such as bars, cables, arches etc. The main propose of a structure is to transmit external loads to the foundation or the adjoining structure. In general, following are the two types of structures :

(a) Statically determinate structures, and

(b) Statically indeterminate structures.

1·18. Statically determinate structures

If the forces in the members of a structure as well as its reactions can be found by the conditions of equilibrium (i.e., $\Sigma H = 0$, $\Sigma V = 0$ and $\Sigma M = 0$) or the principle of statics alone, it is known as a *statically determinate structures*.

1·19. Statically indeterminate structures

If the forces in the members of a structure as well as its reactions cannot be found out by the conditions of equilibrium or the principle of statics alone, it is known as a *statically indeterminate structure* or *redundant structure*. In such cases the equations of equilibrium are supplemented by the equation of compatibility of deformation. The number of unknown forces or reactions, over and above the equilibrium equations, is known as the degree of indeterminateness or degree of redundancy. A statically indeterminate structure may be further classified into the following two types :

(a) Internally indeterminate structures, and

(b) Externally indeterminate structures.

1·20. Internally indeterminate structures

A frame or a truss is called a perfect one or statically determinate, if the number of members and number of joints may be expressed by the relation,

$$n = 2j - 3$$

where n is the number of members and j is the number of joints. But if the number of members in a frame is more than $(2j-3)$, it is known as an internally indeterminate structure. The number of members which are more than $(2j-3)$, is called the degree of redundancy.

1·21. Externally indeterminate structures

A frame or a truss, whose reactions cannot be found out by the conditions of equilibrium alone, is called an externally indeterminate structure. The number of extra or redundant reactions, over and above the equilibrium equations, is called the degree of redundancy.

It may be noted that a truss may be either internally indeterminate, externally indeterminate, or both.

2

ROLLING LOADS

2·1. Introduction

In day to day work, we see rolling loads (*i.e.* travelling or moving loads, like axle-loads of locomotives) crossing bridges. A little consideration will show that when these loads move across the girder, then every cross-section of the girder will be subjected to some shear force as well as bending moment. The magnitude of the shear force and bending moment will go on changing, with the change in position of the loads.

2·2. Effects of rolling loads

In the previous article, we have discussed that the magnitude of shear force as well as bending moment goes on charging with the change in positions of the rolling loads. The rolling loads have the following two effects :—

(i) At a given cross-section of the girder, the loads may occupy such a position, so as to cause maximum S.F. or

        B.M. at that section. Such a position is called *critical position*.

        (*ii*) The loads may occupy such a position anywhere on the girder, so that it may cause maximum positive or negative S.F. and B.M. at any point, on the cross-section of the girder.

The maximum values of S.F. and B.M. may be plotted with the given span as the base. Thus the maximum S.F. and B.M. diagrams may be taken as ordinary S.F. and B.M. diagrams for various positions of the moving loads across the girder. Such diagrams are very essential for designing the girders or beams. In the succeeding pages, we shall discuss the following cases of rolling loads :—

     (1) A single concentrated load,

     (2) A uniformly distributed load, longer than the span,

     (3) A uniformly distributed load, shorter than the span,

     (4) Two concentrated or point loads, and

     (5) Several concentrated or point loads.

2·3. Sign conventions

We find different sign conventions, in different books, regarding S.F. and B.M. at a section. But we shall follow the following sign conventions, which are commonly used all over India and are also internally recognised.

(1) *Shear force*

        Since the S.F. at a section is the unbalanced vertical force, therefore it tends to slide one portion of the beam upward or downward with respect to the other. The S.F. at a section is said to be positive, when the right hand portion tends to slide upwards with respect to left hand portion. Or in other words, all the upward forces to the right of the section cause positive shear, and those acting downwards cause negative shear. On the other hand, all the upward forces to the left of the section cause negative shear, and those acting downwards cause positive shear.

(2) *Bending moment*

        At a section, where B.M. is such that it tends to bend the beam at that point to a curvature having concavity at the top is taken as positive. On the other hand where the B.M. is such that it tends to bend the beam at that point to a curvature having convexity at the top is taken as negative. The positive B.M. is often called *sagging moment* and negative as *hogging moment*.

2·4. A single concentrated load

Consider a concentrated load W rolling over the beam of span l from A to B. We shall discuss the following possible cases of S.F. and B.M.

     (*a*) Positive shear force,

     (*b*) Negative shear force, and

     (*c*) Bending moment.

(a) Positive shear force.

Let us consider an instant, when the load is at any point C at a distance y from A. We know that the positive S.F. at any section X,

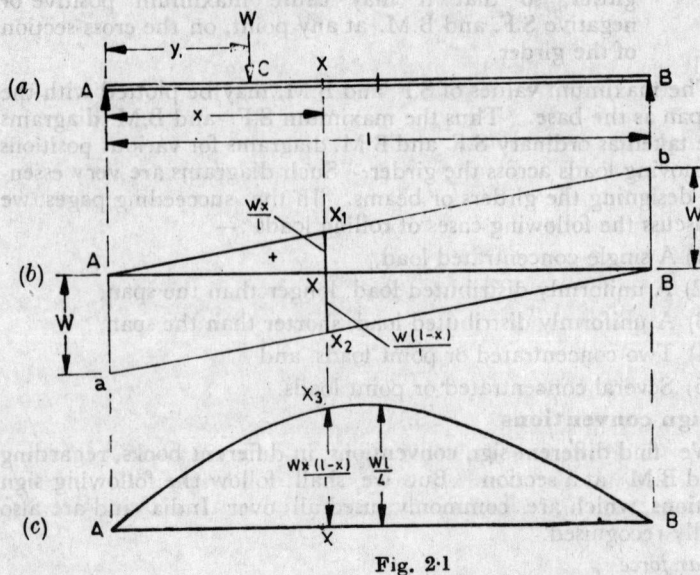

Fig. 2·1

at a distance x from A, is equal to the reaction R_B *minus* any load between X and B. Now, a little consideration will show that the positive S.F. is maximum when (i) the load is between A and X (*i.e.* for no load in between X and B) and (ii) the value of reaction R_B is maximum. Now the above two conditions are possible, only, when the load is on the section itself. Thus the maximum positive S F. at X,

$$F_{max} = +R_B = \frac{Wx}{l}$$

Now, if we want to plot the positive S.F. diagram for all sections *i.e.* from $x=0$ to $x=l$, let us substitute the various values of x in the above equation. We find that when $x=0$, $F_{max}=0$ and when $x=l$

$$F_{max} = \frac{Wl}{l} = W$$

Since the value of x varies linearly in the above equation (because x has a unit power), therefore the positive S.F. diagram will consist of a straight line having zero ordinate A, and W at B shown in Fig. 2·1 (b).

(b) Negative shear force.

Let us again consider an instant, when the load is at a point C at a distance y from A. We know that the negative shear force at any section X, at a distance x from A, is equal to the reaction R_A *minus* any load between A and X. Now, a little consideration will show that the negative S.F. is maximum when (i) the load is between X and B (*i.e.* for no load in between A and X) and (ii) the value of reaction R_A is maximum. Now the above two conditions are possible, only, when the load is on the section itself. Thus the maximum negative S F. at X,

$$F_{max} = -R_A = -\frac{W(l-x)}{l}$$

If we want to plot the negative S.F. diagram for all sections *i.e.* from $x=0$ to $x=l$, let us substitute the various values of x in the above equation. We find that when $x=0$, $F_{max} = -\frac{W(l-0)}{l} = W$ and when $x=l$ $F_{max} = -\frac{W(l-l)}{l} = 0$.

Since the value of x varies linearly in the above equation (because x has a unit power), therefore the negative S.F. diagram will consist of a straight line having W ordinate at A and zero at B as shown in Fig. 2·1 (b).

From the above discussion, we find that the *maximum positive and negative S.F. at a section occurs, when the load is on the section itself*. At such a stage, there will be maximum positive S.F. in the portion AX and maximum negative S.F. in the portion XB. Thus the S.F. diagram will consist of two parallel straight lines, one for positive and the other for negative, having end ordinates equal to W as shown in Fig. 2·1 (b)

(c) *Bending moment*

We know that when the load W is just at A, the reaction R_A will be equal to W and the reaction R_B will be zero ; thus the bending moment at A will also be equal to zero. As the load rolls towards B, the reaction R_A will go on decreasing, whereas the reaction, R_B will go on increasing.

Now let us consider an instant, when the load is at any point C at a distance y from A. We know that B.M.* at any section X, at a distance x from A is equal to the moment of the reaction R_A about X, *minus* the moment of any load between A and X. Now a little consideration will show that the B.M.** is maximum when (i) there is no load between A and X and (ii) the value of the reaction R_A is maximum. Now the above two conditions are possible, only, when the load is on the section itself. Therefore maximum bending moment,

$$M_{max} = R_A.x = \frac{Wx(l-x)}{l} = Wx - \frac{Wx^2}{l}$$

It is thus obvious, that maximum B.M. diagram for all the sections (from $x=0$ to $x=l$) will be a parabola Its value will be zero, when $x=0$ and $x=l$

---

*It is also equal to the moment of the reaction R_B about X *minus* the moment of any load between X and B.

**If the B.M. is to be found out from the reaction R_B, then the B.M. is maximum when (i) there is no load between X and B and (ii) the value of the reaction R_B is maximum.

The absolute maximum moment will occur at a point, where

$$\frac{d}{dx}\left(Wx - \frac{Wx^2}{l}\right) = 0$$

$$\frac{d}{dx}\left(x - \frac{x^2}{l}\right) = 0$$

$$\therefore \qquad 2x = l$$

or $\qquad\qquad\qquad\qquad\qquad x = \dfrac{l}{2}$

Therefore absolute maximum bending moment will occur at the centre of the girder.   Thus

$$M_{max\ max} = W \cdot \frac{l}{2} - \frac{W\left(\dfrac{l}{2}\right)^2}{l} = \frac{Wl}{4}$$

Example 2·1.   *A load of 20 tonnes crosses a bridge AB of span 25 m. Find the values of maximum S.F and B.M. at a section 10 m from the left end support.*

Solution.

Given.   Load, $W = 20$ t

Span, $l = 25$ m

Distance of the section from the left end support,

$$x = 10 \text{ m}$$

Maximum positive shear force

We know that the maximum positive S.F takes place, when the load is at the section itself.

$$\therefore \qquad F_{max} = +R_B = +\frac{W \cdot x}{l} = \frac{20 \times 10}{25} \text{ t}$$

$$= \mathbf{8\ t\ \ Ans.}$$

Maximum negative shear force

We know that the maximum negative S.F. also takes place when the load is at the section itself.

$$\therefore \qquad F_{max} = -R_A = -\frac{W(l-x)}{l} = -\frac{20(25-10)}{25} \text{ t}$$

$$= \mathbf{-12\ t\ \ Ans.}$$

Maximum bending moment

We also know that the maximum B.M. also takes place, when the load is at the section itself.

$$\therefore \qquad M_{max} = R_A \cdot x = \frac{Wx(l-x)}{l} = \frac{20 \times 10(25-10)}{25} \text{ t-m}$$

$$= \mathbf{120\ t\text{-}m\ \ Ans.}$$

2·5. A uniformly distributed load, longer than the span

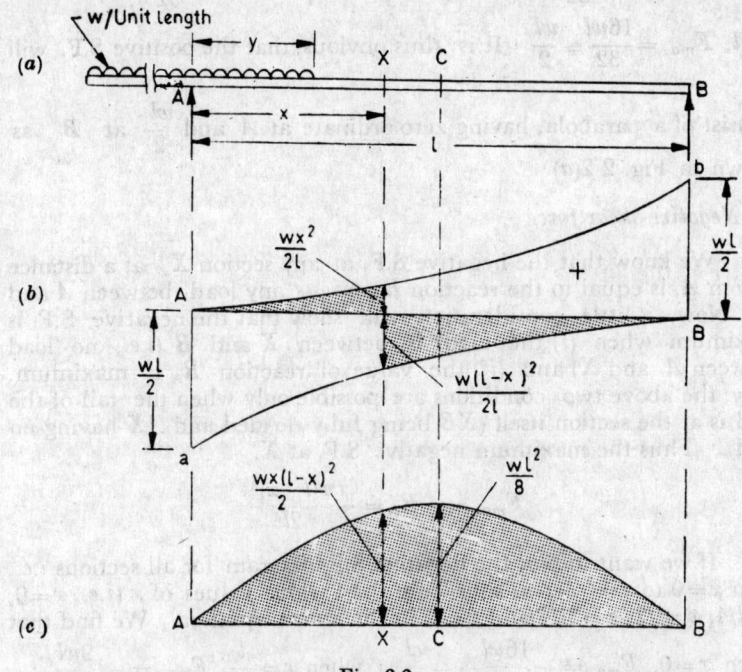

Fig. 2·2

Consider a uniformly distributed load of w/unit length, *longer* than the span l, rolling from A to B. We shall discuss the following possible cases of S.F. and B.M.

(*a*) Positive shear force,

(*b*) Negative shear force, and

(*c*) Bending moment

(*a*) *Positive shear force*

Let us consider an instant, when the head of the load is at any point C at a distance y from A. We know that the positive S.F. at any section X, at a distance x from A is equal to the reaction R_B *minus* any load between X and B. Now, a little consideration will show that the positive S.F. is maximum when (*i*) the load is between A and X (*i.e.*, no load between X and B) and (*ii*) the value of reaction R_B is maximum. Now, the above two conditions are possible, only, when the head of the load is on the section itself. Thus the maximum positive S.F. at X,

$$F_{max} = +R_B = +\frac{wx^2}{2l}$$

Now, if we want to plot the positive S.F. diagram for all sections *i.e.*, from $x=0$ to $x=l$, let us substitute the various values of x (*i.e.*, $x=0$, $x=l/4$, $x=l/2$, $x=3l/4$ and $x=l$) in the above equation. We find that when $x=0$, $F_{max}=0$; when $x=l/4$, $F_{max}=wl/32$; when

$x=l/2$, $F_{max}=\dfrac{4wl}{32}=\dfrac{wl}{8}$; when $x=\dfrac{3l}{4}$, $F_{max}=\dfrac{9wl}{32}$ and when

$x=l$, $F_{max}=\dfrac{16wl}{32}=\dfrac{wl}{2}$. It is thus obvious, that the positive S.F. will

consist of a parabola, having zero ordinate at A and $\dfrac{wl}{2}$ at B as shown in Fig. 2·2(a).

(b) Negative shear force

We know that the negative S.F. at any section X, at a distance x from A, is equal to the reaction R_A *minus* any load between A and X. Now, a little consideration will show that the negative S.F. is maximum when (i) the load* is between X and B (i.e., no load between A and X) and (ii) the value of reaction R_A is maximum. Now, the above two conditions are possible only when the tail of the load is at the section itself (XB being fully loaded and AX having no load). Thus the maximum negative S.F. at X,

$$F_{max}=-R_A=-\frac{w(l-x)^2}{2l}$$

If we want to plot the negative S.F. diagram for all sections i.e., from $x=0$ to $x=l$, let us substitute the various values of x (i.e., $x=0$, $x=l/4$, $x=l/2$, $x=3l/4$ and $x=l$) in the above equation. We find that

when $x=0$, $F_{max}=-\dfrac{16wl}{32}=-\dfrac{wl}{2}$; when $x=\dfrac{l}{4}$, $F_{max}=-\dfrac{9wl}{32}$;

when $x=\dfrac{l}{2}$, $F_{max}=-\dfrac{4wl}{32}=-\dfrac{wl}{8}$ and when $x=l$, $F_{max}=0$. It is thus

obvious, that the negative S.F will also consist of a parabola having $-\dfrac{wl}{2}$ ordinate at A and zero at B as shown in Fig. 2·2 (b).

From the above discussion, we find that the *maximum positive S.F. at a section occurs when the head of the load is on the section itself, whereas maximum negative S.F. at a section occurs when the tail of the load is at the section itself.* The S.F. diagram will consist of two parabolas, one for positive and the other for negative, having end ordinates as $wl/2$ as shown in Fig. 2·2 (b).

(c) Bending moment

We know that when the head of the load is just at A, the reaction R_B will be zero ; thus the bending moment at A will also be equal to zero. As the head of the load rolls towards B, the reactions R_A and R_B will go on increasing.

Now let us consider an instant, when the head of the load is

---

*This will happen when the head of the load crosses the section X and continues to advance further. After some time the load will cover the entire span (as the load is longer than the span). As the load still continues to advance, the tail of the load will first reach at A, and then at the section X.

at any point C at a distance y from A. We know that the B.M.* at any section X, at a distance x from A is equal to the moment of reaction R_A, about X *minus* the moment of any load between A and X about X. Now a little consideration will show that the B.M. at the section X may be maximum under any one of the following three conditions :—

(1) When the head of the load at X (*i.e.*, for no load in section XB.

(2) When the entire span is fully loaded (*i.e.*, for the maximum values of R_A and R_B).

(3) When tail of the load is at X (*i.e.*, for no load in section AX).

Now we shall consider the above three conditions, one by one. In first case, when the head of the load is at X, the B.M. at X,

$$M_X = R_B(l-x) = \frac{wx^2}{2l}\ (l-x)$$

$$= \frac{wx(l-x)}{2l} \times x \qquad\qquad ...(i)$$

In the second case, when the entire span is fully loaded, the B.M. at X,

$$M_X = R_A.x - wx.\frac{x}{2} = \frac{wl}{2}\ x - \frac{wx^2}{2} = \frac{wx}{2}\ (l-x)$$

$$= \frac{wx(l-x)}{2l} \times l = \frac{wx}{2}\ (l-x) \qquad\qquad ...(ii)$$

In the third case, when the tail of the load is X, the B.M. at X,

$$M_X = R_A.x = \frac{w(l-x)^2 x}{2l}$$

$$= \frac{wx(l-x)}{2l} \times (l-x) \qquad\qquad ...(iii)$$

Now, from the above three equations, we find that the maximum** B.M. at any section X will take place when the entire span is fully loaded.

$$M_{max} = \frac{wx}{2}\ (l-x)$$

The absolute maximum B.M. will occur at the centre of the span, when the entire span is loaded (just as in the case of static loading).

$$M_{max} = \frac{wl^2}{8}$$

---

* It is also equal to the moment of the reaction R_B about X *minus* the moment of any load between X and B about X.

**The simple reason for the maximum B.M. to be in equation (*ii*) is that in all three equations there is a common factor $\frac{wx(l-x)}{2l}$. In equation (*i*) it is multiplied by x, in equation (*ii*) it is multiplied by l whereas in equation (*iii*) it is multiplied by $(l-x)$. Since the value of l is maximum of all the three values, therefore the value of equation (*ii*) is maximum of all the three.

Example 2·2. *A girder of span 30 m is simply supported at its ends. A uniformly distributed load of 1·5 tonne/metre and 40 metres long is made to roll over the girder from one end to another. Determine the magnitude of maximum S.F. and B.M at a section 10 m from the left end support.* (Gujarat University, 1972)

Solution.

Given. Span, $l = 30$ m

Load, $w = 1·5$ t/m

Length of the load $= 40$ m

Distance of the section from the left end support,

$$x = 10 \text{ m}$$

Maximum positive S.F.

We know that the maximum positive S.F. takes place, when the head of the load is on the section itself.

$$\therefore \qquad F_{max} = +R_B = +\frac{wx^2}{2l} = \frac{1·5 \times 10^2}{2 \times 30} = 2·5 \text{ t Ans}$$

Maximum negative S.F.

We know that the maximum negative S.F. takes place, when the tail of the load is on the section itself.

$$\therefore \qquad F_{max} = -R_A = -\frac{w(l-x)^2}{2l} = -\frac{1·5 \times (30-10)^2}{2 \times 30}\text{T}$$

$$= -10 \text{ t Ans.}$$

Maximum bending moment

We know that the maximum B.M. takes place when the entire span is fully loaded.

Using the relation,

$$M_{max} = \frac{wx}{2}(l-x) \text{ with usual notations.}$$

$$= \frac{1·5 \times 10}{2}(30-10) = 150 \text{ t·m Ans.}$$

Example 2·3. *A uniformly distributed load of 2 t/m and 20 m long crosses a girder of span 16 m. Calculate the maximum S.F. and B.M. at 0, 4, 8, 12 and 16 m from the left end support and construct the diagrams.* (Nagpur University

Solution.

Given. Load, $w = 2$ t/m

Span, $l = 16$ m

Maximum positive S.F. diagram

We know that the maximum positive S.F. takes place, when the head of the load is on the section itself.

i.e., $F_{max} = +R_B = \dfrac{wx^2}{2l}$

$\therefore \qquad 0\ F_{max} = 0$

$\qquad\qquad 4\ F_{max} = \dfrac{2 \times 4^2}{2 \times 16} = 1 \ \text{t}$

$\qquad\qquad 8\ F_{max} = \dfrac{2 \times 8^2}{2 \times 16} = 4 \ \text{t}$

$\qquad\qquad 12\ F_{max} = \dfrac{2 \times 12^2}{2 \times 16} = 9 \ \text{t}$

$\qquad\qquad 16\ F_{max} = \dfrac{2 \times 16^2}{2 \times 16} = 16 \ \text{t}$

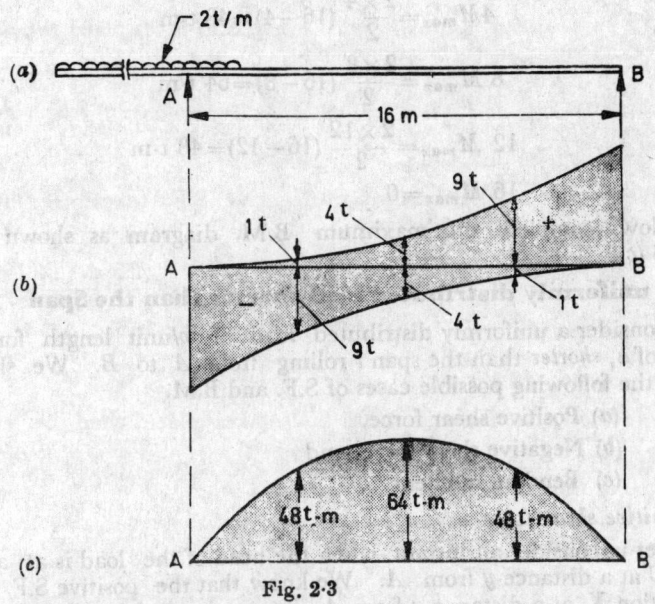

Fig. 2·3

Now complete the maximum positive S.F. diagram as shown in Fig. 2·3 (b).

Maximum negative S.F. diagram

We know that the maximum negative S.F. takes place when the tail of the load is on the section itself.

i.e., $F_{max} = -R_A = -\dfrac{w(l-x)^2}{2l}$

$\therefore \qquad 0\ F_{max} = -\dfrac{2(16-0)^2}{2 \times 16} = -16 \ \text{t}$

$\qquad\qquad 4\ F_{max} = -\dfrac{2(16-4)^2}{2 \times 16} = -9 \ \text{t}$

$\qquad\qquad 8\ F_{max} = -\dfrac{2(16-8)^2}{2 \times 16} = -4 \ \text{t}$

$$12\,F_{max} = -\frac{2(16-12)^2}{2 \times 16} = -1 \text{ t}$$

$$16\,F_{max} = 0$$

Now complete the maximum negative S.F. diagram as shown in Fig. 2·3 (b)

Maximum B.M. diagram

We know that the maximum B.M. occurs, when the entire span is fully loaded

i.e.

$$M_{max} = \frac{wx}{2}\,(l-x)$$

$$\therefore \qquad 0\,M_{max} = 0$$

$$4\,M_{max} = \frac{2 \times 4}{2}\,(16-4) = 48 \text{ t-m}$$

$$8\,M_{max} = \frac{2 \times 8}{2}\,(16-8) = 64 \text{ t-m}$$

$$12\,M_{max} = \frac{2 \times 12}{2}\,(16-12) = 48 \text{ t-m}$$

$$16\,M_{max} = 0$$

Now complete the maximum B.M. diagram as shown in Fig. 2·3 (c).

2·6 A uniformly distributed load shorter than the Span

Consider a uniformly distributed load of w/unit length for a length of a, *shorter* than the span l rolling from A to B. We shall discuss the following possible cases of S.F. and B.M.

    (a) Positive shear force,

    (b) Negative shear force, and

    (c) Bending moment.

(a) Positive shear force

Let us consider an instant, when the head of the load is at any point C at a distance y from A. We know that the positive S.F. at any section X, at a distance x from A, is equal to the reaction R_B *minus* any load between X and B. Now, a little consideration will show that the positive S.F. is maximum when (*i*) the load is between A and X (*i.e.* no load between X and B) and the value of reaction R_B is maximum. Now the above two conditions are possible. only, when the head of the load is on the section itself. Now, considering the length of the load a to be less than x (*i.e.*, when head of the load is at X, the tail being between A and X) the maximum positive S.F. at X,

$$*F_{max} = +R_B = +\frac{wa}{l}\left(x - \frac{a}{2}\right)$$

---

*If the length of the load a is greater than x (*i.e.*, when the head of the load is at x, its tail is still behind A or at A), then the maximum S.F. at X,

$$F_{max} = +R_B = \frac{wx^2}{2l}$$

Now, if we want to plot the positive S.F. diagram for all sections *i.e.* from $x=0$ to $x=l$, let us substitute the various values of x in the above equation. We find that when $x=0$, $F_A=0$; when $x=\dfrac{a}{2}$, $F_{max}=0$; when $x=l$, $F_{max}=\dfrac{wa}{l}\left(l-\dfrac{a}{2}\right)$; and when $x=\left(l+\dfrac{a}{2}\right)$ $F_{max}=wa$.

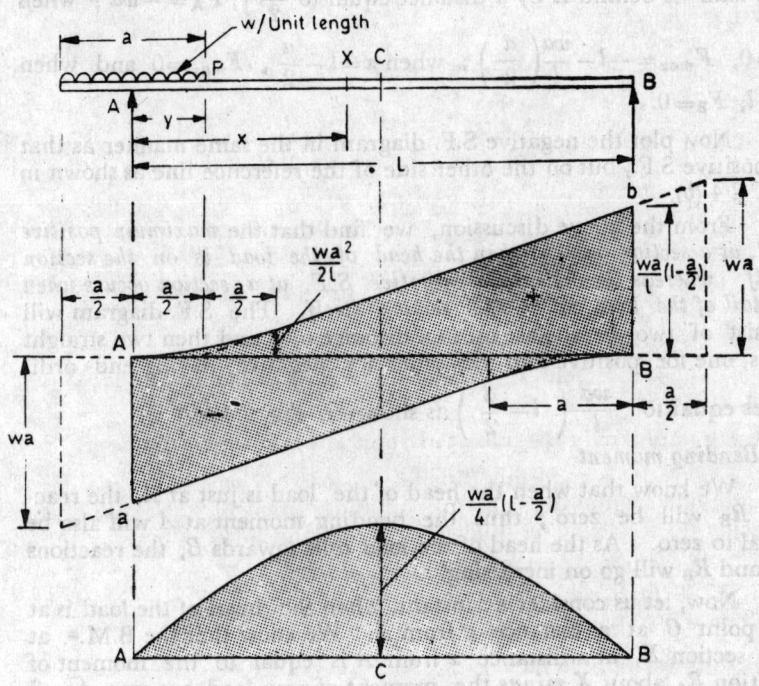

Fig. 2·4

Now, to plot the positive S.F. diagram, mark the ordinate at $x=\dfrac{a}{2}$, $F_{max}=0$, and at $x=l+\dfrac{a}{2}$, $F_{max}=wa$. Join these two points by a straight line. Now reject the portion for which the straight line law does not hold good *i.e.* from $x=0$ to $x=a$. In this position the S.F. increases by a parabolic law. Therefore S.F. at a distance a from the support A is equal to $\dfrac{wa^2}{2l}$

(b) Negative shear force

We know that the negative S.F. at any section X, at a distance x from A is equal to the reaction R_A *minus* any load between A and X. Now, a little consideration will show that the negative S.F. is maximum when (*i*) the load is between X and B (*i.e.* no load between A and X) and (*ii*) value of reaction R_A is maximum. Now, the above two conditions are possible, only, when the tail of the load is at the section itself. Thus the negative S.F. at X,

$$F_{max} = -R_A = -\frac{wa}{l}\left(l - x - \frac{a}{2}\right)$$

If we want to plot negative S.F. diagram for all sections *i.e.* from $x=0$ to $x=l$, let us substitute the various values of x in the above equation. We find that when $x = -\frac{a}{2}\left(i.e.\text{ when the head of}\right.$ the load is behind A by a distance equal to $\frac{a}{2}\right)$, $F_A = -wa$; when $x = 0$, $F_{max} = -l - \frac{wa}{l}\left(\frac{a}{2}\right)$; when $x = l - \frac{a}{2}$, $F_{max} = 0$ and when $x = l$, $F_B = 0$.

Now plot the negative S.F. diagram in the same manner as that of positive S.F., but on the other side of the reference line as shown in Fig. 2·4 (*b*).

From the above discussion, we find that the *maximum positive S.F. at a section occurs, when the head of the load is on the section itself, whereas the maximum negative S.F. at a section occurs when the tail of the load is at the section itself.* The S.F. diagram will consist of two parabolas up to a distance of a and then two straight lines, one for positive and the other for negative, having end ordinates equal to $\frac{wa}{l}\left(l - \frac{a}{2}\right)$ as shown in Fig. 2·4(*b*).

(c) Bending moment

We know that when the head of the load is just at A, the reaction R_B will be zero ; thus the bending moment at A will also be equal to zero. As the head of the load rolls towards B, the reactions R_A and R_B will go on increasing.

Now, let us consider an instant, when the head of the load is at any point C at a distance y from A. We know that the B.M.* at any section X, at a distance x from A is equal to the moment of reaction R_A about X *minus* the moment of any load between A and X about X.

Now a little consideration will show that the B.M. at section X may be maximum under any one of the following three conditions:—

(1) When the head of the load is at X (*i.e.* for no load in section XB).

(2) When the tail of the load is at X (*i.e.* for no load in section AX).

(3) When the load is on the section (*i.e.* partly in AX) and partly in XB (*i.e.* for maximum values of R_A and R_B).

Now we shall discuss the above three conditions, one by one. In the first case, when the head of the load is at X and tail of the load is a distance $(x-a)$ from A (considering the length of the load a to be less than x), at the B.M. at X.

---

*It is also equal to the moment of the reaction R_B about X *minus* the moment of any load between X and B about X.

$$*M_X = R_B\,(l-x) = \frac{wa}{l}\left(x - \frac{a}{2}\right)(l-x) \qquad \ldots(i)$$

In the second case, when the tail of the load is at X,

$$M_X = R_A \cdot x \; \frac{wax}{l}\left(l - x - \frac{a}{2}\right) \qquad \ldots(ii)$$

In the third case, when the load is partly in AX and partly in XB, first of all we have to find out the position of the load for maximum B.M. on the section. Let us assume the position of the load as shown in Fig. 2·5.

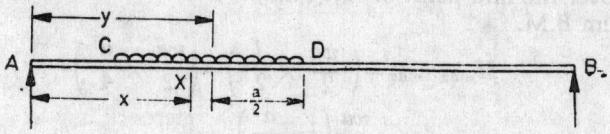

Fig. 2·5

Let y be the distance between the centre of gravity of the load from A and part CX of the load be behind the section X. From the geometry of the figure, we find that

$$CX = x - \left(y - \frac{a}{2}\right) = x - y + \frac{a}{2}$$

At this instant the reaction at A,

$$R_A = \frac{wa}{l}(l - y)$$

$\therefore$ B.M. at X at this instant,

$$M_X = R_A \cdot x - \frac{w}{2}\left(x - y + \frac{a}{2}\right)^2$$

$$= \frac{wax}{l}(l - y) - \frac{w}{2}\left(x - y + \frac{a}{2}\right)^2 \qquad \ldots(iii)$$

The moment at X will be maximum, for the value of y, when $\dfrac{dM_X}{dy} = 0$. Therefore differentiating the equation (iii) and equating it to zero,

i.e.
$$\frac{dM_X}{dy}\left[\frac{wax}{l}(l - y) - \frac{w}{2}\left(x - y + \frac{a}{2}\right)^2\right] = 0$$

$\therefore$
$$\frac{ax}{l} - \left(x + y + \frac{a}{2}\right) = 0$$

or
$$\frac{ax}{l} = x - y + \frac{a}{2}$$

or
$$\frac{x - y + \frac{a}{2}}{a} = \frac{x}{l}$$

___
*If the length of the load a is greater than x (i.e. when the head of the load is at x, its tail is still behind A or at A) the B.M. at x,

$$M_X = R_B\,(l - x) = \frac{wa^2}{2l}\,(l - x)$$

$$\frac{CX}{CD} = \frac{x}{l}$$

It means that *the B.M. at the section X is maximum, when the position of the load is such that the section X divides the load in the same ratio, as it divides the span l.* It is an important relation, which will hold good for point loads and distributed loads also.

We know that the absolute maximum B.M. will occur at the centre of the span. As per above deduction, we find it will occur when the load is also centrally placed (*i.e.*, when the mid point of the load is over the mid-point of the span). At this instant the absolute maximum B.M.

$$M_{max. \ max} = \left(\frac{wa}{2} \times \frac{l}{2} \right) - \left(\frac{wa}{2} \times \frac{a}{4} \right)$$

$$= \frac{wa}{4} \left(l - \frac{a}{2} \right)$$

Example 2·4. *A simply supported beam has a span of 20 metres. A uniformly distributed load of 2 tonnes/m and 5 metres long crosses the span. Find the maximum B.M. produced at a point 8 metres from the left support.*

Solution.

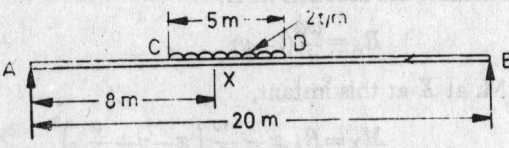

Fig 2·6

Given. Span, $l = 20$ m
Load, $w = 2$ t/m
Length of load, $a = 5$ m
Distance of the section from the left support
                       $x = 8$ m

We know that the **maximum bending moment at the section** takes place, when the position of the load is such that the section X divides the load in the same ratio, as it divides the span as shown in Fig. 2·6.

$$\therefore \qquad \frac{CX}{CD} = \frac{x}{l}$$

$$\frac{CX}{5} = \frac{8}{20}$$

or $\qquad\qquad CX = \frac{8 \times 5}{20} = 2$ m

Now taking moments about A,

$$R_B \times 20 = 5 \times 2 \times 8 \cdot 5 = 85$$

$$\therefore \qquad\qquad R_B = \frac{85}{20} = 4 \cdot 25 \text{ T}$$

and maximum B.M. at X,

$$M_{max} = R_B \times 12 - \frac{2 \times 3^2}{2} = 4\cdot25 \times 12 - 9 \text{ t-m}$$

$$= 42 \text{ t-m} \quad \textbf{Ans}.$$

Example 2·5. *A uniformly distributed load of $\frac{1}{2}$ tonne/m, covering a length of 15 m, crosses a girder of span 50 m. Find the values of maximum S.F. and B.M. at a section 10 m from the left hand support*

Solution.

Given. Load, $w = \frac{1}{2}$ t/m
Length of load, $a = 15$ m
Span, $l = 50$ m
Distance of the section from the left hand support,

$$x = 10 \text{ m}$$

Maximum positive S.F.

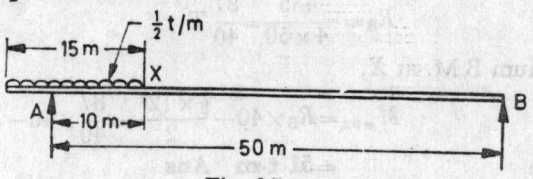

Fig. 2·7

We know that the maximum positive S.F. will take place, when the head of the load is on the section itself as shown in Fig. 2·7.

$$\therefore \quad F_{max} = + R_B = + \frac{wx^2}{2l} = \frac{\frac{1}{2} \times 10^2}{2 \times 50} = 0\cdot5 \text{ T} \quad \textbf{Ans}.$$

Maximum negative S.F.

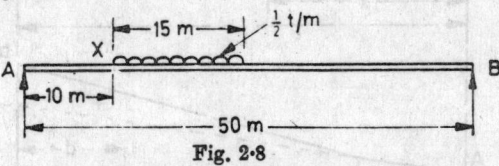

Fig. 2·8

We know that the maximum negative S.F. will take place, when the tail of the load is on the section itself as shown in Fig. 2·8.

Now taking moments about B,

$$R_B \times 50 = \frac{1}{2} \times 15 \times \frac{65}{2} = 243\cdot75$$

or

$$R_A = \frac{243\cdot75}{50} = 4\cdot875 \text{ T}$$

$$\therefore \quad F_{max} = -R_A = -4\cdot875 \text{ t} \quad \textbf{Ans}.$$

Maximum bending moment

We also know that the maximum B.M. at the section will take place, when the position of the load is such that the section X, divides the load in the same ratio, as it divides the span as shown in Fig. 2·9.

$$\therefore \quad \frac{CX}{CD} = \frac{x}{l}$$

$$\frac{CX}{15} = \frac{10}{50}$$

or $$CX = \frac{10 \times 15}{50} = 3 \text{ m}$$

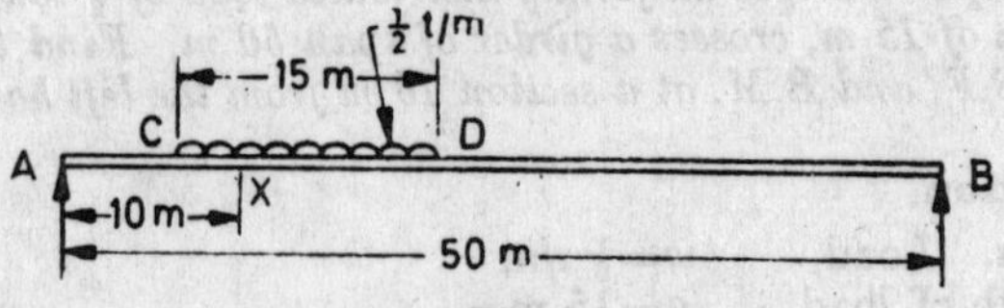

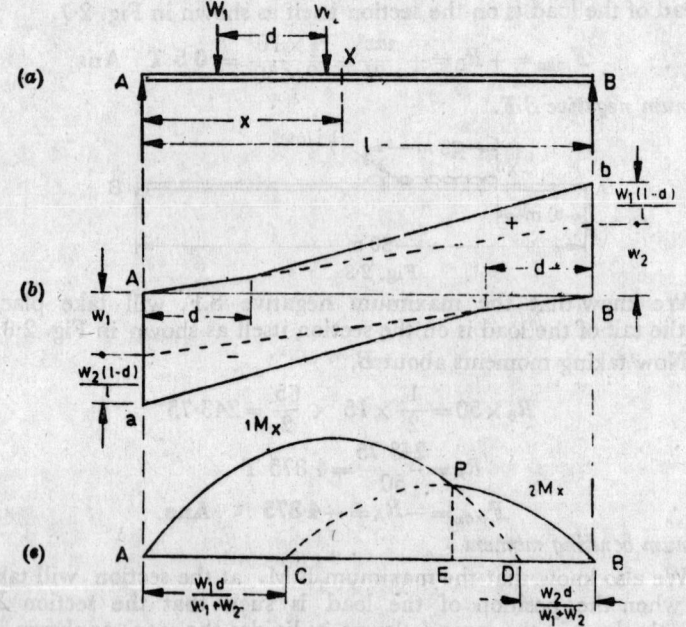

Fig. 2·9

Now taking moments about A,

$$R_B \times 50 = \frac{1}{2} \times 15 \times \frac{29}{2} = \frac{435}{4}$$

or $$R_B = \frac{435}{4 \times 50} = \frac{87}{40} \text{ T}$$

and maximum B.M. at X,

$$M_{max} = R_B \times 40 - \frac{\frac{1}{2} \times 12^2}{2} = \frac{87}{40} \times 40 - \frac{144}{4} \text{ t·m}$$

$$= 51 \text{ t·m} \quad \textbf{Ans.}$$

2·7. Two concentrated loads

Fig. 2·10.

Consider two concentrated or point loads, W_1 and W_2 spaced d apart rolling over the beam of span l from A to B. Let the leading load W_2 be lighter than the trailing load W_1. In practical life, such a situation takes place, when a four wheeled vehicle with two axles crosses a bridge. We shall discuss the following possible cases of S.F. and B.M.

    (a) Positive shear force,

    (b) Negative shear force, and

    (c) Bending moment.

(a) Positive shear force

Let us consider an instant, when the leading load W_2 be at any point C. We know that the positive S.F. at any section X, at a distance x from A is equal to the reaction R_A *minus* any load between X and B. Now, a little consideration will show that the positive S.F. is maximum when (i) both the loads are between A and X (*i.e.*, no load between X and B) and (ii) the value of reaction R_B is maximum.

Now, we see that the positive S.F. at section X may be maximum under any one of the following two conditions :—

    (1) When the leading load W_2 is at X, and the trailing load W_1 is in AX (*i.e.*, for no load in section XB)

    (2) When leading load W_2 is in section XB and the trailing load W_1 is at X (*i.e.*, for the maximum value of R_B)

Now, we shall discuss the above two conditions, one by one.

In the first case, when the leading load W_2 is at X and the trailing load W_1 is in section AX at a distance d behind the load W_2 (considering the distance d between the two loads less than x) the maximum positive S.F. at X,

$$*F_{max} = + R_B = \frac{W_2 x + W_1 (x-d)}{l} \qquad \ldots (i)$$

In the second case, when the load W_2 is in section XB and the trailing load W_1 is at X, then the maximum positive S.F. at X,

$$**F_{max} = + R_B - W_2 = \frac{W_1 x + W_2 (x+d)}{l} - W_2 \qquad \ldots (ii)$$

It is thus obvious, that the maximum positive S.F. at the section is greater of the two values obtained in equations (i) and (ii) depending upon the magnitudes of W_1, W_2, d, l and x. If the S.F. obtained in equation (i) is to be greater than that obtained in equation (ii), then

$$\frac{W_2 x + W_1 (x-d)}{l} > \frac{W_1 x + W_2 (x+d)}{l} - W_2$$

$$\therefore \quad W_2 x + W_1 x - W_1 d > W_1 x + W_2 x + W_2 d - W_2 l$$

$$W_1 d + W_2 d < W_2 l$$

---

*If the distance d between the loads is more than x, then the load W_1 will be still behind A, when the load W_2 has reached at X,

∴. Maximum positive S.F. at X,

$$F_{max} = + R_B = \frac{W_2 x}{l}$$

**If the distance $(l-x)$ is less than d, then the load W_2 will cross B, when the load W_1 will be at X. At this moment, the maximum positive S.F. at X,

$$F_{max} = + R_B = \frac{W_1 x}{l}$$

or
$$d(W_1+W_2) < W_2l$$
$$\therefore \quad d < \frac{W_2l}{W_1+W_2}$$

This is a standard case, for which the maximum positive S.F. diagram has been drawn.

Now, if we want to plot the positive S.F. diagram for all sections *i.e.*, from $x=0$ to $x=l$, let us first divide the span into two portions *i.e.* 0 to d and d to l (the idea of dividing the span from 0 to d is that first of all only the leading load W_2 will be on the beam, till it rolls beyond d and then the load W_1 will follow). Let us substitute the various values of x in equation (*i*) first from 0 to d and then d to l. We find that when $x=0$. $F_{max}=0$ (at this instant W_I is to be neglected) when $x=d$, $F_{max}=\dfrac{W_2d}{l}$ and when $x=l$, $F_{max}=W_2+\dfrac{W_1(l-d)}{l}$.

(b) *Negative shear force*

We know that the negative S.F. at any section X, at a distance x from A, is equal to the reaction R_A *minus* any load between A and X. Now, a little consideration will show that the negative S.F. is maximum when both the loads are between X and B (*i.e.*, for no load between A and X) and (*ii*) the value of reaction R_A is maximum. Now we see that the negative S.F. at a section may be maximum under any one of the following two conditions:—

(1) When the trailing load W_1 is at X and the leading load W_2 is in XB (*i.e.*, for no load in section AX.)

(2) When the leading load W_2 is at X and the trailing load W_1 is in AX at a distance d behind X (*i.e.* for the maximum value of R_A).

Now, we shall discuss the above two conditions, one by one.

In the first case when the trailing load W_1 is at X and the leading load W_2 is in section XB, then the maximum negative S.F at X,

$$*F_{max}=-R_A=-\frac{W_1(l-x)+W_2(l-x-d)}{l} \qquad ...(iii)$$

In the second case, when the leading load W_2 is at X and the trailing load W_1 is in section AX at a distance d behind X, then the maximum negative S.F. at X,

$$**F_{max}=-(R_A-W_A)=-\frac{W_2(l-x)+W_1(l-x+d)}{l}+W_1 \qquad ...(iv)$$

It is thus obvious, that the maximum negative S.F. at the section is greater of the two values obtained in equations (*iii*) and (*iv*)

----

*If the distance d between the two loads is more than $(l-x)$, the leading load W_2 will cross B when the load W_1 has reached at X.
∴ maximum negative S.F. at X,

$$F_{max}=-R_A=-\frac{W_1(l-x)}{l}$$

**If the distance d between the two loads is more than x, then the load W_1 will be still behind A, when the load W_2 has reached at X. At this instant the maximum negative S.F. at X,

$$F_{max}=-R_A=-\frac{W_2(l-x)}{l}$$

depending upon the magnitudes of W_1, W_2, d, l and x. If the S.F. obtained in the equation (*iii*) is to be greater than that obtained in equation (*iv*), then

$$\frac{W_1(l-x)+W_2(l-x-d)}{l} > \frac{W_2(l-x)+W_1(l-x+d)}{l} - W_1$$

or $W_1l-W_1x+W_2l-W_2x-W_2d > W_2l-W_2x+W_1l-W_1x+W_1d-W_1l$

$$W_1d+W_2d < W_1l$$

$$\therefore \qquad d(W_1+W_2) < W_1l$$

or $$d < \frac{W_1l}{(W_1+W_2)}$$

This is a standard case, for which the maximum negative S.F. diagram has been drawn.

If we want to plot the negative S.F. diagram for all section *i.e.*, from $x=0$ to $x=l$, let us first divide the span into two portions *i.e.* 0 to $(l-d)$ and $(l-d)$ to l [the idea of dividing the span to $(l-d)$ to l is that when the trailing load W_1 reaches at a distance $(l-d)$ from A, the leading load W_2 will cross B]. Let us substitute the various values of x in equation (*iii*) first of all 0 to $(l-d)$ and then $(l-d)$ to l. We find that when $x=0$, $F_{max}=W_1+\dfrac{W_2(l-d)}{l}$; when $x=(l-d)$, $F_{max}=\dfrac{W_1d}{l}$ and when $x=l$ $F_{max}=0$ (at this instant, W_2 is to be neglected).

(c) Bending moment

We also know that when the leading load is just at A the reaction R_B will be zero : thus the B.M. at A will also be equal to zero. As the loads roll towards B the reactions R_A and R_B will go on increasing.

Now, let us consider an instant, when the leading load is at any point C at a distance y from A. We know that the B.M.* at any section X, at a distance x from A is equal to the moment of reaction R_A about X *minus* the moment of any load between A and X about X.

Now a little consideration will show that the B.M. at section X may be maximum under any one of the following two conditions :—

    (1) When the leading load W_2 is at X (*i.e.* for no load in section XB).

    (2) When the trailing load W_1 is at X (*i.e.* for no load in section AX).

Now we shall discuss the above two conditions one by one.

In the first case, when the leading load W_2 is at X and the trailing load is at a distance $(x-d)$ from A (considering the distance d between the two loads to be less than x) the B.M. at X,

---

*It is also equal to the moment of the reaction R_B about X *minus* the moment of any load between X and B about X.

$$*2M_X = R_B(l-x) = \frac{[W_2 x + W_1(x-d)](l-x)}{l} \qquad \ldots(v)$$

In the second case, when the trailing load W_1 is at X and the leading load W_2 is in the section XB, then the B.M. at X,

$$**1M_X = R_A . x = \frac{[W_1(l-x) + W_2(l-x-d)]x}{l} \qquad \ldots(vi)$$

It is thus obvious, that the maximum B.M. at the section is greater of the two values obtained in equations (v) and (vi) depending upon the magnitudes of W_1, W_2, d, l and x. If we plot the B.M. diagrams, we find that the two diagrams will be parabolas. The first B.M. diagram will be zero at $x=0$ and $x = \frac{W_2 d}{W_1 + W_2}$ from the end B and the second B.M. diagram will be zero at $x=l$ and $x = \frac{W_1 d}{W_1 + W_2}$ from the end A as shown in Fig. 2·10(c) we also find from the figure that the $1M_X$ is equal to $2M_X$ at P. Therefore, let us equate the two bending moments in order to locate the point P i.e.

$$\frac{[W_2 x + W_1(x-d)](l-x)}{l} = \frac{[W_1(l-x) + W_2(l-x-d)]x}{l}$$

Simplifying the above equation, we get

$$x = \frac{W_1 l}{W_1 + W_2}$$

From the above equation, we find that the point P divides the span AB in the ratio of $W_2 : W_1$. It will be seen from the B.M. diagram that for all sections from A to E, the maximum B.M. is given by the equation (vi) when the load W_1 is on the section and W_2 in front of it and for all sections from E to B, the maximum B.M. is given by the equation (v) when the load W_2 is on the section and the load W_1 behind it.

The absolute maximum B.M. anywhere in the girder occurs in the range $1M_X$ at $x = \frac{1}{2}AD$ or in the range $2M_X$ at $x = AC + \frac{1}{2}CB$, depending upon the greater value of $1M_X$ or $2M_X$.

Example 2·6. *Two point loads of 12 tonnes and 16 tonnes, spaced 5 metres apart, cross a girder of 25 metres span from left to right with the 12 tonnes load leading. Construct the maximum S.F. and B.M. diagrams stating the absolute maximum values.*

(*A.M.I.E. Summer, 1979*

---

*If the distance d between the two loads is more than x, then the load W_1 will be still behind A, when the load W_2 has reached at X, the B.M. at X,

$$M_X = R_B(l-x) = \frac{W_2 x}{l}(l-x)$$

**If the distance d between the two loads is more than $(l-x)$ than the leading load W_2 will cross B, when the load W_1 has reached at X. Thus B.M. at X,

$$M_X = R_A . x = \frac{W_1(l-x)x}{l}$$

Solution.

Given.   Leading load, $W_2 = 12$ t
Trailing load, $W_1 = 16$ t
Spacing of loads, $d = 5$ m
Span, $l = 25$ m

Positive shear force diagram

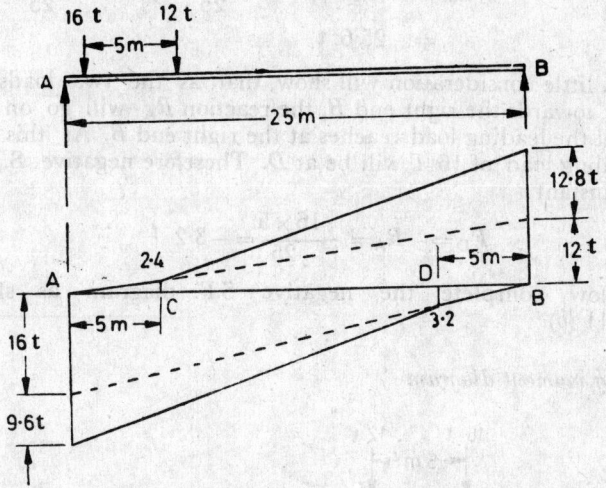

Fig. 2·11

Now for drawing the positive S F. diagram, let us divide the span AB into two sections AC and CB, such that AC is equal to 5 m (*i.e.* spacing of the loads). We know that when the leading load of 12 T is at A, then the reaction R_B is equal to zero. Therefore the positive S.F. is also equal to zero. But when the leading load is between A and C, the positive S.F. is equal to R_B. Therefore positive S.F., when the leading load is at C,

$$F_C = + R_B = \frac{12 \times 5}{25} = 2 \cdot 4 \text{ t} \qquad\qquad ...(i)$$

Now, when the leading load enters into the section CB, the trailing load of 16 T also enters into the section AC. A little consideration will show that as the two loads go on moving, towards the right end B, the reaction R_B will also go on increasing till the leading load of 12 T reaches the right end B. At this moment the value of reaction R_B will be maximum. Therefore maximum positive S.F.,

$$F_{max} = + R_B = + \frac{12 \times 25 + 16 \times 20}{25} = \frac{620}{25} \text{ t}$$

$$= + 24 \cdot 8 \text{ t}$$

Now complete the positive S.F. diagram as shown in Fig. 2·11 (b).

Negative shear force diagram

For drawing the negative S.F. diagram, let us divide the span AB into two sections AD and DB, such that DB is equal to 5 m (*i.e.* spacing of the loads). We know that the reaction at A *i.e.* R_A will be maximum, when the leading load of 12 T is in section AD and the trailing load of 16 T is at A. Therefore maximum negative S.F.,

$$F_{max} = -R_A = -\frac{16 \times 25 + 12 \times 20}{25} = -\frac{640}{25} \text{ t}$$

$$= -25 \cdot 6 \text{ t}$$

A little consideration will show, that as the two loads go on moving towards the right end B, the reaction R_A will go on decreasing, till the leading load reaches at the right end B. At this moment the trailing load of 16 Γ will be at D. Therefore negative S.F. at D, at this instant

$$F_D = -R_A = -\frac{16 \times 5}{25} = -3 \cdot 2 \text{ t}$$

Now complete the negative S.F. diagram as shown in Fig. 2.11 (*b*).

Bending moment diagram

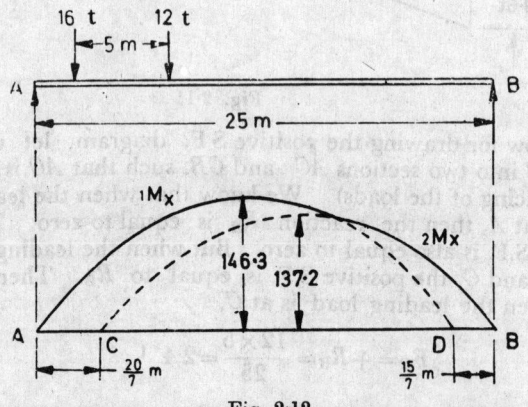

Fig. 2·12

For drawing the B.M. diagram, let us divide the span AB into three sections AC, CD and DB, such that $AC = \dfrac{W_1 d}{W_1 + W_2} = \dfrac{16 \times 5}{16 + 12}$

$= \dfrac{80}{28} = \dfrac{20}{7}$ m, and $DB = \dfrac{W_2 d}{W_1 + W_2} = \dfrac{12 \times 5}{16 + 12} = \dfrac{60}{28} = \dfrac{15}{7}$ m as shown in

Fig 2·12 (*b*). We know that the B.M. at any section X, at a distance x from A, may be maximum under any one of the following two conditions :—

    (1) when the leading load of 12 T is at X.

    (2) when the trailing load of 16 T is at X.

In the first case, the B.M. at X, at a distance x from A,

$$2M_X = \frac{[W_2 x + W_1 (x-d)](l-x)}{l}$$

If we plot the above B.M. diagram, we shall find that the diagram is a parabola having zero ordinates at C and B as shown in Fig. 2·12 (b). The maximum ordinate of $2M_X$ will be at the mid-point of CB, i.e., $\dfrac{20}{7} + \dfrac{1}{2}\left(25 - \dfrac{20}{7}\right) = \dfrac{20}{7} + \dfrac{155}{14} = \dfrac{195}{14}$

$= 13·93$ m from A. Therefore maximum B.M.

$$2M_{max} = \frac{[(12 \times 13·93) + 16(13·93 - 5)](25 - 13·93)}{25} \text{ t-m}$$

$$= 137·2 \text{ t-m}$$

Similarly, in the second case, the B.M. at X, at a distance x from A,

$$1M_X = \frac{[W_1(l-x) + W_2(l-x-d)]x}{l}$$

If we plot the above B.M. diagram, we find that the diagram is a parabola having zero ordinates at A and D as shown in Fig. 2·12 (b). The maximum ordinates of $1M_X$ will be at the mid-point of AD, i.e., $\dfrac{1}{2}\left(25 - \dfrac{15}{7}\right) = \dfrac{160}{1_4} = \dfrac{80}{7} = 11·43$ m from A. Therefore maximum B.M.

$$1M_{max} = \frac{[16(25 - 11·43) + 12(25 - 11·43 - 5)]11·43}{25} \text{ t-m}$$

$$= 146·3 \text{ t-m}$$

Now complete the $1M_X$ and $2M_X$ diagrams as shown in Fig. 2·12 (b).

Exercise 2-1

1. A load of 15 tonnes crosses a beam of 20 m span, Find the values of maximum positive and negative S.F. and B.M. at a section 8 m from the left end support. **[Ans. +6 T; −9 T; 72 t-m]**

2. A uniformly distributed load of 2 t/m, and 30 m long crosses a girder of span 25 m. Determine the values of maximum positive and negative S. F. and B.M. at a section 10 m from the left end of the girder. *(Punjab University)* **[Ans. +21 T; −16 T; 156·25 t-m]**

3. A uniformly distributed load of 2·5 t/m and 6 m long crosses a beam of 30 m span. Calculate the maximum S.F. and B.M. at a section 10 m from the left hand support. *(Calcutta University)* **[Ans. +3·5 T; −8·5 T; 90 t-m]**

4. A uniformly distributed load of 0·5 t/m and 5 m long rolls across a beam 15 m long simply supported at its both ends. Find the maximum S.F. and B.M. at a section 6 m from the left hand support. Also find the absolute maximum B. M. on the span. *(Patna University)* **[Ans. +0·58 T; −1·08 T; 7·5 t- m; 7·81 t-m]**

5. Two point loads of 4 T and 6 T, spaced 6 m apart, cross a girder of 16 m span, with the 4 T load leading from left to right. Calculate the values of maximum S. F. and B.M. *(Jadavpur University)* **[Ans. +7·75 T− ; 8·5 T; 28·8 t-m]**

2.8.   Several concentrated loads

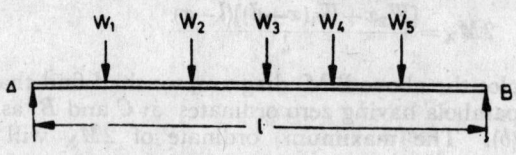

Fig. 2·13.

Consider several concentrated or point loads, W_1, W_2......rolling from A to B over span l. We shall discuss the following possible cases of S.F. and B.M.

(a) Positive shear force,

(b) Negative shear force, and

(c) Bending moment.

(a) Positive shear force

The value of maximum positive S.F. may be obtained by trial and error. However, one of the point loads is to be on the section itself.

(b) Negative shear force

Like the value of maximum positive S.F., the value of negative S.F. may also be obtained by trial and error. In this case also one of the point loads is to be on the section itself.

(c) Bending moment

In order to obtain the position and amount of maximum B.M. over the span, the following two propositions are indispensable.

(i) Proposition for the maximum B.M. under any given load,

(ii) Proposition for the maximum B.M. at any section on the span.

2·9.   Proposition for the maximum bending moment under any given load

According to this proposition, *"when a series of point loads crosses a girder, simply supported at its ends, the maximum B.M. under any given load occurs, when the centre of the span is mid-way between the c.g. of the load system and the load under which the maximum B.M. is required to be found out."*

Proof.

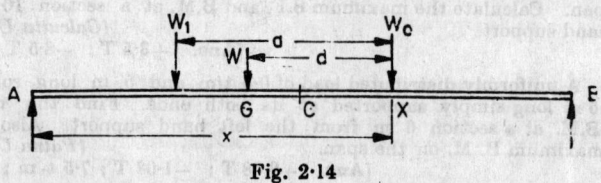

Fig. 2·14

Consider several point loads W_1, W_2, W_3......rolling from A to B. Let the given load be W_0, under which the maximum B.M. is required to be found out.

Let the position of the load W_0, for the maximum B.M. under it, be at X at a distance x from A.

Let $\quad\quad\quad\quad W =$ Total load on the span AB,

$\quad\quad\quad\quad\quad G =$ Centre of gravity of the total load on the span,

$\quad\quad\quad\quad\quad W_1 =$ Position and magnitude of the total load in section AX,

$\quad\quad\quad\quad\quad d =$ Distance between the c.g. of the total load on the span and the section X,

$\quad\quad\quad\quad\quad d_1 =$ Distance between the c.g. of the total load in section AX and the section X.

We know that the reaction at A,

$$R_A = \frac{W.GB}{AB} = \frac{W}{l} \; (l - x + d)$$

and the B.M. at X,

$$M_X = R_A.x - W_1 d_1$$

$$= \frac{Wx}{l} \; (l - x + d) - W_1 d_1$$

$$= \frac{W}{l} \; (lx - x^2 + dx) - W_1 d_1$$

Now, for the maximum B.M. at x, let us differentiate the above equation with respect to x and equate it to zero,

$$\frac{dM_X}{dl} \left[\frac{W}{l} \; (lx - x^2 + dx) - W_1 d_1 \right] = 0$$

$$l - 2x + d = 0$$

$$x - \frac{l}{2} = \frac{d}{2}$$

Expressing the above equation geometrically, we get

$$AX - AC = \frac{GX}{2}$$

$$CX = GC$$

It is thus obvious, that the centre of the span is midway between the c.g. of the load system and the load under which the maximum B.M. is required to be found out.

Example 2·7. *Two point loads of 4 tonnes and 6 tonnes, spaced 6 metres apart, cross a girder of 16 metres span, with the 4 tonnes load leading. Determine the magnitude of maximum B.M. under the 6 tonne load.*

Solution.

Given. Leading load, $W_2 = 4$ t

Trailing load, $\quad\quad\quad W_1 = 6$ t

Spacing of loads $d = 5$ m
Span, $l = 16$ m

We know that for maximum B.M. under the 6 tonne load, the load should occupy such a position on the girder that the centre of the span is mid-way between the *c.g.* of the two loads and the 6 tonne load. From the geometry of the load system, we find that the distance between the *c.g.* of the two loads and the 6 tonne load,

$$x = \frac{4 \times 5}{4+6} = \frac{20}{10} = 2 \text{ m}$$

Therefore the 6 tonne load should be at a distance of $8 - \frac{1}{2}(2) = 7$ m from A. The 4 tonne load will be at a distance $7 + 5 = 12$ m from A as shown in Fig. 2·15.

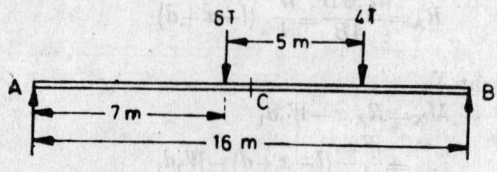

Fig. 2·15

Taking moments about B,

$$R_A \times 16 = 6 \times 9 + 4 \times 4 = 70$$

$$\therefore \quad R_A = \frac{70}{16} \ t$$

$\therefore$ Maximum B.M. under the 6 tonne load,

$$M_{max} = R_A \times 7 = \frac{70}{16} \times 7 = \frac{490}{16} \text{ t-m}$$

$$= 30·625 \text{ t-m} \quad \textbf{Ans}.$$

Example 2·8. *The load system shown in Fig. 2·16. crosses a beam simply supported over a span 24 m.*

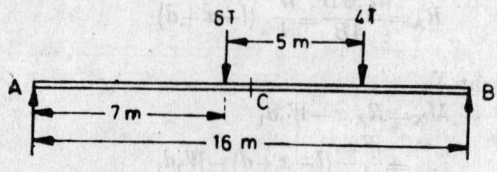

Fig. 2·16

Determine the maximum bending moment under 25 tonnes load.

Solution.

Given span, $l = 24$ m

We know that for maximum B.M. under the 25 tonnes load, the load should occupy such a position on the girder that the span is mid-way between the *c. g.* of the load system and the 25 tonnes load. From the geometry of the load system, we find that the distance between the *c.g.* of the load system from the 10 tonnes load,

$$x=\frac{(20\times3)+(20\times5)+(25\times7)+(15\times10)}{10+20+20+25+15}=5{\cdot}39 \text{ m}$$

Therefore 15 tonnes load should be at a distance of $12{\cdot}0+\frac{1}{2}(7-5{\cdot}39)=12{\cdot}805$ m from the left support A of the beam as shown in Fig. 2·17.

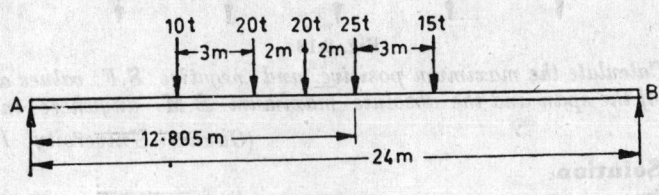

Fig. 2·17.

Taking moments about A,

$$R_B\times24=(10\times5{\cdot}805)+(20\times8{\cdot}805)+(20\times10{\cdot}805)$$
$$+(15\times12{\cdot}805)$$
$$=1{,}007{\cdot}5$$

or $$R_B=\frac{1{,}007{\cdot}5}{24}=42{\cdot}0 \text{ t}$$

$$M_{max}=42\times11{\cdot}195-15\times3=\textbf{425·2 t-m Ans.}$$

2·10. Absolute maximum bending moment

In the previous article, we have discussed the proposition for the position of the load system, for finding out the maximum bending moment under any given load. This proposition may also be used for finding out the absolute maximum bending moment (briefly written as $(M_{max\ max})$ anywhere on the span. For doing so, several trials are to be made. This is done first by selecting a wheel load and then arranging suitably the load system (as discussed in Art 2·9). Finally the value of maximum B. M. is obtained. After this another wheel load is selected and the same proceedure is repeated to get the value of maximum bending moment. Two or three such trials give the value of absolute maximum bending moment. However, the following points help in reducing the number of trials to a minimum.

1. The absolute maximum bending moment always occur *under* a wheel load, and not between the two wheel loads.

2. The absolute maximum bending moment always occur under a wheel load, when it is near the centre of the span.

It does not occur at the centre of the span, unless *c. g.* of the load system coincides with some heavy load.

3.   The absolute maximum bending moment generally occurs under the heavier wheel load, specially that which is very near the *c. g.* of the load system.

Example 2·9. *A train of 5 wheel-loads as shown in Fig. 2·18 crosses a simply supported beam of span 22·5 metres.*

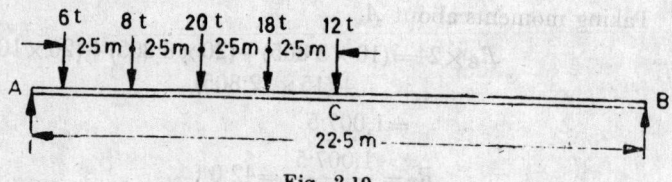

Fig. 2·18

Calculate the maximum positive and negative *S.F.* values at the centre of the span and the absolute maximum *B.M.* anywhere in the span.

(*Gauhati University, 1974*)

Solution.

Given.   Span, $l = 22·5$ m

Maximum positive S.F.

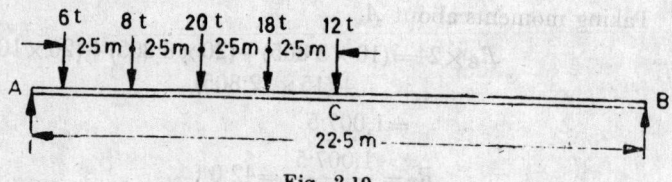

Fig. 2·19

We know that the maximum positive S.F. at the centre of the span will take place, when the leading load is at the centre of the span as shown in Fig. 2·19.

Taking moments about *A*,

$$R_B \times 22·5 = 6 \times 1·25 + 8 \times 3·75 + 20 \times 6·25 + 18 \times 8·75 + 12 \times 11·25$$
$$= 455·0$$

or $R_B = \dfrac{455·0}{22·5} = 20·22$ t

$\therefore$ $F_{max} = + R_B = \mathbf{20·22}$ **t   Ans.**

Maximum negative S.F.

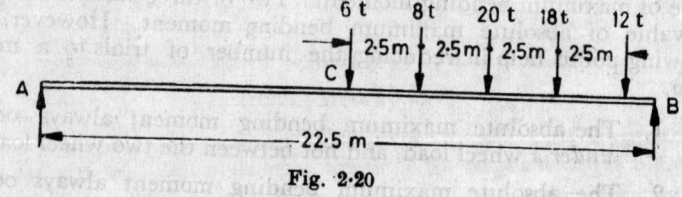

Fig. 2·20

We know that the maximum negative S.F. at the centre of the span will take place, when the trailing load is at the centre of the span as shown in Fig. 2·20.

Taking moments about B,

$$R_A \times 22 \cdot 5 = 12 \times 1 \cdot 25 + 18 \times 3 \cdot 75 + 20 \times 6 \cdot 25 + 8 \times 8 \cdot 75 + 6 \times 11 \cdot 25$$
$$= 345 \cdot 0$$

or $$R_A = \frac{345 \cdot 0}{22 \cdot 5} = 15 \cdot 33 \text{ t}$$

$$\therefore \quad F_{max} = -R_A = -15 \cdot 33 \text{ t} \quad \textbf{Ans}.$$

Absolute Maximum B.M.

First of all let us find out the *c.g.* of the load system. From the geometry of the load system, we find that the distance between the *c.g.* of the load system and 6 T load,

$$x = \frac{8 \times 2 \cdot 5 + 20 \times 5 + 18 \times 7 \cdot 5 + 12 \times 10}{6 + 8 + 20 + 18 + 12} = \frac{375}{64} = 5 \cdot 86 \text{ m}$$

Now, let us find out the load under which the absolute maximum B.M. will take place.* By inspection, we find that the absolute maximum B.M. will take place under the 20 T load. We know that for maximum B.M. under the 20 T load, the load should occupy such a position on the beam that the centre of the span is mid-way between the *c.g.* of the load system and the 20 T load.

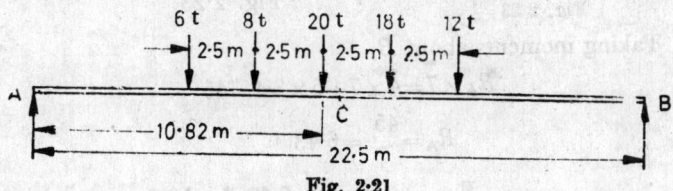

Fig. 2·21

Therefore 20 t load should be at a distance of $11 \cdot 25 - \frac{1}{2}(5 \cdot 86 - 5 \cdot 0) = 10 \cdot 82$ m from A as shown in Fig. 2·21.

Taking moments about B,

$$R_A \times 22 \cdot 5 = 6 \times 16 \cdot 68 + 8 \times 14 \cdot 18 + 20 \times 11 \cdot 68$$
$$+ 18 \times 9 \cdot 18 + 12 \times 6 \cdot 68$$
$$= 689 \cdot 6$$

or $$R_A = \frac{689 \cdot 6}{22 \cdot 5} = 30 \cdot 65 \text{ m}$$

$$\therefore \quad M_{max \; max} = 30 \cdot 65 \times 10 \cdot 82 - 6 \times 5 \cdot 0 - 8 \times 2 \cdot 50 \text{ t-m}$$

$$= 281 \cdot 64 \text{ t-m} \quad \textbf{Ans}.$$

Example 2·10. *The axle loads ABC of a 6-wheel truck travelling over a simply supported span of 7 metres are : Axle A=2 tonnes, axle*

---

*The absolute maximum B.M. generally occurs under a heavier load, which is very near the *c.g.* of the load system.

$B=4$ tonnes, Axle $C=5$ tonnes with 5 tonnes load leading. The axle distances are $AB=2\cdot4$ metres and $BC=3\cdot6$ metres. Determine :

    (a) *the maximum positive and negative shearing forces, and*

    (b) *the maximum bending moment developed as the truck crosses the span.*

                                         (*Cambridge University*)

Solution.

Given. Span $l=7$ m

Maximum positive S.F.

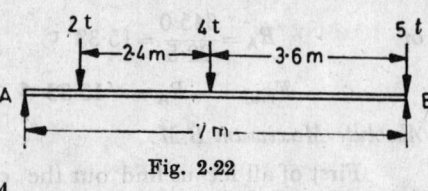

Fig. 2·22

    We know that the maximum positive S.F. will take place at B, when the leading load is at B as shown in Fig. 2·22.

Taking moments about A,

$$R_B \times 7 = 2\times1 + 4\times3\cdot4 + 5\times7 = 50\cdot6$$

or

$$R_B = \frac{50\cdot6}{7} = 7\cdot23 \text{ t}$$

$$F_{max} = +R_B = 7\cdot23 \text{ t} \quad \textbf{Ans.}$$

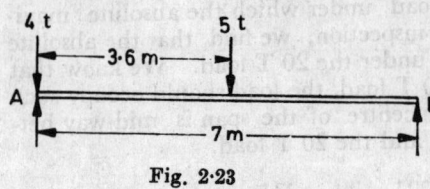

Fig. 2·23

Maximum negative S.F.

    We know that the maximum negative S.F. will take place at A, when the *4 t load is at A as shown in Fig. 2·23.

Taking moments about B,

$$R_A \times 7 = 4\times7 + 5\times3\cdot4 = 45$$

or

$$R_A = \frac{45}{7} = 6\cdot43 \text{ t}$$

$$\therefore \qquad F_{max} = -R_A = -\textbf{6·43 t} \quad \textbf{Ans.}$$

Maximum B.M.

    First of all let us find out the *c.g.* of the load system. From the geometry of the load system, we find that the distance between the *c.g.* of the load system and 2 t load,

$$x = \frac{4\times2\cdot4 + 5\times6}{2+4+5} = \frac{39\cdot6}{11} = 3\cdot6 \text{ m}$$

---

    *The other possibility of maximum negative B.M. is when the 2 t load is at A. In this case R_A will be equal to 5·34 t. Therefore the maximum negative B.M. will be equal to —5·34 t, which is less than that obtained when the 4 t load is at A.

Now, let us find out the load, under which the absolute maximum B.M. will take place. By inspection, we find that the absolute maximum B.M. will take place either under the 4 t load (because it is very near the *c.g.* of the load system) or under the 5 t load (because it is the heavier load near the *c.g.* of the load system). Here we shall find out the maximum B.M. under both the loads, one by one.

Now, for maximum B.M. under the 4 T load, the load should occupy such a position on the girder that the centre of the span is mid-way between the *c.g.* of the load system and the 4 T load.

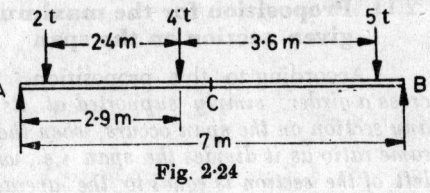

Fig. 2·24

Therefore the 4 T load should be at a distance of $3 \cdot 5 - \frac{1}{2}(3 \cdot 6 - 2 \cdot 4) = 2 \cdot 9$ m from A as shown in Fig. 2·24.

Taking moments about A,

$$R_B \times 7 = 2 \times 0 \cdot 5 + 4 \times 2 \cdot 9 + 5 \times 6 \cdot 5 = 45 \cdot 1$$

or

$$R_B = \frac{45 \cdot 1}{7} = 6 \cdot 44 \text{ t}$$

$\therefore$

$$M_{max} = 6 \cdot 44 \times 4 \cdot 1 - 5 \times 3 \cdot 6 = 8 \cdot 4 \text{ t-m} \qquad \ldots(i)$$

Now, for maximum B.M. under 5 t load, the load should occupy such a position on the girder that the centre of the span is mid-way between the *c.g.* of the load system and 5 t load.

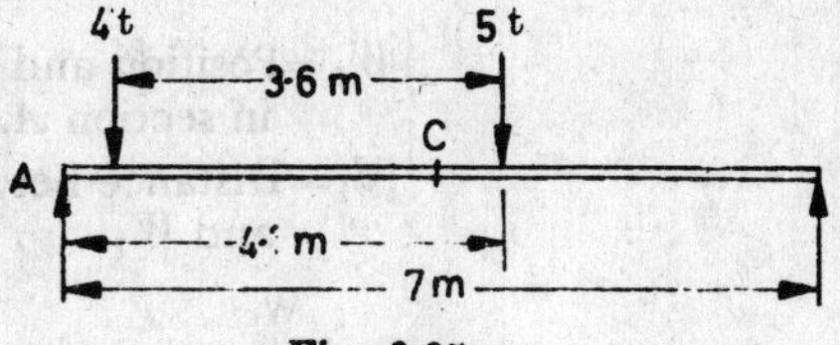

Therefore the 5 t load should be at a distance of $3 \cdot 5 + \frac{1}{2}(3 \cdot 6 - 2 \cdot 4) = 4 \cdot 1$ m from A as shown in Fig. 2·25.

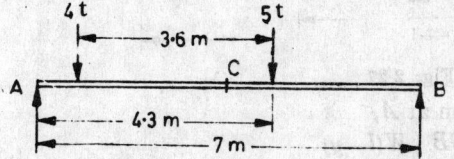

Fig. 2·25

Since the 2 t load will be out of the span, therefore we have to calculate the *c.g.* of the load system once again. From the geometry of the two loads, we find that the distance between the *c.g.* of the load system and 4 T load,

$$x = \frac{5 \times 3 \cdot 6}{4 + 5} = \frac{18}{9} = 2 \text{ m}$$

Fig. 2·26

Therefore the 5 t load should be at a distance of $3 \cdot 5 + \frac{1}{2}(3 \cdot 6 - 2 \cdot 0) = 4 \cdot 3$ m from A as shown in Fig. 2·26. It may be noted that even now the 2 t load is out of the span.

Taking moments about A,

$$R_B \times 7 = 4 \times 0.7 + 5 \times 4.3 = 24.3$$

or
$$R_B = \frac{24.3}{7} = 3.47 \text{ t}$$

$\therefore$
$$M_{max} = 3.47 \times 2.7 = 9.37 \text{ t-m} \qquad \qquad ...(ii)$$

Thus the absolute maximum B.M. equal to 9·37 t-m will take place under the 5 t load, when it is 4·3 m from A. **Ans.**

2·11. Proposition for the maximum bending moment at any given section on the span

According to this proposition, *"When a series of point loads cross a girder, simply supported at its ends, the maximum B.M. at any section on the span occurs, when the section divides the load in the same ratio as it divides the span i.e., when the average loading on the left of the section is equal to the average loading on the right of the section."*

Proof

Now consider several point loads W_1, W_2......rolling from A to B. Let the given section be X, at a distance x from A, where maximum bending moment is required to be found out.

Let
$W = $ Total load on the span AB,

$G = $ Centre of gravity of the total load on the span at a distance y from A,

$W_1 = $ Position and magnitude of the total load in section AX, and

$d_1 = $ Distance between the c.g. of the total load and W_1.

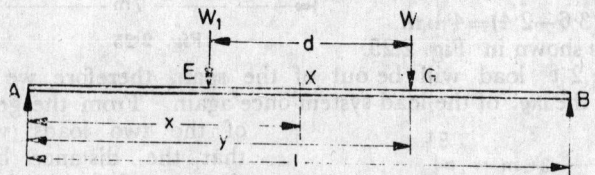

Fig. 2·27

We know that the reaction at A,

$$R_A = \frac{W.GB}{AB} = \frac{W(l-y)}{l}$$

and   B.M. at X, $M_X = R_A . x - W_1[x-(y-d)]$

$$= \frac{W(l-y)x}{l} - W_1[x-(y-d)]$$

$$= \frac{W}{l}(lx-xy) - W_1[x-(y-d)]$$

$$= Wx - \frac{Wxy}{l} - W_1x + W_1y - W_1d$$

Now, for maximum B.M. at X, let us differentiate the above equation with respect to x and equate it to zero.

i.e.
$$\frac{dM_x}{dy}\left[Wx - \frac{Wxy}{l} \times W_1x + W_1y - W_1d \right] = 0$$

$$-\frac{Wx}{l} + W_1 = 0$$

or
$$W_1 = \frac{Wx}{l}$$

or
$$\frac{W_1}{x} = \frac{W}{l} = \frac{W - W_1}{l - x}$$

Expressing the above equation geometrically, we get

$$\frac{W_1}{AX} = \frac{W}{AB} = \frac{W - W_1}{XB}$$

It is thus obvious, that the average loading on the left of the section is equal to the average loading on the right of the section. Or in other words, *the section divides the load in the same ratio as it divides the span.*

In actual practice with point loads, it is rarely possible that the average loading in section AX is exactly equal to the average loading in section XB. Generally, the loading in the two sections will be such that it is either heavier in AX and lighter in XB or *vice versa.* With the passage of a load over the section X the above condition may alter *i.e.* the loading in section AX may become lighter from heavier, and consequently the loading in section XB may become heavier from lighter. In such a case, the maximum B.M. on the section will occur, when this load is on the section.

Example 2·11. *Fig. 2·28 shows a train of wheel loads crossing a beam AB of 12 m span from left to right with 20 T load leading.*

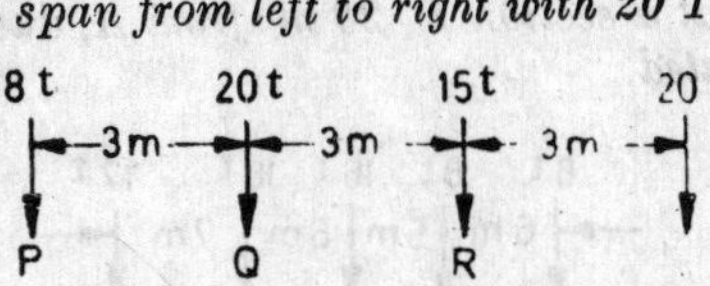

Fig. 2·28.

Determine the magnitude of maximum B.M. at a section X, 4·5 m from the left support.

Solution.

Given. Span, $\quad l = 12$ m

Distance of the section X and the left end support,
$$x = 4·5 \text{ m}$$

First of all let us find out the load, which should be placed at the section for maximum B.M. For doing so, we can assume the loads to cross the section, one be one, and study the average loading on both sides of the section as discussed below :

Load crossing the section	Average loading in AX	Average loading in XB	Remarks
S(20 T)	$\dfrac{43 \times 2}{9}$	$\dfrac{20 \times 2}{15}$	Average loading in AX is heavier
R(15 T)	$\dfrac{28 \times 2}{9}$	$\dfrac{35 \times 2}{15}$	Average loading in AX is heavier
Q(20 T)	$\dfrac{8 \times 2}{9}$	$\dfrac{55 \times 2}{15}$	Average loading in AX is lighter

It is thus obvious, that when the load Q of 20 T will be at the section, it will cause maximum B.M. at the section X. Therefore the load system should occupy a position as shown in Fig. 2·29.

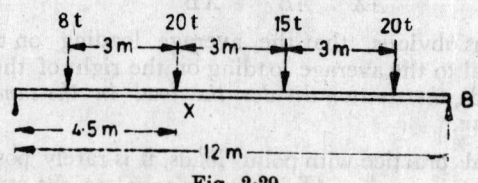

Fig. 2·29.

Taking moments about B,

$$R_A \times 12 = 8 \times 10\cdot5 + 20 \times 7\cdot5 + 15 \times 4\cdot5 + 20 \times 1\cdot5 = 331\cdot5$$

or $$R_A = \frac{331\cdot5}{12} = 27\cdot625 \text{ T}$$

∴ Maximum B.M. at X,
$$M_{max} = 27\cdot625 \times 4\cdot5 - 8 \times 3 = \mathbf{100\cdot31 \text{ t-m}} \quad \mathbf{Ans.}$$

Example 2·12. *For the span shown in the sketch below, obtain the bending moment at a section P, 20 m from A, due to the given loads in the position indicated.*

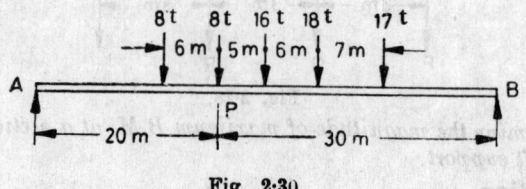

Fig. 2·30

Also determine the position of the loads, for maximum bending moment at section P and the value of the maximum moment.

Solution.

Given.   Span $l = 50$ m

Distance of the section P and the left support
$$x = 20 \text{ m}$$

Bending moment at P with the loads in position

Taking moments about B,

$$R_A \times 50 = 8 \times 36 + 8 \times 30 + 16 \times 25 + 18 \times 19 + 17 \times 12$$
$$= 1,474$$

$$R_A = \frac{1,474}{50} = 29 \cdot 48 \text{ T}$$

$\therefore$ B.M. at P, $M_P = R_A \times 20 - 8 \times 6 = 29 \cdot 48 \times 20 - 8 \times 6 \text{ t-m}$

or $= \textbf{541·6 t-m Ans}.$

Maximum bending moment at P

First of all, let us find out the load, which should be placed at the section for maximum B.M. For doing so, we can assume the loads to cross the section P, one by one, and study average loading on both the sides of the section as discussed below :

Load crossing the section	Average loading in AP	Average loading in PB	Remarks
17 T	$\dfrac{50}{20}$	$\dfrac{17}{30}$	Average loading in AP is heavier
18 T	$\dfrac{32}{20}$	$\dfrac{35}{30}$	Average loading in AP is heavier
16 T	$\dfrac{16}{20}$	$\dfrac{51}{30}$	Average loading in AP is lighter

It is thus obvious, that when the 16 T load will be at the section, it will cause maximum B M. at the section P. Therefore the load system should occupy a position as shown in Fig $2 \cdot 31$.

Taking moments about B,
$$R_A \times 50 = 8 \times 41 + 8 \times 35 + 16 \times 30 \times 18 \times 24 \times 17 \times 17$$
$$= 1,809$$

$\therefore$ $R_A = \dfrac{1,809}{50} = 36 \cdot 18 \text{ T}$

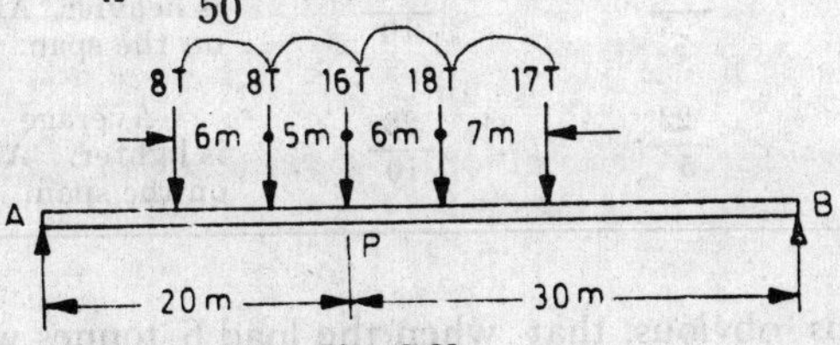

Fig. 2·31

$\therefore$ Maximum B.M. at P,
$$M_{max} = R_A \times 20 - (8 \times 11 + 8 \times 5) \text{ t-m}$$

$$= 36 \cdot 18 \times 20 - (88 + 40) \text{ t·m}$$
$$= 595 \cdot 6 \text{ t·m Ans.}$$

Example 2·13. *A simply supported girder of 15 m span is traversed by a train of loads as shown in Fig. 2·32.*

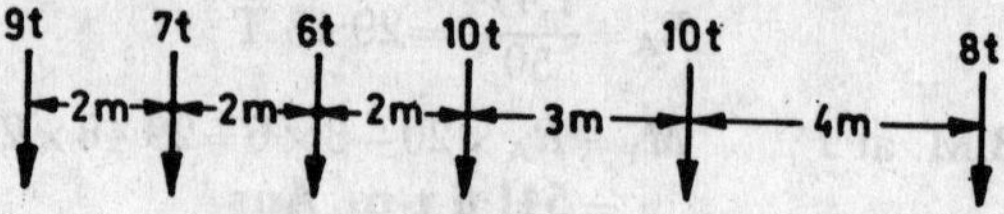

Fig. 2·32.

Determine the maximum bending moment at a section 5 m from the left support. Also find the magnitude of the absolute maximum bending moment anywhere in the girder.

Solution.

Given. Span, $l = 15$ m

Maximum bending moment at a section 5 m from the left support.

Distance of section X and the left support

$$x = 5 \text{ m}$$

First of all, let us find out the load, which should be placed at the section for the maximum B. M. For doing so, we can assume the loads to cross the section, one by one, and study the average loading on both sides of the section as discussed below :

Load crossing the section	Average loading gAX	Average loading in XB	Remarks
9 t	$\dfrac{9}{5}$	$\dfrac{33}{10}$	Average loading in XB is heavier. The 8 tonnes load is out of the span.
7 t	$\dfrac{16}{5}$	$\dfrac{34}{10}$	Average loading in XB is heavier. All the loads are on the span.
6 t	$\dfrac{22}{5}$	$\dfrac{28}{10}$	Average loading in XB is lighter. All the loads are on the span.

thus obvious, that when the load 6 tonnes will be at the secti   t will cause maximum B. M. at the section.

Therefore the load system should occupy a position as shown in Fig. 2·33.

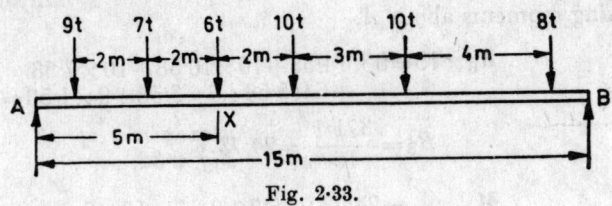

Fig. 2·33.

Taking moments about B,

$$R_A \times 15 = (9 \times 14) + (7 \times 12) + (6 \times 10) + (10 \times 8)$$
$$+ (10 \times 5) + (8 \times 1) = 408$$

or

$$R_A = \frac{408}{15} = 27·2 \text{ t}$$

∴   Maximum B. M. at X,

$$M_{max} = 27·2 \times 5 - 9 \times 4 - 7 \times 2 = 86·0 \text{ t-m} \quad \textbf{Ans.}$$

Absolute maximum bending moment

First of all, let us find out the *c. g.* of the load system. From the geometry of the load system, we find that the distance between the *c. g.* of the load system and the 9 tonnes load,

$$x = \frac{(7 \times 2) + (6 \times 4) + (10 \times 6) + (10 \times 9) + (8 \times 13)}{9 + 7 + 6 + 10 + 10 + 8} \text{ m}$$

$$= \frac{292}{50} = 5·84 \text{ m}$$

Now let us find out the load under which the absolute maximum B. M. will take place. By inspection, we find that the absolute maximum B. M. will take place under the first 10 tonnes load (because it is a heavier load very near the *c. g.* of the load system). We know that for maximum B. M. under this 10 tonnes load, the load should occupy such a position on the girder that the centre of the girder is mid way between the *c. g.* of the load system and the 10 tonnes load.

Therefore 10 tonnes load should be at a distance of
$$7·50 + \tfrac{1}{2}(6·0 - 5·84) = 7·58 \text{ m as shown in Fig. } 2·34.$$

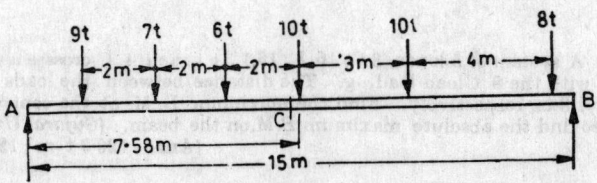

Fig. 2·34.

Taking moments about A,

$$R_B \times 15 = 8 \times 14 \cdot 58 + 10 \times 10 \cdot 58 + 10 \times 7 \cdot 58$$
$$+ 6 \times 5 \cdot 58 + 7 \times 3 \cdot 58 + 9 \times 1 \cdot 58 = 371 \cdot 0$$

or

$$R_B = \frac{371 \cdot 1}{15} = 24 \cdot 73 \text{ t}$$

$\therefore$

$$M_{maxmax} = 24 \cdot 73 \times 7 \cdot 42 - 8 \times 7 - 10 \times 3 \text{ t-m}$$
$$= 97 \cdot 5 \text{ t-m Ans.}$$

Exercise 2·2

1. A train of wheel loads as shown in Fig. 2·35 cross a simply supported beam of span 25 m left to right with the 20 t' load leading.

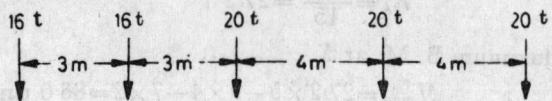

16 t   16 t   20 t   20 t   20 t

|← 3 m →|← 3 m →|← 4 m →|← 4 m →|

Fig. 2·35

Determine the maximum B. M. under the central load.

(*Calcutta University*)
[**Ans.** 384 t-m)

2. A load system shown in Fig. 2·36 crosses a girder of span 24 m from the left to right with the 15 t load leading.

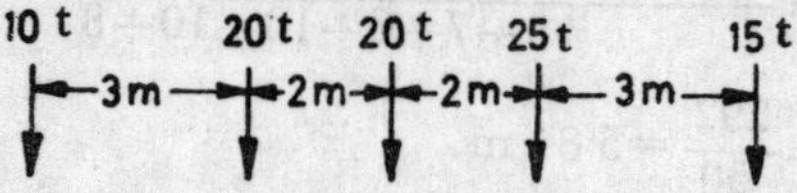

10 t   20 t   20 t   25 t   15 t

|←3 m→|←2 m→|←2 m→|←3 m→|

Fig. 2·36

Find the value of maximum B.M. under the 25 T load. (*Delhi University*)
[**Ans** 494 t-m]

3. The following load system crosses a girder of 25 m span with the 12 t load leading.

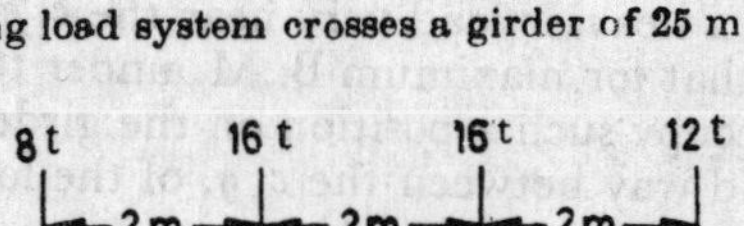

8 t   16 t   16 t   12 t

|←2 m→|←2 m→|←2 m→|

Fig. 2·37

Determine the value of (*i*) maximum B.M. at a section 8 m from left end of the girder and (*ii*) absolute maximum B.M. on the girder.

4. A system of 5 loads, 8 t, 16 t, 16 t, 6 t and 4 t crosses a beam of 15 m span with the 8 t load leading. The distance between the loads are 2·4, 3·0, 2·4 or 1·8 m respectively. Find the maximum B. M. at the centre of the span. Also find the absolute maximum B.M. on the beam. (*Gujarat University*)
[**Ans.** 126·3 t-m ; 127·5 t-m]

2·12. Equivalent uniformly distributed load

We have seen in the preceding articles, that when a system of loading rolls over a girder, it causes S.F. and B.M. over the span. The given system of loading can always be replaced by a static uniformly distributed load equal to the length of the span, such that the B.M. or S.F. at any section under this static load is the same (or not less than) as obtained due to the system of rolling loads. Such a static load is called *equivalent uniformly distributed load.*

As a matter of fact, the equivalent uniformly distributed load will have different values for B.M. and S.F. for the same given system of loading. We know, that the B.M. diagram for a uniformly distributed load is a parabola and the S.F. diagram is a straight line. This equivalent uniformly distributed should be such that the parabolic B.M. diagram and the straight line S.F. diagram, due to it, entirely envelops the actual B.M. and S.F. diagrams due to the given system of rolling loads. Once the B.M. and S.F. diagrams for the given system of rolling loads have been obtained, it is easy to obtain the equivalent uniformly distributed load. This is done, by drawing a parabola and the straight lines so as to envelop the respective diagrams, under the actual system of loading. Then the uniformly distributed load is calculated, which will have the enveloping figures as their B.M and S.F. diagrams respectively.

The equivalent uniformly distributed load is very useful in the design offices. The design procedure is very much simplified, if the actual loading is replaced by an equivalent uniformly distributed load.

Though there are many methods for finding out the equivalent uniformly distributed loads for various types of load systems, yet the following simple method is that which is widely used, and internationally recognised.

Let $M_{max\ max}$=Absolute maximum B.M. in the span,

l=Length of the span,

w'=Equivalent uniformly distributed load per unit length.

Now equating the B.M. produced by the equivalent uniformly distributed load to the absolute maximum B.M.

i.e. $$\frac{w'l^2}{8} = M_{max\ max}$$

$$\therefore \quad w' = \frac{8 \times M_{max\ max}}{l^2}$$

Example 2·14. *A uniformly distributed load of 2 t/m and 4 metres long rolls over a girder of 16 m span. Find the equivalent uniformly distributed load.*

Solution.

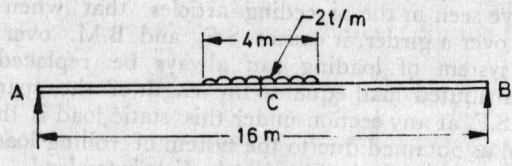

Fig. 2·38

Given.   Load, $w = 2$ t/m
Length of load, $a = 4$ m
Span, $l = 16$ m
Let $w' =$ Equivalent uniformly distributed load.

We know that the absolute maximum B.M. in the span will occur, when the load is occupying central position over the girder as shown in Fig. 2·35.   Therefore reaction at A,

$$R_A = R_B = \frac{4 \times 2}{2} = 4 \text{ T}$$

∴   Absolute maximum B.M.

$$M_{max\ max} = 4 \times 8 - \frac{2 \times 2^2}{2} = 28 \text{ t-m}$$

Using the relation,

$$w' = \frac{8 \times M_{max\ max}}{l^2} \quad \text{with usual notations.}$$

$$= \frac{8 \times 28}{16 \times 16} = \mathbf{0·875 \text{ t/m}} \quad \mathbf{Ans}.$$

Example 2·15.   *A girder having a span of 18 metres is simply supported at the ends.   It is traversed by train of loads as shown in Fig. 2·36, the 10 T load is leading.*

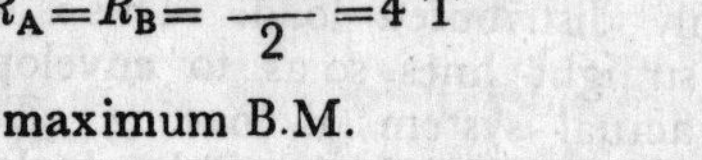

15 t   15 t   25 t   10 t
├──3 m──┼──3 m──┼─3 m─┤

Fig. 2·39

Find (a) maximum S.F. at 6 m and 9 m from the left support,

(b) the maximum B.M. in the girder, and

(c) the E.U.D. load to give the maximum B.M.

Solution.

Given.   Span, $l = 18$ m

Maximum positive S.F. at 6 m from the left support

We know that the maximum positive S.F will take place, when the leading load of 10 T is at the section itself.   But, in this case there is every* possibility that the maximum positive S.F. may take place, when the second load of 25 t is at the section itself. Here we shall find out the maximum, S.F. in both cases, one by one.

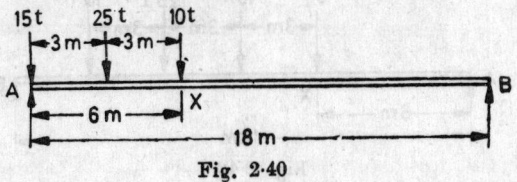

Fig. 2·40

Now consider the leading load of 10 ^{t} to be on the section as shown in Fig. 2·37.

Taking moments about A,

$$R_B \times 18 = 10 \times 6 + 25 \times 3 = 135$$

or

$$R_B = \frac{135}{18} = 7 \cdot 5 \ ^t$$

∴

$$F_{max} = + R_B = 7 \cdot 5 \ t \qquad \qquad ...(i)$$

Similarly, consider the second load of 25 t to be on the section as shown in Fig. 2·41.

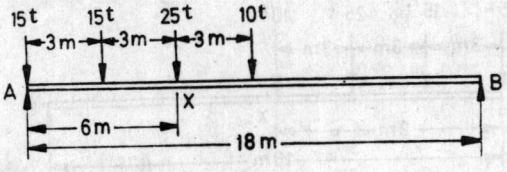

Fig. 2·41

Taking moment about A,

$$R_B \times 18 = 10 \times 9 + 25 \times 6 + 15 \times 3 = 285$$

or

$$R_B = \frac{285}{18} = 15 \cdot 83 \ t$$

∴

$$F_{max} = 15 \cdot 83 - 10 = 5 \cdot 83 \ t \qquad \qquad ...(ii)$$

Thus the maximum positive S.F. equal to 7·5 t will take place when the leading load of 10 ^{t} is at the section. **Ans.**

Maximum negative S.F. at 6 m from the left support

We know that the maximum negative S.F. will take place, when the trailing load of 15 t is at the section itself as shown in Fig. 2·42

Taking moments about B,

$$R_A \times 18 = 10 \times 3 + 25 \times 6 + 15 \times 9 + 15 \times 12 = 495$$

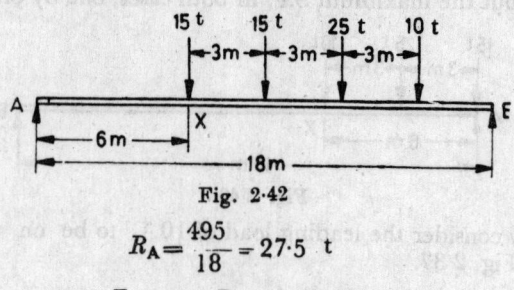

Fig. 2·42

or
$$R_A = \frac{495}{18} = 27\cdot5 \text{ t}$$

∴
$$F_{max} = -R_A = -27\cdot5 \text{ t} \quad \textbf{Ans}.$$

Maximum positive S.F. at 9 m from the left support

We know that the maximum positive S.F. will take place, when the leading load of 10 t is at the section itself. But in this case there is every possibility that the maximum positive S.F. may take place, when the second load of 25 t is at the section itself (because the heavier loads will come near the right support). Here we shall find out the maximum S.F. in both cases, one by one.

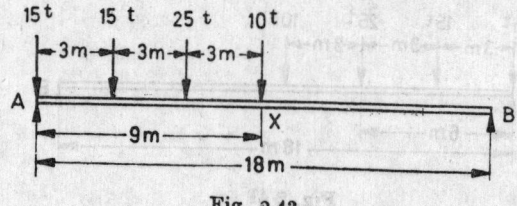

Fig. 2·43

Now consider the leading load of 10 t to be on the section as shown in Fig. 2·43

Taking moments about A,

$$R_B \times 18 = 10 \times 9 + 25 \times 6 + 15 \times 3 = 285$$

or
$$R_B = \frac{285}{18} = 15\cdot83 \text{ t}$$

∴
$$F_{max} = R_B = 15\cdot83 \text{ t}$$

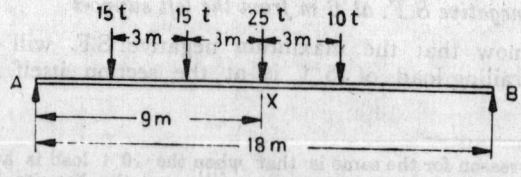

Fig. 2·44

Similarly consider the second load of 25 t to be on the section as shown in Fig. 2·44.

Taking moments about A,

$$R_B \times 18 = 10 \times 12 + 25 \times 9 + 15 \times 6 + 15 \times 3 = 480$$

or

$$R_B = \frac{480}{18} = 26 \cdot 67 \text{ }^t$$

$$\therefore \quad F_{max} = 26 \cdot 67 - 10 = 16 \cdot 67 \text{ }^t$$

Thus the maximum positive S.F. equal to 16·67 ^{t} will take place, when the second load of 25 ^{t} is at the section. **Ans.**

Maximum negative S.F. at 9 m from the left support

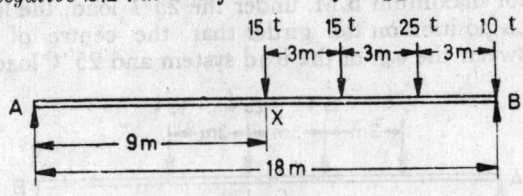

Fig. 2·45

We know that the maximum negative S.F. will take place, when the trailing load is at the section itself as shown in Fig. 2·45

Taking moments about B,

$$R_A \times 18 = 15 \times 9 + 15 \times 3 = 300$$

or

$$R_A = \frac{300}{18} = 16 \cdot 67 \text{ }^t$$

$$\therefore \quad F_{max} = -R_A = -\mathbf{16 \cdot 67 \text{ t Ans.}}$$

Maximum B.M. in the girder

First of all, let us find out the *c.g.* of the load system. From the geometry of the load system, we find that the distance between the *c.g.* of the load system and the trailing load of 15 ^{t},

$$x = \frac{15 \times 3 + 25 \times 6 + 10 \times 9}{15 + 15 + 25 + 10} = \frac{285}{65} = \frac{57}{13} = 4 \cdot 38 \text{ m}$$

Now let us find out the load, under which the maximum B.M. will take place. By inspection, we find that the absolute maximum B.M. will take place either under the last but one 15 $^{t'}$ load (because it is very near the *c.g.* of the load system) or under the 25 ^{t} load (because it is the heavier load near the *c.g.* of the load system). Here we shall find out the maximum B.M. under both the loads, one by one.

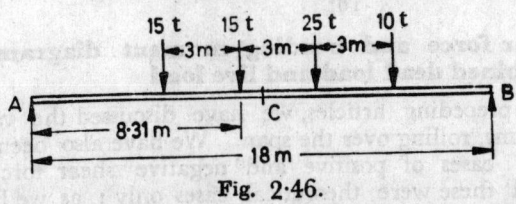

Fig. 2·46.

Now, for maximum B.M. under the last but one 15 '^{t}' load, the load should occupy such a position on the girder that the centre of the span is mid-way between the *c.g.* of the load system and the 15 't' load. Therefore the last but one 15 t load will be at a distance of $9-\frac{1}{2}(4\cdot38-3\cdot0)=8\cdot31$ m from A as shown in Fig. 2·43.

Taking moments about B,

$$R_A \times 18 = 15 \times 12\cdot69 + 15 \times 9\cdot69 + 25 \times 6\cdot69 + 10 \times 3\cdot69 = 539\cdot9$$

or

$$R_A = \frac{539\cdot9}{18} = 30 \text{ 't'}$$

$$\therefore \qquad M_{max} = 30 \times 8\cdot31 - 15 \times 3 = 204\cdot3 \text{ t-m} \qquad \text{...}(i)$$

Now, for maximum B.M. under the 25 't' load, the load should occupy such a position on the girder that the centre of the span is mid-way between the *c.g.* of the load system and 25 't' load.

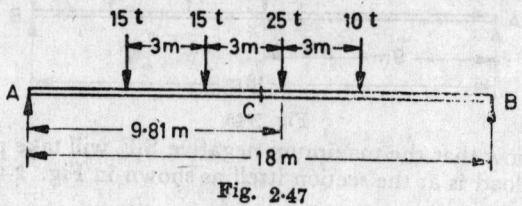

Fig. 2·47

Therefore the 25 T load will be at a distance of $9+\frac{1}{2}(6\cdot0-4\cdot38)$ $=9\cdot81$ m from A as shown in Fig. 2·47

Taking moments about A,

$$R_B \times 18 = 15 \times 3\cdot81 + 15 \times 6\cdot81 + 25 \times 9\cdot81 + 10 \times 12\cdot81 = 532\cdot65$$

or

$$R_B = \frac{532\cdot65}{18} = 29\cdot6 \text{ t}$$

$$\therefore \quad M_{max} = 29\cdot6 \times 8\cdot19 - 10 \times 3 = 212\cdot4 \text{ t-m} \qquad \text{...}(ii)$$

Thus absolute maximum B.M. *i.e.* $M_{max\ max}$ equal to 212·4 t-m will take place under the 25 t load, when it is 9·81 m from A. **Ans.**

Equivalent uniformly distributed load to give the maximum B.M.

Let $\qquad w' =$ Equivalent uniformly distributed load.

Using the relation,

$$w' = \frac{8 \times M_{max\ max}}{l^2} \text{ with usual notations.}$$

$$= \frac{8 \times 212\cdot4}{18^2} = 5\cdot24 \text{ t/m} \quad \text{Ans.}$$

2·13 Shear force and bending moment diagrams for the combined dead load and live load

In the preceding articles, we have discussed the various types of load systems, rolling over the span. We have also been discussing the possible cases of positive and negative shear forces. Strictly speaking, all these were theoretical cases only; as we have always neglected the self weight of the girder, which will also act as a uniformly distributed load over the entire span.

As a matter of fact, while designing bridge girders and trusses, both dead and live loads are taken into consideration to get the value of maximum B.M. and S.F. at any point. This is done by treating the dead and moving loads separately. The maximum B.M. and S.F. diagrams are drawn separately for the dead and moving loads. The combined diagrams are then obtained by adding the ordinates of the two sets of diagram algebraically. In the case of B.M. diagrams, both the dead and live loads give positive values at every section of a simply supported girder. Thus the combined diagram will also have positive ordinates at all points. But in the case of S.F. diagrams, there will be *positive as well as negative* S.F. due to moving load at every section, whereas there will be *either positive or negative* S.F. due to dead load at the section. It is thus obvious, that the combined S.F. diagram at any section may or may not have the same sign, as the loads move.

Now consider a simply supported beam AB subjected to a combined dead and moving loads.

Let $\qquad$ $l =$ Length of the span,

$\qquad$ $w =$ Uniformly distributed load due to self weight of the girder (*i.e.* dead load) per unit length

$\qquad$ $w' =$ Equivalent uniformly distributed load (*i.e.* live load) per unit length.

We know that the S.F. due to dead load at any section X at a distance x from A,

$$F_1 = -\frac{wl}{2} + wx \qquad \qquad ...(i)$$

We have also seen in art. 2·5 that the maximum positive S.F. due to live load at section X,

$$F_2 = +\frac{w'x^2}{2l} \qquad \qquad ...(ii)$$

and maximum negative S.F. due to live load at section X,

$$F_3 = -\frac{w'(l-x)^2}{2l} \qquad \qquad ...(iii)$$

Combined S.F. at section X due to dead and live load,

$$F_X = F_1 + F_2 = \left(-\frac{wl}{2} + wx \right) + \frac{w'x^2}{2l} \qquad ...(+ve\ S.F.)$$

$$= F_1 + F_3 = \left(-\frac{wl}{2} + wx \right) - \frac{w'(l-x)^2}{2l} \qquad ...(-ve\ S.F.)$$

We know that the B. M. at any section due to dead load,

$$\therefore \qquad M_1 = \frac{wlx}{2} - \frac{wx^2}{2} = \frac{wx}{2}(l-x)$$

We have already discussed in Art 2·5 that the B. M. at any

section, due to live load, will be maximum when the entire span is fully loaded. Therefore B. M. at any section due to live load,

$$M_2 = \frac{w' x}{2}(l-x)$$

∴ Combined B. M. at section X, due to dead load and live load,

$$M_x = M_1 + M_2 = \frac{wx}{2}(l-x) + \frac{w'x}{2}(l-x)$$

Example 2·16   *A beam of 20 m span has a dead load of 2 t/m. If a live load of 5 t/m, longer than the span, crosses the beam, find out the value of maximum positive as well as negative S.F. at a section 12 m from the left end support.*

Solution.

Given.   Span, $l = 20$ m
Dead load, $w = 2·0$ t|m
Live load, $w' = 5$ t/m

Distance of section X from the left end support,
         $x = 12$ m

Maximum positive S.F.

et $F_1 =$ S.F. due to dead load at the section,
            $F_2 =$ Positive S.F. due to live load.

Using the relation,

$$F_1 = -\frac{wl}{2} + wx \quad \text{with usual notations.}$$

$$= -\frac{2 \times 20}{2} + 2 \times 12 = +4 \; \text{t}$$

Now using the relation,

$$F_2 = \frac{w'x^2}{2l} \quad \text{with usual notations.}$$

$$= \frac{5 \times 12^2}{2 \times 20} = 18 \; \text{t}$$

∴ Combined maximum positive S.F.,

$$F_X = F_1 + F_2 = 4 + 18 = \textbf{22 t Ans.}$$

Maximum negative S. F.

Let $F_3 =$ Negative S.F. due to live load.

Using the relation,

$$F_3 = -\frac{w'(l-x)^2}{2l} \quad \text{with usual notations.}$$

$$= -\frac{5(20-12)^2}{2 \times 20} = -8·0 \; \text{t}$$

∴ Combined maximum negative S.F.,

$$F_X = F_1 + F_3 = 4 - 8 \cdot 0 = -4 \cdot 0 \text{ t Ans.}$$

2·14 Focal length due to combined dead load and live load

We have already discussed in Art. 2·13. that whenever a simply supported beam AB is subjected to a combined dead load and live load, the S.F. at a section will be positive as well as negative due to live load, whereas it will be either positive or negative due to the dead load. Now let us look into a practical aspect of this statement.

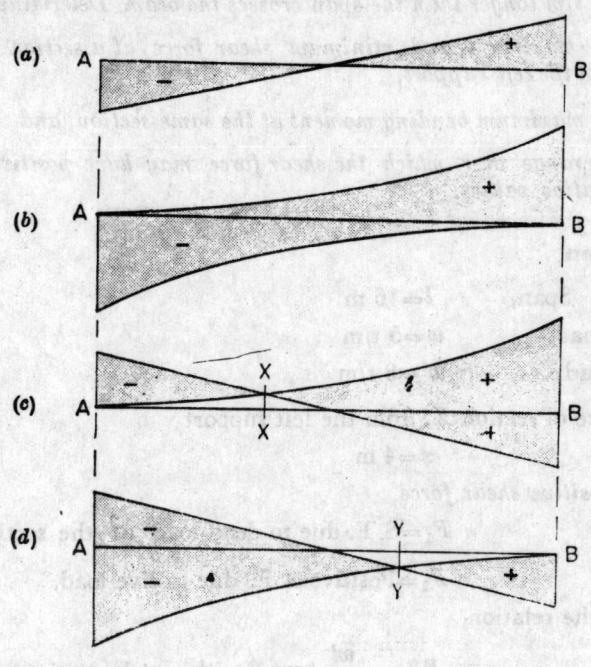

Fig. 245.

Consider a simply supported beam AB subjected to a combined uniformly distributed dead load and moving loads. In Fig. 2·48 (a) is shown the S.F. diagram due to uniformly distributed dead load. In Fig. 2·48(b) is shown the S.F. diagram due to moving loads. In Fig. 2·45 (c) is shown the combined S.F. diagram by algebraically adding the positive S.F. ordinates. We see that the maximum positive S.F. occurs for a length of BX of the beam. In Fig. 2·48 (d) is shown the combined S.F. diagram by algebraically adding the negative S.F. ordinates. We see that maximum negative S.F. occurs for a length AY of the beam. Thus we see that the beam AB can be divided into three zones namely AX, XY and YB. In zone AX, there will be always negative S.F. whereas in zone YB there will be always positive S.F. The central zone XY, in which the S.F. changes sign with the movement of the load, is known as *focal length* of the beam.

A little consideration will show that both the outer zones *i.e.*

AX and *BY* are of equal lengths. Therefore the focal length of a beam may be easily found out by locating the points *X* and *Y*. This can be done by equating the total S.F. due to dead load and the S.F. due to live load to zero.

It may be noted that in the case of trussed girders, the focal length plays an important role, the girder in the focal length needs counterbracing, so as to resist the changing S.F. more efficiently.

Example 2·17. *A simply supported beam of 16 m span is subjected to a uniform dead load of 5 t/m. In addition a uniform live load of 8 t/m longer than the span crosses the beam. Determine*

(a) *the maximum and minimum shear force at a section 4 m from the left support,*

(b) *the maximum bending moment at the same section, and*

(c) *the range over which the shear force may have positive or negative values.*

Solution.

Given.   Span, $l = 16$ m

Dead load, $w = 5$ t/m

Live load, $w' = 8$ t/m

Distance of section *X*, from the left support

$$x = 4 \text{ m}$$

Maximum positive shear force

Let $F_1 = $ S. F. due to dead load at the section,

and $F_2 = $ Positive S. F. due to live load.

Using the relation,

$$F_1 = -\frac{wl}{2} + wx \quad \text{with usual notations.}$$

$$= -\frac{5 \times 16}{2} + 5 \times 4 = -20 \text{ t}$$

Now using the relation,

$$F_2 = \frac{w'x^2}{2l} \quad \text{with usual relations.}$$

$$= \frac{8 \times 4^2}{2 \times 16} = 4 \text{ t}$$

∴ Combined maximum positive S. F.,

$$F_x = F_1 + F_2 = -20 + 4 = -16 \text{ t}$$

Thus the section will never be subjected to positive S. F. **Ans**.

Maximum negative shear force.

$$F_3 = \text{Negative S. F. due to live load.}$$

Using the relation,

$$F_3 = -\frac{w'\,(l-x)^2}{2l} \text{ with usual notations.}$$

$$= -\frac{8(16-4)^2}{2 \times 16} = -36 \text{ t}$$

$\therefore$ Combined maximum negative S. F.

$$F_x = F_1 + F_3 = -20 - 36 = \textbf{-56 t Ans}.$$

Minimum shear force

The shear force at the section will be minimum, when there is a live load on the beam, and is equal to shear force due to dead load only *i.e.*, -20 t **Ans**.

Maximum bending moment

Let $\qquad M_1 = $ B. M. due to dead load at the section

and $\qquad M_2 = $ B. M. due to live load,

Using the relation,

$$M_1 = \frac{wx}{2}(l-x) \text{ with usual notations.}$$

$$= \frac{5 \times 4}{2}\,(16-4) = 120 \text{ t-m}$$

Now using the relation,

$$M_2 = \frac{w'\,x}{2}\,(l-x) \text{ with usual notations}.$$

$$= \frac{8 \times 4}{2}\,(16-4) = 192 \text{ t-m}$$

$\therefore$ Combined maximum B. M.,

$$M_x = M_1 + M_2 = 120 + 192 = \textbf{312 t-m \quad Ans}.$$

The range over which the shear force may have positive or negative values.

Let $\qquad x = $ Distance from the left support, where the S. F. changes sign.

Again using the relation,

$$F_1 = -\frac{Wl}{2} + wx \text{ with usual notations.}$$

$$F_1 = -\frac{5 \times 16}{2} + 5x = -40 + 5x \qquad \ldots(i)$$

Now again using the relation.

$$F_2 = \frac{w'\, x^2}{2l} \text{ with usual notations.}$$

$$= \frac{8x^2}{2 \times 16} = \frac{x^2}{4} \qquad \ldots(ii)$$

Now adding equations (i) and (ii) and equating the same to zero,

$$-40 + 5x + \frac{x^2}{4} = 0$$

$$\therefore \qquad x^2 + 20x = 160$$

$$x^2 + 20x + 100 = (160 + 100) \qquad \text{(Adding 100 to both sides)}$$

$$(x+10)^2 = 260$$

$$\therefore \qquad x + 10 = 16 \cdot 12$$

or $$x = 6 \cdot 12$$

Therefore the range over which the shear force may have positive or negative values is from $6 \cdot 12$ m from the left support to $(16 - 6 \cdot 12) = 9 \cdot 88$ m from the left support. **Ans.**

Exercise 2-3

1. A load of 1 t/m for a length of 4 m rolls over a girder of 10 m span. Find the equivalent uniform distributed load. *(Rajasthan University)*
[**Ans.** 0·19 t/m]

2. A girder of span 16 m is subjected to a dead load of 3 t/m. Calculate the portion of the girder, for which S.F. changes sign, when an equivalent uniformly distributed load of 6 t/m crosses the girder. *(Agra University)*
[**Ans.** 4·3 m]

3. Determine the focal length of a girder of 16 m span subjected to a dead load of 3 t/m, when a uniformly distributed load of 2 t/m for a length of 4 m travels from left to right. *(Punjab University)*
[**Ans.** 1·72 m]

HIGHLIGHTS

1. A single concentrated load W rolling over the beam of span l will cause the maximum B.M. and S.F., on a section X at a distance x from A, when the load is on the section itself. Therefore

$$+F_{max} = +R_B = \frac{Wx}{l}$$

$$-F_{max} = -R_A = -\frac{W(l-x)}{l}$$

and $$M_{max} = \frac{Wx(l-x)}{l}$$

2. A uniformly distributed load of w/unit length, longer than the span, will cause maximum positive S.F. when the *head* of the load is at the section, and maximum negaive S.F. when the *tail* of the load is at the section. The load will cause maximum B.M., when the entire span is loaded.

3. A uniformly distributed load of w/unit length shorter than the span, will cause maximum positive S.F. when the *head* of the load is at the section and maximum negative S.F. when the *tail* of the load is at the section. The load will cause maximum B.M., when it occupies such as a position that the section divides the load in the same ratio, as it divides the span.

4. Two concentrated loads, spaced a distance d apart, will cause maximum positive S.F. when the leading load is at the section and maximum negative S.F. when the trailing load at the section. The maximum B.M. will take place under the heavier load, when it occupies such a position on the span that the centre of the span is mid-way between the *c.g.* of the two loads and the heavier load.

5. When a series of point loads crosses a girder, simply supported at its ends, the maximum B.M. under any given load occurs when the centre of the span is mid-way between the *c.g.* of the load system and the load under which the maximum B.M. is required to be found out.

6. When a series of point loads crosses a girder, simply supported at its ends, the maximum B.M. at any section on the span occurs, when the average loading on the left of the section is equal to the average loading on the right of the section.

7. Equivalent uniformly distributed load is that imaginary static load, which causes approximately the same B.M. or S.F. at any section as obtained due to the system of rolling loads. The equivalent uniformly distributed load,

$$w' = \frac{8 \times M_{max\ max}}{l^2}$$

where $M_{max\ max}$=Absolute maximum B.M., over the span,
               l=Length of the span.

8. The focal length of a girder is its that central portion, in which the S.F. changes sign due to rolling of the loads.

Do You Know ?

1. What do you understand by the term 'rolling loads ? Discuss their effects on a simply supported girder.

2. Derive an expression for the maximum S.F. and B.M. at a section produced by a single load, when it rolls over a girder.

3. State the position of a uniformly distributed load for maximum B.M. and S.F., when it crosses a girder of (*i*) smaller and (*ii*) longer length than that of the load.

4. Two point loads W_1 and W_2 at a distance d apart travel across a simply supported girder. Find the position, where the greatest B.M. anywhere in the girder will occur.

5. State and prove the propositions for (i) the maximum B.M. under any given load, and (ii) the maximum B.M. at any given section on the span.

6. What is meant by 'equivalent uniformly distributed load' ? How is it determined ?

7. Explain the term equivalent uniformly distributed load and indicate its practical application.

8. Discuss the combined effect of dead load and live load on the S.F., and B.M. diagrams of a girder.

9. Define 'focal length'. Derive an expression for the same and discuss its importance in the design of trussed girders.

10. What is meant by counterbracing and why it is done ?

3

INFLUENCE LINES

1. Introduction. 2. Uses of influence lines. 3. Influence lines for a single concentrated load. 4. Influence lines for a uniformly distributed load longer than the span. 5. Influence lines for a uniformly distributed load shorter than the span. 6. Influence lines for two concentrated loads. 7. Influence lines for several concentrated loads.

3·1. Introduction

In the last chapter, we have seen, that whenever a system of loading crosses a beam, the shear force (negative as well as positive) and bending moment at any section of the beam goes on changing, as the system of loading moves from one end to the other. We have also been drawing the S.F. and B.M. diagrams, showing the maximum values of these functions

In the modern design offices, it is the usual practice to plot curves showing the variation of S.F. and B.M. for one particular section. These curves are plotted on the span of a beam, as the base and the variable function as ordinates at various points along the span. The variable function may be shear force, bending moment deflection etc. Since these curves indicate the extent, to which a particular function at the given section is effected or influenced during the passage of the load system across the span, these are known as *influence lines.*

3·2. Uses of influence lines.

The study of influence lines is very interesting and extremely useful as well, as it enables speedily in determining the value of a function at the given section, when any system of loading rolls over the beam. It also enables us in determining the position of the load system, which will cause the maximum values of the function at that section. In the succeeding pages, we shall discuss the influence lines for the following cases of rolling loads :—

(1) A single concentrated load,
(2) A uniformly distributed load, longer than the span,
(3) A uniformly distributed load shorter than the span,
(4) Two concentrated or point loads, and
(5) General concentrated or point loads.

3·3 Influence lines for a single concentrated load

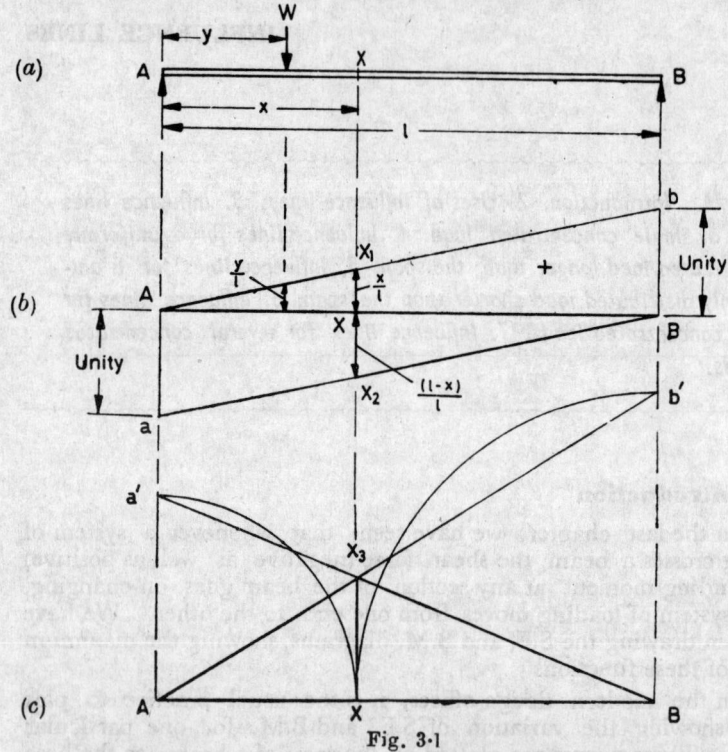

Fig. 3·1

Consider a concentrated load W rolling over the beam of span l from A to B. Let us consider a section X, at a distance x from A, where S.F. and B.M. is required to be found out.

(a) Shear force

We have already discussed in Art. 2·4 that the maximum positive and negative S.F. occurs, when the load W is on the section itself, and the values of maximum positive and negative S.F. are :

$$F_{max} = + R_B = \frac{Wx}{l} \qquad \ldots (+ve \text{ S.F.})$$

$$= - R_A = - \frac{W(l-x)}{l} \qquad \ldots (-ve \text{ S.F.})$$

We can also find out the values of maximum positive and negative S.F., from the influence line diagram as discussed as follows :

In order to plot the influence line diagram for the S.F., take *AB* as the base. Now erect perpendiculars *Aa* and *Bb* each equal to unity, to some suitable scale, on opposite sides of the base. Join *Ab* and *Ba*. Through the given section *X*, draw an ordinate cutting the parallel lines *Ab* and *Ba* at X_1 and X_2. Now AX_1X_2B is the required influence line for S.F. for the given section *X* as shown in Fig. 3·1 (*b*).

Now, the magnitude of S.F. (positive or negative) will be given by ordinate under the given load multiplied by the load itself. It is thus obvious, that the maximum positive and negative S.F. occurs, when the load is on the section itself. Now, from the influence line diagram, we see that the positive S.F. at section *X*, when the load is on the section.

$$F_{max} = +W \times XX_1 = W\frac{x}{l} = \frac{W.x}{l}$$

$$= -W \times XX_2 = -W\frac{l-x}{l} = -\frac{W(l-x)}{l}$$

This influence line diagram for S.F. can be utilized for finding out the S.F. at section *X*, for any load position.

(*b*) *Bending moment*

We have also seen in Art. 2·4 that the maximum B.M. at section *X* occurs when the load *W* is on the section itself and its value,

$$M_X = \frac{Wx(l-x)}{l}$$

Now plot the influence line diagram for the B.M. Take *AB* as the base and erect perpendiculars *Aa'* and *Bb'* on the same side of the base, such that *Aa'* is equal to *AX* and *Bb'* is equal to *XB*. Join *Ab'* and *Ba'* to intersect at X_3. Now AX_3B is the required influence line for the B.M., for the given section *X* as shown in Fig. 3·1 (*c*).

Now the magnitude of B.M. will be given by ordinate *XX_3 under the given load multiplied by the load itself. Thus B.M. at *X*,

$$M_X = W. XX_3 = \frac{Wx(l-x)}{l}$$

Example 3·1. *A single point load of 8 t crosses a girder of 12 m span. Using influence lines, find the maximum positive and negative S.F. and B.M. at a point 4 m from the left end.*

Solution.

Given. Load, $W = 8$ t

Span, $l = 12$ m

Distance between the left end of the girder and the section *X*, $x = 4$ m

---

*The length of ordinate XX_3 may be found out from the similar traingles $AX_3 X$ and $Ab' B$, in which

$$\frac{XX_3}{Bb'} = \frac{AX}{AB}$$

or $$XX_3 = \frac{AX . Bb'}{AB} = \frac{AX . XB}{AB} = \frac{x(l-x)}{l}$$

Maximum positive S.F.

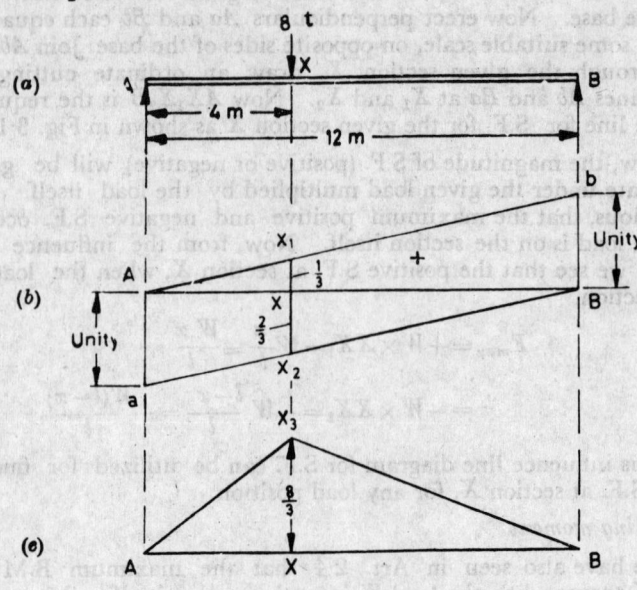

Fig. 3·2

First of all, draw the influence lines for the positive S.F. on the base AB with ordinate XX_1 equal to $\dfrac{x}{l} = \dfrac{4}{12} = \dfrac{1}{3}$ as shown in Fig. 3·2 (b).

We know that the maximum positive S.F. at X takes place, when the load is on the section itself.

$$\therefore \quad *F_{max} = +W \times XX_1 = 8 \times \frac{1}{3} \text{ t}$$

$$= 2·67 \text{ t Ans.}$$

Maximum negative S.F.

Now draw the influence lines for the negative S.F. on the base AB with ordinate XX_2, equal to $\dfrac{(l-x)}{l} = \dfrac{12-4}{12} = \dfrac{8}{12} = \dfrac{2}{3}$ as shown in Fig. 3·2 (b).

We know that the maximum negative S.F. at X takes place when the load is on the section itself.

$$\therefore \quad *F_{max} = -W \times XX_2 = 8 \times \frac{2}{3} \text{ t}$$

$$= -5·33 \text{ t Ans.}$$

---

*As per Art. 2·4 we know that maximum positive S.F. at X,

$$F_{max} = +\frac{Wx}{l} = 8 \times \frac{4}{12} = 2·67 \text{ t Ans.}$$

and the maximum negative S.F. at X,

$$F_{max} = -\frac{W(l-x)}{l} = -\frac{8(12-4)}{12} = -5·33 \text{ t Ans.}$$

Maximum B.M.

Now draw the influence lines for the B.M. on the base AB with ordinate XX_3 equal to $\dfrac{x(l-x)}{l} = \dfrac{4(12-4)}{12} = \dfrac{4 \times 8}{12} = \dfrac{8}{3}$ as shown in Fig. 3·2 (c). We know that maximum B.M. at X takes place when the load is on the section itself.

$$\therefore \qquad *M_{max} = W \times XX_3 = 8 \times \frac{8}{3} \text{ t-m}$$

$$= 21·33 \text{ t-m } \textbf{Ans.}$$

3·4. Influence lines for a uniformly distributed load longer than the span.

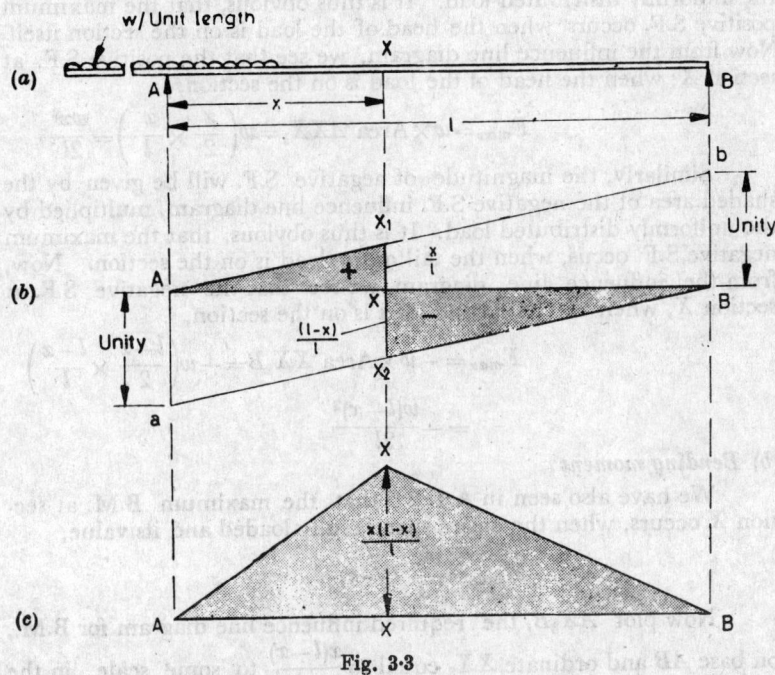

Fig. 3·3

Consider a uniformly distributed load of w/unit length, *longer* than the span l rolling from A to B. Let us consider a section X, at a distance x from A, where S.F. and B.M. is required to be found out.

(a) *Shear force*

We have already seen in Art. 2·5 that the maximum positive S.F. occurs, when the head of the load is on the section itself and the value of maximum positive S.F.

---

*As per Art. 2·4, we know that the maximum B.M. at X,

$$M_{max} = \frac{Wx(l-x)}{l} = \frac{8 \times 4(12-4)}{12} = \frac{8 \times 4 \times 8}{12} = 21·33 \text{ t-m Ans.}$$

$$F_{max} = +R_B = \frac{wx^2}{2l}$$

We have also discussed that the maximum negative S.F. occurs, when the tail of the load is on the section and the value of maximum negative S.F.

$$F_{max} = -R_A = -\frac{w(l-x)^2}{2l}$$

We can also find out the values of maximum positive and negative S.F., from the influence line diagram. Now draw AX_1X_2B the required influence line for S.F. for the given section X in the same way as discussed in Art. 3·3 as shown in Fig. 3·3 (b).

Now the magnitude of the positive S.F. will be given by the shaded area of the positive S.F. influence line diagram, multiplied by the uniformly distributed load. It is thus obvious, that the maximum positive S.F. occurs, when the head of the load is on the section itself. Now from the influence line diagram, we see that the positive S.F. at section X, when the head of the load is on the section,

$$F_{max} = w \times \text{Area } AXX_1 = w\left(\frac{x}{2} \times \frac{x}{l}\right) = \frac{wx^2}{2l}$$

Similarly, the magnitude of negative S.F. will be given by the shaded area of the negative S.F. influence line diagram, multiplied by the uniformly distributed load. It is thus obvious, that the maximum negative S.F. occus, when the tail of the load is on the section. Now, from the influence line diagram, we see that the negative S.F. at section X, when the tail of the load is on the section,

$$F_{max} = -w \times \text{Area } XX_2B = -w\left(\frac{l-x}{2} \times \frac{l-x}{l}\right)$$

$$= -\frac{w(l-x)^2}{2l}$$

(b) Bending moment

We have also seen in Art. 2·5 that the maximum B.M. at section X occurs, when the entire span is fully loaded and its value,

$$M_{max} = \frac{wx}{2}(l-x)$$

Now plot AX_3B, the required influence line diagram for B.M., on base AB and ordinate XX_3 equal to $\frac{x(l-x)}{l}$ to some scale, in the same way as discussed in Art. 3·3.

The magnitude of B.M. will be given by the shaded area of the B.M. influence line diagram multiplied by the uniformly distributed load. Now from the influence line diagram, we see that the maxmum B.M. at section X, when the entire span is fully loaded,

$$M_{max} = w \times \text{Area } AX_3B = w \times \frac{l}{2} \times \frac{x(l-x)}{l} = \frac{wx}{2}(l-x)$$

Example 3·2. *A uniformly distributed load of 5 t/m, longer than span, rolls over a beam of 25 m span. Using influence lines determine the maximum S.F. and B.M at a section 10 m from the left end support.*

Solution.

Given. Load, $w = 5$ t/m

Span, $l = 25$ m

Distance between left end of the beam and section X,

$$x = 10 \text{ m}$$

Maximum positive S.F.

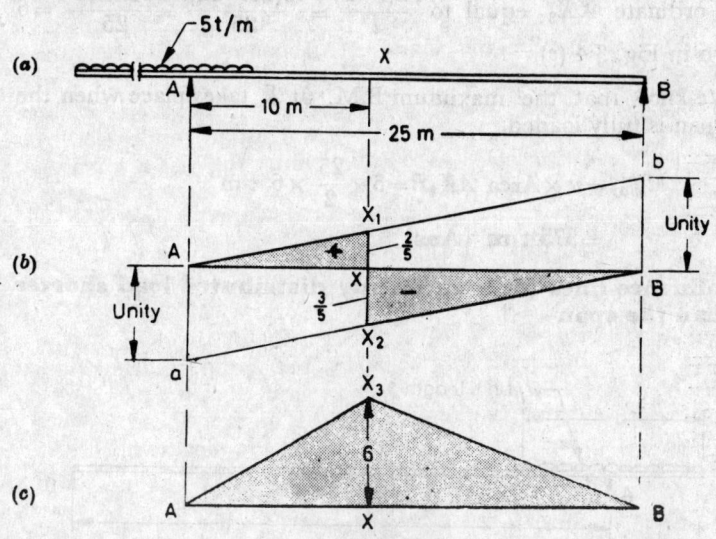

Fig. 3·4

First of all, draw the influence lines for the positive S.F. on the base AB with ordinate XX_1 equal to $\dfrac{x}{l} = \dfrac{10}{25} = \dfrac{2}{5}$ as shown in Fig. 3·4 (b).

We know that the maximum positive S.F. at X takes place, when the head of the load is on the section.

$$\therefore \qquad {}^*F_{max} = w \times \text{Area } AXX_1 = 5 \times \frac{10}{2} \times \frac{2}{5} \text{ t}$$

$$= 10 \text{ t Ans.}$$

Maximum negative S.F.

Now draw the influence lines for the negative S.F. on the base AB with ordinate XX_2 equal to $\dfrac{(l-x)}{l} = \dfrac{25-10}{25} = \dfrac{15}{25} = \dfrac{3}{5}$ as shown in Fig. 3·4(b).

We know that the maximum negative S.F. at X takes place, when the tail of the load is on the section.

---

*As per Art. 2·5 we know that the maximum positive S.F. at X,

$$F_{max} = +\frac{wx^2}{2l} = \frac{5 \times 10^2}{2 \times 25} \quad 10 \text{ t Ans.}$$

and maximum negative S.F. at X,

$$F_{max} = -\frac{w(l-x)^2}{2l} = -\frac{5(25-10)^2}{2 \times 25} = -\frac{5 \times 15^2}{2 \times 25} = -22·5 \text{ t Ans.}$$

$$\therefore \quad *F_{max} = -w \times \text{Area } XX_2B = -5 \times \frac{15}{2} \times \frac{3}{5} \text{ t}$$

$$= -22 \cdot 5 \text{ t} \quad \textbf{Ans}.$$

Maximum B.M.

Now draw the influence lines for the B.M. on the base AB with central ordinate XX_3 equal to $\dfrac{x(l-x)}{l} = \dfrac{10(25-10)}{25} = \dfrac{10 \times 15}{25} = 6$ as shown in Fig. 3·4 (c).

We know that the maximum B.M. at X takes place when the entire span is fully loaded.

$$\therefore \quad *M_{max} = w \times \text{Area } AX_3B = 5 \times \frac{25}{2} \times 6 \text{ t-m}$$

$$= 375 \text{ t-m} \quad \textbf{Ans}.$$

3·5. Influence lines for a uniformly distributed load shorter than the span

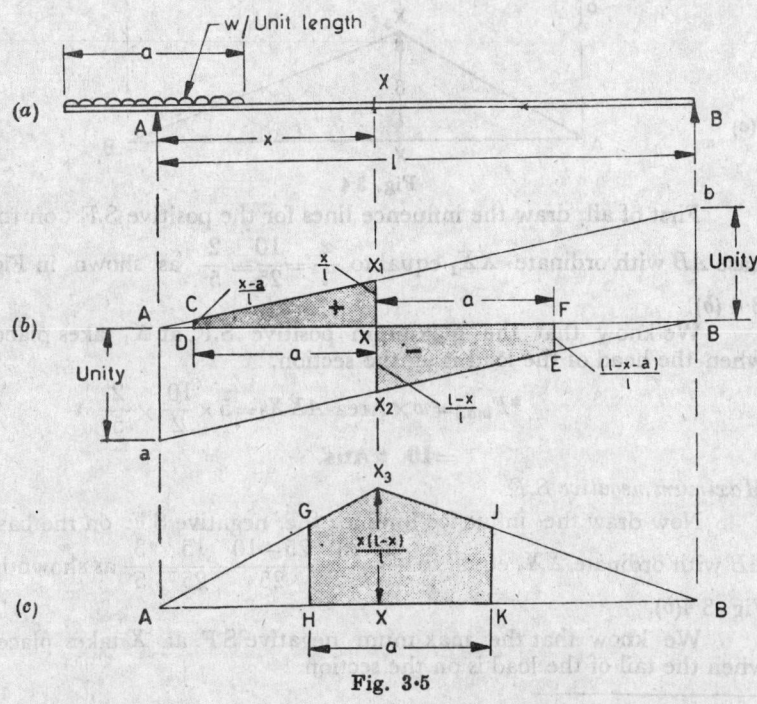

Fig. 3·5

---

*As per Art. 2·5, we know that the maximum B.M. at X,

$$M_{max} = \frac{wx}{2}(l-x) = \frac{5 \times 10}{2}(25-10) = \frac{5 \times 10 \times 15}{2} \text{ t-m}$$

$$= 375 \text{ t-m} \quad \textbf{Ans}.$$

Consider a uniformly distributed load of w/unit length, *shorter* than the span l rolling from A to B. Let us consider a section X, at a distance x from A, where S.F. and B.M. is required to be found out.

(a) *Shear force*

We have already seen in Art. 2·6 that the maximum positive S.F. occurs, when the head of the load is on the section itself and the value of maximum positive S.F.,

$$F_{max} = +R_B = +\frac{wa}{l}\left(x - \frac{a}{2}\right)$$

We have also discussed that the maximum negative S.F. occurs, when the tail of the load is on the section and the value of maximum negative S.F.

$$F_{max} = -R_A = -\frac{wa}{l}\left(l - x - \frac{a}{2}\right)$$

We can also find out the values of maximum positive and negative S.F. from the influence diagram.

Now draw AX_1X_2B the required influence lines for S.F. for the given section X as shown in Fig. 3·5 (b).

Now the magnitude of positive S.F. will be given by the shaded area of the positive S.F. influence line diagram multiplied by the uniformly distributed load. It is thus obvious, that the maximum positive S.F. occurs, when the head of the load is on the section itself. Now from the influence line diagram we see that the positive S.F. at section X, when the head of the load is on the section,

$$F_{max} = w \times \text{Area } CDXX_1 = w \times \left(\frac{x}{l} + \frac{x-a}{l}\right)$$

$$= \frac{wa}{l}\left(x - \frac{a}{2}\right)$$

Similarly, the magnitude of negative S.F. will be given by the shaded area of the negative S.F. influence line diagram multiplied by the uniformly distributed load. It is thus obvious, that the maximum negative S.F. occurs when the tail of the load is on the section. Now from the influence line diagram we see that the negative S.F. at section X, when the tail of the load on the section,

$$F_{max} = -w \times \text{Area } XX_2EF = w \times \frac{a}{2}\left(\frac{l-x}{l} + \frac{l-x-a}{l}\right)$$

$$= \frac{wa}{l}\left(l - x - \frac{a}{2}\right)$$

(b) *Bending moment*

We have also seen in Art. 2·6 that the maximum B.M. at the section X occurs, when the position of the load is such that the section X divides the load in the same ratio, as it divides the span l. This can also be proved from the influence line diagram for the

B.M. Now plot AX_3B the required influence line diagram for B.M. on base AB and ordinate XX_3 equal to $\dfrac{x(l-x)}{l}$ to same scale in the same way as discussed in Art. 3·3.

The magnitude of B.M. will be given by the shaded area of the B.M. influence line diagram multiplied by the uniformly distributed load. A little consideration will show that the area of the shaded portion will be maximum, when the load occupies such a position that the end ordinates GH and JK are equal.* Now from similar triangles AGH and AXX_3, we find that

$$\frac{GH}{XX_3}=\frac{AH}{AX}$$

or
$$GH=\frac{AH.XX_3}{AX}$$

and from similar triangles BJK and BXX_3, we find that

$$\frac{JK}{XX_3}=\frac{KB}{XB}$$

or
$$JK=\frac{KB.XX_3}{XB}$$

Since the ordinates GH and JK are equal, therefore

$$\frac{AH.XX_3}{AX}=\frac{KB.XX_3}{XB}$$

$$\therefore \quad \frac{AH}{A_X}=\frac{KB}{XB}$$

or
$$\frac{AX}{XB}=\frac{AH}{KB}$$

$$\therefore \quad \frac{AX}{XB}=\frac{AX-AH}{XB-KB}=\frac{HX}{XK}$$

i.e., the given cross section X divides the load in the same ratio, as it divides the span.

Example 3·3. *A simply supported beam has a span of 20 metres. A uniformly distributed load of 2 tonnes/m and 5 metres long crosses the span. Find the maximum B.M. produced at a point 8 metres from the left support.*

Solution

Given.   Span, $l=20$ m

Load, $w=2$ t/m

---

*If the load moves forward or backward from this position, the area of the shaded portion will decrease. This will happen, because the area added will be less than the area subtracted.

** We have already solved this example on page 22 (example 2·4). Here we shall solve the problem with the help of influence lines.

Length of load, $a = 5$ m

Distance of the section from the left support,

$x = 8$ m

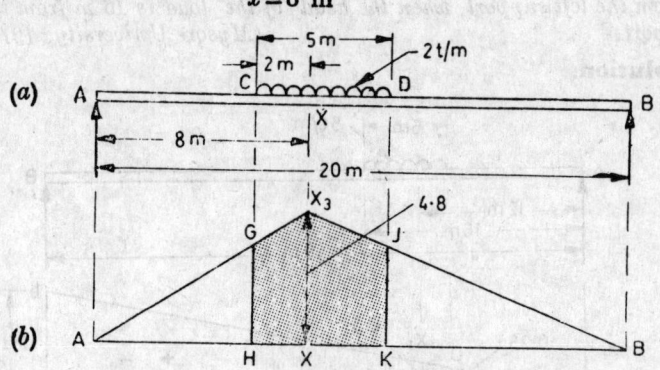

Fig. 3·6

We know that the maximum B.M. at the section takes place, when the position of the load is such that the section X divides the load in the same ratio, as it divides the span as shown in Fig. 3·6 (a)

$$\therefore \quad \frac{CX}{CD} = \frac{x}{l}$$

$$\frac{CX}{5} = \frac{8}{20}$$

or $$CX = \frac{8 \times 5}{20} = 2 \text{ m}$$

Now draw the influence lines for B.M. on the base AB with ordinate XX_3 equal to $\frac{x(l-x)}{l} = \frac{8(20-8)}{20} = \frac{8 \times 12}{20} = \frac{24}{5} = 4·8$ as shown in Fig. 3·6 (b). From the geometry of the figure, we find that

$$\frac{GH}{\frac{24}{5}} = \frac{8-2}{8} = \frac{6}{8}$$

$$\therefore \quad GH = \frac{6}{8} \times \frac{24}{5} = \frac{18}{5} = 3·6$$

We know that the maximum B.M. at X,

$$M_{max} = w \times \text{Area } GHKJ$$
$$= w(\text{Area } GHXX_3 + \text{Area } JKXX_3)$$
$$= 2\left[2 \times \frac{1}{2}\left(4·8 + 3·6 \right) + 3 \times \frac{1}{2}\left(4·8 + 3·6 \right) \right]$$
$$\left(\because \quad GH = JK = 3·6 \right)$$
$$= 2[8·4 + 12·6] = \textbf{42 t-m Ans.}$$

Example 3·4. *A uniformly distributed load of 5 t/m of 6 m length crosses a girder of span 40 m from left to right. With the help of influence lines, determine the values of S.F. and B.M. at a point 12 m from the left support, when the head of the load is 16 m from the left support.*   (*Mysore University, 1978*)

Solution.

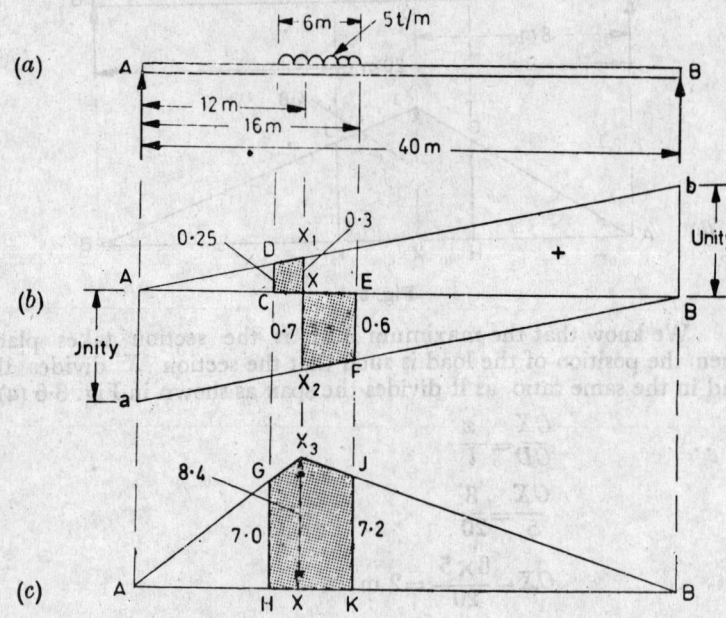

Fig. 3·7

Given.   Load, $w = 5$ t/m

Length of the load, $a = 6$ m

Span, $l = 40$ m

Distance of the point from the left support,

$$x = 12 \text{ m}$$

Distance of the head of the load from the left support

$$= 16 \text{ m}$$

Shear force

First of all, draw the influence lines for the positive as well as negative S.F. on base AB with ordinate XX_1 equal to $\dfrac{x}{l} = \dfrac{12}{40} = 0.3$

and ordinate XX_2 equal to $\dfrac{(l-x)}{40} + \dfrac{(40-12)}{40} = \dfrac{28}{40} = 0.7$ as shown in Fig. 3·7.

From the geometry of the figure, we find that

$$CD = \frac{3}{10} \times \frac{10}{12} = \frac{1}{4} = 0.25$$

and
$$EF = \frac{7}{10} \times \frac{24}{28} = \frac{6}{10} = 0.6$$

We know that the S.F. at X,

$$F_X = w[\text{Area } CDXX_1 - \text{Area } EFXX_2]$$
$$= 5[2 \times \tfrac{1}{2}(0.3 + 0.25) - 4 \times \tfrac{1}{2}(0.7 + 0.6)] \text{ t}$$
$$= 5[0.55 - 2.6] = -5 \times 2.05 \text{ t}$$
$$= \mathbf{-10.25 \text{ t}} \quad \textbf{Ans.}$$

Bending moment

Now draw the influence lines for the B.M. on the base AB with

ordinate XX_3 equal to $\dfrac{x(l-x)}{l} = \dfrac{12(40-12)}{40} = \dfrac{12 \times 28}{40} = 8.4$ as shown

in Fig. 3.7 (c).

From the geometry of the figure, we find that

$$GH = 8.4 \times \frac{10}{12} = 7.0$$

and
$$JK = 8.4 \times \frac{24}{28} = 7.2$$

We know that the B.M. at X,

$$M_X = w[\text{Area } GHXX_3 + \text{Area } JKXX_3]$$
$$= 5[2 \times \tfrac{1}{2}(7.0 + 8.4) + 4 \times \tfrac{1}{2}(8.4 + 7.2)] \text{ t-m}$$
$$= 5[15.4 + 31.2] = 5 \times 46.6 \text{ t-m}$$
$$= \mathbf{233 \text{ t-m}} \quad \textbf{Ans.}$$

3.6. Influence lines for two concentrated loads

Consider two concentrated loads W_1 and W_2, spaced d apart, rolling over the span l from A to B. Let us consider a section X, at a distance x from A, where S.F. and B.M. is required to be found out.

(a) Shear force

We have already seen in Art. 2.7 that the maximum positive S.F. occurs, when the leading load W_2 is on the section and the trailing load W_1 is in section AX at a distance d behind the load W_2. At this instant the value of maximum positive S.F.

$$F_{max} = +R_B = \frac{W_2 x + W_1 (x - d)}{l}$$

We have also discussed that the negative S.F. occurs, when the trailing load W_1 is on the section and leading load W_2 is in the section XB. At this instant the value of maximum S.F.

$$F_{max} = R_A = -\frac{W_1(l-x) + W_2(l-x-d)}{l}$$

We can also find out the values of maximum positive and negative S.F. from the influence line diagram. Now draw AX_1X_2B

the required influence line diagram for S.F. for the given section X in the same way as discussed in Art. 2·4 as shown in Fig. 3·3. The

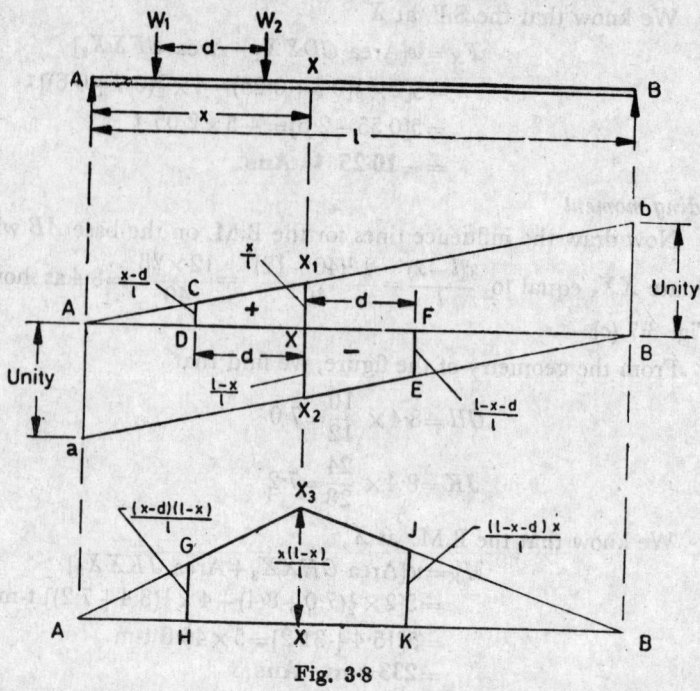

Fig. 3·8

magnitude of positive S.F. will be given by the sum of multiples of loads and the lengths of corresponding ordinates under them in the positive S.F. influence line diagram. It is thus obvious, that the maximum positive S.F. occurs, when the leading load W_2 is on the section and the trailing load W_1 is in section AX at a distance d behind the load W_2. Now, from the influence line diagram, we see that the maximum positive S.F. at section and X, when the leading load is at X,

$$F_{max} = W_2 \times XX_1 + W_1 \times CD$$

$$= W_2 \times \frac{x}{l} + W_1 \frac{(x-d)}{l}$$

$$= \frac{W_2 x + W_1 (x-d)}{l}.$$

Similarly, the magnitude of the negative S.F. will be given by the sum of multiples of loads and the lengths of corresponding ordinates, under them in the negative S.F. influence line diagram. It is thus obvious, that the maximum negative S.F. occurs when the trailing load W_1 is on the section and the leading load W_2 is in the section XB. Now, from the influence line diagram, we see that the negative S.F. at section X, when the trailing load is at X,

$$F_{max} = -(W_1 \times XX_2 + W_2 \times FE)$$

$$= -\left[W_1 \times \frac{(l-x)}{l} + W_2 \times \frac{(l-x-d)}{l} \right]$$

$$= -\frac{W_1(l-x) + W_2(l-x-d)}{l}$$

(b) Bending moment

We have also seen in Art. 2·7 that the maximum B.M. at section X occurs, under any one of the following two conditions :

(1) When the leading load W_2 is at X and the trailing load W_1 is in section AX at a distance d behind W_2. At this instant the B.M. at X,

$$2M_X = \frac{[W_2 x + W_1(x-d)](l-x)}{l} \qquad \text{...(i)}$$

(2) When the trailing load W_1 is at X and the leading load is in section XB. As this instant the B.M. at X,

$$1M_X = \frac{[W_1(l-x) + W_2(l-x-d)]x}{l} \qquad \text{...(ii)}$$

Now plot AX_3B the required influence line diagram for B.M. on base AB and ordinate XX_3 equal to $\frac{x(l-x)}{l}$ to some suitable scale, in the same way as discussed in Art. 3·3. The magnitude of B.M. will be given by the sum of multiple of loads and the lengths of corresponding ordinates under them, in the B.M. influence line diagram.

The B.M. at section X, when the leading load W_1 is at X and the trailing load W_1 is in section AX,

$$2M_X = W_2 \times XX_3 + W_1 \times GH$$

$$= W_2 \times \frac{x(l-x)}{l} + W_1 \times \frac{(x-d)(l-x)}{l}$$

$$= \frac{[W_2 x + W_1(x-d)](l-x)}{l} \qquad \text{...(iii)}$$

and the B.M. at section X, when the trailing load W_1 is on the section and the leading load is in section XB,

$$1M_X = W_1 \times XX_3 + W_2 \times JK$$

$$= W_1 \times \frac{x(l-x)}{l} + W_2 \times \frac{(l-x-d)x}{l}$$

$$= \frac{[W_1(l-x) + W_2(l-x-d)]x}{l} \qquad \text{...(iv)}$$

Example 3·5. *Two point loads of 8 t and 16 t spaced 2 m apart, cross a girder of span 10 m with the 8 t load leading from left to right. Draw the influence lines for S.F. and B.M. and find the value of maximum S.F. and B.M. at a section 4 m from the left end support.*

Solution.

Given. Leading load, $W_2 = 8$ t

Trailing load, $W_1 = 16$ t

Spacing of loads, $d = 2$ m

Span, $l = 10$ m

Distance between the left end of the girder and section X,

$$x = 4 \text{ m}$$

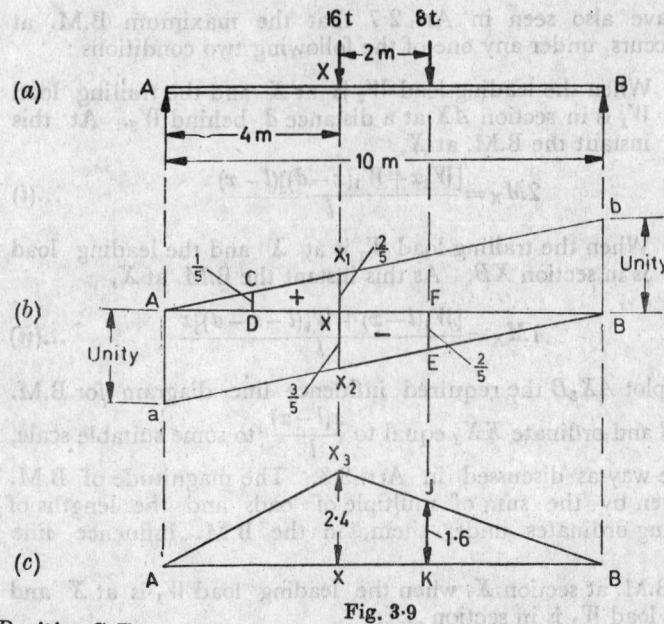

Fig. 3·9

Positive S.F.

First of all, draw the influence lines for the positive S.F. on the base AB, with ordinate XX_1 equal to $\dfrac{x}{l} = \dfrac{4}{10} = 0 \cdot 4$ as shown in Fig. 3·9 (b). From the geometry of the figure, we find that $CD = 0 \cdot 2$.

We know that the maximum positive S.F. as X takes place, when the leading load is X.

$$\therefore \qquad F_{max} = W_1 \times CD + W_2 \times XX_1 = 16 \times 0 \cdot 2 + 8 \times 0 \cdot 4 \text{ t}$$
$$= 6 \cdot 4 \text{ t} \quad \textbf{Ans.}$$

Negative S.F.

Now draw the influence lines for the negative S.F. on the base AB, with ordinate XX_2 equal to $\dfrac{(l-x)}{l} = \dfrac{10-4}{10} = \dfrac{6}{10} = 0 \cdot 6$ as shown in Fig. 3·9(b). From the geometry of the figure, we find that $EF = 0 \cdot 4$.

We know that the maximum negative S.F. at X takes place, when the trailing load is on the section.

$$\therefore \quad F_{max} = -[W_1 \times XX_2 + W_2 \times FE] = -[16 \times 0 \cdot 6 + 8 \times 0 \cdot 4] \text{ t}$$
$$= -12 \cdot 8 \text{ t} \quad \textbf{Ans.}$$

Bending moment

Now draw the influence lines for the B.M. on the base AB with ordinate XX_3 equal to $\dfrac{x(l-x)}{l} = \dfrac{4(10-4)}{10} = \dfrac{4 \times 6}{10} = 2 \cdot 4$ as shown in Fig. 3·9 (c). From the geometry of the figure, we find that $JK = 1 \cdot 6$.

We know that the maximum B.M. at X, takes place, when the trailing load is at X and the leading load is in XB.

$$\therefore \quad M_{max} = 16 \times XX_3 + 8 \times JK = 16 \times 2 \cdot 4 + 8 \times 1 \cdot 6 \text{ t·m}$$
$$= 51 \cdot 2 \text{ t·m} \quad \textbf{Ans.}$$

Exercise 3–1

1. A point load of 5 tonnes rolls over a girder of 20 m span. Draw the influence lines and find the values of maximum positive and negative S.F. and B.M. at a point 8 m from the left hand end. [**Ans.** $+2$ t ; -3 t ; 72 t-m]

2. A uniformly distributed load of 2 t/m longer than the span rolls over a beam of 15 m span. Using influence lines, determine the values of maximum positive and negative S.F. and B.M. at a point 5 m from the left end support.
[**Ans.** $+1 \cdot 67$ t ; $-6 \cdot 67$ t ; 50 t-m]

3. A uniformly distributed load of $9 \cdot 5$ t/m and 5 m long rolls across a beam of 15 m long simply supported at its both ends. Find the maximum S.F. and B.M. at a section 6 m from the left hand support. Also find the absolute maximum B.M. on the beam. (*Patna University*)
[**Ans.** $+0 \cdot 58$ t ; $-1 \cdot 08$ t ; $7 \cdot 5$ t·m ; $7 \cdot 81$ t-m]

3·7. Influence lines for several concentrated loads

Consider several concentrated or point loads W_1, W_2......rolling from A to B over a span l. Let us consider section X, at a distance x from A, where S.F. and B.M. is required to be found out.

(a) Shear force

We have already seen in Art. 2·8, that the maximum positive and negative S.F. is to be found out by trial and error only. However, one of the point loads is to be on the section itself, for the maximum positive and negative shear force.

We can also confirm it from the influence line diagram. Now draw AX_1X_2B the required influence line for S.F. for the given section X in the same way as discussed in Art. 2·4 as shown in Fig. 3·10(b). Now draw the ordinates through the lines of action of the loads W_1, W_2......Let the lengths of these ordinates be $f_1 f_1'$, $f_2 f_2'$ ......The magnitude of positive as well as negative S.F. will be given by the algebraic* sum of multiples of loads and the lengths of corresponding ordinates under them e.g., as per Fig. 3·10(a). The S.F. at X,

$$F_X = W_1 \cdot f_1 f_1' + W_2 \cdot f_2 f_2' - W_3 \cdot f_3 f_3' - W_4 \cdot f_4 f_4' - W_5 f_5 f_5'$$

---

*The lengths of ordinates in the *positive* S.F. influence line diagram are taken *positive*, whereas the lengths of ordinates in the *negative* S.F. influence line diagram are taken as *negative*.

(b) Bending moment

Now plot AX_3B the required influence line diagram for B.M. on base AB and ordinate XX_3 equal to $\dfrac{x(l-x)}{l}$ to some suitable scale

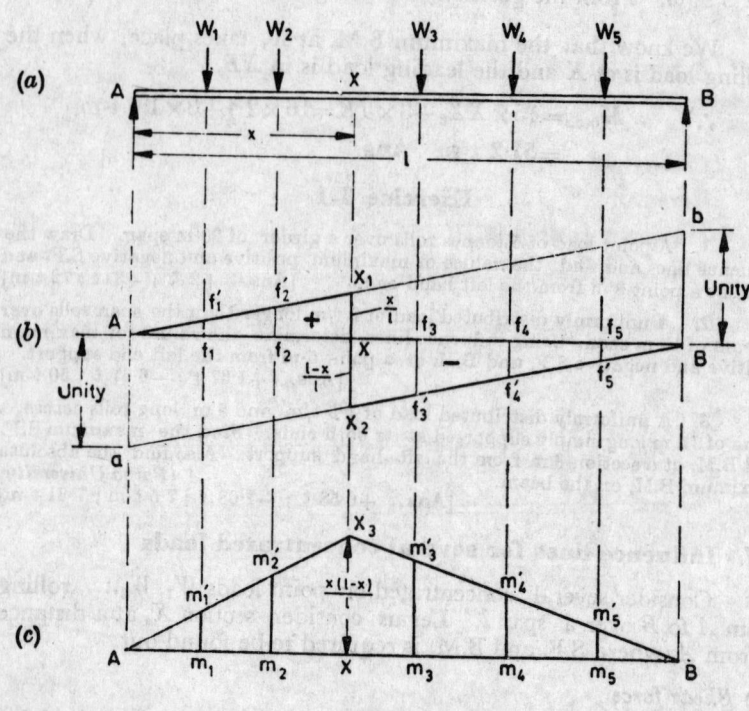

Fig. 3·10

in the same way as discussed in Art. 3·3. The magnitude of B.M. will be given by the sum of multiples of loads and the lengths of corresponding ordinates under them, e.g. as per Fig. 3·10(b) the B.M. at X,

$$M_X = W_1.m_1m_1' + W_2.m_2m_2' + W_3.m_3m_3' + W_4.m_4m_4' + W_5.m_5m_5$$

Note. The following two propositions (as already discussed in Arts. 2·9 and 2·10) hold good for influence lines also :

1. When a series of point loads crosses a girder simply supported at its ends, the maximum B.M. under any given load occurs, when the centre of the span is mid-way between the c.g. of the load system and the load under which the maximum B.M. is required to be found out.

2. When a series of point loads crosses a girder simply supported at its ends, the maximum B.M. at any section on the span occurs, when the section divides the load in the same ratio as it divides the span, i.e., when the average loading on the left of the section is equal to the average loading on the right of the section.

Examples 3·6. *A train of 5 wheel-loads as shown in Fig. 3·11 crosses a simply supported beam of span 22·5 metres.*

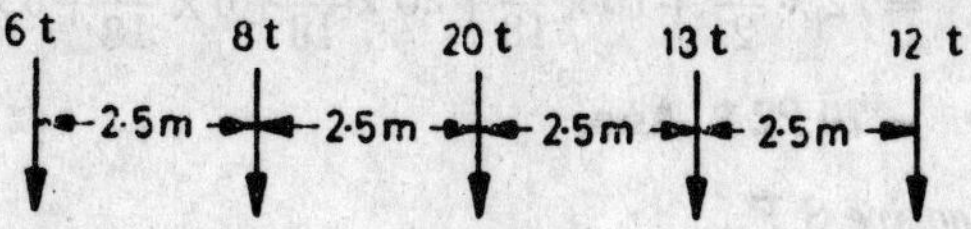

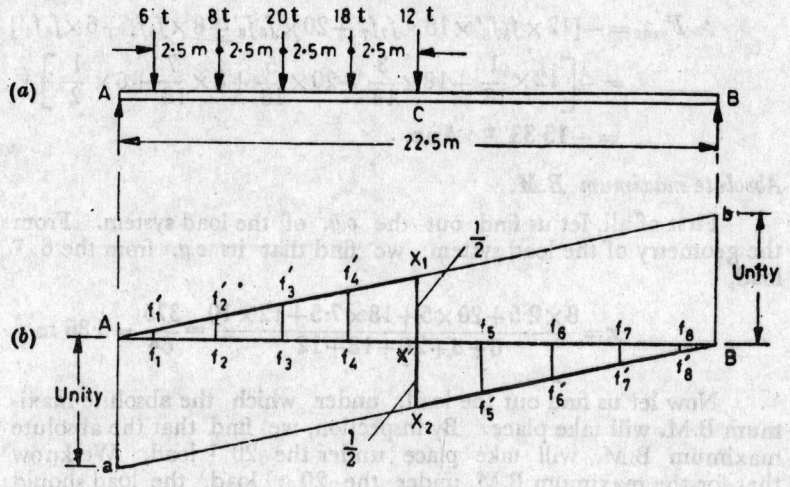

Fig. 3·11

Calculate the maximum positive and negative S.F. values at the centre of the span, and the absolute maximum B.M. anywhere in the span.

*Solution.

Given.   Span, $l = 22·5$ m

Maximum positive S.F.

First of all, draw the influence lines for the positive S.F. on the base AB, with ordinate XX_1 equal to $\dfrac{x}{l} = \dfrac{11·25}{22·5} = \dfrac{1}{2}$ as shown in Fig. 3·12 (b).

Fig. 3·12

From the geometry of the figure AXX_1, we find that $f_1f_1' = \dfrac{1}{18}$; $f_2f_2' = \dfrac{3}{18}$; $f_3f_3' = \dfrac{5}{18}$ and $f_4f'_4 = \dfrac{7}{18}$.

We know that the maximum positive S.F. at C occurs, when the leading load is at C.

*We have already solved this example on page 35 (example 2·9). Here we shall solve this problem with the help of influence lines.

$$\therefore\ F_{max} = 12 \times XX_1 + 18 \times f_4 f_4' + 20 \times f_3 f_3' + 8 \times f_2 f_2' + 6 \times f_1 f_1'$$

$$= 12 \times \frac{1}{2} + 18 \times \frac{7}{18} + 20 \times \frac{5}{18} + 8 \times \frac{3}{18} + 6 \times \frac{1}{18}\ \ t'$$

$$= 20 \cdot 22\ ^t\ \mathbf{Ans.}$$

Maximum negative S.F.

Now draw the influence lines for the negative S.F. on the base AB, with ordinate XX_2 equal to $\dfrac{(l-x)}{l} = \dfrac{(22 \cdot 5 - 11 \cdot 25)}{22 \cdot 5} = \dfrac{11 \cdot 25}{22 \cdot 5} = \dfrac{1}{2}$ as shown in Fig. 3·12(b).

From the geometry of the figure BXX_2, we find that $f_5 f_5' = \dfrac{7}{18}$;

$f_6 f_6' = \dfrac{5}{18}$; $f_7 f_7' = \dfrac{3}{18}$ or $f_8 f_8' = \dfrac{1}{18}$.

We know that the maximum negative S.F. at C occurs, when the trailing load is at C.

$$\therefore\ F_{max} = -[12 \times f_8 f_8' \times 18 \times f_7 f_7' + 20 \times f_6 f_6' + 8 \times f_5 f_5' + 6 \times f_4 f_4']$$

$$= -\left[12 \times \frac{1}{18} + 18 \times \frac{3}{18} + 20 \times \frac{5}{18} + 8 \times \frac{7}{18} + 6 \times \frac{1}{2}\right]\ t$$

$$= -15 \cdot 33\ ^t\ \mathbf{Ans.}$$

Absolute maximum B.M.

First of all, let us find out the *c.g.* of the load system. From the geometry of the load system, we find that its *c.g.* from the 6 ^{t} load,

$$x = \frac{8 \times 2 \cdot 5 + 20 \times 5 + 18 \times 7 \cdot 5 + 12 \times 10}{6 + 8 + 20 + 18 + 12} = \frac{375}{64} = 5 \cdot 86\ m$$

Now let us find out the load, under which the absolute maximum B.M. will take place. By inspection, we find that the absolute maximum B.M. will take place under the 20 ^{t} load. We know that for the maximum B.M. under the 20 t load, the load should occupy such a position on the beam that the centre of the span is mid-way between the *c.g.* of the load and the 20 t load.

Therefore 20 T load should be at a distance $11 \cdot 25 - \frac{1}{2}(5 \cdot 86 - 5 \cdot 0)$ $= 10 \cdot 82\ m$ from A as shown in Fig. 3·13.

Now draw the influence lines for the B.M. on the base AB, with ordinate XX_3 equal to $\dfrac{x(l-x)}{l} = \dfrac{11 \cdot 25(22 \cdot 50 - 11 \cdot 25)}{52 \cdot 50}$

---

*The absolute maximum B.M. generally occurs under a heavier load, which is very near the *c.g.* of the load system.

$$= \frac{11 \cdot 25 \times 21 \cdot 25}{22 \cdot 50} = 5 \cdot 625 \text{ as shown in Fig. 3·13 (b).}$$

From the geometry of figure AX_3B, we find that $m_1 m_1' = 2·91$; $m_2 m_2' + 4·16$; $m_3 m_3' = 5·41$; $m_4 m_4' = 4·59$ and $m_5 m_5' = 3·34$.

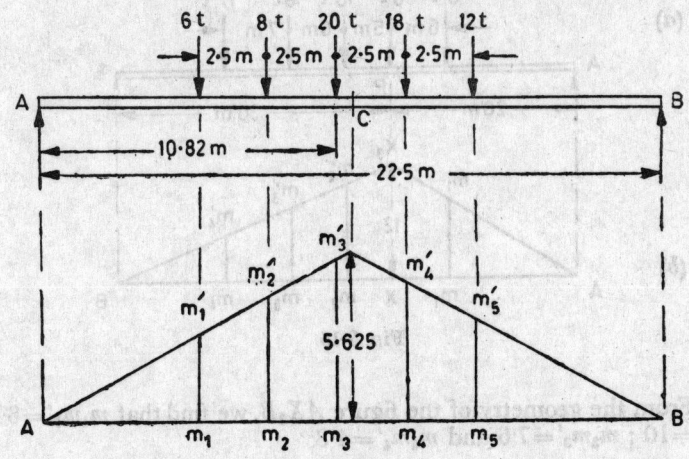

Fig. 3·13

$$\therefore \quad M_{max\ max} = 6 \times m_1 m_1' + 8 \times m_2 m_2' + 20 \times m_3 m_3$$
$$+ 18 \times m_4 m_4' + 12 m_5 m_5'$$
$$= 6 \times 2·91 + 8 \times 4·16 + 20 \times 5·41 + 18 \times 4·59$$
$$+ 12 \times 3·34 \text{ t-m}$$
$$= 281·64 \text{ t-m} \quad \textbf{Ans.}$$

Example 3·7. *For the span shown in the sketch below, obtain the bending moment at a section P, 20 m from A due to the loads in the position indicated.*

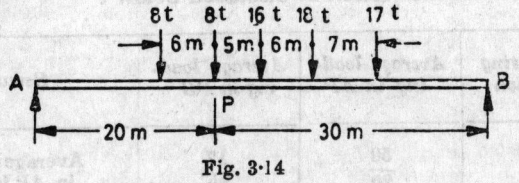

Fig. 3·14

Also determine the position of the loads for maximum bending moment of section P and the value of maximum moment.

*Solution.

Given. Span, $l = 50$ m

B.M. at P with the loads in position

First of all, draw the influence lines for the B.M. on the base

---

*We have already solved this example. on page 42 (example 2·12). Here we shall solve this problem with the help of influence lines.

AB, with ordinate XX_2 equal to $\dfrac{x(l-x)}{l} = \dfrac{20(50-20)}{50} = \dfrac{20 \times 30}{50} = 12$ as shown in Fig. 3·15 (b).

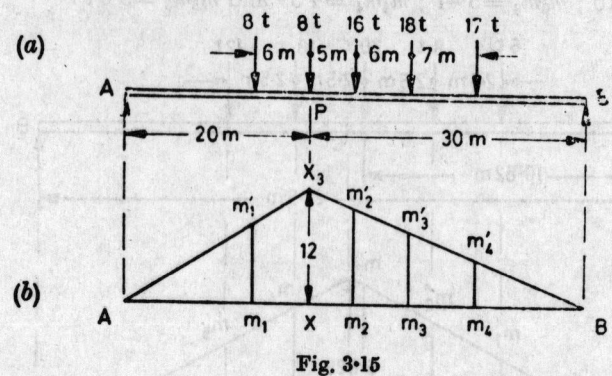

Fig. 3·15

From the geometry of the figure AX_3B, we find that $m_1m_1' = 8\cdot4$; $m_2m_2' = 10$; $m_3m_3' = 7\cdot6$ and $m_4m_4' = 4\cdot8$

∴ B.M. at P,

$M_P = 8 \times m_1m_1' + 8 \times XX_3 + 16 \times m_2m_2' + 18 \times m_3m_3' + 17 \times m_4m_4'$

$= 8 \times 8\cdot4 + 8 \times 12 + 16 \times 10 + 18 \times 7\cdot6 + 17 \times 4\cdot8$ t-m

$= \mathbf{541\cdot6\ t\text{-}m}$ **Ans**.

Maximum BM at P

First of all, let us find out the load, which should be placed at the section, for maximum B.M. For doing so, we can assume the loads to cross the section P one by one and, study the average loading on both the sides of the section as discussed below :

Load crossing the section	Average loading in AP	Average loading in PB	Remarks
17 t	$\dfrac{50}{20}$	$\dfrac{17}{30}$	Average loading in AP is heavier
18 t	$\dfrac{32}{20}$	$\dfrac{35}{30}$	Average loading in AP is heavier
16 t	$\dfrac{16}{20}$	$\dfrac{51}{30}$	Average loading in AP is lighter

It is thus obvious, that when the 16 t load will be at the section, it will cause maximum B.M. at the section P. Therefore the load system should occupy a position as shown in Fig. 3·16

From the geometry of the figure AX_3B, we find that $m_1m_1' = 5\cdot4$; $m_2m_2' = 9\cdot0$; $m_3m_3' = 9\cdot6$ and $m_4m_4' = 6\cdot8$.

∴ Maximum B.M. at P,

$$M_{max} = 8 \times m_1 m_1' + 8 \times m_2 m_2' + 16 \times XX_3 + 18 \times m_3 m_3' + 17 \times m_4 m_4'$$

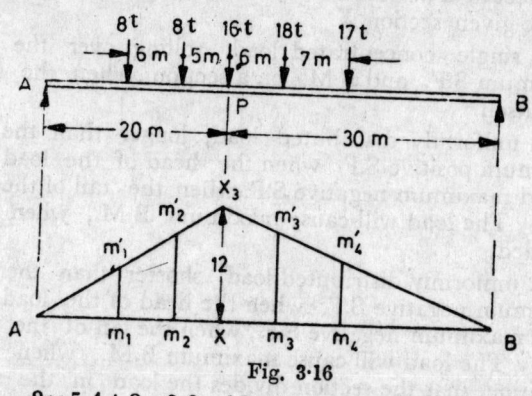

Fig. 3·16

$$= 8 \times 5·4 + 8 \times 9·0 + 16 \times 12 + 18 \times 9·6 + 17 \times 6·8 \text{ t-m}$$
$$= 595·6 \text{ t-m Ans.}$$

Exercise 3·2

1. Two point loads of 5 t or 15 t, spaced 4 m apart cross a girder of 12 m span with 5 T load leading. Find the value of maximum B.M. under the 15 T load on the girder.

(Nagpur University)
[Ans. 35·16 t-m]

2. A train of wheel loads as shown in Fig. 3-17 crosses a simply supported beam of span 25 m from left to right with the 20 T load leading.

```
16 t       16 t       20 t           20 t            20t
|←— 3m —→|←— 3m —→|←——— 4 m ——→|←—— 4 m ——→|
↓          ↓          ↓               ↓                ↓
```

Fig. 3.17

Using influence lines, determine the maximum B.M. under the central load.

(Calcutta University)
[Ans. 384 t-m]

3. A system of 5 loads of 8 t, 16 t, 16 t', 6 t and 4 t crosses a beam of 15 m span with the 8 T load leading. The distance between the loads are 2·4, 3·0, 2·4, and 1·8 m respectively. Find the maximum B.M. at the centre of the span. Also find the absolute maximum B.M. on the beam.

(Gujarat University)
[Ans. 126·3 t-m ; 127·5 t-m]

HIGHLIGHTS

1. The influence line diagram for S.F. is drawn first by drawing the base AB equal to the span to some scale, and then by erecting perpendiculars Aa and Bb, each equal to unity, on opposite sides of the base. Join Ab and Ba. Now through the given section X, draw ordinate cutting the parallel lines Ab and Ba at X_1 and X_2. Now AX_1X_2B is the required influence line for S.F. for the given section X.

2. The influence line diagram for B.M. is drawn first by drawing the base AB equal to the span to some suitable scale and then by erecting perpendiculars Aa' and Bb' on the same side of the base such that Aa' is equal to AX and Bb' is equal to XB. Join Ab' and Ba' to intersect at X_3. Now AX_3B is the required influence line for B.M. for the given section X.

3. A single concentrated load, rolling over the beam will cause maximum S.F. and B.M. on a section, when the load is on the section itself.

4. A uniformly distributed load, longer than the span will cause maximum positive S.F. when the head of the load is at the section ; and maximum negative S.F. when the tail of the load is at the section. The load will cause maximum B.M., when the entire span is loaded.

5. A uniformly distributed load, shorter than the span, will cause maximum negative S.F. when the head of the load is at the section and maximum negative S.F., when the tail of the load is at the section. The load will cause maximum B.M., when it occupies such a position, that the section divides the load in the same ratio, as it divides the span.

6. Two concentrated loads, spaced a distance d apart, will cause maximum positive S.F., when the leading load is at the section, and maximum negative S.F. when the trailing load is at the section. The maximum B.M. will take place, under the heavier load, when it occupies such a position on the span, that the centre of the span is mid-way between the $c.g.$ of the two loads and the heavier load.

7. When a series of point loads crosses a girder, simply supported at its ends, the maximum B.M. under any given load occurs, when the centre of the span is mid-way between the $c.g.$ of the load system as the load under which the maximum B.M. is required to be found out.

8. When a series of a point loads crosses a girder, simply supported at its ends, the maximum B.M. at any section on the span occurs, when the average loading on the left of the section is equal to the average loading on the right of the section.

Do You Know ?

1. Explain the term influence lines.

2. Discuss the uses and principles of influence lines.

3. Draw influence lines for (i) the reaction at a support, (ii) shear force and (iii) bending moment at any section of a simply supported beam of span l. In each case explain clearly how you obtain this influence line. Indicate salient points.

4. With the help of influence lines, locate the position of a uniformly distributed load shorter than the span for maximum B.M. rolling over a beam.

5. Draw the influence lines for the S.F. and B.M., when two concentrated loads cross over a beam.

4

INFLUENCE LINES FOR TRUSSED BRIDGES

1. Introduction. 2. Through type trusses. 3. Deck type trusses. 4. Principles for the influence lines for forces in the members of trussed bridges. 5. Influence lines for a Pratt truss with parallel chords. 6. Influence lines for an inclined Pratt truss. 7. Influence lines for a deck type Warren girder. 8. Influence lines for a composite truss.

4·1. Introduction

In the last chapter, we have discussed the influence lines for the simply supported beams and girders. But in actual practice, simply supported beams and girders are used for small spans only. The trussed bridges are used for long spans, specially for heavier loads. The bridge trusses may be classified into the following two types, depending upon the position of loading :

(a) Through type trusses, and

(b) Deck type trusses.

4·2. Through type trusses

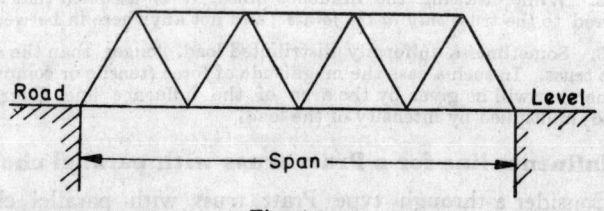

Fig. 4·1

A truss, which receives load at its bottom chord joints, is called a *through type truss* as shown in Fig. 4·1.

85

4·3. Deck type trusses

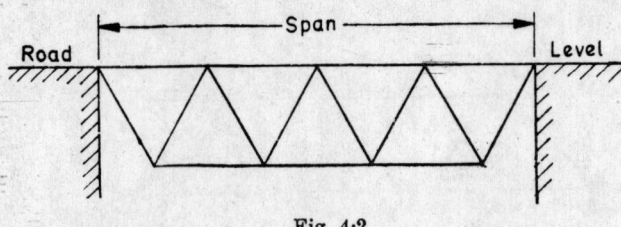

Fig. 4·2

A truss, which receives load at its top chord joints, is called a *deck type truss* as shown in Fig. 4·2.

4·4. Principles for the influence lines for forces in the members of the trussed bridges

A trussed bridge consists of two long trusses, separated by a distance equal to the width of the roadway. The two trusses are connected together by cross girders, which transfer the load to the truss at its joints. The influence line diagrams, for the forces in the members of a truss, are drawn by assuming a load of 2 units moving across the bridge ; so that a load of 1 unit is transferred to the each main truss.

The influence line diagrams for the forces in the top and bottom chord members are drawn for the B.M., whereas the influence line diagrams for diagonal and vertical members are drawn for S.F. in the panel. Though there are several types of bridge trusses in use, yet the following are very common these days :

1. Pratt truss with parallel chords,
2. Pratt truss with inclined chords,
3. Warren truss, and
4. Composite truss.

Note : 1. There are different notations used for drawing the influence lines for the forces in the various members of the trusses. But in this book the influence line diagrams for tensile force will be drawn above the reference line, whereas those, for compressive forces, below it. These notations are widely used and internationally recognised.

2. While drawing the influence lines, it is assumed that the load is transferred to the truss *only at the joints* ; and not anywhere in between them.

3. Sometimes a uniformly distributed load, longer than the span, rolls over the truss. In such a case the magnitude of force (tensile or comprehensive) in the member will be given by the area of the influence line diagram (for a unit load) multiplied by intensity of the load.

4·5. Influence line for a Pratt truss with parallel chords

Consider a through type Pratt truss with parallel chords of panels each of length a and of height h as shown in Fig. 4·3. Here we shall draw the influence lines for half the members of the truss, when a unit load rolls over the bottom chord. In a through type Pratt truss, the top chord members (*i.e.*, $L_1 U_2$, $U_2 U_3$, $U_3 U_4$...) are in compression, and the bottom chord members (*i.e.* L_1L_2, L_2L_3, L_3L_4...) are in tension.

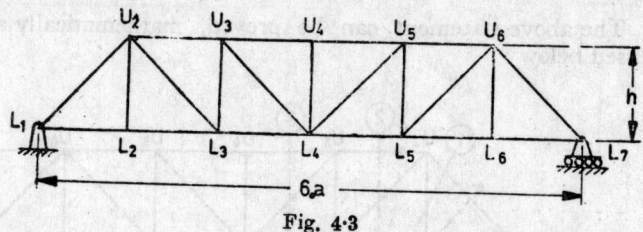

Fig. 4·3

(i) *Influence lines for forces in members L_1U_2, U_2U_3 and U_3U_4*

First of all, pass section 1-1 cutting the member L_1U_2 as shown in Fig. 4·4 (a). Now, though L_2 draw the perpendicular to L_1U_2. Let the unit load move from the joint L_1 to L_7. Now consider the equilibrium of the left part of the truss and assume the direction of force in member L_1U_2, in the left part of the truss as shown in Fig. 4·4(a), thus causing compression in the member. Now taking moments about the *opposite joint L_2, and equating the same,

$$f_{L_1U_2} \times AL_2 = V_1 \times a \qquad \text{(where } V_1 \text{ is vertical reaction at } L_1 \text{)}$$

or
$$f_{L_1U_2} = \frac{V_1 \times a}{AL_2} = \frac{V_1 \times a}{\dfrac{a \times h}{\sqrt{a^2+h^2}}} \qquad \left(\because AL_2 = \frac{a \times h}{\sqrt{a^2+h^2}} \right)$$

From the above equation, we find that $V_1 \times a$ is equal to the moment M_{L2} about the joint L_2.

$$f_{L_1U_2} = \frac{M_{L_2}}{\dfrac{a \times h}{\sqrt{a^2+h^2}}} = M_{L_2} \times \frac{\sqrt{a^2+h^2}}{ah}$$

Therefore the influence line for the force in the member L_1U_2 is equal to $\dfrac{\sqrt{a^2+h^2}}{ah}$ times the influence line for M_{L_2}. We know that the influence line for M_{L_2} is a triangle with ordinate equal to $\dfrac{x\,(l-x)}{l}$ $= \dfrac{a\,(6a-a)}{6a} = \dfrac{a \times 5a}{6a} = \dfrac{5a}{6}$. It is thus obvious, that the influence line for the force in member L_1U_2 is also a triangle with ordinate equal to $\dfrac{5a}{6} \times \dfrac{\sqrt{a^2+h^2}}{a \times h} = \dfrac{5}{6h}\sqrt{a^2+h^2}$ under the joint L_2 as shown in Fig. 4·4 (b).

---

*It is the joint in the opposite chords (*i.e.* bottom chord in this case) at which *vertical member* through the joints L_1 or U_2 or meets *e.g.*, opposite joints for the members U_2U_3 or or U_2U_4 are L_3 and L_4 respectively. Similarly, opposite joints for the members L_1L_2 and L_2L_3 is U_2, and opposite joint for the member L_3L_4 is U_3.

Thus the influence line for the member L_1U_2 is equal to the influence lines for B.M. at the opposite joint L_2, divided by AL, i.e. vertical distance between the member L_1U_2 and the opposite joint L_2.

The above statement can be proved, mathematically also, as discussed below :

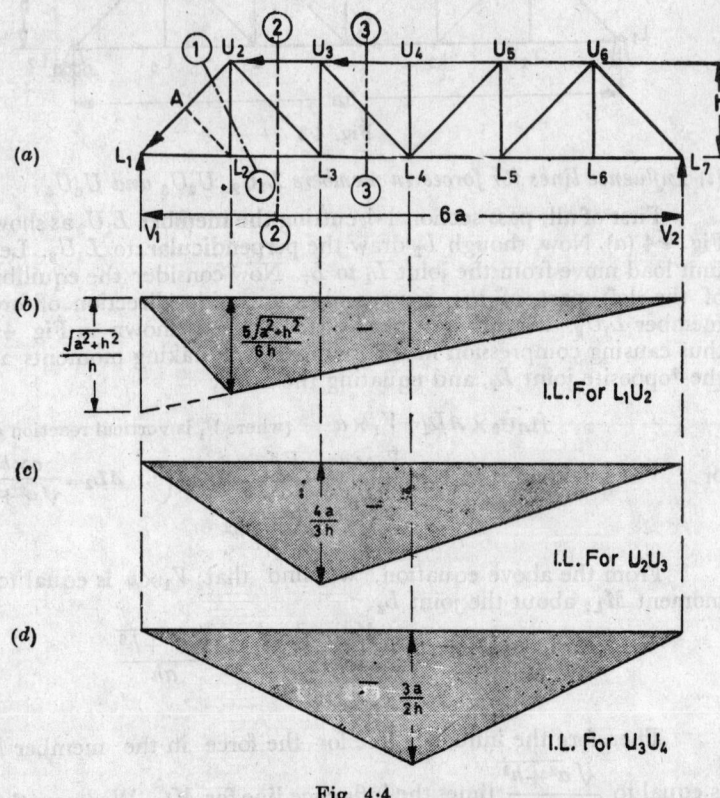

Fig. 4·4

First of all, pass section 1-1 cutting the member L_1U_2 as shown in Fig. 4·4 (a). Now consider the equilibrium of the left part of the truss and the unit load to roll from L_1 to L_2. Let the unit load be at a distance x from the joint L_1. At this instant, the vertical reaction at L_1,

$$V_1 = \frac{1(6a-x)}{6a} = \frac{(6a-x)}{6a}$$

Taking moments about the joint L_2 and equating the same,

$$f_{L_1U_2} \times AL = \frac{(6a-x) \times a}{6a} - 1(a-x) = \frac{(6a-x)}{6} - (a-x)$$

$$\therefore \quad f_{L_1U_2} = \frac{\dfrac{(6a-x)}{6} - (a-x)}{AL} \qquad \text{...(i)}$$

From the above equation, we find that the influence line for force in member L_1U_2 is a straight line, as the value of x varies linearly (because x has a unit power). We also find that when $x=0$ (i.e., when the load is at L_1) $F_{L_1U_2}=0$ and when $x=a$ (i.e., when the load is at L_2),

$$f_{L_1U_2}=\frac{\dfrac{(6a-a)}{6}-(a-a)}{AL}=\frac{5a}{6\times AL}$$

$$=\frac{5a\sqrt{a^2+h^2}}{6ah}=\frac{5\sqrt{a^2+h^2}}{6h}$$

Now consider the unit load to roll from L_1 to L_7. Let the load be at a distance of x from the joint L_1. At this instant, the vertical reaction at L_1,

$$V_1=\frac{1(6a-x)}{6x}=\frac{(6a-x)}{6a}$$

Taking moments about the joint L_2 and equating the same,

$$f_{L_1U_2}\times AL=\frac{(6a-x)}{6a}\times a=\frac{(6a-x)}{6}$$

$$\therefore \qquad\qquad f_{L_1U_2}=\frac{(6a-x)}{6AL} \qquad\qquad ...(ii)$$

From the above equation, we find that the influence line for force in member U_1L_2 is a straight line, as the value of x varies linearly (because x has a unit power). We also find that when $x=a$ (i.e. when the load is at L_2),

$$f_{L_1U_2}=\frac{(6a-a)}{6AL}=\frac{5a\sqrt{a^2+h^2}}{6ah}=\frac{5\sqrt{a^2+h^2}}{6h}$$

and when $x=6a$ (i.e. when the load is at L_7) $f_{L_1U_2}=0$.

Note : Similarly draw the influence lines for the members U_2U_3 and U_3U_4 by passing section (2—2) and (3—3) and taking moments about the respective opposite joints L_3 and L_4 as shown in Fig. 4·4 (c) and (d).

(ii) *Influence lines for forces in members* L_1L_2, L_2L_3 *and* L_3L_4

First of all, pass section 4·4 cutting the member L_1L_2 as shown in Fig. 4·5 (a). Let the unit load move from the joint L_1 to L_7. Now consider the equilibrium of the left part of the truss and assume the direction of force in member L_1L_2 in the left part of the truss, thus causing tension in the member as shown in Fig. 4·5 (a). Now taking moments about the opposite joint U_2 and equating the same,

$$f_{L_1L_2}\times h=V_1\times a \text{ (where } V_1 \text{ is vertical reaction at } L_1)$$

or $\qquad\qquad f_{L_1L_2}=\dfrac{V_1\times a}{h}$

From the above equation, we find that $V_1\times a$ is equal to the moment M_{U_2}, about the joint U_2.

$$\therefore \qquad\qquad f_{L_1L_2}=\frac{M_{U_2}}{h}$$

Therefore the influence line for the force in member L_1L_2 is equal to $1/h$ times the influence line for M_{U_2}. We know that the influence line for M_{U_2} is a triangle with ordinate equal to

$$\frac{x(l-x)}{t}=\frac{a(6a-a)}{6a}=\frac{5a}{6}$$

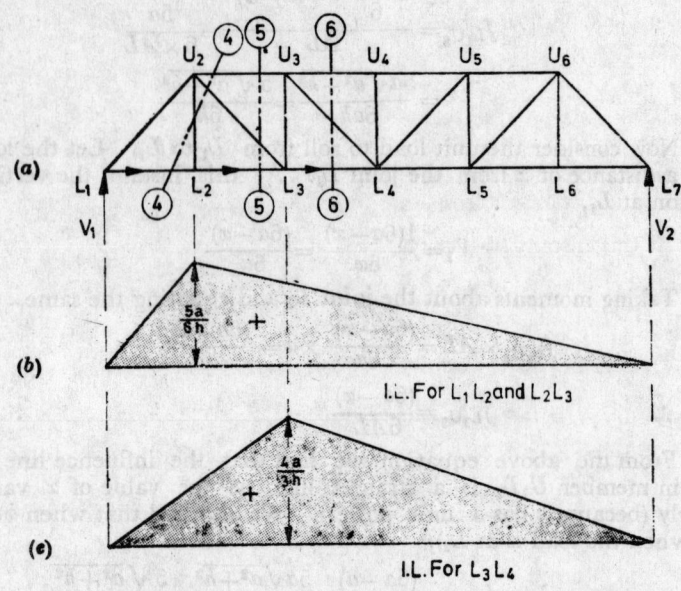

Fig. 4·5

It is thus obvious, that the influence line for the force in member L_1L_2 is also a triangle with ordinate equal to $\dfrac{5a}{6}\times\dfrac{1}{h}=\dfrac{5a}{6h}$ under the joint U_2 as shown in Fig. 4·5 (b). *Thus the influence line for force in the member L_1L_2 is equal to the influence line for B.M. at the opposite joint U_2 divided by h i.e., vertical distance between the member L_1L_2 and the opposite joint U_2.*

The above statement can be proved, mathematically also, as discussed below :

First of all pass section $4-4$ cutting the member L_1L_2 as shown in Fig. 4·5 (a). Now consider the equilibrium of the left part of the truss and the unit load to roll from L_1 to L_2. Let the load be at a distance x from the joint L_1. At this instant, the vertical reaction at L_1,

$$V_1=\frac{1(6a-x)}{6a}=\frac{(6a-x)}{6a}$$

Taking moments about the joint U_2, and equating the same,

$$f_{L_1L_2}\times h=\frac{(6a-x)\times a}{6a}-1(a-x)=\frac{(6a-x)}{6}-(a-x)$$

$$\therefore \qquad f_{\mathrm{L_1L_2}} = \frac{\frac{(6a-x)}{6}-(a-x)}{h} \qquad \ldots(i)$$

From the above equation, we find that the influence line for L_1L_2 is a straight line as the value of x varies linearly (because x has a unit power). We also find that when $x=0$ (*i.e.*, when the load is at L_1) $f_{\mathrm{L_1L_2}}=0$ and when $x=a$ (*i.e.*, when the load is at L_2)

$$f_{\mathrm{L_1L_2}} = \frac{\frac{(6a-a)}{6}-(a-a)}{h} = \frac{5a}{6h}$$

Now, consider the unit load to roll from L_2 to L_7. Let the load be at a distance x from the joint L_1.

At this instant, the vertical reaction at L_1,

$$V_1 = \frac{1(6a-x)}{6a} = \frac{(6a-x)}{6a}$$

Taking moments about the joint U_2, and equating the same,

$$f_{\mathrm{L_1L_2}} \times h = \frac{(6a-x)}{6a} \times a = \frac{(6a-x)}{6}$$

$$\therefore \qquad f_{\mathrm{L_1L_2}} = \frac{(6a-x)}{6h} \qquad \ldots(ii)$$

From the above equation, we find that the influence line for L_1L_2 is a straight line, as the value of x varies linearly (because x has a unit power). We also find that when $x=a$ (*i.e.* when the load is at L_2) $f_{\mathrm{L_1L_2}} = \frac{(6a-a)}{6h} = \frac{5a}{6h}$ and when $x=6a$ (*i.e.* when the load is at L_7) $f_{\mathrm{L1L2}}=0$.

Note : 1. The influence line for L_2L_3 is similar to that for L_1L_2. This can be easily proved by passing a section 5-5 cutting the member L_2L_3 as shown in Fig. **4·5(a)**.

2. Similarly, draw the influence line for the members L_3L_4 by passing section 6-6 and taking moments about the opposite joint U_3 as shown in Fig. **4·5(c)**.

(iii) Influence line for force in member U_2L_2

We see that when the unit load is at L_1, the force in member U_2L_2 is zero. As the load rolls towards the joint L_2, the tensile force in the member U_2L_2 will go on increasing, till the load reaches the joint L_2. At this moment, the tensile force in the member U_2L_2 will be equal to unity. As the load moves further, towards the joint L_3, the force in the member U_2L_2 will go on decreasing till the load reaches the joint L_3. At this moment, the force in U_2L_2 will also be equal to zero. A little consideration will show, that as the load moves away from L_3 the force in the member U_2L_2 will continue to be equal to zero.

Thus the influence line for the force in member U_2L_2 will be a triangle with base equal to L_1U_3 and ordinate equal to unity under the joint L_2 as shown in Fig. 4·6 (b).

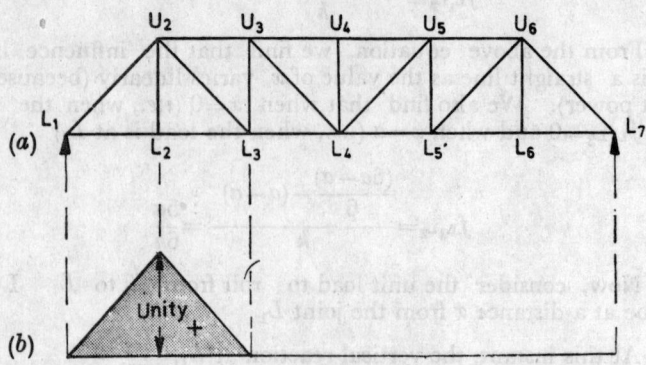

Fig. 4·6

(iv) Influence lines for forces in members U_2L_3 and U_3L_4

First of all pass section 7-7 cutting the member U_2L_3 as shown in Fig. 4·7 (a). We see that when the unit load is at the joint L_1, the force in the member U_2L_3 is zero. Let the load be between L_1 and L_2. Now consider the equilibrium of the right* part of the truss. Since the vertical reaction V_2 is acting upwards, therefore direction of the force in member U_2L_3 near the joint L_3 will be downwards, thus causing compression in the member as shown in Fig. 4·7 (a).

$$f_{U_2L_3} \sin \theta = V_2$$

or $$f_{U_2L_3} = \frac{V_2}{\sin \theta} = V_2 \cosec \theta \text{ (Compn)}$$

Therefore the influence line for the force in member U_2L_3 is equal to $\cosec \theta$ times the influence line for V_2. We know that the influence line for the reaction V_2 is a triangle, with a unit ordinate at L_7 and zero at L_1. Therefore the influence line for the force in member U_2L_3 between L_1 and L_2 will be a straight line with ordinate equal to $\frac{1}{6} \cosec \theta$ under the joint L_2 as shown in Fig. 4·7 (b).

Now let the unit load be between L_2 and L_7, consider the equilibrium left** part of the truss. Since the reaction V_1 is acting upwards, therefore direction of the force in the member U_2L_3 near the

---

*If the equilibrium of the left part is considered then the direction of the force in the member U_2L_3 will be upward near the joint L_3

∴ $$f_{U_2L_3} \sin \theta = (1 - V_1) = V_2$$

or $$f_{U_2L_3} = V_2 \cosec \theta \text{ (Compn)}$$

**If the equilibrium of the right part is considered, then the direction of the force in the member U_2L_3 near the joint L_3 will be upwards.

∴ $$f_{U_2L_3} \sin \theta = 1 - V_2 = V_1$$

or $$f_{U_2L_3} = V_1 \cosec \theta \text{ (Tension)}$$

joint U_2 will be downwards, thus causing tension in the member, as shown in Fig. 4·7 (a).

$$\therefore \qquad f_{U_2L_3} \sin \theta = V_1$$

or
$$f_{U_2L_3} = \frac{V_1}{\sin \theta} = V_1 \operatorname{cosec} \theta \text{ (Tension)}$$

Therefore the influence line for the force in the member U_2L_3 is equal to cosec θ times the influence line for V_1. We know that the influence line for the reaction V_1 is a triangle with a unit ordinate at L_1, and zero at L_7. Thus the influence line for the force in member U_2L_3 between L_3 and L_7 will be a straight line with ordinate equal to $\frac{2}{3}$ cosec θ under the joint L_3, and zero under the joint L_7 as shown in Fig. 4·7 (b). The influence line between the joints L_2 and L_3 will also be a straight line joining the ordinates under the joints L_2 and L_3 as shown in Fig. 4·7 (b).

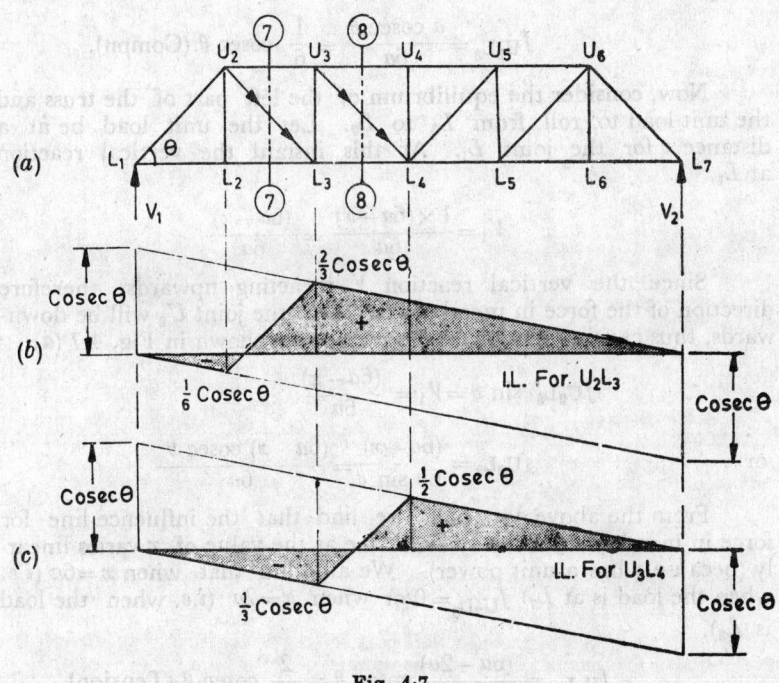

Fig. 4·7

The above statement can be proved, mathematically also, as discussed below :

First of all, pass section 7-7 cutting the member U_2L_3 as shown in Fig. 4·7 (a). Consider the equilibrium of the right part of the truss and the unit load to roll from L_1 to L_2. Let the load be at a distance x from the joint L_1. At this instant, the vertical reaction at L_7,

$$V_2 = \frac{1 \times x}{6a} = \frac{x}{6a}$$

Since the vertical reaction V_2 is acting upwards, therefore direction of the force in member $U_2 L_3$ near the joint L_3 will be downwards, thus causing compression in the member, as shown in Fig. 4·7 (a).

$$\therefore \qquad f_{U_2 L_3} \sin \theta = V_2 = \frac{x}{6a}$$

or

$$f_{U_2 L_3} = \frac{x}{6a \sin \theta} = \frac{x \operatorname{cosec} \theta}{6a}$$

From the above equation, we find that the influence line for force in member $U_2 L_3$ is a straight line, as the value of x varies linearly (because x has a unit power). We also find that when $x=0$ (i.e. when the load is at L_1) $f_{U_2 L_3} = 0$ and when $x=a$ (i.e., when the load is at L_2)

$$f_{U_2 L_3} = \frac{a \operatorname{cosec} \theta}{6a} = \frac{1}{6} \operatorname{cosec} \theta \text{ (Compn)}.$$

Now, consider the equilibrium of the left part of the truss and the unit load to roll from L_3 to L_7. Let the unit load be at a distance x for the joint L_1. At this instant the vertical reaction at L_1,

$$V_1 = \frac{1 \times (6a - x)}{6a} = \frac{(6a - x)}{6a}$$

Since the vertical reaction V_1 is acting upwards, therefore direction of the force in member $U_2 L_3$ near the joint U_2 will be downwards, thus causing tension in the member as shown in Fig. 4·7 (a).

$$\therefore \qquad f_{U_2 L_3} \sin \theta = V_1 = \frac{(6a - x)}{6a}$$

or

$$f_{U_2 L_3} = \frac{(6a - x)}{6a \sin \theta} = \frac{(6a - x) \operatorname{cosec} \theta}{6a}$$

From the above equation, we find that the influence line for force in member $U_2 L_3$ is a straight line as the value of x varies linearly (because x has a unit power). We also find that when $x=6a$ (i.e. when the load is at L_7) $f_{U_2 L_3} = 0$ or when $x=2a$ (i.e. when the load is L_3).

$$f_{U_2 L_3} = \frac{(6a - 2a)}{6a} \operatorname{cosec} \theta = \frac{2}{3} \operatorname{cosec} \theta \text{ (Tension)}$$

Now, consider the equilibrium of the left part and the unit load to roll from L_2 to L_3. Since the unit load is rolling in the panel, which has been cut by the section (7-7), therefore, for some time the unit load will be on the left of section and for some time it will be on the right of the section. We have already discussed in Art. 4·4 that the load is assumed to be transferred to the truss, at the joints only, therefore vertical load transferred to the joint L_2, when the unit load is at a distance of x from the joint L_1 as shown in Fig. 4·8,

$$V_{L_2} = \frac{(2a-x)}{a}$$

and vertical reaction at the joint L_1

$$V_1 = \frac{(6a-x)}{6a}$$

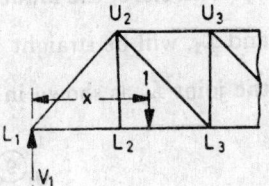

Fig. 4·8

$$\therefore \quad f_{U_2L_3} \sin \theta = V_1 - V_{L_2}$$

$$= \frac{(6a-x)}{6a} - \frac{(2a-x)}{a}$$

or $\qquad f_{U_2L_3} = \left[\frac{(6a-x)}{6a} - \frac{(2a-x)}{a} \right] \operatorname{cosec} \theta$

From the above equation we find that the influence line for force in member U_2L_3 is a straight line as the value of x varies linearly (because x has a unit power). We also find that when $x=a$ (i.e. when the load is at L_2),

$$f_{U_2L_3} = \left[\frac{(6a-a)}{6a} - \frac{(2a-a)}{a} \right] \operatorname{cosec} \theta$$

$$= \left(\frac{5}{6} - 1 \right) \operatorname{cosec} \theta$$

(Minus sign means compn.)

$$= - \frac{1}{6} \operatorname{cosec} \theta$$

and when $x=2a$ (i.e. when the load is at L_3),

$$f_{U_2L_3} = \left[\frac{(6a-2a)}{6a} = \frac{2a-2a}{a} \right] \operatorname{cosec} \theta$$

$$= \frac{2}{3} \operatorname{cosec} \theta \qquad \text{(Plus sign means tension)}$$

Note. Similarly draw the influence line for the force in member U_3L_3 by passing section (8—8) as shown in Fig. 4·7 (c).

(v) *Influence-lines for force in member U_3L_3*

First of all pass section* 9-9 cutting the member U_3L_3 as shown in Fig. 4·9 (a). We see that when the load is at the joint L_1, the force in member U_3L_2 is zero. Let the load be between L_1 and L_3. Now consider the equilibrium of the right** part. Since the vertical reaction V_2 is acting upwards, therefore direction of the force in member U_3L_3 near the joint U_3 will be downwards, thus causing tension in the member as shown in Fig. 4·9 (a).

$$\therefore \qquad f_{U_3L_3} = V_2 \text{ (Tension)}$$

---

*It may be noted that the section has to cut the three members i.e., U_2U_3, U_3L_3 and L_3L_4. There cannot be any other section, which will cut three members only, with U_3L_3 as one of the members.

**If the equilibrium of the left part is considered, then the direction of the force in the member U_3L_3 will be upward near the joint L_3.

$$\therefore \qquad f_{U_3L_3} = 1 - V_1 = V_2 \text{ (Tension)}.$$

Therefore, the influence line for the force in member U_3L_3 is equal to the influence line for V_2. We know that the influence line for the reaction V_2 is a triangle with a unit ordinate at L_1 and zero at L_1. Therefore, the influence line for force in member U_3L_3 between L_1 and L_2, will be straight line with ordinate equal to $\frac{2}{6}=\frac{1}{3}$ under the joint L_3 as shown in Fig. 4·9 (b).

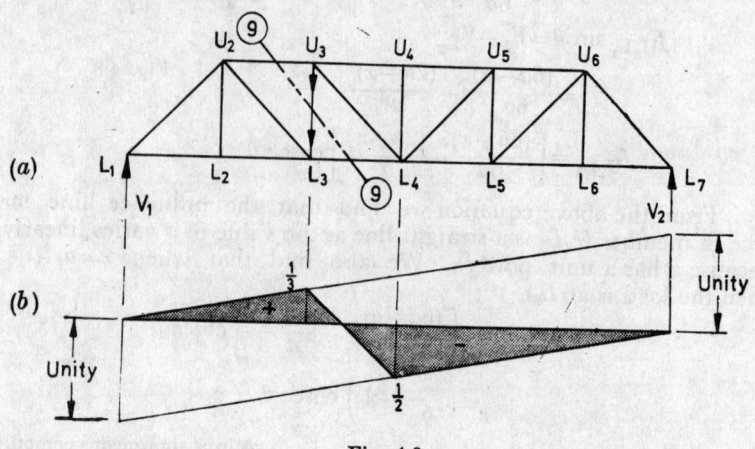

Fig. 4·9

Now, let the unit load be between L_4 and L_7. Consider the equilibrium of the *left part of the truss. Since the reaction V_1 is acting upwards, therefore direction of the force in member U_3L_3 near the joint L_3 will be downwards, thus causing compression in the member as shown in Fig. 4·9 (a).

$$\therefore \qquad f_{U3L3}=V_1$$

Therefore the influence line for the force in the member U_3L_3 is equal to the influence line for V_1. We know that the influence line for the reaction V_1 is a traingle with a unit ordinate at L_1 and zero at L_7. Therefore, the influence line for force in member U_3L_3, between L_4 and L_7, will be a straight line with ordinate equal to $\frac{3}{6}=\frac{1}{2}$ as shown in Fig. 4·9 (b). The influence line between the joints L_3 and L_4 will also be a straight line joining the ordinates under the joints L_3 and L_4 as shown in Fig. 4·9 (b).

The above statement can be proved, mathematically also, as discussed below :

First of all pass section 9-9 cutting the member U_3L_3 as shown in Fig. 4·9 (a). Consider the equilibrium of the right part of the

_____

*If the equilibrium of the right part is considered, then direction of force in member U_3L_3 near the joint U_3 will be upwards.

$$\therefore \qquad f_{U3L3}=1-V_2=V_1 \text{ (Compn.)}$$

truss and the unit load to roll from L_1 to L_3. Let the load be at a distance x from the joint L_1. At this instant, the vertical reaction L_7,

$$V_2 = \frac{1 \times x}{6a} = \frac{x}{6a}$$

Since the vertical reaction V_2 is acting upwards, therefore, direction of the force in member U_3L_3 near the joint U_3 will be downwards, thus causing tension in the member, as shown in Fig. 4·9 (a).

$$\therefore \qquad fU_3L_3 = V_2 = \frac{x}{6a}$$

From the above equation, we find that the influence line for force in member U_3L_3 is a straight line, as the value of x varies linearly (because x has a unit power). We also find that when $x=0$ (i.e. when the load is at L_1) $fU_3L_3 = 0$ and when $x=2a$ (i.e. when the load is L_3)

$$fU_3L_3 = \frac{2a}{6a} = \frac{1}{3}$$

Now consider the equilibrium of the left part of the truss and the unit load to roll from L_4 to L_7. Let the unit load be at a distance of x from the joint L_1. At this instant, vertical reaction at L_1,

$$V_1 = \frac{1(6a-x)}{6a} = \frac{(6a-x)}{6a}$$

Since the vertical reaction V_1 is acting upwards, therefore, direction of the force in member U_3L_3 near the joint L_3 will be downwards, thus causing compression in the member, as shown in Fig. 4·9 (a).

$$\therefore \qquad fU_3L_3 = V_1 = \frac{(6a-x)}{6a}$$

From the above equation, we find that the influence line for force in member U_3L_3 is a straight line, as the value of x varies linearly (because x has a unit power). We also find that when $x=6a$ (i.e. when the load is L_7) $f_{U3L3}=0$ and when $x=3a$ (i.e. when the load is at L_4,

$$fU_3L_3 = \frac{(6a-3a)}{6a} = \frac{3a}{6a} = \frac{1}{2}$$

Now, consider the equilibrium of the left part of the truss and the unit load to be rolling from L_3 to L_4.

Since the load is rolling in the panel, which has been cut by the section 9-9, therefore, for some distance the unit load will be on the left of the section and for some time, it will be on the right of the section. Now the vertical load transferred to the joint L_3,

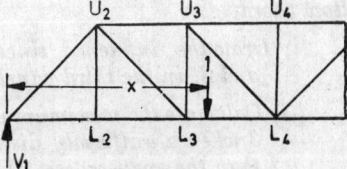

Fig. 4·10

when the unit load is at a distance of x from the joint L_1,

$$V_{L3} = \frac{(3a-x)}{a}$$

and the vertical reaction at the joint L_1,

$$V_1 = \frac{(6a-x)}{6a}$$

$$\therefore \quad f_{U3L3} = V_{L3} - V_1 = \frac{(3a-x)}{a} - \frac{(6a-x)}{6a}$$

From the above equation, we find that the influence line for force in member U_3L_3 is a straight line as the value of x varies linearly (because x has a unit power). We also find that when $x=2l$ (i.e., the load is at L_3),

$$f_{U3L3} = \frac{3a-2a}{a} - \frac{6a-2a}{6a}$$

$$= 1 - \frac{4}{6} = \frac{1}{3} \qquad \text{(Plus sign means tension)}$$

and when $x=3a$ (i.e when the load is at L_4)

$$f_{U3L3} = \frac{3a-3a}{a} - \frac{6a-3a}{6a}$$

$$= -\frac{1}{2} \qquad \text{(Minus sign means compn)}$$

(vi) *Influence line for force in member* U_4L_4

Since no force is induced in the member U_4L_4 at the joint U_4 when the unit load moves from L_1 to L_7, therefore, the influence line for force in member U_4L_4 will be a straight line, coinciding with the base.

Example 4·1. *A Pratt truss consists of 6 panels, each of 6 m, its height being 8 m as shown in Fig. 4·11.*

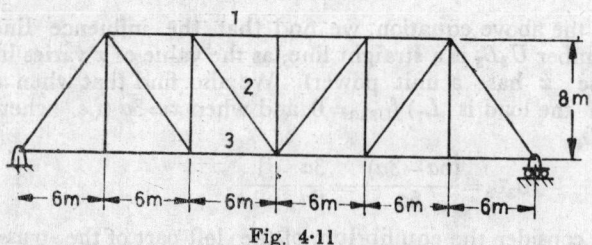

Fig. 4·11

It is simply supported over a span of 36 m and is loaded over the bottom chord.

(i) *Draw the influence lines for force in member serialled 1, 2 and 3, in the third panel from the left, giving principal values.*

(ii) *Calculate the maximum values of forces in members 1, 2 and 3 when a uniformly distributed load of intensity 6 t/m longer than the span crosses the structure.* Bihar University, 1974)

Solution.

Given.   Length of each panel, $a = 6$ m
Height of truss, $h = 8$ m
Span of truss, $l = 36$ m
Intensity of load, $w = 6$ t/m

Maximum force in member 1

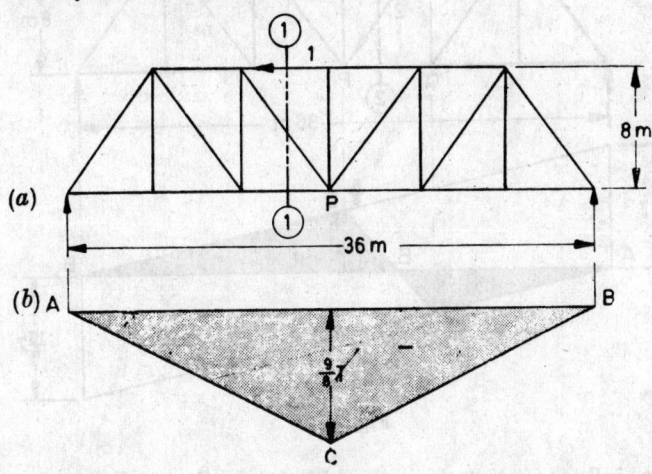

Fig. 4·12

First of all, pass section 1·1 cutting the member 1 as shown in Fig. 4·12 (a). We know that the influence line for the force in member 1 will be given by the influence line for B.M. at the opposite joint P, divided by the vertical distance between the member 1 and the opposite joint P. We know that the influence line for B.M. at P is a triangle having ordinate equal to $\dfrac{18(36-18)}{36} = 9$. Therefore, influence line for member 1 will also be a triangle, having ordinate equal to $9 \times \dfrac{1}{8} = \dfrac{9}{8}$ T (Compression) under the joint P as shown in Fig. 4·12 (b).

From the geometry of Fig. 4·12 (b), we find that the maximum compression force in the member 1 will occur, when the U.D.L. covers the entire span,

$$\therefore \quad F_{max} = 6 \times \text{Area of } \triangle ABC = 6 \times \frac{36}{2} \times \frac{9}{8} = \frac{243}{2} \text{ t}$$

$$= 121·5 \text{ t (Compression)   Ans.}$$

Maximum force in member 2

First of all, pass section 2-2 cutting the member 2 as shown in Fig. 4·13 (a). We know that the influence line for the force in member 2, in the first two panels will be a straight line with ordinate equal to $\dfrac{2}{6} \operatorname{cosec} \theta = \dfrac{1}{3} \times \dfrac{10}{8} = \dfrac{5}{12}$ t (Compression) under the joint Q and in the last 3 panels, it will also be a straight line with ordinate equal to $\frac{3}{6}$ cosec $\theta = \frac{1}{2} \times \frac{10}{8} = \frac{5}{8}$ t (Tension) under the joint

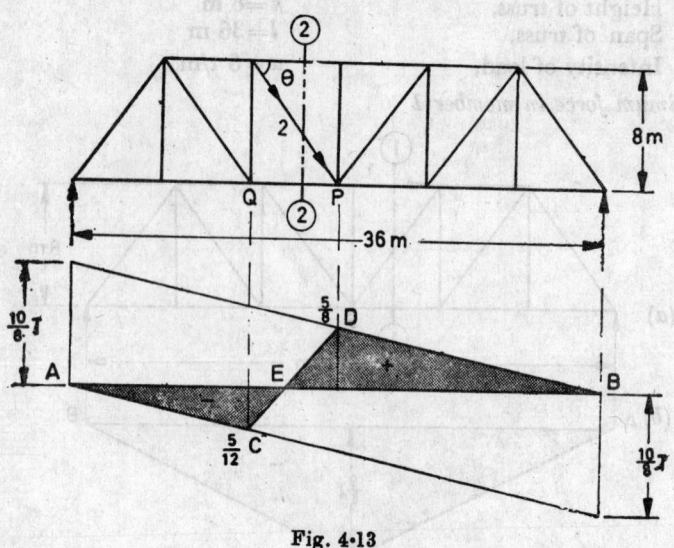

Fig. 4·13

P. The influence line between the joints Q and P will also be a straight line joining the ordinates under the joints Q and P as shown in Fig. 4·13 (b). From the geometry of Fig. 4·13 (b), we find that the point E cuts the base line AB at a distance of $\frac{12}{5}$ m from the joint Q and $\frac{18}{5}$ m from the joint P. We also find that the maximum compressive force in the member 2 will occur, when the U.D.L. covers the section AE of the span.

$$\therefore \qquad F_{max} = 6 \times \text{Area } ACE = 6 \times \frac{1}{2} \left(12 + \frac{12}{5} \right) \times \frac{5}{12} \text{ t}$$

$$= \mathbf{18 \cdot 0} \text{ t (Compn.) Ans}.$$

Similarly, the maximum tensile force in the member 2 will occur, when the U.D.L. covers the section EB of the span.

$$\therefore \qquad F_{max} = 6 \times \text{Area } BDE = 6 \times \frac{1}{2} \left(18 + \frac{18}{5} \right) \times \frac{5}{8} \text{ t}$$

$$= \mathbf{40 \cdot 5} \text{ t (Tension) Ans}.$$

Maximum force in member 3

First of all pass section 3-3 cutting the member 3 as shown in Fig. 4·14(a). We know that the influence line for the force in member 3 will be given by influence line for B.M. at the opposite joint R divided by the vertical distance between the member 3 and the opposite joint R. We know that the influence line for B.M. at R is a triangle having ordinate equal to

$$\frac{12(36-12)}{36} = \frac{12 \times 24}{36} = 8.$$

Therefore the influence line or force in member 3 will also be a triangle having ordinate equal to $8 \times \dfrac{1}{8} = 1$ T under the joint R as shown in Fig. 4·14 (b).

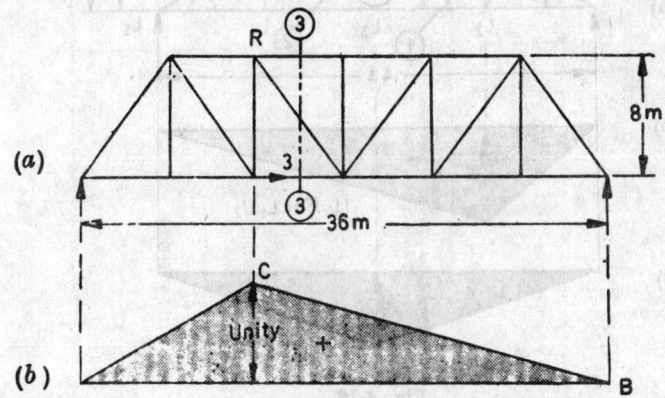

Fig. 4·14

From the geometry of Fig. 4·14 (b) we find that the maximum tensile force in member 3 will occur, when the U.D.L. covers the entire span.

$$\therefore \quad F_{max} = 6 \times \text{Area } ABC = 6 \times \frac{36}{2} \times 1 \ t$$

$$= 108 \ t \ \text{(Tensile)} \quad \text{Ans.}$$

4·6. Influence lines for an inclined Pratt truss

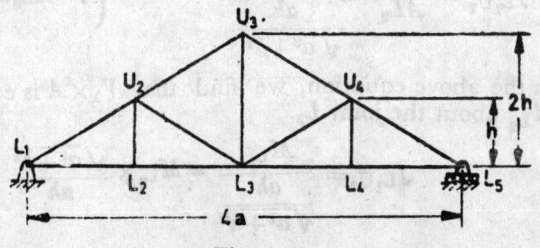

Fig. 4·15

Consider a through type inclined Pratt truss of 4 panels, each of length a and of height $2h$ as shown in Fig. 4·15. From the geometry of the figure, we find that the height of member U_2L_2 is equal to h. Here we shall draw the influence lines for half the members of the truss, when a unit load rolls over the bottom chord. In this type of truss the top chord members (*i.e.* L_1U_2, U_2U_3...) are in compression and the bottom chord members (*i.e.* L_1L_2, L_2L_3......) are in tension.

(*i*) *Influence lines for force in members* L_1U_2 *and* U_2U_3

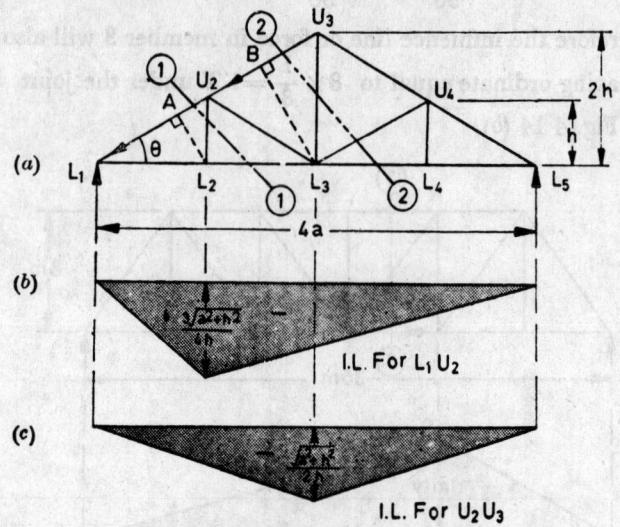

Fig. 4·16

First of all, pass section 1-1 cutting the member L_1U_2 as shown in Fig. 4·16 (*a*). Now through L_2, draw AL_2 perpendicular to L_1U_2. Let the unit load move from the joint L_1 to L_2, Now, consider the equilibrium of the left part of the truss. Assume the direction of the force in member L_1U_2 near the joint L_1 in the left part of truss thus causing compression in the member, as shown in Fig. 4·16 (*a*). Now, taking moments about the opposite joint L_2 and equating the same,

$$f_{L_1U_2} \times AL_2 = V_1 \times a \qquad \text{(where } V_1 \text{ is the vertical reaction at } L_1\text{)}$$

$$\therefore \quad f_{L_1U_2} = \frac{V_1 \times a}{AL_2} = \frac{V_1 \times a}{\dfrac{ah}{\sqrt{a^2+h^2}}} \qquad \left(\because \ AL_2 = \frac{ah}{\sqrt{a^2+h^2}} \right)$$

From the above equation, we find that $V_1 \times A$ is equal to the moment M_{L_2} about the joint L_2.

$$\therefore \qquad\qquad f_{L_1U_2} = \frac{M_{L_2}}{\dfrac{ah}{\sqrt{a^2+h^2}}} = M_{L_2} \times \frac{\sqrt{a^2+h^2}}{ah}$$

Therefore the influence* line for the force in member L_1U_2 is equal to $\dfrac{\sqrt{a^2+h^2}}{ah}$ times the influence line for M_{L_2}. We know that the influence line for M_{L_2} is a triangle with ordinate equal to

———————————————————————
*For mathematical proof please refer to page 84 of this book.

$$\frac{x(l-x)}{l} = \frac{a(4a-a)}{4a} = \frac{a \times 3a}{4a} = \frac{3a}{4}$$

It is thus obvious, that the influence line for the force in member L_1U_2 is a triangle having ordinate equal to

$$\frac{3a}{4} \times \frac{\sqrt{a^2 \times h^2}}{ah} = \frac{3\sqrt{a^2+h^2}}{4h}$$

under the joint L_2 as shown in Fig. 4·16 (b). *Thus the influence line for the force in member L_1U_2 is equal to the influence line for B.M. at the opposite joint L_2, divided by AL_2 i.e., vertical distance between the member L_1U_2 and the opposite joint L_2.*

Note. Similarly, draw the influence lines for the force in member U_2U_3 by passing section and taking moments about the opposite joint L_3, as shown in Fig. 4·16 (c).

(ii) Influence lines for force in members L_1L_2 and L_2L_3

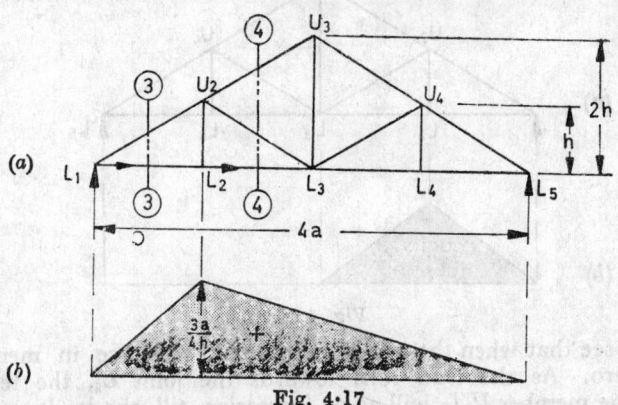

Fig. 4·17

First of all, pass section 3-3 cutting the member L_1L_2 as shown in Fig 4·17 (a). Let the unit load move from the joint L_1 to L_5. Now consider the equilibrium of the left part of the truss. Assume the direction of the force in the member L_1L_2 in the left part of the truss, as shown in Fig. 4·17 (a), thus causing tension in the member. Now taking moments about the opposite joint U_2 and equating the same,

$$f_{L_1L_2} \times h = V_1 \times a \qquad \text{(where } V_1 \text{ is vertical reaction at } L_1\text{)}$$

or

$$f_{L_1L_2} = \frac{V_1 \times a}{h}$$

From the above equation, we find that $V_1 \times a$ is equal to the moment M_{U_2} about the joint U_2.

$$\therefore \qquad f_{L_1L_2} = \frac{M_{U_2}}{h}$$

Therefore the influence[*] line for the force in member L_1L_2 is equal to $1/h$ times the influence line for M_{U_2}. We know that

*For mathematical proof, please refer to page 86 of this book.

the influence line for M_{U_2} is a triangle with the ordinate equal to $\dfrac{x(l-x)}{l} = \dfrac{a(4a-a)}{4a} = \dfrac{3a}{4}$. It is thus obvious, that the influence line for the force in member L_1L_2 is a triangle having ordinate equal to $\dfrac{3a}{4} \times \dfrac{1}{h} = \dfrac{3a}{4h}$ under the joint U_2 as shown in Fig. 4·17 (b). *Thus the influence line for the force in member L_1L_2 is equal to the influence line for B.M. at the opposite joint U_2 divided by h i.e., vertical distance between the member L_1L_2 and the opposite joint U_2.*

Note. The influence line for force in member L_2L_3 is similar to that for L_1L_2. This can be easily proved by passing a section 4-4 cutting the member L_2L_3 as shown in Fig. 4·17 (a).

(iii) *Influence line for force in member U_2L_2*

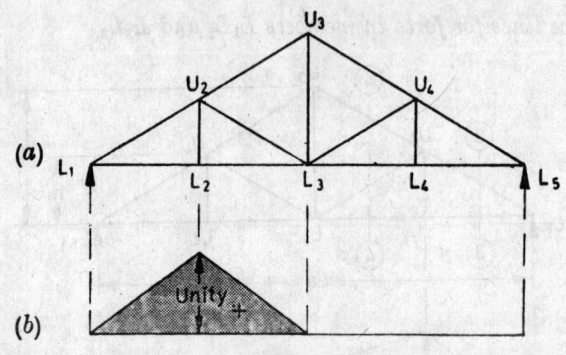

Fig. 4·18

We see that when the unit load is at L_1, the force in member U_2L_2 is zero. As the load rolls towards the joint L_2, the tensile force in the member U_2L_2 will go on increasing, till the load reaches the joint L_2. At this moment, the tensile force in the member U_2L_2 will be equal to unity. As the load moves further, towards the joint L_3, the force in the member U_2L_2 will go on decreasing till the load reaches the joint L_3. At this moment the force in U_2L_2 will also be equal to zero. A little consideration will show, that as the load moves away from L_3, the force in member U_2L_2 will continue to be equal to zero.

Thus the influence line for the force in member U_2L_2 will be a triangle with base equal to L_1L_3 and ordinate equal to unity under the joint L_2 as shown in Fig. 4·18 (b).

(iv) *Influence lines for force in member U_2L_3*

First of all, pass section 5-5 cutting the member U_2L_3 as shown in Fig. 4·19 (a). Now produce the member L_3U_2 and draw a perpendicular L_1C from the joint L_1. From the geometry of the figure, we find that L_1C is equal to $\dfrac{2ah}{\sqrt{a^2+h^2}}$. Let the unit load be between L_1 and L_2. Now consider the equilibrium of the right part of the truss. Assume the direction of force in member U_2L_3

near the joint L_3 in the right part of the truss, thus causing compression in the member, as shown in Fig. 4·19 (a). Now taking moments about the joint L_1 and equating the same,

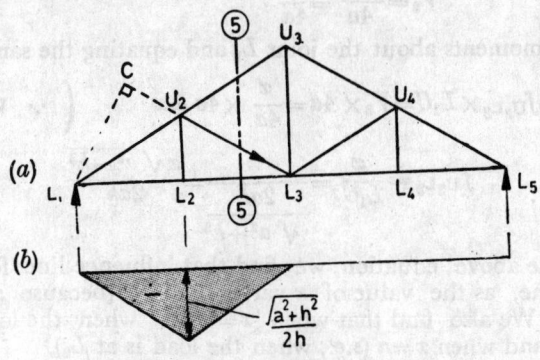

(a)

(b)

$$\frac{\sqrt{a^2+h^2}}{2h}$$

Fig. 4·19

$$f_{U_2L_3} \times L_1C = V_2 \times 4a \qquad \text{(where } V_2 \text{ is the vertical reaction at } L_5\text{)}$$

or $$f_{U_2L_2} = \frac{V_2 \times 4a}{L_1C} = \frac{V_2 \times 4a}{\dfrac{2ah}{\sqrt{a^2+h^2}}} = V_2 \times \frac{2\sqrt{a^2+h^2}}{h}$$

Therefore the influence line for the force in the member U_2L_3 is equal to $\dfrac{2\sqrt{a^2+h^2}}{h}$ times the influence line for V_2. We know that the influence line for the vertical reaction V_2 is a triangle with a unit ordinate at L_5, and zero at L_1. Therefore the influence line for force in member U_2L_3 between L_1 and L_2 will be a straight line with ordinate equal to $\dfrac{1}{4} \times \dfrac{2\sqrt{a^2+h^2}}{h} = \dfrac{\sqrt{a^2+h^2}}{2h}$ under the joint L_2 as shown in Fig. 4·19 (b).

Now let the unit load be between L_3 and L_5. Now consider the equilibrium of the left part of the truss. A little consideration will show, that if we take moments about the joint L_1 there is no moment due to external force (since the moment of vertical reaction at L_1, about the joint L_1 will be zero). It is thus obvious, that the influence line for U_2L_3, when the load is between L_3 and L_5 will coincide with the reference line as shown in Fig. 4·19 (b). The influence line between the joints L_2 and L_3 will also be a straight line joining the ordinates under the joints L_2 and L_3 as shown in Fig. 4·19 (b).

The above statement can be proved, mathematically also, as discussed below :

First of all, pass section 5-5 cutting the member U_2L_3 as shown in Fig. 4·19 (a). Consider the equilibrium of right part of the truss, and the unit load to roll from L_1 to L_2. Let the unit load be at a distance x from the joint L_1. At this instant the vertical reaction at L_5,

$$V_2 = \frac{1 \times x}{4a} = \frac{x}{4a}$$

Taking moments about the joint L_1 and equating the same,

$$f_{U_2L_3} \times L_1C = V_2 \times 4a = \frac{x}{4a} \times 4a = x \qquad \left(\because V_2 = \frac{x}{4a} \right)$$

$$\therefore \qquad f_{U_2L_3} = \frac{x}{L_1C} = \frac{x}{\dfrac{2ah}{\sqrt{a^2+h^2}}} = \frac{x\sqrt{a^2-h^2}}{2ah}$$

From the above equation, we find that influence line for U_2L_3 is a straight line, as the value of x varies linearly (because x has a unit power). We also find that when $x=0$ (*i.e.* when the load is at L_1), $f_{U_2L_3}=0$ and when $x=a$ (*i.e.*, when the load is at L_2),

$$f_{U_2L_3} = \frac{a\sqrt{a^2+h^2}}{2ah} = \frac{\sqrt{a^2+h^2}}{2h}$$

Now consider the equilibrium of the *left part of the truss and the unit load to roll from L_3 to L_5. Now taking moments about the joint L_1,

$$f_{U_2L_3}.\frac{2ah}{\sqrt{a^2 \times h^2}} = 0$$

The right hand side of the equation will be zero, because there is no moment due to external force (the moment of vertical reaction at L_1, about the joint L_1 will be zero). It is thus obvious, that the influence line for U_2L_3, when the load is between L_3 and L_5 will coincide with the reference line.

Now consider the equilibrium of the left part of the truss and the unit load to roll from L_2 to L_3. Since the unit load is rolling in the panel, which has been cut by the section 5-5, therefore, for some time the unit load will be on the left of the section and for some time it will be on the right of the section. Since the load is assumed to be transferred to the truss at the joints only, therefore vertical load transferred to the joint L_2, when the unit load is at a distance x from the joint L_1 (considering the panel $L_2 L_3$ only),

$$V_{L_2} = \frac{1 \times (2a-x)}{a} = \frac{(2a-x)}{a}$$

*If the equilibrium of the right part is considered, then vertical reaction at L_5 when the unit load is at a distance x from the joint L_1

$$V_2 = \frac{x}{4a}$$

Now taking moments about the joint L_1, and equating the same,

$$f_{U_2L_3} \times \frac{2ah}{\sqrt{a^2+h^2}} = V_2 \times 4a - 1 \times x = \frac{x}{4a} \times 4a - x = 0$$
$$f_{U_2L_3} = 0$$

Now taking moments about the joint L_1 and equating the same,

$$f_{U_2 L_3} \times \frac{2ah}{\sqrt{a^2 + h^2}} = \frac{(2a - x)}{a} \times a = (2a - x)$$

or

$$f_{U_2 L_3} = \frac{(2a - x)\sqrt{a^2 + h^2}}{2ah}$$

From the above equation, we find that the influence line for $U_2 L_3$ is a straight line as the value of x varies linearly (because x has a unit power). We also find that when $x = 2a$ (*i.e.* when the load is at L_3) $f_{U_2 L_3} = 0$ and when $x = a$ (*i.e.* when the load is at L_2)

$$f_{U_2 L_3} = \frac{(2a - a)\sqrt{a^2 + h^2}}{2ah} = \frac{\sqrt{a^2 + h^2}}{2h}$$

(ii) Influence line for force in member $U_3 L_3$

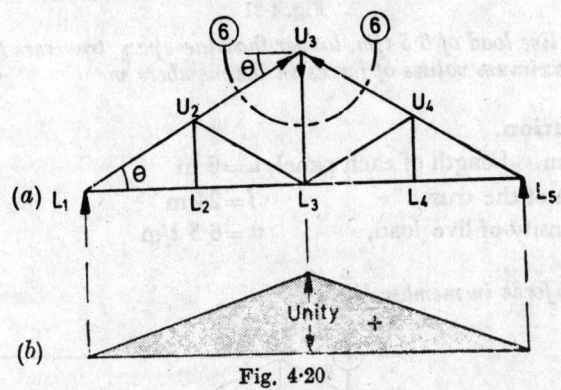

Fig. 4·20

First of all pass section 6-6 cutting the member $U_3 L_3$ as shown in Fig. 4·20 (a). A little consideration will show that the force* in member $U_2 U_3$ will be equal to the force in member $U_3 U_4$. Now resolving the forces vertically at the joint U_3,

$$f_{U_3 L_3} = 2 f_{U_2 U_3} \sin \theta = f_{U_2 U_3} \times 2 \sin \theta$$

$$= f_{U_2 U_3} \times \frac{2h}{\sqrt{a^2 + h^2}} \qquad \left(\because \sin \theta = \frac{h}{\sqrt{a^2 + h^2}} \right)$$

It is thus obvious, that the influence line for the force in member $U_3 L_3$ will be $\dfrac{2h}{\sqrt{a^2 + h^2}}$ times the influence line for $U_2 U_3$. We know that the influence line for force in member $U_2 U_3$ is a triangle with ordinate equal to $\dfrac{\sqrt{a^2 + h^2}}{2h}$ under the joint L_3. Therefore, the influence line for force in member $U_2 U_3$ will also be a triangle with

---

*This can be easily proved by resolving the forces horizontally at the joint U_3 and equating the same.

ordinate equal to $\dfrac{\sqrt{a^2+h^2}}{2h} \times \dfrac{2h}{\sqrt{a^2+h^2}} = 1$ (*i.e.* unity) under the joint L_3 as shown in Fig. 4·20 (*b*).

Example 4·2. *Draw the influence lines for forces in the members U_2U_3, U_2L_3 and L_2L_3 of the truss shown in Fig. 4·21.*

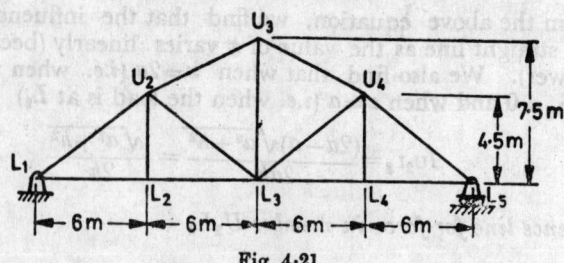

Fig. 4·21

If a live load of 6·5 t/m, longer than the span, traverses the girder, find the maximum values of forces in the members mentioned above.

Solution.

Given.   Length of each panel, $a=6$ m

Span of the truss, $l=24$ m

Intensity of live load, $w=6·5$ t/m

Maximum force in member U_2U_3

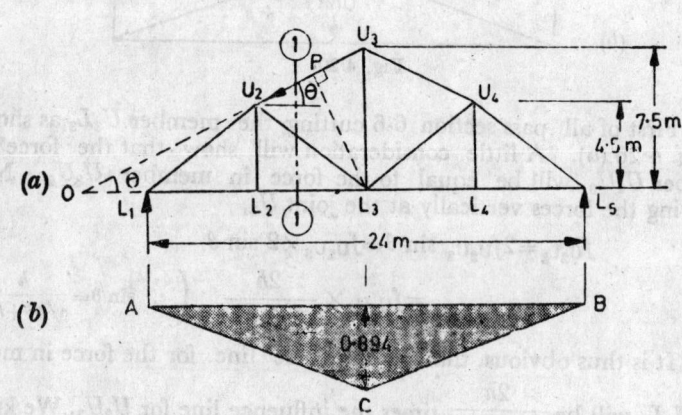

Fig. 4·22

First of all, pass section 1—1 cutting the member U_2U_3 as shown in Fig. 4·22(*a*). Now extend the member U_3U_2 and L_2L_1 meeting at O, and at an angle θ. From the geometry of Fig. 4·22 (*a*), we find that

$$\tan \theta = \frac{7·5-4·5}{6} = \frac{3}{6} = 0·5$$

$$\therefore \qquad \frac{U_3 L_3}{OL_3} = 0.5$$

or
$$OL_3 = \frac{U_3 L_3}{0.5} = \frac{7.5}{0.5} = 15 \text{ m}$$

or
$$OL_1 = 15 - 12 = 3 \text{ m}$$

From the joint L_3, draw PL_3 perpendicular to OU_3. Now from the geometry of the figure, we find that,

$$\sin\theta = \frac{PL_3}{OL_3}$$

or
$$PL_3 = OL_3 \times \sin\theta = 15 \times \frac{1}{\sqrt{5}} = \frac{15}{2.236} = 6.71 \text{ m}$$

We know that the influence line for the force in member $U_2 U_3$ will be given by the influence line for B.M. at the opposite joint L_3 divided by the vertical distance between the member $U_2 U_3$ and the opposite joint L_3 i.e., the distance PL_3. We know that the influence line for B.M. at L_3 is a triangle having ordinate equal to

$$\frac{12(24-12)}{24} = \frac{12 \times 12}{24} = 6$$

Therefore influence line for force in member $U_2 U_3$ will also be a triangle having ordinate $= 6 \times \dfrac{1}{6.71} = 0.894$ T (Compression) under the joint L_3 as shown in Fig. 4·22 (b).

From the geometry of Fig. 4·22 (b), we find that the maximum compressive force in member $U_2 U_3$ will occur, when the U.D.L. covers the entire span.

$$\therefore \qquad F_{max} = 6.5 \times \text{Area of } ABC = 6.5 \times \frac{24}{2} \times 0.894 \text{ t}$$

$$= 69.73 \text{ t (Compn.) Ans.}$$

Maximum force in member $U_2 L_3$

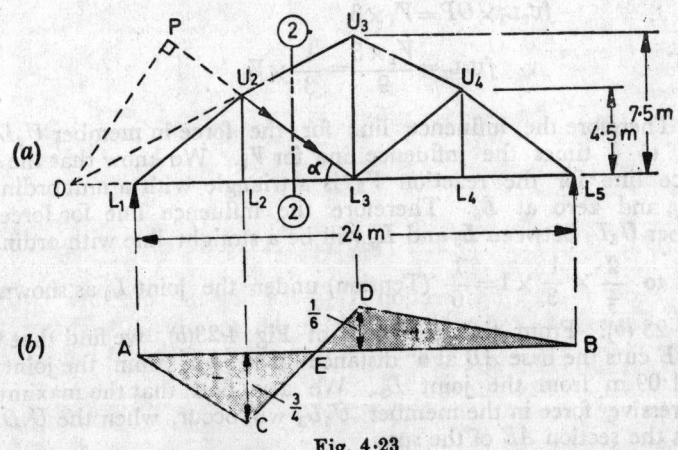

Fig. 4·23

First of all, pass section 2—2 cutting the member U_2L_3 as shown in Fig. 4·23(a). Now extend the number, U_3U_2 and L_3L_1 meeting at O. We know that $OL_1=3$ m and $OL_3=12+3=15$ m. Let $\angle U_2L_3L_2=\alpha$. From the geometry of Fig. 4·23(a), we find that

$$\tan \alpha=\frac{4\cdot5}{6}=\frac{3}{4}$$

$\therefore$ $$\sin \alpha=\frac{3}{5}$$

Through O, draw OP perpendicular to L_3U_2 produced at P.

$\therefore$ $$OP=OL_3 \sin \alpha=15\times\frac{3}{5}=9 \text{ m}$$

Now consider the equilibrium of the right part of the truss and the unit load to roll from L_1 to L_2. Assume the direction of the force in member U_2L_3 near the joint L_3, thus causing compression in the member as shown in Fig. 4·23(a). Now taking moments about O and equating the same,

$$f_{U_2L_3}\times OP=V_2\times(24+3)$$

or $$f_{U_2L_3}=\frac{V_2\times 27}{9}=3V_2$$

Therefore, the influence line for the force in member U_2L_3 is equal to 3 times the influence line for V_2. We know that the influence line for the reaction V_2 is a triangle with a unit ordinate at L_5 and zero at L_1. Therefore the influence for force in member U_2L_3 between L_1 and L_2 will be a straight line with ordinate equal to $\frac{1}{4}\times3\times1$ $=\frac{3}{4}$ (Compn.) under the joint L_2 as shown in Fig. 4·23 (b).

Now, let the unit load be between L_3 and L_5. Now consider the equilibrium of the left part of the truss and the unit load to roll from L_3 to L_5. Assume the direction of force in member U_2L_3 near the joint U_2, thus causing tension in the member as shown in Fig. 4·23 (b). Now taking moments about O and equating the same,

$\therefore$ $$f_{U_2L_3}\times OP=V_1\times 3$$

or $$f_{U_2L_3}=\frac{V_1\times 3}{9}=\frac{1}{3}\times V_1$$

Therefore the influence line for the force in member U_2L_3 is equal to $\frac{1}{3}$ times the influence line for V_1. We know that the influence line for the reaction V_2 is a triangle with a unit ordinate at L_1, and zero at L_5. Therefore the influence line for force in member U_2L_3 between L_3 and L_5 will be a straight line with ordinate equal to $\frac{2}{4}\times\frac{1}{3}\times1=\frac{1}{6}$ (Tension) under the joint L_3 as shown in Fig. 4·23 (b). From the geometry of Fig. 4·23(b), we find that the point E cuts the base AB at a distance of 4·91 m from the joint L_2 and 1·09 m from the joint L_3. We also find that the maximum compressive force in the member U_2L_3 will occur, when the U.D.L. covers the section AE of the span.

$$\therefore \quad F_{max} = 6 \cdot 5 \times \text{Area } ACE = 6 \cdot 5 \times \frac{1}{2} \left(6 + 4 \cdot 91 \right) \times \frac{3}{4} \, t$$

$$= 26 \cdot 6 \quad ^t \quad \textbf{Ans.}$$

Similarly, the maximum tensile force in the member $U_2 L_3$ will occur, when U.D.L. covers the section, EB of the span.

$$\therefore \quad F_{max} = 6 \cdot 5 \times \text{Area } BDE = 6 \cdot 5 \times \frac{1}{2} \, (12 + 1 \cdot 09) \times \frac{1}{6} \quad t$$

$$= 7 \cdot 09 \quad ^t \quad \textbf{Ans.}$$

Maximum force in member $L_2 L_3$

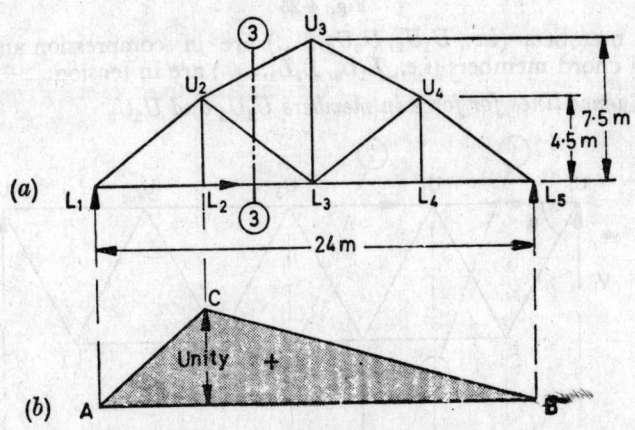

Fig. 4·24

First of all, pass section 3-3 cutting the member $L_2 L_3$ as shown in Fig. 4·24 (a). We know that the influence line for the force in member $L_2 L_3$ will be given by the influence line for B.M. at the opposite joint U_2, divided by the vertical distance between the member $L_2 L_3$ and the opposite joint U_2. We know that the influence line for B.M. at the joint U_2 is a triangle having ordinate equal to $\frac{6(24-6)}{24} = \frac{6 \times 18}{24} = 4 \cdot 5$. Therefore the influence line for force in member $L_2 L_3$ will also be a triangle having ordinate equal to $4 \cdot 5 \times \frac{1}{4 \cdot 5} = 1$ under the joint U_2 as shown in Fig. 4·24 (b).

From the geometry of Fig. 4·24 (b), we find that the maximum tensile force in the member $L_2 L_3$ will occur, when the U.D.L. covers the entire span.

$$\therefore \quad F_{max} = 6 \cdot 5 \times \text{Area } ABC = 6 \cdot 5 \times \tfrac{1}{2} \times 24 \times 1 \quad t$$

$$= 78 \quad ^t \quad \textbf{Ans.}$$

4·7. Influence lines for a deck type Warren truss

Consider a deck type warren truss of 4 panles each of length a and of height h as shown in Fig. 4·25. Here we shall draw the influence lines for half the members of the truss, when a unit load rolls over the top chord. In a deck type Warren truss, the top

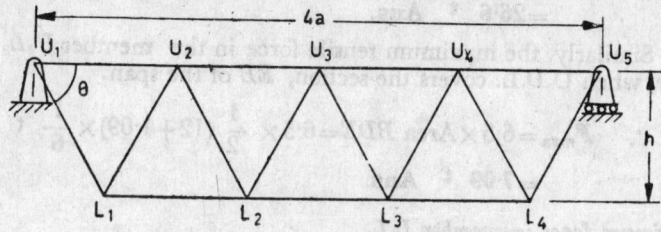

Fig. 4·25

chord members (*i.e.*, U_1U_2, U_2U_3......) are in compression and the bottom chord members (*i.e.*, L_1L_2, L_2L_3......) are in tension.

(*i*) *Influence lines for force in members U_1U_2 and U_2U_3*

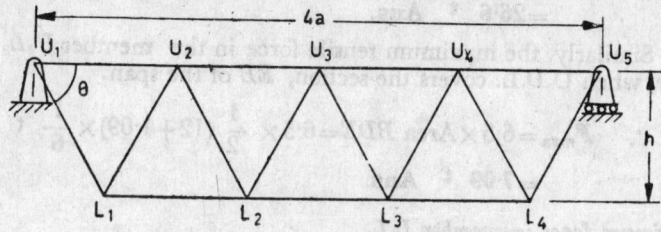

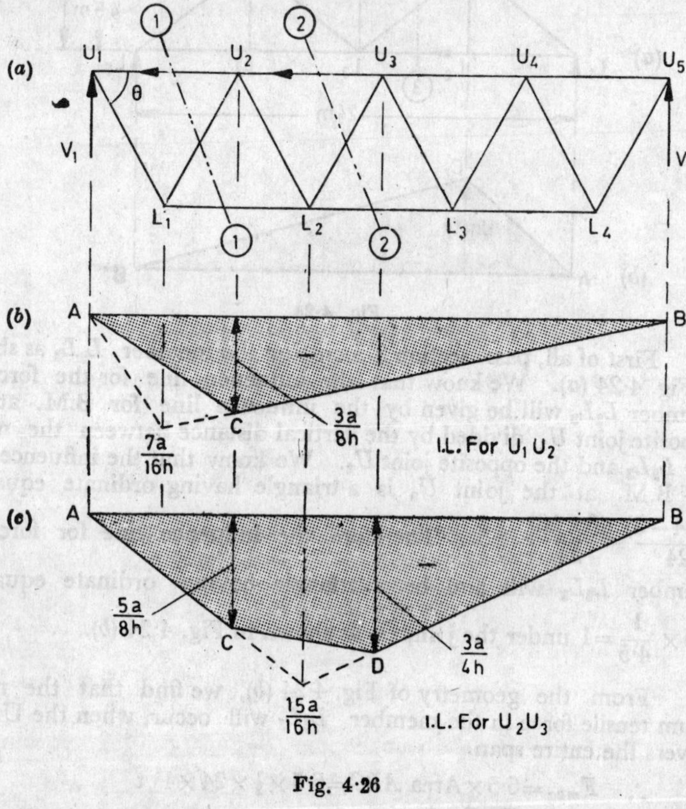

Fig. 4·26

First of all, pass section 1·1 cutting the member U_1U_2 as shown in Fig. 4·26 (*a*). Let the unit load move from the joint U_1 to U_5. Now consider the equilibrium of the left part of the truss and assume the direction of force in member U_1U_2 in the left part of the truss, thus causing compression in the member as shown in

Fig. 4·26 (a). Now taking moments about the *joint L_1 and equating the same,

$$f_{U_1U_2} \times h = V_1 \times \frac{a}{2}$$

(where V_1 is the vertical reaction at U_1)

or

$$f_{U_1U_2} = V_1 \times \frac{a}{2} \times \frac{1}{h}$$

From the above equation, we find that $V_1 \times \frac{a}{2}$ is equal to the moment M_{L_1} about the joint L_1.

$$\therefore \quad f_{U_1U_2} = \frac{M_{L_1}}{h}$$

Therefore the influence line for the force in member U_1U_2 is equal to $1/h$ times the influence line for M_{L_1}. We know that the influence line for M_{L_1} is a triangle with ordinate equal to

$$\frac{x(l-x)}{l} = \frac{\frac{a}{2}\left(4a - \frac{a}{2}\right)}{4a} = \frac{\frac{a}{2} \times \frac{7a}{2}}{4a} = \frac{7a}{16}$$

Thus for influence line for U_1U_2 between the joints U_2 and U_5 will be a straight line, such that the ordinate under the joint L_1 is equal to $\frac{7a}{16h}$ as shown in Fig. 4·26 (b).

Now draw the vertical ordinates from the joints U_1 and U_2 on the influence line. From the geometry of the figure, we find that the ordinate under the joint U_1 is zero and that under the joint U_2 is $\frac{3a}{8h}$. Join these ordinates. Now ACB is the required influence line as shown in Fig. 4·26 (b).

The above statement can be proved, mathematically also, as discussed below :

First of all, pass section 1-1 cutting the member U_1U_2 as shown in Fig. 4·26 (a). Consider the equilibrium of the left part of the truss and the unit load to roll from U_2 to U_5. Let the load be at a distance x from the joint U_1. At this instant, the vertical reaction at L_1,

$$V_1 = \frac{4a - x}{4a}$$

Taking moments about the joint L_1 and equating the same,

$$f_{U_1U_2} \times h = \frac{(4a-x)}{4a} \times \frac{a}{2} = \frac{(4a-x)}{8}$$

*Since the joint L_1 is not the opposite joint to member U_1L_2 (because it is not vertically below either of the joint L_1 or L_2) therefore the relation for influence line holds good only from the joints U_2 to U_5 i.e. in all the other spans, except that which has been cut by the section.

$$\therefore \qquad f_{U_1U_2} = \frac{(4a-x)}{8h}$$

From the above equation, we find that the influence line for U_1U_2 is a straight line, as the value of x varies linearly (because x has a unit power). We also find that when $x=4a$ (i.e. when the load is U_5) $F_{U_1U_2}=0$ and when $x=a$ (i.e. when the load is U_2),

$$f_{U_1U_2} = \frac{4a-a}{8h} = \frac{3a}{8h}$$

Now consider the equilibrium of the left part and the unit load to roll from U_1 to U_2. Let the unit load be at a distance x from the joint U_1. Since the unit load is rolling in the panel, which has been cut by the section 1-1 therefore, for some time, the unit load will be on the left of the section and for some time it will be on the right of the section. We know that the load is assumed to be transferred to the truss at the joints only, therefore vertical load transferred to the joint U_1, when the unit load is at a distance x from the joint U_1 (considering the panel U_1U_2 only),

$$V_{U_1} = \frac{a-x}{a}$$

and the vertical reaction of the joint U_1 (considering the truss as a whole),

$$V_1 = \frac{4a-x}{4a}$$

Now taking moments about the joint L_1 and equating the same,

$$f_{U_1U_2} \times h = V_1 \times \frac{a}{2} - V_{U_1} \times \frac{a}{2}$$

$$= \frac{(4a-x)}{4a} \times \frac{a}{2} - \frac{(a-x)}{a} \times \frac{a}{2}$$

$$= \frac{(4a-x)}{8} - \frac{(a-x)}{2}$$

or $\qquad f_{U_1U_2} = \dfrac{\dfrac{(4a-x)}{8} - \dfrac{(a-x)}{2}}{h}$

From the above equation, we find that the influence line for force in member U_1U_2 is a straight line, as the value of x varies linearly (because x has a unit power) We also find that when $x=0$ (i.e. when the load is at U_1) $F_{U_1U_2}=0$ and when $x=a$ (i.e. when the load is at U_2).

$$f_{U_1U_2} = \frac{\dfrac{(4a-a)}{8} - \dfrac{(a-a)}{2}}{h} = \frac{3a}{8h}$$

Note : Similarly, draw the influence lines for the member U_2U_3 by passing section 2-2 and taking moments about the joint L_1.

(ii) *Influence lines for forces in members U_1L_1, L_1L_2 and L_2L_3*

First of all, pass section 3-3 cutting the member U_1L_1 as shown in Fig. 4·27 (a). Now through U_2 draw AU_2 perpendicular to U_1L_1.

Now consider the equilibrium of the left part of the truss and assume direction of the force in member U_1L_1 in the left part of the truss, as shown in Fig. 4·27 (a), thus causing tension in the member. Now taking moments about the *joint U_2 and equating the same.

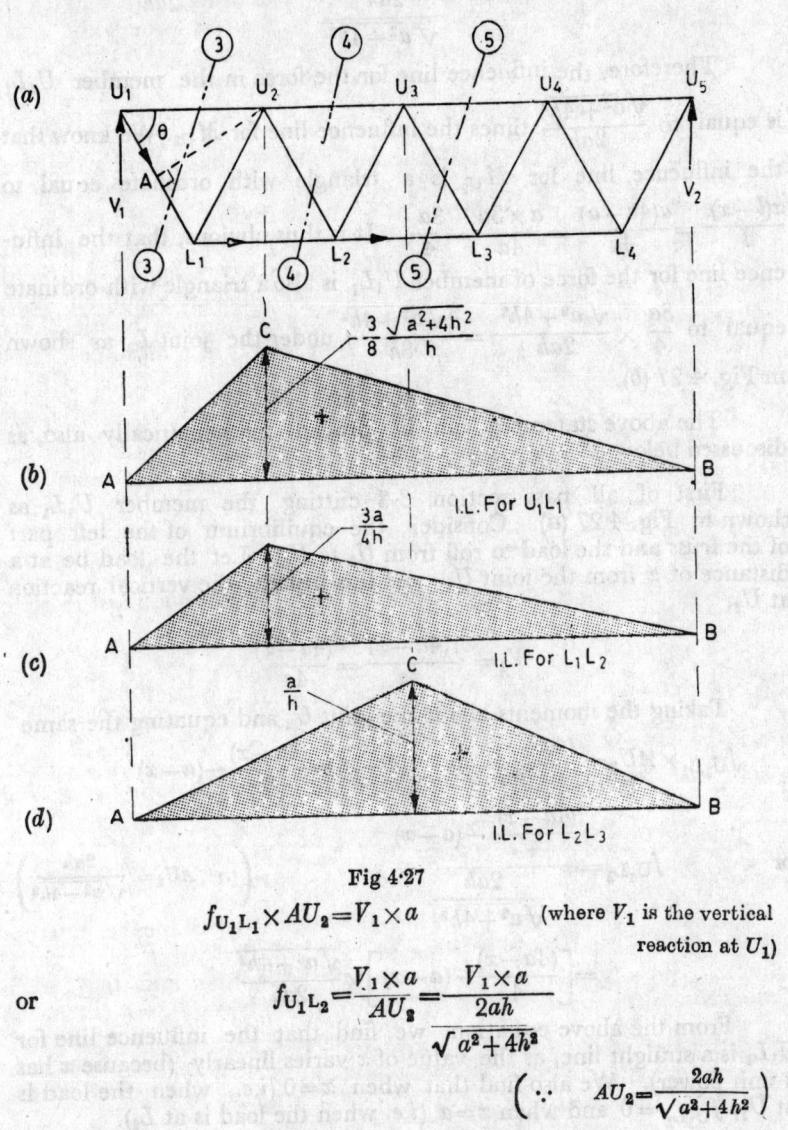

Fig 4·27

$$f_{U_1L_1} \times AU_2 = V_1 \times a \qquad \text{(where } V_1 \text{ is the vertical}$$
$$\text{reaction at } U_1)$$

or

$$f_{U_1L_2} = \frac{V_1 \times a}{AU_2} = \frac{V_1 \times a}{\dfrac{2ah}{\sqrt{a^2+4h^2}}}$$

$$\left(\because \quad AU_2 = \frac{2ah}{\sqrt{a^2+4h^2}} \right)$$

---

*The joint U_2 is not the opposite joint to the member U_1L_1 (because it is not vertically above the joints U_1 or L_1). Since the load is not rolling over the member U_1L_1, therefore for all practical purposes, the joint U_2 will behave like an opposite joint.

From the above equation, we find that $V_1 \times a$ is equal to the moment M_{U_2} about the joint U_2

$$\therefore \qquad f_{U_1L_1} = \frac{M_{U_2}}{\dfrac{2ah}{\sqrt{a^2+4h^2}}} = M_{U_2} \times \frac{\sqrt{a^2+4h^2}}{2ah}$$

Therefore, the influence line for the force in the member U_1L_1 is equal to $\dfrac{\sqrt{a^2+4h^2}}{2ah}$ times the influence line for M_{U_2}. We know that the influence line for M_{U_2} is a triangle with ordinate equal to $\dfrac{x(l-x)}{l} = \dfrac{a(4a-a)}{4a} = \dfrac{a \times 3a}{4a} = \dfrac{3a}{4}$. It is thus obvious, that the influence line for the force of member U_1L_1 is also a triangle with ordinate equal to $\dfrac{3a}{4} \times \dfrac{\sqrt{a^2+4h^2}}{2ah} = \dfrac{3\sqrt{a^2+4h^2}}{8h}$ under the joint L_2 as shown in Fig. 4·27 (b).

The above statement can be proved, mathematically also, as discussed below :

First of all pass section 3-3 cutting the member U_1L_1 as shown in Fig. 4·27 (a). Consider the equilibrium of the left part of the truss and the load to roll from U_1 to U_2. Let the load be at a distance of x from the joint U_1. At this instant, the vertical reaction at U_1,

$$V_1 = \frac{1(4a-x)}{4a} = \frac{(4a-x)}{4a}$$

Taking the moments about the joint U_2 and equating the same

$$f_{U_1L_1} \times AU_2 = \frac{(4a-x) \times a}{4a} - 1(a-x) = \frac{(4a-x)}{4} - (a-x)$$

or $\qquad f_{U_1L_1} = \dfrac{\dfrac{(4a-x)}{4} - (a-x)}{\dfrac{2ah}{\sqrt{a^2+4h^2}}}$ $\qquad \left(\because AU_2 = \dfrac{2ah}{\sqrt{a^2-4h^2}} \right)$

$$= \left[\frac{(4a-x)}{4} - (a-x) \right] \times \frac{\sqrt{a^2+4h^2}}{2ah}$$

From the above equation, we find that the influence line for U_1L_1 is a straight line, as the value of x varies linearly (because x has a unit power). We also find that when $x=0$ (i.e., when the load is at U_1) $f_{U_1L_1} = 0$ and when $x=a$ (i.e. when the load is at L_2).

$$f_{U_1L_1} = \left[\frac{(4a-a)}{4} - (a-a) \right] \times \frac{\sqrt{a^2+4h^2}}{2ah}$$

$$= \frac{3\sqrt{a^2+4h^2}}{8h}$$

Now consider the unit load to roll from U_2 to U_5. Let the load be at a distance x from the joint U_1. At this instant, the vertical reaction at U_1,

$$V_1 = \frac{1(4a-x)}{4a} = \frac{(4a-x)}{4a}$$

Taking moments about the joint U_2 and equating the same,

$$f_{U_1L_1} \times AU_2 = \frac{(4a-x)}{4a} \times a = \frac{(4a-x)}{4}$$

or
$$f_{U_1L_1} = \frac{(4a-x)}{4AU_2} = \frac{(4a-x)}{\dfrac{4 \times 2ah}{\sqrt{a^2+4h^2}}} \qquad \left(\because AU_2 = \frac{2ah}{\sqrt{a^2+4h^2}} \right)$$

$$= \frac{(4a-x)\sqrt{a^2+4h^2}}{8ah}$$

From the above equation, we find that the influence line for U_1L_1 is a straight line, as the value of x varies linearly (because x has a unit power). We also find that when $x=a$ (i.e. when the load is at U_2)

$$f_{U_1L_1} = \frac{(4a-a)\sqrt{a^2+4h^2}}{8ah} = \frac{3\sqrt{a^2+4h^2}}{8h}$$

and when $x=4a$ (i.e. when the load is at U_5) $f_{U_1L_1}=0$.

Note. Similarly, draw the influence lines for the members L_1L_2 and L_2L_3 by passing sections (4—4) and (5—5) and taking moments about the joints U_2 and U_3 as shown in Fig. 4·27 (c) and (d).

(iii) Influence lines for forces in members L_1U_2 and L_2U_3

First of all pass section 6—6 cutting the member L_1U_2 as shown in Fig. 4·28 (a). Let the load be between U_2 and U_5. Now consider the equilibrium of the left *part of the truss. Since the reaction V_1 is acting upwards, therefore, the direction of the force in member L_1U_2 near the joint L_1 will be downwards, thus causing compression in the member, as shown in Fig. 4·28 (a).

$$\therefore \quad f_{L_1U_2} \sin\theta = V_1$$

or
$$f_{L_1U_2} = \frac{V_1}{\sin\theta} = V_1 \operatorname{cosec}\theta \quad (\text{Compression})$$

Therefore, the influence line for the force in the member L_1U_2 is equal to cosec θ times the influence line for V_1. We know that the influence line for the reaction V_1 is a triangle with a unit ordinate at U_1 and zero at U_5. Thus the influence line for force in member

---

*If the equilibrium of the right part is considered, then the direction of the force in the member L_1U_2 near the joint U_2 will be upwards.

$$\therefore \quad f_{L_1U_2} \sin\theta = 1-V_2 = V_1$$

or
$$f_{L_1U_2} = V_1 \operatorname{cosec}\theta \quad (\text{Compression}).$$

L_1U_2 between U_2 and U_5 will be a straight line with ordinate equal to $\dfrac{3}{4}$ cosec θ under the joint U_2 as shown in Fig. 4·28 (b). The in-

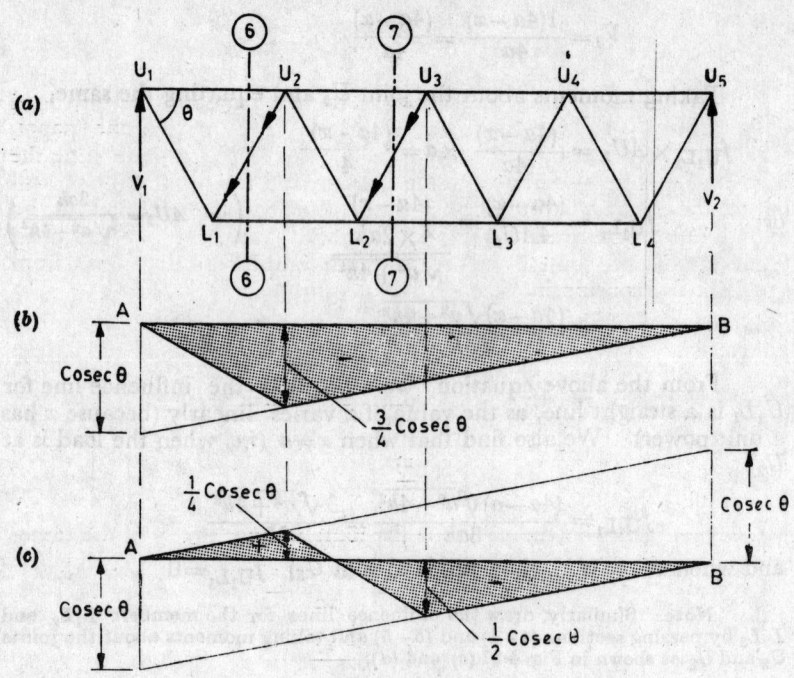

Fig. 4·28

fluence line between the joints U_1 and U_2 will also be a straight line joining the ordinates under the joint U_2 to the zero ordinate under the joint U_1 as shown in Fig. 4·28 (b).

The above statement can be proved, mathematically also, as discussed below :

First of all, pass section 6-6 cutting the member L_1U_2 as shown in Fig. 4·28(a). Consider the equilibrium of left part of the truss and the unit load to roll from U_2 to U_5. Let the unit load be at a distance x from the joint U_1. At this instant, the vertical reaction at U_1,

$$V_1 = \frac{1 \times (4a-x)}{4a} = \frac{(4a-x)}{4a}$$

Since the vertical reaction V_1 is acting upwards, therefore direction of the force in member L_1U_2 near the joint U_2 will be downwards, causing compression in the member as shown in Fig. 4·28(a).

$$\therefore \qquad f_{L_1U_2} \sin \theta = V_1 = \frac{(4a-x)}{4a}$$

or

$$f_{L_1U_2} = \frac{4a-x}{4a \sin \theta} = \frac{(4a-x) \ cosec \ \theta}{4a}$$

From the above equation, we find that the influence line for L_1U_2 is a straight line, as the value of x varies linearly (because x has a unit power). We also find that when $x=4a$ (i.e., when the load is at U_5) $f_{L_1U_2}=0$ and when $x=a$ (i.e., when the load is at U_2),

$$f_{L_1U_2} = \frac{(4a-a)\ \text{cosec}\ \theta}{4a} = \frac{3}{4}\ \text{cosec}\ \theta$$

Now, consider the equilibrium of the left part and the unit load to roll from U_1 to U_2. Since the unit load is rolling in the panel, which has been cut by the section 6—6, therefore for some time the unit load will be on the left of the section and for some time it will be on the right of the section. Since the load is assumed to be transferred to the truss at the joints only, therefore vertical load transferred to the joint U_1, when the unit load is at a distance x from the joint U_1 (considering the panel U_1U_2 only)

$$V_{U_1} = \frac{(a-x)}{a}$$

and the vertical reaction at the joint U_1 (considering the truss as a whole),

$$V_1 = \frac{(4a-x)}{4a}$$

Now taking moments about the joint L_1 and equating the same,

$$f_{L_1U_2} \sin\ \theta = V_1 - U_{U1} = \frac{(4a-x)}{4a} - \frac{(a-x)}{a}$$

or

$$f_{L_1U_2} = \frac{\dfrac{(4a-x)}{4a} - \dfrac{(a-x)}{a}}{\sin\ \theta}$$

$$= \left[\frac{(4a-x)}{4a} - \frac{(a-x)}{a}\right]\ \text{cosec}\ \theta$$

From the above equation, we find that the influence line for L_1U_2 is a straight line as the value of x varies linearly (because x has a unit power). We also find that when $x=0$ (i.e. when the load is at U_1) $f_{L_1U_2}=0$ and when $x=a$ (i.e. when the load is at U_2),

$$f_{L_1U_2} = \left[\frac{(4a-a)}{4a} - \frac{(a-a)}{a}\right]\ \text{cosec}\ \theta = \frac{3}{4}\ \text{cosec}\ \theta$$

Note : Similarly, draw the influence line for the member L_2U_3 by section (7—7) as shown in Fig. 4·28 (c).

(iv) Influence lines for force in member U_2L_2

First of all, pass section 8-8 cutting the member U_2L_2 as shown in Fig. 4·29 (a). Let the load be between U_1 and U_2. Now consider the equilibrium of the right part of the truss. Since the vertical reaction V_2 is acting upwards, therefore direction of the force in member U_2L_2 near the joint L_2 will be downwards, thus causing compression in the member, as shown in Fig. 4·29 (a).

$$\therefore \qquad f_{U_2L_2} \sin\ \theta = V_2$$

or

$$f_{U_2L_2} = \frac{V_2}{\sin\ \theta} = V_2\ \text{cosec}\ \theta\ \text{(Compn)}$$

Therefore, the influence line for the force in member U_2L_2 is equal to cosec θ times the influence line for V_2. We know that

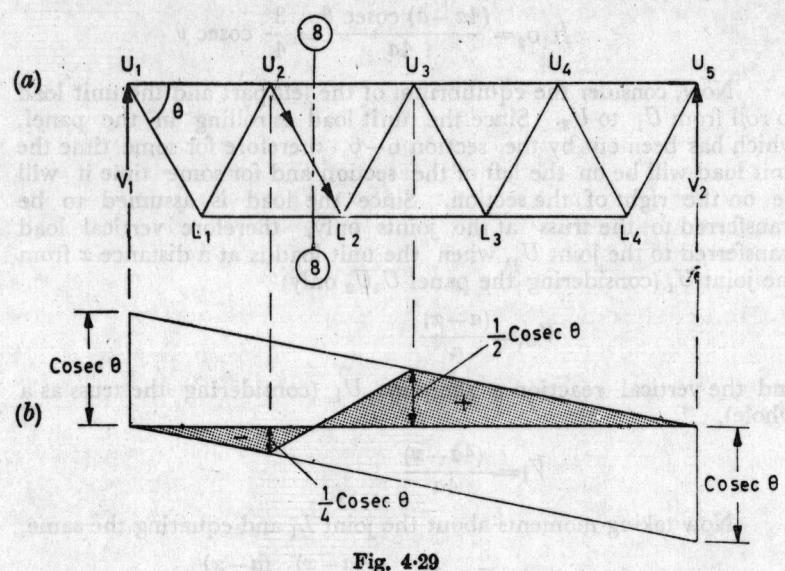

Fig. 4·29

the influence line for the reaction V_2 is a triangle with a unit ordinate at U_5, and zero at U_1. Therefore the influence line for force in member U_2L_2 between U_1 and U_2 will be a straight line with ordinate equal to $\frac{1}{4}$ cosec θ under the joint U_2 and zero under the joint U_1 as shown in Fig. 4·29 (b).

Now let the unit load be between U_3 and U_5. Consider the equilibrium of the left part of the truss. Since the vertical reaction V_1 is acting upwards, therefore direction of the force in member U_2L_2 near the joint U_5 will be downwards, thus causing tension in the member, as shown in Fig. 4·29 (a).

$$\therefore \qquad f_{U_2L_2} \sin \theta = V_1$$

or $\qquad\qquad f_{U_2L_2} = \dfrac{V_1}{\sin \theta} = V_1 \text{ cosec } \theta \text{ (Tension)}$

Therefore, the influence line for the force in the member U_2L_2 is equal to cosec θ times the influence line for V_1. We know that the influence line for the reaction V_1 is a triangle with a unit ordinate at U_1, and zero at U_5. Thus the influence line for force in member U_2L_2 between U_1 and U_2 will be a straight line with ordinate equal to $\frac{1}{2}$ cosec θ under the joint U_3 and zero under the joint U_5 as shown in Fig. 4·29 (b). The influence line between the joints U_2 and U_3 will also be a straight line joining the ordinates under the joints U_2 and U_3 as shown in Fig. 4·29(b).

The above statement can be proved, mathematically also, as discussed below :

First of all, pass section 8—8 cutting the member U_2L_2 as shown in Fig. 4·29 (a). Consider the equilibrium of the right part of the truss and the unit load to roll from U_1 to U_2. Let the load be at a distance x from the joint U_1. At this instant, vertical reaction at U_5,

$$V_2 = \frac{1 \times x}{4a} = \frac{x}{4a}$$

Since the vertical reaction V_2 is acting upwards, therefore direction of the force in member U_2L_2 near the joint L_2 will be downwards, thus causing compression in the number, as shown in Fig. 4·29 (a).

$$\therefore \qquad f_{U_2L_2} \ \sin \theta = V_2 = \frac{x}{4a}$$

or
$$f_{U_2L_2} = \frac{x}{4a \sin \theta} = \frac{x \ \mathrm{cosec} \ \theta}{4a}$$

From the above equation, we find that the influence line for U_2L_2 is a straight line, as the value of x varies linearly (because x has a unit power). We also find that when $x=0$ (i.e. when the load is at U_1) $f_{U_2L_2}=0$ and when $x=a$ (i.e. when the load is at U_2)

$$f_{U_2L_2} = \frac{a \ \mathrm{cosec} \ \theta}{4a} = \frac{1}{4} \ \mathrm{cosec} \ \theta$$

Now consider the equilibrium of the left part of the truss and the unit load to roll from U_3 to U_5. Let the unit load be at a distance x from the joint U_1. At this instant the vertical reaction at U_1,

$$V_1 = \frac{1 \times (4a-x)}{4a} = \frac{(4a-x)}{4a}$$

Since the vertical reaction V_1 is acting upwards, therefore direction of the force in member U_2L_2 near the joint U_2 will be downwards, thus causing tension in the member as shown in Fig. 4·29(a).

$$\therefore \qquad f_{U_2L_2} \ \sin \theta = V_1 = \frac{(4a-x)}{4a}$$

or
$$f_{U_2L_2} = \frac{(4a-x)}{4a \sin \theta} = \frac{(4a-x) \ \mathrm{cosec} \ \theta}{4a}$$

From the above equation, we find that the influence line for U_2L_2 is a straight line as the value of x varies linearly (because x has a unit power). We also find that when $x=4a$ (i.e. when the load is at U_5) $f_{U_2L_2}=0$ and when $x=2a$ (i.e. when the load is at U_3),

$$f_{U_2L_2} = \frac{(4a-2a) \ \mathrm{cosec} \ \theta}{4a} = \frac{1}{2} \ \mathrm{cosec} \ \theta$$

Now consider the equilibrium of the left part and the unit load to be rolling from U_2 to U_3. Since the unit load is rolling in the panel, which has been cut by the section 8-8, therefore for some time the unit load will be on the left of the section and for some time it will be on the right of the section. We know that the load is assumed to be transferred to the truss at the joints only, therefore, vertical load transferred to the joint U_2 when the unit load is at a distance of x from the joint U_1,

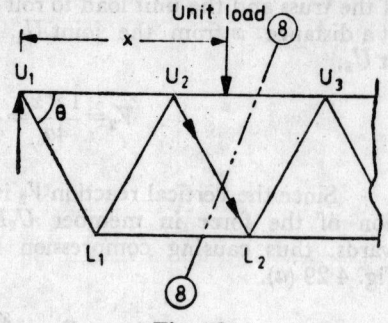

Fig. 4·30

$$V_{U_2} = \frac{(2a-x)}{a}$$

and vertical reaction at the joint U_1 (considering the truss as a whole),

$$V_1 = \frac{(4a-x)}{4a}$$

$$\therefore \quad f_{U_2L_2} \sin\theta = V_1 - V_{U_2} = \frac{(4a-x)}{4a} - \frac{(2a-x)}{a}$$

or

$$f_{U_2L_2} = \left[\frac{(4a-x)}{4a} - \frac{(2a-x)}{a} \right] \operatorname{cosec}\theta$$

From the above equation, we find that the influence line for U_2L_2 is a straight line as the value of x varies linearity (because x has a unit power). We also find that when $x=a$ (*i.e.*, when the load is at U_2)

$$f_{U_2L_2} = \left[\frac{(4a-a)}{4a} - \frac{(2a-a)}{a} \right] \operatorname{cosec}\theta = \left(\frac{3}{4} - 1 \right) \operatorname{cosec}\theta$$

(Minus sign means compn.)

$$= -\frac{1}{4} \operatorname{cosec}\theta$$

and when $x=2a$ (*i.e.* when the load is at U_3)

$$f_{U_2L_2} = \left[\frac{(4a-2a)}{4a} - \frac{(2a-2a)}{a} \right] \operatorname{cosec}\theta$$

$$= \frac{1}{2} \operatorname{cosec}\theta$$

(Plus sign means tension)

4·8. Influence lines for a composite truss

Consider a through type composite truss of 8 panels, each of length a and height $2h$ as shown in Fig. 4·31. A little consideration will show, that the truss consists of (*i*) a primary truss of panels each length of $2a$ and of height $2h$ and (*ii*) 4 secondary trusses of 2 panels, each of length a and of height h as shown in Fig. 4·32 (a) and (b). Here we shall draw the influence lines for all the members in one

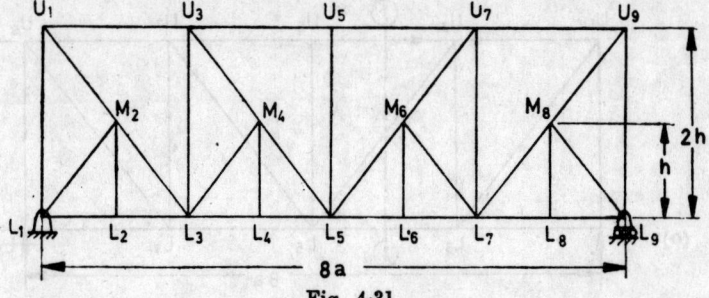

Fig. 4·31

panel of the primary truss (i.e. U_3U_5, L_3L_4, L_4L_5, U_3M_4, M_4L_5, M_4L_4, L_3M_4 and U_3L_3). Now, let us draw the two trusses independently as shown in Fig. 4·32 (a) and (b).

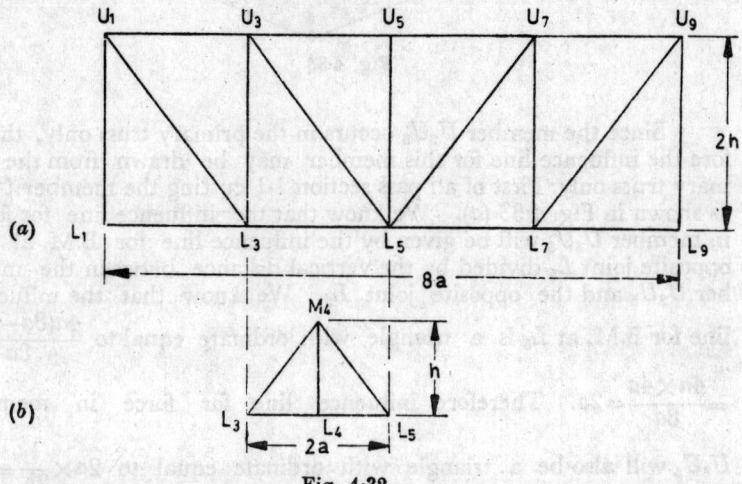

Fig. 4·32

A little consideration will show, that some of the members in the second panel of the primary truss occur in the primary truss only (e.g. U_3U_5, U_3M_4, and U_3L_3). Some of the members occur in the secondary truss only (e.g. M_4L_4 and L_3M_4). But some of the members occur in both the primary as well as secondary trusses (e.g. L_3L_4, L_4L_5, M_4L_5). The influence lines for the members, which occur in the primary truss only, will be given by the influence line for the corresponding member of the primary truss only. Similarly, the influence lines for the members, which occur in the secondary truss only, will be given by the influence line for the corresponding member of the secondary truss only. But the influence lines for the members which occur in both the primary and secondary trusses, may be drawn by joining the points obtained by algebraically adding the ordinates & the influence lines for the corresponding members in the primary as well as the secondary truss.

(i) Influence lines for force in members U_3U_5

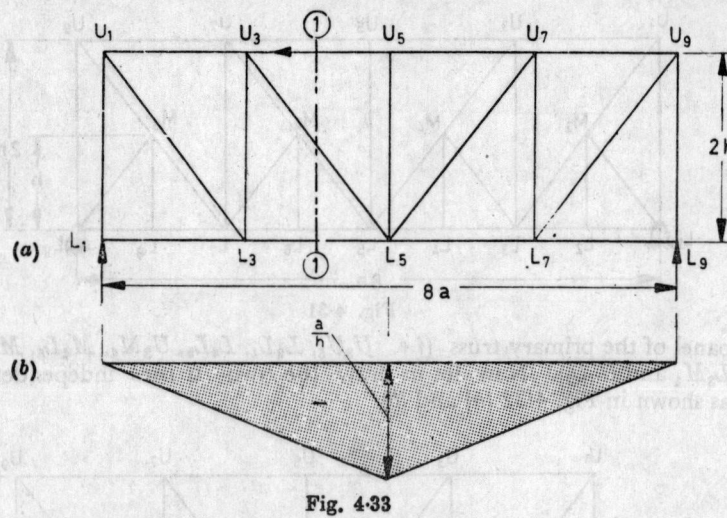

Fig. 4·33

Since the member U_3U_5 occurs in the primary truss only, there-fore the influence line for this member may be drawn from the pri-mary truss only. First of all pass section 1-1 cutting the member U_3U_5 as shown in Fig. 4·33 (a). We know that the influence line for force in member U_3U_5 will be given by the influence line for B.M. at the opposite joint L_5 divided by the vertical distance between the mem-ber U_3U_5 and the opposite joint L_5. We know that the influence line for B.M. at L_5 is a triangle with ordinate equal to $\dfrac{4a(8a-4a)}{8a}$

$=\dfrac{4a \times 4a}{8a}=2a$. Therefore influence line for force in member

U_3U_5 will also be a triangle with ordinate equal to $2a \times \dfrac{1}{2h}=\dfrac{a}{h}$

(Compression) under the joint L_5 as shown in Fig. 4·33 (b).

(ii) Influence line for force in member U_3M_4

Since the member U_3M_4 occurs in the primary truss only, therefore the influence line for this member will be the same as for the member U_3L_5 in the primary truss. First of all pass section 2—2 cutting the member U_3L_5 as shown in Fig. 4·34 (a) We know that the influence* line for U_3L_5 between the joints L_1 and L_3 will be a straight line with ordinate equal to $\dfrac{2}{8}$ cosec $\theta=\dfrac{1}{4}$ cosec θ (Comp-ression) under the joint L_3, and between the joints L_5 and L_9, it will also be a straight line with ordinate equal to $\frac{1}{2}$ cosec θ (Tension)

---

*For details, please refer to page 91 of this book.

under the joint L_5. The influence line between the joints L_3 and L_5 will also be a straight line joining the ordinate under the joints L_3 and L_5 as shown in Fig. 4·34 (b).

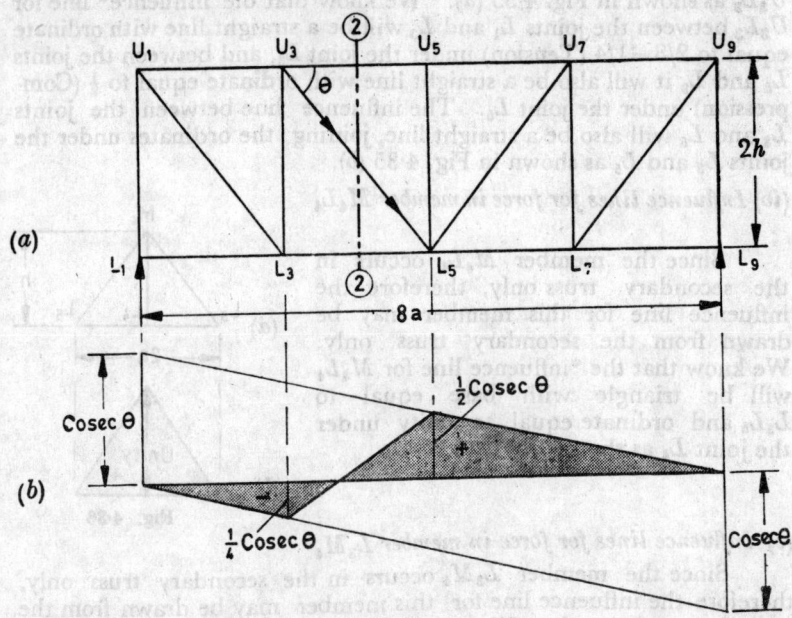

Fig. 4·34

(iii) Influence line for force in member U_3L_3

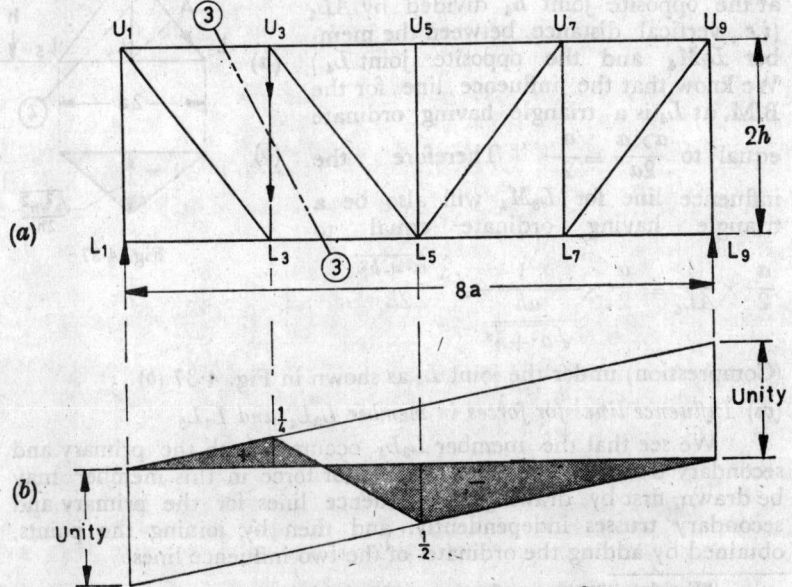

Fig. 4·35

Since the member U_3L_3 occurs in the primary truss only, therefore the influence line for this member may be drawn from the primary truss only. First of all pass section 3-3 cutting the member U_3L_3 as shown in Fig. 4·35 (a). We know that the influence* line for U_3L_3 between the joints L_1 and L_3 will be a straight line with ordinate equal to $2/8=1/4$ (Tension) under the joint L_3, and between the joints L_5 and L_9 it will also be a straight line with ordinate equal to $\frac{1}{2}$ (Compression) under the joint L_5. The influence line between the joints L_3 and L_5 will also be a straight line, joining the ordinates under the joints L_3 and L_5 as shown in Fig. 4·35 (b).

(iv) Influence lines for force in member M_4L_4

Since the member M_4L_4 occurs in the secondary truss only, therefore the influence line for this member may be drawn from the secondary truss only. We know that the *influence line for M_4L_4 will be triangle with base equal to L_3L_5 and ordinate equal to unity under the joint L_4 as shown in Fig. 4·36 (b).

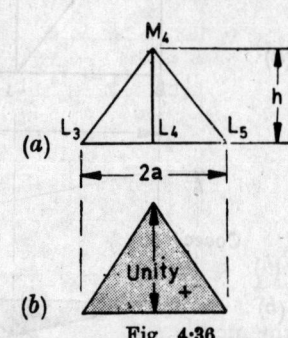

Fig. 4·36

(v) Influence lines for force in member L_3M_4

Since the member L_3M_4 occurs in the secondary truss only, therefore the influence line for this member may be drawn from the secondary truss only. We know that the influence line for L_3M_4 will be given by the influence line for B.M. at the opposite joint L_4 divided by AL_4 (i.e., vertical distance between the member L_3M_4 and the opposite joint L_4.) We know that the influence line for the B.M. at L_4 is a triangle having ordinate equal to $\dfrac{a \times a}{2a} = \dfrac{a}{2}$. Therefore the influence line for L_3M_4 will also be a triangle having ordinate equal to

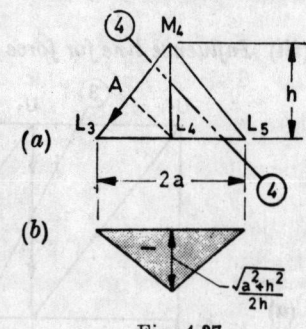

Fig. 4·37

$$\frac{a}{2} \times \frac{1}{AL_4} = \frac{a}{2} \times \frac{1}{\dfrac{ah}{\sqrt{a^2+h^4}}} = \frac{\sqrt{a^2+h^2}}{2h}$$

(Compression) under the joint L_4 as shown in Fig. 4·37 (b).

(vi) Influence lines for forces in member L_3L_4 and L_4L_5

We see that the member L_3L_4 occurs in both the primary and secondary trusses. The influence line for force in this member may be drawn, first by drawing the influence lines for the primary and secondary trusses independently, and then by joining the points, obtained by adding the ordinates of the two influence lines.

*For details, please refer to page 87 of this book.

The influence line for L_3L_4 may be first drawn for the member L_3L_5 from the primary truss. First of all pass section 5—5 cutting

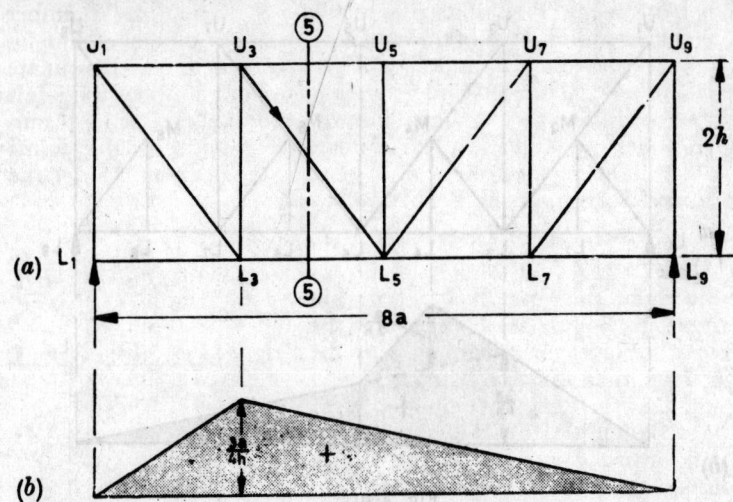

(a)

(b)

Fig. 4·38

the member L_3L_5 as shown in Fig. 4·38 (a). We know that the influence* line for force in member L_3L_5 will be given by the influence line for B.M. at the opposite joint U_3 divided by the vertical distance between the member L_3L_5 and the opposite joint U_3 of the primary truss. We know that the influence line for B.M. at U_3 is a triangle having ordinate equal to $\dfrac{2a(8a-2a)}{8a}=\dfrac{2a\times6a}{8a}=\dfrac{3a}{2}$ under the joint L_3. Therefore the influence line for force in member L_3L_5 will also be a triangle with ordinate equal to $\dfrac{3a}{2}\times\dfrac{1}{2h}=\dfrac{3a}{4h}$ (Tension) under the joint U_3 as shown in Fig. 4·38 (b).

We also know that the influence line for force in member L_3L_4 in the secondary truss will be given by the influence line for B.M. at the opposite joint M_4 divided by the vertical distance between the member L_3L_4 and opposite joint M_4 of the secondary truss. We know that the influence line for B.M. at M_4 is a triangle with ordinate equal to $\dfrac{a\times(2a-a)}{2a}=\dfrac{a\times a}{2a}=\dfrac{a}{2}$ under the joint L_4. Therefore the influence line for force in member L_3L_4 will also be a triangle with ordinate equal to $\dfrac{a}{2}\times\dfrac{1}{h}=\dfrac{a}{2h}$ (Tension) under the joint L_4 as shown in Fig. 4·39 (b)

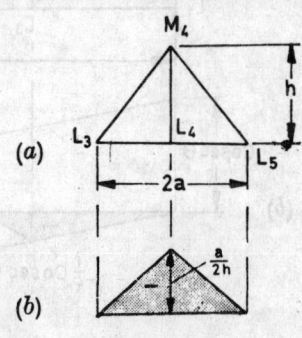

(a)

(b)

Fig. 4·39

We see that the member M_4L_4 occurs in both the primary and secondary trusses. The influence line for this member may be drawn

Now complete the influence line for force in member L_3L_4, by superimposing the influence line for force in L_3L_4 in the secondary

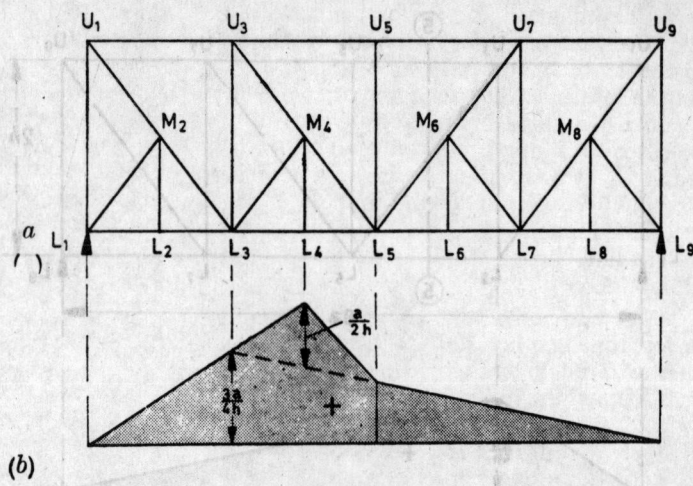

(b)

Fig. 4·40

truss to the influence line for force in L_3L_5, in the primary truss as shown in Fig. 4·40 (b).

Note : The influence line for L_4L_5 is similar to that for L_3L_4.

(vii) Influence line for force in member M_4L_5.

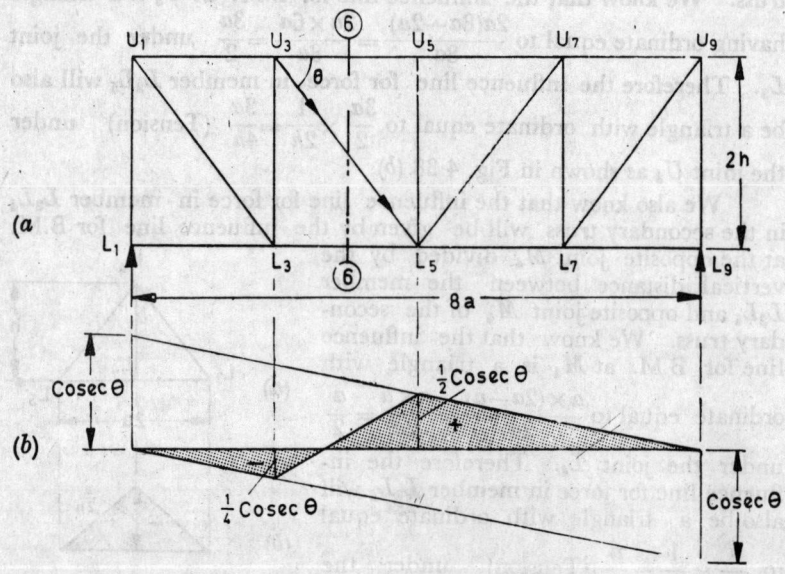

Fig. 4·41

We see that the member M_4L_5 occurs in both the primary and secondary trusses. The influence line for this member may be drawn first by drawing the influence lines for the primary and secondary trusses independently, and then by joining the points, obtained by adding the ordinates of the two influence lines.

The influence line for force in member M_4L_5 will be the same as for the member U_3L_5 in the primary truss. First of all, pass section 6-6 cutting the member U_3L_5 as shown in Fig. 4·41 (a). We know that the influence* line for force in member U_3L_5 between the joints L_1 and L_3 will be a straight line with ordinate equal to $\frac{2}{8}$ cosec θ $=\frac{1}{4}$ cosec θ (Compression) under the joint L_3, and between the joints L_5 and L_9, it will also be a straight line with ordinate equal to $\frac{1}{2}$ cosec θ (Tension) under the joint L_5.

The influence line between the joints L_3 and L_5 will also be a straight line joining the ordinate under the joints L_3 and L_5 as shown in Fig. 4·41 (b).

We also know that the influence line for force in member M_4L_5 in the secondary truss will be given by the influence line for B.M. at the opposite joint L_4 divided (a) by BL_4 i.e. vertical distance between the member M_4L_5 and the opposite joint L_4. We know that the influence line for the (b) B.M. at L_4 is a triangle with ordinate

equal to $\dfrac{a \times a}{2a} = \dfrac{a}{2}$. Therefore the in-

fluence line for force in member M_4L_5 will also be a triangle with ordinate

equal to $\dfrac{a}{2} \times \dfrac{1}{BL_4} = \dfrac{a}{2} \times \dfrac{1}{\dfrac{ah}{\sqrt{a^2+h^2}}} = \dfrac{\sqrt{a^2+h^2}}{ah} =$ (Compression)

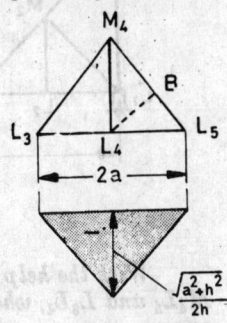

Fig. 4·42

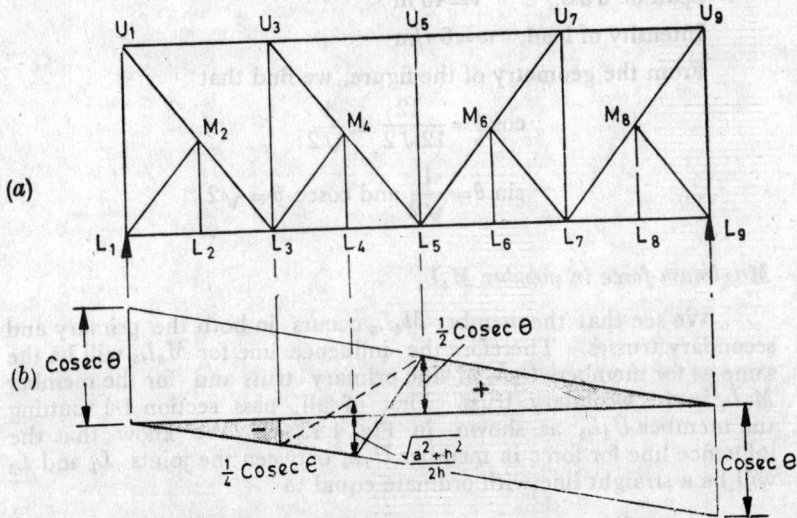

Fig. 4·43

under the joint L_4 as shown in Fig. 4·42 (b).

Now the complete influence line for force in member is M_4L_5 by superimposing the influence line for force in M_4L_5 in the secondary truss, to the influence line for force in U_3L_5 in the primary truss as shown in Fig. 4·43 (b).

Example 4·3. *Draw the influence lines for the forces in the members M_4L_5 and L_3L_4 for the truss shown in Fig. 4·44.*

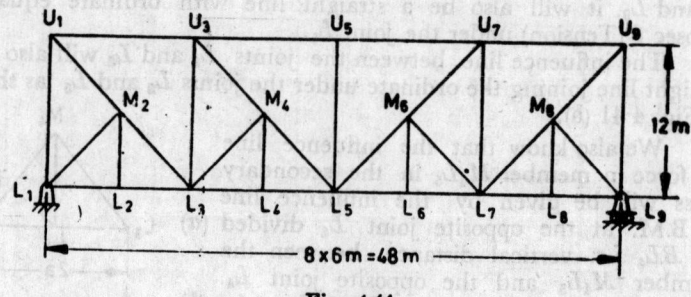

Fig. 4·44

With the help of the above, find the maximum force in the members M_4L_5 and L_3L_4, when there is a moving load of 6 t/m longer than the span.

(*London University*),

Solution.

Given. Length of each panel,

$$a = 6 \text{ m}$$

Span of truss, $l = 48 \text{ m}$

Intensity of load, $w = 6$ t/m

From the geometry of the figure, we find that

$$\cos \theta = \frac{12}{12\sqrt{2}} = \frac{1}{\sqrt{2}}$$

$\therefore$ $\sin \theta = \dfrac{1}{\sqrt{2}}$ and $\operatorname{cosec} \theta = \sqrt{2}$

Maximum force in member M_4L_5

We see that the member M_4L_5 occurs in both the primary and secondary trusses. Therefore the influence line for M_4L_5 will be the same as for members U_3L_5 of the primary truss and for the member M_4L_5 in the secondary truss. First of all, pass section 1-1 cutting the member U_3L_5 as shown in Fig. 4·45 (a). We know that the influence line for force in member U_3L_5 between the joints L_1 and L_3 will be a straight line with ordinate equal to

$$\frac{2}{8} \operatorname{cosec} \theta = \frac{1}{4} \times \sqrt{2} = \frac{1}{2\sqrt{2}} \text{ (Compression)}$$

under the joint L_3 and between the joints L_5 and L_9, it will also be a straight line with ordinate equal to

$$\frac{1}{2}\ \mathrm{cosec}\ \theta = \frac{1}{2} \times \sqrt{2} = \frac{1}{\sqrt{2}}\ \text{(Tension)}$$

under the joint L_5. The influence line between the joints L_3 and L_5 will also be a straight line joining the ordinates under the joints L_3

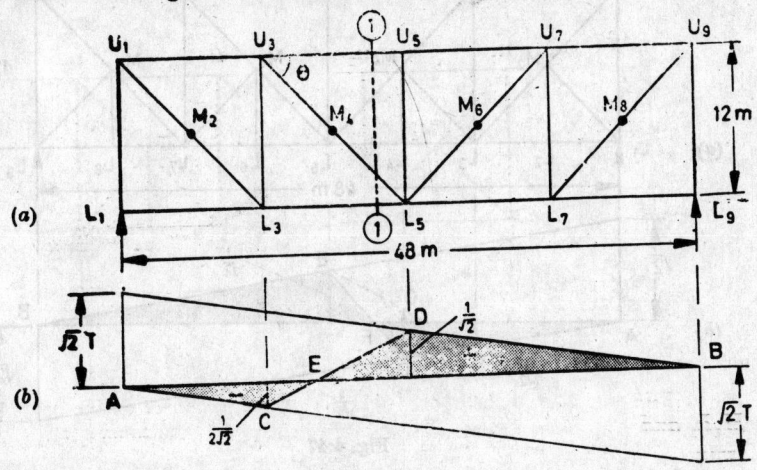

Fig. 4·45

and L_5 as shown in Fig. 4·45 (b). From the geometry of Fig. 4·45 (b), we find that the point E cuts the base line AB at a distance of 4 m from the joint L_3 and 8 m from the joint L_5.

$$\therefore\ \text{Ordinate of the influence line under the joint } M_4 = \frac{1}{4\sqrt{2}}$$

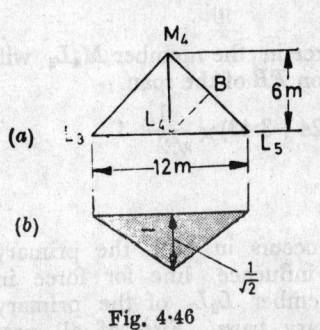

Fig. 4·46

We also know that the influence line for force in member M_4L_5 in the secondary truss will be given by the influence line for B.M. at the opposite joint L_4 divided by the vertical distance between the member M_4L_5 and the opposite joint distance L_4 i.e. BL_4. We know that the influence line for B.M. at L_4 is a triangle with ordinate equal to $\frac{6 \times 6}{12} = 3$. Therefore influence line for M_4L_5 will also be a triangle with

ordinate equal to $3 \times \frac{1}{3\sqrt{2}} = \frac{1}{\sqrt{2}}$ (Compression) under the joint L_4 as shown in Fig. 4·46 (b).

Now complete the influence line for force in member M_4L_5 by superimposing the influence line for force in member M_4L_5 of the secondary truss, to the influence line for force in member U_3L_5 of the primary truss as shown in Fig. 4·47 (b). Therefore ordinate of the complete diagram under the joint M_4

$$= \frac{1}{4\sqrt{2}} = \frac{1}{\sqrt{2}} = -\frac{3}{4\sqrt{2}} = \frac{3}{4\sqrt{2}} \text{ (Compn.)}$$

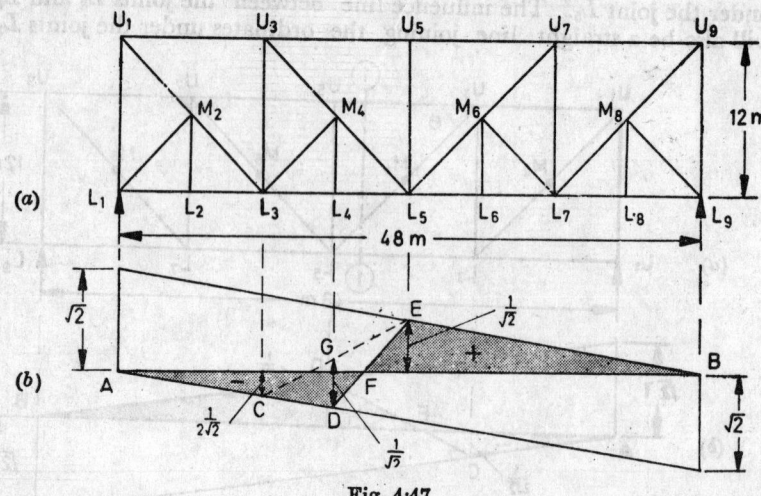

Fig. 4·47

It is thus obvious, that ACD will be a straight line. From the geometry of Fig. 4·47 (b), we find that the point F cuts the base AB at a distance of 3·43 m from the joint L_5 and 2·57 m from the joint L_4. We also find that the maximum compressive force in the member M_4L_5 will occur, when the U.D.L. covers the section AF of the span.

$$\therefore \quad F_{max} = 6 \times \text{Area } AFD = 6 \times \tfrac{1}{2} (18 + 2 \cdot 57) \times \frac{3}{4\sqrt{2}} \text{ t}$$

$$= 32 \cdot 74 \text{ t} \quad \textbf{Ans.}$$

Similarly, the maximum tensile force in the member M_4L_5 will occur, when the U.D.L. covers the section FB of the span.

$$\therefore \quad F_{max} = 6 \times \text{Area } FBE = 6 \times \tfrac{1}{2} (24 + 3 \cdot 43) \times \frac{1}{\sqrt{2}} \text{ t}$$

$$= 58 \cdot 20 \text{ t} \quad \textbf{Ans.}$$

Maximum force in member L_3L_4

We see that the member L_3L_4 occurs in both the primary and secondary truss. Therefore the influence line for force in member L_3L_4 will be the same as for member L_3L_5 of the primary truss and for member L_3L_4 of the secondary truss. First of all pass section 2-2 cutting the members L_3L_5 as shown in Fig. 4·48(a). We know that the influence line for the force in member L_3L_5 will be given by the influence line for B.M. at the opposite joint U_3 divided by the vertical distance between the member L_3L_5 and the opposite joint U_3. We know that the influence line for B.M. at U_3 is a triangle having ordinate equal to $\frac{12 \times 36}{48} = 9$. Therefore influence line for the force in member L_3L_5 will also be a triangle having ordinate equal to $\dfrac{9}{12}=\dfrac{3}{4}$ (Tension) under the joint U_3 as shown in Fig. 4·48 (b).

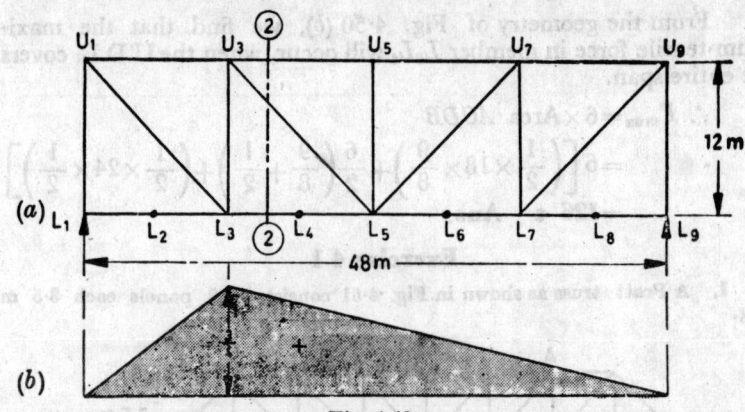

(a)

(b)

Fig. 4·48

We also know that the influence line for the force in member L_3L_4 of the secondary truss will be given by the influence line for B.M. at the opposite joint M_4 divided by the vertical distance between the member L_3L_4 and the opposite joint M_4. We know that the influence line for B.M. at M_4 is a triangle having ordinate equal to $\dfrac{6\times6}{12}=3$. (a)

Therefore the influence line for the force in member L_3L_4 will also be a triangle having ordinate equal to $3\times\frac{1}{6}=\frac{1}{2}$ (Tension) under the joint M_4 as shown in Fig. 4·49 (b). (b)

Now complete the influence line for force in member L_3L_4 by superimposing the

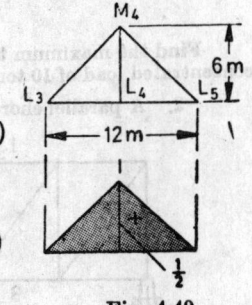

Fig. 4·49

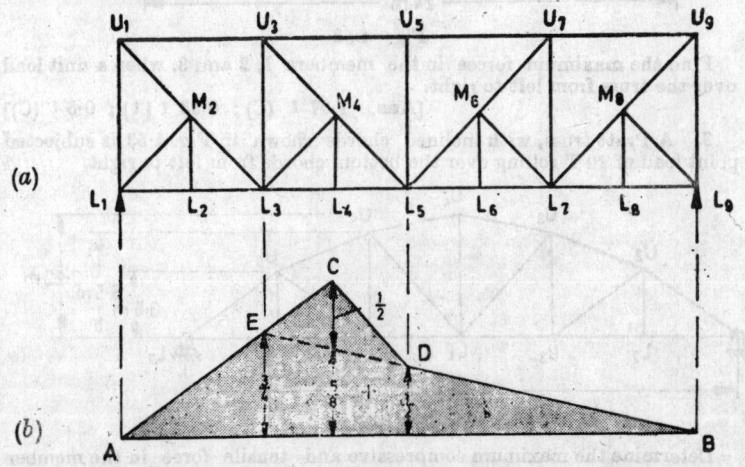

(a)

(b)

Fig. 4·50

influence line for the force in member L_3L_4 of the secondary truss to the influence line for the force in member L_3L_5 of the primary truss as shown in Fig. 4·50 (b).

From the geometry of Fig. 4·50 (b), we find that the maximum tensile force in member L_3L_4 will occur, when the U.D.L. covers the entire span.

$$\therefore F_{max} = 6 \times \text{Area } ACDB$$
$$= 6\left[\left(\frac{1}{2} \times 18 \times \frac{9}{8}\right) + \frac{6}{2}\left(\frac{9}{8} + \frac{1}{2}\right) + \left(\frac{1}{2} \times 24 \times \frac{1}{2}\right)\right]$$
$$= 126 \text{ t   Ans.}$$

Exercise 4·1

1. A Pratt truss as shown in Fig. 4·51 consists of 8 panels each 3·5 m long.

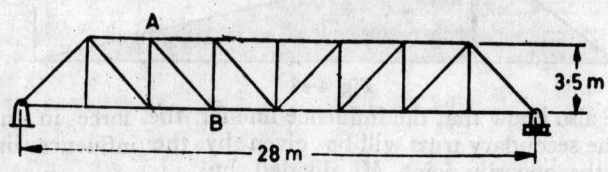

Fig. 4·51

Find the maximum tension and compression in the member AB when a concentrated load of 10 tonnes rolls over the truss. [**Ans.** $2·5\sqrt{2}$ t ; $6·25\sqrt{2}$ t]

2. A parallel chord through type N-truss is shown in Fig. 4·52.

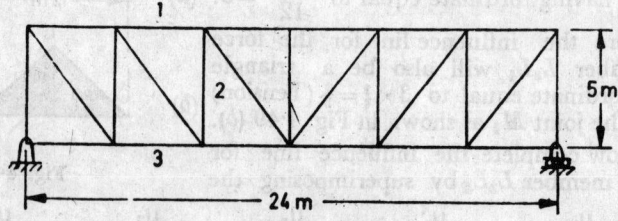

Fig. 4·52

Find the maximum forces in the members 1, 2 and 3, when a unit load rolls over the truss from left to right.

[**Ans.** $1·07$ t (C) ; $0·33$ t (t) ; $0·5$ t (C)]

3. A Pratt truss, with inclined chords shown in Fig. 4·53 is subjected to a point load of 10 T rolling over the bottom chords from left to right.

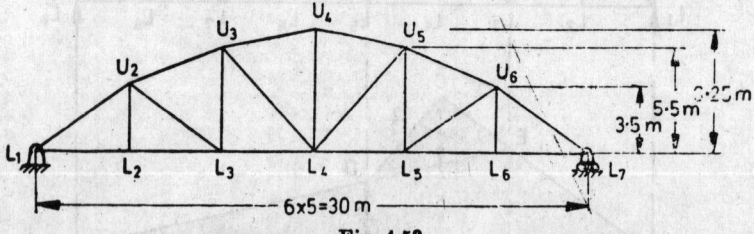

Fig. 4·53

Determine the maximum compressive and tensile force in the member U_2L_3.   [**Ans.** $7·13$ t ; $3·17$ t]

4. Fig. 4·54 shows a Warren truss with inclined top chords.

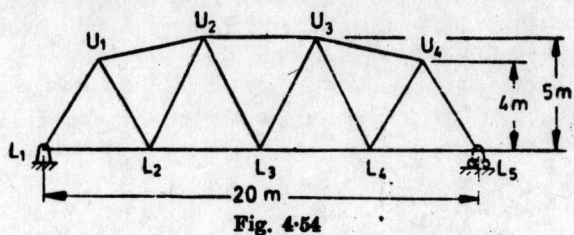

Fig. 4·54

Find the force in member L_2L_3 when a uniformly distributed load of 6 t/m rolls over the bottom chord of the truss.   [**Ans.**   52·5 *T* (Tension)]

HIGHLIGHTS

1. A truss, which receives load at its bottom chord joints, is called a through type truss.

2. A truss, which receives load at its top chord joints, is called a deck type truss.

3. While drawing the influence lines, it is assumed that the load is transferred to the truss, only at its joints.

4. The influence line for a force in parallel chord members is equal to the influence line for B. M. at the opposite joint divided by the vertical distance between the member and the vertical joint.

5. The influence lines for force in the members of a composite truss is drawn, first by drawing the influence lines for the primary and secondary trusses independently, and then by joining the points obtained by adding the ordinates of the two influence lines.

Do You Know ?

1. What do you understand by the terms 'through type trusses and 'deck type trusses' ?

2. Discuss the principles and assumptions, on which the influence lines for the trussed bridges are drawn.

3. Define the term 'opposite joint'. What important rule does it play in drawing the influence lines of a truss.

4. Describe the procedure for drawing the influence lines for the forces in the vertical and diagonal members of a truss. How does it differ from that for the bottom chord horizontal members ?

5. Describe, in detail, the procedure for drawing the influence lines for the force in the members of a composite truss.

4. Fig. 4·61 shows a Warren truss with inclined top chords.

Find the force in member L_2L_3, when a uniformly distributed load of 6 t/m rolls over the bottom chord of the truss. (Ans. 65·5 T (Tension))

5

DIRECT AND BENDING STRESSES

1. Introduction. 2. Eccentric loading. 3. Columns with eccentric loading. 4. Symmetrical columns with eccentric loading about one axis. 5. Symmetrical columns with eccentric loading about two axes. 6. Unsymmetrical columns with eccentric loading. 7. Limit of eccentricity. 8. Wind pressure on walls and chimneys.

5·1. Introduction

We know, that whenever a body is subjected to an axial tension or compression, then a direct stress comes into play at every section of the body. We also know that whenever a body is subjected to a bending moment, then a bending stress comes into play. It is thus obvious, that if a member is subjected to an axial loading, along with a transverse bending, a direct stress as well as a bending stress comes into play. The magnitude and nature of these stresses may be easily found out from the magnitude and nature of the load and the moment. A little consideration will show, that since both these stresses act normal to a cross-section, therefore the two stresses may be algebraically added into a single resultant stress.

5·2. Eccentric loading

A load, whose line of action does not coincide with the axis of a column or a strut, is known as an *eccentric load*. A bucket full of water, carried by a person, in his hand, is an excellent example of an eccentric load. A little consideration will show, that the man will feel this load as more severe than the same load, if he had carried the same bucket over his head. The simple reason for the same is that if he carries the bucket in his hand, then in addition to his carrying bucket, he has also to lean or bend on the other side of the bucket, so as to counteract any possibility of his falling towards the bucket. Thus we say that he is subjected to :

1. Direct load, due to weight of the bucket, and
2. Moment due to eccentricity of the load.

5·3. Columns with eccentric loading

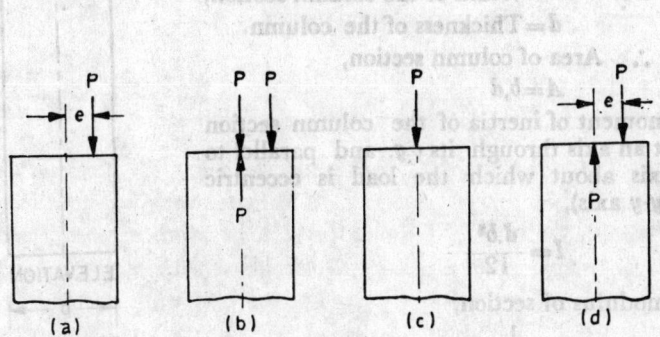

Fig. 5·1

Consider a column subjected to an eccentric loading. The eccentric load may be easily analysed as shown in Fig. 5·1.

(1) The given load P, acting at an eccentricity of e, is shown in Fig. 5·1 (*a*).

(2) Let us introduce, along the axis of the strut, two equal and opposite forces, P as shown in Fig. 5·1 (*b*).

(3) The forces, thus acting, may be split up into three forces.

(4) One of these forces will be acting along the axis of the strut. (This force will cause a direct stress) as shown in Fig. 5·1 (*c*).

(5) The other two forces will form a couple, as shown in Fig. 5·1(*d*). The moment of this couple will be equal to $P \times e$. (This couple will cause a bending stress).

Note. A column may be of symmetrical or unsymmetrical section and subjected to an eccentric load, with eccentricity about one of the axis, or both the axes. In the succeeding pages, we shall discuss these cases one by one,

19·4. Symmetrical columns with eccentric loading about one axis

Consider a column *ABCD* subjected to an eccentric load about one axis (*i.e.* about *y-y* axis) as shown in Fig. 5·2.

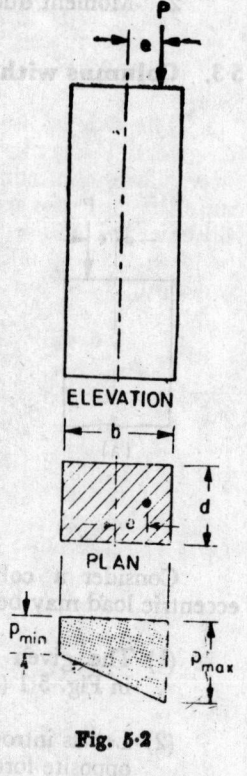

ELEVATION

Fig. 5·2

Let P=Load acting on the column,

          e=Eccentricity of the load

∴   Moment,

     $M=P.e$

Let b=Width of the column section,

          d=Thickness of the column.

∴   Area of column section,

     $A=b.d$

and moment of inertia of the column section about an axis through its *c.g.* and parallel to the axis about which the load is eccentric (*i.e.* *y-y* axis),

$$I=\frac{d.b^3}{12}$$

and modulus of section,

$$Z=\frac{1}{6}\,db^2$$

We know that the direct stress on the column due to the load,

$$p_0=\frac{P}{A}$$

and bending stress at any point of the column section at a distance y from y-y axis

$$p_b=\frac{M.y}{I}=\frac{M}{Z} \qquad \left(\because\ Z=\frac{I}{y}\right)$$

Now for the bending stress at the extreme, let us substitute $y=b/2$ in the above equation,

$$p_b=\frac{M\cdot\dfrac{b}{2}}{I}=\frac{M\cdot\dfrac{b}{2}}{\dfrac{db^3}{12}} \qquad \left(\text{Substituting } I=\frac{db^3}{12}\right)$$

$$=\frac{6.M}{db^2}=\frac{6P\cdot e}{db^2} \qquad\qquad (\text{Substituting } M=P.e)$$

$$=\frac{6\,P.e}{A\cdot b} \qquad\qquad\qquad (\text{Substituting } b.d=A)$$

We have already discussed in the previous article, that an eccentric load causes a direct stress as well as bending stress. It is thus obvious, that the total stress at the extreme fibre,

$$= p_0 \pm p_b = \frac{P}{A} \pm \frac{6P.e}{Ab} \qquad \text{(In terms of eccentricity)}$$

$$= \frac{P}{A} \pm \frac{M}{Z} \qquad \text{(In terms of modulus of section)}$$

The $+$ve or $-$ve sign will depend . upon the position of the fibre, with respect to the eccentric load. A little consideration will show, that the maximum stress will be at the corners B and C (because these corners are near the load), whereas the maximum stress will be at the corners A and D (because these corners are away from the load). The total stress along the width of the column will vary by a straight line law. The maximum stress,

$$p_{max} = \frac{P}{A} + \frac{6P.e}{Ab} = \frac{P}{A}\left(1 + \frac{6e}{b}\right)$$

$$= \frac{P}{A} + \frac{M}{Z}$$

and $\qquad p_{min} = \frac{P}{A} - \frac{6P.e}{Ab} = \frac{P}{A}\left(1 - \frac{6e}{b}\right)$

$$= \frac{P}{A} - \frac{M}{Z}$$

Note. From the above equations, we find that

1. If p_0 is *greater* than p_b, the stress, throughout the section, will be of the same nature (*i.e.* compressive).

2. If p_0 is *equal* to p_b, even then the stress throughout the section will be of the same nature. The minimum stress will be equal to zero, whereas the maximum stress will be equal to $2 \times p_0$.

3. If p_0 is *less* than p_b, the stress will change its sign (partly compressive and partly tensile).

Example 5·1. *A rectangular strut is 15 cm wide and 12 cm thick. It carries a load of 18,000 kg at an eccentricity of 1 cm in a plane bisecting the thickness. Find the maximum and minimum intensities of stress in the section.*

Solution.

Given.   Width, $b = 15$ cm

Thickness, $d = 12$ cm

∴ Area, $A = 15 \times 12 = 180$ cm²

Load, $P = 18,000$ kg

Eccentricity $e = 1$ cm

Maximum intensity of stress in the section

Let $p_{max} = $ Max. intensity of
stress in the section.

Using the relation,

$$p_{max} = \frac{P}{A}\left(1 + \frac{6e}{b}\right) \text{ with usual notation}$$

$$= \frac{18,000}{180}\left(1 + \frac{6 \times 1}{15}\right) \text{ kg/cm}^2$$

$$= 140 \text{ kg/cm}^2 \quad \textbf{Ans.}$$

Minimum intensity of stress in the section

Let $p_{min.} = $ Minimum intensity of stress
in the section

Now using the relation,

$$p_{min} = \frac{P}{A}\left(1 - \frac{6e}{6}\right) \text{ with usual notations.}$$

$$= \frac{18,000}{180}\left(1 - \frac{6 \times 1}{15}\right) \text{ kg/cm}^2$$

$$= 60 \text{ kg/cm}^2 \quad \textbf{Ans.}$$

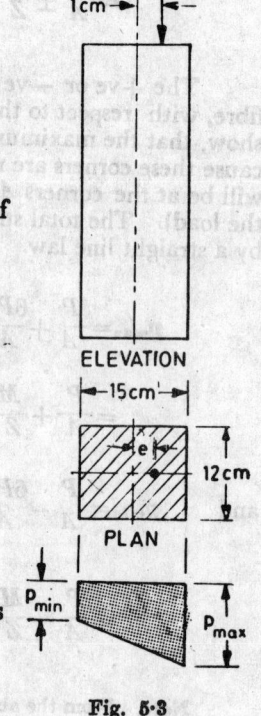

ELEVATION

PLAN

Fig. 5·3

Example 5·2. *A rectangular column 20 cm wide and 15 cm thick is carrying a vertical load of 1,000 kg at an eccentricity of 5 cm in a plane bisecting the thickness. Determine the maximum and minimum intensities of stress in the section.*

Solution.

Given. Width, $b=20$ cm

Thickness, $d=15$ cm

∴   Area, $A=20\times15=300$ cm²

Load, $P=1,000$ kg

Eccentricity $e=5$ cm

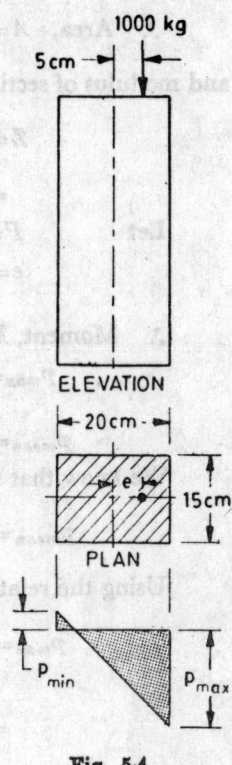

Maximum intensity of stress in the section

Let

$p_{max}=$ Max. intensity of stress.

Using the relation,

$$p_{max}=\frac{P}{A}\left(1+\frac{be}{b}\right) \text{ with usual notations.}$$

$$=\frac{1,000}{300}\left(1+\frac{6\times5}{20}\right) \text{ kg/cm}^2$$

$$=8\cdot38 \text{ kg/cm}^2 \text{ Ans.}$$

Minimum intensity of stress in the section

Let

$p_{min}=$ Minimum intensity of stress.

Now using the relation,

$$p_{min}=\frac{P}{A}\left(1-\frac{6e}{b}\right) \text{ with usual notations.}$$

$$=\frac{1,000}{300}\left(1-\frac{6\times5}{20}\right) \text{ kg/cm}^2$$

$$=-1\cdot67 \text{ kg/cm}^2$$

$$=1\cdot67 \text{ kg/cm}^2 \text{ (tension) Ans.}$$

Fig. 5·4

Example 5·3. *In a tension specimen 1·3 cm in diameter the line of pull is parallel to the axis of the specimen but is displaced from it. Determine the distance of the line of pull from the axis, when the maximum stress is 15 per cent greater than the mean stress on a section normal to the axis.*

Solution.

Given. Dia. of specimen = 1·3 cm

$$\therefore \text{ Area, } A = \frac{\pi}{4} \times 1 \cdot 3^2 = 1 \cdot 33 \text{ cm}^2$$

and modulus of section,

$$Z = \frac{\pi}{32} d^3 = \frac{\pi}{32} \times 1 \cdot 3^3 \text{ cm}^3$$

$$= 0 \cdot 217 \text{ cm}^3$$

Let P = Pull on the specimen in kg

        e = Distance of the line of pull from the axis.

$\therefore$ Moment, $M = P.e$

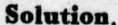

     p_{max} = Maximum stress in the section,

     p_{mean} = Mean stress in the section

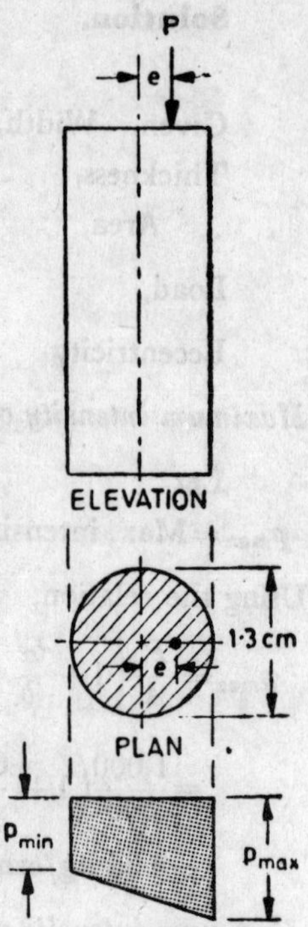

We know that the mean stress,

$$p_{mean} = \frac{P}{A} = \frac{P}{1 \cdot 33} \text{ kg/cm}^2 \quad ...(i)$$

Using the relation,

$$p_{max} = \frac{P}{A} + \frac{M}{Z} \text{ with usual notations.}$$

$$= \frac{P}{1 \cdot 33} + \frac{P.e}{0 \cdot 217} \quad ...(ii)$$

Fig. 5·5

Now $p_{max} = p_{mean} \times \dfrac{115}{100}$ ...(given)

$$\therefore \quad \frac{P}{1 \cdot 33} + \frac{Pe}{0 \cdot 217} = \frac{P}{1 \cdot 33} \times \frac{115}{100}$$

or $\dfrac{1}{1 \cdot 33} + \dfrac{e}{0 \cdot 217} = \dfrac{115}{133}$

$$\therefore \quad e = \left(\frac{115}{133} - \frac{1}{1 \cdot 33} \right) \times 0 \cdot 217 \text{ cm}$$

$$= 0 \cdot 0245 \text{ cm} \quad \textbf{Ans.}$$

Example 5·4. *A hollow rectangular masonry pier is 120 cm ×80 cm wide and 15 cm thick. A vertical load of 20 tonnes is transmitted in the vertical plane bisecting 120 cm side and at an eccentricity of 10 cm from the geometric axis of the section.*

Calculate the maximum and minimum stress intensities in the section.

Solution.

Given. Outer width,

$B = 120$ cm

Outer thickness,

$D = 80$ cm

Inner width,

$b = 120 - (15 \times 2) = 90$ cm

Inner thickness,

$d = 80 - (15 \times 2) = 50$ cm

∴ Area,

$A = (120 \times 80) - (90 \times 50)$ cm²

$= 5,100$ cm²

and section modulus,

$$Z = \frac{1}{6}(BD^2 - bd)^2$$

$$= \frac{1}{6}[120 \times 80^2 - 90 \times 50^2] \text{ cm}^3$$

$$= 90,500 \text{ cm}^3$$

Load, $P = 20$ t $= 20,000$ kg

Eccentricity,

$e = 10$ cm

∴ Moment,

$M = Pe = 20,000 \times 10$ kg-cm

$= 2,00,000$ kg-cm

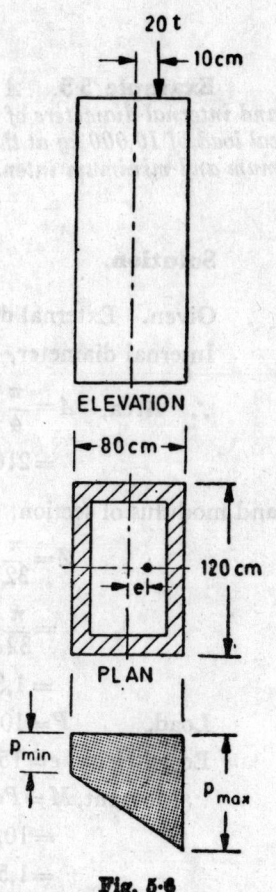

Fig. 5·6

Maximum stress intensity in the section

Let $p_{max} = $ Max. stress intensity in the section.

Using the relation,

$$p_{max} = \frac{P}{A} + \frac{M}{Z} \text{ with usual notations.}$$

$$= \frac{20,000}{5,100} + \frac{2,00,000}{90,500} \text{ kg/cm}^2$$

$$= 5,828 \text{ kg/cm}^2 \text{ Ans.}$$

Minimum stress intensity in the section

Let $p_{min} = $ Minimum stress intensity in the section.

Now using the relation,

$$p_{min} = \frac{P}{A} - \frac{M}{Z} \text{ with usual notations.}$$

$$\therefore \qquad p_{min} = \frac{20,000}{5,100} - \frac{2,00,000}{90,500} \text{ kg/cm}^2$$

$$= 1,996 \text{ kg/cm}^2 \quad \textbf{Ans.}$$

Example 5·5. *A hollow circular column having external and internal diameters of 30 cm and 25 cm respectively carries a vertical load of 10,000 kg at the outer edge of the column. Calculate the maximum and minimum intensities of stress in the section.*

Solution.

Given. External diameter, $D = 30$ cm

Internal diameter, $\qquad d = 25$ cm

$$\therefore \text{ Area, } \quad A = \frac{\pi}{4}(30^2 - 25^2) \text{ cm}^3$$

$$= 216 \text{ cm}^2$$

and modulus of section,

$$Z = \frac{\pi}{32} \times \frac{(D^4 - d^4)}{D}$$

$$= \frac{\pi}{32} \times \frac{(30^4 - 25^4)}{30} \text{ cm}^3$$

$$= 1,372 \text{ cm}^3$$

Load, $\qquad P = 10,000$ kg

Eccentricity, $e = 15$ cm

$\therefore$ Moment, $M = Pe$

$$= 10,000 \times 15 \text{ kg-cm}$$

$$= 1,50,000 \text{ kg-cm}$$

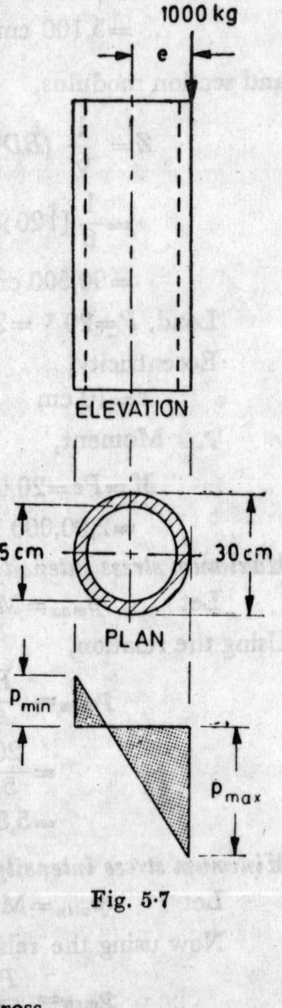

Maximum intensity of stress in the section

Let $p_{max} = $ Max. intensity of stress.

Using the relation,

$$p_{max} = \frac{P}{A} + \frac{M}{Z} \quad \text{with usual notations.}$$

$$= \frac{10,000}{216} + \frac{1,50,000}{1,372}$$

$$= 46\cdot3 + 109\cdot3 \text{ kg/cm}^2$$

$$= \textbf{155·6 kg/cm}^2 \quad \textbf{Ans.}$$

Fig. 5·7

Minimum intensity of stress in the section

Let $\qquad p_{min} = $ Minimum intensity of stress.

Now using the relation,

$$y_{min} = \frac{P}{A} - \frac{M}{Z} \quad \text{with usual notations.}$$

$$= \frac{10,000}{216} - \frac{1,50,000}{1,372} = -63 \cdot 0 \text{ kg/cm}^2$$

$$= 63 \cdot 0 \text{ kg/cm}^2 \text{ (tension) Ans.}$$

Exercise 5·1

1. A short column 10 cm × 10 cm is subjected to an eccentric load of 6,000 kg at an eccentricity of 4 cm in the plane bisecting the two opposite faces. Find the maximum and minimum intensities of stress at the base section.

(*Nagpur University*)

[**Ans.** 204 kg/cm² (compn.) ; 84 kg/cm² (tension)]

2. A hollow circular column of 25 cm external and 20 cm internal diameter respectively carries an axial load of 20,000 kg. It also carries a load of 10,000 kg on a bracket whose line of action is 20 cm from the axis of the column. Determine the maximum and minimum intensities of stress at the base section.

(*Madras University*)

[**Ans.** 390 kg/cm² (compn.) ; 51·3 kg/cm² tension)]

3. A concrete block 2m × 2 m weighing 9 tonnes is subjected to an eccentric load of 2 tonnes. Determine the eccentricity of the load, if the maximum stress is equal to twice the minimum stress. (*Ranchi University*)

[**Ans.** 0·61 m]

4. A hollow rectangular column is having external and internal dimensions as 2·4 cm × 1·8 cm and 1·2 cm × 1·2 cm respectively. Calculate the safe load, that can be placed at an eccentricity of 0·9 cm on a plane bisecting the thickness, if the maximum compression stress is not to exceed 556 kg/cm².

(*Jadavpur University*)

[**Ans.** 6,055 kg]

5·5. Symmetrical columns with eccentric loading about two axes

Consider a column *ABCD* subjected to a load with eccentricity about two axes as is shown in Fig. 5·8.

Let P = Load acting on the column *ABCD*,

A = Cross-sectional area of the column

e_X = Eccentricity of the load about X-X axis

∴ Moment of the load X-X

axis,

$$M_X = P.e_X$$

I_{XX} = Moment of inertia of the column section about X-X axis

e_y, M_Y, I_{YY} = corresponding values for Y-Y axis.

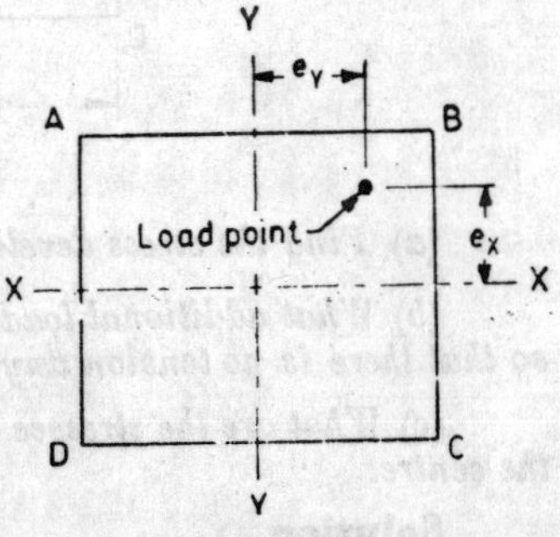

Fig. 5·8

The effect of such a load may be split up into the following three parts :

1. Direct stress on the column, due to the load,

$$p_0 = \frac{P}{A} \qquad \qquad \qquad ...(i)$$

2. Bending stress due to eccentricity e_X,

$$p_{bX} = \frac{M_X \cdot y}{I_{XX}} = \frac{P \cdot e_X \cdot y}{I_{XX}} \qquad \qquad ...(ii)$$

3. Bending stress due to eccentricity e_Y,

$$p_{bY} = \frac{M_Y \cdot x}{I_{YY}} = \frac{P \cdot e_Y \cdot x}{I_{YY}} \qquad \qquad ...(iii)$$

The total stress at the extreme fibre

$$= p_0 \pm p_{bX} \pm p_{bY} = \frac{P}{A} \pm \frac{M_X \cdot y}{I_{XX}} \pm \frac{M_Y \cdot x}{I_{YY}}$$

The +ve or −ve sign depends upon the position of the fibre with respect to the load. A little consideration will show that the stress will be maximum at B, where both the +ve signs are to be adopted. The stress will be minimum at D, where both the −ve signs are to be adopted. While calculating the stress at A, the value of M_Y is to be taken as +ve, whereas the value of M_X as −ve. Similarly for the stress at C, the value of M_X is to be taken as +ve, whereas the value of M_Y as −ve.

Example 5·6. *A masonry pier of 3 m × 4 m supports a vertical load of 80 tonnes as shown in Fig. 5·9.*

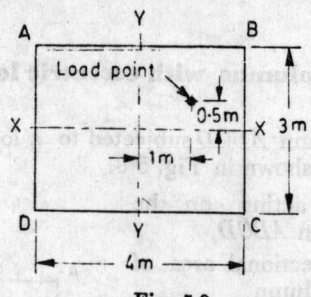

Fig. 5·9

(a) *Find the stress developed at each corner of the pier.*

(b) *What additional load should be placed at the centre of the pier, so that there is no tension anywhere in the pier section ?*

(c) *What are the stresses at the corners with the additional load in the centre.*

Solution.

Given. Width, $b = 4$ m

Thickness, $d = 3$ m

∴ Area, $A = 4 \times 3 = 12$ cm²

Moment of inertia about X-X axis,

$$I_{XX} = \frac{1}{12} \times 4 \times 3^3 = 9 \text{ m}^4$$

and

$$I_{YY} = \frac{1}{12} \times 3 \times 4^3 = 16 \text{ m}^4$$

Load, $P = 80$ t

Eccentricity about X-X axis,

$$e_X = 0.5 \text{ m}$$

and

$$e_Y = 1.0 \text{ m}$$

$\therefore$ Moment, $M_X = P.e_X = 80 \times 0.5 = 40$ t-m

Similarly $M_Y = P.e_Y = 80 \times 1.0 = 80$ t-m

Distance between Y-Y axis and A as well as B,

$$x = 2 \text{ m}$$

Distance between X-X axis and A as well as D,

$$y = 1.5 \text{ m}$$

(a) Stresses developed at each corner

From the geometry of the figure, we find that the stress at A,

$$p_A = \frac{P}{A} + \frac{M_X.y}{I_{XX}} - \frac{M_Y.x}{I_{YY}}$$

$$= \frac{80}{12} + \frac{40 \times 1.5}{9} - \frac{80 \times 2}{16} \quad \text{t/m}^2$$

$$= 3.34 \text{ t/m}^2 \text{ Ans.}$$

Similarly

$$p_B = \frac{P}{A} + \frac{M_X.y}{I_{XX}} + \frac{M_Y.x}{I_{YY}}$$

$$= \frac{80}{12} + \frac{40 \times 1.5}{9} + \frac{80 \times 2}{16} \quad \text{t/m}^2$$

$$= 23.34 \text{ t/m}^2 \text{ Ans.}$$

$$p_C = \frac{P}{A} - \frac{M_X.y}{I_{XX}} + \frac{M_Y.x}{I_{YY}}$$

$$= \frac{80}{12} - \frac{40 \times 1.5}{9} + \frac{80 \times 2}{16} = 10.0 \text{ t/m}^2 \text{ Ans.}$$

and
$$p_D = \frac{P}{A} - \frac{M_X \cdot y}{I_{XX}} - \frac{M_Y \cdot x}{I_{YY}}$$

$$= \frac{80}{12} - \frac{40 \times 1 \cdot 5}{9} - \frac{80 \times 2}{16} = -10 \cdot 0 \text{ t/m}^2$$

$$= 10 \cdot 0 \text{ t/m}^2 \text{ (tension) Ans.}$$

(b) *Additional load at the centre for no tension in the pier section*

Let $W =$ Additional load that should be placed at
the centre for no tension in the pier
section.

We know that the compressive stress due to the load

$$= \frac{W}{A} = \frac{W}{12}$$

We also know that for no tension, in the pier section, the compressive stress due to the load W should be equal to the tensile stress at D *i.e.* $10 \cdot 0$ t/m².

$$\therefore \qquad\qquad \frac{W}{12} = 10 \cdot 0$$

or $W = 10 \cdot 0 \times 12 = 120$ tonnes **Ans.**

Stresses at the corners with the additional load in the centre

We find that the stress due to the additional load

$$= \frac{W}{A} = \frac{120}{12} = 10 \cdot 0 \text{ t/m}^2$$

$\therefore$ Stress at A, $p_A = 3 \cdot 34 + 10 \cdot 0 = 13 \cdot 34 \text{ t/m}^2$ **Ans.**

Similarly $p_B = 23 \cdot 34 + 10 \cdot 0 = 33 \cdot 34$ t/m² **Ans.**

$$p_C = 10 \cdot 0 + 10 \cdot 0 = 20 \cdot 0 \text{ t/m}^2 \text{ Ans.}$$

and $p_D = -10 \cdot 0 + 10 \cdot 0 = 0$ **Ans.**

5·6. Unsymmetrical columns with eccentric loading

In the previous articles, we have discussed the symmetrical column sections subjected to eccentric loading. In an unsymmetrical column first *c.g.* and then moment of inertia of the section is found out. After that the distances between the *c.g.* of the section, and its corners are calculated. The stresses on the corners are then found out, as usual, by using the respective values of moment of inertia and distance of the corner from the *c.g.* of the section.

Example 5·7. *A hollow cylindrical shaft of 20 cm external dia-meter has got eccentric bore of 14 cm diameter, such that the thickness varies from 2 cm at one end to 4 cm at the other. Calculate the extreme stress intensities, if the shaft is subjected to a load of 40 tonnes along the axis of the bore.*

Solution.

Given. External diameter
$$D = 20 \text{ cm}$$
Internal diameter,
$$d = 14 \text{ cm}$$

∴ Area, $A = \dfrac{\pi}{4}(20^2 - 14^2) \text{ cm}^2$

$$= 160 \cdot 28 \text{ cm}^2$$

Load, $P = 40$ T

Let $p_{max} =$ Maximum stress at B, and

$p_{min} =$ Minimum stress at A,

First of all let us find out the c.g. of the section. Let $\bar{x}$ be the distance between the c.g. of the section from left end A.

(i) main circle,

$$a_1 = \frac{\pi}{4} \times 20^2 = 100\,\pi \text{ cm}^2$$

$$x_1 = 10 \text{ cm}$$

(ii) bore $a_2 = -\dfrac{\pi}{4} \times 14^2 \text{ cm}^2$

$$= -49\pi \text{ cm}^2$$

$$x_2 = 4 + \frac{14}{2} = 11 \text{ cm}$$

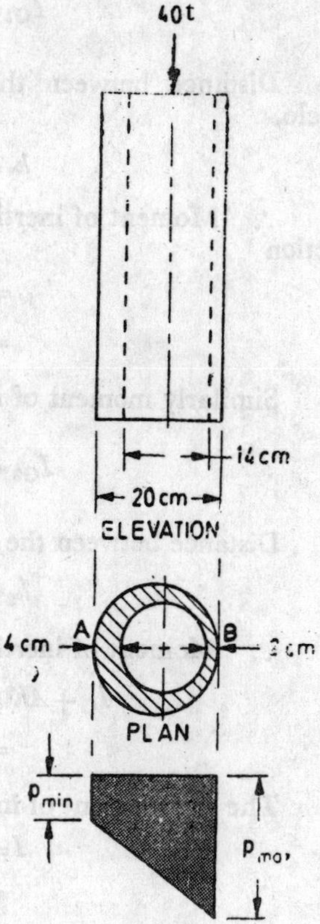

We know that

$$\bar{x} = \frac{a_1 x_1 + a_2 x_2}{a_1 + a_2}$$

$$= \frac{100\pi \times 10 - 49\pi \times 11}{100\pi - 49\pi} \text{ cm}$$

$$= 9 \cdot 04 \text{ cm}$$

Fig. 5·10

From the geometry of the figure, we find that the eccentricity of the load,

$$e = 11 - 9 \cdot 04 = 1 \cdot 96 \text{ cm}$$

∴ Moment, $M = P \cdot e = 40 \times 1 \cdot 96 = 78 \cdot 4$ t-cm

Distance of corner A from the c.g. of the section,

$$y_A = 9 \cdot 04 \text{ cm}$$

Similarly $y_B = 20 - 9 \cdot 04 = 10 \cdot 96 \text{ cm}$

We know that the moment of inertia of the main circle about its *c.g.*

$$I_{G1} = \frac{\pi}{64} \times 20^4 = 2,500 \ \pi \ \text{cm}^4$$

Distance between the *c.g.* of the section and *c.g.* of the main circle,

$$h_1 = 10 - 9\cdot04 = 0\cdot96 \ \text{cm}$$

∴ Moment of inertia of the main circle about the *c.g.* of the section

$$= I_G + Ah^2 = 2,500 \ \pi + 100 \ \pi \times 0\cdot96^2 \ \text{cm}^4$$
$$= 2,592\cdot16 \ \pi \ \text{cm}^4$$

Similarly moment of inertia of the bore about its *c.g.*

$$I_{G2} = -\frac{\pi}{64} \times 14^4 = -600\cdot25 \ \pi \ \text{cm}^4$$

Distance between the *c.g.* of the section and the *c.g.* of the bore,

$$h_2 = 11 - 9\cdot04 = 1\cdot96 \ \text{cm}$$

∴ Moment of inertia of the bore about the *c.g.* of the section

$$I_G + Ah^2 = -600\cdot25 \ \pi - 49 \ \pi \times 1\cdot96^2 \ \text{cm}^4$$
$$= -788\cdot5 \ \pi \ \text{cm}^4$$

The net moment of inertia of the section about its *c.g.*

$$I = 2,592\cdot16 \ \pi - 788\cdot5 \ \pi \ \text{cm}^4$$
$$= 5,666\cdot7 \ \text{cm}^4$$

Using the relation,

$$p_{max} = \frac{P}{A} + \frac{M \cdot y_B}{I} \ \text{with usual notations.}$$

$$= \frac{40}{160\cdot28} + \frac{78\cdot4 \times 10\cdot96}{5,666\cdot7} \ \text{t/cm}^2$$

$$= 0\cdot4 \ \text{t/cm}^2 \quad \textbf{Ans.}$$

Now using the relation,

$$p_{min} = \frac{P}{A} - \frac{M \cdot y_A}{I} \ \text{with usual notations.}$$

$$= \frac{40}{160\cdot28} - \frac{78\cdot4 \times 9\cdot04}{5,666\cdot7} \ \text{t/cm}^2$$

$$= 0\cdot125 \ \text{t/cm}^2 \quad \textbf{Ans.}$$

Example 5·8. *A short C.I. column has a rectangular section 16 cm × 20 cm with a circular hole of 8 cm diameter as shown in Fig. 5·11.*

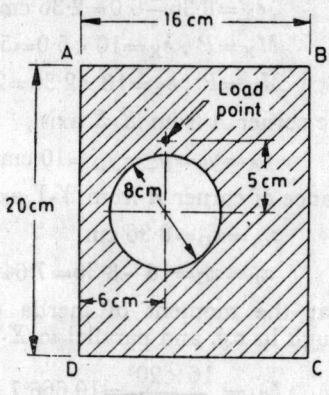

Fig. 5·11

It carries an eccentric load of 10 tonnes, located as shown in the figure. Determine the values of the stresses at the four corners of the section.

(*Cambridge University*)

Solution

Given.   Outer width, $B = 16$ cm

Outer Depth, $D = 20$ cm

Hole radius $= 4$ cm

∴ Hole diameter, $d = 8$ cm

and   Area of column, $A = 20 \times 16 - \dfrac{\pi}{4} \times 8^2 = 269 \cdot 7$ cm²

Load, $P = 10$ t

First of all, let us find out the *c.g.* of the section. Let $\bar{x}$ be the distance between the *c.g.* of the section from the respective line *A.D.*

(*i*) Outer rectangle

$$a_1 = 20 \times 16 = 320 \text{ cm}^2$$

$$x_1 = 8 \text{ cm}$$

(*ii*) Circular hole

$$a_2 = -\frac{\pi}{4} \times 8^2 = -50 \cdot 3 \text{ cm}^2$$

$$x_2 = 6 \text{ cm}$$

We know that, $\bar{x}_1 = \dfrac{a_1 x_1 + a_2 x_2}{a_1 + a_2}$

$$= \frac{320 \times 8 - 50 \cdot 3 \times 6}{320 - 50 \cdot 3} = 8 \cdot 36 \text{ cm}$$

From the geometry of the figure, we find that the eccentricity of load about X-X axis,

$$e_X = 5 \cdot 0 \text{ cm}$$

and

$$e_Y = 8 \cdot 36 - 6 \cdot 0 = 2 \cdot 36 \text{ cm}$$

$\therefore$ Moment $\quad M_X = P \cdot e_X = 10 \times 5 \cdot 0 = 50 \cdot 0$ t-cm

Similarly $\quad M_Y = P \cdot e_Y = 10 \times 2 \cdot 36 = 23 \cdot 6$ t-m

Distance of the corner A from X-X axis,

$$y_A = y_B = y_C = y_D = 10 \text{ cm}$$

Similarly distance of corner A from Y-Y axis.

$$x_A = x_D = 8 \cdot 36 \text{ cm}$$

and

$$x_B = x_C = 16 - 8 \cdot 36 = 7 \cdot 64 \text{ cm}$$

We know that the moment of inertia of the main rectangle $ABCD$, passing through its $c.g.$ and parallel to X-X axis,

$$I_{G1} = \frac{16 \times 20^3}{12} = 10,666 \cdot 7 \text{ cm}^4$$

and moment of inertia of the circular hole, passing through its $c.g.$ and parallel to X-X axis,

$$I_{G2} = \frac{\pi}{64} \times 8^4 = 201 \cdot 1 \text{ cm}^4$$

Since the $c.g.$ of the rectangle and the circular hole coincides with the X-X axis, therefore moment of inertia of the section about X-X axis,

$$I_{XX} = 10,666 \cdot 7 - 201 \cdot 1 = 10,465 \cdot 6 \text{ cm}^4 \qquad \ldots(i)$$

We also know that the moment of inertia of the main rectangle $ABCD$, passing through its $c.g.$ and parallel to Y-Y axis,

$$I_{G3} = \frac{20 \times 16^3}{12} = 6,826 \cdot 7 \text{ cm}^4$$

Distance between the $c.g.$ of the rectangle from Y-Y axis,

$$h_1 = 8 \cdot 36 - 8 \cdot 0 = 0 \cdot 36 \text{ cm}$$

$\therefore$ Moment of inertia of the rectangle about Y-Y axis,

$$= I_G + Ah^2 = 6,826 \cdot 7 + 320 \times 0 \cdot 36^2 \text{ cm}^4$$
$$= 6,868 \cdot 2 \text{ cm}^4$$

and moment of inertia of the circular hole passing through its $c.g.$ and parallel to Y-Y axis,

$$I_{G4} = \frac{\pi}{64} \times 8^4 = 201 \cdot 1 \text{ cm}^4$$

Distance between the $c.g.$ of the circular section from Y-Y axis,

$$h_2 = 8 \cdot 36 - 6 \cdot 0 = 2 \cdot 36 \text{ cm}$$

$\therefore$ Moment of inertia of the circular hole about Y-Y axis

$$=Ig+Ah^2=201\cdot1+50\cdot3\times2\cdot36^2 \text{ cm}^4$$
$$=481\cdot2 \text{ cm}^4$$

The net moment of inertia of the section about $Y-Y$ axis,

$$I_{YY}=6{,}868\cdot2-481\cdot2=6{,}387\cdot0 \text{ cm}^4 \qquad \ldots(ii)$$

Now from the geometry of the figure, we find that the stress at A,

$$p_A=\frac{P}{A}+\frac{M_X.y_A}{I_{XX}}+\frac{M_Y.x_A}{I_{YY}}$$

$$=\frac{10}{269\cdot7}+\frac{50\cdot0\times10}{10{,}465\cdot6}+\frac{23\cdot6\times8\cdot36}{6{,}387\cdot0} \text{ t/cm}^2$$

$$=0\cdot1158 \text{ t/cm}^2 \quad \textbf{Ans.}$$

Similarly $\quad p_B=\dfrac{P}{A}+\dfrac{M_X.y_B}{I_{XX}}-\dfrac{M_Y.x_B}{I_{YY}}$

$$=\frac{10}{269\cdot7}+\frac{50\cdot0\times10}{10{,}465\cdot6}-\frac{23\cdot6\times7\cdot64}{6{,}387\cdot0} \text{ t/cm}^2$$

$$=0\cdot0565 \text{ t/cm}^2 \quad \textbf{Ans.}$$

$$p_C=\frac{P}{A}-\frac{M_X.y_C}{I_{XX}}-\frac{M_Y.y_C}{I_{YY}}$$

$$=\frac{10}{269\cdot7}-\frac{50\cdot0\times10}{10{,}465\cdot6}-\frac{23\cdot6\times7\cdot64}{6{,}387} \text{ t/cm}^2$$

$$=-0\cdot0391 \text{ t/cm}^2$$

$$=0\cdot0391 \text{ t/cm}^2 \text{ (tensile)} \quad \textbf{Ans.}$$

and $\quad p_D=\dfrac{P}{A}-\dfrac{M_X.y_D}{I_{XX}}+\dfrac{M_Y.x_D}{I_{YY}}$

$$=\frac{10}{269\cdot7}-\frac{50\cdot0\times10}{10{,}465\cdot6}+\frac{23\cdot6\times8\cdot36}{6{,}387\cdot0} \text{ t/cm}^2$$

$$=0\cdot0202 \text{ t/cm}^2 \quad \textbf{Ans.}$$

5·7. Limit of eccentricity

In Art. 5·2 and 5·3, we have seen that when an eccentric load is acting on a column, it produces direct stress as well as bending stress. On one side of the neutral axis there is a maximum stress (equal to the *sum* of direct and bending stress) ; and on the other side of the neutral axis there is a minimum stress (equal to direct stress *minus* bending stress). A little consideration will show that so long as the bending stress remains less than the direct stress, the resultant stress is compressive. If the bending stress is equal to the direct stress, then there will be a zero stress on one side. But if the bending stress exceeds the direct stress, then there will be a tensile stress on one side. Though cement concrete can take up some tensile stress, yet it is desirable that no tensile stress should come into play.

We have seen, that if the tensile stress is not to be permitted to come into play, then bending stress should be less than the direct stress ; or, maximum, it may be equal to the direct stress, *i.e.*,

$$p_b \lesseqgtr p_o$$

$$\frac{M}{Z} \lesseqgtr \frac{P}{A}$$

$$\frac{P.e}{Z} \lesseqgtr \frac{P}{A} \qquad (\because \quad M=P.e)$$

or

$$e \lesseqgtr \frac{Z}{A}$$

It means that for no tension condition, e should be less than $\frac{Z}{A}$ or equal to $\frac{Z}{A}$. Now we shall discuss the limit for e in the following cases :—

   (*a*) For a rectangular section,

   (*b*) For a hollow rectangular section,

   (*c*) For a circular section,

   (*d*) For a hollow rectangular section.

(*a*) *Limit of eccentricity for a rectangular section*

Consider a rectangular section of width b and thickness d as shown in Fig. 5·12. We know that the modulus of section

$$Z=\frac{1}{6} bd^2 \qquad ...(i)$$

and Area, $A=bd$ $\qquad ...(ii)$

We also know that for no tension condition,

$$e \lesseqgtr \frac{Z}{A}$$

$$\lesseqgtr \frac{\frac{1}{6} bd^2}{bd}$$

$$\lesseqgtr \frac{1}{6} d$$

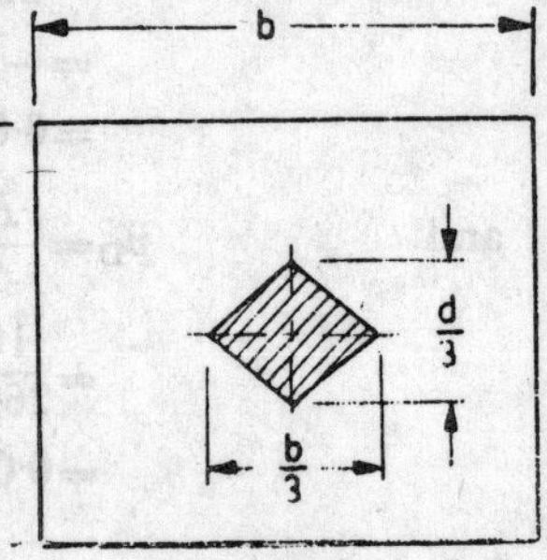

Fig. 5·12

It means that the load can be eccentric, on either side of the geometrical axes, by an amount equal to $d/6$. Thus if the line of action of the load is *within the middle third*, as shown by the dotted area in Fig. 5·12, then the stress will be compressive throughout.

(*b*) *Limit of eccentricity for a hollow rectangular section*

Consider a hollow rectangular section with B and D as outer width and thickness and b and d internal dimensions respectively. We know that the modulus of section,

$$Z=\frac{(BD^3-bd^3)}{6D} \qquad ...(i)$$

and   Area $A = BD - bd$ ...(ii)

We also know that for no tension condition,

$$e \leqq \frac{Z}{A}$$

$$\leqq \frac{\dfrac{(BD^3 - bd^3)}{6D}}{BD - bd}$$

$$\leqq \frac{BD^3 - bd^3}{6D(BD - bd)}.$$

It means that the load can be eccentric, on either side of the geometrical axis, by an amount equal to $\dfrac{BD^3 - bd^3}{6D(BD - bd)}$.

(c) *Limit of eccentricity for a circular section*

Consider a circular section of diameter d as shown in Fig. 5·13. We know that the modulus of section,

$$Z = \frac{\pi}{32} d^3 \qquad \qquad ...(ii)$$

and   Area, $A = \dfrac{\pi}{4} d^2$

We also know that for no tension condition,

$$e \leqq \frac{Z}{A}$$

$$\leqq \frac{\dfrac{\pi}{32} d^3}{\dfrac{\pi}{4} d^2}$$

$$\leqq \frac{d}{8}$$

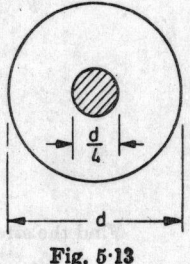

Fig. 5·13

It means that the load can be eccentric, on any side of the geometrical centre, by an amount equal to $d/8$. Thus, if the line of action of the load is within a circle of diameter equal to one-fourth of the main circle as shown by the dotted area in Fig. 5·13, then the stress will be compressive throughout.

(d) *Limit of eccentricity for a hollow circular section*

Consider a hollow circular section of external and internal diameters as D and d respectively. We know that the modulus of section,

$$Z = \frac{\pi}{32} \times \frac{(D^4 - d^4)}{D} \qquad \qquad ...(i)$$

and   Area $A = \dfrac{\pi}{4} (D^2 - d^2)$ ...(ii)

We also know that for no tension condition,

$$e \leqq \frac{Z}{A}$$

$$e \leqq \frac{\frac{\pi}{32} \times \frac{(D^4 - d^4)}{D}}{\frac{\pi}{4}(D^2 - d^2)}$$

$$\leqq \frac{(D^2 + d^2)}{8D} \qquad [(D^4 - d^4) = (D^2 + d^2)(D^2 - d^2)]$$

It means, that the load can be eccentric, on any side of the geometrical centre, by an amount equal to $\frac{(D^2 + d^2)}{8D}$.

Exercise 5-2

1. A rectangular pier is subjected to a compressive load of 45 tonnes as shown in Fig. 5·14.

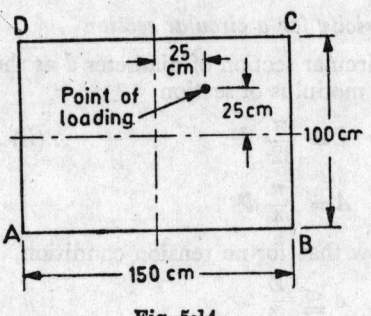

Fig. 5·14

Find the stress intensities on all the four corners of the pier.

(*Allahabad University*)

[Ans.　$p_A = -4\cdot5$ kg/cm^2 ; $p_B = +1\cdot5$ kg/cm^2 ; $p_C = +10\cdot5$ kg/cm^2]
$p_D = +4\cdot5$ kg/cm^2;

2. A hollow square column has 1·25 m outside length and 75 cm inside length. The column is subjected to a load of 12 tonnes located on a diagonal and at a distance of 70 cm from the vertical axis of the pier. Determine the stress intensities on the outside corners of the column.　(*Bombay University*)

[Ans.　$+53\cdot95$ t/m^2 ; $12\cdot0$ t/m^2; $29\cdot95$ t/m^2; $12\cdot0$ t/m^2]

5·8. Wind pressure on walls and chimneys

The walls and chimneys are always subjected to some wind pressure, whose effect is to cause some bending stress at its base. Consider a vertical wall subjected to wind pressure on one side as shown in Fig. 5·15. A little consideration will show, that as a result of this wind pressure, the wall will tend to fall on the right side (since the direction of the wind is from left to right). It may be noted, that before the wall may fall down, it will tend to bend, which will cause some bending stress at the base of the wall. A little consideration will show that the stress at B will be more than that at A.

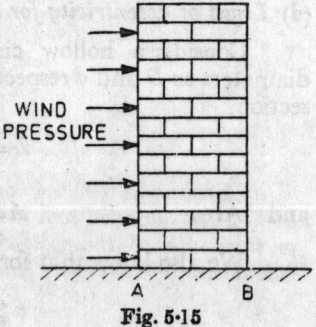

Fig. 5·15

Let $p=$Intensity of wind pressure,

           $b=$Width of the wall subjected to wind pressure, and

           $h=$Height of the wall.

The total wind pressure on the wall,

$$P=pbh$$

This pressure will act at height of $h/2$ from the base.

$\therefore$ Moment of this wind pressure about the base,

$$M=P.\ \frac{h}{2}=\frac{Ph}{2}$$

and bending stress due to this wind pressure,

$$p_b=\frac{M}{Z}=\frac{Ph}{2Z}$$

(where Z is the modulus of base section)

Let $W=$Weight of the wall, and

           $A=$Area of the base section.

$\therefore$ Direct stress,

$$p_0=\frac{W}{A}=\frac{\rho hA}{A}$$

(where ρ is the weight of wall material per m³)

$$=ph$$

$\therefore$ Maximum stress,

$$p_{max}=p_0+p_b=\frac{W}{A}+\frac{M}{Z}$$

and minimum stress,

$$p_{min}=p_0-p_b=\frac{W}{A}-\frac{M}{Z}$$

Example 5·9. *A masonry wall 4 m wide, 1·5 m thick is 6 m high. It is subjected to a horizontal wind pressure of 100 kg/m², acting on the width of the wall. Determine the maximum and minimum intensities of stress at the base of the wall, if the masonry weighs 2,240 kg/m³.*

Solution.

Given. Width, $\quad b = 4$ m

Thickness, $\quad d = 1.5$ m

Height, $\quad h = 6$ m

Wind pressure, $p = 100$ kg/m^3

Density, $\quad \rho = 2,240$ kg/m^3

Let $p_{max} = $ Maximum intensity of stress, and

$p_{min} = $ Minimum intensity of stress.

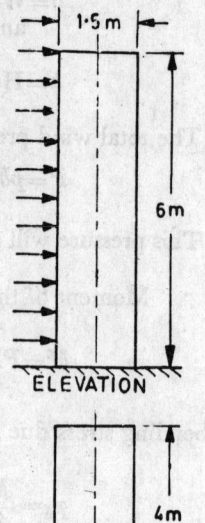

We know that the direct stress,

$$p_0 = \rho h = 2,240 \times 6$$
$$= 13,440 \text{ kg/m}^2$$

Total wind pressure,

$$P = pbh = 100 \times 4 \times 6 \text{ kg}$$
$$= 2,400 \text{ kg}$$

Moment of wind pressure, about the base,

$$M = \frac{Ph}{2} = \frac{2,400 \times 6}{2} \text{ kg-m}$$
$$= 7,200 \text{ kg-m}$$

Modulus of base section,

$$Z = \frac{1}{6} bd^2$$
$$= \frac{1}{6} \times 4 \times 1.5^2 \text{ m}^3$$
$$= 1.5 \text{ m}^3$$

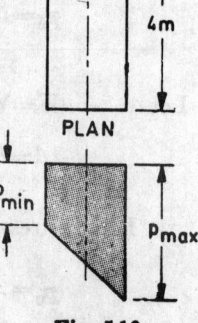

Fig. 5·16

∴ Bending stress,

$$p_b = \frac{M}{Z} = \frac{7,200}{1.5} = 4,800 \text{ kg/m}^2$$

Using the relation,

$$p_{max} = p_0 + p_b \quad \text{with usual notations.}$$
$$= 13,440 + 4,800 = \textbf{18,240 kg/m}^2 \quad \textbf{Ans.}$$

Now using the relation,

$$p_{min} = p_0 - p_b \quad \text{with usual notations.}$$
$$= 13,440 - 4,800 = \textbf{8,640 kg/m}^2 \quad \textbf{Ans.}$$

Example 5·10. *A masonry chimney of a hollow circular section has external and internal diameters of 2 m and 1·5 m respectively. It* *is subjected to a horizontal wind pressure of 160 kg/m². Determine the greatest height of the chimney, if no tension is allowed to occur at the base. Take specific weight of the masonry as 1,920 kg/m³.*

Solution.

Given. External dia.,

$$D=2 \text{ m}$$

Internal dia.,

$$d=1.5 \text{ m}$$

Wind pressure,

$$p=160 \text{ kg/m}^2$$

Specific weight,

$$\rho=1,920 \text{ kg/m}^3$$

Let h=height of the chimney.

We know that the direct stress,

$$p_0=ph=1,920\ h$$

Total wind pressure,

$$p=pdh=160\times 2h=320\ h$$

Moment of wind pressure about the base,

$$M=\frac{Ph}{2}=\frac{320\ h\times h}{2}=160\ h^2$$

Modulus of base section,

$$Z=\frac{\pi}{32}\times\frac{(D^4-d^4)}{D}=\frac{\pi}{32}\times\frac{2^4-1.5^4}{2}$$

$$=\frac{10.9375\ \pi}{64} \text{ m}^3$$

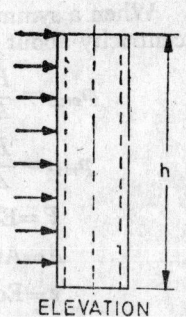

ELEVATION

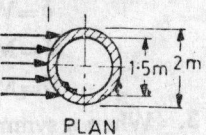

PLAN

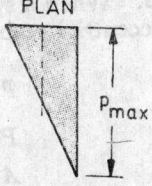

Fig. 5·17

∴ Bending stress,

$$p_b=\frac{M}{Z}=\frac{160\ h^2}{\dfrac{10.9375\ \pi}{64}}$$

Now, for no tension at the base, the bending stress should be equal to the direct stress.

or

$$\frac{160\ h^2}{\dfrac{10.9375\ \pi}{64}}=1,920\ h$$

∴

$$h=\frac{1920\times\dfrac{10.9375\ \pi}{64}}{160}=\textbf{6·45 m Ans.}$$

HIGHLIGHTS

1. Whenever a column section is subjected to an eccentric loading, it causes a direct stress, as well as the bending stress to come into play. Since both these stresses act normal to the cross-section, these can be algebraically added into a single resultant stress.

2. When a symmetrical column section is subjected to a load with eccentricity about one axis, then maximum stress,

$$p_{max} = \frac{P}{A}\left(1 + \frac{6e}{b}\right) = \frac{P}{A} + \frac{M}{Z}$$

and

$$p_{min} = \frac{P}{A}\left(1 - \frac{6e}{b}\right) = \frac{P}{A} - \frac{M}{Z}$$

where

P = Eccentric load

A = Area of the cross-section

e = Eccentricity of the load

b = Width of the column

M = Moment of the eccentric load (*i.e.* $P.e$)

Z = Modulus of section

3. When a symmetrical column section is subjected to a load with eccentricity about the axes, the total stress at the extreme fibre,

$$p = \frac{P}{A} \pm \frac{M_X.y}{I_{XX}} \pm \frac{M_Y.x}{I_{YY}}$$

where

P = Eccentric load

A = Area of the cross-section

M_X = Moment of load about X-X axis (*i.e.*, $P.e_X$)

y = Distance of the extreme fibre from Y-Y axis,

I_{XX} = Moment of inertia of the column section about X-X axis,

M_Y, x, I_{YY} = corresponding value about Y-Y axis

The +ve or −ve sign depends upon the position of the fibre with respect to the load.

4. When an unsymmetrical column is subjected to an eccentric load, first of all find out the *c.g.* and then the moment of inertia of the section is found out. After that the distances between the *c.g.* of the section and its corners are calculated. The stresses on the corners are then found out by using the respective values of moment of inertia and distance of the corners from the *c.g.* of the section.

5. Limit of eccentricity is the maximum distance between geometrical axis of a column section and the point of loading, such that no tensile stress comes into play at any colour of the section.

6. Whenever a wall or chimney is subjected to a wind pressure,

$$p_{max}=\frac{W}{A}+\frac{M}{Z} \quad \text{and} \quad p_{min}=\frac{W}{A}-\frac{M}{Z}$$

where

$W=$ The weight of the unit or chimney

$M=$ Moment due to wind

Do You Know ?

1. Distinguish clearly between direct stress and bending stress.

2. What is meant by eccentric loading ? Explain its effects on a short column.

3. Obtain a relation for the maximum and minimum stresses at the base of a symmetrical column, when it is subjected to

(a) an eccentric load about one axis, and

(b) an eccentric load about two axes.

4. Show that for no tension in the base of a short column, the line of action of the load should be within the middle third.

5. Define the term limit of eccentricity. How will you find out this limit in case of a hollow circular section ?

6

DAMS AND RETAINING WALLS

6·1. Introduction

A dam* is constructed to store large quantities of water, which is used for the purposes of irrigation and power generation. A dam may be of any cross-section, but the dams of trapezoidal cross-section are very popular these days. A retaining wall is generally constructed to retain earth in hilly areas. Though there are many types of dams, yet the following are important from the subject point of view.

(1) Rectangular dams.

---

*A dam constructed with earth is called an *earthen dam* ; whereas a dam constructed with cement concrete is called a *gravity dam*.

(2) Trapezoidal dams having water face vertical, and

(3) Trapezoidal dams having water face inclined.

We shall discuss the above three type of dams, one by one.

6·2. Rectangular dams

Consider a unit length of a rectangular dam, retaining water on one of its vertical sides, as shown in Fig. 6·1.

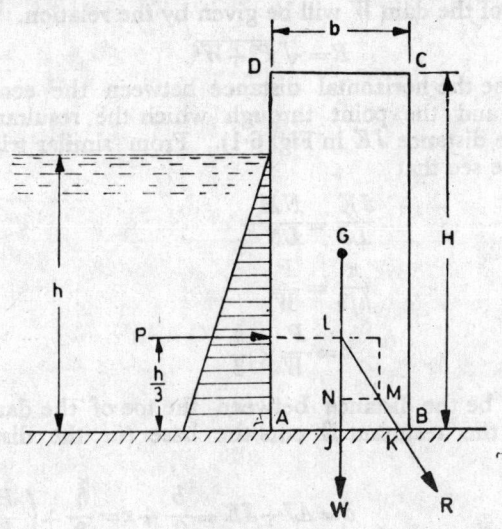

Fig. 6·1

Let
b = width of the dam,

H = Height of the dam,

ρ = Specific weight of the dam masonry,

h = Height of water retained by the dam, and

*w = Specific weight of the water.

The weight of dam per unit length,

$$W = \rho.b.h.$$

This weight W will act through the centre of gravity of the dam section.

We know that the intensity of water pressure will be zero at the water surface and will increase** by a straight line law to wh at the bottom. Thus the average intensity of water pressure on the

---

*Unless mentioned, otherwise the specific weight of the water is taken as 1,000 kg/m³.

**Sometimes, the dam is subjected to two kinds of liquids (e.g. some insoluble oil over water). In such a case, the pressure diagram will be zero at the top and will increase by a straight line law to w_1h_1 upto the depth of oil. It will further increase by a straight line law to $w_1h_1 + w_2h_2$ at the bottom of the water ; where w_1, w_2 and h_1, h_2 are the specific weights and heights of the oil and water respectively. The total pressure on the dam may be found out as usual.

exposed face of the dam is equal to $wh/2$: Therefore the total pressure on a unit length of the dam

$$P = \frac{wh}{2} \times h = \frac{wh^2}{2}$$

This water pressure P will act a height of $h/3$ from the bottom of the dam as shown in Fig. 6·1. Now complete the parallelogram with P and W as adjacent sides. The resultant of water pressure P and weight of the dam W will be given by the relation,

$$R = \sqrt{P^2 + W^2}$$

Let x be the horizontal distance between the centre of gravity of the dam and the point through which the resultant R cuts the base (i.e. the distance JK in Fig. 6·1). From similar triangles LMN and LJK, we see that

$$\frac{JK}{LJ} = \frac{NM}{LN}$$

$$\therefore \quad \frac{x}{h/3} = \frac{P}{W}$$

or

$$x = \frac{P}{W} \times \frac{h}{3}$$

Let d^* be the distance between the toe of the dam A and the point where the resultant R cuts the base (i.e. the distance AK in Fig. 6·1)

$$\therefore \quad d = AJ + JK = \frac{b}{2} + x = \frac{b}{2} + \left(\frac{P}{W} \times \frac{h}{3} \right)$$

and the eccentricity of the resultant,

$$e = d - \frac{b}{2} \qquad \text{(x in the case)}$$

A little consideration will show, that as a result of the eccentricity, some moment will come into play, which will cause some bending stress at the base section of the dam. The magnitude of this moment,

$$M = \text{Weight of the dam} \times \text{eccentricity}$$
$$= W.e.$$

We know that the moment of the inertia of the base section, about its c.g.,

$$I = \frac{1 \times b^3}{12} = \frac{b^3}{12}$$

*The distance d may also be found out by taking moments of (i) water pressure (ii) weight of dam and (iii) resultant force about A and equating the same, i.e.

$$W.d = P. \frac{h}{3} + W. \frac{b}{2}$$

(∵ Vertical component of the resultant force is W and is acting at a distance d from A and its horizontal component is acting through A.)

$$\therefore \quad d = \frac{b}{2} + \left(\frac{P}{W} \times \frac{h}{3} \right)$$

Let $\qquad$ $y=$ Distance between the centre of gravity of the base section and extreme fibre of the base ($b/2$ in this case).

$p_b=$ Bending stress in the fibre at a distance y from the centre of gravity of the base section.

We also know that

$$\frac{M}{I}=\frac{p_b}{y}$$

$$\therefore \qquad p_b=\frac{M.y}{I}=\frac{W.e \times \dfrac{b}{2}}{\dfrac{b^3}{12}}=\frac{6W.e}{b^2}$$

Now the distribution of direct stress at the base,

$$p_0=\frac{\text{Weight of dam}}{\text{Width of dam}}=\frac{W}{b}$$

Now, a little consideration will show that the stress across the base at B will be maximum, whereas the stress across the base at A will be minimum.

$$\therefore \qquad p_{max}=p_0+p_b=\frac{W}{b}+\frac{6W.e}{b^2}=\frac{W}{b}\left(1+\frac{6e}{b}\right)$$

and

$$p_{min}=p_0-p_b=\frac{W}{b}-\frac{6W.e}{b^2}=\frac{W}{b}\left(1-\frac{6e}{b}\right)$$

Note 1. When the reservoir is empty, there will be no water pressure on the dam. In this case, there will be no eccentricity, and thus the weight of the dam W will act through the *c.g.* of the base section, which will cause direct stress only.

2. Sometimes, the value of p_{min} comes out to be negative. In such a case, there will be a tensile stress at the base of the dam.

Example 6·1. *A tank is filled with a liquid of sp. gr. 1 up to a depth of 0·5 m over another liquid of sp. gr. 2. Find the total pressure per metre length and its line of action, on the 1 m deep side of the tank.*

Solution

Given. Sp. gr. of upper liquid

$$=1$$

$\therefore$ Specific weight, $\qquad w_1=1{,}000 \text{ kg/m}^3$

Depth of the upper liquid, $h=0·5$ m

Sp. gr. of the lower liquid $\quad =2$

Sp. weight $\qquad w_2=2{,}000 \text{ kg/m}^3$

Depth of lower liquid $\qquad h_2=1-0·5=0·5$ m

Total pressure per metre length of the dam

The pressure diagram on the tank is shown in Fig. 6·2.

Let $P_1=$ total pressure of the upper liquid per metre length, and

$P_2=$ total pressure of the lower liquid per metre length.

We know that pressure BD
$$= w_1 h_1 = 1,000 \times 0.5 = 500 \text{ kg/m}^2$$

Similarly, pressure EF
$$= w_2 h_2 = 2,000 \times 0.5 = 1,000 \text{ kg/m}^2$$

∴ Total pressure of the upper liquid per metre length,
$$P_1 = \text{Area of triangle } ABD \times \text{length of wall}$$
$$= (\tfrac{1}{2} \times 500 \times 0.5) \times 1 = 125 \text{ kg}$$

Similarly, total pressure of the lower liquid per metre length,
$$P_2 = \text{Area of figure } BDFC \times \text{length of wall}$$
$$= (\text{Area of rectangle } BCDE + \text{Area of triangle } DEF) \times \text{length of wall}$$
$$= [(500 \times 0.5) + (\tfrac{1}{2} \times 1,000 \times 0.5)] \times 1 = 500 \text{ kg}$$

∴ Total pressure,
$$P = P_1 + P_2 = 125 + 500 \text{ kg}$$
$$= 625 \text{ kg} \quad \textbf{Ans.}$$

Line of action of the total force

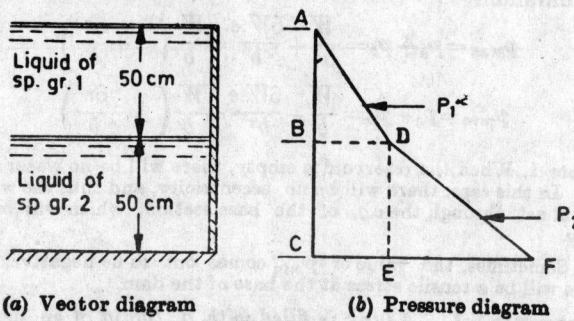

(a) Vector diagram   (b) Pressure diagram
Fig. 6·2

Let $h =$ Depth of the line of action of the total pressure from A.

Taking moments of all pressures about A, and equating the same,

$$P \times h = \left[\text{Pressure } ABD \times \frac{2 \times 0.5}{3} \right]$$
$$+ \left[\text{Pressure } BCDE \times \left(0.5 + \frac{0.5}{2} \right) \right]$$
$$+ \left[\text{Pressure } DEF \times \left(0.5 + \frac{2 \times 0.5}{3} \right) \right]$$
$$625\, h = \left(125 \times \frac{1}{3} \right) + \left(250 \times \frac{3}{4} \right) + \left(250 \times \frac{5}{6} \right) = \frac{2,625}{6}$$

∴ $h = \dfrac{2,625}{6} \times \dfrac{1}{625} = 0.7 \text{ m} \quad \textbf{Ans.}$

Example 6·2. *A concrete dam of rectangular section 15 m high and 6 m wide contains water up a height of 13 m.   Find*

(a) *total pressure on 1 m length of the dam,*

(b) *the point, where the resultant cuts the base,*

(c) *maximum and minimum intensities of stores at the base.*

Assume weight of concrete as 2,530 kg/m².

Solution

Given. Height of dam,

$$H=15 \text{ m}$$

Width of dam, $b=6$ m

Height of water, $h=13$ m

Weight of concrete,

$$\rho = 2{,}530 \text{ kg/m}^3$$

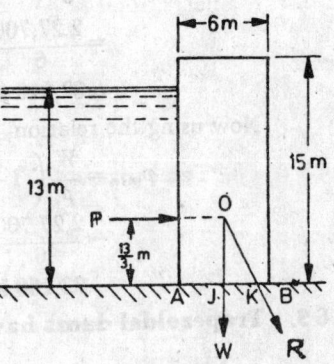

Fig. 6·3

Total pressure on 1 m length of the dam

Let

$P=$Total pressure on 1 m length of the dam.

Using the relation,

$$P=\frac{wh^2}{2} \text{ with usual notations.}$$

$$=\frac{1{,}000 \times 13^2}{2}=\textbf{84,500 kg Ans.}$$

The point, where the resultant cuts the base

Let the resultant cut the base at K as shown in Fig. 6·3.

Now let   *$x=$Horizontal distance between *c.g.* of the dam section and the point where the resultant cuts the base (*i.e. JK*).

Weight of the concrete per m length of the dam,

$$W=2{,}530 \times 6 \times 15 = 2{,}27{,}700 \text{ kg}$$

Using the relation,

$$x=\frac{P}{W} \times \frac{h}{3} \quad \text{with usual notations.}$$

$$=\frac{84{,}500}{2{,}27{,}700} \times \frac{13}{3}=\textbf{1·61 m} \quad \text{Ans.}$$

---

*The eccentricity (*e*) may also be found out by taking moments about *A*. Let *d* be the distance *AK*.

$$\therefore \qquad W.d = P \times \frac{h}{3} + W \times \frac{b}{2}$$

or $\qquad d = \frac{b}{2} + \frac{P}{W} \times \frac{h}{3} = \frac{6}{2} + \frac{84{,}500}{2{,}27{,}700} \times \frac{13}{3} = 4·61 \text{ m}$

$\therefore$ eccentricity, $e = d - \dfrac{b}{2} = 4·61 - 3·0 = 1·61$ m

Maximum and minimum intensities of stress at the base

Let p_{max}=Maximum intensity of stress across the base at B,

        p_{min}=Minimum intensity of stress across the base at A.

We know that the eccentricity of the resultant,

$$e=x=1\cdot61 \text{ m}$$

Using the relation,

$$p_{max}= \frac{W}{b}\left(1+\frac{6e}{b} \right) \text{ with usual notations.}$$

$$=\frac{2,27,700}{6}\left(1+\frac{6\times1\cdot61}{6} \right) \text{ kg/m}^3$$

$$=\textbf{82,010 kg/m}^2 \textbf{ (Compression) Ans.}$$

Now using the relation,

$$p_{min}= \frac{W}{\rho}\left(1-\frac{6e}{b} \right) \text{ with usual notations.}$$

$$-\frac{2,27,700}{6}\left(1-\frac{6\times1\cdot61}{6} \right)=-23,150 \text{ kg/m}^2$$

$$=\textbf{23,150 kg/m}^2 \textbf{ (Tension) Ans.}$$

6·3. Trapezoidal dams having water face vertical

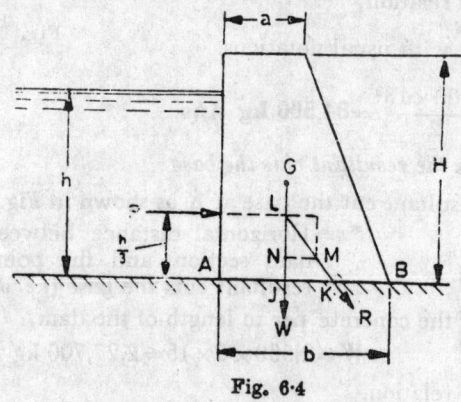

Fig. 6·4

Consider a unit length of a trapezoidal dam having its water face vertical as shown in Fig. 6·4.

Let a=Top width of the dam,

                            b=Bottom width of the dam,

                            H=Height of the dam,

                            ρ=Specific weight of the dam masonry,

                            h=Height of water retained by the dam, and

                            w=Specific weight of the water.

Now know that the weight of the dam per unit length,

$$W=\rho \times \frac{(a+b)}{2}\times H$$

Like a rectangular dam, the total pressure on a unit length of the trapezoidal dam,

$$P = \frac{wh^2}{2}$$

and the horizontal distance between the c.g. of the dam and the point where the resultant R cuts the base,

$$x = \frac{P}{W} \times \frac{h}{3}$$

The distance between the toe of the dam A and the point where the resultant R cuts the base (i.e. distance AK in Fig. 6·4),

$$d = AJ + JK = AJ + \left(\frac{P}{W} \times \frac{h}{3} \right)$$

Now the distance AJ* may be found out by splitting the dam section into a rectangle and a triangle. Now taking their moments about A and equating the same with the moment of the dam section about A,

$\therefore$ eccentricity, $\quad e = d - \dfrac{b}{2}$

The stress across the base at B, will be maximum, whereas the stress across the base at A will be minimum, such that

$$p_{max} = \frac{W}{b} \left(1 + \frac{6e}{b} \right)$$

and

$$p_{min} = \frac{W}{b} \left(1 - \frac{6e}{b} \right)$$

Note. When the reservoir is empty, there will be no water pressure on the dam. In this case, the eccentricity of the weight of the dam,

$$e = \frac{b}{2} - AJ$$

Since the eccentricity in this case will be *minus*, therefore the total stress across the base at B will be minimum whereas the stress across the base at A, will be maximum, such that

$$p_{min} = \frac{W}{b} \left(1 - \frac{6e}{b} \right)$$

and

$$p_{max} = \frac{W}{b} \left(1 + \frac{6e}{b} \right)$$

Example 6·3. *A concrete dam of trapezoidal section having water in vertical face is 16 m high. The base of the dam is 8 m wide and top 3 m wide. Find*

(a) *the resultant thrust on the base per metre length of the dam,*

(b) *the point, where the resultant thrust cuts the base, and*

(c) *intensities of maximum and minimum stresses across the base.*

---

*The distance AJ may also be found out from the relation,

$$AJ = \frac{a^2 + ab + b^2}{3(a+b)}$$

Take weight of the concrete as 2,400 kg/m³ and the water level coinciding with the top of the dam.

Solution

Given.   Height of dam,
$$H = 16 \text{ m}$$

Height of water, $h = 16$ m

Base width $b = 8$ m

Top width $a = 3$ m

Weight of masonry,
$$\rho = 2,400 \text{ kg/m}^3$$

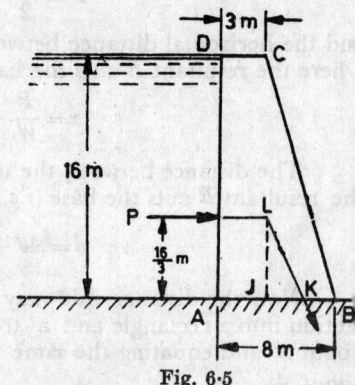

Fig. 6·5

Resultant thrust on the base per metre length.

Let P = Total pressure of water per metre length.

Using the relation,

$$P = \frac{wh^2}{2} \text{ with usual notations.}$$

$$= \frac{1000 \times 16^2}{2} = 1,28,000 \text{ kg}$$

We know that the weight of dam per metre length,

$$W = \rho \times \left(\frac{a+b}{2}\right) \times H = 2,400 \times \left(\frac{3+8}{2}\right) \times 16 \text{ kg}$$
$$= 2,11,200 \text{ kg}$$

∴ Resultant thrust,
$$R = \sqrt{P^2 + W^2} = \sqrt{(1,28,000)^2 + (2,11,200)^2} \text{ kg}$$
$$= 246,780 \text{ kg   Ans.}$$

The point, where the resultant cuts the base

Let the resultant cut the base at K as shown in Fig. 6·5.

Now let x = Horizontal distance between the *c.g.* of the dam section and K (*i.e.* JK).

First of all, let us find out the position* of the *c.g.* of the dam section.   Now taking moment of the area about A and equating the same,

$$\left(16 \times 3 + \frac{16 \times 5}{2}\right) AJ = \left(16 \times 3 \times \frac{3}{2}\right) + \left[16 \times \frac{5}{2}\left(3 + \frac{5}{3}\right)\right]$$
$$88 \times AJ = 72 + 186\cdot7 = 258\cdot7$$

or $*AJ = \dfrac{258\cdot7}{88} = 2\cdot94$ m

---

*The distance AJ may also be found out from the following relation,
$$AJ = \frac{a^2 + ab + b^2}{3(a+b)} = \frac{3^2 + 3 \times 8 + 8^2}{3(3+8)} = \frac{97}{33} = 2\cdot94 \text{ m}$$

Now using the relation,

$$x = \frac{P}{W} \times \frac{h}{3} \text{ with usual notations.}$$

$$= \frac{1,28,000}{2,11,200} \times \frac{16}{3} = 3 \cdot 235 \text{ m}$$

∴   horizontal distance AK, $d^* = AJ + x = 2 \cdot 94 + 3 \cdot 235$ m
                                $= 6 \cdot 175$ m   **Ans.**

Intensities of maximum and minimum stresses across the base

Let p_{max} = Intensity of maximum stress across **the**
                    base at B, and

         p_{min} = Intensity of minimum stress across **the**
                    base at A.

We know that the eccentricity of the resultant,

$$e = d - \frac{b}{2} = 6 \cdot 175 - \frac{8}{2} = 2 \cdot 175 \text{ m}$$

Using the relation,

$$p_{max} = \frac{W}{b} \left(1 + \frac{6e}{b} \right) \text{ with usual notations.}$$

$$= \frac{2,11,200}{8} \left(1 + \frac{6 \times 2 \cdot 175}{8} \right) \text{ kg/m}^2$$

$$= \mathbf{69,430 \text{ kg/m}^2 \text{ (Compression)}} \text{ Ans.}$$

Now using the relation,

$$p_{min} = \frac{W}{b} \left(1 - \frac{be}{b} \right) \text{ with usual notations.}$$

$$= \frac{2,11,200}{8} \left(1 - \frac{6 \times 2 \cdot 175}{8} \right) \text{ kg/cm}^2$$

$$= -16,630 \text{ kg/m}^2$$

$$= \mathbf{16,630 \text{ kg/m}^2 \text{ (Tension)}} \text{ Ans.}$$

Example 6·4. *A masonry trapezoidal dam 4 m high, 1′ m wide at its top and 3 m wide at its bottom retains water on its vertical face. Determine the maximum and minimum stresses, at the base, (i) when the reservoir is full and (ii) when the reservoir is empty. Take weight of the masonry as 2,000 kg/m³.*   (*Mysore University 1977*)

Solution

Given. Height of dam, $H = 4$ m

---

*The horizontal distance d may also be found out by taking moment about A and equating the same *i.e.*

$$W \cdot d = P \times \frac{h}{3} + W \times AJ$$

to $$d = AJ + \frac{P}{W} \times \frac{h}{3} = 2 \cdot 94 + \frac{1,28,000}{2,11,200} \times \frac{16}{3} = 6 \cdot 175 \text{ m}$$

Top width $a = 1$ m
Bottom width $b = 3$ m
Weight of masonry, $\rho = 2{,}000$ kg/m³
Let $p_{max} =$ Maximum stress at the base B, and
                          $p_{min} =$ Minimum stress at the base A.

Stresses at the base when the reservoir is full

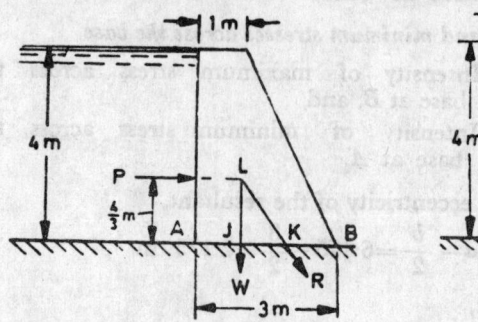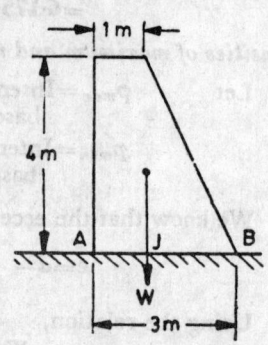

(a) When the reservoir is full.   (b) When the reservoir is empty.

Fig. 6·6

Let the resultant cut the base at K as shown in Fig. 6·6 (a).

Now let $x =$ Horizontal distance between the c.g. of the
             dam section and K (i.e. JK).
             $P =$ Total pressure of water per metre length.

Using the relation,

$$P = \frac{wh^2}{2} \text{ with usual notations.}$$

$$= \frac{100 \times 4^2}{2} = 8{,}000 \text{ kg}$$

We know that the weight of the dam per metre length,

$$W = \rho \times \frac{(a+b)}{2} \times H = 2{,}000 \left(\frac{1+3}{2} \right) \times 4 \text{ kg}$$

$$= 16{,}000 \text{ kg}$$

Now let us find out the position* of the c.g. of the dam section.
Taking moments of the area about A and equating the same,

$$\left(4 \times 1 + \frac{4 \times 2}{2} \right) AJ = \left(4 \times 1 \times \frac{1}{2} \right) + \left[\frac{4 \times 2}{2} \left(1 + \frac{2}{3} \right) \right]$$

$$[8AJ = 2 + 6 \cdot 67 = 8 \cdot 67$$

or $$AJ = \frac{8 \cdot 67}{8} = 1 \cdot 08 \text{ m}$$

---

*The distance AJ may also be found out from the following relation :
$$AJ = \frac{a^2 + ab + b^2}{3(a+b)} = \frac{1^2 + 1 \times 3 + 3^2}{3(1+3)} = \frac{13}{12} = 1 \cdot 08 \text{ m}$$

Using the relation,

$$x = \frac{P}{W} \times \frac{h}{3} \text{ with usual notations.}$$

$$= \frac{8,000}{16,000} \times \frac{4}{3} = 0.67 \text{ m}$$

∴ Horizontal distance AK,

$$d = AJ + x = 1.08 + 0.67 = 1.75 \text{ m}$$

and eccentricity, $$e = d - \frac{b}{2} = 1.75 - \frac{3}{2} = 0.25 \text{ m}$$

Using the relation,

$$p_{max} = \frac{W}{b}\left(1 + \frac{6e}{b} \right) \text{ with usual notations}$$

$$= \frac{16,000}{3}\left(1 + \frac{6 \times 0.25}{3} \right) \text{ kg/m}^2$$

$$= 8,000 \text{ kg/m}^2 \quad \text{Ans.}$$

Now using the relation,

$$p_{min} = \frac{W}{b}\left(1 - \frac{6e}{b} \right) \text{ with usual notations.}$$

$$= \frac{16,000}{3}\left(1 - \frac{6 \times 0.25}{3} \right) \text{ kg/m}^2$$

$$= 2,667 \text{ kg/m}^2 \quad \text{Ans.}$$

Stresses at the base when the reservoir is empty

We know that the distance AJ,

$$d = 1.08 \text{ m}$$

and eccentricity, $$e = d - \frac{b}{2} = 1.08 - 1.5 \text{ m} = -0.42 \text{ m}$$

(Minus sign indicates that the stress at A will be more than that at B).

Using the relation,

$$p_{max} = \frac{W}{b}\left(1 + \frac{6e}{b} \right) \text{ with usual notations.}$$

∴ $$p_A = \frac{16,000}{3}\left(1 + \frac{6 \times 0.42}{3} \right) \text{ kg/m}^2$$

$$= 9,813 \text{ kg/m}^2 \quad \text{Ans.}$$

Now using the relation,

$$p_{min} = \frac{W}{b}\left(1 - \frac{6e}{b} \right) \text{ with usual notations.}$$

∴ $$p_B = \frac{16,000}{3}\left(1 - \frac{6 \times 0.42}{3} \right) \text{ kg/m}^2$$

$$= 853 \text{ kg/m}^2 \quad \text{Ans.}$$

Example 6·5. *A masonry dam as shown in Fig. 6·7 has a total height of 20 m with a top width of 5 m and a free board of 2 m. Its upstream face is vertical while the downstream face has a batter of 0·66 horizontal to 1·0 vertical. The specific gravity of masonry may be taken as 2·4.*

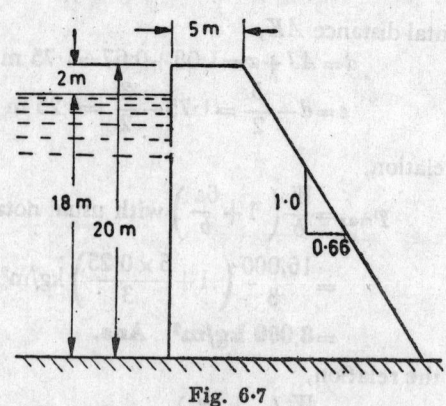

Fig. 6·7

In addition to the hydrostatic pressure on the upstream face, there is an uplift pressure at the foundation, which may be taken to vary linearly from a value equal to the full hydrostatic pressure at the upstream end, to zero at the downstream end. Calculate the extreme values of the normal stresses on the foundation, when the reservoir is full.

Solution

Given.   Height of dam,
$$H = 20 \text{ m}$$

Top width, $a = 5$ m

Free board $= 2$ m

∴ Height of water $h = 20 - 2 = 18$ m

Slope of downstream face
$$= 0·66 \text{ horizontal to } 1·0 \text{ vertical}$$

∴   Bottom width, $b = 5 + \dfrac{20}{1·0} \times 0·66 = 18·2$ m

Specific gravity of masonry
$$= 2·4$$

∴ Specific weight, $\rho = 2·4 \times 1,000 = 2,400$ kg/m³

Uplift pressure at the upstream
$$= wh = 1,000 \times 18 = 18,000 \text{ kg}$$

Uplift pressure at the downstream
$$= 0.$$

Let $p_{max} = $ Maximum stress at the base B, and

                $p_{min} = $ Minimum stress at the base A,

Let the resultant cut the base at K as shown in Fig 6·8.

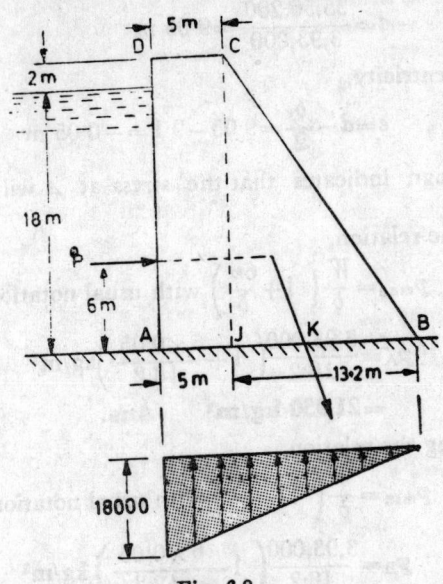

Fig. 6·8

Now let P = Total water pressure per metre length

x = Horizontal distance between the *c.g.* of
the dam section and K (*i.e.* JK).

Using the relation,

$$P = \frac{wh^2}{2} \text{ with usual notations.}$$

$$= \frac{1,000 \times 18^2}{2} = 1,62,000 \text{ kg}$$

We know that the weight of the dam per metre length,

W = Weight of dam section − Uplift pressure

$$= \left(2,400 \times \frac{5+18·2}{2} \times 20 \right) - \left(\frac{18,000 \times 18·2}{2} \right) \text{kg}$$

$$= 5,56,800 - 1,63,800 = 3,93,000 \text{ kg}$$

Now let us find out the point K, where the resultant cuts the base. Taking moments of the dam section about A, and equating the same,

$$3,93,000 \ d = (1,62,000 \times 6) + (2,400 \times 20 \times 5 \times 2·5)$$

$$+ \left[2,400 \times \frac{20 \times 13·2}{2} \left(5 + \frac{13·2}{3} \right) \right]$$

$$- \left(\frac{18,000 \times 18·2}{2} \times \frac{18·2}{3} \right) \text{ kg}$$

$$3,93,000 \ d = 9,72,000 + 6,00,000 + 29,77,920 - 9,93,720$$
$$= 35,56,200 \text{ kg}$$

or
$$d = \frac{35,56,200}{3,93,000} = 9 \cdot 05 \text{ m}$$

The eccentricity,

$$e = d - \frac{b}{2} = 9 \cdot 05 - 9 \cdot 1 = -0 \cdot 05 \text{ m}$$

(Minus sign indicates that the stress at A will be more than that at B).

Using the relation,

$$p_{max} = \frac{W}{b}\left(1 + \frac{6e}{b}\right) \text{ with usual notations.}$$

$$p_A = \frac{3,93,000}{18 \cdot 2}\left(1 + \frac{6 \times 0 \cdot 05}{18 \cdot 2}\right) \text{kg/m}^2$$
$$= 21,950 \text{ kg/m}^2 \qquad \textbf{Ans.}$$

Now using the relation,

$$p_{min} = \frac{W}{b}\left(1 - \frac{6e}{b}\right) \text{ with usual notations.}$$

$$p_B = \frac{3,93,000}{18 \cdot 2}\left(1 - \frac{6 \times 0 \cdot 05}{18 \cdot 2}\right) \text{kg/m}^2$$
$$= 21,230 \text{ kg/m}^2 \quad \textbf{Ans.}$$

Alternate method,

Let the resultant cut the base at K as shown in Fig. 6·8.

Let $\qquad x =$ Horizontal distance between the $c.g.$ of the dam section and K (i.e. JK).

We know that the weight of the dam per metre length,

$$W = \rho \times \frac{(a+b)}{2} \times H = 2,400 \times \frac{5 + 18 \cdot 2}{2} \times 20$$
$$= 5,56,800 \text{ kg}$$

Now let us find out the $c.g.$ of the dam section. Taking moment of the dam section about A, and equating the same, i.e.

$$\left[20 \times 5 + \frac{20 \times 13 \cdot 2}{2}\right] AJ = \left[20 \times 5 \times \frac{5}{2}\right] + \left[\frac{20 \times 13 \cdot 2}{2}\left(5 + \frac{13 \cdot 2}{3}\right)\right]$$
$$232 \ AJ = 250 + 1,240 \cdot 8 = 1,490 \cdot 8$$
$$AJ = \frac{1,490 \cdot 8}{232} = 6 \cdot 42 \text{ m}$$

Using the relation,

$$x = \frac{P}{W} \times \frac{h}{3} \text{ with usual notations.}$$
$$= \frac{1,62,000}{5,56,800} \times \frac{18}{3} = 1 \cdot 75 \text{ m}$$

∴ Horizontal distance AK,

$$d = AJ + x = 6 \cdot 42 + 1 \cdot 75 = 8 \cdot 17 \text{ m}$$

and eccentricity, $e = d - \dfrac{b}{2} = 8 \cdot 17 - 9 \cdot 1 = -0 \cdot 93$ m

(Minus sign indicates that the stress at A will be more than that at B).

Using the relation

$$p_{max} = \frac{W}{b} \left(1 + \frac{6e}{b} \right) \text{ with usual notations.}$$

$$= \frac{5,56,800}{18 \cdot 2} \left(1 + \frac{6 \times 0 \cdot 93}{18 \cdot 2} \right) = 39,950 \text{ kg/m}^2$$

Since there is hydrostatic pressure at A, therefore the actual stress at A,

$$p_A = 39,950 - 18,000 = \textbf{21,950 kg/m}^2 \quad \textbf{Ans.}$$

Now using the relation,

$$p_{min} = \frac{W}{b} \left(1 - \frac{6e}{b} \right) \text{ with usual notations}$$

$$= \frac{5,56,800}{18 \cdot 2} \left(1 - \frac{6 \times 0 \cdot 93}{18 \cdot 2} \right) = 21,230 \text{ kg/m}^2$$

Since there is zero hydrostatic pressure at B, therefore actual stress at B,

$$p_B = \textbf{21,230 kg/m}^2 \quad \textbf{Ans.}$$

6·4. Trapezoidal dams having water face inclined

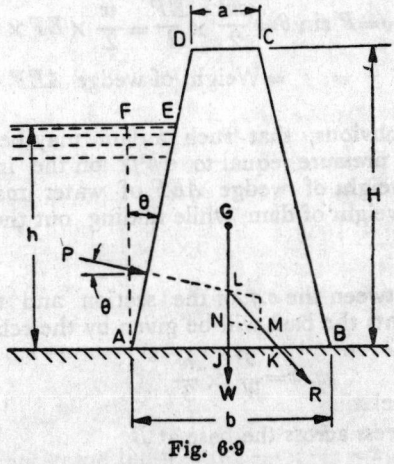

Fig. 6·9

Consider a unit length of a trapezoidal dam, having its water surface inclined as shown in Fig. 6·9.

Let $a =$ Top width of the dam,

$b =$ Bottom width of the dam,

$H =$ Height of the dam,

$\rho =$ Specific weight of the dam masonry,

$h =$ Height of water retained by the dam,

$w =$ Specific weight of the water, and

$\theta =$ Inclination of the water face with the vertical.

Length of the sloping side AE, which is subjected to water pressure,

$$l = \frac{h}{\cos \theta}$$

Now we see that the weight of the dam per unit length,

$$W = \rho \times \frac{(a+b)}{2} \times H$$

The intensity of water pressure will be zero at the water surface and will increase by a straight line law to wh at the bottom. Therefore the total pressure on a unit length of the dam,

$$P = \frac{wh}{2} \times l = \frac{whl}{2}$$

This water pressure P will act at a height of $h/3$ from the bottom of the dam as shown in Fig. 6·9.

The horizontal component of this water pressure,

$$P_H = P \cos \theta = \frac{whl}{2} \times \frac{h}{l} = \frac{wh^2}{2}$$

and vertical component of this water pressure,

$$P_V = P \sin \theta = \frac{whl}{2} \times \frac{EF}{l} = \frac{w}{2} \times EF \times h$$

$$= \text{Weight of wedge } AEF \text{ of water}$$

It is thus, obvious, that such a dam may be taken to have a horizontal water pressure equal to $wh^2/2$ on the imaginary vertical face AF. The weight of wedge AEF of water may be considered as a part of the weight of dam, while finding out the c.g. of the dam section.

Distance between the c.g. of the section and the point where the resultant R cuts the base will be given by the relation

$$x = \frac{P}{W} \times \frac{h}{3} \qquad \text{(As usual)}$$

The total stress across the base at B,

$$p_{max} = \frac{W}{b} \left(1 + \frac{6e}{b} \right) \qquad \text{(As usual)}$$

and the total stress across the base at A,

$$p_{min} = \frac{W}{b}\left(1 - \frac{6e}{b}\right) \qquad \text{(As usual)}$$

Note. When the reservoir is empty there will be neither water pressure on the dam, nor there will be the weight of wedge AEF of water. In this case the eccentricity of the weight of the dam,

$$e = \frac{b}{2} - AJ$$

Since the eccentricity will be *minus*, therefore the total stress across the base at B,

$$p_{min} = \frac{w}{b}\left(1 - \frac{6e}{b}\right)$$

and the total stress across the base at A,

$$p_{max} = \frac{W}{b}\left(1 + \frac{6e}{b}\right)$$

Example 6·6. *A masonry dam of trapezoidal section is 10 m high. It has top width of 1 m and bottom width 7 m. The face exposed to water has a slope of 1 horizontal to 10 vertical.*

Calculate the maximum and minimum stresses on the base, when the water level coincides with the top of the dam. Take weight of the masonry of 2,000 kg/m³. ...ersity, 1979

Solution

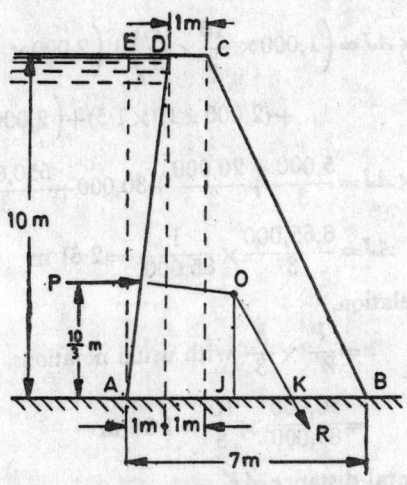

Fig. 6·10

Given. Height of dam, $H = 10$ m
Top width of dam, $a = 1$ m
Bottom width of dam, $b = 7$ m
Height of water, $h = 10$ m
Weight of masonry, $\rho = 2{,}000$ kg/m³

Let the resultant cut the base at K as shown in Fig. 6·10.
Now let p_{max} = Maximum stress on the base at B,

p_{min} = Minimum stress on the base at A,

P = Total water* pressure per metre length of the dam,

x = Horizontal distance between the $c.g.$ of the dam section and K (i.e. JK).

Using the relation,

$$P = \frac{wh^2}{2} \text{ with usual notations.}$$

$$= \frac{1,000 \times 10^2}{2} = 50,000 \text{ kg}$$

We know that the weight of the dam,

$$W = \left(w \times \frac{10}{2} \times 1\right) + \rho\left(\frac{a+b}{2}\right) \times H \text{ kg}$$

$$= (1,000 \times 5) + 2,000 \times \left(\frac{1+7}{2}\right) \times 10 \text{ kg}$$

$$= 85,000 \text{ kg}$$

Now let us find out the position of the $c.g.$ of the dam section. Taking moments of the weight of the dam section about A and equating the same,

$$W \times AJ = \left(1,000 \times \frac{10}{2} \times \frac{1}{3}\right) + \left(2,000 \times \frac{10}{2} \times \frac{2}{3}\right)$$

$$+ (2,000 \times 10 \times 1\cdot5) + \left(2,000 \times \frac{10 \times 5}{2} \times \frac{11}{3}\right)$$

$$85,000 \times AJ = \frac{5,000}{3} + \frac{20,000}{3} + 30,000 + \frac{550,000}{3} = \frac{6,65,000}{3}$$

$$AJ = \frac{6,65,000}{3} \times \frac{1}{85,000} = 2\cdot61 \text{ m}$$

Using the relation,

$$x = \frac{P}{W} \times \frac{h}{3} \text{ with usual notations.}$$

$$= \frac{50,000}{85,000} \times \frac{10}{3} = 1\cdot96 \text{ m}$$

∴ Horizontal distance AK,

$$d = AJ + x = 2\cdot61 + 1\cdot96 = 4\cdot57 \text{ m}$$

We know that the eccentricity of the resultant,

$$e = d - \frac{b}{2} = 4\cdot57 - \frac{7}{3} = 1\cdot07 \text{ m}$$

*Strictly speaking, the total pressure is acting normally to the face AD of the dam. But here we shall assume the pressure to act normally to the imaginary vertical plane AE as discussed in Art. 6·4.

Using the relation,

$$p_{max} = \frac{W}{b}\left(1 + \frac{6e}{b}\right) \text{ with usual notations.}$$

$$= \frac{85,000}{7}\left(1 + \frac{6 \times 1 \cdot 07}{7}\right) \text{ kg/m}^2$$

$$= 23,420 \text{ kg/m}^2 \quad \text{Ans.}$$

Now using the relation,

$$p_{min} = \frac{W}{b}\left(1 - \frac{6e}{b}\right) \text{ with usual notations.}$$

$$= \frac{85,000}{7}\left(1 - \frac{6 \times 1 \cdot 07}{7}\right) \text{ kg/m}^2$$

$$= 10,060 \text{ kg/m}^2 \quad \text{Ans.}$$

Exercise 6-1

1. A masonry dam 6 m high and 3 m wide has water level with its top. Find (i) total pressure per metre length of the dam, (ii) point at which the resultant cuts the base and (iii) maximum and minimum intensities of stress at the bottom of the dam.

[**Ans.** 18,000 kg ; 2·5 m ; 36,000 kg/m^2 ; —12,000 kg/m^2]

2. A masonry trapezoidal dam 1 m wide at top, 4 m at its base and 6 m high is retaining water on its vertical face to a height equal to the top of the dam. Determine the maximum and minimum stress intensities. Take density of masonry as 2,250 kg/m^3. *(Aligarh University)*

[**Ans.** 16,030 kg/m^2 ; 843·7 kg/m^2]

3. A trapezoidal masonry dam 2 m wide at top, 8 m wide at its bottom is 12 m high. The face exposed to water has a slope of 1 horizontal to 12 vertical. Determine the maximum and minimum stress intensities, when the water rises to the top of the dam. Take weight of the masonry as 2·4 tonnes/m^3. *(U.P.S.C. Engg. Service)*

[**Ans.** 32,800 kg/m^2 ; 4,690 kg/m^2]

4. A trapezoidal masonry dam 1 m wide at its bottom 7 m wide at its top is 10 high. The face exposed to water has a slope of 1 in 10. Find the maximum and minimum intensities of stress at the base, where the water level rises up to the top of the dam. The masonry weighs 2,000 kg/m^3. *(Calcutta University)*

[**Ans.** 23,200 kg/m^2 ; 1,090 kg/m^2]

6·5. Conditions for the stability of a dam

In the previous articles, we used to derive a relation for the position of a point, through which the resultant R (of the water pressure P and the weight of dam W) cuts the base. The position of this point helps us in finding out the total stresses across the base, at toe and heel of the dam. Apart from finding out the total stresses, this point helps us in checking the stability of the dam. In general, a dam is checked for the following conditions of stability :—

1. To avoid tension in the masonry at the base of the dam
2. To safeguard the dam from overturning
3. To prevent the sliding of dam, and
4. To prevent the crushing of masonry at the base of the dam.

Now we shall discuss, all the above conditions of stability, one by one.

6·6. Condition to avoid tension in the masonry of the dam at its base

We have discussed in Art. 6·2 that the water pressure, acting on one side of the dam, produces bending stress ; whereas the weight of the dam produces direct stress at the bottom of the dam. We have also seen, that on one side of the dam there is a maximum stress (equal to *sum* of the direct and bending stress) ; whereas on other side of the dam, there is a minimum stress (equal to direct stress *minus* bending stress). A little consideration will show. that so long as the bending stress remains less than the direct stress, the resultant stress is compressive But, when the bending stress is equal to the direct stress, there will be zero stress on one side. But, when the bending stress exceeds the direct stress, there will be a tensile stress on one side. Though cement concrete can take up a small amount of tensile stress, yet it is desirable to avoid tension in the masonry of the dam at its base.

It is this obvious, that in order to avoid the tension in the masonry of the dam, at its base, the bending stress should be less than the direct stress or it may be equal to the direct stress, *i.e.*,

$$p_b \leqq p_0$$

$$\frac{6W.e}{b^2} \leqq \frac{W}{b}$$

or
$$e \leqq \frac{b}{6}$$

It means that the eccentricity of the resultant can be equal to $d/6$ on either side of geometrical axis of base section. Thus the resultant must lie *within the middle third* of the base width, in order to avoid tension.

6·7. Condition to prevent the overturning of the dam

We have already discussed that when a dam is retaining water, it is subjected to some water pressure. We can easily find out the resultant R of the water pressure P and the weight of dam W. Since the dam is in equilibrium, therefore the resultant R must be balanced by equal and opposite reaction acting at K. This reaction may be split up at K into two components *viz*., horizontal and vertical. The horizontal component must be equal to the water pressure P ; whereas the vertical component must be equal to the weight W. Thus the following four forces, acting on the dam, keep it in equilibrium :

  (a) Water pressure P,
  (b) Horizontal component of the reaction,
  (c) Weight of the dam W,
  (d) Vertical component of the reaction.

These four forces may be grouped into two sets or couples. The moment of a couple consisting of the first two forces,

$$M_1 = \text{force} \times \text{arm} = P \times \frac{h}{3} \qquad \qquad ...(i)$$

Similarly, moment of a couple consisting of the last two forces,
$$M_2 = W \times JK \qquad \qquad ...(ii)$$

A little consideration will show that the moment of the first two forces will tend to overturn the dam about B ; whereas the moment of the last two forces will tend to restore the dam. Since the dam is in equilibrium and a couple can only be balanced by a couple, therefore overturning moment must be equal to the restoring moment, i.e.

$$P \times \frac{h}{3} = W \times JK$$

or

$$JK = \frac{P}{W} \times \frac{h}{3}$$

Incidentally, this equation is the same which we derived in Art. 6·2 and gives the position of the point K, where the resultant cuts the base. Since the dam will tend to overturn about B, therefore balancing moment about B,

$$M_3 = W \times JB$$

Now, we see that the dam is safe against overturning, so long as the balancing moment is more than the overturning moment (or restoring moment, which is equal to overturning moment) i.e.

$$W \times JB > W \times JK$$

or

$$JB > JK$$

It is thus obvious, that the condition to prevent the dam from overturning is that the point K should be between J and B or more precisely between A and B.

As a matter of fact, this is a superfluous condition. We know that to avoid tension in the masonry of a dam at its bottom, the resultant must lie within the middle third of the base width. Since we have to check the stability of a dam for tension in the base masonry, therefore the stability of the dam for overturning is automatically checked.

6·8 Condition to prevent the sliding of dam

We have already discussed in Art. 6·7 that there are four forces which act on a dam and keep it in equilibrium. Out of these four forces, two are vertical and the following two are horizontal :

(a) Water pressure P, and

(b) Horizontal component of the reaction.

A little consideration will show, that the horizontal component of the reaction will be given by the frictional force at the base of the dam.

Let μ = Coefficient of friction between the base of dam and the soil.

We know that the maximum available force of friction,

$$F_{max} = \mu W$$

It is thus obvious, that so long as F_{max} is more* than the water pressure P, the dam is safe against sliding.

6·9. Condition to prevent the crushing of masonry, at the base of the dam

We have already discussed in Art. 6·2 that whenever a dam is retaining water, the masonry of dam at its bottom is subjected to some stress. This stress varies from p_{max} to p_{min} by a straight line law. A little consideration will show, that the condition to prevent the crushing of masonry at the base of the dam, is that the maximum stress p_{max} should be less than the permissible stress in the masonry.

Example 6·7. *A trapezoidal masonry dam having 3 m top width, 8 m bottom width and 12 m high is retaining water as shown in Fig. 6·11.*

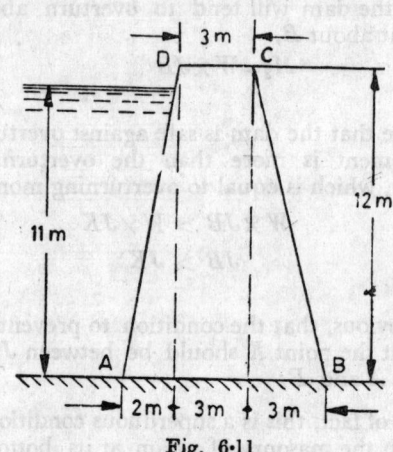

Fig. 6·11

Check the stability of the dam, when it is retaining water to a height of 11 m. The masonry weighs 2,000 kg/m³, and coefficient of friction between the dam masonry and soil is 0·6. Take the allowable compressive stress as 40,000 kg/m².

(*London University*)

Solution

Given. Top width of dam, $a = 3$ m
Bottom width of dam, $b = 8$ m
Height of dam, $H = 12$ m
Height of water, $h = 11$ m
Weight of masonry, $\rho = 2,000$ kg/m³
Coefficient of friction, $\mu = 0·6$
Allowable compression stress $= 40,000$ kg/m²

---

*Some authorities feel that the dam will be safe, when the force of friction is at least 1·5 times the total water per metre length.

$$\frac{\mu W}{P} = 1·5$$

Check for tension in the masonry

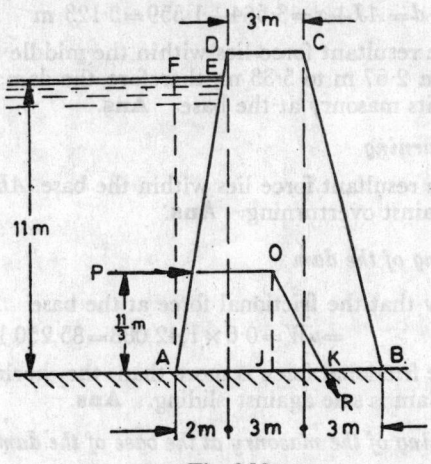

Fig. 6·12

Let P = water pressure per metre length of the dam, and

           x = Horizontal distance between the *c.g.* of the dam
                 section and K (*i.e. JK*).

We know that the water pressure per metre length,

$$P = \frac{wh^2}{2} = \frac{1000 \times 11^2}{2} = 60,500 \text{ kg}$$

We also know that the weight of dam per metre length,

$$W = \left(1,000 \times \frac{1}{2} \times 11 \times \frac{11}{6}\right) + 2,000\left(\frac{3+8}{2}\right) \times 12$$

$$= 1,42,083 \text{ kg}$$

Now let us find out the position of the *c.g.* of the dam section.
Taking moments of the weight of the dam section about A and equating the same,

$$W \times AJ = \left(1,000 \times \frac{11}{2} \times \frac{11}{6} \times \frac{2}{3}\right) + \left(2,000 \times \frac{12 \times 2}{2} \times \frac{4}{3}\right)$$

$$+ \left(2,000 \times 12 \times 3 \times \frac{7}{2}\right) + \left(2,000 \times \frac{12 \times 3}{2} \times 6\right)$$

$$1,42,083 \times AJ = 506,733$$

$$\therefore \qquad AJ = \frac{5,06,733}{1,42,083} = 3\cdot564 \text{ m}$$

Using the relation,

$$x = \frac{P}{W} \times \frac{h}{3} \text{ with usual notations.}$$

$$= \frac{60,500}{1,42,083} \times \frac{11}{3} = 1\cdot559 \text{ m}$$

∴ Horizontal distance AK,

$$d = AJ + x = 3.564 + 1.559 = 5.123 \text{ m}$$

Since the resultant force lies within the middle third of the base width *i.e.* from 2·67 m to 5·33 m, therefore the dam is safe against the tension in its masonry at the base. **Ans**.

Check for overturning

Since the resultant force lies within the base AB, therefore the dam is safe against overturning. **Ans**.

Check for sliding of the dam

We know that the frictional force at the base

$$= \mu W = 0.6 \times 1,42,083 = 85,250 \text{ kg}$$

Since the frictional force is more than the horizontal pressure, therefore the dam is safe against sliding. **Ans**.

Check for crushing of the masonry at the base of the dam

Let p_{max} = Maximum intensity of stress at the base of the dam.

We know that the eccentricity of the resultant,

$$e = d - \frac{b}{2} = 5.123 - \frac{8}{2} = 1.23 \text{ m}$$

Using the relation,

$$p_{max} = \frac{W}{b} \left(1 + \frac{6e}{b} \right) \text{ with usual notations.}$$

$$= \frac{1,42,083}{8} \left(1 + \frac{6 \times 1.123}{8} \right) \text{ kg/m}^2$$

$$= 32,710 \text{ kg/m}^2$$

Since the maximum stress is less than the allowable stress, therefore masonry of the dam is safe against crushing. **Ans**.

6·10. Minimum base width of a dam

We have already discussed in Arts. 6·6 to 6·9, the general conditions for the stability of a dam, when the section is given. But sometimes, while designing a dam, we have to calculate its necessary base width. This can be easily found out by studying the conditions for the stability of a dam. Thus the base width (b) of a dam may be obtained from the following conditions : —

1.   To avoid tension in the masonry at the base of the dam, the eccentricity, $e = \dfrac{b}{6}$. In this case, the maximum stress, $p_{max} = \dfrac{2W}{b}$ and the minimum stress, $p_{min} = 0$. The stress diagram at the base will be a triangle.

2.   To avoid the sliding of dam, the force of friction between the dam and soil, is at least 1·5 times the total water pressure metre length, *i.e.*,

$$\frac{\mu W}{P} = 1 \cdot 5$$

3.   To prevent the crushing of masonry at the base of the dam, the maximum stress should be less than the permissible stress of the soil.

Note. If complete data of a dam is given, then the base width for all the above three conditions should be found out separately. The maximum value of the base width from the above three conditions will give the necessary base width of the dam. But sometimes, sufficient data is not given to find out the values of base width for all the above mentioned three conditions. In such a case, the value of minimum base width may be found out, for any one of the above three conditions.

Example 6·8. *A mass concrete dam shown in Fig. 6·13 has a trapezoidal cross-section. The height above the foundation is 61·5 m and its water face is vertical. The width at the top is 4·5 m.*

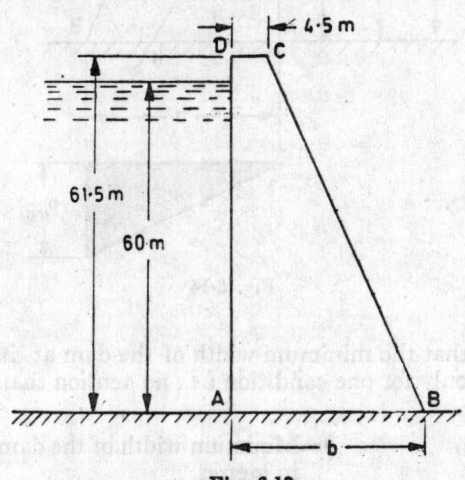

Fig. 6·13

Calculate the necessary minimum width of the dam at its bottom, to ensure that no tension shall be developed when water is stored up to 60 metres. Draw the pressure diagram at the base of the dam, for this condition, and indicate the maximum pressure developed.

Take density of concrete as 2,400 kg/m³ and density of water as 1,000 kg/m³.

Solution

Given. Height of dam,
$$H = 61 \cdot 5 \text{ m}$$

Top width of dam, $a = 4 \cdot 5$ m

Height of water, $h = 60$ m

Density of concrete, $\rho = 2,400$ kg/m^3

Density of water, $w = 1,000$ kg/m^3

Minimum width of the dam at its bottom

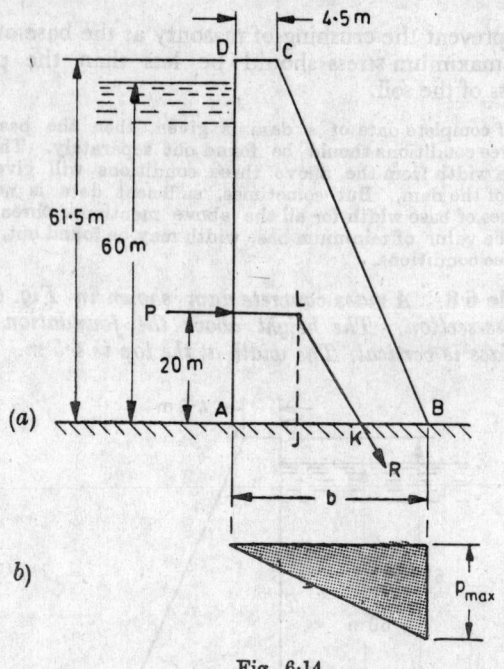

Fig. 6·14

We see that the minimum width of the dam at its bottom is to be found out only for one condition *i.e.*, no tension shall be developed at the base.

Let
      b = Minimum width of the dam at its bottom in metres,

      P = Total water pressure on the dam per metre length,

      x = Horizontal distance between the *c.g.* of the dam section and K (*i.e.* JK).

Using the relation,

$$P = \frac{wh^2}{2} \text{ with usual notations.}$$

$$= \frac{1,000 \times 60^2}{2} = 18,00,000 \text{ kg} \qquad \ldots(i)$$

We know that the weight of dam per metre length,

$$W = \rho \times \frac{(a+b)}{2} \times H$$

$$W = 2,400 \times \frac{4 \cdot 5 + b}{2} \times 61 \cdot 5 \text{ kg}$$

$$= 73,800 \, (4 \cdot 5 + b) \text{ kg} \qquad \qquad ...(ii)$$

Now let us find out the position of the *c.g.* of the dam section. We know that the distance AJ,

$$AJ = \frac{a^2 + ab + b^2}{3(a+b)} = \frac{4 \cdot 5^2 + 4 \cdot 5 b + b^2}{3(4 \cdot 5 + b)}$$

$$= \frac{20 \cdot 25 + 4 \cdot 5 b + b^2}{3(4 \cdot 5 + b)}$$

Now using the relation,

$$x = \frac{P}{W} \cdot \frac{h}{3} \text{ with usual notations.}$$

$$= \frac{18,00,000}{73,800(4 \cdot 5 + b)} \times \frac{60}{3} = \frac{488}{(4 \cdot 5 + b)}$$

$\therefore$ Horizontal distance AK,

$$d = AJ + x = \frac{20 \cdot 25 + 4 \cdot 5 \, b + b^2}{3(4 \cdot 5 + b)} + \frac{488}{4 \cdot 5 + b}$$

$$= \frac{20 \cdot 25 + 4 \cdot 5 \, b + b^2 + 1,464}{3(4 \cdot 5 + b)}$$

$$= \frac{1,484 \cdot 25 + 4 \cdot 5 \, b + b^2}{3(4 \cdot 5 + b)}$$

$\therefore$ Eccentricity of the resultant,

$$e = d - \frac{b}{2} = \frac{1484 \cdot 25 + 4 \cdot 5 \, b + b^2}{3(4 \cdot 5 - b)} - \frac{b}{2}$$

We know, that in order to avoid tension in the masonry at the base of the dam, the eccentricity,

$$e = \frac{b}{6}$$

$\therefore$
$$\frac{1484 \cdot 25 + 4 \cdot 5 \, b + b^2}{3(4 \cdot 5 + b)} - \frac{b}{2} = \frac{b}{6}$$

$$\frac{1,484 \cdot 25 + 4 \cdot 5 \, b + b^2}{3(4 \cdot 5 + b)} = \frac{b}{6} + \frac{b}{2} = \frac{2b}{3}$$

$$1,484 \cdot 25 + 4 \cdot 5 \, b + b^2 = 2b(4 \cdot 5 + b) = 9b + 2b^2$$

or $\qquad b^2 + 4 \cdot 5b - 1,484 \cdot 25 = 0$

Solving this equation, as a quadratic equation for b,

$\therefore$
$$b = \frac{-4 \cdot 5 + \sqrt{4 \cdot 5^2 + 4 \times 1,484 \cdot 25}}{2} \text{ m}$$

$$= \frac{-4 \cdot 5 + 77 \cdot 05}{2} = \textbf{36 \cdot 275 m \quad Ans.}$$

Pressure diagram at the base of the dam

Let $\qquad p_{max} = $ Maximum stress across the base at B.

Substituting the value of b in equation (ii),

$$W = 73,800(4 \cdot 5 + b) = 73,800(4 \cdot 5 + 33 \cdot 275)$$

$$= 30,09,200 \text{ kg}$$

$$\therefore \qquad p_{max} = \frac{2W}{b} = \frac{2 \times 30,09,200}{36 \cdot 275} = 1,65,900 \text{ kg/m}^2$$

$$= 165 \cdot 9 \text{ t/m}^2$$

and $p_{min} = 0$

The pressure diagram at the base of the dam is shown in Fig. 6·14 (b). **Ans.**

Example 6·9. *A concrete dam has its upstream face vertical and a top width of 3 m. Its downstream face has a uniform batter. It stores water to a depth of 15 m with a free board of 2 m as shown in Fig. 6·15.*

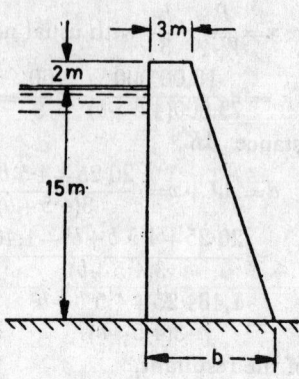

Fig. 6·15

The weights of water and concrete may be taken as 1,000 kg/m³ and 2,570 kg/m³. Calculate

   (a) *the minimum dam width at the bottom for no tension in concrete. Neglect uplift.*

   (b) *the extreme intensities of pressure on the foundation, when the reservoir is empty.* (*Andhra University, 1975*)

Solution

Given. Top width, $a = 3$ m
Height of water, $h = 15$ m
Height of dam, $H = 15 + 2 = 17$ m
Weight of water, $w = 1,000$ kg/m³
Weight of concrete, $\rho = 2,570$ kg/m³

Minimum dam width at the bottom

We see that the minimum dam width at the bottom is to be found out only for one condition *i.e.*, no tension should be developed at the base.

Let $b =$ Minimum dam width at the bottom,

$P =$ Total pressure on the dam per metre length.

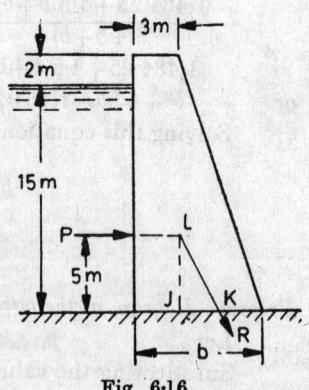

Fig. 6·16

$x=$ Horizontal distance between the *c.g.* of the dam section and K (*i.e.* JK)

Using the relation,

$$P=\frac{wh^2}{2} \text{ with usual notations.}$$

$$=\frac{1,000 \times 15^2}{2}=1,12,500 \text{ kg} \qquad \dots(i)$$

We know that the weight of dam per metre length,

$$W=\rho \times \frac{(a+b)}{2} \times H=2,570 \times \frac{(3+b)}{2} \times 17 \text{ kg}$$

$$=21,850 \ (3+b) \text{ kg} \qquad \dots(ii)$$

Now, let us find out the position of the *c.g.* of the dam section. We know that the distance AJ

$$=\frac{a^2+ab+b^2}{3(a+b)}=\frac{3^2+3b+b^2}{3(3+b)}=\frac{9+3b+b^2}{3(3+b)}$$

Now using the relation,

$$x=\frac{P}{W} \times \frac{h}{3} \text{ with usual notations.}$$

$$=\frac{1,12,500}{21,850(3+b)} \times \frac{15}{3}=\frac{25.9}{(3+b)}$$

$\therefore$ Horizontal distance AK,

$$d=AJ+x=\frac{9+3b+b^2}{3(3+b)}+\frac{25.9}{(3+b)}$$

$$=\frac{9+3b+b^2+77.7}{3(3+b)}=\frac{86.7+3b+b^2}{3(3+b)}$$

$\therefore$ Eccentricity of the resultant,

$$e=d-\frac{b}{2}=\frac{86.7+3b+b^2}{3(3+b)}-\frac{b}{2}$$

We know, that in order to avoid tension in the concrete, at the dam base, the eccentricity,

$$e=\frac{b}{6}$$

or

$$\frac{86.7+3b+b^2}{3(3+b)}-\frac{b}{2}=\frac{b}{6}$$

$\therefore$

$$\frac{86.7+3b+b^2}{3(3+b)}=\frac{b}{6}+\frac{b}{2}=\frac{2b}{3}$$

$$86.7+3b+b^2=2b(3+b)=6b+2b^2$$

or

$$b^2+3b-86.7=0$$

Solving this equation, as a quadratic equation for b,

$$b=\frac{-3 \pm \sqrt{3^2+4 \times 86.7}}{2}=\frac{-3+18.6}{2} \ m$$

$$=7 \cdot 8 \text{ m Ans.}$$

Extreme intensities of pressure on the foundation, when the reservoir is empty

Let p_{max} = Maximum intensity of pressure on the foundation,

p_{min} = Minimum intensity of pressure on the foundation.

We know that the weight of dam per metre length,

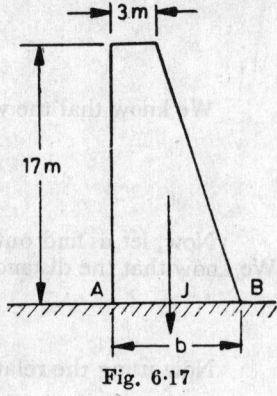

$$W = \rho \times \frac{a+b}{2} \times H$$

$$= 2{,}570 \times \frac{3+7{\cdot}8}{2} \times 17 \text{ kg}$$

$$= 2{,}35{,}900 \text{ kg}$$

We also know that distance AJ,

$$d = \frac{a^2 + ab + b^2}{3(a+b)}$$

$$= \frac{3^2 + 3 \times 7{\cdot}8 + 7{\cdot}8^2}{3(3+7{\cdot}8)} \text{ m}$$

$$= 2{\cdot}85 \text{ m}$$

Fig. 6·17

and eccentricity, $e = d - \dfrac{b}{2} = 2{\cdot}85 - \dfrac{7{\cdot}8}{2}$ m

$$= -1{\cdot}05 \text{ m}$$

(Minus sign indicates that the pressure at A will be more than that at B).

Using the relation,

$$p_{max} = \frac{W}{b} \left(1 + \frac{6e}{b} \right) \text{ with usual notations.}$$

$\therefore$ $p_A = \dfrac{2{,}35{,}900}{7{\cdot}8} \left(1 + \dfrac{6 \times 1{\cdot}05}{7{\cdot}8} \right) = 54{,}700 \text{ kg/m}^2$

$$= 54{\cdot}70 \text{ t/m}^2 \quad \textbf{Ans.}$$

Now using the relation,

$$p_{min} = \frac{W}{b} \left(1 - \frac{6e}{b} \right) \text{ with usual notations.}$$

$\therefore$ $p_B = \dfrac{2{,}35{,}900}{7{\cdot}8} \left(1 - \dfrac{6 \times 1{\cdot}05}{7{\cdot}8} \right) = 5{,}820 \text{ kg/m}^2$

$$\textbf{5·82 t/m}^2 \quad \textbf{Ans.}$$

6·11. Maximum height of a dam

We have already discussed in Art. 6·5, the various conditions for the minimum base width of a dam. The same conditions also hold good for the maximum height of a dam.

Example 6·10. *Assuming uniformly varying stress across the base, find the limit of height of a triangular masonry dam, with water upto the top of the vertical face, in order that the vertical compressive stress across the base shall not exceed 10 kg/cm². The masonry weighs 2,000 kg/m³.*

(*Bombay University, 1973*)

Solution

Given.   Maximum compressive stress at the base,
$$p_{max} = 10 \text{ kg/cm}^2 = 1,00,000 \text{ kg/m}^2$$

Weight of masonry, $\rho = 2,000 \text{ kg/m}^3$

Let e = Eccentricity of the resultant,
H = Height of the dam in metres,
b = bottom width of the dam in metres.

$\therefore$ Height of water = h m
and weight of dam per metre length,

$$W = \rho \times \frac{b}{2} \times H = 2,000 \times \frac{bH}{2} = 1,000 \ bH$$

We know that if the compressive stress is not to exceed 10 kg/cm^2 (*i.e.* 1,00,000 kg/m^2) the eccentricity,

$$e = \frac{b}{6}$$

Using the relation,

$$p_{max} = \frac{W}{b}\left(1 + \frac{6e}{b}\right) \text{ with usual notations.}$$

$$1,00,000 = \frac{1,000 \ bH}{b}\left(1 + \frac{6 \times \dfrac{b}{6}}{6}\right) = 1,000 \ H(1+1)$$

$$= 2,000 \ H$$

$$\therefore \qquad H = \frac{1,00,000}{2,000} = \textbf{50 m} \quad \textbf{Ans.}$$

Exercise 6–2

1. A masonry rectangular dam 5 m high, 2 m wide is retaining water to a height of 4 m. If the coefficient of friction between the wall and the soil is 0·6, check the stability of the dam. Take the weight of masonry as 2,000 kg/m^3.   [**Ans.** (*i*) Unsafe against tension in masonry ;
(*ii*) Safe against overturning ;
(*iii*) Safe against sliding]

2. A trapezoidal concrete dam 10 m high has top width 1 m and bottom width 6 m. The face exposed to water has a slope of 1 horizontal to 10 vertical.

Check the stability of the dam, when the water level coincides with the top of the dam, if the coefficient of friction between the bottom of the dam and the soil is 0·6. Take weight of the concrete and its allowable stress as 2,300 kg/m^3 and 40,000 kg/m^2.   (*Bombay University*)
[**Ans.** (*i*) Unsafe against tension in concrete;
(*ii*) Safe against overturning ;
(*iii*) Safe against sliding ;
(*iv*) Safe against crushing of masonry]

3. A trapezoidal dam 4 m high has top width of 1 m with vertical face exposed to water. Find minimum bottom width of the dam, if no tension is to develop at the base.   (*Patna University*)
[**Ans.** 2·55 m]

6·12. Retaining walls

We have already discussed in Art. 6·1 that a retaining wall is, generally, constructed to retain earth in hilly areas. The analysis of a retaining wall is, somewhat, like a dam. The retaining wall is subjected to pressure, produced by the retained earth in a similar manner, as the dam is subjected to water pressure.

6·13. Earth pressure on a retaining wall

It has been established since long, that the earth particles lack in cohesion, and hence have a definite *angle of repose. These earth particles always exert some lateral pressure on the walls, which retain or support them. The magnitude of this lateral pressure depends upon type of earth particles and the manner, in which they have been deposited on the back of the retaining wall. It has been experimentally found that the lateral pressure is minimum, when the earth particles have been loosely dumped, whereas the pressure is relatively high, when the same particles have been compacted by tamping or rolling. The earth pressures may be classified into the following two types :

(i) Active earth pressure, and (ii) Passive earth pressure.

6·14. Active earth pressure

The pressure, exerted by the retained material, called backfill, on the retaining wall is known as *active earth pressure*. As a result of the active pressure, the retaining wall tends to slide away from the retained earth. It has been observed, that the active pressure of the retained earth, acts on the retaining wall, in the same way as the pressure of the stored water on the dam.

6·15. Passive earth pressure

Sometimes, the retaining wall moves laterally against the retained earth, which gets compressed. As a result of the moments of the retaining wall, the compressed earth is subjected to a pressure (which is in the opposite direction of the active pressure) known as *passive earth pressure*.

It may be noted that the active pressure is the practical pressure, which acts on the retaining walls ; whereas the passive earth pressure is a theoretical pressure, which rarely comes into play.

6·16. Theories for active earth pressure

There are many theories and hypothesis for the active earth pressure, on the retaining walls. But none of these gives the exact value of the active pressure. The following two theories are considered to give a fairly reliable values

(a) Rankine's theory, and

(b) Coulomb's wedge theory.

---

*It may be defined as the maximum natural slope, at which the soil particles will rest due to their internal friction, if left unsupported for a sufficient length of time.

6·17.   *Rankine's theory for active earth pressure

It is one of the most acceptable theories, for the determination of active earth pressure on the retaining wall. This theory is based on the following assumptions :

1. The retained material is homogeneous and cohesionless.
2. The back of the wall is smooth, i.e., the frictional resistance between the retaining wall and the retained material is neglected.
3. The failure of the retained material takes place along a plane, called rupture plane.

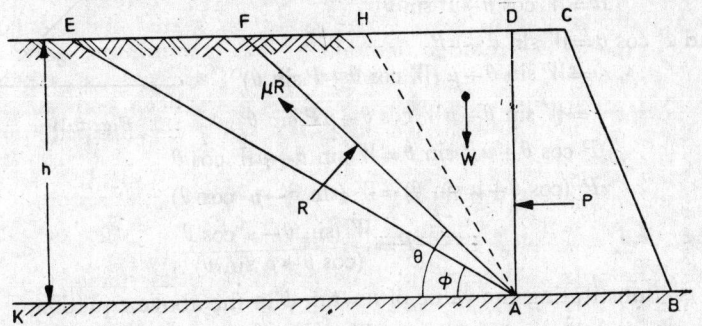

Fig. 6·18

Consider a trapezoidal retaining wall ABCD, retaining earth up to a height of h on its vertical face AD. Let the retained earth be levelled with the top of the wall DC. Draw AE at an angle φ with AK (where φ is the angle of repose of the retained earth). A little consideration will show, that if retaining wall is removed, the retained earth will be subjected to tension and will slide down along certain plane, whose inclination will be more than that of the angle of repose with AK. Let such a plane AF be inclined at an angle θ with AK as shown in Fig. 6·18. Now consider a horizontal force P offered by the retaining wall, which will keep the wedge AFD of the retained earth in equilibrium.

We see that, the wedge AFD of the retained earth is in equilibrium, under the action of the following forces**.

(a) Weight of the wedge AFD,

$$w = \frac{1}{2} \, w \times AD \times DE = \frac{wh^2}{2} \cot \theta$$

where w=specific weight of the material

(b) Horizontal thrust P offered by the retaining wall on the retained material.

---

*This theory was given by Prof. W.J. Rankine, a British Engineer in 1857.

**The frictional force, along the face AD of the retaining wall, is neglected.

(c) Normal reaction R acting at right angle to the plane AF.

(d) The frictional force, $F=\mu R$ acting on the opposite direction of the motion of the retained earth (where μ is the co-efficient of friction of the retained material).

The above condition is similar to the equilibrium of a body of weight W on a rough inclined plane, when it is subjected to a horizontal force P as shown in Fig. 6·19. From the geometry of the figure we find that

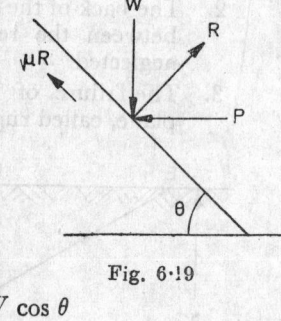

Fig. 6·19

$$R=W\cos\theta+P\sin\theta$$

and $P\cos\theta=W\sin\theta-\mu R$

$$=W\sin\theta-\mu(W\cos\theta+P\sin\theta)$$

$$=W\sin\theta-\mu W\cos\theta-\mu P\sin\theta$$

or $\quad P\cos\theta+\mu P\sin\theta=W\sin\theta-\mu W\cos\theta$

$$P(\cos\theta+\mu\sin\theta)=W(\sin\theta-\mu\cos\theta)$$

$$\therefore\quad P=\frac{W(\sin\theta-\mu\cos\theta)}{(\cos\theta+\mu\sin\theta)}$$

Substituting the value of $\mu=\tan\phi$ in the above equation,

$$P=\frac{W(\sin\theta-\tan\phi\cos\theta)}{(\cos\theta+\tan\phi\sin\theta)}$$

Multiplying the numerator and denominator by $\cos\phi$,

$$P=\frac{W(\sin\theta\cos\phi-\sin\phi\cos\theta)}{(\cos\theta\cos\phi+\sin\phi\sin\theta)}$$

$$=W\frac{\sin(\theta-\phi)}{\cos(\theta-\phi)}=W\tan(\theta-\phi)$$

$$=\frac{wh^2}{2}\cot\theta\cdot\tan(\theta-\phi)$$

$$\left(\because W=\frac{wh^2}{2}\cot\theta\right)$$

A little consideration will show, that if the retaining wall is removed, the retained earth will immediately slide down across a plane, where the tendency for the material to slide down is greatest. Let such a plane be AH. Therefore maximum value of the force P is required to retain the wedge AHD of the earth. In order to locate the plane AH (i.e. the plane of rupture) differentiate the equation for P and equal to zero i.e.

$$\frac{dp}{d\theta}\left[\frac{wh^2}{2}\left[\cot\theta\tan(\theta-\phi)\right]\right]=0$$

or $\quad\dfrac{wh^2}{2}\left[\cot\theta\sec^2(\theta-\phi)-\operatorname{cosec}^2\theta\tan(\theta-\phi)\right]=0$

$$\therefore\quad\cot\theta\sec^2(\theta-\phi)-\operatorname{cosec}^2\theta\tan(\theta-\phi)=0$$

Substituting $\tan \theta = t$ and $\tan (\theta - \phi) = t_1$ in the above equation

$$\frac{1}{t} \times (1+t_1^2) - \left(1 + \frac{1}{t^2} \right) \times t_1 = 0$$

$$\frac{1+t_1^2}{t} - t_1 \left(1 + \frac{1}{t^2} \right) = 0$$

or

$$t \, (1+t_1^2) - t_1 \, (1+t^2) = 0$$

$$t + tt_1^2 - t_1 - t_1 t^2 = 0$$

$$(t - t_1) - tt_1 \, (t - t_1) = 0$$

$$(t - t_1)(1 - tt_1) = 0$$

therefore either $t = t_1$ or $1 - tt_1 = 0$. Since $\tan \theta$ can not be equal to $\tan (\theta - \phi)$ therefore $1 - tt_1 = 0$

or

$$1 - \tan \theta \, . \, \tan (\theta - \phi) = 0$$

this statement is possible, only if $\theta + (\theta - \phi) = \dfrac{\pi}{2}$

or

$$\theta = \frac{\pi}{4} + \frac{\phi}{2}$$

Thus the plane of rupture is inclined at $\left(\dfrac{\pi}{4} + \dfrac{\phi}{2} \right)$ with the horizontal.

We also see that

$$\angle HAE = \angle HAK - \angle KAE = \left(\frac{\pi}{4} + \frac{\phi}{2} \right) - \phi$$

$$= \left(\frac{\pi}{4} - \frac{\phi}{2} \right) = \frac{1}{2} \left(\frac{\pi}{2} - \phi \right)$$

$$= \tfrac{1}{2} \angle DAE$$

Now substituting the values in the equation for P,

$$P = \frac{wh^2}{2} \cot \theta \tan (\theta - \phi)$$

$$= \frac{wh^2}{2} \cot \left(\frac{\pi}{4} + \frac{\phi}{2} \right) \tan\left(\frac{\pi}{4} - \frac{\phi}{2} \right)$$

$$= \frac{wh^2}{2} \times \frac{\tan \left(\dfrac{\pi}{4} - \dfrac{\phi}{2} \right)}{\tan \left(\dfrac{\pi}{4} + \dfrac{\phi}{2} \right)}$$

$$= \frac{wh^2}{2} \times \frac{1 - \sin \phi}{1 + \sin \phi}.$$

Note 1. Similarly, it can be proved that if the retained material is surcharged (i.e., the angle of surcharge is α with the horizontal), the total pressure on the retaining wall per unit length,

$$P = \frac{wh^2}{2} \cos \alpha \, . \, \frac{\cos \alpha - \sqrt{\cos^2 \alpha - \cos^2 \phi}}{\cos \alpha + \sqrt{\cos^2 \alpha - \cos^2 \phi}}$$

This pressure may now be resolved into horizontal and vertical components.

The horizontal component $P_H = P \cos \alpha$ will act at a height $h/3$ from the base and vertical component $P_V = P \sin \alpha$. It will act along DA.

2, If the retained material is subjected to some superimposed or surcharged load (*i.e.* the pressure due to traffic etc.) it will cause a constant pressure on the retaining wall from top to bottom. The total horizontal pressure due to surcharged load,

$$P = p \times \frac{1 - \sin \phi}{1 + \sin \phi}$$

where p is the intensity of the surcharged load.

Example 6·11. *Find the resultant lateral pressure and the distance of the point of application from the bottom in the case of retaining wall as shown in Fig. 6·20.*

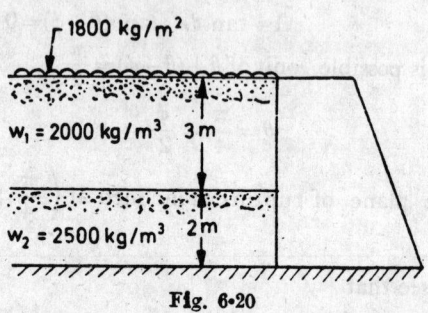

Fig. 6·20

Take weight of upper soil as 2,000 kg/m³ and $\phi = 30°$ and weight of lower soil as 2,400 kg/m³ and $\phi = 30°$.

(*Oxford University,*)

Solution

Given. Surcharge $= 1,800$ kg/m²
Weight of upper soil, $w_1 = 2,000$ kg/m³
Depth of upper soil, $h_1 = 3$ m
Weight of lower soil, $w_2 = 2,400$ kg/m³
Depth of lower soil, $h_2 = 2$ m

Resultant lateral pressure per metre length of the dam

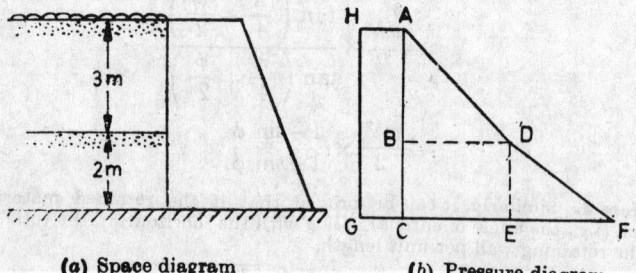

(*a*) Space diagram   (*b*) Pressure diagram

Fig. 6·21

The pressure diagram on the dam is shown in Fig. 6·21(*b*)
Let $P_1 =$ Total pressure due to upper soil, and

P_2=Total pressure due to lower soil,

P_3=Total pressure due to surcharge.

We know that the pressure BD

$$=w_1h_1 \times \frac{1-\sin \phi}{1+\sin \phi}=2,000 \times 3 \times \frac{1-\sin 30°}{1+\sin 30°} \text{ kg/m}^2$$

$$=2,000 \text{ kg/m}^2$$

Similarly, pressure EF

$$=w_2h_2 \times \frac{1-\sin \phi}{1+\sin \phi}=2,400 \times 2 \times \frac{1-\sin 30°}{1+\sin 30°} \text{ kg/m}^2$$

$$=1,600 \text{ kg/m}^2$$

and pressure GC $=1,800 \times \dfrac{1-\sin 30°}{1+\sin 30°}=600 \text{ kg/m}^2$

$\therefore$ Total pressure due to upper soil,

P_1=Area of triangle $ABD \times$ length of the wall

$=(\frac{1}{2} \times 2,000 \times 3)=3,000 \text{ kg}$

Similarly, total pressure due to lower soil,

P_2=Area of figure $BDFC \times$ length of wall

$=$(Area of rectangle $BCDE$+Area of triangle DEF) $\times$ length of wall

$=[(2,000 \times 2)+\frac{1}{2} \times 1,600 \times 2] \times 1=5,600 \text{ kg}$ ...(ii)

and pressure due to surcharge,

P_3=Area of rectangle, $ACGH \times$ length of wall

$=600 \times 5 \times 1=3,000 \text{ kg}$ ...(iii)

$\therefore$ Total pressure,

$P=P_1+P_2+P_3=3,000+5,600+3,000 \text{ kg}$

$=\textbf{11,600 kg Ans.}$

Point of application of the resultant pressure

Let y=Height of the point of application of the resultant pressure from the bottom.

Taking moments of all pressures about G and equating the same, *i.e.*

$$P \times y=\left[\text{Pressure } ABD \times \left(2+\frac{3}{3}\right)\right]+\left[\text{Pressure } BCDE \times \frac{2}{2}\right]$$

$$+\left[\text{Pressure } DEF \times \frac{2}{3}\right]+\left[\text{Pressure } ACGH \times \frac{5}{2}\right]$$

$$11,600 \times y=(3,000 \times 3)+(4,000 \times 1)+\left(1,600 \times \frac{2}{3}\right)+\left(3,000 \times \frac{5}{2}\right)$$

$$=21,567$$

$\therefore$ $y=\dfrac{21,567}{11,600}=\textbf{1·86 m Ans.}$

Example 6·12. *A masonry retaining wall of trapezoidal section with a vertical face on the earth side is 1 m wide at the top, 3 m wide at the bottom and 6 m high. It retains sand over the entire height with an angle of surcharge of 20°. Determine the distribution of pressure at the base of the wall. The sand weighs 1,800 kg/m³ and has an angle of repose of 30°. The masonry weighs 2,400 kg/m³.* (London University)

Solution.

Given. Top width,

   $a = 1$ m

Bottom width,

   $b = 3$ m

Height of wall,

   $h = 6$ m

Angle of surcharge,

   $\alpha = 20°$

Specific weight of sand,

   $w = 1,800$ kg/m³

Angle of repose,

   $\phi = 30°$

Specific weight of masonry

   $l = 2,400$ kg/m³

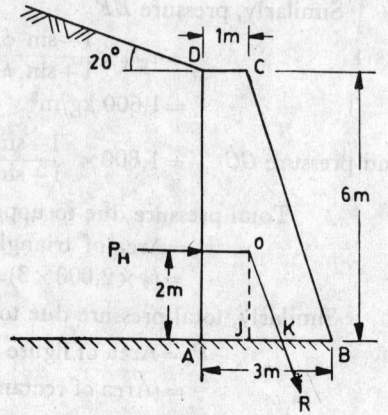

Fig. 6·22

Let P = Total pressure on the wall per metre length,

   W = Weight of the wall per metre length,

   x = Horizontal distance between the *c.g.* of the vertical load of the wall and K (*i.e.*, JK).

   p_{max} = Maximum intensity of pressure at the base, and

   p_{min} = Minimum intensity of pressure at the base.

Using the relation,

$$P = \frac{wh^{'}}{2} \cos \alpha \frac{\cos \alpha - \sqrt{\cos^2 \alpha - \cos^2 \phi}}{\cos \alpha + \sqrt{\cos^2 \alpha - \cos^2 \phi}}$$

with usual notations.

$$= \frac{1,800 \times 6^2}{2} \times \cos 20° \times \frac{\cos 20° - \sqrt{\cos^2 20° - \cos^2 30°}}{\cos 20° + \sqrt{\cos^2 20° - \cos^2 30°}}$$

$$= 32,400 \times 0·9397 \times \frac{0·9397 - \sqrt{0·9397^2 - 0·866^2}}{0·9397 + \sqrt{0·9397^2 - 0·866^2}} \text{ kg}$$

$$= 13,380 \text{ kg}$$

We know that the horizontal component of the pressure,

   $P_H = 13,380 \cos 20° = 13,380 \times 0·9397$ kg

   $= 12,570$ kg

and vertical component of the pressure,

   $P_V = 13,380 \sin 20° = 13,380 \times 0·3420$ kg

   $= 4,576$ kg

Weight of the dam

$$= \rho \times \frac{a+b}{2} \times h = 2,400 \times \frac{1+3}{2} \times 6 \text{ kg}$$

$$= 28,800 \text{ kg}$$

$\therefore$ Total load acting vertically down,

$$W = 28,800 + 4,576 = 33,376 \text{ kg}$$

Now let us find out the position of the *c.g.* of the vertical load. Taking moments of the vertical loads about A and equating the same,

$$W \times AJ = P_V \times 0 + (2,400 \times 1 \times 6 \times 0\cdot5) + \left(2,400 \times \frac{6\times2}{2} \times 2\right)$$

$$33,376 \times AJ = 36,000$$

or

$$AJ = \frac{36,000}{33,376} = 1\cdot08 \text{ m}$$

Now using the relation,

$$x = \frac{P_H}{W} \times \frac{h}{3} \text{ with usual notations.}$$

$$= \frac{12,570}{33,376} \times \frac{6}{3} = 0\cdot75 \text{ m}$$

$\therefore$ Horizontal distance between A and the point K, where the resultant cuts the base,

$$*d = AJ + JK = 1\cdot08 + 0\cdot75 = 1\cdot83 \text{ m}$$

We know that the eccentricity of resultant,

$$e = d - \frac{b}{2} = 1\cdot83 - \frac{3}{2} = 0\cdot33 \text{ m}$$

Using the relation,

$$p_{max} = \frac{W}{b}\left(1 + \frac{6e}{b}\right) \text{ with usual notations.}$$

$$= \frac{33,376}{3}\left(1 + \frac{6\times0\cdot33}{3}\right) = \textbf{18,468 kg/m}^2 \textbf{ Ans.}$$

Now using the relation,

$$p_{min} = \frac{W}{b}\left(1 - \frac{6e}{b}\right) \text{ with usual notations.}$$

$$= \frac{33,376}{3}\left(1 - \frac{6\times0\cdot33}{3}\right) = \textbf{3,783 kg/m}^2 \textbf{ Ans.}$$

---

*The horizontal distance d may also be found out by taking moment about A and equating the same,

$$W.d = P_H \times \frac{h}{3} + (2,400 \times 1 \times 6 \times 0\cdot5) + \left(2,400 \times \frac{6\times2}{2} \times 2\right)$$

$$33,376 \times d = 12,570 \times \frac{6}{3} + 7,200 + 28,800 = 61,140$$

$$\therefore \quad d = \frac{61,140}{33,376} = 1\cdot83 \text{ m} \quad \textbf{Ans.}$$

6·18. *Coulomb's wedge theory for active earth pressure

In Rankine's theory for active earth pressure, we considered the equilibrium of an element within the mass of the retained material. But in this theory, the equilibrium of the whole material supported by the retaining wall is considered, when the wall is at the point of slipping away from the retained material. This theory is based on the concept of sliding wedge, which is torn off from the backfill on the movement of the wall. This theory is based on the following assumptions :

(1) The retained material is homogeneous and cohesionless.
(2) The sliding wedge itself acts as a rigid body and the earth pressure is obtained by considering the limiting equilibrium of the sliding wedge as a whole.
(3) The position and direction of the earth pressure are known *i.e.*, the pressure acts on the back of the wall and at a height of one-third of the wall height from the base. The pressure is inclined at an angle δ (called the angle of wall friction) to the normal to the back.

Consider a trapezoidal retaining wall *ABCD* retaining surcharged earth up to a height of *h* on the inclined face *AD* as shown in Fig. 6·23.

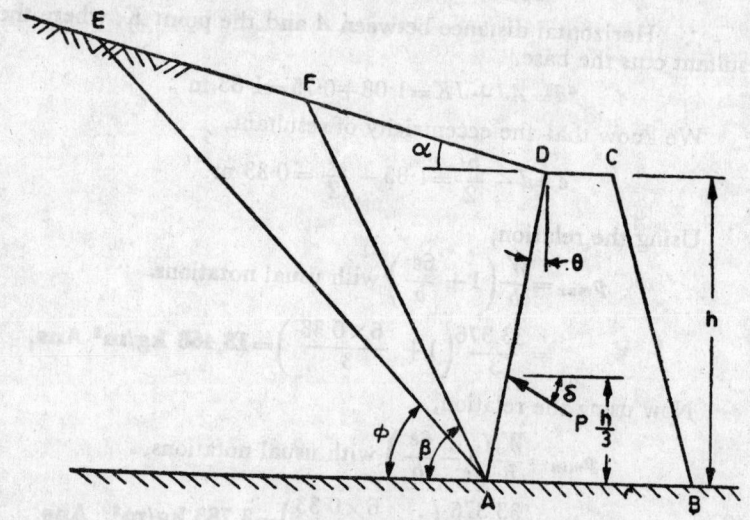

Fig. 6·23

Let   *h* = Height of the wall,
                                      w = Specific weight of the retained earth,
                                      φ = Angle of repose of the retained earth,
                                      α = Angle of surcharge,
                                      θ = Angle, which the inclined face *AB* makes with the vertical, and

---

*This theory was given by Prof. C.A. Coulomb, a French scientist, in 1876.

δ=Angle of friction between the retaining
wall and the retained earth.

In this case the earth pressure* is given by the relation,

$$P = \frac{wh^2}{2} \times \frac{\cos^2(\phi-\theta)}{\cos^2\theta \cos(\delta+\theta)\left[1+\sqrt{\dfrac{\sin(\delta+\phi)\sin(\phi-\alpha)}{\cos(\delta+\theta)\cos(\theta-\alpha)}}\right]^2}$$

6·19. Graphical method for active earth pressure

In the previous articles, we have discussed the analytical method
for the magnitude of the active earth pressure on a retaining wall.
But in this article, we shall discuss the graphical method for the
determination of active earth pressure. The following cases are
important from the subject point of view :

    (1) Graphical method for Rankine's theory and

    (2) Rehbann's graphical method for Coloumb's theory.

6·20. Graphical method for Rankine's theory

The Rankine's theory of active earth pressure may also be
proved graphically as discussed below. Here we shall discuss the
following two important cases :

    (a) *Retaining wall with a vertical soil face, and retaining earth
level with the top of the wall.*

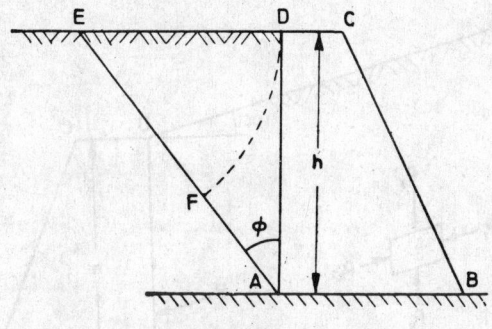

Fig. 6·24

Consider a trapezoidal retaining wall *ABCD*, retaining earth on
its vertical face *AD*. Let the retained earth be levelled with the
top of the wall *DC* as shown in Fig. 6·24.

Let   h=Height of the retained earth,
               w=Specific weight of the retained earth,
               φ=Angle of repose of the retained earth.

First of all, draw *AE* at an angle φ with *AD*. Now with *E* as
centre and radius equal to *ED* draw an arc cutting *AE* at *F* as shown
in Fig. 6·24. Now the active earth pressure (*i.e.* the horizontal thrust

For graphical proof, please refer to page 202.

of the retained earth) on the face AD of the wall will be given by the relation,

$$P = \frac{w}{2} \times AF^2$$

Proof

From the geometry of the figure, we find that

$$EF = ED = h \tan \phi$$

and
$$AF = AE - EF = h \sec \phi - h \tan \phi$$
$$= h(\sec \phi - \tan \phi)$$

$$\therefore \quad AF^2 = h^2(\sec \phi - \tan \phi)^2$$

$$= h^2_1 \times \frac{1 - \sin \phi}{1 + \sin \phi}$$

But
$$P = \frac{wh^2}{2} \times \frac{1 - \sin \phi}{1 + \sin \phi}$$

$$= \frac{w}{2} \times AF^2$$

(b) *Retaining wall with a vertical soil face, and retaining a surcharged earth.*

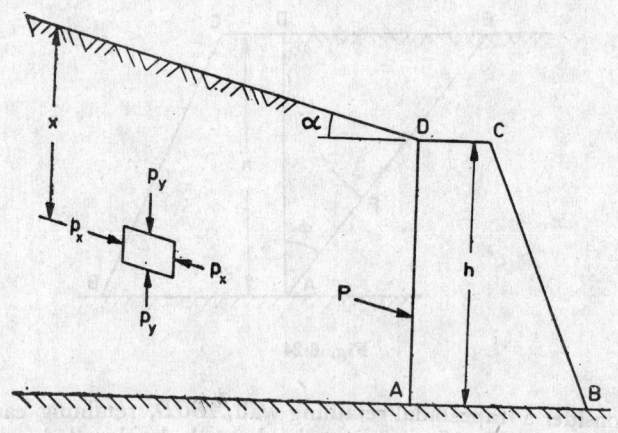

Fig. 6·25

Consider a trapezoidal retaining wall $ABCD$, retaining earth on its vertical face AD. Let the retained earth be surcharged as shown in Fig. 6·25.

Let
$h =$ Height of the retained earth,
$w =$ Specific weight of the retained earth,
$\phi =$ Angle of repose of the retained earth,
$\alpha =$ Angle of surcharge of the retained earth.

Now consider the equilibrium of a small parallelopiped at a depth x. Let p_X and p_Y be the stresses acting on this element along the vertical plane and the plane parallel to the sloping surface of the earth as shown in Fig. 6·25. Now let p_1 and p_2 be the principal stresses on the element, whose values may be easily found out by Mohr's circle as shown in Fig. 6·26.

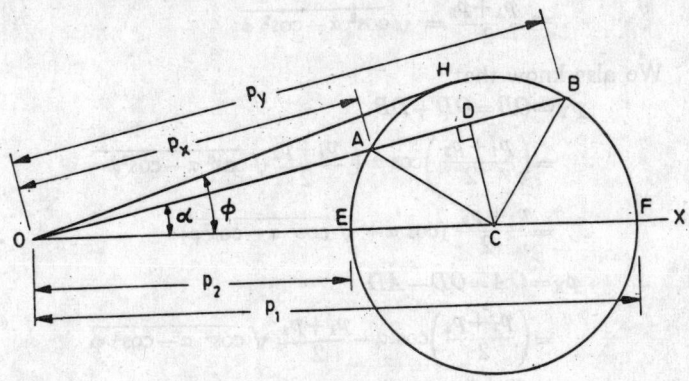

Fig. 6·26

First of all, draw a horizontal line OX. Now draw another line OB at an angle α with OX and cut off OA and OB equal to p_X and p_Y to some scale. Bisect AB at D and draw DC perpendicular to AB meeting OX at C. Now with C as centre and radius equal to CA or CB, draw a circle meeting OX at E and F. Now through O draw OH tangent to the circle. A little consideration will show that the tangent will make an angle ϕ with OX as shown in Fig. 6·26. Now OF and OE will represent the principal stresses to the scale on the element. From the geometry of the figure, we find that

$$OC = \frac{p_1 + p_2}{2}$$

$$AC = BC = \frac{p_1 - p_2}{2}$$

$$OD = OC \cos \alpha = \left(\frac{p_1 + p_2}{2}\right) \cos \alpha$$

$$CD = OC \sin \alpha = \left(\frac{p_1 + p_2}{2}\right) \sin \alpha$$

$$CH = AC = CB = \frac{p_1 - p_2}{2}$$

$$= OC \sin \phi = \frac{p_1 + p_2}{2} \sin \phi$$

$$\therefore \quad p_1 - p_2 = (p_1 + p_2) \sin \phi$$

We know that $\quad AD = DB = \sqrt{AC^2 - CD^2}$

$$= \sqrt{\left(\frac{p_1 - p_2}{2}\right)^2 - \left(\frac{p_1 + p_2}{2}\right)^2 \sin^2 \alpha}$$

$$= \sqrt{\left(\frac{p_1+p_2}{2}\right)^2 \sin^2 \phi - \left(\frac{p_1+p_2}{2}\right)^2 \sin^2 \alpha}$$

$$[\because\ p_1 - p_2 = (p_1+p_2) \sin \phi]$$

$$= \frac{p_1+p_2}{2} \sqrt{\sin^2 \phi - \sin^2 \alpha}$$

$$= \frac{p_1+p_2}{2} = \sqrt{\cos^2 \alpha - \cos^2 \phi}$$

We also know that

$$p_Y = OB = OD + DB$$

$$= \left(\frac{p_1+p_2}{2}\right) \cos \alpha + \frac{p_1+p_2}{2} \sqrt{\cos^2 \alpha - \cos^2 \phi}$$

$$= \frac{p_1+p_2}{2} (\cos \alpha + \sqrt{\cos^2 \alpha - \cos^2 \phi}) \qquad \ldots(i)$$

and

$$p_X = OA = OD - AD$$

$$= \left(\frac{p_1+p_2}{2}\right) \cos \alpha - \frac{p_1+p_2}{2} \sqrt{\cos^2 \alpha - \cos^2 \phi}$$

$$= \frac{p_1+p_2}{2} (\cos u - \sqrt{\cos^2 \alpha - \cos^2 \phi}) \qquad \ldots(ii)$$

Dividing (ii) by (i)

$$\frac{p_Y}{p_X} = \frac{\cos \alpha + \sqrt{\cos^2 \alpha - \cos^2 \phi}}{\cos \alpha - \sqrt{\cos^2 \alpha - \cos^2 \phi}} \qquad \ldots(iii)$$

We also know that the vertical pressure on the element,

$$p_Y = wh \cos \alpha$$

$$\therefore \quad p_X = wh \cos \alpha \times \frac{\cos \alpha - \sqrt{\cos^2 \alpha - \cos^2 \phi}}{\cos \alpha + \sqrt{\cos^2 \alpha - \cos^2 \phi}}$$

The total pressure per unit length of the wall,

$$P = \text{Average intensity} \times \text{Area}$$

$$= \frac{1}{2} wh \cos \alpha \times \frac{\cos \alpha - \sqrt{\cos^2 \alpha - \cos^2 \phi}}{\cos \alpha + \sqrt{\cos^2 \alpha - \cos^2 \phi}} \times h$$

$$= \frac{wh^2}{2} \cos \alpha \times \frac{\cos \alpha - \sqrt{\cos^2 \alpha - \cos^2 \phi}}{\cos \alpha + \sqrt{\cos^2 \alpha - \cos^2 \phi}}$$

Note : The total pressure P may be resolved horizontally as well as vertically. The horizontal component $P_H = P \cos \alpha$ and vertical component $P_V = P \sin \alpha$. The horizontal component will act at a height of one-third of the wall height and the vertical component will act along DA.

6·21 Rehbann's graphical method for Coulomb's theory

The Rehbann's graphical method is very suitable for the determination of active earth pressure, especially when the retaining wall has inclined soil face. Here we shall discuss the general case. Consi-

der a retaining wall $ABCD$, retaining earth on its inclined face AD
as shown in Fig. 6·27.

Let
$h=$ Height of the wall.

$w=$ Specific weight of the retained earth.

$\phi=$ Angle of repose of the retained earth.

$\alpha=$ Angle of surcharge.

$\delta=$ Angle of friction between the retaining wall and the retained earth.

$\theta=$ Angle, which the inclined face AD makes with the vertical.

First of all, through A draw AE at an angle ϕ, and through D
draw DE at an angle α with the horizontal meeting at E. Now draw
a semicircle on AE. Through D draw a line DF at an angle $(\delta+\phi)$

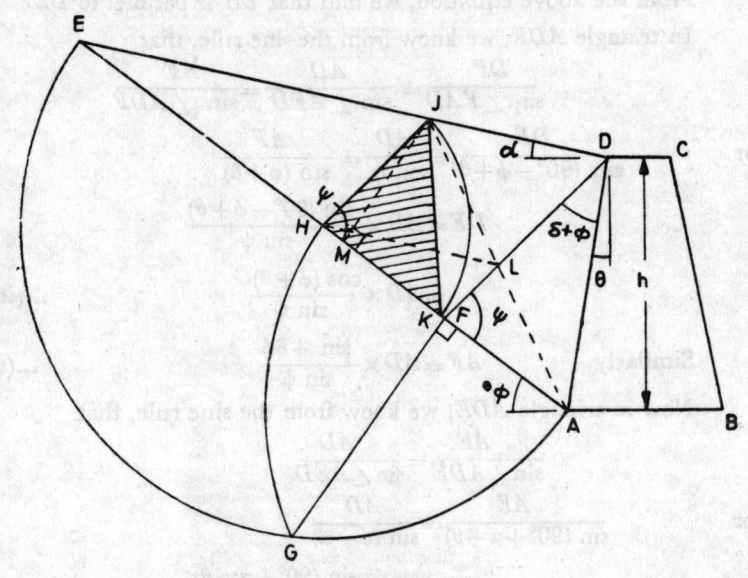

Fig. 6·27

with DA meeting AE at F. Now at F erect a perpendicular FG meeting the semicircle at G. Now with A as centre, and radius equal to
AG draw an arc meeting AE at H. Through H, draw HJ parallel
to FD meeting DE at J. Now with H as centre, and radius equal
to HJ, draw an arc meeting AE at K. Join JK. Now the active
earth pressure on the retaining wall will be given by the relation,

$$P=w\times \text{ Area of triangle } HJK.$$

Proof

Join AJ meeting DF at L. Join HL.

Now from J draw JM perpendicular to AE. Join AG.

We know from the properties of a circle that

$$AF \times FE = FG^2$$

$$AF \times FE + AF^2 = FG^2 + AF^2 \qquad \text{(Adding } AF^2 \text{ on both sides)}$$

$$AF(FE + AF) = AG^2 \qquad \text{(In } \triangle AFG, \ FG^2 + AF^2 = AG^2\text{)}$$

$$AF \times AE = AH^2 \qquad (\because \ AG = AH) \qquad ...(i)$$

or

$$\frac{AF}{AH} = \frac{AH}{AE} \qquad ...(ii)$$

Since DF is parallel to JH, therefore

$$\frac{AF}{AH} = \frac{AL}{AJ} \qquad ...(iii)$$

From equations (ii) and (iii) we find that

$$\frac{AH}{AE} = \frac{AL}{AJ}$$

From the above equation, we find that LH is parallel to DE.

In triangle ADF, we know from the sine rule, that

$$\frac{DF}{\sin \angle FAD} = \frac{AD}{\sin \angle AFD} = \frac{AF}{\sin \angle ADF}$$

or

$$\frac{DF}{\sin (90° - \phi + \theta)} = \frac{AD}{\sin \psi} = \frac{AF}{\sin (\phi + \delta)}$$

$$\therefore \qquad DF = AD \times \frac{\sin (90° - \phi + \theta)}{\sin \psi}$$

$$= AD \times \frac{\cos (\phi + \theta)}{\sin \psi} \qquad ...(iv)$$

Similarly

$$AF = AD \times \frac{\sin + \delta\phi}{\sin \psi} \qquad ...(v)$$

Now in triangle ADE, we know from the sine rule, that

$$\frac{AE}{\sin \angle ADE} = \frac{AD}{\sin \angle AED}$$

or

$$\frac{AE}{\sin (90° + \alpha - \theta)} = \frac{AD}{\sin (\phi - \alpha)}$$

$$\therefore \qquad AE = AD \times \frac{\sin (90° + \alpha - \theta)}{\sin (\phi - \alpha)}$$

$$= AD \times \frac{\cos (\alpha - \theta)}{\sin (\phi - \alpha)} \qquad ...(vi)$$

Since LH is parallel to DE, therefore the two triangles ALH and AJE are similar. Therefore

$$\frac{LH}{JE} = \frac{AH}{AE} \qquad ...(vii)$$

Now in triangle ADE, we know that

$$\frac{DE}{JE} = \frac{DJ + JE}{JE} = \frac{DJ}{JE} + 1$$

$$\frac{DE}{JE} = \frac{LH}{JE} + 1 \qquad (\because DJ = LH)$$

$$= \frac{AH}{AE} + 1 \left(\text{From eqn. } (vii) \frac{LH}{JE} = \frac{AH}{AE} \right) \quad (viii)$$

We have already seen in equation (i) that
$$AF \times AE = AH^2$$

$\therefore \qquad \frac{AH}{AE} = \sqrt{\frac{AF}{AE}}$

Substituting the above value of $\frac{AH}{AE}$ in equation $(viii)$ we get

$$\frac{DE}{JE} = \sqrt{\frac{AF}{AE}} + 1 \qquad \qquad ...(ix)$$

Now from the geometry of the figure, we find that the two triangles DFE and JHE are similar. Therefore

$$\frac{JH}{JE} = \frac{DF}{DE}$$

$\therefore \qquad JH = \frac{DF}{DE} \times JE = \frac{JE}{DE} \times DF$

or $\qquad JH^2 = \left(\frac{JE}{DE} \times DF \right)^2 \qquad \qquad ...(x)$

Now we know that the area of triangle HJK
$$= \tfrac{1}{2} \times HK \times JM = \tfrac{1}{2} \times HK \times JH \sin \psi$$
$$= \tfrac{1}{2} JH^2 \sin \psi \qquad (\because HK = JH)$$
$$= \frac{1}{2} \left(\frac{JE}{DE} \times DF \right)^2 \sin \psi$$

[Substituting the value of JH^2 from eqn. (x)]

Now substituting the values of DF from equation (iv) and $\frac{JE}{DE}$ from equation (ix) along with the values of AF and AE from equations (v) and (vi) and simplifying we get the area of triangle HJK.

$$= \frac{h^2}{2} \times \frac{\cos^2 (\phi - \theta)}{\cos^2 \theta \cos (\delta + \theta) \left[1 + \sqrt{\dfrac{\sin (\delta + \phi) \sin (\phi - \alpha)}{\cos (\delta + \theta) \cos (\theta - \alpha)}} \right]^2}$$

Therefore active earth pressure on the retaining wall,

$$P = \frac{wh^2}{2} \times \frac{\cos^2 (\phi - \theta)}{\cos^2 \theta \cos (\delta + \theta) \left[1 + \sqrt{\dfrac{\sin (\delta + \phi) \sin (\theta - \alpha)}{\cos (\delta + \theta) \cos (\theta - \alpha)}} \right]^2}$$

Example 6·13. *Determine, by Rehbaun's construction, the active pressure on a retaining wall shown in Fig. 6·28.*

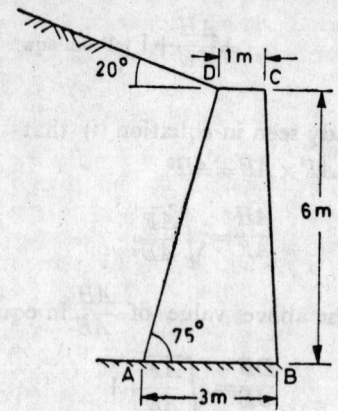

Fig. 6·28

Take angle of repose equal to 35°, unit weight of soil as 1·6 tonnes/m³ and angle of friction between the wall and soil as 20°.

(London University)

Solution.

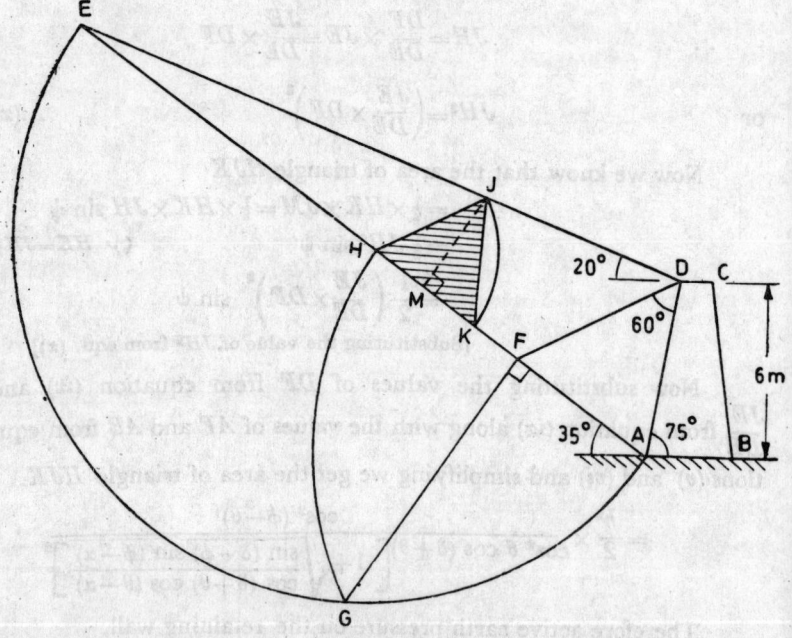

Fig. 6·29

Given.   Top width, $a = 1·0$ m
            Bottom width, $b = 3$ m
            Angle of surcharge, $\alpha = 20°$

Angle which the inclined face AD makes with the vertical,
$$\theta = 90° - 75° = 15°$$
Angle of repose, $\phi = 35°$

Unit weight of soil, $w = 1·6$ tonnes/m³

Angle of friction between the wall and soil,
$$\delta = 25°.$$

Let $P =$ Total active earth pressure per metre length of the wall.

First of all, draw the space diagram of the wall to the scale. Through A, draw AE at an angle ϕ (35°) and through D draw DE at an angle α (20°) with the horizontal meeting at E. Now draw a semicircle on AE. Through D draw a line DF at angle $\delta + \phi$ (25° + 35° = 60°) with DA meeting AE at F. Now at F erect a perpendicular FG meeting the semicircle at G. Now with A as centre, and radius equal to AG draw an arc meeting AE at H. Through H, draw HJ parallel to FD meeting DE at J. Now with H as centre, and radius equal to HJ, draw an arc meeting AE at K. Join JK. Now from J draw JM perpendicular to AE.

From the geometry of the figure, we find that $HK = 4·95$ and $JM = 3·79$

$$\therefore \quad P = \tfrac{1}{2} \times 1·6 \times HK \times JM$$
$$= \tfrac{1}{2} \times 1·6 \times 4·95 \times 3·79 \text{ t}$$
$$= 15 \text{ t} \quad \textbf{Ans.}$$

6·22. Conditions for the stability of a retaining wall

The conditions, for the stability of a retaining wall, are the same as those for the stability of a dam. In general, a retaining wall is checked for the following conditions of stability :—

1. To avoid tension in the masonry at the base of the wall.
2. To safeguard the wall from overturning.
3. To prevent the sliding of wall.
4. To prevent the crushing of masonry at the base of the wall.

Example 6·14. *Examine graphically or otherwise the stability of the retaining wall shown in Fig. 6·30.*

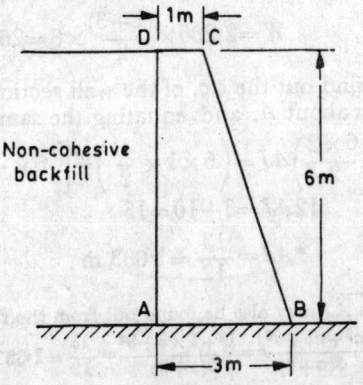

Fig. 6·30

Find the extreme stresses at the base of the wall, taking the densities of soil retained and masonry of the wall as 1·60 gm/cc and 2·20 gm/cc respectively. Assume angle of internal friction of the soil as 30°.

Solution.

Given. Top width,

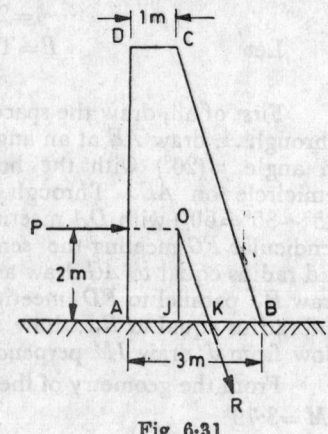

Fig. 6·31

$$a = 1 \text{ m}$$

Bottom width, $b = 3$ m

Height of wall, $= 6$ m

Density of soil, $w = 1·6$ gm/cc

$$= 1,600 \text{ kg/m}^3$$

Density of masonry,

$$\rho = 2·2 \text{ gm/cc}$$

$$= 2,200 \text{ kg/m}^3$$

Angle of internal friction,

$$\phi = 30°$$

Check for tension in the masonry

Let $P =$ Earth pressure per metre length of the wall, and

$x =$ Horizontal distance between the c.g. of the dam section and K (i.e. JK).

We know that the active earth pressure on the wall per metre length,

$$P = \frac{wh^2}{2} \times \frac{1 - \sin \phi}{1 + \sin \phi}$$

$$= \frac{1,600 \times 6^2}{2} \times \frac{1 - \sin 30°}{1 + \sin 30°} \text{ kg}$$

$$= 28,800 \times \tfrac{1}{3} = 9,600 \text{ kg}$$

We also know that the weight of the wall per metre length

$$W = 2,200 \times \frac{(1+3)}{2} \times 6 = 26,400 \text{ kg}$$

Now let us find out the c.g. of the wall section. Taking moments of the wall section about A, and equating the same, i.e.

$$\left(6 \times 1 + \frac{6 \times 2}{2} \right) AJ = \left(6 \times 1 \times \frac{1}{2} \right) + \left[6 \times \frac{2}{2} \left(1 + \frac{2}{3} \right) \right]$$

$$12AJ = 3 + 10 = 13$$

$$\therefore \qquad *AJ = \frac{13}{12} = 1·083 \text{ m}$$

———————————

*The distance AJ may also be found out from the following relation

$$AJ = \frac{a^2 + ab + b^2}{3(a+b)} = \frac{1^2 + 1 \times 3 + 3^2}{3(1+3)} = \frac{13}{12} = 1·083 \text{ m}$$

Using the relation,

$$x = \frac{P}{W} \times \frac{h}{3} \quad \text{with usual notations.}$$

$$= \frac{9,600}{26,400} \times \frac{6}{3} = 0.727 \text{ m}$$

∴ Horizontal distance AK,

$$d = AJ + x = 1.083 + 0.727 = 1.81 \text{ m}$$

Since the resultant force lies within the middle third of the base width *i.e.* from 1.0 m to 2.0 m, therefore the wall is safe against the tension in its masonry at the base. **Ans.**

Check for overturning

Since the resultant force lies within the base AB, therefore the wall is safe against overturning **Ans.**

Check for sliding of the dam

Let coefficient of friction,

$$\mu = 0.6$$

We know that the frictional force at the base

$$= \mu W = 0.6 \times 26,400 = 15,840 \text{ kg}$$

Since the frictional force is more than the horizontal pressure, therefore the wall is safe against sliding **Ans.**

Extreme stresses at the base of the wall

Let p_{max} = Maximum stress at the base of the wall,

p_{min} = Minimum stress at the base of the wall.

We know that the eccentricity of the resultant,

$$e = d - \frac{b}{2} = 1.81 - 1.5 = 0.31 \text{ m}$$

Using the relation,

$$p_{max} = \frac{W}{b}\left(1 + \frac{6e}{b}\right) \quad \text{with usual notations.}$$

$$= \frac{26,400}{3}\left(1 + \frac{6 \times 0.31}{3}\right) \text{ kg/m}^2$$

$$= 14,256 \text{ kg/m}^2 \quad \text{Ans.}$$

Now using the relation,

$$p_{min} = \frac{W}{b}\left(1 - \frac{6e}{b}\right) \quad \text{with usual notations.}$$

$$= \frac{26,400}{3}\left(1 - \frac{6 \times 0.31}{3}\right) \text{ kg/cm}^2$$

$$= 3,344 \text{ kg/m}^2 \quad \text{Ans.}$$

Example 6·15. *A masonry retaining wall 4 m high above ground level as shown in Fig. 6·32 sustains earth with a positive surcharge of 10°. The width of the wall at top is 0·75 m and at the base 2·5 m. The earth face of the wall makes an angle of 20° with the vertical.*

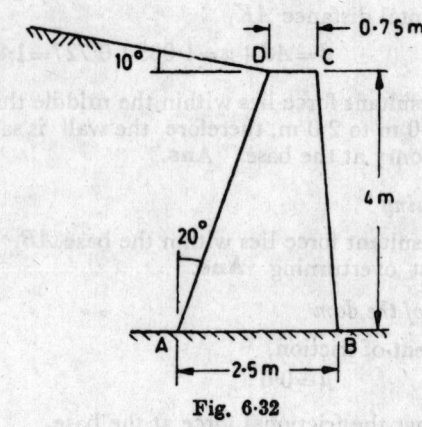

Fig. 6·32

Determine the thrust on the wall by graphical construction or otherwise, and examine the safety of the wall for no tension, overturning and sliding. Given the weight of earth =1,600 kg/m³, masonry =2,000 kg/m³. Maximum pressure allowable on soil 10,000 kg/m² ; angle of repose of the soil =30° ; angle of friction between the soil and wall =20° ; and angle of surcharge=10°. (U.P.S.C. Engg. Services, 1971)

Solution.

Given. Height of wall, $h=4$ m
Angle of surcharge, $\alpha=10°$
Top width, $a=0·75$ m
Bottom width, $b=2·5$ m
Angle of AD with vertical,
                                      $\theta=20°$
Weight of earth, $w=1,600$ kg/m³
Weight of masonry, $\rho=2,000$ kg/m³
Maximum allowable pressure,
                                      $=10,000$ kg/m²
Angle of repose, $\phi=30°$
Angle of friction, $\delta=20°$

Check for tension

Let $P=$Earth pressure per metre length of the wall.
      $x=$Horizontal distance between *c.g.* of the dam and K (*i.e. JK*).

We know that the active earth pressure on the wall per metre length,

$$P = \frac{wh^2}{2} \times \frac{\cos^2(\phi - \theta)}{\cos^2\theta \cos(\delta + \theta)\left[1 + \sqrt{\dfrac{\sin(\delta + \phi)\sin(\phi - \alpha)}{\cos(\delta + \theta)\cos(\theta - \alpha)}}\right]^2}$$

$$= \frac{1,600 \times 4^2}{2} \times \frac{\cos^2 10^\circ}{\cos^2 20^\circ \times \cos 40^\circ\left[1 + \sqrt{\dfrac{\sin 50^\circ \times \sin 20^\circ}{\cos 40^\circ \times \cos 10^\circ}}\right]^2} \text{ kg}$$

$$= 12,800 \times \frac{0 \cdot 9848^2}{0 \cdot 9397^2 \times 0 \cdot 766\left[1 + \sqrt{\dfrac{0 \cdot 766 \times 0 \cdot 342}{0 \cdot 766 \times 0 \cdot 9848}}\right]^2} \text{ kg}$$

$$= 12,800 \times \frac{0 \cdot 9848^2}{0 \cdot 9397^2 \times 0 \cdot 766 \times 1 \cdot 5892^2} \text{ kg}$$

$$= 7,263 \text{ kg}$$

Horizontal component of the pressure,

$$P_H = P \cos\theta = 7,263 \cos 10^\circ = 7,263 \times 0 \cdot 9848 \text{ kg}$$
$$= 7,163 \text{ kg}$$

and vertical component of the pressure,

$$P_V = P \sin\theta = 7,263 \sin 10^\circ = 7,263 \times 0 \cdot 1736 \text{ kg}$$
$$= 1,263 \text{ kg}$$

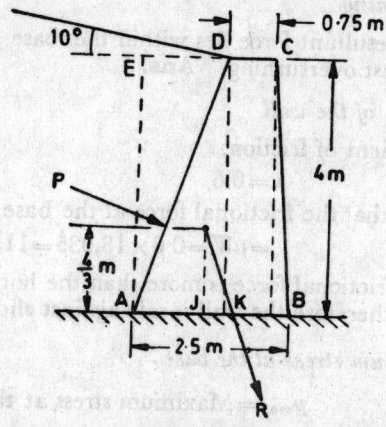

Fig. 6·33

We also know that the weight of the wall per metre length,

$$= \text{Weight of wall } ABCD + \text{Weight of wedge } ADE \text{ of earth}$$
$$= 2,000 \times \frac{(0 \cdot 75 + 2 \cdot 5)}{2} \times 4 + 1,600 \times \frac{1 \cdot 46}{2} \times 4 \text{ kg}$$
$$= 17,672 \text{ kg}$$

Therefore total downward weight of the wall per metre length,
$$W = 17,672 + 1,263 = 18,935 \text{ kg}$$

Now let us find out the *c.g.* of the wall section. Taking moments of the wall section about A, and equating the same

$$W \times AJ = \left(1{,}600 \times \frac{1{\cdot}46 \times 4}{2} \times \frac{1{\cdot}46}{3}\right) + \left(2{,}000 \times \frac{1{\cdot}46 \times 4}{2} \times \frac{1{\cdot}46 \times}{} \right)$$

$$+ \left[2{,}000 \times 0{\cdot}75 \times 4\left(1{\cdot}46 + \frac{0{\cdot}75}{2}\right)\right]$$

$$+ \left[2{,}000 \times \frac{0{\cdot}29 \times 4}{2}\left(2{\cdot}21 + \frac{0{\cdot}29}{3}\right)\right] + 1{,}263 \times \frac{1{\cdot}46}{3}$$

$$18{,}935 \; AJ = 2{,}274 + 5684 + 11{,}010 + 2{,}676 + 615 = 22{,}259$$

$$\therefore \qquad AJ = \frac{22{,}259}{18{,}935} = 1{\cdot}17 \text{ m}$$

Using the relation,

$$x = \frac{P_H}{W} \times \frac{h}{3} \text{ with usual notations.}$$

$$= \frac{7{,}163}{18{,}935} \times \frac{4}{3} = 0{\cdot}5 \text{ m}$$

$$\therefore \quad \text{Horizontal distance } AK,$$

$$d = AJ + x = 1{\cdot}17 + 0{\cdot}5 = 1{\cdot}67 \text{ m}$$

Since the resultant force lies at a point, which is at a distance of 2/3 from A, therefore the wall is safe against the tension in its masonry at the base. **Ans.**

Check for overturning

Since the resultant force lies within the base AB, therefore the wall is safe against overturning. **Ans.**

Check for sliding of the wall

Let coefficient of friction,

$$\mu = 0{\cdot}6$$

We know that the frictional force at the base

$$= \mu W = 0{\cdot}6 \times 18{,}935 = 11{,}361 \text{ kg}$$

Since the frictional force is more than the horizontal component of the pressure, therefore the wall is safe against sliding. **Ans.**

Check for maximum stress at the base

Let $\qquad p_{max} =$ Maximum stress at the base of the wall.

We know that the eccentricity of the resultant,

$$e = d - \frac{b}{2} = 1{\cdot}67 - \frac{2{\cdot}5}{2} = 0{\cdot}42 \text{ m}$$

Now using the relation,

$$p_{max} = \frac{W}{b}\left(1 + \frac{6e}{b}\right) \text{ with usual notations.}$$

$$= \frac{18{,}935}{2{\cdot}5}\left(1 + \frac{6 \times 0{\cdot}42}{2{\cdot}5}\right) = 15{,}148 \text{ kg/m}^2$$

Since the maximum stress is more than the permissible, therefore the wall is not safe against maximum stress at the base. **Ans.**

Exercise 6-3

1. A trapezoidal masonry retaining wall 1 m wide at top, 3 m wide at its bottom is 8 m high. It is retaining earth having level with the top of the wall on its vertical face. Find the maximum and minimum intensities of stress at the base of the wall. if the weight of masonry and earth is 2,400 kg/m³ and 1,800 kg/m³. Angle of repose of the earth is 40°. *(Bihar University)*
[**Ans.** 24,400 kg/m² ; 1,203 kg/m²]

2. A trapezoidal masonry retaining wall 1 m wide at top, 3 m wide at its bottom is 6 m high. The vertical face is retaining earth with angle of repose 30° at surcharge of 20° with the horizontal. Determine the maximum and minimum intensities of stress at the base of the dam. Take the densities of earth and masonry as 2,000 kg/m³ and 2,400 kg/m³. *(Ranchi University)*
[**Ans.** 16,950 kg/m² ; 5,695 kg/m²]

HIGHLIGHTS

1. The water pressure on a dam per metre length,

$$P = \frac{wh^2}{2}$$

where
$w =$ Specific weight of water (generally taken as 1,000 kg/m³) and

$h =$ Height of water.

2. The horizontal distance between the *c.g.* of the dam section and the point, where the resultant cuts the base,

$$x = \frac{P}{W} \times \frac{h}{3}$$

where
$P =$ Water pressure on the dam per metre length,

$W =$ Weight of the dam per metre length, and

$h =$ Height of water.

3. The maximum and minimum intensities of stress at the base of the dam,

$$p_{max} = \frac{W}{b}\left(1 + \frac{6e}{b}\right)$$

and
$$p_{min} = \frac{W}{b}\left(1 - \frac{6e}{b}\right)$$

where
$W =$ Weight of the dam per metre length,

$b =$ Bottom width of the dam, and

$e =$ Eccentricity of the dam.

4. Following are the conditions from the stability of a dam.

  (a) To avoid tension in the masonry of the dam,

  (b) To prevent the overturning of the dam,

  (c) To prevent the sliding of the dam.

  (d) To prevent the crushing of masonry at the base of the dam.

5. The minimum base width of a dam is found out by studying the conditions for the stability of a dam.

6. The pressure exerted by the retained earth on the retaining wall is called as an active earth pressure.

7. The pressure exerted by a retaining wall on the retained earth is called as a passive earth pressure.

8. The Rankine's active earth pressure on the retaining wall,

$$P = \frac{wh^2}{2} \times \frac{1 - \sin \phi}{1 + \sin \phi} \qquad \text{...(when the earth retained in level)}$$

$$= \frac{wh^2}{2} . \cos \alpha \times \frac{\cos \alpha - \sqrt{\cos^2 \alpha - \cos^2 \phi}}{\cos \alpha + \sqrt{\cos^2 \alpha - \cos^2 \phi}}$$

(when the earth retained is surcharged)

where w = Specific weight of the earth,

h = Height of the wall

ϕ = Angle of repose for the earth.

α = Angle of surcharge.

9. The Coulomb's active earth pressure on a retaining wall,

$$P = \frac{wh^2}{2} \times \frac{\cos^2 (\phi - \theta)}{\cos^2 \theta \cos (\delta + \theta) \left[1 + \sqrt{\frac{\sin (\delta + \phi)}{\cos (\delta + \theta)} \frac{\sin (\theta - \alpha)}{\cos (\theta - \alpha)}} \right]}$$

where w = Specific weight of the earth,

h = Height of the wall,

ϕ = Angle of repose for the earth,

θ = Angle, which the inclined face AD, makes with the vertical,

δ = Angle of friction between the retaining wall and the retained earth, and

α = Angle of surcharge.

10. The conditions for the stability of a retaining wall is checked in the same way, as that of a dam.

Do You Know ?

1. What do you understand by the term dam ? Name the various types of dams commonly used these days.

2. Derive an equation for the maximum and minimum intensities of stress at the base of a trapezoidal dam.

3. Name the various conditions for the stability of a dam. Describe any two of them.

4. How will you find out the (i) minimum base width and (ii) maximum height of a dam ?

5. What is a retaining wall ? Discuss its uses.

6. Explain what do you understand by active and passive earth pressures of soil ?

7. What are the assumptions made in Rankine's theory for calculating the magnitude of earth pressure behind retaining walls.

8. State and explain Rankine's theory of earth pressure.

7

DEFLECTION OF CANTILEVERS AND BEAMS

7·1. Introduction

We see, that whenever a cantilever or a beam is loaded, it deflects from its original position. The amount, by which a beam deflects, depends upon its cross-section and the bending. In modern design offices, the design criteria for a cantilever or a beam is for its (*i*) strength and (*ii*) stiffness. As per the strength criterion of the beam design, it should be strong enough to resist bending moment and shear force. Or in other words, the beam should be strong enough to resist the bending stresses and shear stresses. But as per the stiffness criterion of the beam design, which is equally important, it should be stiff enough to resist the deflection of the beam. Or in other words the beam should be stiff enough not to deflect more than the given limit* under the action of the loading. In actual practice, some specifications are always laid, to limit the maximum deflection of a cantilever or a beam, to a small fraction of its span.

In this chapter, we shall discuss the slope and deflection of the *centre line* of cantilevers and beams under the different types of loadings.

7·2. Curvature of the bending beam

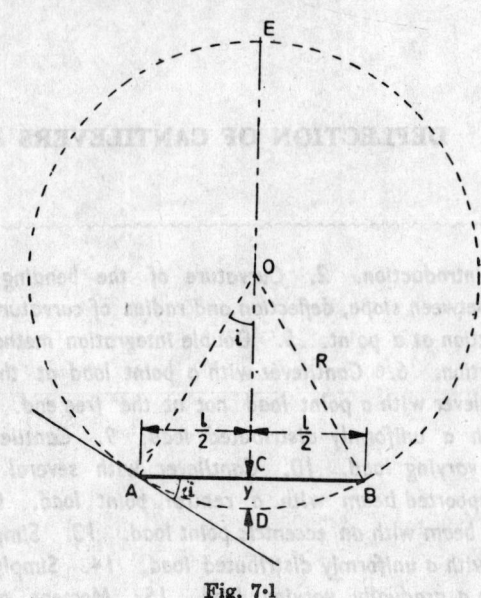

Fig. 7·1

Consider a beam *AB* subjected to a bending moment. Let the beam deflect from *ACB* to *ADB* into a circular arc as shown in Fig. 7·1.

Let

l = Length of the beam *AB*,

M = Bending moment,

R = Radius of curvature, of the bent up beam,

I = Moment of inertia of the beam section,

E = Modulus of elasticity of beam material,

y = Deflection of the beam (*i.e.*, *CD*) and

i = Slope of the beam (*i.e.*, angle which the tangent at *A* makes with *AB*).

From the geometry of a circle, we find that

$$AC \times CB = EC \times CD$$

*As per Indian standard specifications, this limit is span/325.

or
$$\frac{l}{2} \times \frac{l}{2} = (2R-y) \times y$$

$$\therefore \qquad \frac{l^2}{4} = 2Ry - y^2 = 2Ry \qquad \text{(neglecting } y^2\text{)}$$

or
$$y = \frac{l^2}{8R} \qquad \qquad ...(i)$$

We know that for a loaded beam,

$$\frac{M}{I} = \frac{E}{R}$$

or
$$R = \frac{EI}{M}$$

Now substituting this value of R in equation (i) we get the value of deflection,

$$y = \frac{l^2}{8 \times \dfrac{EI}{M}} = \frac{Ml^2}{8EI}$$

Now from the geometry of the figure, we find that the slope of the beam i at A or B is also equal to angle AOC.

$$\therefore \qquad \sin i = \frac{AC}{OA} = \frac{l}{2R}$$

Since the angle i is very small, therefore, $\sin i$ may be taken equal to i (in radians)

$$\therefore \qquad i = \frac{l}{2R} \text{ radians} \qquad \qquad ...(ii)$$

Again substituting the value of R in equation (ii) we get the value of slope,

$$i = \frac{l}{2R} = \frac{l}{2 \times \dfrac{EI}{M}} = \frac{Ml}{2EI} \text{ radians} \qquad ...(iii)$$

7·3. Relation between slope, deflection and radius of curvature

Consider a small portion PQ of a beam, bent up in arc, as shown in Fig. 7·2.

Let
  $ds =$ Length of the beam PQ,

  $R =$ Radius of the arc, into which the beam has been bent,

  $C =$ Centre of the arc,

  $\psi =$ Angle, which the tangent at P makes with X-axis.

$\psi + d\psi =$ Angle, which the tangent Q makes with X-axis.

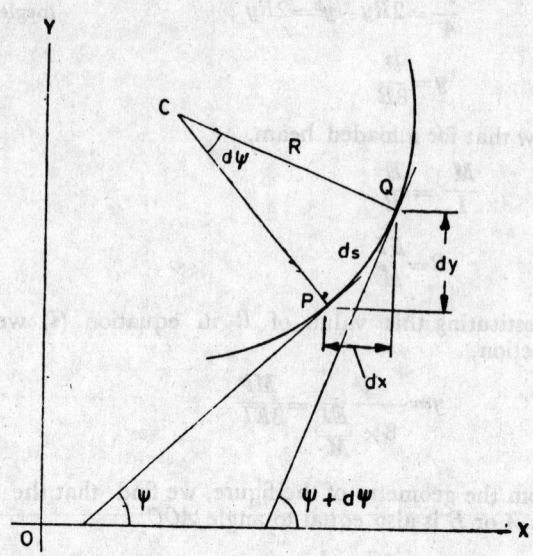

Fig. 7·2

from the geometry of the figure, we find that the angle $PCQ = d\psi$.

and $\qquad\qquad ds = R.d\psi$ (∵ Length of arc = Radius × angle subtended at the centre in radians)

or $\qquad\qquad R = \dfrac{ds}{d\psi} = \dfrac{dx}{d\psi}$ (Considering $ds = dx$)

or $\qquad\qquad \dfrac{1}{R} = \dfrac{d\psi}{dx}$ ...(i)

We know that if x and y be the co-ordinates of point P, then

$$\tan \psi = \dfrac{dy}{dx}$$

Since ψ is a very small angle, therefore, taking $\tan \psi = \psi$, we get

$$\psi = \dfrac{dy}{dx}$$

Differentiating the above equation with respect to x,

$$\dfrac{d\psi}{dx} = \dfrac{d^2y}{dx^2}$$

Now, we know from equation (i) that $\dfrac{1}{R} = \dfrac{d\psi}{dx}$

∴ $\qquad\qquad \dfrac{1}{R} = \dfrac{d^2y}{dx^2}$ ...(ii)

We also know

$$\frac{M}{I} = \frac{E}{R} \quad \text{or} \quad M = EI \times \frac{1}{R}$$

or

$$M = EI \frac{d^2y}{dx^2} \qquad \left(\text{Substituting value of } \frac{1}{R} \right)$$

A little consideration will show, that the above equation is based only on the bending moment. The effect of shear force, being very small as compared to the bending moment, is neglected.

7·4. Slope and deflection at a point

Though there are many methods to find out the slope and deflection at a point in a loaded cantilever or beam, yet the following are important from the subject point of view :

(1) Double integration method,

(2) Moment area method,

(3) Macaulay's method, and

(4) Conjugate beam method.

The first two methods are suitable for a single load, whereas the third one is suitable for several loads and the last one is suitable for cantilevers or beams with varying moment of inertia.

7·5. Double integration method for slope and deflection

We have already discussed in Art. 7·4 that the bending moment at a point,

$$M = EI \frac{d^2y}{dx^2}$$

Integrating the above equation,

$$EI \frac{dy}{dx} = \int M \qquad \qquad ...(i)$$

Integrating the above equation once again,

$$EI.y = \int\int M \qquad \qquad ...(ii)$$

It is thus obvious, that on integrating the original differential equation once, we get the value of slope at any point and on further integrating, we get the value of deflection at any point.

Note. While integrating twice the original differential equation, we will get two constants C_1 and C_2. The value of these constants may be easily determined by using the end conditions.

7·6. Cantilever with a point load at the free end

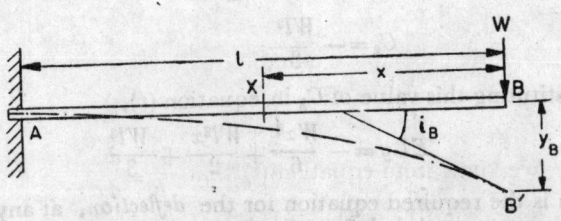

Fig. 7·3

Consider a cantilever AB of length l and carrying a point load W at the free end as shown in Fig. 7·3.

Now consider a section X at a distance x, from the free end B.

Bending moment at this section,

$$M_X = -W.x \qquad \text{(Minus sign due to hogging)}$$

$$\therefore \qquad EI\frac{d^2y}{dx^2} = \times W.x \qquad \qquad \text{...(i)}$$

Integrating the above equation,

$$EI\frac{dy}{dx} = -\frac{Wx^2}{2} + C_1 \qquad \qquad \text{...(ii)}$$

where C_1 is the constant of integration. We know that when $x=l$, $\frac{dy}{dx}=0$. Substituting these values in the above equation, we get

$$0 = -\frac{Wl^2}{2} + C_1 \quad \text{or} \quad C_1 = \frac{Wl^2}{2}$$

Substituting this value of C_1 in equation (ii),

$$EI\frac{dy}{dx} = -\frac{Wx^2}{2} + \frac{Wl^2}{2} \qquad \qquad \text{...(iii)}$$

This is the required equation for the *slope*, at any section by which we can get the slope at any point on the cantilever. A little consideration will show that the maximum slope occurs at the free end. Now using the abbreviation i for the angle of inclination (in radians) and considering $i = \tan i$, for very small angles.

For maximum slope, substituting $x=0$ in equation (iii),

$$EI.i_B = \frac{Wl^2}{2}$$

or

$$i_B = \frac{Wl^2}{2EI} \text{ radians} \qquad \qquad \text{...(iv)}$$

Integrating the equation (iii) once again,

$$EI.y = -\frac{Wx^3}{6} + \frac{Wl^2x}{2} + C_2 \qquad \qquad \text{...(v)}$$

where C_2 is the constant of integration. We know that when $x=l$, $y=0$. Substituting these values in the above equation, we get

$$0 = -\frac{Wl^3}{6} + \frac{Wl^3}{2} + C_2$$

or

$$C_2 = -\frac{Wl^3}{3}$$

Substituting this value of C_2 in equation (v),

$$EI.y = -\frac{Wx^3}{6} + \frac{Wl^2x}{2} + \frac{Wl^3}{3} \qquad \qquad \text{...(vi)}$$

This is the required equation for the *deflection*, at any section, by which we can get the deflection at any point on the cantilever.

A little consideration will show that the maximum deflection occurs at the free end.

For maximum deflection, substituting $x=0$ in equation (*vi*),

$$EI.y_B = -\frac{Wl^3}{3}$$

or

$$y_B = -\frac{Wl^3}{3EI} \qquad \text{(Minus sign means that the deflection is downwards)}$$

$$= \frac{Wl^3}{3EI}$$

7·7. Cantilever with a point load not at the free end

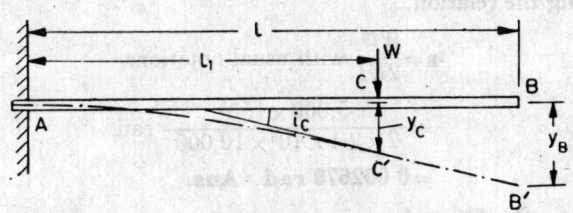

Fig. 7·4

Consider a cantilever AB of length l and carrying a point load W at C at a distance l_1 from the fixed end as shown in Fig. 7·4.

A little consideration will show that the portion AC of the cantilever will bend into AC', while the portion CB will remain straight, and displaced to $C'B'$ as shown in Fig. 7·4.

The portion AC of the cantilever may be taken as similar to a cantilever in Art. 7·6 (*i.e.*, load at the free end).

$$\therefore \qquad i_C = \frac{Wl_1^2}{2EI}$$

Since the portion CB of the cantilever is straight, therefore

$$i_B = i_C = \frac{Wl_1^2}{2EI}$$

and

$$y_C = \frac{Wl_1^3}{3EI}$$

From the geometry of the figure, we find that

$$y_B = y_C + i_C(l-l_1) = \frac{Wl_1^3}{3EI} + \frac{Wl_1^2}{2EI}(l-l_1)$$

Cor. If

$$l_1 = \frac{l}{2}$$

then

$$y_B = \frac{W}{3EI}\left(\frac{l}{2}\right)^2 + \frac{W}{2EI}\left(\frac{l}{2}\right)^2 \times \frac{l}{2} = \frac{5Wl^3}{48EI}$$

Example 7·1. *A cantilever 15 cm wide and 20 cm deep projects 1·5 m out of a wall and is carrying a point load at 5,000 kg at the free end. Find the slope and deflection of the cantilever at the free end. Take $E = 2·1 \times 10^6$ kg/cm².* (Bombay University, 1972)

Solution.

Given. Width $b=15$ cm

Depth, $d=20$ cm

$\therefore$ $I=\dfrac{bd^3}{12}=\dfrac{15\times20^3}{12}=10,000$ cm^4

Length $l=1\cdot5$ m$=150$ cm

Load, $W=5,000$ kg

Young's modulus, $E=2\cdot1\times10^6$ kg/cm^2

Slope at the free end

Let $i_B=$ Slope at the free end.

Using the relation,

$$i_B=\frac{Wl^2}{2EI}\text{ with usual notations.}$$

$$=\frac{5,000\times150^2}{2\times2\cdot1\times10^6\times10,000}\text{ rad}$$

$$=0\cdot002678\text{ rad}\quad\textbf{Ans.}$$

Deflection at the free end

Let $y_B=$ Deflection at the free end.

Using the relation,

$$y_B=\frac{Wl^3}{3EI}\text{ with usual notations.}$$

$$=\frac{5,000\times150^3}{3\times2\cdot1\times10^6\times10,000}=0\cdot2678\text{ cm}\quad\textbf{Ans.}$$

Example 7·2. *A cantilever beam 4 m long, carries a point load of 10 T at a distance of 3 m from the fixed end. Determine the slope and deflection at the free end. Take $E=2\cdot0\times10^6$ kg/cm^2 and $I=40,000$ cm^4.*

Solution.

Given. Beam length,

$$l=4\text{ m}=400\text{ cm}$$

Load, $W=10$ T$=10,000$ kg

Distance between the load and the fixed end,

$$l_1=3\text{ m}=300\text{ cm}$$

$$E=2\cdot0\times10^6\text{ kg/cm}^3$$

$$I=40,000\text{ cm}^4$$

Slope at the free end

Let $i_B=$ Slope at the free end.

Using the relation,

$$i_B=\frac{Wl_1^2}{2EI}\text{ with usual notations.}$$

$$=\frac{10,000\times300^2}{2\times2\cdot0\times10^6\times40,000}=0\cdot005625\text{ rad}\quad\textbf{Ans.}$$

Deflection at the free end

Let y_B=Deflection at the free and.

Using the relation,

$$y_B = \frac{Wl_1{}^3}{3EI} + \frac{Wl_1{}^2}{2EI}(l-l_1) \text{ with usual notations.}$$

$$= \frac{10,000 \times 300^3}{3 \times 2 \cdot 0 \times 10^6 \times 40 \cdot 000}$$

$$+ \frac{10,000 \times 300^2}{2 \times 2 \cdot 0 \times 10^6 \times 40,000}(400-300) \text{ cm}$$

$$= 1 \cdot 6875 \text{ cm} \quad \textbf{Ans}.$$

7·8. Cantilever with a uniformly distributed load

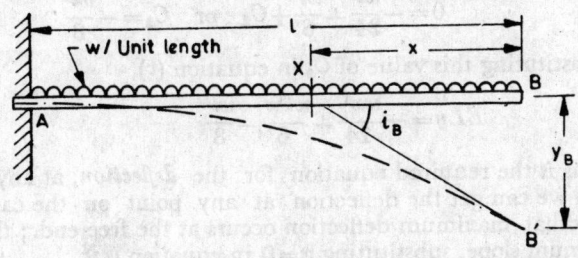

Fig. 7·5

Consider a cantilever AB of length l, and carrying a uniformly distributed load of w per unit length as shown in Fig. 7·5.

Now consider a section X, at a distance x from the free end B.

We know that bending moment at the section,

$$M_X = -\frac{wx^2}{2} \qquad \text{(Minus sign due to hogging)}$$

$$\therefore \qquad EI\frac{d^2y}{dx^2} = -\frac{wx^2}{2} \qquad \qquad ...(i)$$

Integating the above equation,

$$EI\frac{dy}{dx} = -\frac{wx^3}{6} + C_1 \qquad \qquad ...(ii)$$

where C_1 is the constant of integration. We know that when $x=l$, $\frac{dy}{dx}=0$. Substituting these values in equation (ii), we get

$$0 = -\frac{wl^3}{6} + C_1 \qquad \therefore \qquad C_1 = \frac{wl^3}{6}$$

Substituting this value of C_1 in equation (ii), we get

$$EI\frac{dy}{dx} = -\frac{wx^3}{6} + \frac{wl^3}{6} \qquad \qquad ...(iii)$$

This is the required equation for the *slope,* at any section, by which we can get the slope at any section on the cantilever. We know that maximum slope occurs at the free end ; therefore for maximum slope substituting $x=0$ in equation (*iii*),

$$EI.i_B = \frac{wl^3}{6}$$

or

$$i_B = \frac{wl^3}{6EI} \text{ radians} \qquad \qquad ...(iv)$$

Integrating the equation (*iii*) once again,

$$EI.y = -\frac{wx^4}{24} + \frac{wl^3 x}{6} + C_2 \qquad \qquad ...(v)$$

where C_2 is the constant of integration. We know that when $x=l$, $y=0$. Substituting these values in the above equation, we get

$$0 = -\frac{wl^4}{24} + \frac{wl^4}{6} + C_2 \quad \text{or} \quad C_2 = -\frac{wl^4}{8}$$

Substituting this value of C_2 in equation (*v*),

$$EI.y = -\frac{wx^4}{24} + \frac{wl^3 x}{6} - \frac{wl^4}{8} \qquad \qquad ...(vi)$$

This is the required equation for the *deflection,* at any section by which we can get the deflection at any point on the cantilever. We know that maximum deflection occurs at the free end ; therefore for maximum slope, substituting $x=0$ in equation (*vi*),

$$EI.y_B = -\frac{wl^4}{8}$$

or

$$y_B = -\frac{wl^4}{8EI} \qquad \text{(Minus sign means that the deflection is downwards)}$$

$$= \frac{wl^4}{8EI}$$

Note. The above expression for slope and deflection may also be expressed in terms of total load. Such that $W=wl$.

∴

$$i_B = i_A = \frac{wl^3}{6EI} = \frac{Wl^2}{6EI}$$

and

$$y_B = \frac{wl^4}{8EI} = \frac{Wl^3}{8EI}$$

Example 7·3. *A cantilever projecting 2·5 metres from a wall is loaded with a uniformly distributed load of 8,000 kg. Determine the moment of the inertia of the beam section, if the deflection of the beam at the free end be 1 cm. Take $E = 2·05 \times 10^6$ kg/cm.*

Solution.

Given. Length, $l = 2·5$ m $= 250$ cm

Total load $W = 8,000$ kg

Deflection at free end,

$$y_B = 1 \text{ cm}$$

Let I = Moment of inertia of the beam section.

Using the relation,

$$y_B = \frac{Wl^3}{8EI} \quad \text{with usual notations.}$$

$$1 = \frac{8,000 \times 250^3}{8 \times 2 \cdot 05 \times 10^6 \times I}$$

$$\therefore \quad I = \frac{8,000 \times 250^3}{1 \times 8 \times 2 \cdot 05 \times 10^6} = \textbf{7,760 cm}^4 \quad \textbf{Ans.}$$

Example 7·4. *A cantilever 12 cm wide and 20 cm deep is 2·5 metres long. What uniformly distributed load should the beam carry to produce a deflection of 0·5 cm at the free end ?* Take $E = 2 \cdot 0 \times 10 \text{ kg/cm}^2$.   hi University, 1978,

Solution.

Given. Width, $b = 12$ cm

Depth, $d = 20$ cm

We know that moment of inertia

$$I = \frac{bd^3}{12} = \frac{12 \times 20^3}{12} = 8,000 \text{ cm}^4$$

Length, $l = 2 \cdot 5 \text{ m} = 250 \text{ cm}$

Deflection at the free end,

$$y_B = 0 \cdot 5 \text{ cm}$$

$$E = 20 \times 10^6 \text{ kg/cm}^2$$

Let W = Total uniformly distributed load.

Using the relation,

$$y_B = \frac{Wl^3}{8EI} \quad \text{with usual notations.}$$

$$0 \cdot 5 = \frac{W \times 250^3}{8 \times 2 \cdot 0 \times 10^6 \times 8,000}$$

or $$W = \frac{0 \cdot 5 \times 8 \times 2 \cdot 0 \times 10^6 \times 8,000}{250^3} = \textbf{4,096 kg Ans.}$$

Example 7·5. *A cantilever beam AB is 6 m long and is to be subjected to a uniformly distributed load of w t/m, spread over the entire span.*

Assuming rectangular cross-section with depth equal to twice the width, determine the dimensions of the beam, so that the vertical deflection at the free end does not exceed 1·5 cm. The maximum stress due to bending does not exceed 1,000 kg/cm³. Take $E = 2 \cdot 0 \times 10^6 \text{ kg/cm}^2$

Solution.

Given. Length of beam,
$$l = 6 \text{ m} = 600 \text{ cm}$$
$$\text{Load} = w \text{ t/m} = 1{,}000 \; w \text{ kg/m} = 10 \; w \text{ kg/cm}$$

Deflection at the free end,
$$y_B = 1 \cdot 5 \text{ cm}$$

Binding stress $\quad f = 1{,}000 \text{ kg/cm}^2$
$$E = 2 \cdot 0 \times 10^6 \text{ kg/cm}^2$$

Let $\qquad b = $ Width of the cantilever beam in cm, and
$\qquad d = $ Depth of the cantilever beam in cm.

∴ Moment of inertia of the beam section,
$$I = \frac{bd^3}{12} = \frac{b}{12}(2b)^3 = \frac{2b^4}{3} \text{ cm}^4$$

Using the relation,
$$y_B = \frac{wl^4}{8 \, EI} \quad \text{with usual notations.}$$

$$1 \cdot 5 = \frac{10w \times 600^4}{8 \times 2 \cdot 0 \times 10^6 \times \dfrac{2b^4}{3}}$$

$$= \frac{10w \times 600 \times 600 \times 600 \times 600 \times 3}{8 \times 2 \cdot 0 \times 10^6 \times 2b^4}$$

or $\qquad \dfrac{3}{2} = \dfrac{1{,}21{,}500}{b^4}$

∴ $\qquad b^4 = 81{,}000 \qquad \qquad \text{...(i)}$

We know that the moment at the fixed end of the cantilever
$$M = \frac{wl^2}{2} = \frac{10}{2} \times 600^2 = 18{,}00{,}000 \text{ kg-cm}$$

Now using the relation,
$$\frac{M}{I} = \frac{f}{y} \quad \text{with usual notations.}$$

$$\frac{18{,}00{,}000}{\dfrac{2b^4}{3}} = \frac{1{,}000}{b} \qquad \left(\because \; y = \frac{d}{2} = \frac{2b}{2} = b \right)$$

or $\qquad 2{,}700 = b^3$

Substituting the value of b^3 in equation (i),
$$2{,}700 \; b = 81{,}000$$

∴ $\qquad b = \dfrac{81{,}000}{2{,}700} = \textbf{30 cm} \quad \textbf{Ans.}$

and $\qquad d = 2b = 2 \times 30 = \textbf{60 cm} \quad \textbf{Ans.}$

7·9. Cantilever with a gradually varying load

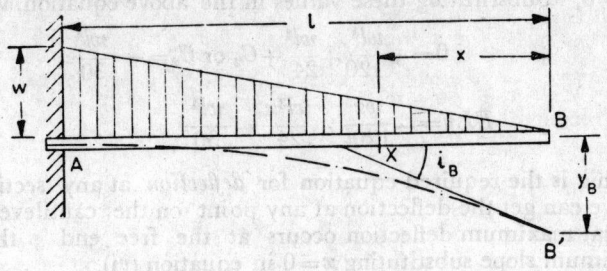

Fig. 7·6

Consider a cantilever AB of length l and carrying a gradually varying load, from zero at B to w per unit length at A, as shown in Fig. 7·6.

Now consider a section X, at a distance x from the free end B. We know that the bending moment at the section,

$$M_X = -\frac{1}{2} \times \frac{wx}{l} \times x \times \frac{x}{3} = -\frac{wx^3}{6l}$$

(Minus sign due to hogging)

$$\therefore \quad EI \frac{d^2y}{dx^2} = -\frac{wx^3}{6l} \qquad ...(i)$$

Integrating the above equation.

$$EI \cdot \frac{dy}{dx} = -\frac{wx^4}{24l} + C_1 \qquad ...(ii)$$

where C_1 is the constant of integration. We know that when $x=l$, $\frac{dy}{dx}=0$. Substituting these values in equation (ii), we get

$$0 = -\frac{wl^4}{24l} + C_1 \quad \text{or} \quad C_1 = \frac{wl^3}{24}$$

$$\therefore \quad EI.\frac{dy}{dx} = -\frac{wx^4}{24l} + \frac{wl^3}{24} \qquad ...(iii)$$

This is the required equation for the *slope* at any section, by which we can get the slope at any section on the cantilever. We know that the maximum slope occurs at the free end ; therefore for maximum slope, substituting $x=0$ in equation (iii),

$$EI.i_B = \frac{wl^3}{24}$$

$$i_B = \frac{wl^3}{24EI} \text{ radians} \qquad ...(iv)$$

Integrating the equation once again,

or $\qquad EI.\, y = -\frac{wx^5}{120l} + \frac{wl^3x}{24} + C_2 \qquad ...(v)$

where C_2 is the constant of integration. We know that when $x=l$, $y=0$, Substituting these values in the above equation, we get

$$0=-\frac{wl^4}{120}+\frac{wl^4}{24}+C_2 \text{ or } C_2=-\frac{wl^4}{30}$$

$$\therefore \quad EI.y=-\frac{wx^5}{120l}+\frac{wl^3x}{24}-\frac{wl^4}{30} \qquad ...(vi)$$

This is the required equation for *deflection*, at any section, by which we can get the deflection at any point on the cantilever. We know that maximum deflection occurs at the free end ; therefore for maximum slope substituting $x=0$ in equation (vi),

$$EI.y_B=-\frac{wl^4}{30EI}$$

$$\therefore \quad y_B=-\frac{wl^4}{30EI} \qquad \text{(Minus sign means that the deflection is downwards)}$$

$$=\frac{wl^4}{30EI}$$

Example 7·6. *A cantilever of 2 m span carries a triangular load of zero intensity at the free end, and 10 t/m at the fixed end. Determine the slope and deflection at the free end. Take $I=10,000$ cm⁴ and $E=2·0\times 10^6$ kg/cm².*

Solution.

Given. Span, $l=2$ m$=200$ cm

Load at the fixed end,

$$w=10 \text{ t/m}=10,000 \text{ kg/m}=100 \text{ kg/cm}$$

Moment of inertia, $I=10,000$ cm⁴

Young's modulus $E=2·0\times 10^6$ kg/cm²

Slope at the free end

Let $i_B=$Slope at the free end.

Using the relation,

$$i_B=\frac{wl^3}{24EI} \text{ with usual notations.}$$

$$=\frac{100\times 200^3}{24\times 2·0\times 10^6\times 10,000}\text{rad}$$

$$=0·00667 \text{ rad Ans.}$$

Deflection at the free end

Let $y_B=$Deflection at the free end.

Using the relation,

$$y_B=\frac{wl^4}{30EI} \text{ with usual notations.}$$

$$=\frac{100\times 200^4}{30\times 2·0\times 10^6\times 10,000}=0·267 \text{ cm Ans.}$$

7·10. Cantilever with several loads

If a cantilever is loaded with a several point or uniformly distributed loads, the slope as well as the deflection, at any point on the cantilever, is equal to the algebraic sum of the slopes and deflections at that point due to various loads, acting individually.

Example 7·7. *A cantilever 2 m long is carrying a load of 2,000 kg at free end and 3,000 kg at a distance 1 m from the free end. Find the slope and deflection at the free end. Take $E = 2·0 \times 10^6$ kg/cm² and $I = 15,000$ cm⁴.*

Solution.

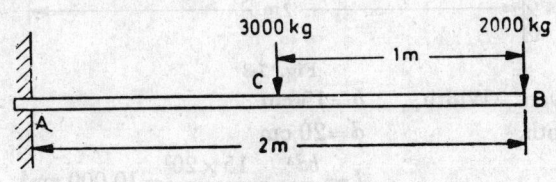

Fig. 7·7

Given. Length, $l = 2$ m $= 200$ cm
Load 1, $W_1 = 2,000$ kg
Load 2, $W_2 = 3,000$ kg
Length of AC, $l_1 = 1$ m $= 100$ cm
Young's modulus, $E = 2·0 \times 10^6$ kg/cm²
Moment of inertia, $I = 15,000$ cm⁴

Slope at the free end

Let $i_B =$ Slope at the free end B.

Using the relation,

$$i_B = \left(\frac{W_1 l^2}{2EI}\right) + \left(\frac{W_2 l_1^2}{2EI}\right) \text{ with usual notations.}$$

$$= \frac{2,000 \times 200^2}{2 \times 2·0 \times 10^6 \times 15,000} + \frac{3,000 \times 100^2}{2 \times 2·0 \times 10^6 \times 15,000} \text{ rad}$$

$= 0·001833$ rad   **Ans.**

Deflection at the free end

Let $y_B =$ Deflection at the free end B.

Using the relation,

$$y_B = \left(\frac{W_1 l^3}{3EI}\right) + \left[\frac{W_2 l_1^3}{3EI} + \frac{W_2 l_1^2}{2EI}(l - l_1)\right] \text{with usual notations.}$$

$$= \frac{2,000 \times 200^3}{3 \times 2 \times 10^6 \times 15,000} + \frac{3,000 \times 100^3}{3 \times 2 \times 10^6 \times 15,000}$$

$$+ \frac{3,000 \times 100^2}{2 \times 2 \times 10^6 \times 15,000}(200 - 100) \text{ cm}$$

$= 0·2611$ cm   **Ans.**

Example 7·8. *A metallic cantilever 15 cm wide 10 cm deep and of 2 m span carries a uniformly varying load of 5 t/m at the free end to 15 t/m at the fixed end. Find the slope and deflection of the cantilever at the free end. Take* $E = 10 \times 10^5$ *kg/cm².*

(Cambridge University)

Solution.

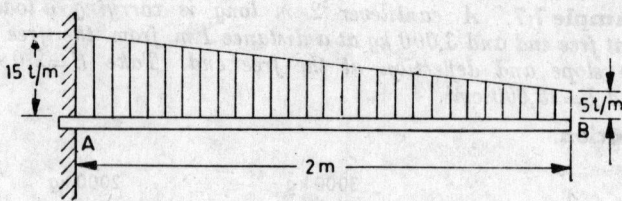

Fig. 7·8

Given.   Width, $b = 15$ cm

Depth, $d = 20$ cm

∴ $I = \dfrac{bd^3}{12} = \dfrac{15 \times 20^3}{12} = 10{,}000$ cm⁴

Span, $l = 2$ m $= 200$ cm

*Load 1 (*u.d.l.*),

$$w_1 = 5 \text{ t/m} = 5{,}000 \text{ kg/m} = 50 \text{ kg/cm}$$

Load 2 (triangular),

$$w_2 = 10 \text{ t/m} = 10{,}000 \text{ kg/m} = 100 \text{ kg/cm}$$

Young's modulus,

$$E = 10 \times 10^6 \text{ kg/cm}^2$$

Slope at the free end

Let $i_B =$ Slope at the free end.

Using the relation,

$$i_B = \frac{w_1 l^3}{6EI} + \frac{w_2 l^3}{24EI} \text{ with usual notations.}$$

$$= \frac{50 \times 200^3}{6 \times 10 \times 10^6 \times 10{,}000}$$

$$+ \frac{100 \times 200^3}{24 \times 10 \times 10^6 \times 10{,}000} = \mathbf{0·01 \text{ rad} \text{ Ans.}}$$

Deflection at the free end

Let $y_B =$ Deflection at the free end.

Using the relation,

$$y_B = \frac{w_1 l^4}{8EI} + \frac{w_2 l^4}{30EI} \text{ with usual notations.}$$

$$= \frac{50 \times 200^4}{8 \times 10 \times 10^6 \times 10{,}000}$$

$$+ \frac{100 \times 200^4}{30 \times 10 \times 10^6 \times 10{,}000} = \mathbf{1·533 \text{ cm} \text{ Ans.}}$$

---

*The trapezoidal load has beam assumed to be consisting of *u.d.l.* of 5 t/m and a triangular load of zero at B and 10 t/m at A.

Example 7·9. *A cantilever of length 2a is carrying a load of W at the free end, and another load of W at its centre. Determine from first principles, the slope and deflection of the cantilever at the free end.*

Solution.

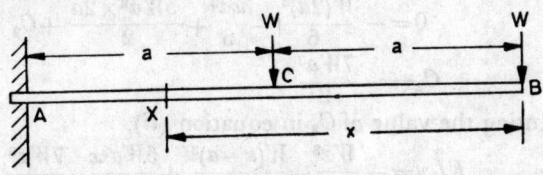

Fig. 7·9

The cantilever AB of length $2a$ and carrying point loads of W each at C and B is shown in Fig. 7·9.

Slope at the free end

Let i_B = slope at the free end.

Consider a section X, at a distance x from B. Therefore B.M. at X,

$$M_X = -Wx - W(x-a) \qquad \text{(Minus sign due to hogging)}$$

$$\therefore \quad EI\frac{d^2y}{dx^2} = -Wx - W(x-a) \qquad \qquad ...(i)$$

Integrating the above equation,

$$EI\frac{dy}{dx} = -\frac{Wx^2}{2} - \frac{W(x-a)^2}{2} + C_1 \qquad \qquad ...(ii)$$

where C_1 is the constant of integration. We know that when $x = 2a$, $\frac{dy}{dx} = 0$. Substituting these values in the above equation, we get

$$0 = -\frac{W}{2}(2a)^2 - \frac{w}{2}(2a-a)^2 + C_1$$

$$\therefore \quad C_1 = +\frac{5Wa^2}{2}$$

Substituting this value of C_1 in equation (ii),

$$EI\frac{dy}{dx} = -\frac{Wx^2}{2} - \frac{W(x-a)^2}{2} + \frac{5Wa^2}{2}$$

Now for the slope at the free end, substitute $x = 0$ in equation (iii) (neglecting the negative term in the bracket).

$$EI.i_B = \frac{5Wa^2}{2}$$

or

$$i_B = \frac{5Wa^2}{2EI} \quad \textbf{Ans.}$$

Deflection at the free end

Let y_B = Deflection at the free end.

Integrating the equation (iii) once again,

$$EI.y = -\frac{Wx^3}{6} - \frac{W(x-a)^3}{6} + \frac{5Wa^2x}{2} + C_2$$

where C_2 is the constant of integration. We know that when $x=2a$, $y=0$. Substituting these values in the above equation, we get

$$0 = -\frac{W(2a)^3}{6} - \frac{wa^3}{6} + \frac{5Wa^2 \times 2a}{2} + C_2$$

$$\therefore \qquad C_2 = -\frac{7Wa^3}{2}$$

Substituting the value of C_2 in equation (ii),

$$EI.y = -\frac{Wx^3}{6} - \frac{W(x-a)^3}{6} + \frac{5Wa^2x}{2} - \frac{7Wa^3}{2}$$

Now for the deflection at B, substituting $x=0$ in the above equation (neglecting the $-$ve term in the bracket).

$$y_B = -\frac{7Wa^3}{2EI} \qquad \text{(Minus sign means that the deflection is downwards)}$$

$$= \frac{7Wa^3}{2EI} \quad \textbf{Ans.}$$

Exercise 7·1

1. A cantilever 2 m long and 10 cm×25 cm cross-section carries a concentrated load of 2,000 kg at its free end. Determine the deflection at its free end. Take $E = 2·0 \times 10^6$ kg/cm².   (*Punjab University*)
[**Ans.**   0·205 cm]

2. A cantilever 4 m long carries a load of 2,000 kg at its free end. Find the slope and deflection at the free end. Take $E = 2·0 \times 10^6$ kg/cm² and $I = 21,000$ cm⁴.   ((*Ranchi University*)
[**Ans.**   0·0038 radian ; 1·01 cm]

3. A cantilever 2 m long carries a point load of 900 kg at the free end and a load of 800 kg uniformly distributed over a length of 1 m from the fixed end. Determine the deflection at the free end, if $E = 2·2 \times 10^6$ kg/cm². Take $I = 2,250$ cm⁴.   (*Bihar University*)
[**Ans.**   0·5322 cm]

4. A cantilever 2 m long carries a point load of 100 kg at the free end and a uniformly distributed load of 200 kg/m over a length of 1·25 m from the fixed end. Find the deflection at the free end, if $E = 2·0 \times 10^6$ kg/cm². Take $I = 13,824$ cm⁴.   (*Patna University*)
[**Ans.**   0·146 cm]

5. A horizontal cantilever of uniform section and length L carries a load W at a distance $L/4$ from the free end. Derive from the first principles the deflection at the free end in terms of W, L, E and I.   (*London University*)
$$\left[\textbf{Ans.} \quad \frac{27WL^3}{128EI} \right]$$

7·11. Simply supported beam with a central point load

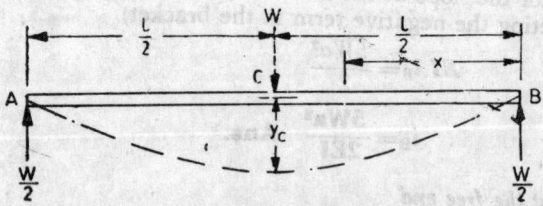

Fig, 7·10

DEFLECTION OF CANTILEVERS AND BEAMS

Consider a simply supported beam AB of length l and carrying a point load W at the centre of beam C as shown in Fig. 7·10. From the geometry of the figure, we find that the reaction at A,

$$R_A = R_B = \frac{W}{2}$$

Consider a section X, at a distance x from A. We know that the bending moment at this section,

$$M_X = R_B . x = \frac{W}{2} x$$

$$\therefore \quad EI \frac{d^2y}{dx^2} = \frac{Wx}{2} \qquad ...(i)$$

Integrating the above equation,

$$EI \frac{dy}{dx} = \frac{Wx^2}{4} + C_1 \qquad ...(ii)$$

where C_1 is the constant of integration. We know when $x = \frac{l}{2}$, $\frac{dy}{dx} = 0$. Substituting these values in equation (ii), we get

$$0 = \frac{Wl^2}{16} + C_1 \text{ or } C_1 = -\frac{Wl^2}{16}$$

Substituting this value of C_1 in equation (ii),

$$EI \frac{dy}{dx} = \frac{Wx^2}{4} - \frac{Wl^2}{16} \qquad ...(iii)$$

This is the required equation for the *slope*, at any section, by which we can get the slope at any point on the beam. A little consideration will show that the maximum slope occurs at A and B. Thus for maximum slope at B, substituting $x = 0$ in equation (iii),

$$EI . i_B = \frac{Wl^2}{16}$$

or $\qquad i_B = -\frac{Wl^2}{16EI}$ (Minus sign means that the tangent at B makes an angle with AB in the negative or anti-clockwise direction)

or $\qquad i_B = \frac{Wl^2}{16EI}$ radians

By symmetry, $\quad i_A = \frac{Wl^2}{16EI}$ radians

Integrating the equation (iii) once again,

$$EI . y = \frac{Wx^3}{12} - \frac{Wl^2}{16} x + C_2 \qquad ...(iv)$$

where C_2 is the constant of integration. We know that when $x = 0$, $y = 0$. Substituting these values in equation (iv), we get $C_2 = 0$.

$$\therefore \quad EI . y = \frac{Wx^3}{12} - \frac{Wl^2}{16} x \qquad ...(v)$$

This is the required equation for the *deflection*, at any section, by which we can get the deflection at any point on the beam. A little consideration will show, that maximum deflection occurs at the mid-point C. Thus, for maximum deflection, substituting $x=l/2$ in equation (v),

$$EI.y_C = \frac{W}{12}\left(\frac{l}{2}\right)^3 - \frac{Wl^2}{16}\left(\frac{l}{2}\right) = \frac{Wl^3}{96} - \frac{Wl^3}{32} = -\frac{Wl^3}{48}$$

or
$$y_C = -\frac{Wl^3}{48EI}$$

(Minus sign means that the deflection is downwards)

$$= \frac{Wl^3}{48EI}$$

Example 7·10. *A uniform beam $(I=7,800 \text{ cm}^4)$ is 6 metres long and carries a central point load of 5 tonnes.*

Taking $E=2\cdot1\times10^6$, kg/cm², calculate the deflection under the load, if the beam is simply supported at its ends.

Solution.

Given. Moment of inertia,
$$I=7,800 \text{ cm}^4$$
Length of beam $l=6 \text{ m}=600 \text{ cm}$
Central load, $W=5 \text{ t}=5,000 \text{ kg}$
$$E=2\cdot1\times10^6 \text{ kg/cm}^2$$
Let $y_C=$Deflection under the load.
Using the relation,
$$y_C=\frac{Wl^3}{48EI} \text{ with usual notations.}$$
$$=\frac{5,000\times600^3}{48\times2\cdot1\times10^6\times7,800}=1\cdot38 \text{ cm } \textbf{ Ans.}$$

Example 7·11. *A beam 3 metres long, simply supported at its ends, is carrying a point load at its centre. If the slope at the ends of the beam is not to exceed 1°, find the deflection at the centre of the beam.*   (Oxford University)

Solution.

Given. Length, $l=3 \text{ m}$
Slope at A, $i_A=1°=0\cdot01745$ radian
We know that slope at an end,
$$i_A=\frac{Wl^2}{16EI}$$
and deflection at the centre,
$$y_C=\frac{Wl^3}{48EI}=\frac{Wl^2}{16EI}\times\frac{l}{3}$$
$$=i_A\times\frac{l}{3} \qquad \left(\because \frac{Wl^2}{16EI}=i_A\right)$$
$$=0\cdot01745\times\frac{3}{3}=0\cdot01745 \text{ m}$$
$$=1\cdot745 \text{ cm } \textbf{ Ans.}$$

Example 7·12. *A beam AB of span 5 m is simply supported at A and B. A cantilever DC of length 3 m, which is fixed at D, meets the beam AB at the mid-point C, thereby forming a rigid joint at C as shown in Fig. 7·11.*

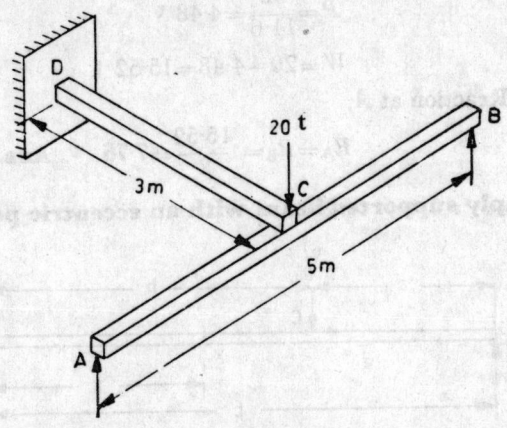

Fig. 7·11

A vertical load of 20 ᵗ *is applied vertically at the common joint C. Find out the reaction at the ends of the simply supported beam. Take $I_{AB}=I_{CD}$.*   (Cambridge University)

Solution.

Given.   Length of beam AB,

$$l_{AB}=5 \text{ m}$$

Length of cantilever CD,

$$l_{CD}=3 \text{ m}$$

Load at C $=20$ ᵗ

$$I_{AB}=I_{CD}$$

Let $P=$ Load shared by the cantilever.

∴ Load shared by the beam AB

$$W=20-P \text{ tonnes}$$

Using the relation,

$$y_C \text{ (cantilever)} = \frac{Wl^3}{3EI} \text{ with usual notations.}$$

$$= \frac{P \times 3^3}{3EI} = \frac{9P}{EI} \qquad \qquad ...(i)$$

Again using the relation,

$$y_C \text{ (beam)} = \frac{Wl^3}{48EI} \text{ with usual notations.}$$

$$= \frac{(20-P) \times 5^3}{48EI} = \frac{125(20-P)}{48EI} \qquad ...(ii)$$

Since the deflection of the beam AB and cantilever CD is the same, therefore equating (i) and (ii)

$$\frac{9P}{EI}=\frac{125\,(20-P)}{48EI}$$

$$\therefore \qquad 9P=2\cdot6\,(20-P)=52-2\cdot6\,P$$

or
$$P=\frac{52}{11\cdot6}=4\cdot48 \text{ t}$$

$$\therefore \qquad W=20-4\cdot48=15\cdot52 \text{ t}$$

$\therefore$ Reaction at A,

$$R_A=R_B=\frac{15\cdot52}{2}=7\cdot76 \text{ t Ans.}$$

7·12. Simply supported beam with an eccentric point load

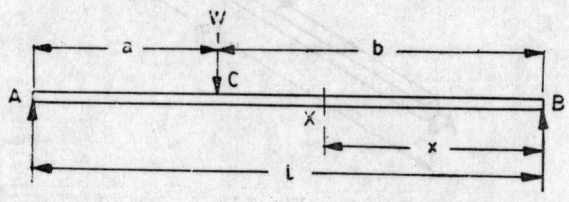

Fig. 7·12

Consider a simply supported beam AB of length l and carrying an eccentric point load W at C, as shown in Fig. 7·12. From the geometry of the figure, we find that the reaction $R_A=\dfrac{Wb}{l}$ and reaction $R_B=\dfrac{Wa}{l}$.

Consider a section X, in CB at a distance x from B, such that x is less than b (i.e., $x<b$). We know that the bending moment at this section,

$$M_X=R_B.x=\frac{Wax}{l}$$

$$\therefore \qquad EI\frac{d^2y}{dx^2}=\frac{Wax}{l} \qquad \qquad ...(i)$$

Integrating the above equation,

$$EI\frac{dy}{dx}=\frac{Wax^2}{2l}+C_1 \qquad \qquad ...(ii)$$

where C_1 is the constant of integration. We know that at C, $x=b$ and $\dfrac{dy}{dx}=i_C$.

$$\therefore \qquad EI.i_C=\frac{Wab^2}{2l}+C_1$$

or
$$C_1=EI.i_C-\frac{Wab^2}{2l}$$

Substituting this value of C_1 in equation (ii),

$$EI\frac{dy}{dx} = \frac{Wax^2}{2l} + EI.i_C - \frac{Wab^2}{2l} \qquad ...(iii)$$

Integrating the above equation once again,

$$EI.y = \frac{Wax^3}{6l} + EI.i_C x - \frac{Wab^2x}{2l} + C_2 \qquad ...(iv)$$

where C_2 is the constant of integration. We know that when $x=0$, $y=0$. Therefore $C_2=0$

$$\therefore \qquad EI\,y = \frac{Wax^3}{6l} + EI.i_C x - \frac{Wab^2x}{2l} \qquad ...(v)$$

The equations (iii) and (iv) are the required equations for slope and deflection at any point in the section AC. A little consideration will show that these equations are useful, only, if the value of i_C is known.

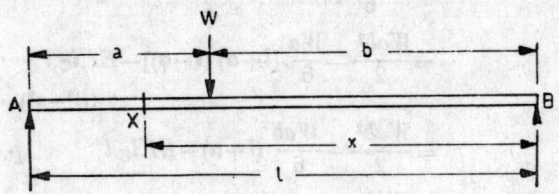

Fig. 7·13

Now consider a section X, in AB, at a distance x from B such that x is greater than a (i.e., $x > b$) as shown in Fig. 7·13.

Bending moment at this section,

$$M_X = \frac{Wax}{l} - W(x-b)$$

$$\therefore \qquad EI\frac{d^2y}{dx^2} = \frac{Wax}{l} - W(x-b) \qquad ...(vi)$$

Integrating the above equation,

$$EI\frac{dy}{dx} = \frac{Wax^2}{2l} - \frac{W(x-b)^2}{2} + C_3 \qquad ...(vii)$$

where C_3 is the constant of integration. We know that at C, $x=b$ and $\frac{dy}{dx} = i_C$.

$$\therefore \qquad EI.i_C = \frac{Wab^2}{2l} + C_3$$

or $$\qquad C_3 = EI.i_C - \frac{Wab^2}{2l}$$

Substituting this value of C_3 in equation (vii),

$$EI\cdot\frac{dy}{dx} = \frac{Wax^2}{2l} - \frac{W(x-b)^2}{2} + EI.i_C - \frac{Wab^2}{2l} \qquad ...(viii)$$

Integrating the above equation once again,

$$EI.y = \frac{Wax^3}{6l} - \frac{W(x-b)^3}{6} + EI.i_C x - \frac{Wab^2 x}{2l} + C_4 \quad \ldots(ix)$$

where C_4 is the constant of integration. We know that when $x = l$, $y = 0$.

Substituting these values in the above equation,

$$0 = \frac{Wal^2}{6} - \frac{Wa^3}{6} + EI.i_C.l - \frac{Wab^2}{2} + C_4 \quad [\because (x-b) = a]$$

$$\therefore \quad C_4 = \frac{Wa^3}{6} - \frac{Wal^2}{6} + \frac{Wab^2}{2} - EI.i_C.l$$

$$= \frac{Wa}{6}(a^2 - l^2) + \frac{Wab^2}{2} - EI.i_C.l$$

$$= -\frac{Wa}{6}(l^2 - a^2) + \frac{Wab^2}{2} - EI.i_C.l$$

$$= \frac{Wab^2}{2} - \frac{Wa}{6}[l+a)(l-a)] - EI.i_C.l$$

$$[(l^2-a^2) = (l+a)(l-a)]$$

$$= \frac{Wab^2}{2} - \frac{Wab}{6}(l+a) - EI.i_C.l \quad [\because (l-a) = b]$$

$$= \frac{Wab}{6}[3b - (l+a)] - EI.i_C.l$$

$$= \frac{Wab}{6}[3b - (a+b+a)] - EI.i_C.l \quad [\because l = a+b)$$

$$= \frac{Wab}{6}(2b - 2a) - EI.i_C.l$$

$$= \frac{Wab}{3}(b-a) - EI.i_C.l$$

Substituting this value of C_4 in equation (ix),

$$EI.y = \frac{Wax^3}{6l} - \frac{W(x-b)^3}{6} + EI.i_C.x - \frac{Wab^2 x}{2l}$$

$$\frac{Wab}{3}(b-a) - EI.i_C.l \quad \ldots(x)$$

The equations (ix) and (x) are the required equations for the slope and deflection at any point in the section AC. A little consideration will show that these equations are useful, only, if the value of i_C is known.

Now, to obtain the value of i_C, let us first find out the deflection at C from both the equations i.e., from equations for the sections AC and CB.

Now substituting $x = b$ in equations (v) and (x) and equating the same.

$$\frac{Wab^3}{6l} + EI.i_C.b - \frac{Wab^3}{2l}$$

$$= \frac{Wab^3}{6l} - \frac{W(b-b)^3}{6} + EI.i_C\, b - \frac{Wab^3}{2l}$$

$$+ \frac{Wab}{3}(b-a) - EI.i_C.l$$

$$\therefore \qquad EI.i_C = \frac{Wab}{3l}(b-a)$$

Substituting the value of $EI.i_C$ in equation (iii),

$$EI\,\frac{dy}{dx} = \frac{Wax^2}{2l} + \frac{Wab}{3l}(b-a) - \frac{Wab^2}{2l}$$

$$= \frac{Wa}{6l}[3x^2 + 2b(b-a) - 3b^2]$$

$$= \frac{Wa}{6l}(3x^2 - b^2 - 2ab) \qquad\qquad ...(xi)$$

This is the required equation for *slope* at any section in AC.

We know that the slope is maximum at B. Thus for maximum slope, substituting $x = 0$ in equation (xi),

$$EI.i_B = \frac{Wa}{6l}(-b^2 - 2ab) = -\frac{Wa}{6l}(b^2 + 2ab)$$

$$= -\frac{Wab}{6l}(b + 2a)$$

$$= -\frac{Wa}{6l}(l-a)(l+a) \qquad \left(\because\; \begin{array}{l}a = l - b\text{ and}\\ a + b = l\end{array}\right)$$

$$= -\frac{Wa}{6l}(l^2 - a^2)$$

$$\text{or} \qquad i_B = -\frac{Wa}{6lEI}(l^2 - a^2) \qquad \begin{array}{l}\text{[Minus sign means that the tangent}\\ \text{at } B \text{ makes an angle with } AB \text{ in the}\\ \text{negative or anti-clockwise direction.]}\end{array}$$

$$= \frac{Wa}{6lEI}(l^2 - a^2) \qquad\qquad ...(xii)$$

Similarly $\quad i_A = \dfrac{Wb}{6lEI}(l^2 - b^2) \qquad$ (Substituting a for b)

For deflection at any point in AC, substitute the value of $EI.i_C$ in equation (v),

$$EI.y = \frac{Wax^3}{6l} + \frac{Wab}{3l}(b-a)\,x - \frac{Wab^2x}{2l}$$

$$= \frac{Wax}{6l}[x^2 + 2b(b-a) - 3b^2]$$

$$= \frac{Wax}{6l}(x^2 + 2b^2 - 2ab - 3b^2)$$

$$= \frac{Wax}{6l}(x^2 - b^2 - 2ab) = -\frac{Wax}{6l}[b(b+2a) - x^2]$$

$$= -\frac{Wax}{6l}[(l-a)(l+a) - x^2] \qquad \left(\because\; \begin{array}{l}b = l - a\text{ and}\\ a + b = l\end{array}\right)$$

$$= -\frac{Wax}{6l}\,[l^2-a^2-x^2]$$

$$\therefore \qquad y = -\frac{Wax}{6lEI}\,(l^2-a^2-x^2) \qquad \text{(Minus sign means that the deflection is downwards)}$$

$$\therefore \qquad = \frac{Wax}{6lEI}\,(l^2-a^2-x^2) \qquad \qquad \ldots(xiii)$$

For deflection at C (i.e., under the load) substitute $x=b$ in the above equation

$$y_C = \frac{Wab}{6lEI}\,(l^2-a^2-b^2) \qquad \qquad \ldots(xiv)$$

Maximum deflection will occur in CB, since $b > a$. For maximum deflection, put $\dfrac{dy}{dx}=0$. Therefore equating the equation (xi) to zero,

$$\frac{Wa}{6l}(3x^2-b^2-2ab)=0$$

or

$$3x^2-b(b+2a)=0$$

$$3x^2-(l-a)(l+a)=0 \qquad (\because \ b=l-a \text{ and } a+b=l)$$

$$3x^2-(l^2-a^2)=0$$

$$3x^2=l^2-a^2$$

or

$$x=\sqrt{\frac{l^2-a^2}{3}}$$

For maximum deflection, substituting this value of x in equation $(xiii)$,

$$y_{max} = \frac{Wa}{6lEI}\sqrt{\frac{l^2-a^2}{3}}\times\left[l^2-a^2-\left(\frac{l^2-a^2}{3}\right)\right]$$

$$= \frac{Wa}{6lEI}\sqrt{\frac{l^2-a^2}{3}}\times\left[\frac{2}{3}(l^2-a^2)\right]$$

$$= \frac{Wa}{9\sqrt{3}\,.\,lEI}(l^2-a^2)^{\frac{3}{2}}$$

Example 7·13. *A beam of 5 metres span is carrying a point load of 3,000 kg at a distance 3·75 metres from the left end. Calculate the slopes at the two supports and deflection under the load. Take $EI=2·6\times10^{10}$ kg-cm².*

ersity, 1977)

Solution.

Given. Span, $l=5$ m $=500$ cm

Load, $W=3,000$ kg

Distance between the load and left end,

$$a=3·75 \text{ m}=375 \text{ cm}$$

$$\therefore \qquad b=(5-3·75)=1·25 \text{ m}=125 \text{ cm}$$

$$EI=2\,6\times10^{10} \text{ kg-cm}^2$$

Slope at A

Let $i_A=$ Slope at A.

Using the relation,

$$i_A = \frac{Wb}{6EIl}(l^2 - b^2) \text{ with usual notations.}$$

$$= \frac{3{,}000 \times 125\,[500^2 - 125^2]}{6 \times 2 \cdot 6 \times 10^{10} \times 500} \text{ rad}$$

$$= 0 \cdot 001125 \text{ rad } \textbf{Ans.}$$

Slope at B

Let i_B = Slope at B.

Using the relation,

$$i_B = \frac{Wa}{6EIl}(l^2 - a^2) \text{ with usual notations.}$$

$$= \frac{3{,}000 \times 375\,[500^2 - 375^2]}{6 \times 2 \cdot 6 \times 10^{10} \times 500} \text{ rad}$$

$$= 0 \cdot 001575 \text{ rad } \textbf{Ans.}$$

Deflection under the load

Let y_C = Deflection under the load.

Now using the relation,

$$y_C = \frac{Wab}{6lEI}(l^2 - a^2 - b^2) \text{ with usual notations.}$$

$$= \frac{3{,}000 \times 375 \times 125}{6 \times 500 \times 2 \cdot 6 \times 10^6}[500^2 - 375^2 - 125^2] \text{ cm}$$

$$= 0 \cdot 169 \text{ cm } \textbf{Ans.}$$

Example 7·14. *A simply supported beam of span 10 metres is carrying a point load of 1,000 kg at a distance 6 metres from the left end. If $I = 2 \cdot 0 \times 10^6$ kg/cm^2 and $I = 1{,}00{,}000$ cm^4, determine.*

(1) Slope at the left end,

(2) Deflection under the load, and

(3) Maximum deflection of the beam.

Solution.

Given. Span, $l = 10$ m $= 1{,}000$ cm

Load, $W = 1{,}000$ kg

Distance between load and left end,

$a = 6$ m $= 600$ cm

∴ $b = (10 - 6) = 4$ m $= 400$ cm

$E = 2 \cdot 0 \times 10^6$ kg/cm^2

$I = 1{,}00{,}000$ cm^4

Slop at A

Let i_A = Slope at the left end A.

Using the relation

$$i_A = \frac{Wb}{6EIl}(l^2 - b^2) \text{ with usual notations.}$$

$$= \frac{1{,}000 \times 400\,[1{,}000^2 - 400^2]}{6 \times 2 \cdot 0 \times 10^6 \times 1{,}00{,}000 \times 1{,}000} \text{ rad}$$

$$= 0 \cdot 0028 \text{ rad } \textbf{Ans.}$$

Deflection under the load

Let $\quad y_C$=Deflection under the load.

Using the relation,

$$y_C = \frac{Wab}{6lEI} \; (l^2 - a^2 - b^2) \text{ with usual notations.}$$

$$= \frac{1,000 \times 600 \times 400}{6 \times 1,000 \times 2 \cdot 0 \times 10^6 \times 1,00,000}$$
$$[1,000^2 - 600^2 - 400^2] \text{ cm}$$

$$= 0 \cdot 096 \text{ cm} \quad \text{Ans.}$$

Maximum deflection

Let $\quad y_{max}$=Maximum deflection of the beam.

Using the relation,

$$y_{max} = \frac{Wa}{9\sqrt{3} \times lEI} \; (l^2 - a^2)^{\frac{3}{2}} \text{ with usual notations.}$$

$$= \frac{1,000 \times 600}{9\sqrt{3} \times 1,000 \times 2 \cdot 0 \times 10^6 \times 1,00,000} \; [1,000^2 - 600^2]^{\frac{3}{2}}$$

$$= 0 \cdot 09875 \text{ cm} \quad \text{Ans.}$$

7·13. Simply supported beam with a uniformly distributed load

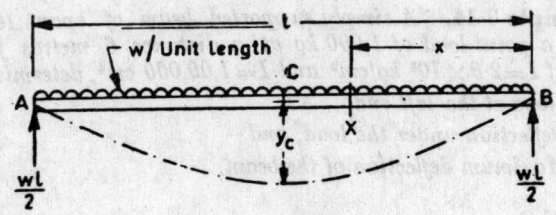

Fig. 7·14

Consider a simply supported beam AB of length l, and carrying a uniformly distributed load of w per unit length as shown in Fig. 7·14. From the geometry of the figure, we find that the reaction,

$$R_A = R_B = \frac{wl}{2}$$

Consider a section X, at a distance x from B. We know that the bending moment at this section,

$$M_X = \frac{wlx}{2} - \frac{wx^2}{2}$$

$$\therefore \quad EI \frac{d^2y}{dx^2} = \frac{wlx}{2} - \frac{wx^2}{2} \qquad \qquad \text{...(i)}$$

Integrating the above equation,

$$EI \frac{dy}{dx} = \frac{wlx^2}{4} - \frac{wx^3}{6} + C_1 \qquad \qquad \text{...(ii)}$$

where C_1 is the constant of integration. We know when $x = \dfrac{l}{2}$

$\dfrac{dy}{dx} = 0$.

Substituting these values in the above equation,

$$0 = \frac{wl^3}{16} - \frac{wl^3}{48} + C_1 \quad \text{or} \quad C_1 = -\frac{wl^3}{24}$$

Substituting this value of C_1 equation (*ii*),

$$EI\,\frac{dy}{dx} = \frac{wlx^2}{4} - \frac{wx^3}{6} - \frac{wl^3}{24} \qquad \qquad \text{...(iii)}$$

This is the required equation for the *slope* at any section, by which we can get the slope at any point on the beam. We know that maximum slope occurs at A and B. Thus for maximum slope, substituting $x=0$ in equation (*iii*),

$$EI.i_B = -\frac{wl^2}{24}$$

or $\qquad \qquad \qquad i_B = -\dfrac{wl^3}{24EI}$

(Minus sign means that the tangent at A makes an angle with AB in the negative or anticlock- wise direction)

or $\qquad \qquad \qquad i_B = \dfrac{wl^3}{24EI}$

By symmetry, $\qquad i_A = \dfrac{wl^3}{24EI}$

Integrating the equation (*iii*) once again,

$$EI.y = \frac{wlx^3}{12} - \frac{wx^4}{24} - \frac{wl^3x}{24} + C_2 \qquad \qquad \text{...(iv)}$$

where C_2 is the constant of integration. We know when $x=0$, $y=0$. Substituting these values in equation (*iv*) we get $C_2 = 0$.

$\therefore \qquad \qquad EI.y = \dfrac{wlx^3}{12} - \dfrac{wx^4}{24} - \dfrac{wl^3x}{24} \qquad \qquad \text{...(v)}$

This is the required equation for the *deflection* at any section, by which we can get the deflection at any point on the beam. We know that maximum deflection occurs at the mid-point C. Thus for maximum deflection, substituting $x=l/2$ in equation (*v*),

$$EI.y_C = \frac{wl}{12}\left(\frac{l}{2}\right)^3 - \frac{w}{24}\left(\frac{l}{2}\right)^4 - \frac{wl^3}{24}\left(\frac{l}{2}\right)$$

$$= \frac{wl^4}{96} - \frac{wl^4}{384} - \frac{wl^4}{48}$$

$$= -\frac{5wl^4}{384}$$

or

$$y_C = -\frac{5wl^4}{384EI}$$

(Minus sign means that the deflection is downwards)

$$= \frac{5wl^4}{384EI}$$

Note. The above expressions for slope and deflection may also be expressed in terms of total load. Such that $W = wl$.

$$\therefore \qquad i_B = i_A = \frac{wl^3}{24EI} = \frac{Wl^2}{24EI}$$

and

$$y_C = \frac{5wl^4}{384EI} = \frac{5Wl^3}{384EI}$$

Example 7·15. *A steel beam simply supported over a span of 4 metres is carrying a uniformly distributed load of 10 kg/cm. Find the deflection of the beam at the centre, if $E = 2·0 \times 10^6$ kg/cm² and $I = 2,000$ cm⁴.*

Solution.

Given.   Span, $l = 4$ m $= 400$ cm

Load $w = 10$ kg/cm

$$E = 2·0 \times 10^6 \text{ kg/cm}^2$$

$$I = 2,000 \text{ cm}^4$$

Let $y_C = $ Deflection at the centre.

Using the relation,

$$y_C = \frac{5wl^4}{384EI} \text{ with usual notations.}$$

$$= \frac{5 \times 10 \times 400^4}{384 \times 2·0 \times 10^6 \times 2,000} = \textbf{0·89 cm} \text{ Ans.}$$

Example 7·16. *A timber beam of rectangular section has a span of 4·8 metres and is simply supported at its ends. It is required to carry a total load of 4,500 kg uniformly distributed over the whole span.*

Find the maximum values for the breadth b and depth d of the beam, if maximum bending stress is not to exceed 70 kg/cm² and the maximum deflection is limited to 9·5 mm. Take E for timber as $1·05 \times 10^5$ kg/cm².

Solution.

Given.   Span, $l = 4·8$ m $= 480$ cm

Total uniformly distributed load,

$$W = wl = 4,500 \text{ kg}$$

Bending stress, $f = 70$ kg/cm²

Maximum deflection,

$$y_C = 9·5 \text{ mm} = 0·95 \text{ cm}$$

$$E = 1·05 \times 10^5 \text{ kg/cm}^2$$

Let $b = $ Breadth of the beam, and

$d = $ Depth of the beam.

We know that the maximum bending moment in a simply supported, beam carrying a uniformly distributed load, is at the centre, *i.e.*

$$M = \frac{wl^2}{8} = \frac{Wl}{8} = \frac{4,500 \times 4\cdot8}{8} \text{ kg-m } \quad (\because W = wl)$$

$$= 2,700 \text{ kg-m} = 2,70,000 \text{ kg-cm}$$

We know that moment of intertia of a rectangular section,

$$I = \frac{bd^3}{12}$$

Using the relation,

$$\frac{M}{I} = \frac{f}{y} \text{ with usual notations.}$$

$$\frac{2,70,000}{\frac{bd^3}{12}} = \frac{70}{\frac{d}{2}}$$

$$bd^2 = \frac{2,70,000 \times 6}{70} = 23,143 \text{ cm}^3 \qquad \ldots(i)$$

Now using the relation,

$$y_C = \frac{5Wl^3}{384EI} \text{ with usual notations.}$$

$$0\cdot95 = \frac{5 \times 4,500 \times 480^3}{384 \times 1\cdot05 \times 10^5 \times \frac{bd^3}{12}}$$

$$\therefore \qquad bd^3 = \frac{5 \times 4,500 \times 480^3}{384 \times 1\cdot05 \times 10^5 \times 0\cdot95} = 7,79,500 \text{ cm}^4$$

$$\ldots(ii)$$

Dividing equation (ii) by (i),

$$d = \frac{7,79,500}{23,143} \text{ cm} \qquad \left(\because \frac{bd^3}{bd^2} = d \right)$$

$$= 33\cdot68 \text{ cm} \quad \text{Ans.}$$

Substituting this value of d in equation (i),

$$b \times 33\cdot68^2 = 23,143$$

$$\therefore \qquad b = \frac{23,143}{33\cdot68^2} = 20\cdot4 \text{ cm} \quad \text{Ans.}$$

7·14. Simply supported beam with a gradually varying load

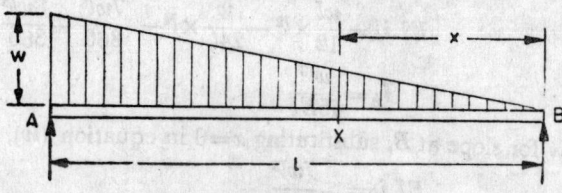

Fig. 7·15

Consider a simply supported beam AB of length l and carrying a gradually varying load from zero at B to w per unit length at A as shown in Fig. 7·15. From the geometry of the figure, we find that the reaction $R_A = \dfrac{wl}{3}$ and $R_B = \dfrac{wl}{6}$.

Now consider a section X, at a distance x from A. We know that the bending moment at this section,

$$M_X = R_B . x - \frac{wx}{l} \times \frac{x}{2} \times \frac{x}{3}$$

$$= \frac{wlx}{6} - \frac{wx^3}{6l} \qquad \left(\because R_B = \frac{wl}{6} \right)$$

$$\therefore \qquad EI \frac{d^2y}{dx^2} = \frac{wlx}{6} - \frac{wx^3}{6l} \qquad \qquad ...(i)$$

Integrating the above equation,

$$EI \frac{dy}{dx} = \frac{wlx^2}{12} - \frac{wx^4}{24l} + C_1 \qquad \qquad ...(ii)$$

where C_1 is the first constant of integration.

Integrating the equation (ii) once again,

$$EI.y = \frac{wlx^3}{36} - \frac{wx^5}{120l} + C_1 x + C_2 \qquad \qquad ...(iii)$$

where C_2 is the second constant of integration. We know that when $x = 0$, $y = 0$. Therefore $C_2 = 0$. We also know that when $x = l$, $y = 0$. Substituting these values in equation (iii),

$$0 = \frac{wl}{36} \times l^3 - \frac{w}{120l} \times l^5 + C_1 l$$

$$= \frac{wl^4}{36} - \frac{wl^4}{120} + C_1 l$$

$$\therefore \qquad C_1 = -\frac{wl^3}{36} + \frac{wl^3}{120} = -\frac{7wl^3}{360}$$

Now substituting this value of C_1 in equation (ii),

$$EI \frac{dy}{dx} = \frac{wlx^2}{12} - \frac{wx^4}{24l} - \frac{7wl^3}{360} \qquad \qquad ...(iv)$$

This is the required equation for *slope* at any section, by which we can get the slope at any section on the beam. A little consideration will show that the maximum slope will be either at the support A or B. Thus, for slope at A, substituting $x = l$ in equation (iv),

$$EI.i_A = \frac{wl}{12} \times l^2 - \frac{w}{24l} \times l^4 - \frac{7wl^3}{360} = \frac{8wl^3}{360} = \frac{wl^3}{45}$$

$$\therefore \qquad i_A = \frac{wl^3}{45EI}$$

Now for slope at B, substituting $x = 0$ in equation (iv),

$$EI.i_B = -\frac{7wl^3}{360}$$

$$\therefore \quad i_B = -\frac{7wl^3}{360EI} \qquad \text{(Minus sign means that the tangent at } B \text{ makes an angle with } AB \text{ in the negative or anti-clockwise direction)}$$

$$= \frac{7wl^3}{360EI} \text{ radians}$$

Now substituting the value of C_1 in equation (iii),

$$EI.y = \frac{wlx^3}{36} - \frac{wx^5}{120l} - \frac{7wl^3x}{360}$$

$$\therefore \quad y = \frac{1}{EI}\left(\frac{wlx^3}{36} - \frac{wx^5}{120l} - \frac{7wl^3x}{360}\right) \qquad \qquad ...(v)$$

This is the required equation for the *deflection* at any section, by which we can get the deflection at any section on the beam. For deflection at the centre of the beam, substituting $x = l/2$ in equation (v),

$$y_C = \frac{1}{EI}\left[\frac{wl}{36}\left(\frac{l}{2}\right)^3 - \frac{w}{120l}\left(\frac{l}{2}\right)^5 - \frac{7wl^3}{360}\left(\frac{l}{2}\right)\right]$$

$$= -\frac{0.00651wl^4}{EI} \qquad \text{Minus sign means that the deflection is downwards)}$$

$$= \frac{0.00651wl^4}{EI}$$

We know that the maximum deflection will occur where slope of the beam is zero. Therefore equating the equation (iv) to zero,

$$\frac{wlx^2}{12} - \frac{wx^4}{24l} - \frac{7wl^3}{360} = 0$$

$$\therefore \quad x = 0.519 \, l$$

Now substituting this value of x in equation (v),

$$y_{max} = \frac{1}{EI}\left[\frac{wl}{36}(0.519l)^3 - \frac{w}{120l}(0.519l)^5 - \frac{7wl^3}{360}(0.519l)\right]$$

$$= \frac{0.00652wl^4}{EI}$$

Example 7·17. *A simply supported beam AB of pan 4 metres is carrying a triangular load rarying from zero at A to 500 kg/m at B. Determine the maximum deflection of the beam. Take rigidity of the beam as 1.2×10^{10} kg-cm^2.* (Oxford University)

Solution.

Given. Span, $\qquad l = 4 \text{ m} = 400 \text{ cm}$

Load, $\qquad\qquad w = 500 \text{ kg/m} = 5 \text{ kg/cm}$

Rigidity of the beam, $EI = 1.2 \times 10^{10}$ kg-cm^2

Let $\qquad\qquad y_{max} = $ Maximum deflection of the beam

Using the relation

$$y_{max} = \frac{0.00652 \; wl^4}{EI} \text{ with usual notations.}$$

$$= \frac{0.00652 \times 5 \times 400^4}{1.2 \times 10^{10}} = 0.0695 \text{ cm} \quad \textbf{Ans.}$$

7·15. Moment area method for slope and deflection

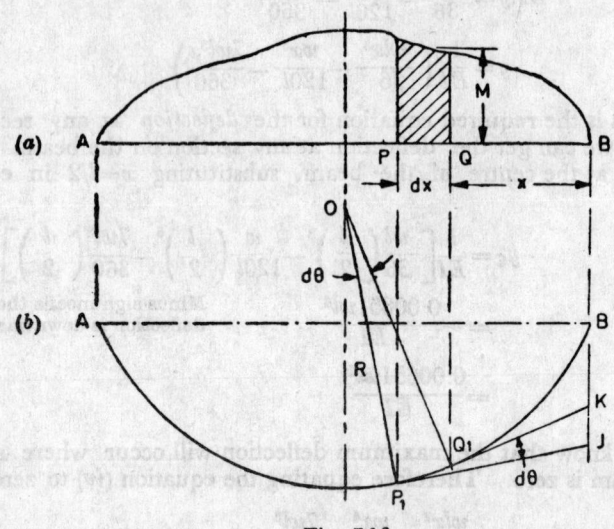

Fig. 7·16

Consider a beam AB carrying some load. As a result of loading, let the beam be subjected to bending moment as shown to Fig. 7·16 (a). Let the beam bent up to AP_1Q_1B as shown in Fig. 7·16 (b).

Now consider an element of small length PQ of the beam at a distance x from B as shown in Fig. 7·16 (b).

Let M = Bending Moment between P and Q,
               dx = Length of the beam PQ,
               R = Radius of the bent up beam,
               $d\theta$ = Angle subtended by the arc P_1Q_1 at the centre, in radians.

From the geometry of the bent up beam, we find that

$$P_1Q_1 = R.d\theta$$

or $dx = R.d\theta$ (Substituting $P_1Q_1 = dx$)

∴ $d\theta = \dfrac{dx}{R}$ ...(i)

We know that for a loaded beam,

$$\frac{M}{I} = \frac{E}{R}$$

$$\therefore \qquad R = \frac{EI}{M}$$

Substituting this value of R in equation (i),

$$d\theta = dx \cdot \frac{M}{EI} = \frac{M.dx}{EI} \qquad \qquad ...(ii)$$

Now the total change of slope from A to B, may be found out by integrating the equation between the limits zero and l,

$$\therefore \qquad i = \int_0^l \frac{M.dx}{EI} = \frac{1}{EI} \int_0^l M.dx$$

$$= \frac{\text{Area of B.M. diagram between } A \text{ and } B}{EI}$$

$$= \frac{A}{EI}$$

Now draw tangents at P_1 and Q_1. Let these two tangents meet at J and K as shown in Fig. 7·16 (b).

From the geometry of the figure, we find that the tangents at P_1 and Q_1 meet at an angle $d\theta$.

and

$$JK = x.d\theta = \frac{x.M.dx}{EI} = \frac{M.x.dx}{EI} \qquad \qquad ...(iii)$$

The total deflection may be found out by integrating the above equation between the limits zero and l.

$$\therefore \qquad y = \int_0^l \frac{M.x.dx}{EI} = \frac{1}{EI} \int_0^l M.x.dx$$

$$= \frac{\text{Area of the B.M. diagram between } A \text{ and}}{\underline{} B \times \text{Distance of } c.g. \text{ of B.M. from the end}}$$
$$= \frac{}{EI}$$

$$= \frac{A\bar{x}}{EI}$$

The above two relations of slope and deflection are also known as Mohr's theorems I and II, which state as follows :

Mohr's theorem I. It states, *"The change of slope between any two points on an elastic curve is equal to the area of B.M. diagram between these points divided by EI."*

Mohr's theorem II. It states, *"The intercept taken on a vertical reference line of tangents at any two points on an elastic curve, is equal to the moment the B.M. diagram between these points about the reference line divided by EI."*

These two Mohr's theorems are also known as moment area theorems. These theorems enable us quicker solutions than the double integration method. In the following pages, we shall discuss the application of these theorems on cantilevers and beams.

(i) Cantilever with a point load at the free end

Consider a cantilever AB of length l and carrying a point load W at the free end as shown in Fig. 7·17 (a).

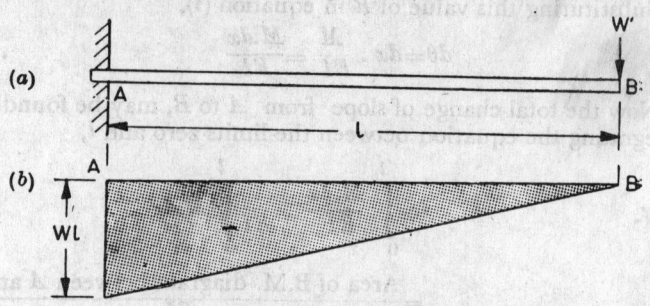

Fig. 7·17

We know that the B.M. will be zero at B, and will increase by a straight line law to Wl at A as shown in Fig. 7·17 (b).

Therefore area of B.M. diagram,

$$A = \frac{1}{2} \, Wl.l = \frac{Wl^2}{2}$$

and distance between the c.g. of B.M. diagram and B,

$$\bar{x} = \frac{2l}{3}$$

$$\therefore \qquad i_B = \frac{A}{EI} = \frac{Wl^2}{2EI} \text{ radians}$$

and

$$y_B = \frac{A.\bar{x}}{EI} = \frac{\dfrac{Wl^2}{2} \times \dfrac{2l}{3}}{EI}$$

$$= \frac{Wl^3}{3EI}$$

(ii) Cantilever with a point load not at the free end

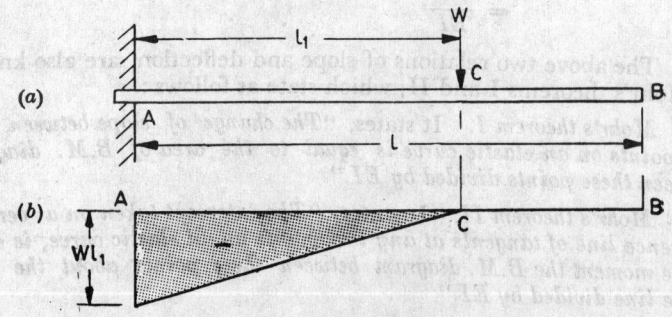

Fig. 7·18

Consider a cantilever AB of length l and carrying a point load W at a distance l_1 from the fixed end as shown in Fig. 7·18 (a).

We know that the B.M. will be zero at B and C, and will increase by a straight line to Wl_1 at A as shown in Fig. 7·18(b).

Therefore area of B.M. diagram,

$$A = \frac{1}{2} \times Wl_1 \times l_1 = \frac{Wl_1^2}{2}$$

and distance between the *c.g.* of the B.M. diagram and B,

$$\bar{x} = \frac{2l_1}{3} + (l - l_1)$$

$$\therefore \quad i_B = \frac{A}{EI} = \frac{Wl_1^2}{2EI} \text{ radians}$$

and

$$y_B = \frac{A\bar{x}}{EI} = \frac{\dfrac{Wl_1^2}{2}\left[\dfrac{2l_1}{3} + (l - l_1)\right]}{EI}$$

$$= \frac{Wl_1^3}{3EI} + \frac{Wl_1^2}{2EI}(l - l_1)$$

(iii) Cantilever with a uniformly distributed load

Consider a cantilever AB of length l, and carrying a uniformly distributed load of w per unit length as shown in Fig. 7·19 (a).

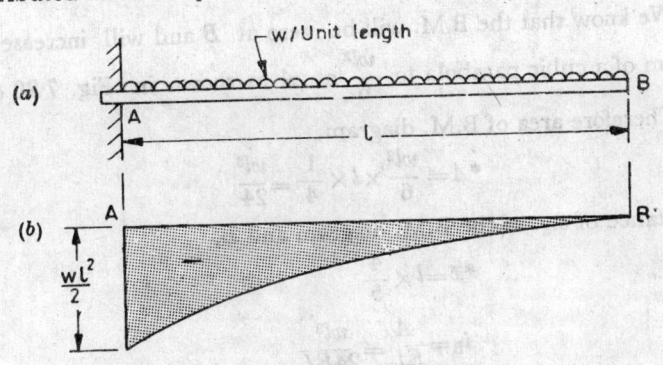

Fig. 7·19

We know that the B.M. will be zero at B, and will increase in the from of a parabola to $\dfrac{wl^2}{2}$ at A, as shown in Fig. 7·19 (b) .

Therefore area of B.M. diagram,

$$A = \frac{wl^2}{2} \times l \times \frac{1}{3} = \frac{wl^3}{6}$$

and distance between the *c.g.* of the B.M. diagram and B,

$$\bar{x} = \frac{3l}{4}$$

$$\therefore \quad i_B = \frac{A}{EI} = \frac{wl^3}{6EI} \text{ radians}$$

and
$$y_B = \frac{A\bar{x}}{EI} = \frac{\frac{wl^3}{6} \times \frac{3l}{4}}{EI} = \frac{wl^4}{8EI}$$

(iv) Cantilever with a gradually varying load

Consider a cantilever AB of length l, and carrying a gradually varying load from zero at B to w per unit length at A as shown in Fig. 7·20 (a).

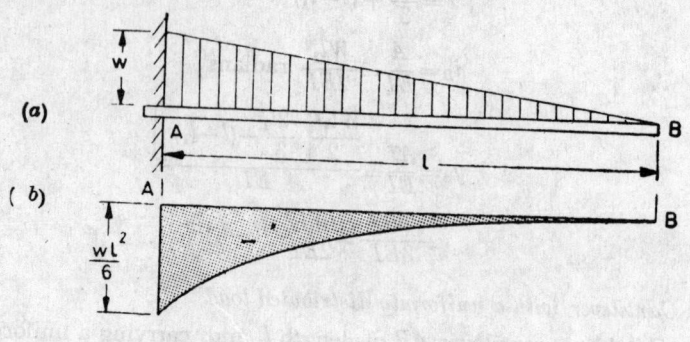

Fig. 7·20

We know that the B.M. will be zero at B and will increase in the form of a cubic parabola to $\frac{wl^2}{6}$ at A as shown in Fig. 7·20 (b).

Therefore area of B.M. diagram,
$$*A = \frac{wl^2}{6} \times l \times \frac{1}{4} = \frac{wl^3}{24}$$

and distance of c.g. of the B.M. diagram and B,
$$*\bar{x} = l \times \frac{4}{5}$$

$$\therefore \qquad i_B = \frac{A}{EI} = \frac{wl^3}{24EI}$$

---

*In general, the area of a parabola [having concave side as shown in Fig. 7·20 (b)],

$$A = l \times h \times \frac{1}{n+1}$$

and distance of c.g. from B,

$$\bar{x} = l \times \frac{n+1}{n+2}$$

where l is the length of the base of parabola, and n is the degree of the parabolic curve. In this case, $n=3$,

$$A = l \times h \times \frac{1}{3+1} = l \times h \times \frac{1}{4}$$

and
$$\bar{x} = l \times \frac{3+1}{3+2} = l \times \frac{4}{5}$$

and
$$y_B = \frac{A\bar{x}}{EI} = \frac{\dfrac{wl^3}{24} \times l \times \dfrac{4}{5}}{EI}$$

$$= \frac{wl^4}{30EI}$$

(v) *Simply supported beam with a central point load*

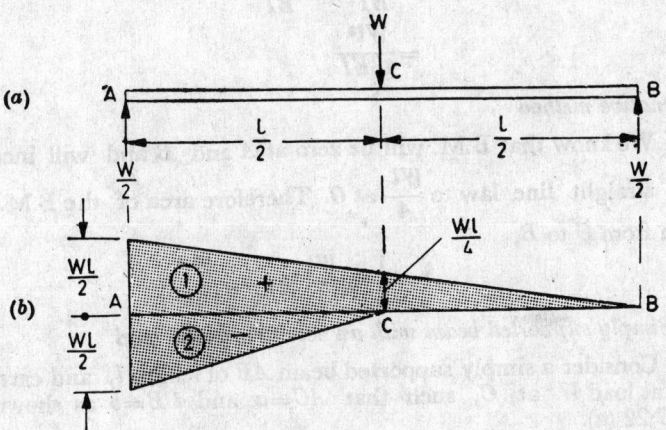

Fig. 7·21

Consider a simply supported beam AB of length l and carrying a point load W at C i.e., the centre of the beam as shown in Fig. 7·21 (a).

We know that the reaction at A,
$$R_A = R_B = \frac{W}{2}$$

B.M. at A, due to reaction R_B,
$$M_1 = + \frac{Wl}{2}$$

Similarly the B.M. at A, due to the load W,
$$M_2 = -W \times \frac{l}{2} = -\frac{Wl}{2}$$

Now draw B.M. diagram with M_1 and M_2 as shown in Fig. 7·21 (b). Such a B.M. diagram is called component B.M. diagram.

∴   Area of B.M. diagram from C to B,
$$*A = \frac{1}{2} \times \frac{Wl}{4} \times \frac{l}{2} = \frac{Wl^2}{16}$$

and distance of *c.g.* of the B.M. diagram on CB from B,
$$\bar{x} = \frac{2}{3} \times \frac{l}{2} = \frac{l}{3}$$

---

*Here the slope of the beam is to be found out from C to B. The deflection of the beam at C, is the intercept made by the tangents at C and B on a vertical line through B.

$$\therefore \qquad i_B = \frac{A}{EI} = \frac{Wl^2}{16EI}$$

*By symmetry, $\quad i_A = \dfrac{Wl^2}{16EI}$

and
$$y_C = \frac{A\bar{x}}{EI} = \frac{\dfrac{Wl^2}{16} \times \dfrac{l}{3}}{EI}$$

$$= \frac{Wl^3}{48EI}$$

Alternative method

We know that B.M. will be zero at A and B and will increase by a straight line law to $\dfrac{Wl}{4}$ at C. Therefore area of the B.M. diagram from C to B,

$$A = \frac{1}{2} \times \frac{Wl}{4} \times \frac{l}{2} - \frac{Wl^2}{16}$$

(vi) Simply supported beam with an eccentric point load

Consider a simply supported beam AB of length l, and carrying a point load W at C, such that $AC = a$ and $CB = b$ as shown in Fig. 7·22 (a).

We know that the reaction $R_A = \dfrac{Wb}{l}$ and reaction $R_B = \dfrac{Wa}{l}$

Therefore the B.M. at A due to reaction R_B,

$$M_1 = +\frac{Wa}{l} \times l = +Wa$$

Similarly B.M. at A due to the load W,
$$M_2 = -Wa.$$

Now draw the component B.M. diagram as shown in Fig. 7·22 (b).

Area of B.M. diagram 1,

$$A_1 = \frac{1}{2} \times Wa \times l = \frac{Wal}{2}$$

Similarly $\qquad A_2 = \dfrac{1}{2} \times Wa \times a = \dfrac{Wa^2}{2}$

---

*It may also be found out by studying the component B.M. diagram from A to C. Therefore area of B.M. diagram from A to C,

$$A = \frac{1}{2}\left(\frac{Wl}{2} + \frac{Wl}{2}\right) \times \frac{l}{2} - \frac{1}{2} \times \frac{Wl}{2} \times \frac{l}{2} = \frac{Wl^2}{16}$$

and
$$A\bar{x} = A_1\bar{x}_1 - A_2\bar{x}_2 = \left(\frac{3Wl^2}{16} \times \frac{2l}{9}\right) - \left(\frac{Wl^3}{8} \times \frac{l}{6}\right)$$

$$= \frac{Wl^3}{24} - \frac{Wl^3}{48} = \frac{Wl^3}{48}$$

From the geometry of the loading, we see that the slope at any section is not known. It is thus obvious, that the slope and deflec-

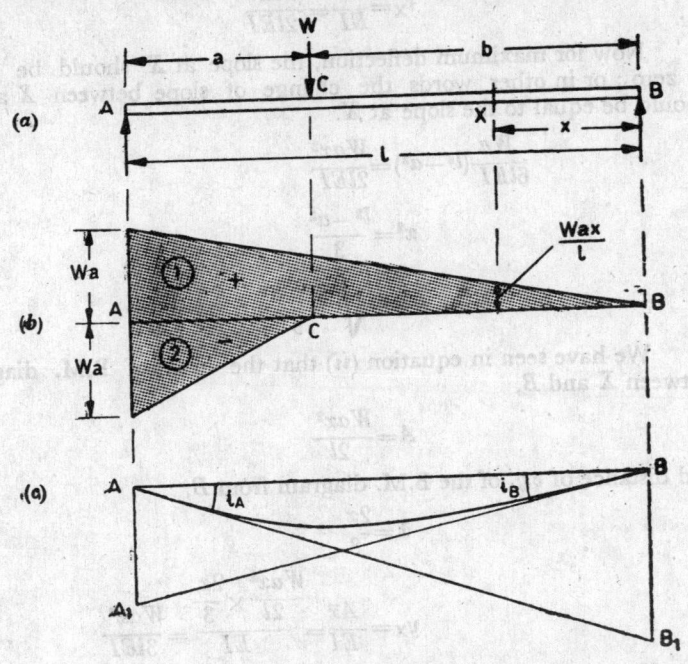

Fig. 7·22

tion can not be found out directly. Now draw vertical lines through A and B. Let AA_1 and BB_1 be equal to intercept of the tangents at A and B as shown in Fig. 7·22 (c).

We see that
$$AA_1 = i_B \times l$$

But
$$AA_1 = \frac{A_1 x_1 - A_2 x_2}{EI}$$
$$= \frac{1}{EI}\left[\left(\frac{Wa}{2} \times \frac{l}{3}\right) - \frac{Wa^2}{2} \times \frac{a}{3}\right]$$
$$= \frac{Wa}{6EI}(l^2 - a^2)$$

$\therefore$
$$i_B = \frac{Wa}{6EIl}(l^2 - a^2)$$

Similarly
$$i_A = \frac{Wb}{6EIl}(l^2 - b^2) \qquad \text{(substituting } b \text{ for } a)$$

Now consider any section X, at a distance x from B. We find that the area of B.M. diagram between X and B,
$$A = \frac{1}{2} \times \frac{Wax}{l} \times x = \frac{Wax^2}{2l} \qquad \dots(ii)$$

$\therefore$ Change of slope between X and B,

$$i_X = \frac{A}{EI} = \frac{Wax^2}{2lEI} \qquad \qquad ...(iii)$$

Now for maximum deflection, the slope at X should be equal to zero ; or in other words the change of slope between X and B should be equal to the slope at X.

$$\therefore \qquad \frac{Wa}{6lEI}(l^2 - a^2) = \frac{Wax^2}{2lEI}$$

or

$$x^2 = \frac{l^2 - a^2}{3}$$

$$\therefore \qquad x = \sqrt{\frac{l^2 - a^2}{3}}$$

We have seen in equation (ii) that the area of B.M. diagram between X and B,

$$A = \frac{Wax^2}{2l}$$

and distance of $c.g.$ of the B.M. diagram from B,

$$\bar{x} = \frac{2x}{3}$$

$$y_X = \frac{A\bar{x}}{EI} = \frac{\dfrac{Wax^2}{2l} \times \dfrac{2x}{3}}{EI} = \frac{Wax^3}{3lEI} \qquad \qquad ...(iv)$$

Now for maximum deflection, subtituting the value of $x = \sqrt{\dfrac{l^2 - a^2}{3}}$ in equation (iv),

$$y_{max} = \frac{Wa}{3lEI}\left(\sqrt{\frac{l^2 - a^2}{3}}\right)^3$$

$$= \frac{Wa}{9\sqrt{3}lEI}(l^2 - a^2)^{\frac{3}{2}}$$

(vii) *Simply supported beam with a uniformly distributed load*

Consider a simply supported beam AB length l, and carrying a uniformly distributed load of w per unit length as shown in Fig 7·23 (a).

We know that the reaction at A,

$$R_A = R_B = \frac{wl}{2}$$

Therefore the B.M. at A due to rection R_B,

$$M_1 = \frac{wl}{2} \times l = \frac{wl^2}{2}$$

Similarly B.M. at A due to load w,

$$M_2 = -wl \times \frac{l}{2} = -\frac{wl^2}{2}$$

Now draw the component B.M. diagram as shown in Fig. 7·23 (b).

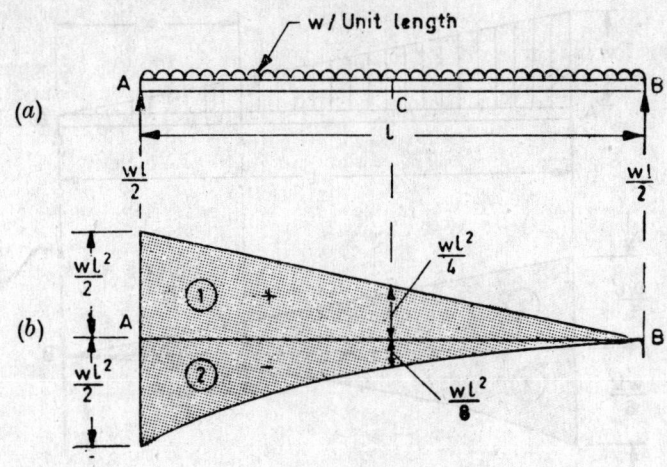

Fig. 7·23

∴ Area of B.M. diagram from C to B,

$$A_1 = +\frac{1}{2} \times \frac{wl^2}{4} \times \frac{l}{2} = \frac{wl^3}{16} \qquad (+\text{ve B.M.})$$

$$A_2 = -\frac{1}{3} \times \frac{wl^2}{8} \times \frac{l}{2} = \frac{wl^3}{48} \qquad (-\text{ve B.M.})$$

∴ Net area of B.M. diagram from C to B,

$$A = \frac{wl^3}{16} - \frac{wl^3}{48} = \frac{wl^3}{24}$$

and distance of *c.g.* of the B.M. diagram on CB from B,

$$\overline{x}_1 = \frac{2}{3} \times \frac{l}{2} = \frac{l}{3} \qquad (\text{for } +\text{ve B.M.})$$

Similarly $\qquad \overline{x}_2 = \frac{3}{4} \times \frac{l}{2} = \frac{3l}{8} \qquad (\text{for } -\text{ve B.M.})$

∴ $\qquad i_B = \frac{A}{EI} = \frac{wl^3}{24EI}$

By symmetry $\quad i_A = \frac{wl^3}{24EI}$

and $\qquad y_C = \frac{A\overline{x}}{EI} = \frac{A_1\overline{x}_1 + A_2\overline{x}_2}{EI} = \frac{\left(\dfrac{wl^3}{16} \times \dfrac{l}{3}\right) + \left(\dfrac{wl^3}{48} \times \dfrac{3l}{8}\right)}{EI}$

$$= \frac{5wl^4}{384EI} \qquad\qquad (\text{As before})$$

(viii) Simply supported beam with a gradually varying load

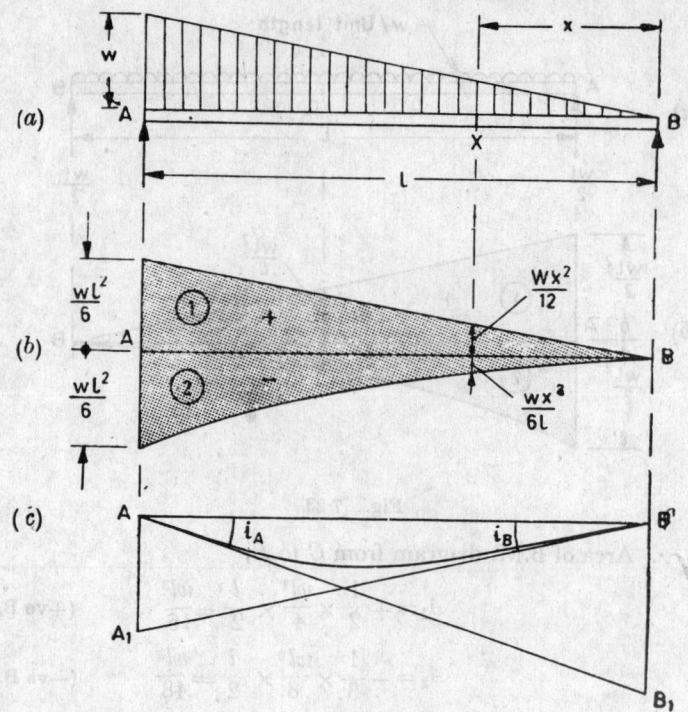

Fig. 7·24

Consider a simply supported beam AB of length l, and carrying a gradually varying load from zero at B to w per unit length at A as shown in Fig. 7·24.

We know that the reaction at A,

$$R_A = \frac{wl}{3}$$

and reaction $R_B = \frac{wl}{6}$

Therefore B.M. diagram at A due to reaction R_B,

$$M_1 = \frac{wl}{6} \times l = +\frac{wl^2}{6}$$

Similarly B.M. at A due to the load,

$$M_2 = -\frac{w \times l}{2} \times \frac{l}{3} = -\frac{wl^2}{6}$$

Now draw the component B.M. diagram as shown in Fig. 7·24 (b).

Area of B.M. diagram 1,

$$A_1 = \frac{1}{2} \times \frac{wl^2}{6} \times l = \frac{wl^3}{12}$$

Similarly $\qquad A_2 = \dfrac{1}{4} \times \dfrac{wl^2}{6} \times l = \dfrac{wl^3}{24}$

From the geometry of the loading, we see that the slope at any section is not known. It is thus obvious, that the slope and deflection cannot be found out directly. Now draw vertical lines through A and B. Let AA_1 and BB_1 be equal to the intercepts of the tangents at A and B as shown in Fig. 7·24 (c).

We see that $AA_1 = i_B \times l$ and $BB_1 = i_A \times l$

But $\qquad AA_1 = \dfrac{A_1 \bar{x}_1 - A_2 \bar{t}_2}{EI} = \dfrac{1}{EI}\left[\left(\dfrac{wl^3}{12} \times \dfrac{l}{3}\right) - \left(\dfrac{wl^3}{24} \times \dfrac{l}{5}\right)\right]$

$\qquad\qquad = \dfrac{7wl^4}{360EI}$

$\therefore \qquad i_B = \dfrac{7wl^4}{360EI}$ $\qquad\qquad$ (As before)

Similarly $\qquad BB_1 = \dfrac{A_1 x_1 - A_2 x_2}{EI} = \dfrac{1}{EI}\left[\left(\dfrac{wl^3}{12} \times \dfrac{2l}{3}\right) - \left(\dfrac{wl^3}{24} \times \dfrac{4l}{5}\right)\right]$

$\qquad\qquad = \dfrac{wl^4}{45EI}$

$\therefore \qquad i_A = \dfrac{wl^3}{45EI}$ $\qquad\qquad$ (As before)

Now consider any section X, at a distance x from B. We find that the area of B.M. diagram between X and B,

$$A = \left(\dfrac{1}{2} \times \dfrac{wlx}{6} \times x\right) - \left(\dfrac{1}{4} \times \dfrac{wx^3}{6l} \times x\right)$$

$$= \dfrac{wlx^2}{12} - \dfrac{wx^4}{24l}$$

$\therefore$ Slope at x,

$$i_X = \dfrac{A}{EI} = \dfrac{1}{EI}\left(\dfrac{wlx^2}{12} - \dfrac{wx^4}{24l}\right)$$

Now for maximum deflection, the slope at X should be equal to zero, or in other words the change of slope between B and x should be equal to the slope at X.

$\therefore \qquad \dfrac{7wl^4}{360EI} = \dfrac{1}{EI}\left(\dfrac{wlx^2}{12} - \dfrac{wx^4}{24l}\right)$

$\qquad\qquad 7l^4 = 30\, lx^2 - \dfrac{15\, x^4}{l}$

or $\qquad\qquad x = 0·519\, l$

We know that the deflection of the beam at X (considering the portion XB of the beam),

$$y_X = \dfrac{A_1 x_1 - A_2 x_2}{EI} = \dfrac{1}{EI}\left[\left(\dfrac{wlx^2}{12} \times \dfrac{2x}{3}\right) - \left(\dfrac{wx^4}{24l} \times \dfrac{4x}{5}\right)\right]$$

$$= \dfrac{1}{EI}\left[\dfrac{wlx^3}{18} - \dfrac{wx^5}{30l}\right]$$

Now for the deflection at the centre, substituting $x = \dfrac{l}{2}$ in the above equation,

$$y_C = \frac{0.00651 wl^4}{EI}$$

(As before)

For maximum deflection, substituting the value of $x = 0.519\, l$ and the above equation,

$$y_{max} = \frac{0.00652 wl^4}{EI}$$

(As before)

Example 7·18. *A cantilever of length 2a is carrying a load of W at the free end, and another load of W at its centre. Determine from first principles, the slope and deflection of the cantilever at the free end.*

*Solution.

The cantilever AB of length $2a$ and carrying point loads of W each at C and B is shown in Fig. 7·25 (a).

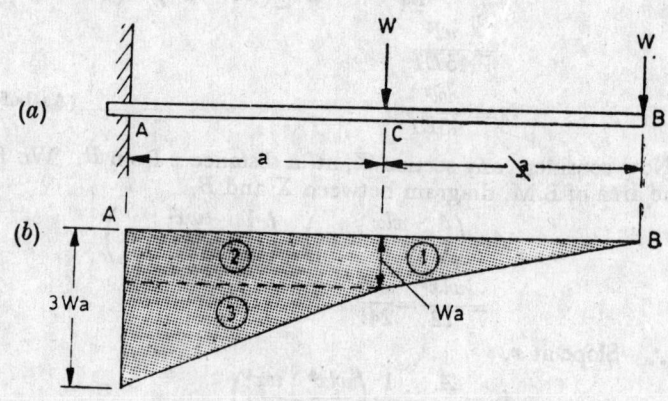

Fig. 7·25

Slope at the free end

Let $\qquad i_B =$ Slope at the free end.

We know that the B.M. at B,

$$M_B = 0$$
$$M_C = -Wa$$
$$M_A = -W \times 2a - Wa = -3Wa$$

Now plot the B.M. diagram as shown in Fig. 7·25(b).

Area of B.M. diagram 1,

$$A_1 = \frac{1}{2} \times Wa \times a = \frac{Wa^2}{2}$$

---

*We have already solved the problem by the double integration method

Similarly $A_2 = Wa \times a = Wa^2$

nd $A_3 = \dfrac{1}{2} \times 2Wa \times a = Wa^2$

∴   Total area of B.M. diagram,

$$A = \frac{Wa^2}{2} + Wa^2 + Wa^2 = \frac{5Wa^2}{2}$$

∴ $i_B = \dfrac{A}{EI} = \dfrac{5Wa^2}{2EI}$ **Ans.**

Deflection at the free end

Let y_B = Deflection at the free end.

Now total moment of the B.M. diagram about B,

$$A\bar{x} = A_1\bar{x}_1 + A_2\bar{x}_2 + A_3\bar{x}_3$$
$$= \frac{Wa^2}{2} \times \frac{2a}{3} + Wa^2 \times \frac{3a}{2} + Wa^2 \times \frac{5a}{3} = \frac{7Wa^3}{2}$$

∴ $y_B = \dfrac{A\bar{x}}{EI} \times \dfrac{7Wa^3}{2EI}$ **Ans.**

Example 7·19. *A beam AB of length l is loaded with a uni-ormly distributed load as shown in Fig. 7·26.*

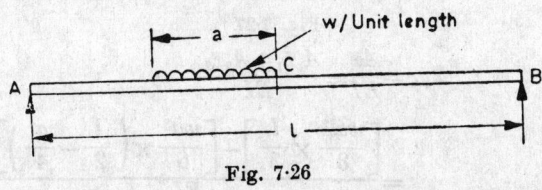

Fig. 7·26

Determine, from first principles, the central deflection of the bean

Solution.

For the sake of convenience, let us assume another load of w per unit length to act for a length of a in CB as shown in Fig. 7·27 (a).

We know that the reaction $R_A = R_B = wa$. Therefore B.M. due to reaction R_B, $M_1 = wal$

Similarly B.M. due to load,

$$M_2 = -w \times 2a \times \frac{l}{2} = -wal$$

Now draw the component B.M. diagram as shown in Fig. 7·27 (b).

∴   Area of B.M. diagram from C to B,

$$A_1 = \frac{1}{2} \times \frac{wal}{2} \times \frac{l}{2} = \frac{wal^2}{8} \qquad (+ve \text{ B.M.})$$

and

$$A_2 = -\frac{1}{3} \times \frac{wa^2}{2} \times a = -\frac{wa^3}{3} \qquad (-ve \text{ B.M.})$$

and distance of c.g. of the B.M. diagram on CB from B,

$$\bar{x}_1 = \frac{2}{3} \times \frac{l}{2} = \frac{l}{3}$$

Similarly $\quad \bar{x}_2 = \frac{l}{2} - a + \frac{3a}{4} = \left(\frac{l}{2} - \frac{a}{4} \right)$

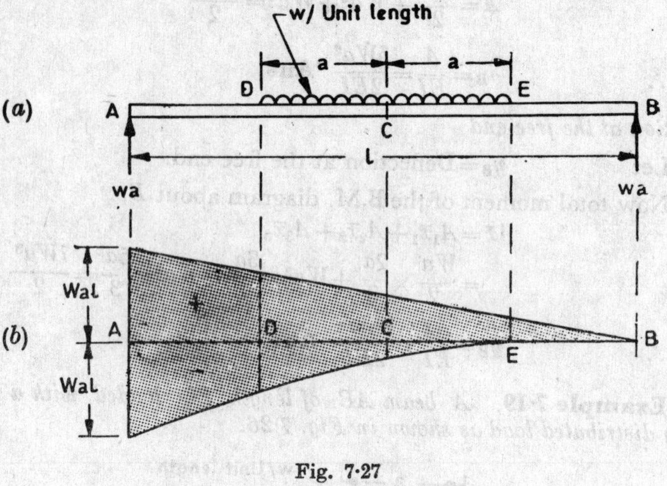

Fig. 7·27

$$\therefore \quad 2y_C = \frac{A\bar{x}}{EI} = \frac{A_1\bar{x}_1 + A_2\bar{x}_2}{EI}$$

$$= \frac{\left[\dfrac{wal^2}{8} \times \dfrac{l}{3} \right] - \left[\dfrac{wa^3}{6} \times \left(\dfrac{l}{2} - \dfrac{a}{4} \right) \right]}{EI}$$

$$= \frac{\dfrac{wal^3}{24} - \dfrac{wa^3}{24}(2l-a)}{EI}$$

$$= \frac{wa}{24EI}(l^3 - 2la^2 + a^3)$$

or $\quad y_C = \dfrac{wa}{48EI}(l - 2la^2 + a^2)$ **Ans.**

7·16. Macaulay's method*

We have seen in the previous articles and examples, that the problems of deflections in beams are bit tedious and laborious, specially when the beam is carrying some point loads. Mr. W.H. Macaulay devised a method, of continuous expression, for bending moment and integrating it in such a way, that the constants of integration are valid for all sections of the beam ; even though the law of bending moment varies from section to section. This method will be best understood by the following examples.

---

*The method was originally proposed by Mr. A. Clebsch, which was further developed by Mr. W.H. Macaulay.

(i) *Simply supported beam with a central point load.*

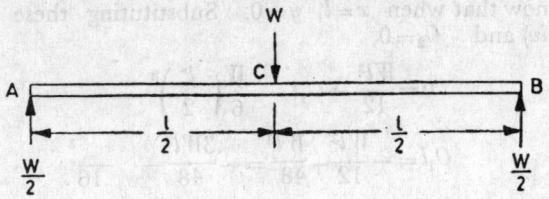

Fig. 7·28

Consider a simply supported beam AB of length l and carrying a point load W at the centre of the beam C as shown in Fig. 7·28.

Take A as the origin. The bending moment at any point in section AC at a distance x from A,

$$M_X = \frac{W}{2} x$$

The bending moment at any section CB, at a distance x from A,

$$M_X = \frac{W}{2} x - W \left(x - \frac{l}{2} \right) \qquad \text{...(i)}$$

Thus we can express the bending moment, for all the sections of the beam, in a single equation, *i.e.*

$$M_X = \frac{W}{2} x \ \bigg| \ -W \left(x - \frac{l}{2} \right)$$

For any point in section AC, stop at the dotted line, and for any point in section CB add the expression beyond the dotted line also.

Re-writing the above equation,

$$EI \frac{d^2y}{dx^2} = \frac{W}{2} x \ \bigg| \ -W \left(x - \frac{l}{2} \right) \qquad \text{...(ii)}$$

Integrating the above equation,

$$EI \frac{dy}{dx} = \frac{Wx^2}{4} + C_1 \ \bigg| \ -\frac{W}{2} \left(x - \frac{l}{2} \right)^2 \qquad \text{...(iii)}$$

It may be noted, that the integration of $\left(x - \frac{l}{2} \right)$ has been made as a whole, and not for individual terms for the expression. This is only, due to this simple integration, that the Macaulay's method is more effective. This type of integration is also justified as the constant of integration C_1 is not only valid for the section AC, but also for section CB.

Integrating the equation (iii) once again,

$$EI.y = \frac{Wx^3}{12} + C_1 x + C_2 \ \bigg| \ -\frac{W}{6} \left(x - \frac{l}{2} \right)^3 \qquad \text{...(iv)}$$

It may be noted that the integration of $\left(x - \frac{l}{2} \right)^2$ has again made as a whole, and not for individual terms. We know that when

$x=0$, $y=0$. Substituting these values in equation (iv) we find $C_2=0$. We also know that when $x=l$, $y=0$. Substituting these values in equation (iv) and $C_2=0$,

$$0=\frac{Wl^3}{12}+C_1 l-\frac{W}{6}\left(\frac{l}{2}\right)^3$$

or
$$C_1 l=-\frac{Wl^3}{12}+\frac{Wl^3}{48}=-\frac{3Wl^3}{48}-\frac{3}{16},$$

or
$$C_1=-\frac{Wl^2}{16}$$

Substituting the value of C_1 in equation (iii),

$$EI\frac{dy}{dx}=\frac{Wx^2}{4}-\frac{Wl^2}{16}\left|-\frac{W}{2}\left(x-\frac{l}{2}\right)^2\right. \qquad \ldots(v)$$

This is the required equation for *slope* at any section. We know that maximum slope occurs at A and B. Thus for maximum slope, substituting $x=0$ in equation (v) upto the dotted line only,

$$EI.i_A=-\frac{Wl^2}{16}$$

$$i_A=-\frac{Wl^2}{16EI} \qquad \text{(As before)}$$

By symmetry, $i_B=-\frac{Wl^2}{16EI}$ (As before)

Substituting the value of C_1 again in equation (iv) and $C_2=0$,

$$EI.y=\frac{Wx^3}{12}-\frac{Wl^2 x}{16}\left|-\frac{W}{6}\left(x-\frac{l}{2}\right)^3\right. \qquad \ldots(vi)$$

This is required equation for *deflection* at any section. We know that maximum deflection occurs at C. Thus for maximum deflection, substituting $x=\frac{l}{2}$ in equation (vi) for the portion AC only (remembering that C lies in AC),

$$EI.y_C=\frac{W}{12}\left(\frac{l}{2}\right)^3-\frac{Wl^2}{16}\left(\frac{l}{2}\right)=-\frac{Wl^3}{48}$$

or
$$y_C=\frac{Wl^3}{48EI} \qquad \text{(As before)}$$

(ii) Simply supported beam with an eccentric point load.

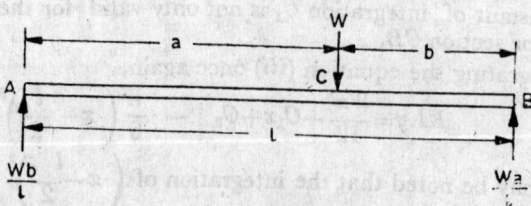

Fig. 7·29

Consider a simply supported beam AB of length l and carrying an eccentric point load W at C such that $AC=a$ and $CB=b$ as shown in Fig. 7·29.

Take A as the origin. The bending moment at any point in section AC at a distance x from A is,

$$M_x = \frac{Wb}{l}.x \qquad \qquad ...(i)$$

The bending moment at any point in section CB at a distance x from A,

$$M_x = \frac{Wbx}{l} - W(x-a)$$

or

$$EI\frac{d^2y}{dx^2} = \frac{Wbx}{l} - W(x-a) \qquad \qquad ...(ii)$$

Integrating the above equation,

$$EI\frac{dy}{dx} = \frac{Wbx^2}{2l} + C_1 \left| -\frac{W(x-a)^2}{2} \right. \qquad \qquad ...(iii)$$

Integrating the above equation once again,

$$EI.y = \frac{Wbx^3}{6l} + C_1x + C_2 \left| -\frac{W(x-a)^3}{6} \right. \qquad \qquad ...(iv)$$

We know that when $x=0$, $y=0$. Substituting these values in equation (iv) upto the dotted line only, we get $C_2=0$.

We also know that when $x=l$, $y=0$. Substituting these values again in equation (iv) and $C_2=0$,

$$EI.y = \frac{Wb}{6l}l^3 + C_1 l \left| -\frac{W(l-a)^3}{6} \right.$$

$$0 = \frac{Wbl^2}{6} + C_1 l \left| -\frac{Wb^3}{6} \right. \qquad \qquad [\because \ (l-a)=b]$$

or

$$C_1 l = \frac{Wb^3}{6} - \frac{Wbl^2}{6} = \frac{Wb}{6}(b^2-l^2)$$

or

$$C_1 = -\frac{Wb}{6l}(l^2-b^2)$$

Substituting this value of C_1 in equation (iii),

$$EI\frac{dy}{dx} = \frac{Wbx^2}{2l} - \frac{Wb}{6l}(l^2-b^2) \left| -\frac{W(x-a)^2}{2} \right. \qquad \qquad ...(v)$$

This is the required equation for *slope at* any point. We know that slope is maximum at A or B. Substituting $x=0$ upto dotted line only.

$$EI.i_A = -\frac{Wb}{6l}(l^2-b^2)$$

or $i_A = \dfrac{Wb}{6EIl}\,(l^2 - b^2)$ (As before)

Similarly $i_B = \dfrac{Wa}{6EIl}\,(l^2 - a^2)$ (Substituting a for b)

Substituting the value of C_1 again in equation (iv) and $C_2 = 0$,

$$EI.y = \frac{Wbx^3}{6l} - \frac{Wbx}{6l}(l^2 - b^2) - \frac{W(x-a)^2}{6}$$

This is the required equation for *deflection* at any point. For deflection in AC, consider the above equation upto dotted line only.

$$EI.y = \frac{Wbx^3}{6l} - \frac{Wbx}{6l}\,(l^2 - b^2)$$

$$= \frac{Wbx}{6l}\,(x^2 - l^2 + b^2)$$

$$= -\frac{Wbx}{6l}\,(l^2 - b^2 - x^2)$$

$\therefore$ $y = \dfrac{Wbx}{6lEI}\,(l^2 - b^2 - x^2)$ (As before)

Example 7·20. *A horizontal girder of steel having uniform section is 14 metres long and is simply supported at its ends. It carries concentrated loads of 12 tonnes and 8 tonnes at two points 3 metres and 4·5 metres from the two ends respectively. I for the section of the girder is 16×10^4 cm⁴ and $E = 2·1 \times 10^6$ kg/cm². Calculate the deflection of the girder at points under the two loads.*

Solution.

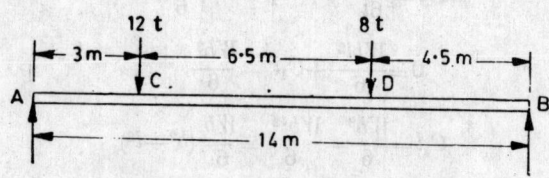

Fig. 7·30

Given.   Span $l = 14$ m

$I = 16 \times 10^4$ cm⁴

$E = 2·1 \times 10^6$ kg/cm²

Taking moments about A,

$$R_B \times 14 = 12 \times 3 + 8 \times 9·5 = 112$$

$\therefore$ $R_B = \dfrac{112}{14} = 8$ t

and $R_A = (12 + 8) - 8 = 12$ t

The B.M. at any section X at a distant x from A,

$$EI\frac{d^2y}{dx^2}=12x\Big|-12(x-3)\Big|-8(x-9\cdot5)$$

Integrating the above equation,

$$EI\frac{dy}{dx}=\frac{12x^2}{2}+C_1-\frac{12(x-3)^2}{2}\Big|-\frac{8(x-9\cdot5)^2}{2}$$
$$=6x^2+C_1-6(x-3)^2-4(x-9\cdot5)^2 \qquad \dots(i)$$

Integrating the above equation once again,

$$EI.y=\frac{6x^3}{3}+C_1x+C_2-\frac{6(x-3)^3}{3}\Big|-\frac{4(x-9\cdot5)^3}{3}$$
$$=2x^3+C_1x+C_2\Big|-2(x-3)^3 \quad \frac{}{3}(x-9\cdot5)^3 \quad\dots(ii)$$

We know that when $x=0$, $y=0$. Therefore $C_2=0$ and when $x=14$, $y=0$.

$$\therefore \qquad 0=2\times14^3+C_1\times14-2(14-3)^2-\frac{4}{3}(14-9\cdot5)^3$$

or $$C_1=-193\cdot18$$

Substituting the values of $C_1=-193\cdot18$ and $C_2=0$ in equation (ii),

$$EI.y=2x^3-193\cdot18x\Big|-2(x-3)^3\Big|-\frac{4}{3}(x-9\cdot5)^3 \quad\dots(iii)$$

For deflection under the load of 12 tonnes, substituting $x=3$ m in equation (iii) upto first dotted line only,

i.e., $$EI.y_C=2(3)^3-193\cdot18\times3=-525\cdot54$$

$$\therefore \qquad y_C=-\frac{525\cdot54\times10^6}{2\cdot1\times10^6\times16\times10^4}\text{ cm} \qquad (1\text{ m}^3=10^6\text{ cm}^3)$$

$$=-0\cdot0015\text{ cm}=\textbf{0·0015 cm (downwards) Ans.}$$

Similarly, for deflection under the load of 8 tonnes, substitute $x=9\cdot5$ m in equation (iii) up to second dotted line only.

i.e., $$EI.y_D=2(9\cdot5)^3-193\cdot18\times9\cdot5-2(9\cdot5-3)^3=-669\cdot71$$

$$\therefore \qquad y_D=-\frac{669\cdot71\times10^6}{2\cdot1\times10^6\times16\times10^4}\text{ cm} \qquad (1\text{ m}^3=10^6\text{ cm}^3)$$

$$=-0\cdot00199=\textbf{0·00199 cm (downwards) Ans.}$$

Example 7·21. *Calculate the net deflection at P and Q for the beam as shown in Fig. 7·31.*

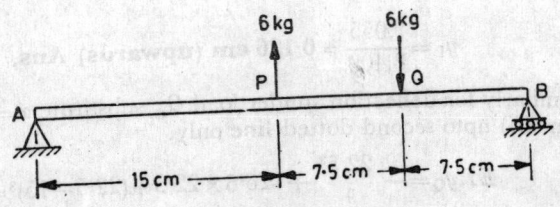

Fig. 7·31

Take the value of EI as 8,400 kg-cm². (U.P.S.C., Engg. Services, 1971)

Solution.

Given. $EI = 8,400$ kg-cm^2

Taking moments about A,

$$R_B \times 30 + 6 \times 15 = 6 \times 22 \cdot 5 = 135$$

$$R_B = \frac{135 - 90}{30} = 1 \cdot 5 \text{ kg (downwards)}$$

$\therefore$ $R_A = 1 \cdot 5$ kg (upwards)

The B.M. at any section, distant x from A,

$$EI \frac{d^2y}{dx^2} = -1 \cdot 5x \ \Big| \ + 6(x-15) \ \Big| \ -6(x-22 \cdot 5)$$

Integrating the above equation,

$$EI \frac{dy}{dx} = -\frac{1 \cdot 5x^2}{2} + C_1 \ \Big| \ + \frac{6(x-15)^2}{2} \ \Big| \ - \frac{6(x-22 \cdot 5)^2}{2}$$

$$= -\frac{1 \cdot 5x^2}{2} + C_1 \ \Big| \ + 3(x-15)^2 \ \Big| \ - 3(x-22 \cdot 5)^2 \quad ...(i)$$

Integrating the above equation once again,

$$EI.y = -\frac{1 \cdot 5x^3}{2 \times 3} + C_1 x + C_2 \ \Big| \ + \frac{3(x-15)^3}{3} \ \Big| \ - \frac{3(x-22 \cdot 5)^3}{3}$$

$$= -\frac{x^3}{4} + C_1 x + C_2 \ \Big| \ + (x-15)^3 \ \Big| \ - (x-22 \cdot 5)^3 \quad ...(ii)$$

We know that when $x = 0$, $y = 0$. Therefore $C_2 = 0$ and when $x = 30$, $y = 0$

or $0 = -\frac{30^3}{4} + C_1 \times 30 \ \Big| \ + (30-15)^3 \ \Big| \ - (30-22 \cdot 5)^3$

$\therefore$ $C_1 = 126 \cdot 6$

Substituting the value of $C_1 = 126 \cdot 6$ and $C_2 = 0$ in equation (ii),

$$EI.y = -\frac{x^3}{4} + 126 \cdot 6x \ \Big| \ + (x-15)^2 \ \Big| \ - (x-22 \cdot 5)^3 \quad ...(iii)$$

For deflection under P, substitute $x = 15$ cm in equation (iii) upto the first dotted line only.

$$EI.y_P = -\frac{15^3}{4} + 126 \cdot 6 \times 15 = 1,055$$

$$y_P = \frac{1,055}{8,400} = 0 \cdot 126 \text{ cm (upwards) Ans.}$$

Similarly for deflection under load Q, substitute $x = 22 \cdot 5$ cm in equation (iii) upto second dotted line only.

$$EI.y_Q = -\frac{22 \cdot 5^3}{4} + 126 \cdot 6 \times 22 \cdot 5 + (22 \cdot 5 - 15)^3 = 422$$

$\therefore$ $y_Q = \frac{422}{8,400} = 0 \cdot 0502 \text{ cm (upwards) Ans.}$

7·17. Conjugate beam method

The conjugate beam method, which was first proposed by Prof. H.F.B. Mueller-Breslau in 1885, is a slightly modified form of moment area method. This method may be conveniently used for finding out the slope and deflection of cantilevers and beams. It is specially useful for cantilevers and beams with varying flexural rigidities.

In this method an imaginary beam, of length equal to, that of original beam and width equal to $1/EI$ and loaded with bending moment diagram, known as *conjugate beam*, is drawn. The slope and deflection is then found out by the following two statements :

1. *The shear force at any section of the conjugate beam is equal to the slope of the elastic curve at the corresponding section of the actual beam.*

2. *The bending moment at any section of the conjugate beam is equal to the deflection of the elastic curve at the corresponding section of the actual beam.*

Note : The following relations exist between the actual beam and the conjugate beam :

S. No.	Actual beam	Conjugate beam	Remarks
1.	Fixed end	Free end	Slope and deflection at fixed end of the actual beam is zero. S.F. and B.M. at the free end of the conjugate beam is also zero.
2.	Free end	Fixed end	Slope and deflection at the free end of the actual beam exist. S.F. and B.M. at the fixed end of the conjugate beam also exist.
3.	Simply supported or roller supported end	Simply supported end	Slope at the free end of the actual beam exists. But deflection is zero. S.F. at the simply supported end of the conjugate beam exists. But the B.M. is zero.

(i) Cantilever with a point load at the free end

Consider a cantilever AB of length l, and carrying a point load W at the free end B as shown in Fig. 7·32(a). We know that the B.M. will be zero at B, and will increase by a straight line law to $W.l$ at A as shown in Fig. 7·32 (b). The shear force at B on the conjugate beam AB (having free end at A and fixed at B),

$$F_B = \frac{1}{2} \times Wl \times l \times \frac{1}{EI} = \frac{Wl^2}{2EI}$$

∴ Slope at B,

$$i_B = F_B = \frac{Wl^2}{2EI} \text{ radians} \qquad \text{(As before)}$$

The bending moment at B on the conjugate beam AB,

$$M_B = \frac{Wl^2}{2EI} \times \frac{2l}{3} = \frac{Wl^3}{3EI}$$

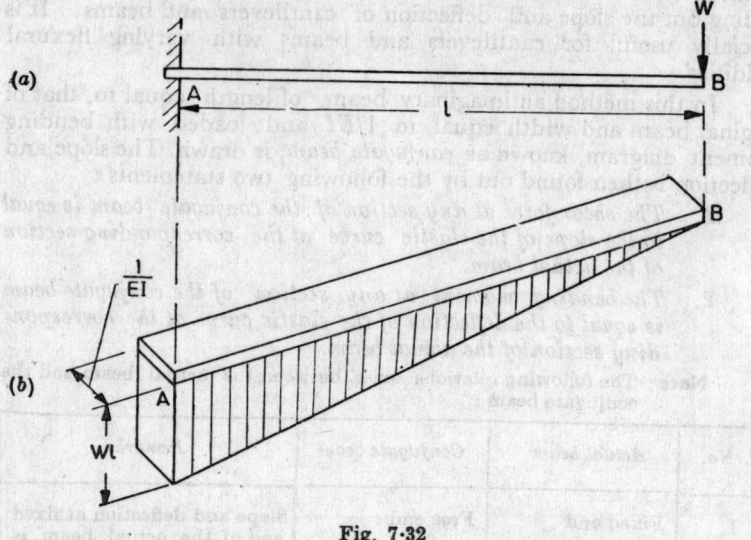

Fig. 7·32

$\therefore$ Deflection at B,

$$y_B = M_B = \frac{Wl^3}{3EI}$$

(As before)

(ii) Cantilever with a uniformly distributed load

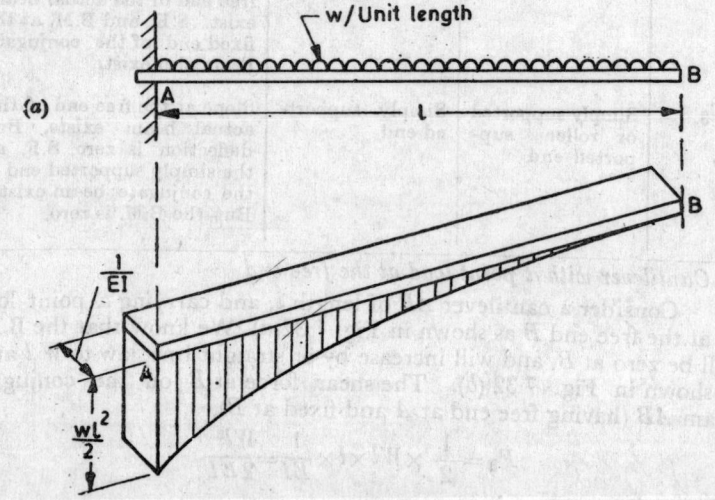

Fig. 7·33

Consider a cantilever AB of length l, and carrying a uniformly distributed load of w per unit length as shown in Fig. 7·33 (a). We know that the B.M. will be zero at B and will increase in the form of parabola to $\dfrac{wl^2}{2}$ at A as shown in Fig. 7.33 (b).

The shear force at B on the conjugate beam AB (having free end at A and fixed at B),

$$F_B = \frac{1}{3} \times \frac{wl^2}{2} \times l \times \frac{1}{EI} = \frac{wl^3}{6EI}$$

∴ Slope at B,

$$i_B = F_B = \frac{wl^3}{6EI} \qquad \text{(As before)}$$

The bending moment at B on the conjugate beam AB,

$$M_B = \frac{wl^3}{6EI} \times \frac{3l}{4} = \frac{wl^4}{8EI}$$

∴ Deflection at B,

$$y_B = M_B = \frac{wl^4}{8EI} \qquad \text{(As before)}$$

(iii) Cantilever with a gradually varying load

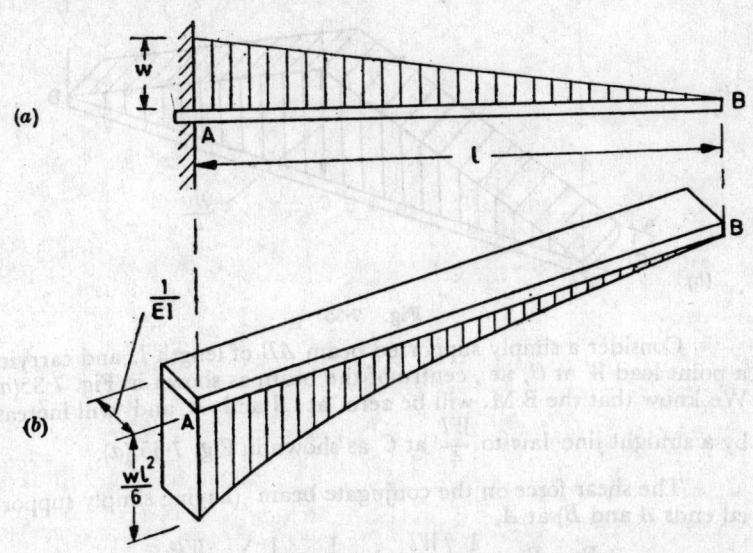

Fig. 7.34

Consider a cantilever AB of length l and carrying a gradually varying load from zero at B to w per unit length at A, as shown in Fig. 7.34 (a). We know that the B.M. will be zero at B and will increase in the form of cubic parabola to $\dfrac{wl^2}{6}$ at A as shown in Fig. 7.34 (b).

The shear force at B on the conjugate beam AB (having free end at A and fixed at B),

$$F_B = \frac{1}{4} \times \frac{wl^2}{6} \times l \times \frac{1}{EI} = \frac{wl^3}{24EI}$$

$\therefore$ Slope at B,

$$i_B = F_B = \frac{wl^3}{24EI} \qquad \text{(As before)}$$

The bending moment at B on the conjugate beam AB,

$$M_B = \frac{wl^3}{24EI} \times \frac{4l}{5} = \frac{wl^4}{30EI}$$

$\therefore$ Deflection at B,

$$y_B = M_B = \frac{wl^4}{30EI} \qquad \text{(As before)}$$

(iv) Simply supported beam with a central point load

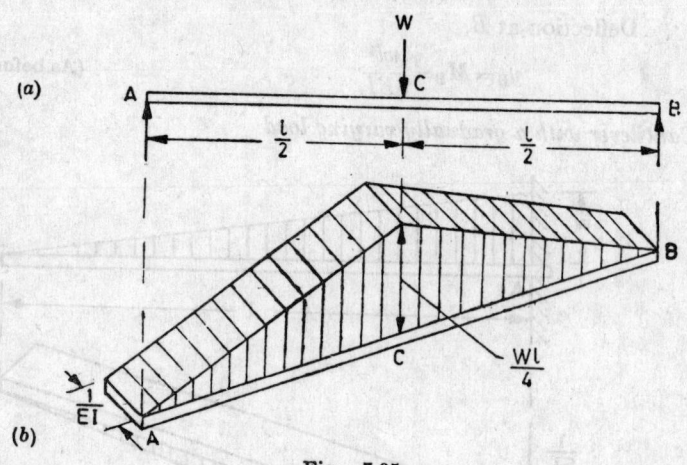

Fig. 7·35

Consider a simply supported beam AB of length l, and carrying a point load W at C, i.e., centre of the beam as shown in Fig. 7·35(a). We know that the B.M. will be zero at A and B, and will increase by a straight line law to $\frac{Wl}{4}$ at C as shown in Fig. 7·35 (a).

The shear force on the conjugate beam (having simply supported ends A and B) at A,

$$F_A = R_A = \frac{1}{2}\left(\frac{Wl}{4} \times l \times \frac{1}{2} \times \frac{1}{EI}\right) = \frac{Wl^2}{16EI}$$

$\therefore$ Slope at A,

$$i_A = F_A = \frac{Wl^2}{16EI} \qquad \text{(As before)}$$

Similarly,

$$i_B = \frac{Wl^2}{16EI} \qquad \text{(By symmetry)}$$

The bending moment at C on the conjugate beam AB,

$$M_C = \left(\frac{Wl^2}{16EI} \times \frac{l}{2} \right) - \left(\frac{Wl^2}{16EI} \times \frac{l}{6} \right) = \frac{Wl^3}{48EI}$$

∴ Deflection at C,

$$y_C = M_C = \frac{Wl^3}{48EI} \qquad \text{(As before)}$$

(v) Simply supported beam with an eccentric point load

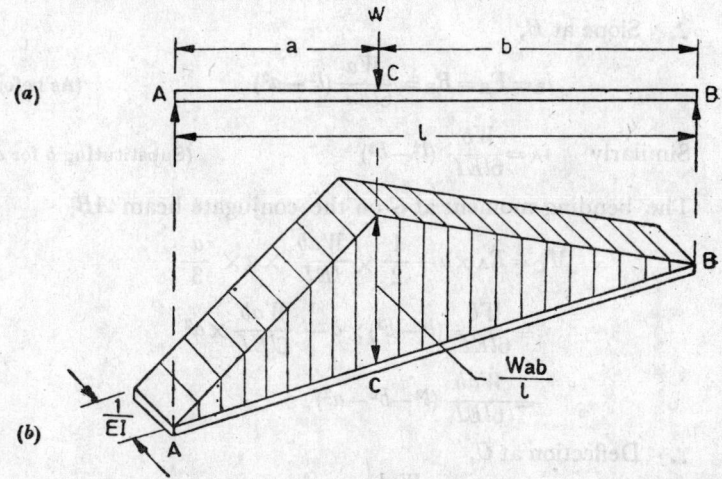

Fig. 7·36

Consider a simply supported beam AB of length l and carrying a point load W at C, such that $AC = a$ and $CB = b$ as shown in Fig. 7·36 (a). We know that the B.M. will be zero at A and B, and will increase by a straight line law to $\dfrac{Wab}{l}$ at C as shown in Fig. 7·36 (b).

The reaction at B on the conjugate beam (having simply supported ends A and B) may be found out by taking the moments of the B.M. diagram about A.

$$R_B \times l = \left[\frac{1}{2} \times \frac{Wab}{l} \times a \times \frac{2a}{3} \times \frac{1}{EI} \right]$$

$$+ \left[\frac{1}{2} \times \frac{Wab}{l} \times b \left(a + \frac{b}{3} \right) \frac{1}{EI} \right]$$

$$= \frac{Wab}{3lEI} \times a^2 + \frac{Wab}{2lEI} \left(ab + \frac{b^2}{3} \right)$$

$$= \frac{Wab}{6lEI} [2a^2 + 3ab + b^2]$$

$$= \frac{Wab}{6lEI} [a^2 + b^2 + 2ab + a^2 + ab]$$

$$R_B \times l = \frac{Wab}{6lEI} [(a+b)^2 + a(a+b)]$$

$$= \frac{Wa(l-a)}{6lEI} [l^2 + al] \qquad (\because a+b=l)$$

$$= \frac{Wal(l-a)}{6lEI} (l+a)$$

or $$R_B = \frac{Wa}{6lEI} (l^2 - a^2)$$

$\therefore$ Slope at B,

$$i_B = F_B = R_B = \frac{Wa}{6lEI} (l^2 - a^2) \qquad \text{(As before)}$$

Similarly $\quad i_A = \frac{Wb}{6lEI} (l^2 - b^2) \qquad$ (Substituting b for a)

The bending moment at C on the conjugate beam AB,

$$M_C = R_A \times a - \frac{1}{2} \times \frac{Wab}{lEI} \times a \times \frac{a}{3}$$

$$= \frac{Wb}{6lEI} (l^2 - b^2) \times a - \frac{Wab}{6lEI} \times a^2$$

$$= \frac{Wab}{6lEI} (l^2 - b^2 - a^2)$$

$\therefore$ Deflection at C,

$$y_C = M_0 = \frac{Wab}{6lEI} (l^2 - b^2 - a^2) \qquad \text{(As before)}$$

From the geometry of the figure, we find that the maximum deflection will take place in CB because $b > a$. We know that the shear force at any section X, at a distance x from B,

$$F_X = R_B - \frac{Wab}{lEI} \times \frac{x}{b} \times \frac{x}{2}$$

$$= \frac{Wa}{6lEI} (l^2 - a^2) - \frac{Wax^2}{2lEI}$$

$$= \frac{Wa}{6lEI} (l^2 - a^2 - 3x^2)$$

We know that the maximum deflection will take place, at a section where the slope of the elastic curve is zero (or in other words maximum B.M. takes place at a section where S.F. is zero) therefore equating the above equation to zero, we get

$$x = \sqrt{\frac{l^2 - a^2}{3}}$$

$\therefore$ Maximum bending moment,

$$M_{max} = R_B . x - \frac{Wax^2}{2lEI} \times \frac{x}{3}$$

$$M_{max} = \frac{Wa}{6lEI}(l^2-a^2)x - \frac{Wax^3}{6lEI}$$

$$= \frac{Wax}{6lEI}(l^2-a^2-x^2)$$

$$= \frac{Wa}{6lEI}\sqrt{\frac{l^2-a^2}{3}}\left[l^2-a^2-\left(\frac{l^2-a^2}{3}\right)\right]$$

$$= \frac{Wa}{6lEI}\sqrt{\frac{l^2-a^2}{3}}\left[\frac{2}{3}(l^2-a^2)\right]$$

$$= \frac{Wa}{9\sqrt{3}\cdot lEI}(l^2-a^2)^{\frac{3}{2}}$$

$\therefore$ Maximum deflection,

$$y_{max} = \frac{Wa}{9\sqrt{3}.lEI}(l^2-a^2)^{\frac{3}{2}} \qquad \text{(As before)}$$

(vi) *Simply supported beam with a uniformly distributed load*

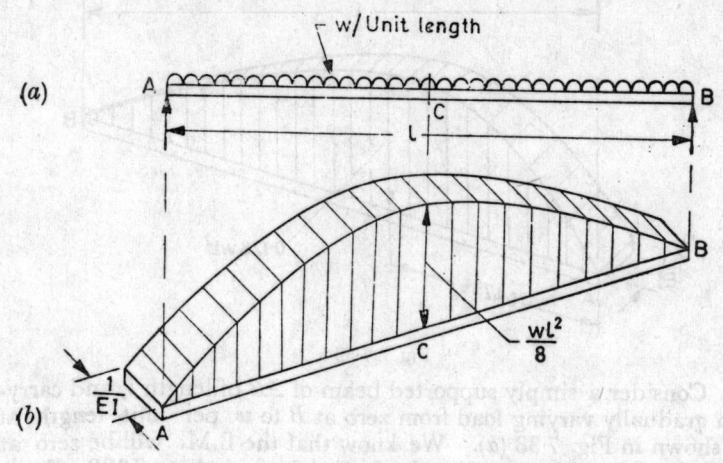

Fig. 7·37

Consider a simply supported beam AB of length l and carrying a uniformly distributed load of w per unit length as shown in Fig. 7·37 (a). We know that the B.M. at A and B will be zero and will increase in the form of parabola to $\frac{wl^2}{8}$ at the centre of the beam as shown in Fig. 7·37 (b).

The shear force on the conjugate beam (having simply supported ends A and B) at A,

$$F_A = R_A = \frac{1}{2}\left(\frac{wl^2}{8}\times l\times\frac{2}{3}\times\frac{1}{EI}\right) = \frac{wl^3}{24EI}$$

$\therefore$ Slope at A,

$$i_A = F_A = \frac{wl^3}{24EI} \qquad \text{(As before)}$$

Similarly $\qquad i_B = \dfrac{wl^3}{24EI}$ (By symmetry)

The bending moment at C on the conjugate beam AB,

$$M_C = \frac{wl^3}{24EI} \times \frac{l}{2} - \frac{wl^3}{24EI} \times \frac{3l}{16} = \frac{5wl^4}{384EI}$$

$\therefore$ Deflection at C,

$$y_C = M_C = \frac{5wl^4}{384EI}$$ (As before)

(vii) *Simply supported beam with a gradually varying load*

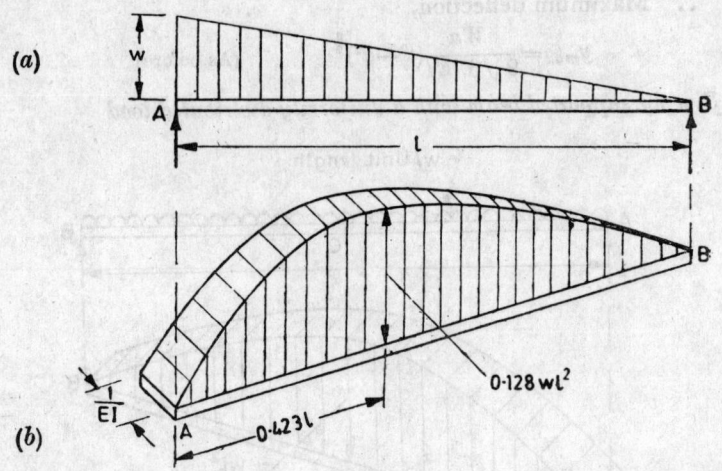

Fig. 7·38

Consider a simply supported beam of AB of length l, and carrying a gradually varying load from zero at B to w per unit length at A is shown in Fig. 7·38 (a). We know that the B.M. will be zero at A and B, and will increase in the form of parabola to $0·128\ wl^2$ at a distance of $0·423\ l$ form A as shown in Fig. 7·38 (b).

The reaction at B on the conjugate beam (having simply supported ends A and B) may be found out by taking moments of the B.M. diagram about A.

$$R_B \times l = \frac{wl^3}{24EI} \times \frac{8l}{15} = \frac{wl^4}{45EI}$$

or $\qquad R_B = \dfrac{wl^3}{45EI}$

$\therefore$ Slope at B,

$$i_B = R_B = \frac{wl^3}{45EI}$$ (As before)

and $\qquad R_A = \dfrac{wl^3}{24EI} - \dfrac{wl^3}{45EI} = \dfrac{7wl^3}{360EI}$

$$\therefore \quad \text{Slope at } A, R_A = \frac{7wl^3}{360EI}$$

(As before)

Example 7·22. *A. rolled steel joist ISMB 250×125 mm, shown in Fig. 7·39, carries a single concentrated load of 2 tonnes at the right third point over a simply supported span of 9 m.*

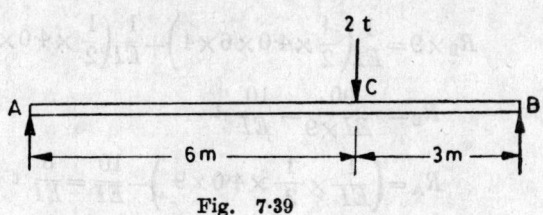

Fig. 7·39

If the value of I_{XX} for the beam is $5,131·6$ cm⁴ and the value of E for the material is $2·0 \times 10^6$ kg/cm², calculate by the use of conjugate beam method (i) the deflection under the load and (ii) the maximum deflection on the span.   *(U 2' (A.M.I.E., Winter, 1981)*

Solution.

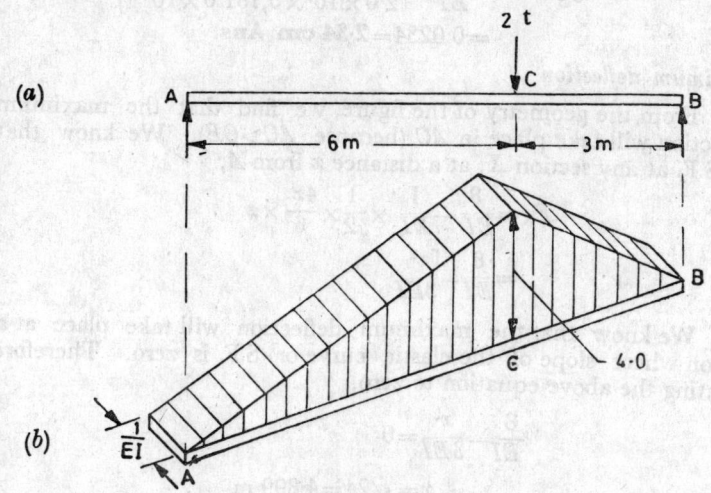

Fig. 7·40

Given   load, $W = 2$ t

Span, $l = 9$ m

Distance AC, $a = 6$ m

Distance CB, $b = 3$ m

Moment of inertia, $I_{XX} = 5,131·6$ cm⁴ $= 5,131·6 \times 10^{-8}$ m⁴

Young's modulus, $E = 2·0 \times 10^6$ kg/cm² $= 2·0 \times 10^{10}$ kg/m²

$$= 2·0 \times 10^7 \text{ t/m}^2$$

Deflection under the load

We know that the B.M. will be zero at A and B, and will increase by a straight line law to $\dfrac{Wab}{l} = \dfrac{2 \times 6 \times 3}{9} = 4 \cdot 0$ t-m.   Now draw the conjugate beam as shown in Fig. 7·40 (*b*). Taking moments about A,

$$R_B \times 9 = \frac{1}{EI}\left(\frac{1}{2} \times 4\cdot 0 \times 6 \times 4\right) + \frac{1}{EI}\left(\frac{1}{2} \times 4\cdot 0 \times 3 \times 7\right) = \frac{90}{EI}$$

or
$$R_B = \frac{90}{EI \times 9} = \frac{10}{EI}\ \text{t}$$

and
$$R_A = \left(\frac{1}{EI} \times \frac{1}{2} \times 4\cdot 0 \times 9\right) - \frac{10}{EI} = \frac{8}{EI}\ \text{t}$$

∴ B.M. at C on the conjugate beam AB,

$$M_C = \frac{10}{EI} \times 3 - \frac{1}{EI} \times \frac{1}{2} \times 3 \times 4 \times 1 = \frac{24}{EI}$$

∴ Deflection at C

$$y_C = M_C = \frac{24}{EI} = \frac{24}{2\cdot 0 \times 10^7 \times 5,131\cdot 6 \times 10^{-8}}\ \text{m}$$
$$= 0\cdot 0234 = \textbf{2·34 cm Ans.}$$

Maximum deflection

From the geometry of the figure, we find that the maximum deflection will take place in AC (because $AC > CB$). We know that the S.F. at any section X, at a distance x from A,

$$F_X = \frac{8}{EI} - \frac{1}{EI} \times \frac{1}{2} \times \frac{4x}{6} \times x$$
$$= \frac{8}{EI} - \frac{x^2}{3EI}$$

We know that the maximum deflection will take place at a section where slope of the elastic curve or S.F. is zero. Therefore equating the above equation to zero,

$$\frac{8}{EI} - \frac{x^2}{3EI} = 0$$

or
$$x = \sqrt{24} = 4\cdot 899\ \text{m}$$

∴ Maximum bending moment,

$$M_{max} = \frac{8}{EI} \times 4\cdot 899 - \frac{1}{EI} \times \frac{1}{2} \times \frac{4 \times 4\cdot 899}{6} \times 4\cdot 899 \times \frac{4\cdot 899}{3}$$
$$= \frac{235\cdot 15}{9EI}$$

∴ Maximum deflection,

$$y_{max} = M_{max} = \frac{235\cdot 15}{9 \times 2\cdot 0 \times 10^7 \times 5\cdot 131\cdot 6 \times 10^{-8}}\ \text{m}$$
$$= 0\cdot 025\ \text{m} = \textbf{2·55 cm   Ans.}$$

Example 7·23. *A simply supported beam AB of span 4 m is carrying a load of 10,000 kg at its mid span C, has cross-sectional moment of inertia 2,400 cm⁴ over the left half of the span and 4,800 cm⁴ over the right half. Find the angles θ_A and θ_B of the end tangents and the deflection δ_C under the load. Take $E = 2\cdot0 \times 10^6$ kg/cm².*

Solution.

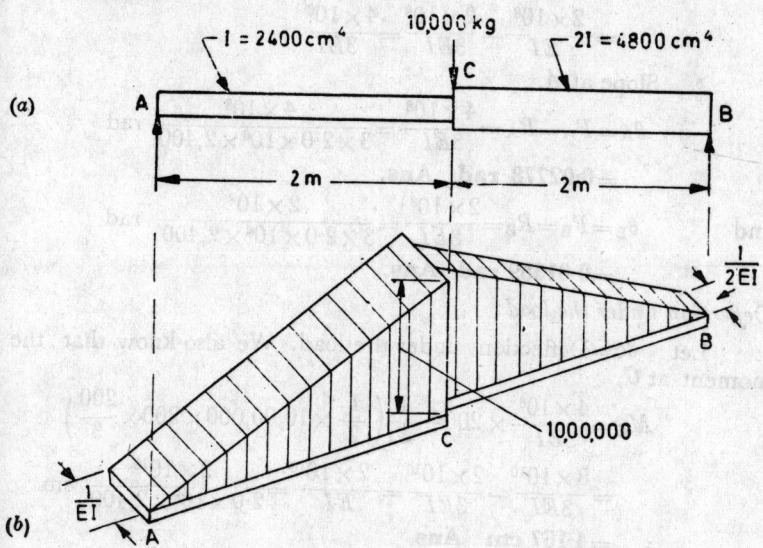

Fig. 7·41

Given.   Length of AC,

$$l_{AC} = l_{CB} = 2 \text{ m} = 200 \text{ cm}$$

Load, $W = 10,000$ kg

M.I. of AC, $I_{AC} = I = 2,400$ cm⁴

M.I. of CB, $I_{CB} = 2I = 4,800$ cm⁴

Young's modulus, $E = 2\cdot0 \times 10^6$ kg/cm²

Slopes at A and B

Let θ_A = Slope at A of the elastic curve, and

               θ_B = Slope at B of the elastic curve.

We know that the B.M. will be zero at A and B, and will increase by a straight line law to $\dfrac{Wab}{l} = \dfrac{10,000 \times 200 \times 200}{400} = 10,00,000$ kg-cm. Now draw the conjugate beam as shown in Fig. 7·41 (b). Taking moment about A,

$$R_B \times 400 = \frac{1}{EI}\left(\frac{1}{2} \times 10,00,000 \times 200 \times \frac{400}{3}\right)$$

$$+ \frac{1}{2EI}\left(\frac{1}{2} \times 10,00,000 \times 200 \times \frac{800}{3}\right)$$

$$= \frac{3 \times 10^{10}}{3EI}$$

or $\qquad R_\mathrm{B} = \dfrac{8 \times 10^{10}}{3EI} \times \dfrac{1}{400} = \dfrac{2 \times 10^8}{3EI}$

and $\qquad R_\mathrm{A} = \left[\dfrac{1}{EI}\left(\dfrac{1}{2} \times 10,00,000 \times 200 \right) \right.$

$$\left. + \dfrac{1}{2EI}\left(\dfrac{1}{2} \times 10,00,000 \times 200 \right) \right] - \dfrac{2 \times 10^8}{3EI}$$

$$= \dfrac{2 \times 10^8}{EI} - \dfrac{2 \times 10^8}{3EI} = \dfrac{4 \times 10^8}{3EI}$$

∴ Slope at A,

$$\theta_\mathrm{A} = F_\mathrm{A} = R_\mathrm{A} = \dfrac{4 \times 10^8}{3EI} = \dfrac{4 \times 10^8}{3 \times 2 \cdot 0 \times 10^6 \times 2,400}\ \text{rad}$$

$$= 0 \cdot 02778 \text{ rad   Ans.}$$

and $\qquad \theta_\mathrm{B} = F_\mathrm{B} = R_\mathrm{B} = \dfrac{2 \times 10^8}{3EI} = \dfrac{2 \times 10^8}{3 \times 2 \cdot 0 \times 10^6 \times 2,400}\ \text{rad}$

$$= 0 \cdot 01389 \text{ rad   Ans.}$$

Deflection under the load

Let δ_C = Deflection under the load. We also know that the moment at C,

$$M_\mathrm{C} = \dfrac{4 \times 10^8}{3EI} \times 200 - \dfrac{1}{EI}\left(\dfrac{1}{2} \times 10,00,000 \times 200 \times \dfrac{200}{3} \right)$$

$$= \dfrac{8 \times 10^{10}}{3EI} - \dfrac{2 \times 10^{10}}{3EI} = \dfrac{2 \times 10^{10}}{EI} = \dfrac{2 \times 10^{10}}{2 \cdot 0 \times 10^8 \times 2,400}\ \text{cm}$$

$$= 4 \cdot 167 \text{ cm   Ans.}$$

Example 7·24: *A beam ABCD is simply supported at its ends A and D over a span of 30 m. It is made up of 3 portions AB, BC and CD each 10 m in length. The moment of inertia of section, I over each of these individual portions is uniform, and the I values for them are I, 3I and 2I respectively, where $I = 2 \times 10^6$ cm⁴. The beam carries a point load of 15 tonnes at B and another load of 30 tonnes at C as shown in Fig. 7·42.*

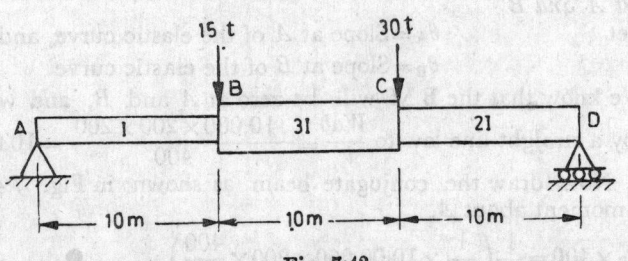

Fig. 7·42

Neglecting the self load of the beam, calculate the deflection at B and the slope at C. Take the values of E, the modulus of material of the beam 2×10^6 kg/cm² uniform throughout the length.

(London University)

Solution.

Given. Length of AB,
$$l_{AB}=l_{BC}=l_{CD}=10 \text{ m}$$

M.I. of AB,	$I_{AB}=I=2\times10^6 \text{ cm}^4=2\times10^{-2} \text{ m}^4$
M.I. of BC,	$I_{BC}=3I=3\times2\cdot0\times10^{-2}=6\times10^{-2} \text{ m}^4$
M.I. of CD,	$I_{CD}=2I=2\times4\times10^{-2}=4\times10^{-2} \text{ m}^4$
Load at B	$=15 \text{ t}$
Load at C	$=30 \text{ t}$

Young's modulus, $E=2\cdot0\times10^6 \text{ kg/cm}^2=2\cdot0\times10^{10} \text{ kg/m}^2$
$$=2\cdot0\times10^7 \text{ t/m}^2$$

Deflection at B

Taking moments about A,
$$V_D\times30=15\times10+30\times20=750$$

$$\therefore \qquad V_D=\frac{750}{30}=25 \text{ t}$$

and $$V_A=30+15-25=20 \text{ t}$$

$$\therefore \qquad M_B=20\times10=200 \text{ t-m}$$

and $$M_C=25\times10=250 \text{ t-m}$$

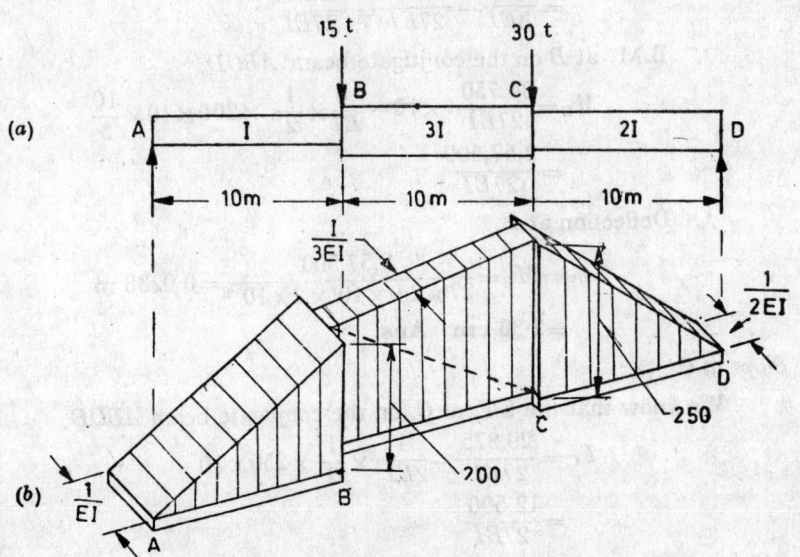

Fig. 7·43

We know that the B.M. will be zero at A and D, and will increase by a straight line law to 200 t-m at B and 250 t-m at C. Now draw the conjugate beam as shown in Fig. 7·43 (b).

Now taking moments about A of the conjugate beam,

$$R_D \times 30 = \frac{1}{EI}\left(\frac{1}{2} \times 200 \times 10 \times \frac{20}{3} \right)$$

$$+ \frac{1}{3EI}\left(\frac{1}{2} \times 200 \times 10 \times \frac{40}{3} \right)$$

$$+ \frac{1}{3EI}\left(\frac{1}{2} \times 250 \times 10 \times \frac{50}{3} \right)$$

$$+ \frac{1}{2EI}\left(\frac{1}{2} \times 250 \times 10 \times \frac{70}{3} \right)$$

$$= \frac{20,000}{3EI} + \frac{40,000}{9EI} + \frac{62,500}{9EI} + \frac{43,750}{3EI}$$

$$= \frac{2,93,750}{9EI}$$

or

$$R_D = \frac{2,93,750}{9EI \times 30} = \frac{29,375}{27EI}$$

and

$$R_A = \frac{1}{EI}\left(\frac{1}{2} \times 200 \times 10 \right) + \frac{1}{3EI}\left(\frac{200+250}{2} \times 10 \right)$$

$$+ \frac{1}{2EI}\left(\frac{1}{2} \times 250 \times 10 \right) - \frac{29,375}{27EI}$$

$$= \frac{7,125}{3EI} - \frac{29,375}{27EI} = \frac{34,750}{27EI}$$

$\therefore$ B.M. at B on the conjugate beam $ABCD$,

$$M_B = \frac{34,750}{27EI} \times 10 - \frac{1}{EI} \times \frac{1}{2} \times 200 \times 10 \times \frac{10}{3}$$

$$= \frac{2,57,500}{27EI}$$

$\therefore$ Deflection at B,

$$y_B = M_B = \frac{2,57,500}{27 \times 2 \cdot 0 \times 10^7 \times 2 \times 10^{-2}} = 0 \cdot 0238 \text{ m}$$

$$= 2 \cdot 38 \text{ cm} \quad \textbf{Ans.}$$

Slope at C

We know that the S.F. at C, on the conjugate beam $ABCD$,

$$F_C = \frac{29,375}{27EI} - \frac{1}{2EI} \times \frac{1}{2} \times 250 \times 10$$

$$= \frac{12,500}{27EI}$$

$\therefore$ Slope at C,

$$i_C = F_C = \frac{12,500}{27EI} = \frac{12,500}{27 \times 2 \cdot 0 \times 10^7 \times 2 \times 10^{-2}} \text{ rad}$$

$$= 0 \cdot 001157 \text{ rad} \quad \textbf{Ans.}$$

Exercise 7–2

1. A wooden beam of 14 cm wide and 24 cm deep has a span of 4 metres. Determine the load, that can be placed at its centre, to cause the beam a deflection of 1 cm. Take $E = 6 \times 10^4$ kg/cm². **[Ans. 725 kg]**

2. A simply supported beam of 10 metres span is carrying a point load of 10,000 kg at 6 metres from the left end. Calculate the deflection under the load. Take $E = 2 \cdot 0 \times 10^6$ kg/cm² and $I = 1,00,000$ cm⁴. (*Baroda University*) **[Ans. 0·96 cm]**

3. Détermine the maximum deflection of the beam in the above example. (*Gujarat University*) **[Ans. 0·9875 cm]**

4. A simply supported beam of span 10 metres is loaded with a uniformly distributed load of 500 kg/m over a length of 3 metres from the left end. Find the maximum deflection of the beam. Take $E = 2 \cdot 0 \times 10^6$ kg/cm² and $I = 3,000$ cm⁴. (*Poona University*) **[Ans. 2 cm]**

5. A simply supported beam of span L is carrying a point load of W at a distance of $L/3$ from the left end. Show, from the first principles, that the deflection under the load is $\dfrac{0 \cdot 0164 \, WL^2}{EI}$. (*Oxford University*)

6. A beam, simply supported over a span of 10 metres, carries point loads 4, 2 and 6 tonnes at 2, 5 and 9 metres from the right hand support. Determine the maximum deflection of the beam. Take $E = 2 \cdot 0 \times 10^6$ kg/cm², and $I = 45,200$ cm⁴. (*Rajasthan University*) **[Ans. 1·4 cm]**

HIGHLIGHTS

1. Slope and deflection at B of a cantilever AB, fixed at A,

$$i_B = \frac{Wl^2}{2EI} \quad \text{and} \quad y_B = \frac{Wl^3}{3EI}$$

(when it carries a point load W at the free end)

$$i_B = \frac{wl^3}{6EI} \quad \text{and} \quad y_B = \frac{wl^4}{8EI}$$

(with it carries a u.d.l. of w/unit length)

$$i_B = \frac{wl^3}{24EI} \quad \text{and} \quad y_B = \frac{wl^4}{30}$$

(when it carries gradually varying load from zero at B to w per unit length at A)

where
- i_B = Slope at the free end B,
- y_B = Deflection at the free end B,
- l = Length of the cantilever AB,
- E = Modulus of elasticity of the cantilever material, and
- I = Moment of inertia of the cantilever cross-section.

2. Slope and deflection of a simply supported beam AB,

$$i_A = i_B = \frac{Wl^2}{16EI} \quad \text{and} \quad y_C = \frac{Wl^3}{48EI}$$

(when it carries a central point load W)

$$i_A = i_B = \frac{wl^3}{24EI} \quad \text{and} \quad y_C = \frac{5wl^4}{384EI}$$

(when it carries a u.d.l. of w/unit length)

$$i_A = \frac{wl^3}{45EI} \quad \text{and} \quad i_B = \frac{7wl^3}{360EI}$$

$$y_C = \frac{0.00651wl^4}{EI} \quad \text{and} \quad y_{max} = \frac{0.0065wl^4}{EI}$$

(when it carries a gradually varying load
from zero at B to w per unit length at A)

3. If a cantilever (or a beam) is loaded with several point or uniformly distributed loads, the slope as well as deflection at any point on the cantilever (or beam) is equal to the algebraic sum of the slopes and deflections, at that point, due to various loads, acting individually.

4. The moment area method for slope and deflection at a section is given by the Mohr's theorems.

Mohr's theorem I. It states, *"The change of slope between any two points, on an elastic curve, is equal to the area of B.M. diagram between these points divided by EI."*

Mohr's theorem II. It states *"The intercept taken on a vertical reference line of tangents drawn at any two points on an elastic curve is equal to the moment of the B.M. diagram between these points about the reference line divided by EI."*

5. The Macaulay's method is most suitable to find out the slope and deflection at any point in a cantilever or a beam, especially when it is carrying some point loads. In this method a continuous expression is integrated in such a way that the constants of integration are valid for all sections even though the law of bending moment varies from section to section.

6. The conjugate beam method is a slightly modified form of moment area method. It is specially useful for cantilevers and beams with varying flexural rigidities. The slope and deflection is found out by the following two statements.

1. *"The shear force at any section of the conjugate beam is equal to the slope of the elastic curve at the corresponding section of the actual beam."*

2. *"The bending moment at any section of the conjugate beam is equal to the deflection of the elastic curve at the corresponding section of the actual beam."*

Do You Know ?

1. What is the relation between slope deflection and radius of curvature of a simply supported beam ?

2. Drive a relation for the slope and deflection of a cantilever, when

    (a) it carries a point load at the free end,

    (b) it carries a uniformly distributed load over the entire span.

A cantilever beam of length l is loaded only one half of its length from the free end, with a uniformly distributed load of w per metre. Derive a formula for the deflection at the free end.

4. How would you find out the slope and deflection at any point on a cantilever, when it carries several loads ?

5. A simply supported beam AB of span l and stiffness EI carries a concentrated load P at its centre. Find the expression for slope of the beam at the support A and deflection of the beam at its centre.

6. A simply supported beam AB of span l and uniformly flexural rigidity EI is subjected to a point load W at a section C, distant a from A and b from B. Prove by any standard method that the deflection at a section in the part AC at a distance x from A is given by the expression :

$$\frac{Wbx}{6lEI}[al+ab-x^2]$$

7. Derive a relation for the slope and deflection of simply supported beam, subjected to a uniformly distributed load of w/m length.

8. The beam AB of span l carries a distributed load of varying intensity from zero at A to w per unit length at B. Measuring x from the end A, establish the equation for the deflection curve of the beam. Hence calculate the deflection at the centre of the beam.

9. What is moment area method ? Explain the two Mohr's theorems as applicable to the slope and deflection of a cantilever or a beam.

10. What is Macaulay's method for finding out the slope and deflection of a beam ? Discuss the cases where it is of a particular use.

11. Explain the conjugate beam method for slope and deflection of cantilevers and beams. Describe the utility of this method.

8

DEFLECTION OF PERFECT FRAMES

8·1. Introduction

A frame may be defined as a structure, made up of several bars riveted or welded together. These bars are made up of angle irons or channel sections, and are called members of a frame or framed structure. Though these members are welded or riveted together at their joints, yet for calculation purposes, these joints are assumed to be hinged, or simply pin-jointed.

8·2. Perfect frames

A perfect frame is that in which the number of members is just sufficient to keep it in equilibrium. This means that the shape of a perfect frame is not distorted, when loaded.

The simplest form of a perfect frame is a triangle, in which there are three members and three joints as shown in Fig. 8·1(a). It may be noted, that if such a frame is loaded, its shape is not distorted.

Thus we see, that for a three jointed frame, there should be three members to prevent distortion. We further see, that if we want to add one more joint to a triangular frame, we require two members as shown in Fig. 8·1(b). Thus, in general, we may say that for every additional joint to a triangular frame, two members are required. The

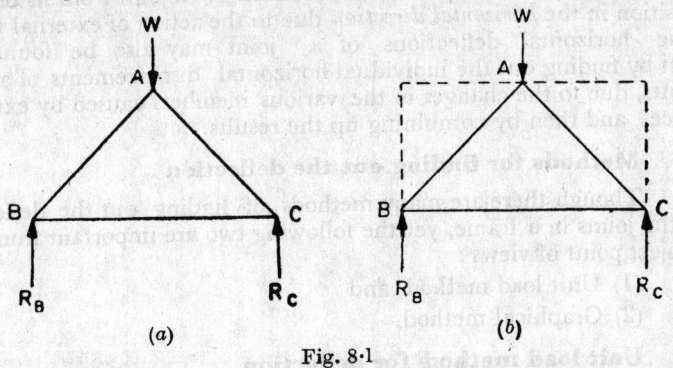

Fig. 8·1

number of members and number of joints, in a perfect frame, may be expressed by the relation :

$$n = 2j - 3$$

where
$n = $ No. of members, and
$j = $ No. of joints.

A frame, which does not satisfy the above equation, is known as an *imperfect frame*.

8·3. Types of deflections

We know that whenever a frame or framed structure carries some external loads, all of its members are subjected to either compressive or tensile forces. We also know that whenever a body is subjected to some external forces, it undergoes some deformation. It is thus obvious, that all the members of a frame, when subjected to some external force, undergo a change in their lengths. This change in lengths, of the members of a frame, will cause some displacement of all its joints ; except those which are rigidly fixed. The net displacement of any joint may be found out by studying the combined effect of changes in all members of the frame. Though a framed joint may suffer deflection in any direction, depending upon the direction of the load system and rigidity of the frame members, yet the following two types of deflections are important from the subject point of view :

(1) Vertical deflection, and

(2) Horizontal deflection.

8·4. Vertical deflection

It is the total displacement, suffered by a joint from its original position, in the *vertical direction* due to the action of external forces. The vertical deflection of a joint may be found out first by finding out the individual vertical displacements of all the joints, due to the changes in the various members caused by external forces, and then by combining up the results.

8·5. Horizontal deflection

It is the total displacement, suffered by a joint from its original position in the *horizontal direction* due to the action of external forces. The horizontal deflections of a joint may also be found out first by finding out the individual horizontal displacements of all the joints, due to the changes in the various members caused by external forces, and then by combining up the results.

8·6. Methods for finding out the deflection

Though there are many methods of finding out the deflection of the joints in a frame, yet the following two are important from the subject point of views :

   (1) Unit load method, and
   (2) Graphical method.

8·7. Unit load method for deflection

The deflection of a joint, by the unit load method, is found out as discussed below :

   1. First of all find out the forces P_1, P_2, P_3...in the various members of the frame, due to the external loading.

   2. Now find out the stresses in the various members of the frame

   i.e. $p_1 = \dfrac{P_1}{A_1}$, $p_2 = \dfrac{P_2}{A_2}$, $p_3 = \dfrac{P_3}{A_3}$ ...where A_1, A_2, A_3 are the cross-sectional areas of the members 1, 2, 3...The type of stress (*i.e.* compression or tension) should be clearly considered. The usual practice is to use +ve for tension and −ve for compression.

   3. Assume all the external loads on the frame to be removed. Now assume a unit vertical load to act at a joint whose deflection is required to be found.

   4. Now, find out the forces v_1, v_2, v_3...in the various members of the frame due to this *unit vertical load*. The type of force (*i.e.* push or pull) should be clearly considered. The usual practice is to use +ve for pull (because it induces tension in the member) and −ve for push (because it induces compression in the member.

   5. Now the vertical deflection of the joint will be given by the equation.

$$\delta_V = \frac{P_1 v_1 l_1}{A_1 E} + \frac{P_2 v_2 l_2}{A_2 E} + \ldots\ldots$$

$$= \sum \frac{Pvl}{AE}$$

$$= \sum \frac{pvl}{E} \qquad \left(\because\ p = \frac{P}{A}\right)$$

where l_1, l_2, l_3...are the lengths of the members of the frame and E is the Young's modulus of elasticity of the frame material.

6. Similarly, the horizontal deflection of the joint will be given by equation,

$$\delta_H = \frac{P_1 u_1 l_1}{A_1 E} + \frac{P_2 u_2 l_2}{A_2 E} + \dots$$

$$= \sum \frac{Pul}{AE}$$

$$= \sum \frac{pul}{E} \qquad \left(\because p = \frac{P}{A} \right)$$

where u_1, u_2, u_3......are the forces in the various members of the frame, due to a *unit horizontal load* at a joint, whose horizontal deflection is required to be found out.

Proof

Consider a perfect frame subjected to a number of forces.

Let P_1, P_2, P_2...= Forces in the members 1, 2, 3...of the frame, due to external loading.

v_1, v_2, v_3...= Forces in the members 1, 2, 3...of the frame, due to a unit vertical load at the joint, whose vertical deflection is required to be found out.

l_1, l_2, l_3...= Lengths of the members 1, 2, 3...of the frame.

A_1, A_2, A_3...= Cross-sectional areas of the members 1, 2, 3...of the frame.

Now consider another small vertical load δW applied gradually on the joint, whose vertical deflection is required to be found out. We see that when a unit vertical load is applied at the joint, the forces in the members are v_1, v_2, v_3 and when a vertical load δW is applied on the same joint, the forces in the members will be $v_1 \delta W$, $v_2 \delta W$, $v_3 \delta W$...Let the vertical weight δW cause a vertical deflection δv at the joint. Therefore work done by the weight δW

$$= \tfrac{1}{2} \times \text{weight} \times \text{deflection}$$
$$= \tfrac{1}{2} \times \delta W \times \delta v \qquad \dots(i)$$

A little consideration will show that the total force acting on member 1,

$$= P_1 + v_1 \delta W$$

$\therefore$ Stress in member $1 = \dfrac{\text{force}}{\text{Area}} = \dfrac{P_1 + v_1 \delta W}{A_1}$

and strain in member $1 = \dfrac{\text{Stress}}{E} = \dfrac{(P_1 + v_1 \delta W)}{A_1 E}$

and deformation of member $1 = \text{Strain} \times \text{length} = \dfrac{(P_1 + v_1 \delta W) l_1}{A_1 E} \qquad \dots(ii)$

Now stress in member 1 due to load δW

$$= \frac{\text{force}}{\text{Area}} = \frac{v_1 \delta W}{A_1}$$

and strain $\qquad = \dfrac{\text{Stress}}{E} = \dfrac{v_1 \delta W}{A_1 E}$

and deformation $\qquad = \text{Strain} \times \text{length} = \dfrac{v_1 \delta W l_1}{A_1 E}$ ...(iii)

∴ Work stored in member 1

$\qquad = \tfrac{1}{2} \times \text{Total force} \times \text{deformation due to } \delta W$

$\qquad = \tfrac{1}{2}(P_1 + v_1 \delta W) \times \left(\dfrac{v_1 \delta W l_1}{A_1 E} \right)$ ...(iv)

$\qquad = \dfrac{P_1 v_1 \delta W l_1}{2 A_1 E}$

(neglecting small quantities of second order)

∴ Total work stored in the frame

$$= \sum \frac{P v \delta W l}{2AE} = \frac{1}{2} \sum \frac{P v \delta W l}{AE}$$

Now equating the total work done and the work stored in the frame,

$$\tfrac{1}{2} \delta W \times \delta_V = \frac{1}{2} \sum \frac{P v \delta W l}{AE}$$

$$\delta_V = \sum \frac{P v l}{AE}$$

$$= \frac{\Sigma p v l}{E} \qquad \left(\because p = \frac{P}{A} \right)$$

Similarly, it can be proved that the horizontal deflection,

$$\delta_H = \frac{\Sigma p u l}{E}$$

Example 8·1. *A crane shown in Fig. 8·2 has cross-sectional areas of tie and jib as 30 cm² and 70 cm² respectively.*

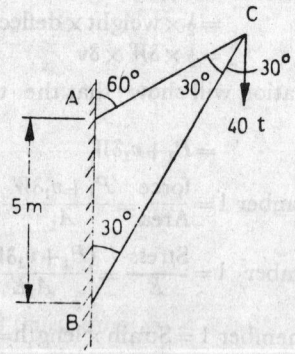

Fig. 8·2

Determine the vertical deflection of the joint C, when a load of 40 t is suspended from it. Take E as $2 \cdot 0 \times 10^6$ kg/cm².

(Patna University, 1972)

Solution.

Given. Area of member AC

$$= 30 \text{ cm}^2 \qquad (\because AC \text{ is tie member})$$

Area of member $BC = 70 \text{ cm}^2 \qquad (\because BC \text{ is jib member})$

Load at $\qquad C = 40 \text{ t}$

Young's modulus, $E = 2 \cdot 0 \times 10^6 \text{ kg/cm}^2 = 2 \cdot 0 \times 10^3 \text{ t/cm}^3$

Let $\qquad \delta_V = $ Vertical deflection of the joint C.

From the geometry of the figure, we find that the length, $AC = 5$ m $= 500$ cm and that of $BC = 8 \cdot 66$ m $= 866$ cm

We know that when a load of a 40 t is suspended from the joint C, the force in member AC,

$$P_{AC} = 40 \text{ t (Tension)}$$

and $\qquad P_{BC} = 69 \cdot 28 \text{ t (Compn)}$

Similarly, when a unit vertical load is suspended from C, the force in member AC,

$$v_{AC} = 1 \text{ t (Tension)}$$

$$v_{BC} = 1 \cdot 732 \text{ t (Compn)}$$

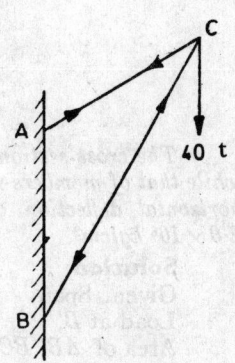

Fig. 8·3

Now complete the following table :

Tension +ve ; Compn. —ve

Member	l cm	P tonnes	a cm²	$p = \dfrac{P}{a}$ t/cm²	v t/cm²	pvl
AC	500	+40·0.	30	$+\dfrac{4}{3}$	+1·0	0·667×10³
BC	866	—69·28	70	$-\dfrac{6·928}{7}$	—1·732	1·485×10³

$$\therefore \quad \Sigma pvl = 2 \cdot 152 \times 10^3$$

Using the relation,

$$\delta_V = \frac{\Sigma pvl}{E} \quad \text{with usual notations.}$$

$$= \frac{2 \cdot 152 \times 10^3}{2 \cdot 0 \times 10^3} = \mathbf{1 \cdot 076 \text{ cm}} \quad \textbf{Ans}$$

Example 8·2. *A symmetrical frame of 1·5 m span is hinged a..
A and is supported on rollers at C. The frame is carrying a load of 1(
tonnes at B as shown in Fig. 8·4.*

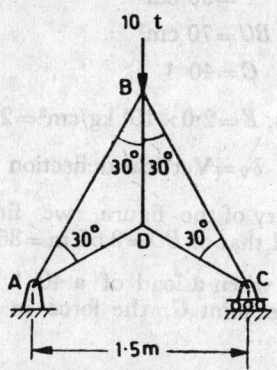

Fig. 8·4

The cross-sectional area of members AB and BC is 10 square cm,
while that of members AD, BD and CD is 5 square cm. Determine the
horizontal deflection of the joint C. Take E for the frame material as
$2·0 \times 10^6$ kg/cm².

Solution.

Given. Span $AC = 1·5$ m $= 150$ cm
Load at B, $W = 10$ t
Area of AB, BC $= 10$ cm²
Area of AD, BD, $CD = 5$ cm²
Young's modulus, $E = 2·0 \times 10^6$ kg/cm² $= 2·0 \times 10^3$ t/cm²
Let $\delta_H =$ Horizontal deflection of C.

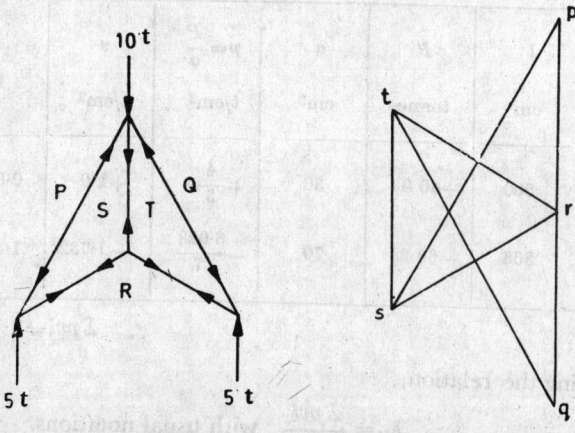

(a) Space diagram   (b) Vector diagram

Fig. 8·5

In order to determine the forces in the various members of the
truss, draw the space and vector diagrams as shown in Fig. 8·5 (a)
and (b).

Now consider a unit horizontal load acting at C. In order to determine the forces in the various members of the truss, draw the space and vector diagrams as shown in Fig. 8·6 (a) and (b).

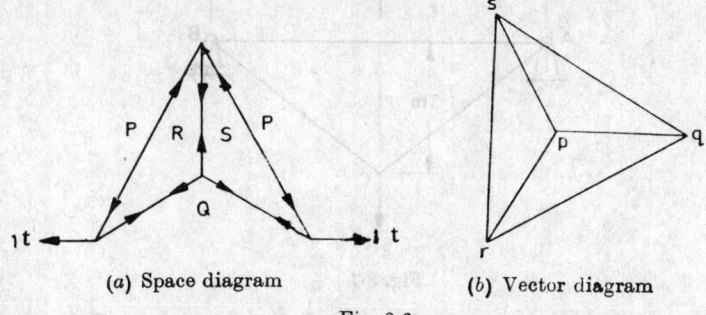

(a) Space diagram   (b) Vector diagram

Fig. 8·6

Now complete the following table by measuring the various sides of the vector diagrams in Figs. 8·5 (b) and 8·6 (b).

Tension +ve ; Compn −ve

Member	l cm	P tonnes	a cm^2	$p = \dfrac{P}{a}$ t/cm^2	u t/cm^2	Pul
AB	150	$-5\sqrt{3}$	10	$\dfrac{\sqrt{3}}{2}$	$-1\cdot0$	$75\sqrt{3}$
AD	$\dfrac{150}{\sqrt{3}}$	$+5$	5	$1\cdot0$	$+\sqrt{3}$	150
BD	$\dfrac{150}{\sqrt{3}}$	$+5$	5	$1\cdot0$	$+\sqrt{3}$	150
BC	150	$-5\sqrt{3}$	10	$\dfrac{\sqrt{3}}{2}$	$-1\cdot0$	$75\sqrt{3}$
DC	$\dfrac{150}{\sqrt{3}}$	$+5$	5	$1\cdot0$	$+\sqrt{3}$	150

$$\therefore \quad \Sigma pul = 450 + 150\sqrt{3}$$

Using the relation,

$$\delta_H = \frac{\Sigma pul}{E} \quad \text{with usual notations.}$$

$$= \frac{450 + 150\sqrt{3}}{2\cdot0 \times 10^3} = \textbf{0·0355 cm} \quad \textbf{Ans.}$$

Example 8·3. *A pin-jointed frame shown in Fig. 8·7 is carrying a load of 6 tonnes at C.*

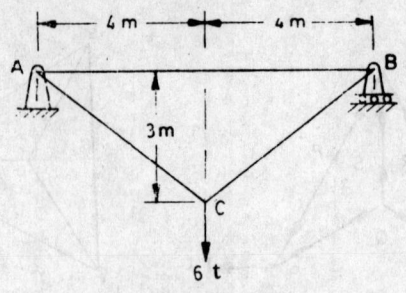

Fig. 8·7

Find the vertical as well as horizontal deflection of C. Take area of member AB as 10 cm², and those of members AC and BC as 15 cm². E = 2·0 × 10³ t/cm².

Solution.

Given. Load $= 6$ t

Area of member $AB = 10$ cm²

Area of member $AC = 15$ cm²

Area of member $BC = 15$ cm²

Young's modulus, $E = 2·0 \times 10^3$ t/m²

From the geometry of the figure, we find that the length of $AB = 8$ m $= 800$ cm and that of AC and $BC = 5$ m $= 500$ cm.

Vertical deflection of C

Let $\delta_V = $ Vertical deflection of joint C.

We know that when the joint C is carrying a load of 6 t, the force in AC,

$P_{AC} = 5$ t (Tension)

$P_{BC} = 5$ t (Tension)

and $P_{AB} = 4$ t (Compn.)

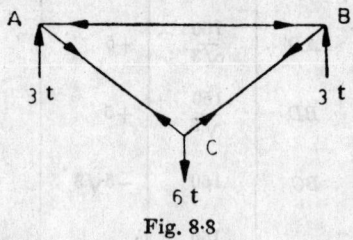

Fig. 8·8

Similarly, when a unit vertical load is attached to the joint C, the force in AC,

$$v_{AC} = \frac{5}{6} \text{ t (Tension)}$$

$$v_{BC} = \frac{5}{6} \text{ t (Tension)}$$

$$v_{AB} = \frac{4}{6} = \frac{2}{3} \text{ t (Compn.)}$$

Now complete the following table :

Tension $+ve$; Compn. $-ve$

Member	l cm	P tonnes	a cm²	$p=\dfrac{P}{a}$ t/cm²	v t/cm²	pvl
AC	500	5	15	$\dfrac{1}{3}$	$\dfrac{5}{6}$	$\dfrac{2,500}{18}$
BC	500	5	15	$\dfrac{1}{3}$	$\dfrac{5}{6}$	$\dfrac{2,500}{18}$
AB	800	-4	10	$-\dfrac{2}{5}$	$-\dfrac{2}{3}$	$\dfrac{3,200}{15}$

$$\therefore \quad \Sigma pvl = \frac{4,420}{9}$$

Using the relation,

$$\delta_V = \frac{\Sigma pvl}{E} \text{ with usual notations.}$$

$$= \frac{4,420}{9} \times \frac{1}{2 \cdot 0 \times 10^3} = 0 \cdot 245 \text{ cm Ans.}$$

Horizontal deflection of C

Let δ_H = Horizontal deflection of C.

We know that when the frame is loaded at C, the member AB is subjected to compression. Since the joint A is hinged, therefore the joints B and C will deflect towards A. It is thus obvious, that the unit horizontal load at C should be applied as shown in Fig. 8·9.

We know that when the joint C is subjected to a unit horizontal load, the force in AC,

$$u_{AC} = \frac{5}{8} \text{ t (Compn.)}$$

$$v_{BC} = \frac{5}{8} \text{ t (Tension)}$$

$$u_{AB} = \frac{1}{2} \text{ t (Compn.)}$$

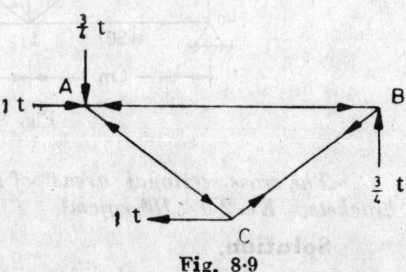

Fig. 8·9

Now complete the following table.

Tension $+ve$; Compn. $-ve$

Member	l cm	P tonnes	a cm²	$p = \dfrac{P}{a}$ t/cm²	u t/cm²	pul
AC	500	5	15	$\dfrac{1}{3}$	$\dfrac{5}{8}$	$+\dfrac{2,500}{24}$
BC	500	5	15	$\dfrac{1}{3}$	$-\dfrac{5}{8}$	$-\dfrac{2,500}{24}$
AB	800	-4	10	$-\dfrac{2}{5}$	$\dfrac{1}{2}$	$+160$

$$\therefore \ \Sigma pul = 160$$

Now using the relation,

$$\delta_H = \frac{\Sigma \ pul}{E} \text{ with usual notations.}$$

$$= \frac{160}{20 \times 10^3} = \mathbf{0\cdot08 \ cm \ Ans.}$$

Example 8·4. *Using castigliano's first theorem, find the vertical deflection component of L_1 in the truss loaded as shown in Fig. 8·10.*

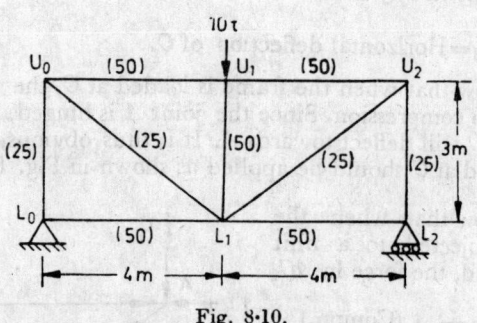

Fig. 8·10.

The cross-sectional areas of the members in cm² are shown in brackets. $E = 2\cdot0 \times 10^6 \ kg/cm^2$.   (*A.M.I.E.*, Winter, 1974).

Solution.

Given. Load at $U_1 = 10$ t

Young' modulus, $E = 2.0 \times 10^6 \text{ kg/cm}^2 = 2.0 \times 10^3 \text{ t/cm}^2$

Let $\delta v = \text{Vertical deflection component of } L_1$

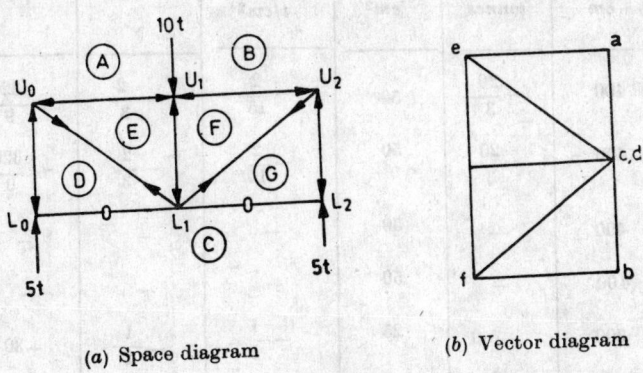

(a) Space diagram   (b) Vector diagram

Fig. 8·11.

In order to determine the forces in the various members of the truss, draw the space diagram and then the vector diagram as shown in Fig. 8·11. (a) and (b).

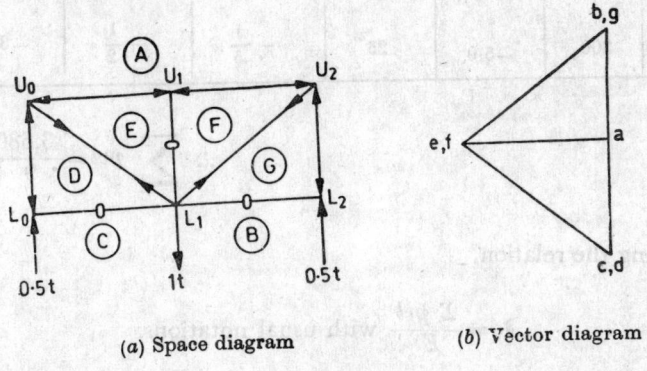

(a) Space diagram   (b) Vector diagram

Fig. 8·12.

Now consider a unit vertical load acting at L_1. In order to determine the forces in the various members of the truss, draw the space and vector diagrams as shown in Fig. 8·12. (a) and (b).

Now complete the following table, by measuring the various sides of the vector diagrams in Fig. 8·12 (b) and (b).

Tension $+ve$; Compn. $-ve$

Member	l cm	P tonnes	a cm²	$p = \dfrac{P}{a}$ t/cm²	v	pvl
$U_0\,U_1$	400	$-\dfrac{20}{3}$	50	$-\dfrac{2}{15}$	$-\dfrac{2}{3}$	$+\dfrac{320}{9}$
$U_1\,U_2$	400	$-\dfrac{20}{3}$	50	$-\dfrac{2}{15}$	$-\dfrac{2}{3}$	$+\dfrac{320}{9}$
$L_0\,L_1$	400	—	50	—	—	—
$L_1\,L_2$	400	—	50	—	—	—
$U_0\,L_0$	300	$-5{\cdot}0$	25	$-\dfrac{1}{5}$	$-\dfrac{1}{2}$	$+30$
$U_0\,L_1$	500	$+\dfrac{25}{3}$	25	$+\dfrac{1}{3}$	$+\dfrac{5}{6}$	$-\dfrac{1,250}{9}$
$U_1\,L_1$	300	$-10{\cdot}0$	50	$-\dfrac{1}{5}$	—	—
$L_1\,U_2$	500	$+\dfrac{25}{3}$	25	$+\dfrac{1}{3}$	$+\dfrac{5}{6}$	$+\dfrac{1,250}{9}$
$U_2\,L_2$	300	$-5{\cdot}0$	25	$-\dfrac{1}{5}$	$-\dfrac{1}{2}$	-30

$$\therefore \sum pvl = \frac{3,680}{9}$$

Using the relation,

$$\delta_V = \frac{\Sigma\ pvl}{E}\ \text{with usual notations.}$$

$$= \frac{3,680}{9 \times 2 \cdot 0 \times 10^3} = 0 \cdot 205\ \text{cm Ans}.$$

Example 8·5 *Determine the vertical deflection of the load in the structure shown in Fig. 8·13*

The tension members are stressed to 1,500 kg/cm² and the compression members to 800 kg/cm². All the inclined members are at 45° with the horizontal. Take $2 \cdot 0 \times 10^6$ kg/cm².

(A.M.I.E., Winter 1977)

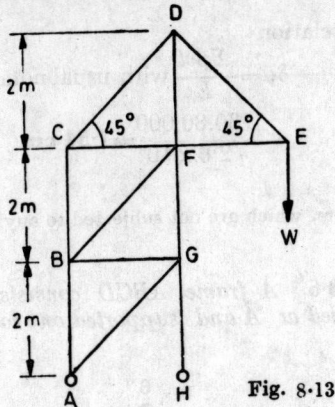

Fig. 8·13

Solution.

Given. Stress in tension members

$$= 1,500 \text{ kg/cm}^2$$

Tension in compression members

$$= 800 \text{ kg/cm}^2$$

Young's modulus, $E = 2 \cdot 0 \times 10^6 \text{ kg/cm}^2$

Let δv = Vertical deflection of the load.

We know that when the joint E is subjected to a load W the members AB, BC, CD, DE are subjected to a tensile force of 1,500 kg/cm², the members EF, FG, GH, CF, DF are subjected to a compressive force of 800 kg/cm² and the members AG, BG and BF are not subjected to any force. Similarly when the joint E is subjected to 1 kg vertical load, the members AB, BC are subjected to 1 kg tensile force, the members CD, DE are subjected to $\sqrt{2}$ kg tensile force, the members EF, CF are subjected to 1 kg compressive force, the members DF, FG, GH are subjected to 2 kg compressive force and the members AG, BG and BF are not subjected to any load.

Now complete the following table.*

Tension $+ve$; Compn. $-ve$

Member	l	$p = \dfrac{P}{a}$	v	pvl
	cm	kg/cm²	kg/cm²	
AB	200	$+1,500$	$+1\cdot0$	3,00,000
BC	200	$+1,500$	$+1\cdot0$	3,00,000
CD	$200\sqrt{2}$	$+1,500$	$+\sqrt{2}$	6,00,000
DE	$200\sqrt{2}$	$+1,500$	$+\sqrt{2}$	6,00,000
EF	200	-800	$-1\cdot0$	1,60,000
FG	200	-800	$-2\cdot0$	3,20,000
GH	200	-800	$-2\cdot0$	3,20,000
CF	200	-800	$-1\cdot0$	1,60,000
DF	200	-800	$-2\cdot0$	3,20,000

$\therefore \ \Sigma\, pvl = 30,80,000$

Using the relation

$$\delta_V = \frac{\Sigma pvl}{E} \text{ with usual notations.}$$

$$= \frac{30,80,000}{2 \cdot 0 \times 10^6} = 1 \cdot 54 \text{ cm} \quad \textbf{Ans.}$$

---

*The members, which are not subjected to any force are **no** entered in the table.

Example 8·6[t] *A frame ABCD consists of two equilateral triangles and is hinged at A and supported on rollers at D as shown in Fig.* 8·14.

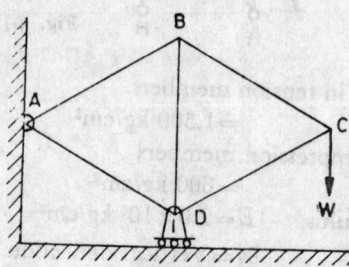

Fig. 8·14

Find the vertical deflection of C due to a load W applied at C. Take all the tension members of area a and compression members of area 2a.   (*Oxford University*)

Solution.

Given. Load at C $= W$

Area of tension members $= a$

Area of compression members $= 2a$

Let $l =$ Length of each member,

              $E =$ Young's modulus for the frame material,

              $\delta_V =$ Vertical deflection of C.

The force in all the members of the frame may be found out by any method. But we shall find out the forces by the method of joints.

First of all consider the joint C. Let the direction of forces in members BC and CD be assumed as shown in Fig. 8·15 . Now resolving the forces horizontally at C,

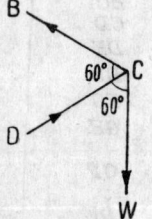

$$P_{BC} \sin 60° = P_{CD} \sin 60°$$

$$\therefore \quad P_{BC} = P_{CD}$$

Now resolving the forces vertically at C,

$$P_{BC} \cos 60° + P_{CD} \cos 60° = W$$

$$2 P_{BC} \times \tfrac{1}{2} = W \qquad (\because P_{BC} = P_{CD})$$

$$\therefore \quad P_{BC} = W \text{ (Tension)}$$

and $P_{CD} = W$ (Compn)

Fig. 8/15

Now consider the joint B. Let the direction of forces in members AB and BD be assumed as shown in Fig. 8·16. Now resolving the forces horizontally at B,

$$P_{AB} \sin 60° = P_{BC} \sin 60°$$

$$\therefore \quad P_{AB} = P_{BC} = W \text{ (Tension)}$$

$$(\because P_{BC} = W)$$

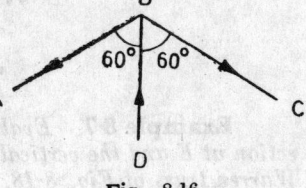

Fig. 8·16

Now resolving the forces vertically at B,

$$P_{BD} = P_{AB} \cos 60° + P_{BC} \cos 60°$$

$$P_{BD} = W \times \tfrac{1}{2} + W \times \tfrac{1}{2} = W \text{ (Compn)}$$

Now consider the joint D.

Let the direction of force in member AD be assumed as shown in Fig. 8·17. Now resolving the forces horizontally at D,

$$P_{AD} \sin 60° = P_{CD} \sin 60°$$

$$\therefore \quad P_{AD} = P_{CD} = W \text{ (Compn)}$$

$$(\because P_{CD} = W)$$

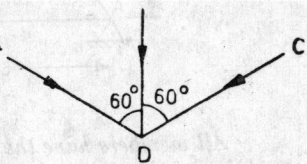

Fig. 8·17

A little consideration will show that when a unit vertical load is applied at C, the forces in all the members will be $\frac{1}{W}$ times the forces due to the load W at C.

Now complete the following table :

Tension $+ve$; Compn $-ve$

Member	l	P	a	$p = \dfrac{P}{a}$	v	pvl
AB	l	$+W$	a	$+\dfrac{W}{a}$	$+1$	$+\dfrac{Wl}{a}$
BC	l	$+W$	a	$+\dfrac{W}{a}$	$+1$	$+\dfrac{Wl}{a}$
CD	l	$-W$	$2a$	$-\dfrac{W}{2a}$	-1	$+\dfrac{Wl}{2a}$
AD	l	$-W$	$2a$	$-\dfrac{W}{2a}$	-1	$+\dfrac{Wl}{2a}$
BD	l	$-W$	$2a$	$-\dfrac{W}{2a}$	-1	$+\dfrac{Wl}{2a}$

$$\therefore \quad \Sigma pvl = \frac{7Wl}{2a}$$

Using the relation,

$$\delta_V = \frac{\Sigma pvl}{E} \text{ with usual notations.}$$

$$= \frac{7Wl}{2aE} \textbf{ Ans.}$$

Example 8·7. *Evaluate the horizontal component of the deflection at E and the vertical component of the deflection at B in the Warren truss of Fig. 8·18.*

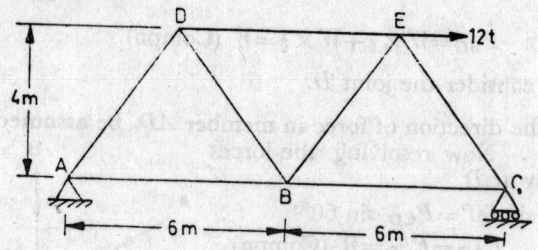

Fig. 8·18.

All members have the area of cross-section of 6·55 cm² and the modulus of elasticity of the material is 2,000 tonnes/cm².

(*A.M.I.E., Summer, 1975*)

Solution.

Given.   Horizontal load at $E = 12$ t

Area of members $= 6\cdot55$ cm²

Young's modulus, $E = 2{,}000$ t/cm²

Horizontal component of deflection at E

Let $\delta_H =$ Horizontal component of deflection at E.

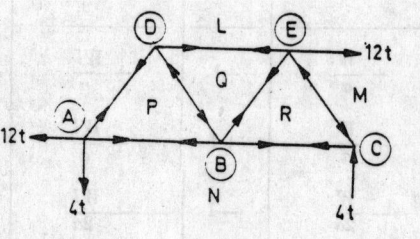

(*a*) Space diagram   (*b*) Vector diagram

Fig. 8·19.

First of all calculate the reactions at A and C. Now in order to determine the forces in the various members of the truss, draw the space diagram and then vector diagram as shown in Fig 8·19. (*a*) and (*b*).

We know that when the truss is carrying a unit horizontal load, at C, the force in all the members of the truss will be 1/12th of the forces obtained from the above vector diagram. Now complete the following table, by measuring the various sides of the vector diagram.

Tension $+ve$; Compn. $-v$.

Member	l cm	p tonnes	a cm²	$p=\dfrac{P}{a}$ t/cm²	u t/cm²	pul
AB	600	$+9\cdot0$	$6\cdot55$	$+\dfrac{9\cdot0}{6\cdot55}$	$+\dfrac{3}{4}$	$+\dfrac{4,050}{6\cdot55}$
BC	600	$+3\cdot0$	$6\cdot55$	$+\dfrac{3\cdot0}{6\cdot55}$	$+\dfrac{1}{4}$	$+\dfrac{450}{6\cdot55}$
AD	500	$+5\cdot0$	$6\cdot55$	$+\dfrac{5\cdot0}{6\cdot55}$	$+\dfrac{5}{12}$	$+\dfrac{12,500}{12\times6\cdot55}$
DE	600	$+6\cdot0$	$6\cdot55$	$+\dfrac{6\cdot0}{6\cdot55}$	$+\dfrac{1}{2}$	$+\dfrac{1,800}{6\cdot55}$
DB	500	$-5\cdot0$	$6\cdot55$	$-\dfrac{5\cdot0}{6\cdot55}$	$-\dfrac{5}{12}$	$+\dfrac{12,500}{12\times6\cdot55}$
BE	500	$+5\cdot0$	$6\cdot55$	$+\dfrac{5\cdot0}{6\cdot55}$	$+\dfrac{5}{12}$	$+\dfrac{12,500}{12\times6\cdot55}$
EC	500	$-5\cdot0$	$6\cdot55$	$-\dfrac{5\cdot0}{6\cdot55}$	$-\dfrac{5}{12}$	$+\dfrac{12,500}{12\times6\cdot55}$

Using the relation, $\therefore$ $\Sigma pvl = 1,600$

$$\delta_H = \frac{pvl}{E} \text{ with usual notations}$$
$$= \frac{1,600}{2,000} = \textbf{0·8 cm } (\rightarrow) \textbf{ Ans}.$$

Vertical deflection at B

Let $\delta_V = $ Vertical deflection at B.

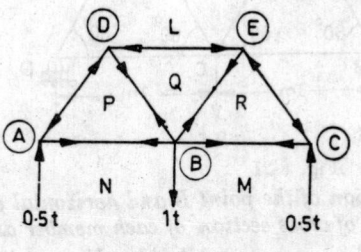
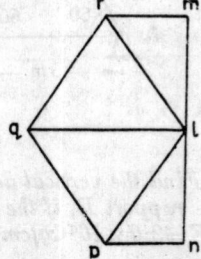

(a) Space diagram   (b) Vector diagram

Fig. 8·20.

In order to determine the forces in the various members of the truss, when subjected to a unit vertical load at B, draw the space diagram and then vector diagram as shown in Fig. 8·20 (a) and (b). Now complete the following table by measuring the various sides of the vector diagram

Tension +ve ; Compn.—ve

Member	l cm	P tonnes	a cm²	$p=\dfrac{P}{a}$ t/cm²	v t/cm²	pvl
AB	600	+9·0	6·55	$+\dfrac{9\cdot0}{6\cdot55}$	$+\dfrac{3}{8}$	$+\dfrac{2,025}{6\cdot55}$
BC	600	+3·0	6·55	$+\dfrac{3\cdot0}{6\cdot55}$	$+\dfrac{3}{8}$	$+\dfrac{675}{6\cdot55}$
AD	500	+5·0	6·55	$+\dfrac{5\cdot0}{6\cdot55}$	$-\dfrac{5}{8}$	$-\dfrac{1,562\cdot5}{6\cdot55}$
DE	600	+6·0	6·55	$+\dfrac{6\cdot0}{6\cdot55}$	$-\dfrac{3}{4}$	$-\dfrac{2,700}{6\cdot55}$
DB	500	−5·0	6·55	$-\dfrac{5\cdot0}{6\cdot55}$	$+\dfrac{5}{8}$	$-\dfrac{1,562\cdot5}{6\cdot55}$
BE	500	+5·0	6·55	$+\dfrac{5\cdot0}{6\cdot55}$	$+\dfrac{5}{8}$	$+\dfrac{1,562\cdot6}{6\cdot55}$
EC	500	−5·0	6·55	$-\dfrac{5\cdot0}{6\cdot55}$	$-\dfrac{5}{8}$	$-\dfrac{1,562\cdot5}{6\cdot55}$

∴ Vertical deflection at B=0 **Ans.**   ∴ $\Sigma\,pvl=0$

Example 8·8. *A truss, shown in Fig. 8·21 is hinged at A, and roller supported at D. It carries point loads at B and C.*

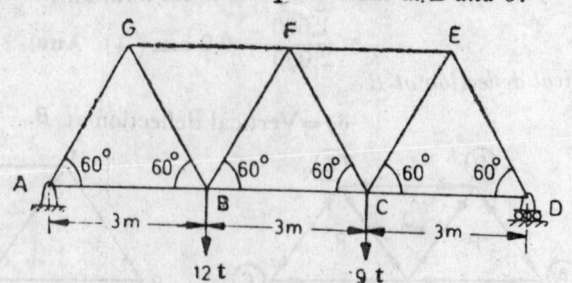

Fig. 8·21

Find the vertical deflection of the point B and horizontal deflection of the support D, if the area of cross section of each member as 50 cm². Take $E=2\cdot0\times10^6$ kg/cm².

(Roorkee University, 1974)

Solution.

Given. Load at B = 12 t

Load at C = 9 t

Area of each member, $a = 50$ cm^2

Young's modulus, $E = 2 \cdot 0 \times 10^6$ kg/cm^2

From the geometry of the figure, we find that the length of each member $= 3$ m $= 300$ cm.

Vertical deflection of B

Let $\delta_V = $ Vertical deflection of B.

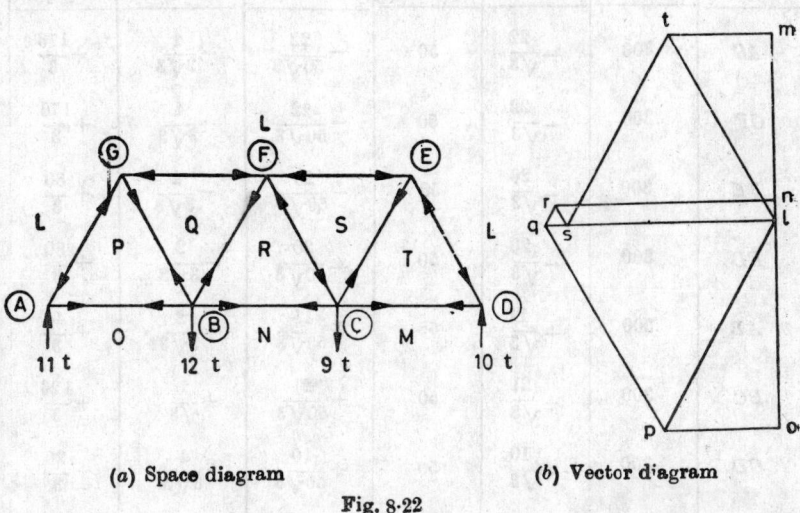

(a) Space diagram   (b) Vector diagram

Fig. 8·22

Now, in order to determine the forces in the various members of the truss, draw the space and vector diagrams as shown in Fig. 8·22 (a) and (b).

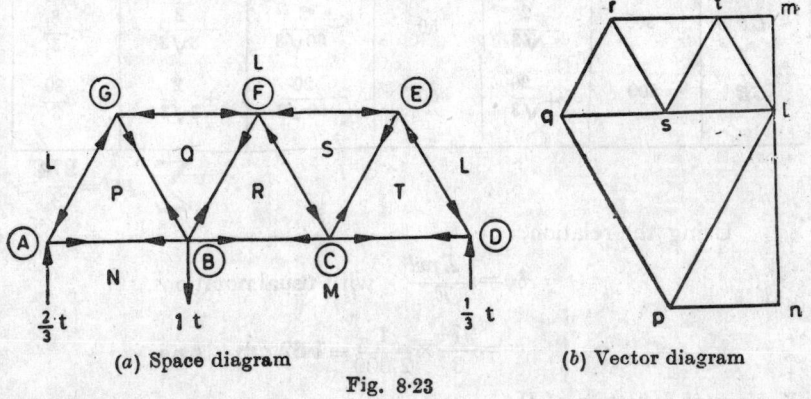

(a) Space diagram   (b) Vector diagram

Fig. 8·23

Now, consider a unit vertical load acting at B. In order to determine the forces in the various members of the truss, draw the space and vector diagrams as shown in Fig. 8·23 (a) and (b).

Now complete the following table, by measuring the various sides of the vector diagrams in Fig. 8·22 (b) and 8·23 (b).

Tension $+ve$; Compn. $-ve$

Member	l cm	P tonnes	a cm²	$p = \dfrac{P}{a}$ t/cm²	v t/cm²	pvl
AG	300	$-\dfrac{22}{\sqrt{3}}$	50	$-\dfrac{22}{50\sqrt{3}}$	$-\dfrac{4}{3\sqrt{3}}$	$+\dfrac{176}{3}$
GF	300	$-\dfrac{22}{\sqrt{3}}$	50	$-\dfrac{22}{50\sqrt{3}}$	$-\dfrac{4}{3\sqrt{3}}$	$+\dfrac{176}{3}$
FE	300	$-\dfrac{20}{\sqrt{3}}$	50	$-\dfrac{20}{50\sqrt{3}}$	$-\dfrac{2}{3\sqrt{3}}$	$+\dfrac{80}{3}$
ED	300	$-\dfrac{20}{\sqrt{3}}$	50	$-\dfrac{20}{50\sqrt{3}}$	$-\dfrac{2}{3\sqrt{3}}$	$+\dfrac{80}{3}$
AB	300	$+\dfrac{11}{\sqrt{3}}$	50	$+\dfrac{11}{50\sqrt{3}}$	$+\dfrac{2}{3\sqrt{3}}$	$+\dfrac{44}{3}$
BC	300	$+\dfrac{21}{\sqrt{3}}$	50	$+\dfrac{21}{50\sqrt{3}}$	$+\dfrac{1}{\sqrt{3}}$	$+\dfrac{126}{3}$
CD	300	$+\dfrac{10}{\sqrt{3}}$	50	$+\dfrac{10}{50\sqrt{3}}$	$+\dfrac{1}{3\sqrt{3}}$	$+\dfrac{20}{3}$
BG	300	$+\dfrac{22}{\sqrt{3}}$	50	$+\dfrac{22}{50\sqrt{3}}$	$+\dfrac{4}{3\sqrt{3}}$	$+\dfrac{176}{3}$
BF	300	$+\dfrac{2}{\sqrt{3}}$	50	$+\dfrac{2}{50\sqrt{3}}$	$+\dfrac{2}{3\sqrt{3}}$	$+\dfrac{8}{3}$
CF	300	$-\dfrac{2}{\sqrt{3}}$	50	$-\dfrac{2}{50\sqrt{3}}$	$-\dfrac{2}{3\sqrt{3}}$	$+\dfrac{8}{3}$
CE	300	$+\dfrac{20}{\sqrt{3}}$	50	$+\dfrac{20}{50\sqrt{3}}$	$+\dfrac{2}{3\sqrt{3}}$	$\dfrac{80}{3}$

$$\therefore \sum pvl = \frac{974}{3}$$

Using the relation,

$$\delta_V = \frac{\Sigma pvl}{E} \quad \text{with usual notations.}$$

$$= \frac{974}{3} \times \frac{1}{2,000} = 1 \cdot 62 \text{ cm} \quad \text{Ans.}$$

Horizontal deflection of D

Let $\delta_H =$ Horizontal deflection of D.

A little consideration will show that when a unit load is applied at C, there will be a force of 1 tonne each in members AB, BC and

CD, whereas there will be no   force in all the other members of the truss as shown in Fig. 8·24.

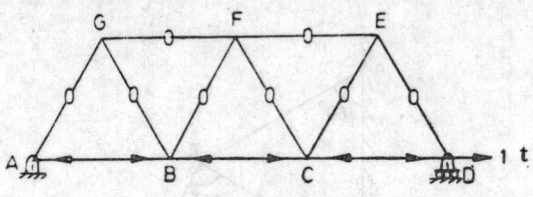

Fig. 8·24

Now complete the following table.

Tension +ve ;   Compn. −ve

Member	l cm	P tonnes	a cm²	$p = \dfrac{P}{a}$ t/cm²	u t/cm²	pul
AB	300	$+\dfrac{11}{\sqrt{3}}$	50	$\dfrac{11}{50\sqrt{3}}$	1	$\dfrac{66}{\sqrt{3}}$
BC	300	$+\dfrac{21}{\sqrt{3}}$	50	$\dfrac{21}{50\sqrt{3}}$	1	$\dfrac{126}{\sqrt{3}}$
CD	300	$+\dfrac{10}{\sqrt{3}}$	50	$\dfrac{10}{50\sqrt{3}}$	1	$\dfrac{60}{\sqrt{3}}$

$$\therefore \sum pul = \frac{252}{\sqrt{3}}$$

Now using the relation,

$$\delta_H = \frac{\Sigma pul}{E} \text{ with usual notations.}$$

$$= \frac{252}{\sqrt{3}} \times \frac{1}{2,000} = 0.073 \text{ cm   Ans.}$$

Exercise 8–1

1.   A crane consists of a jib 7·5 m long cf 15 cm² sectional area and a horizontal tie 6 m long of 10 cm² sectional area.   Determine the vertical and horizontal displacements of the crane head, when a load of 10 tonnes is suspended from it.   Take $E = 2.0 \times 10^6$ kg/cm².   [Ans.   1·28 cm ; 0·4 cm

2.   Fig. 8·22 shows a pin-jointed frame, hinged to a rigid wall at A and B, carries a vertical load of W at D.   The area of each tension member is a and that of compression member is $2a$.

Determine the **vertical and horizontal displacements** of the joint D in terms of W, l, a and E,

(Jiwaji University)

$$\left[\begin{array}{c}\textbf{Ans.} \quad \dfrac{2 \cdot 57\ Wl}{aE}\ ;\ \dfrac{0 \cdot 29\ Wl}{aE}\end{array}\right]$$

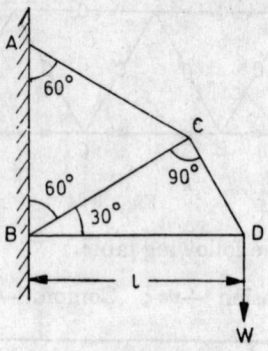

Fig. 8·25

3. A steel truss shown in Fig. 8 26 is hinged at A supported on roller at D.

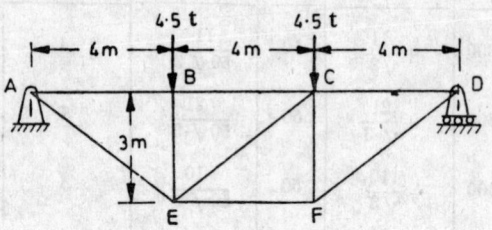

Fig. 8·26

Find the vertical and horizontal deflection of the joint F, if all the tension and compression members are stressed to 1,000 kg/cm² and 300 kg/cm². Take $E = 2 \cdot 0 \times 10^6$ kg/cm².   (*Banaras Hindu University*)

[**Ans.**   0·84 cm ; 0·48 cm]

8·8. Graphical method for deflection

The deflection of a joint may also be found out, by the graphical method, as discussed below :

1.   First of all, find out the forces $P_1, P_2, P_3...$ in the various members of a frame due to the external loading.

2.   Now find out the deformations $\delta l_1, \delta l_2, \delta l_3...$ in the lengths of the various members due to the forces $P_1, P_2, P_3...$ The type of deformation (*i.e.* extension or contraction) should be clearly considered. The usual practice is to use $+ve$ for extensions and $-ve$ for contractions.

3.   Now select a joint in the given frame, which is rigidly fixed and is not subjected to vertical or horizontal displacement. Now start constructing the Williot diagram from this joint, and complete the whole diagram as discussed in the next article.

4. The vertical as well as the horizontal deflection, of any joint, may be found out from the vertical and horizontal components of the line joining the starting point, and the point which represents the final position of the joint in the Williot diagram.

8·9. Williot diagram for the deflection

The Williot diagram is very useful, and interesting as well, for finding out the deflection of any joint in a frame carrying external loads. In general, for drawing the Williot diagrams, the frames may be classified into the following three types.

1. Frames with two joints fixed.

2. Frames with one joint and the direction of the other fixed.

3. Frames with only one joint fixed.

8·10. Williot diagram for the frames, with two joints fixed

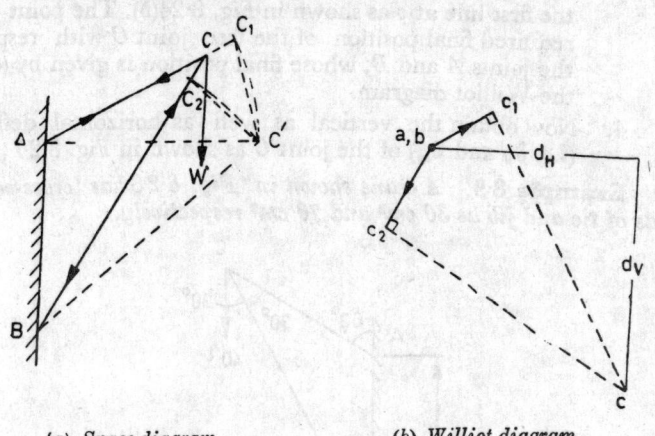

(a) *Space diagram*   (b) *Williot diagram*

Fig. 8·27

Consider a simple triangular cantilever frame ABC fixed in the wall at A and B. Let the free joint C be subjected to a vertical load W. We know that as a result of the load W, the member AC of the frame will be subjected to tension and the member BC will be subjected to compression. First of all, calculate the actual force in the members AC and BC and then the extension δl_1 of the member AC and the contraction δl_2 of the member BC.

Now extend the member AC to AC_1, such that CC_1 is equal to the extensions δl_1 of the member AC. Similarly mark CC_2 equal to the contraction δl_2 of the member BC. A little consideration will show, that final length of the member AC is AC_1 and the length of the member BC is BC_2. Since the joints A and B are fixed, therefore the free joint C has to move to a new position, according to the new lengths AC_1 and BC_2 of the two members. Now, in order to locate the new position of the free joint C, take A as centre and radius equal to AC_1 draw an arc. Again take B as the centre and radius equal to BC_2 draw an arc, meeting the first arc at C'. Now C' is the final position of the free joint C. The vertical as well as horizontal deflections of the free joint C may now be obtained from the vertical and horizontal components of CC'. These deflections may be easily found out, by drawing the Williot diagram as discussed below :

1. Take some suitable point a, b to represent the fixed position of the joints A and B. Through a, draw ac equal to δl_1 and parallel to AC of the space diagram. (The direction of C with respect to a *i.e.* whether upwards or downwards of a, should be decided from the arrow head near the joint A in the member AC of the space diagram).

2. Through b, draw bc equal to δl_2 and parallel to BC of the space diagram. (The direction of C with respect to A, should be decided from the arrow head near the joint B in the member BC of the space diagram).

3. Through c_1 draw a line c_1c at right angles to ac_1. Similarly through c_2, draw a line c_2c at right angles to bc_2, meeting the first line at c as shown in Fig. 8·24(b). The point c is the required final position of the free joint C with respect to the joints A and B, whose final position is given by (a, b) in the Williot diagram.

4. Now obtain the vertical as well as horizontal deflections (*i.e.* δ_V and δ_H) of the joint C as shown in Fig. 8·27 (b).

Example 8·9. *A crane shown in Fig. 8·28 has [cross-sectional areas of tie and jib as 30 cm² and 70 cm² respectively.*

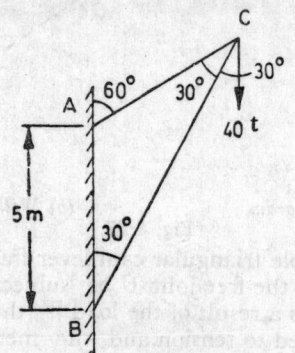

Fig. 8·28

Determine the vertical deflection of the joint C, when a load of 40 T is suspended from it. Take E as $2·0 \times 10^6$ kg/cm².

(*Patna University, 1972*)

*Solution.

Given.   Area of member

$$AC = 30 \text{ cm}^2 \qquad (\because \; AC \text{ is tie member})$$

Area of member $BC = 70$ cm² $(\because$ BC is jib member)

Load at C $= 40$ t

Young's modulus, $E = 2·0 \times 10^6$ kg/cm² $= 2·0 \times 10^3$ t/cm²

From the geometry of the figure, we find that the length of $AC = 5$ m $= 500$ cm and that of $BC = 8·66$ m $= 8·66$ cm.

---

*We have already solved this problem analytically on page 290.

We know that force in member AC,

$$P_{AC} = 40 \text{ t (Tension)}$$
$$P_{BC} = 69 \cdot 28 \text{ t (Compn.)}$$

∴ Increase in the length of member AC,

$$\delta_1 = \frac{Pl}{aE} = \frac{40 \times 500}{30 \times 2 \cdot 0 \times 10^3} = \frac{1}{3} = 0 \cdot 333 \text{ cm}$$

Similarly, decrease in the length of member BC,

$$\delta l_2 = \frac{Pl}{aE} = \frac{69 \cdot 28 \times 866}{70 \times 2 \cdot 0 \times 10^3} = 0 \cdot 43 \text{ cm}$$

Now draw the Williot diagram as discussed below :

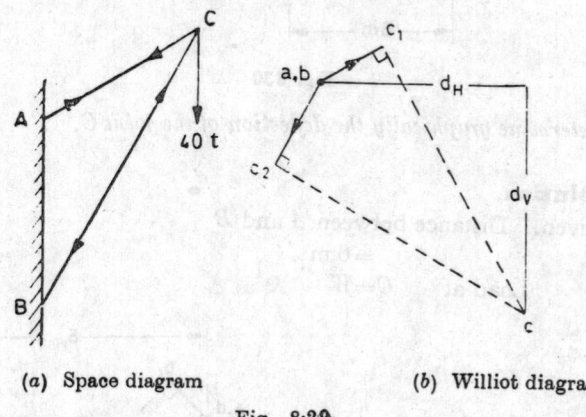

(a) Space diagram   (b) Williot diagram

Fig. 8·29

1. First of all, draw the space diagram of the crane as shown n Fig. 8·29 (a).

2. Now take some suitable point a, b to represent the fixed position of A and B. Through a draw ac, equal to $\delta l_1 = 0 \cdot 333$ cm to some scale and parallel to AC of the space diagram (the direction of c_1 with respect to a, *i.e.* whether upwards or downwards should be decided from the arrow head near the joint A in the member AC of the space diagram).   Similarly through b draw bc_2 equal to $\delta l_2 = 0 \cdot 43$ cm to scale and parallel to BC of the space diagram (the direction of c_2 with respect to b should be decided from the arrow head near the joint B).

3.   Through c_1 and c_2 draw lines c_1c and c_2c at right angles to ac_1 and bc_2 respectively meeting at c.   This is the required position of the joint C with respect to the fixed joints A and B, whose position is given by a and b in the Williot diagram.

By measurement, we find that the vertical deflection of joint C,

$$d_V = 1 \cdot 076 \text{ cm} \textbf{ Ans.}$$

Example 8·10. *A jib crane is fixed at A and D 6 m apart. When a load W is applied at C, the changes in lengths take place, as shown in Fig. 8·30*

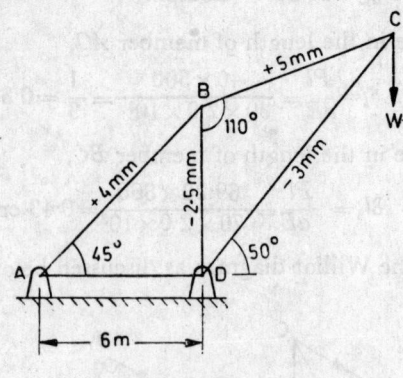

Fig. 8·30

Determine graphically the deflection of the joint C.

Solution.

Given.   Distance between A and B
$$= 6 \text{ m}$$
Load at $C = W$

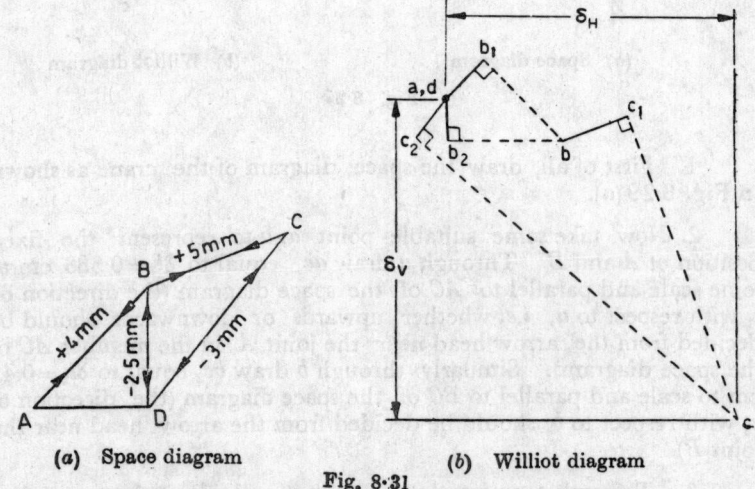

(a) Space diagram   **(b) Williot diagram**

Fig. 8·31

Now draw the Williot diagram as discussed below.

1. First of all, draw the space diagram of the crane as shown in Fig. 8·31 (a).

2. Now take some suitable point a, d to represent the fixed position of A and D. Through a, draw a line equal to 4 mm (given) to

some scale and parallel to AC of the space diagram (the direction of b_1 with respect to a, *i.e.* whether upwards or downwards should be decided from the arrow head near the joint A in the member AB of the space diagram). Similarly through d draw db_2 equal to 2·5 mm (given) to scale and parallel to DB of the space diagram (the direction of b_2 with respect to d should be decided from the arrow head near the joint D).

    3.   Through b_1 and b_2 draw lines b_1b and b_2b at right angles to ab_1 and db_2 respectively meeting at b, which gives the required position of the joint B with respect to the fixed joints A and D.

    4.   Through b, draw bc_1 equal to 5 mm (given) to scale and parallel to BC of the space diagram (the direction of c_1 with respect to b *i.e.* whether upwards or downwards should be decided from the arrow head near the joint B in the member BC of the space diagram). Similarly through d, draw dc_2 equal to 3 mm (given) to scale and parallel to DC of the space diagrams (the direction of c_2 with respect to d should be decided from the arrow head near the joint D).

    5.   Through c_1 and c_2 draw lines c_1c and c_2c at right angles to bc_1 and dc_2 respectively meeting at c. This is the required position of the joint C, with respect to the fixed joints A and D, whose position is given by a and d in the Williot diagram.

    By measurement we find that

$$\delta_V = 2·05 \text{ cm} \quad \textbf{Ans.}$$
$$\delta_H = 2·0 \text{ cm} \quad \textbf{Ans.}$$

8·11. Williot diagram for the frames with one joint, and the direction of the other fixed

    Consider a simple triangular frame ABC, fixed at A and supported on rollers at B. Let the free joint C be subjected to a vertical

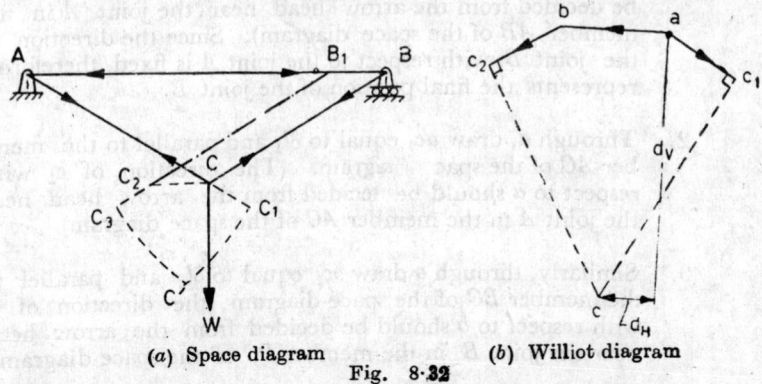

(a) Space diagram   (b) Williot diagram
Fig. 8·32

load W. We know that as a result of this load W, the members AC and BC will be subjected to tension, whereas the member BA will be subjected to compression. First of all calculate the actual forces in the members AB, AC and BC and then the extension δl_1 and δl_2 in the members AC and BC, and the contraction δl_3 in the member AB.

Now, we know that the joint A is fixed and the joint C is supported on rollers, therefore any decrease in the length of member AB will cause the joint B to move horizontally towards left (since the joint A cannot move in any other direction). Now mark the member AB to AB_1, such that BB_1 is equal to the contraction δl_3 of the member AB. Similarly, extend the member AC to CC_1 equal to the extension δl_1 of the member AC. Now a little consideration will show, that if the member AC is not to rotate about the joint A, the final position of C should be C_1. Similarly, if the member BC is neither to rotate about B_1 nor to extend in its length then the final position of C should be C_2 ; such that CC_2 is equal to the contraction δl_3 of the member AB. The member BC would have occupied the position B_1C_2 as shown by the dotted line. Since the member BC had been extended by an amount equal to δl_2, therefore mark C_2C_3 is equal to the extension δl_2. Now we see that if the member BC is not to rotate about the joint B, the final position of C should be C_3. A little consideration will show, that since the joint C has to remain intact, therefore it must move to a new position, according to the new lengths AC_1 and B_1C_3. Now in order to locate the new position of the free joint C, take A as the centre and radius equal to AC_1 draw an arc. Again with B_1 as the centre and radius equal to B_1C_3 draw an arc meeting the first arc at C'. Now C' is the final position of the free joint C. The vertical as well as the horizontal deflections of the free joint C may now be obtained, from the vertical and horizontal components of CC'. The Williot diagram for the deformations of the members AB, AC and BC may be drawn as discussed below :

1. Take some suitable point a to represent the fixed position of the joint A. Through a, draw ab equal to δl_3 and parallel to the member AB of the space diagram (the direction of b with respect to a, *i.e.* whether towards left or right of a, should be decided from the arrow head near the joint A in the member AB of the space diagram). Since the direction of the joint B, with respect to the joint A is fixed, therefore b represents the final position of the joint B.

2. Through a, draw ac_1 equal to δl_1 and parallel to the member AC of the space diagram. (The direction of c_1 with respect to a should be decided from the arrow head near the joint A in the member AC of the space diagram).

3. Similarly, through b draw bc_2 equal to δl_2 and parallel to the member BC of the space diagram (the direction of c_2 with respect to b should be decided from the arrow head near the joint B, in the member BC of the space diagram).

4. Through c_1 draw a line c_1c at right angles to ac_1. Similarly through c_2 draw a line c_2c at right angles to bc_2, meeting first line at c as shown in Fig. 8·29 (*b*). The point c is the required final position of the free joint C with respect to the joints A and B, whose final positions are given by a and b in the Williot diagram.

5.   Now obtain the vertical as well as the horizontal deflections (*i.e.*, dv and d_H) of the joint C as shown in Fig 8·32 (*b*).

Example 8·11. *A pin-jointed frame shown in Fig. 8·33 is carrying a load of 6 tonnes at C.*

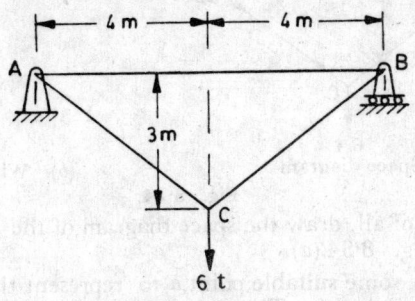

Fig. 8·33

Find the vertical as well as horizontal deflection of C. Take area of member AB as 10 cm² and those of members AC and BC as 15 cm². E=2·0×10³ t/m².   [Bihar University, 1978].

*Solution.

Given.   Load   =6 t
Area of member AB =10 cm²
Area of member AC =15 cm²
Area of member BC =15 cm²
Young's modulus, E =2·0×10³ t/cm²

From the geometry of the figure, we find that the length of $AB=8$ m$=800$ cm and that of AC and $BC=5$ m$=500$ cm.

We know that the force in member AC,
$$P_{AC}=5 \text{ t (Tension)}$$
$$P_{BC}=5 \text{ t (Tension)}$$
$$P_{AB}=4 \text{ t (Compn.)}$$

∴   Increase in the length of member AC,
$$\delta l_1=\frac{Pl}{aE}=\frac{5\times500}{15\times2\cdot0\times10^3}=0\cdot083 \text{ cm}$$

Similarly, increase in the length of member BC,
$$\delta l_2=\frac{Pl}{aE}=\frac{5\times500}{15\times2\cdot0\times10^3}=0\cdot083 \text{ cm}$$

and decrease in the length of member AB,
$$\delta l_3=\frac{Pl}{aE}=\frac{4\times800}{10\times2\cdot0\times10^3}=0\cdot16 \text{ cm}$$

───────────────

*We have already solved the problem analytically

Now draw the Williot diagram as discussed below :

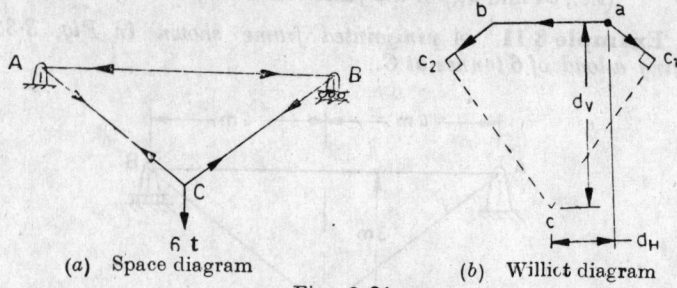

(a) Space diagram   (b) Williot diagram

Fig. 8·34

1. First of all, draw the space diagram of the frame as shown in Fig. 8·34.(a).

2. Take some suitable point a to represent the fixed position of the joint A. Through a draw ab equal to $\delta l_3 = 0.16$ cm to some scale and parallel to AB of the space diagram (the direction of b with respect to a, i.e. whether towards left or right of a should be decided from the arrow head near the joint A in the member AB of the diagram). Since the direction of joint B is fixed, therefore b represents the final position of the joint B.

3. Through a draw ac_1 equal to $\delta l_1 = 0.083$ cm to scale and parallel to AC of the space diagram (the direction of c_1 with respect to a, i.e. whether upwards or downwards should be decided from the arrow head near the joint A in the member AC of the space diagram). Similarly through b, draw bc_2 equal to $\delta l_2 = 0.083$ cm to scale and parallel to BC of the space diagram (the direction of c_2 with respect to b should be decided from the arrow head near the joint B).

4. Through c_1 and c_2 draw lines c_1c and c_2c at right angles to ac_1 and bc_2 respectively meeting at c. This is the required position of the joint C with respect to the fixed joint A, whose position is given by a in the Williot diagram.

By measurement, we find that the vertical deflection of joint C,

$$d_V = 0.245 \text{ cm} \quad \textbf{Ans.}$$

and $$d_H = 0.08 \text{ cm} \quad \textbf{Ans.}$$

8·12. Williot diagram for the frames, with only one joint fixed

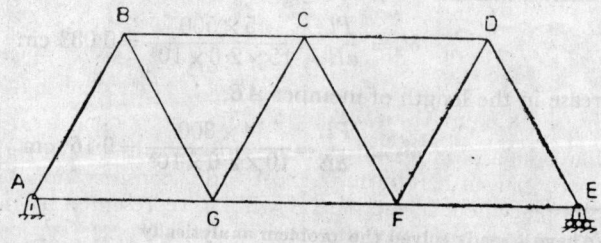

Fig. 8· 35

Sometimes, we are required to find out the vertical as well as horizontal deflection of a joint in a frame, which is fixed at one end, and supported on rollers at the other as shown in Fig. 8·3 5. In such a case, we cannot draw the Williot diagram, until and unless we assume either the direction of joint G or the joint B, with respect to the joint A, to be fixed. Strictly speaking, either of the two assumptions are wrong, since the joints B and G are bound to have vertical as well as horizontal deflection, under the action of the external loading.

In order to start the Williot diagram, we have to make either of two assumptions, stated above. The usual practice, in such cases, is that the direction of G with respect to the joint A, is first assumed to be fixed, and then the Williot diagram for the whole truss is drawn. After completing the Williot diagram we have to apply the necessary correction, for the wrong assumption ; we made at the time of drawing the Williot diagram. This correction is applied by drawing another diagram, called Mohr diagram. The combined Williot and Mohr diagram is known as Williot-Mohr diagram, which is discussed in the succeeding article.

8·13. Mohr diagram

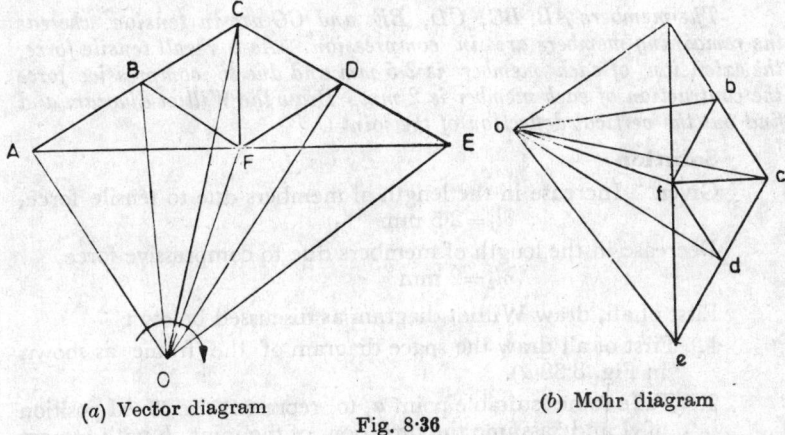

(a) Vector diagram   (b) Mohr diagram

Fig. 8·36

The Mohr diagram may be best understood by studying the effect of a small clockwise rotation, given to a truss $ABCDEF$ about some fixed point (say O) as shown in Fig. 8·36 (a).

If the rays are drawn from the point O to each joint of the truss, then the movement of each joint will be perpendicular to a ray drawn from point O to the joint. The magnitude of the displacement of each joint will be proportional to the length of the ray, drawn from the point O, to the joint.

Now take a point o and through it draw a line oa at right to OA and proportional, in length, to OA. A little consideration will show, that the displacement of the joint A will be represented by the vector oa to some scale. Similarly draw other lines ob, oc, od, oe and of

322   DEFLECTION OF PERFECT FRAMES

representing the displacements of the joints B, C, D, E and F to the scale. If we join ab, bc, cd, de, ef, fa, fb and fd, we shall get a figure, similar to the truss as shown in Fig. 8·36 (b). This new figure, which is similar to the given truss in all respects, except the scale and turned clockwise through 90°, is known as *Mohr diagram*.

The Mohr diagram has got its own importance, as it gives us the ratio, and not the absolute values, of the displacements of the joints. The use of Mohr or Williot-Mohr diagrams may be best understood from the following example.

Example 8·12. *A Warren truss shown in Fig. 8·37 is subjected to some external loads.*

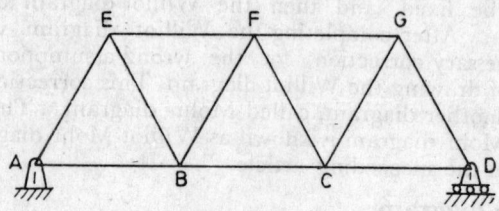

Fig. 8·37

The members AB, BC, CD, BE and CG are in tension, *whereas the remaining members are in compression. As a result tensile force, the extension of each member is 2·5 mm and due to compressive force the contraction of each member is 2 mm. Draw the Williot diagram and find out the vertical deflection of the joint C.*

Solution.

Given. Increase in the length of members due to tensile force,
$$\delta l_1 = 2·5 \text{ mm}$$

Decrease in the length of members due to compressive force,
$$\delta l_2 = 2 \text{ mm}$$

First of all, draw Williot diagram as discussed below :

1. First of all draw the space diagram of the frame as shown in Fig. 8·38(a).

2. Take some suitable point a, to represent the fixed position of A and *assume the direction of the joint B with respect to the joint A to be fixed. Through a draw a line parallel to AB. Now mark the position of b, such that ab is equal to $\delta l_1 = 2·5$ mm to some scale (the direction of b with respect to a, *i.e.* whether towards left or right should be decided from the arrow head near the joint A).

3. Through a draw ae_1 equal to $\delta l_2 = 2$ mm and parallel to AE (the direction should be decided from the arrow head near A). Similarly through b draw be_2 equal to $\delta l_1 = 2·5$ mm and parallel to BE (the direction should be decided from the arrow head near B).

---

*Strictly speaking, this assumption is not correct as the joint B will also undergo some vertical displacement with respect to the joint A.

4. Now through e_1 and e_2, draw e_1e and e_2e at right angles to ae_1 and be_2 respectively meeting at e. The point e gives the final position of the joint E with respect to the fixed joints A and B (assumed to be fixed).

5. Similarly, locate the position f of the joint F from the joints B and E.

6. Now locate the position c of the joint C from the joints B and F.

7. Now locate the position of g of the joint G from the joints F and C.

8. Now locate the position of d of the joint D from the joints C and G.

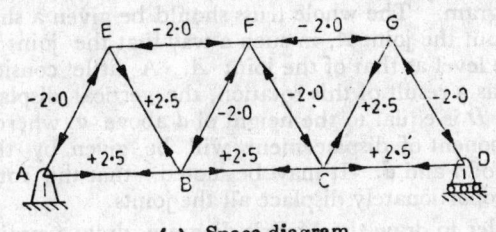

(a) Space diagram

(b) Williot-Mohr diagram
Fig. 8·38

From the Williot diagram (as drawn above), we see that final position of the joint D is d, with respect to joints A and B. Thus the vertical distance between a and d, gives the vertical deflection of the joint D with respect to the joint A. But, from the geometry of the figure, we find that the joint D is supported on rollers and both the joints A and D are at the same level. It is thus obvious, that the vertical deflection of the joint D with respect to the joint A must be zero. This discrepancy has crept in, due to the fact, that we assumed the direction of the joint B to be fixed with respect to the joint A, while drawing the vector diagram. But we know that the direction of the joint B, with respect to the joint A, is not fixed, but rotates about the joint A.

This discrepancy may be easily corrected, with the help of Mohr's diagram. The whole truss should be given a slight clockwise rotation about the joint A, in such a way that the joint D is restored to the same level as that of the joint A. A little consideration will show, that as a result of this rotation, the vertical displacement given to the joint D is equal to the height of d above a, whereas the horizontal component of displacement will be given by the horizontal component of a and d. It may be noted, that the rotation of the truss will proportionately displace all the joints.

In order to draw the Mohr's diagram, draw a straight line dd' parallel to BA, to meet the vertical line through a at d'. The line ad' represents the lower chord AD of the truss. Now with AD as the base, draw a figure $a'\ b'\ c'\ d'\ e'\ f'\ g'$ similar to that of the truss. This figure should be drawn in such a way, that after rotating it through $90°$, it becomes parallel and similar to the given truss. The figure $a'\ b'\ c'\ d'\ e'\ f'g'$ is called Williot diagram and combined diagram is called Williot-Mohr diagram. Now the vertical and horizontal deflection of a joint is given by the respective component of the joint in the Williot diagram and Mohr's diagram (e.g. the vertical and horizontal deflection of a joint C is given by the respective components of c and c').

By measurement, we find that the vertical deflection of the joint C (given by the vertical component of vector cc'),

$$\delta v = 10.5 \text{ mm Ans.}$$

Exercise 8-2

1. Two rods AC and BC are hinged at C and is carrying a load of 8 tonnes at C as shown in Fig. 8·39.

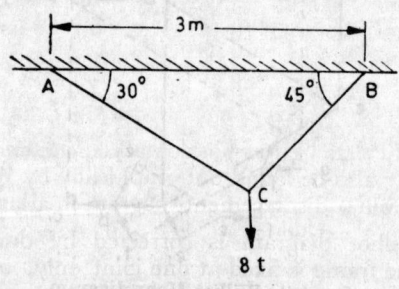

Fig. 8·39

Find graphically, **or** otherwise, the vertical and horizontal deflection of the joint C. Take area of member AB as 9 cm^2 and that of BC as 12 cm^2. $E = 2.02 \times 10^6$ kg/cm^2. *(Baroda University)*

[**Ans.** 0.44 mm ; 0.085 mm]

2. A steel truss of span 15 m is loaded as shown in Fig. 8.40

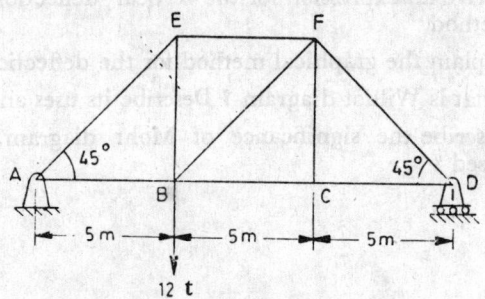

Fig. 8.40

The cross sectional area of the members is such that each member is subjected to a stress of 800 kg/cm^2. Find graphically the vertical and horizontal deflection of the C. Take $E = 2.0 \times 10^6$ kg/cm^2. *(Jabalpur University)*

[**Ans.** 0.73 cm ; 0.4 cm]

HIGHLIGHTS

1. A perfect frame is that in which the number of members is equal to $(2j - 3)$ where j is the number of joints.

2. The vertical deflection of the joint of a perfect frame,

$$\delta_V = \sum \frac{pvl}{E}$$

and horizontal deflection,

$$\delta_H = \sum \frac{pul}{E}$$

where $p_1, p_2, p_3 \ldots$ = Forces in the members 1, 2, 3...

$v_1, v_2, v_3 \ldots$ = Forces in the members 1, 2, 3..., when the external loads are removed, and a unit vertical load is attached to the joint.

$l_1, l_2, l_3 \ldots$ = Length of the members 1, 2, 3...

$u_1, u_2, u_3 \ldots$ = Forces in the members 1, 2, 3..., when the external loads are removed, and a unit horizontal load is acting on the joint.

E = Young's modulus for material.

3. The vertical as well as horizontal deflection of a joint, in a perfect frame, may also be found out graphically by Williot diagram, which is nothing, but a displacement diagram to a larger scale.

4. The Williot diagram is corrected by drawing a Mohr's diagram, when the frame is fixed at one joint only, and direction of any other joint is not fixed.

Do You Know ?

1. What is meant by a perfect frame ? How will you determine, whether the given frame is a perfect one or not ?

2. Derive an expression for the vertical deflection of a joint by unit load method.

3. Explain the graphical method for the deflection.

4. What is Williot diagram ? Describe its uses and importance

5. Describe the significance of Mohr diagram. When and where it is used ?

9

CABLES AND SUSPENSION BRIDGES

·1. Introduction

A suspension bridge consists of two cables, which are stretched ver the span to be bridged. Each cable, passing over two towers, s anchored by backstays to a firm foundation as shown in Fig. 9·1. As the cable is flexible throughout, therefore it cannot resist any moment and can adopt any shape under the loads ; that is why the bending moment at every point of the cable is taken as zero. The roadway is suspended from the cables by means of hangers or sus

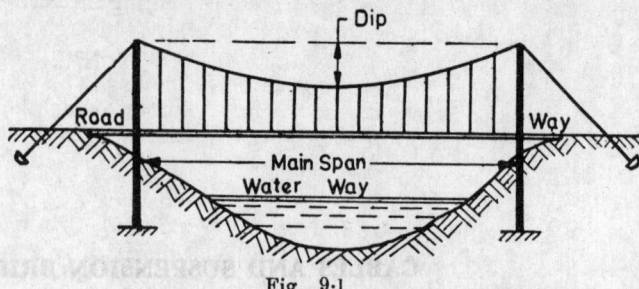

Fig. 9·1

penders. Since the hangers are large in number, therefore the loa transmitted by the hangers, is taken as a uniformly distributed load The central sag or dip of the cable generally varies from 1/10 to 1/1 of the span. It may be noted, that since the cable is in tensio throughout, this type of construction is most economical.

A suspension bridge is generally preferred when its span is mor than 200 metres for a roadway or 300 metres for light traffic way i.e pedestrians, cyclists, motorists etc. Due to oscillations and sag, unde loads, these types of bridges are not suitable for railways or ver heavy traffic.

9·2. Equilibrium of cable under a given system of loading

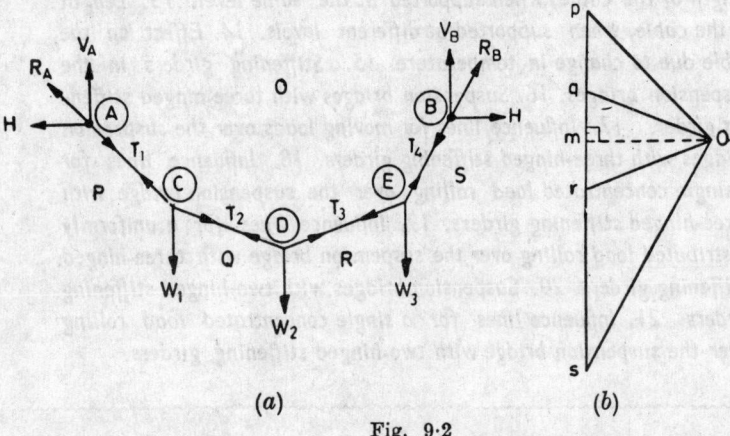

Fig. 9·2

Consider a cable suspended at two points, A and B at the sam level. Let the cable carry point loads W_1, W_2, and W_3 at C, D and E If the weight of the cable is negligible as compared to the loads W_1, W_2, W_3, etc., the cable will take a shape as shown in Fig. 9·2 (a).

Let $T_1 =$ Tension in AC,

           $T_2 =$ Tension in CD,

           $T_3 =$ Tension is DE, and

           $T_4 =$ Tension in EA.

Since all the joints of the cable are in equilibrium, therefore the funicular polygon drawn with the help of loads and tensions in the cable must close. Now draw pq, qr and rs representing the vertical loads W_1, W_2 and W_3 respectively to some suitable scale. Through p draw a line parallel to AC, and through q draw another line parallel to CD, meeting the first line at O. Join or and os. Now $pqrsop$ is the force polygon, corresponding to the system of forces, which keeps the cable in equilibrium. Through o, draw om perpendicular to the load line $pqrs$. Now mp and sm represent the vertical reactions at A and B respectively to the scale.

Now, considering the equilibrium of the point A, we find that the reaction R_A is equal to the tension T_1. Similarly, if we consider the equilibrium of C, we find that the forces acting at this point are T_1, T_2 and W_1. From the law of triangle of forces, we find that W_1 is represented by pq, T_1 is represented by op. Thus T_2 is represented by ro. Similarly by considering the equilibrium of points D and E, we can find that T_3 and T_4 are represented by ro and so respectively. A little consideration will show that the horizontal component of the tensions (i.e. T_1, T_2, T_3 and T_4) will be given by om to the scale. The vertical component of the reactions V_A and V_B will be given by pm and ms to the scale.

9.3. Equation of the cable

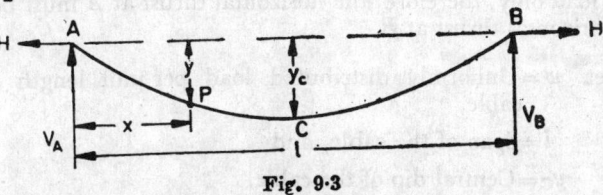

Fig. 9·3

Consider a cable ACB, supported at A and B. Let C be the lowest point of the cable as shown in Fig. 9·3.

Let l = span of the cable, and

                   y_C = central dip of the cable.

Now consider a point P, having coordinates as x and y, with A as the origin as shown in Fig. 9·3.

We know that the equation of a parabola

$$y = kx(l-x) \qquad \qquad \ldots(i)$$

where k is a constant.

When $x = \dfrac{l}{2}$, $y = y_C$. Substituting these values of x and y in equation (i),

$$y_C = k \cdot \frac{l}{2}\left(l - \frac{l}{2}\right) = \frac{kl^2}{4}$$

$$\therefore \qquad k = \frac{4y_C}{l^2}$$

Now substituting the value of k in equation (i),

$$y = \frac{4y_C}{l^2} x(l-x)$$

This is the required equation for the dip y of the cable from its support at a distance x from A or B.

9·4. Horizontal thrust on the cable

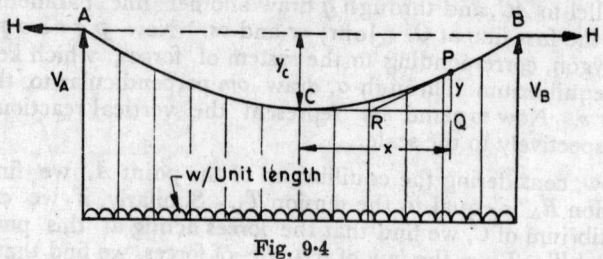

Fig. 9·4

Consider a cable ACB supported at A and B and carrying a uniformly distributed load as shown in Fig. 9·4.

Let C be the lowest point of the cable. A little consideration will show, that as a result of loading, the two supports A and B will tend to come nearer to each other. Since these two supports are in equilibrium, therefore an outward force must act, on both the supports to keep them in equilibrium. As the cable is supporting vertical load only, therefore the horizontal thrust at A must be equal to the horizontal thrust at B.

Let w = Uniformly distributed load per unit length on the cable,

  l = Span of the cable, and

  y_C = Central dip of the cable.

Now consider a point P on the cable, having co-ordinates as x and y as shown in Fig. 9·4. Draw a tangent to the cable at P meeting the horizontal line CQ though C at R. From the geometry of the figure, we know that $CR = RQ = \dfrac{x}{2}$. A little consideration will show that the part CP of the cable is in equilibrium under the action of the following forces :

(1) Horizontal pull (equal to H),

(2) Downward load (equal to wx), and

(3) Tension in the cable at P (equal to T).

Now we see that the triangle PQR represents, to some scale, the above mentioned three forces, which keep the part CP of the cable in equilibrium.

$$\therefore \qquad \frac{PQ}{wx} = \frac{RQ}{H} = \frac{RP}{T}$$

or $\qquad\qquad \dfrac{y}{wx} = \dfrac{x}{2H} \qquad \left(\because\ PQ = y \text{ and } RQ = \dfrac{x}{2} \right)$

$$*H = \frac{wx^2}{2y}$$

We know that when $x = \dfrac{l}{2}$, $y = y_C$, therefore

$$H = \frac{w\left(\dfrac{l}{2}\right)}{2y_C} = \frac{wl^2}{8y_C}$$

This is the required equation for the horizontal thrust on the cable.

Note. The B.M. at the lowest point C of the cable,

$$M_C = \text{B.M. due to loading} - \text{B.M. due to } H$$
$$= \mu_C - H.y_C$$
$$H.y_C = \mu_C \qquad\qquad (\because \text{ B.M. at } C \text{ is zero})$$
$$H = \frac{\mu_C}{y_C}$$

9·5. Tension in the cable

We have already discussed that the cable is always subjected to some tension, whose magnitude can be conveniently found out by the laws of statics. Here we shall discuss the magnitude of tension in the cable in the following two cases :

(a) When supported at the same level, and

(b) When supported at different levels.

9·6. Tension in the cable supported at the same level

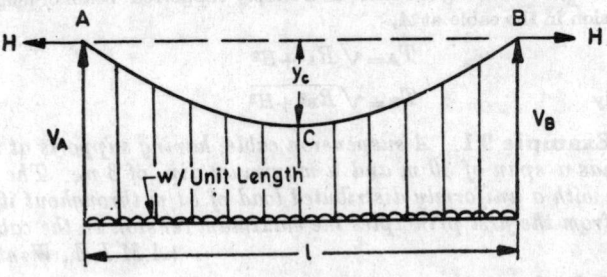

Fig. 9·5

Consider a cable ACB, supported at A and B, at the same level and carrying a uniformly distributed load as shown in Fig. 9·5.

Let $w =$ Uniformly distributed load per unit length, over the cable,

$l =$ Span of the cable,

_____

*This equation may also be written as

$$y = \frac{wx^2}{2H}$$

Since H is a constant quantity, therefore the above equation is that of a parabola. It is thus obvious, that the cable hangs in the form of a parabola.

y_C=Central dip of the cable, and

H=Horizontal thrust at A and B.

Since the cable is symmetrical about the central point C, there fore the vertical reaction at A,

$$R_A=R_B=\frac{wl}{2}$$

We have already discussed in Art. 9·1 that the cable, being flexible, can not resist any moment. Therefore equating the clock wise moment and the anticlockwise moment at C we get,

$$*H.y_C=\left(\frac{wl}{2}\times\frac{l}{2}\right)-\left(\frac{wl}{2}\times\frac{l}{4}\right)=\frac{wl^2}{8}$$

$$\therefore \quad H=\frac{wl^2}{8y_C}=\frac{Wl}{8y_C} \qquad \text{(where } W=\text{)}$$

A little consideration will show, that the maximum tension the cable will be at the supports A and B. Therefore tension in the cable at either of support,

$$T_{max}=\sqrt{R^2+H^2}=\sqrt{\left(\frac{wl}{2}\right)^2+\left(\frac{wl^2}{8y_C}\right)^2}$$

$$=\frac{wl}{2}\sqrt{\left(1+\frac{l^2}{16y_C{}^2}\right)}$$

Note. If the cable is subjected to point loads, with or without uniform distributed load, then the magnitude of tension in the cable will be different the two supports. In such a case, first of all find out the two vertical reactio V_A and V_B, considering the cable as a simply supported beam of length l. No the tension in the cable at A,

$$T_A=\sqrt{R_A{}^2+H^2}$$

similarly

$$T_B=\sqrt{R_B{}^2+H^2}$$

Example 9·1. *A suspension cable, having supports at the sam load, has a span of 30 m and a maximum dip of 3 m. The cable loaded with a uniformly distributed load of 1 t/m throughout its length Find from the first principles the maximum tension in the cable.*

(A.M.I.E., Winter 1974)

Solution.

Given. Span, $l=30$ m

Maximum dip. $y_C=3$ m

Load, $w=1$ t/m

---

*The horizontal thrust may also be found out by considering C as the origin. Now the co-ordinates of the support B are $\frac{l}{2}$ and y_C. We have discuss ed in Art. 9·4 that the horizontal thrust,

$$H=\frac{wx^2}{2y}=\frac{w\left(\frac{l}{2}\right)^2}{2\times y_C}=\frac{wl^2}{8y_C}$$

Let $\qquad\qquad H=$ Horizontal pull in the cable, and

$\qquad\qquad T_{max}=$ Maximum tension in the cable,

We know that the vertical reaction at the support,

$$R=\frac{wl}{2}=\frac{1\times30}{2}=15\text{ t}$$

Using the relation,

$$H=\frac{wl^2}{8y_C}\text{ with usual notations.}$$

$$=\frac{1\times30^2}{8\times3}=37\cdot5\text{ t}$$

Now using the relation,

$$T_{max}=\sqrt{R^2+H^2}\text{ with usual notations.}$$

$$=\sqrt{15^2+37\cdot5^2}=40\cdot39\text{ t }\textbf{Ans.}$$

Example 9·2. *A suspension bridge of 40 m span and 3 m wide platform is subjected to a load of 6,400 kg/m². The bridge is supported by a pair of cables having central dip of 4·5 m. Find the necessary cross sectional area of the cable, if the maximum permissible stress in the cable material, is not to exceed 1,200 kg/cm².*

(*Madras University, 1978*)

Solution.

Given. Span, $\qquad l=40$ m

Width of platform $=3$ m

Load on platform $\quad=6,400$ kg/m²

∴ Total load on the cable,

$$W=6,400\times\frac{3}{2}\times40=38,400\text{ kg}$$

Central dip, $\qquad y_C=4\cdot5$ m

Maximum stress $\quad f=1,200$ kg/cm²

Let $\qquad\qquad H=$ Horizontal thrust on the cable,

$\qquad\qquad T_{max}=$ Maximum tension in the cable, and

$\qquad\qquad A=$ cross sectional area of cable.

Using the relation,

$$H=\frac{Wl}{8y_C}\text{ with usual notations.}$$

$$=\frac{38,400\times40}{8\times4\cdot5}=42,667\text{ kg}$$

We know that the vertical reaction,

$$R=\frac{W}{2}=\frac{38,400}{2}=19,200\text{ kg}$$

Now using the relation,

$$T_{max} = \sqrt{R^2 + H^2} \text{ with usual notations.}$$
$$= \sqrt{19,200^2 + 42,667^2} = 46,790 \text{ kg}$$

Now equating the maximum tension in the cable, to the maximum stress in it,

$$T_{max} = f.A$$
$$46,790 = 1,200 \times A$$

$$\therefore \qquad A = \frac{46,790}{1,200} = \textbf{39 cm}^2 \textbf{ Ans.}$$

Example 9·3. *A cable is used to support six equal and equidistant loads over a span of 14 metres. The central dip of the cable is 1·6 metre and the loads are 2 tonnes each. Find the length of the cable required and its sectional area, if the safe tensile stress is 1·5 t/cm². The distance between the loads is 2 metres.*

Solution.
Given. Span, $l = 14$ m
Central dip, $y_C = 1·6$ m
Safe tensile stress, $f = 1·5$ t/cm²
Let $H = $ Horizontal thrust in the cable.

Length of the cable

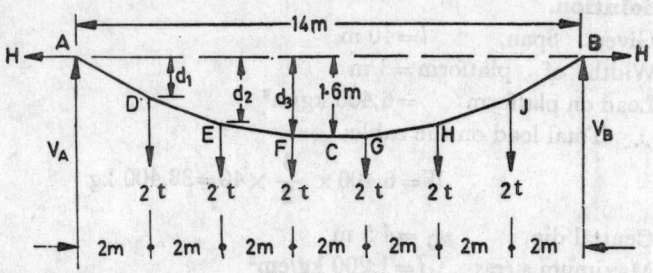

Fig. 9·6

Since the cable, as well as the loading is symmetrical therefore, the vertical reaction at A,

$$R_A = R_B = \frac{6 \times 2}{2} = 6 \text{ t}$$

For the geometry of Fig. 9·6, we find that the moment at C,

$$\mu_C = 6 \times 7 - 2 \times 5 - 2 \times 3 - 2 \times 1 = 24 \text{ t-m}$$

Using the relation,

$$H = \frac{\mu_C}{y_C} \text{ with usual notations.}$$
$$= \frac{24}{1·6} = 15 \text{ t}$$

Let $\quad d_1 =$ Depth of point D from the supports A and B,

$\quad\quad\quad d_2 =$ Depth of point E from the supports A and B,

$\quad\quad\quad d_3 =$ Depth of point F from the supports A and B.

We know that the shape of the cable, will represent the B.M. diagram, to some scale.

Now, taking moments about D and equating the same,

$$H \times d_1 = R_A \times 2$$

or
$$d_1 = \frac{R_A \times 2}{H} = \frac{6 \times 2}{15} = 0.8 \text{ m}$$

Similarly, $\quad H \times d_2 = R_A \times 4 - 2 \times 2 = 6 \times 4 - 2 \times 2 = 20$

$$\therefore \quad d_2 = \frac{20}{15} = 1.33 \text{ m}$$

and $\quad H \times d_3 = R_A \times 6 - 2 \times 4 - 2 \times 2 = 6 \times 6 - 2 \times 4 - 2 \times 2 = 24$

$$\therefore \quad d_2 = \frac{24}{15} = 1.6 \text{ m}$$

A little consideration will show that the length of the cable AD

$$= \sqrt{2^2 + 0.8^2} = \sqrt{4.64} = 2.152 \text{ m}$$

Similarly $\quad DE = \sqrt{2^2 + 0.53^2} = \sqrt{4.28} = 2.069 \text{ m}$

and $\quad BF = \sqrt{2^2 + 0.27^2} = \sqrt{4.0729} = 2.019 \text{ m}$

and $\quad FC = 1 \text{ m}$

$$\therefore \quad \text{Length of the cable}$$
$$= 2[2.152 + 2.069 + 2.019 + 1] \text{ m}$$
$$= \textbf{14.48 m} \quad \textbf{Ans.}$$

Sectional area of the cable

Let $\quad A =$ Sectional area of the cable, and

$\quad\quad\quad T_{max} =$ Maximum tension in the cable.

Using the relation,

$$T_{max} = \sqrt{R^2 + H^2} \quad \text{with usual notations.}$$

$$= \sqrt{6^2 + 15^2} = 16.2 \text{ t}$$

Now equating the maximum tension in the cable to the maximum stress in it,

$$T_{max} = f.A$$

$$16.2 = 1.5 \times A$$

$$A = \frac{16.2}{1.5} = \textbf{10.8 cm}^2 \quad \textbf{Ans.}$$

9·7. Tension in the cable supported at different levels

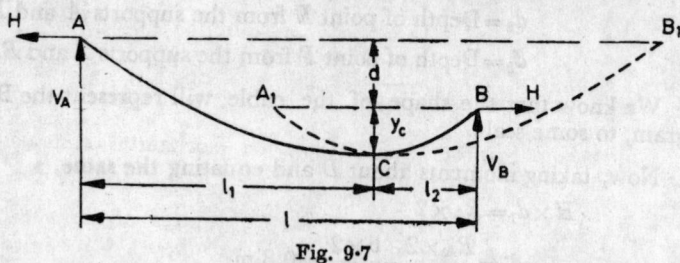

Fig. 9·7

Consider a cable ACB, supported at different levels at A and B and carrying a uniformly distributed load as shown in Fig. 9·7. Let C be the lowest point of the cable.

Let
$\qquad w =$ Uniformly distributed load per unit length of the cable,

$\qquad l =$ Span of the cable,

$\qquad y_C =$ Depth of the lowest point of the cable C, from the support B,

$\qquad d =$ Difference between the levels of the two supports,

$\qquad l_1 =$ Horizontal length between A and C, and

$\qquad l_2 =$ Horizontal length between C and B.

Since the cable is supporting vertical loads only, therefore the horizontal thrust at A, must be equal to the horizontal thrust at B.

In order to locate position of the lowest point of the cable C, let us imagine the portion CB of the cable to be extended to CB_1, such that the new support B_1 is at the same level as that of A. Similarly, imagine the portion AC of the cable to be cut short to A_1C, such that the new support A_1 is at the same level as that of B. From the geometry of the figure, we find that the cable ACB_1 has a span of $2l_1$ and a central dip of (y_C+d); whereas the cable A_1CB has a span of $2l_2$ and a central dip of y_C.

Now in the cable ACB_1 the horizontal thrust,

$$H = \frac{wl^2}{8y_C} = \frac{w(2l_1)^2}{8(y_C+d)} \qquad\qquad\dots(i)$$

Similarly in the cable A_1CB, the horizontal thrust,

$$H = \frac{wl^2}{8y_C} = \frac{w(2l_2)^2}{8y_C} \qquad\qquad\dots(ii)$$

Since the two horizontal thrusts are equal, therefore equating both the equations,

$$\frac{w(2l_1)^2}{8(y_C+d)} = \frac{w(2l_2)^2}{8y_C}$$

or
$$\frac{l_1{}^2}{y_C+d}=\frac{l_2{}^2}{y_C}$$

$$\therefore \qquad \frac{l_1}{l_2}=\sqrt{\frac{y_C+d}{y_C}} \qquad \qquad \text{...}(iii)$$

This is the required equation for the horizontal distance between the support A (or B) and the lowest point of the cable C.

Now take C as the origin. The co-ordinates of the support B are l_2 and y_C, whereas the co-ordinates of the support A are $(-l_1)$ and (y_C+d). We have discussed in Art. 9·4 that the horizontal thrust,

$$H=\frac{wx^2}{2y}=\frac{wl_2{}^2}{2y_C} \qquad \qquad \text{...}(iv)$$

$$=\frac{w(-l_1)^2}{2(y_C+d)}=\frac{wl_1{}^2}{2(y_C+d)} \qquad \qquad \text{...}(v)$$

These are the required equations for the horizontal thrust at A or B.

Now we can find out the vertical reactions at A and B. Taking moments about B and equating the same, i.e.

$$R_A . l=\frac{wl^2}{2}+H.d$$

$$\therefore \qquad R_A=\frac{wl}{2}+\frac{Hd}{l}$$

Similarly, it can be proved that the vertical reaction at B,

$$R_B=\frac{wl}{2}-\frac{Hd}{l}$$

Now the tension in the cable at A,

$$t_A=\sqrt{R_A{}^2+H^2}$$

Similarly $\qquad t'_B=\sqrt{R_B{}^2+H^2}$

Note. Since the value of R_A (the support A being higher than B) is more than R_B, therefore the maximum tension in the cable will be at A.

Example 9·4. *A cable of uniform thickness hangs between two points 120 m apart, with one end 3 m above the other. The cable is loaded with a uniformly distributed load of 1 t/m and the sag of the cable, measured from the higher end, is 5 m. Find the horizontal thrust and maximum tension in the cable.*

Solution.

Given. Span, $\qquad l=120$ m

Difference between the levels of the two supports,

$$d=3 \text{ m}$$

Load, $\qquad w=1$ t/m

Depth of the lowest point of the cable from the lower support,
$$y_C = 5 - 3 = 2 \text{ m}$$

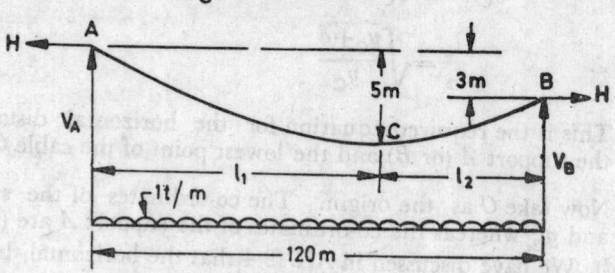

Fig. 9·8.

Horizontal thrust on the cable

        Let
                        $H =$ Horizontal thrust on the cable,

                                $l_1 =$ Horizontal length of AC, and

                                $l_2 =$ Horizontal length of CB.

Using the relation,
$$\frac{l_1}{l_2} = \sqrt{\frac{y_C + d}{y_C}} \quad \text{with usual notations.}$$
$$= \sqrt{\frac{2+3}{2}} = \sqrt{\frac{5}{2}} = \sqrt{2\cdot5} = 1\cdot58$$

$\therefore$ $l_1 = 1\cdot58 \, l_2$

We know that $l_1 + l_2 = 120$

$\therefore$ $1\cdot58 \, l_2 + l_2 = 120$

or $2\cdot58 \, l_2 = 120$

$\therefore$ $l_2 = \dfrac{120}{2\cdot58} = 46\cdot5 \text{ m}$

and $l_1 = 120 - 46\cdot5 = 73\cdot5 \text{ m}$

Now using the relation,
$$H = \frac{w l_1^2}{2(y_C + d)} \quad \text{with usual notations.}$$
$$= \frac{1 \times 73\cdot5^2}{2(2+3)} = 540\cdot25 \text{ t} \qquad \textbf{Ans.}$$

Maximum tension in the cable

        Let $T_{max} =$ Maximum tension in the cable.

We know that the vertical reaction at A,
$$*R_A = \frac{wl}{2} + \frac{Hd}{l} = \frac{1 \times 120}{2} + \frac{540\cdot25 \times 3}{120} \text{ t}$$
$$= 60 + 13\cdot5 = 73\cdot5 \text{ t}$$

---

*This may also be easily found out by considering the load from A to C.

i.e.,

             $R_A = 1 \times 73\cdot5 = 73\cdot5 \text{ t}$

Now using the relation,

$$T_{max} = \sqrt{R_A{}^2 + H^2} \text{ with usual notations.}$$
$$= \sqrt{73 \cdot 5^2 + 540 \cdot 25^2} = 545 \cdot 0 \text{ t} \quad \textbf{Ans.}$$

9·8. Anchor cables

The suspension cable is supported on two towers on both its sides. The cable, after passing over the supporting tower, is anchored down into a huge mass of concrete. The following two arrangements of passing the cable over the supporting towers, are important from the subject point of view :

    (a) Guide pulley support, and

    (b) Roller support.

9·9. Guide pulley support for suspension cable

Sometimes, a suspension cable is passed over a frictionless guide pulley, fixed on the top of supporting tower and then anchored down as shown in Fig. 9·9. A little consideration will show that in such a case,

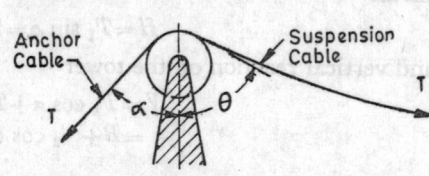

Fig. 9·9

the tension in the suspension cable is equal to that in the anchor cable. This is possible, as the friction of the pulley is neglected.

Let α = Angle, which the anchor cable makes with the vertical,

    θ = Angle, which the tangent to the suspension cable, makes with the vertical, and

    T = Tension in the cable.

It is thus obvious, that the vertical reaction on the tower,

$$V = T(\cos \alpha + \cos \theta)$$
$$= R + T \cos \alpha \qquad \text{(where } R \text{ is the reaction at the support due to load)}$$

and horizontal thrust on the cable,

$$H = T(\sin \alpha - \sin \theta)$$

Note. The value of θ i.e. the angle, which the tangent to the suspension cable makes with the vertical, may be found out by differentiating the equation of the parabola with respect to x.

9·10. Roller support for suspension cable

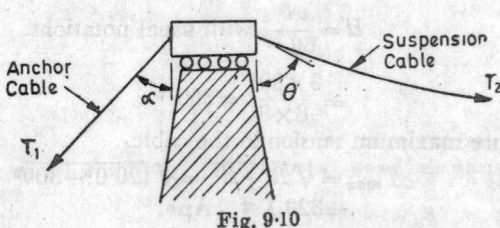

Fig. 9·10

Sometimes, the suspension cable and the anchor cable are attached to a saddle, which is supported on rollers as shown in Fig. 9·10. In such a case, the horizontal components of the tension in the suspension cable and the anchor cable are equal. This is possible, as the sandle is at rest.

Let $\quad$ $\alpha=$ Angle, which the anchor cable makes with the vertical.

$\theta=$ Angle, which the tangent to the suspension cable, makes with the vertical,

$T_1=$ Tension in the anchor cable, and

$T_2=$ Tension in the suspension cable.

It is thus obvious, that the horizontal components of the tension in the anchor cable and suspension cable are equal to the horizontal thrust.

$$\therefore \qquad H=T_1 \sin \alpha = T_2 \sin \theta = H$$

and vertical reaction on the tower

$$V=T_1 \cos \alpha + T_2 \cos \theta$$
$$=R+ t'_2 \cos \theta \qquad \text{(where } R \text{ is the reaction at the support due to load)}$$

Example 9·5. *A cable, supported on piers 80 m apart at the same level has a central dip of 8 m. Calculate the maximum tension in the cable, when it is carrying a load of 3 t/m.*

Also determine the vertical pressure on the pier, if the backstay is inclined at angle of 60° to the vertical when (a) the cable passes over a pulley, and (b) the cable passes over saddles. (*London University*)

Solution.

Given. Span, $\qquad l=80$ m

Central dip, $\qquad y_C = 8$ m

Load, $\qquad w=3$ t/m

Angle, which the backstay makes with the vertical,

$$\alpha = 60°$$

Maximum tension in the cable

Let $\qquad H=$ Horizontal thrust in the cable.

We know that the vertical reaction at the support,

$$R=\frac{wl}{2}=\frac{3 \times 80}{2}=120·0 \text{ t}$$

Using the relation,

$$H=\frac{wl^2}{8y} \text{ with usual notations.}$$

$$=\frac{3 \times 80^2}{8 \times 8}=300 \text{ t}$$

Therefore maximum tension in the cable,

$$T_{max}=\sqrt{R^2+H^2}=\sqrt{120·0^2+300^2} \text{ t}$$
$$=323·1 \text{ t} \quad \textbf{Ans.}$$

Vertical pressure on the pier, when the cable passes over a pulley

Let $\qquad\qquad\qquad$ V = Vertical pressure on the pair

Using the relation,

$$V = R + T \cos \alpha \text{ with usual notations.}$$
$$= 120 \cdot 0 + 323 \cdot 1 \cos 60° \quad t$$
$$= 120 \cdot 0 + 323 \cdot 1 \times 0 \cdot 5 = \textbf{281·55} \quad ^t \quad \textbf{Ans.}$$

Vertical pressure on the pier, when the cable passes over the saddle.

We know that the horizontal component of the tension in the backstay *i.e.* anchor cable is equal to the horizontal thrust in the cable.

$\therefore \qquad\qquad$ $T_1 \sin 60° = 300$

or $\qquad\qquad$ $T_1 = \dfrac{300}{60°} = \dfrac{300}{0 \cdot 866} = 346 \cdot 5 \quad ^t$

Now using the relation,

$$V = R + T_1 \cos \alpha \text{ with usual notations.}$$
$$= 120 \cdot 0 + 346 \cdot 5 \cos 60° \quad 't'$$
$$= 120 \ 0 + 346 \cdot 5 \times 0 \cdot 5 = \textbf{293·25} \quad \textbf{t} \quad \textbf{Ans.}$$

Example 9·6. *An unstiffened suspension cable carries a uniformly distributed load of 6·5 t/m over a span of 30 m as shown in Fig. 9·11.*

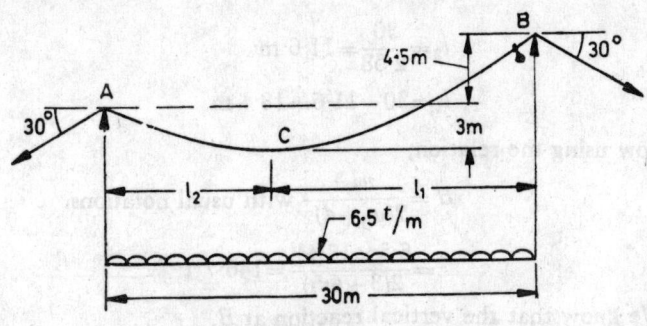

Fig. 9·11

The suspension cable is supported on frictionless rollers, fixed to the piers. The anchor cables are inclined at 30° to the horizontal. One pier is 4·5 m below the other and the maximum dip of the cable at the lowest point is 3 m below the lower pier. Calculate

(a) the maximum and minimum tension in the cable, and

(b) the horizontal and vertical forces at support A.

Solution.

Given $\quad$ Load, $\qquad$ $w = 6 \cdot 5 \text{ t/m}$

$\qquad$ Span, $\qquad\qquad$ $l = 30 \text{ m}$

Angle, which the anchor cables make with the vertical,
$$\alpha = 90° - 30° = 60°$$

Difference of levels between the two supports,
$$d = 4\cdot5 \text{ m}$$

Depth of the lowest point of the cable from the lower pier,
$$y_0 = 3 \text{ m}$$

Maximum tension in the cable

Let T_{max} = Maximum tension in the cable,
            H = Horizontal thrust in the cable,
            l_1 = Horizontal length of CB, and
            l_2 = Horizontal length of AC.

Using the relation,
$$\frac{l_1}{l_2} = \sqrt{\frac{y_0 + d}{y_0}} \text{ with usual notations.}$$
$$= \sqrt{\frac{3\cdot0 + 4\cdot5}{3}} = \sqrt{\frac{7\cdot5}{3}} = \sqrt{2\cdot5} = 1\cdot58$$

$\therefore$ $l_1 = 1\cdot58 \, l_2$

We know that $l_1 + l_2 = 30$

$\therefore$ $1\cdot58 \, l_2 + l_2 = 30$

or $2\cdot58 \, l_2 = 30$

$\therefore$ $l_2 = \dfrac{30}{2\cdot58} = 11\cdot6 \text{ m}$

and $l_1 = 30 - 11\cdot6 = 18\cdot4 \text{ m}$

Now using the relation,
$$H = \frac{wl_1^2}{2(y_0 + d)} \text{ with usual notations.}$$
$$= \frac{6\cdot5 \times 18\cdot4^2}{2(3 + 4\cdot5)} = 146\cdot7 \text{ t}$$

We know that the vertical reaction at B,
$$R_B = \frac{wl}{2} + \frac{H \cdot d}{l} \text{ t}$$
$$= \frac{6\cdot5 \times 30}{2} + \frac{146\cdot7 \times 4\cdot5}{30} = 119\cdot6 \text{ t}$$

and $R_A = 6\cdot5 \times 30 - 119\cdot6 = 75\cdot4 \text{ t}$

Since R_B is more than R_A, therefore maximum tension in the cable will be at B.

Now using the relation,
$$T_{max} = \sqrt{R_B^2 + H^2} \text{ with usual notations.}$$
$$= \sqrt{119\cdot6^2 + 146\cdot7^2} = \mathbf{189\cdot25} \text{ t } \textbf{Ans.}$$

Minimum tension in the cable

The minimum tension in the cable will be at C, *i.e.* the lowest point of the cable and its magnitude will be equal to H,
$$T_{min} = H = 146.7 \text{ t Ans.}$$

Horizontal force at A

Since the cable is supported on frictionless rollers *i.e.* a saddle arrangement, therefore the horizontal force at A will be zero.   **Ans.**

Vertical force at A.

Let T_1 = Tension in the backstay, and
              V_A = Vertical force at A.

We know that the horizontal component of the tension in backstay is equal to the horizontal thrust in the cable at A.

$\therefore$ $T_1 \sin \alpha = H$

or $T_1 \sin 60° = 146.7$

$\therefore$ $T_1 = \dfrac{146.7}{\sin 60°} = \dfrac{146.7}{0.866} = 169.4$ t

Now using the relation,

$V_A = R_A + T_1 \cos \alpha$ with usual notations.
$$= 75.4 + 169.4 \cos 60° = 75.4 + 169.4 \times 0.5 \text{ t}$$
$$= \textbf{160.1 t Ans.}$$

Exercise 9-1

1. A cable of span 40 m has a central dip of 5 m. Find the horizontal pull in the cable, if the cable is loaded with (*i*) a uniformly distributed load of 1.5 t/m and (*ii*) a central point load of 60 T.   [Ans. 60 t-m ; 120 t-m)

2. A cable of span 150 m and central dip 10 m is subjected to a load of 0.75 t/m. Find the maximum tension in the cable.   (*Gujarat University*)
[**Ans.** 218.5 t']

3. A suspension cable of span 160 m and central dip 16 m carries a uniformly distributed load of 0.5 t/m. Find the maximum and minimum tensions in the cable.   (*Aligarh University*)
[**Ans.** 107.7 T ; 100 t]

4. A suspension cable of span 24 m and central dip 2 m is carrying a uniformly distributed load of 2 t/m. Find the sectional area of the cable, if the permissible stress in the cable material is 1,200 kg/cm². (*Osmania University*)
[**Ans.** 31.62 cm²]

5. A suspension cable of span 20 m is supported on piers 2 m and 4 m above the lowest point of the cable. Find the maximum tension in the cable, if it carries a uniformly distributed load 1 t/m over the span.
(*Calcutta University*)
[**Ans.** 20.8 t']

9·11. Length of the cable

It means the actual length of the cable required between the two supports A and B, when it is loaded with a uniformly distributed load, and hangs in the form of a parabola. Here we shall discuss the following two cases :

(*a*) When supported at the same level, and

(*b*) When supported at different levels.

9·12. Length of the cable when supported at the same level

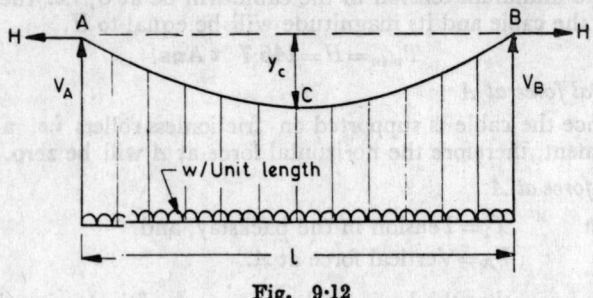

Fig. 9·12

Consider a cable ACB supported at the same level A and B, and carrying a uniformly distributed load as shown in Fig. 9·12. Let O be the lowest point of the cable.

Let $w=$ Uniformly distributed load per unit length on the cable,

$l=$ Span of the cable, and

$y_C=$ Central dip of the cable.

We have already discussed in Art. 9·4 that the cable hangs in the form of a parabola, and the equation of the parabola is given by the relation,

$$y=\frac{wx^2}{2H}$$

Differentiating this equation with respect to x,

i.e. $$\frac{dy}{dx}=\frac{2wx}{2H}=\frac{wx}{H} \qquad \qquad ...(i)$$

Now consider a small length ds of the cable as shown in Fig. 9·13. Taking the length of the arc PQ equal to the length of the chord PQ, we find that

$$ds=\sqrt{dx^2+dy^2}$$

$$=dx\sqrt{\left(1+\frac{dy}{dx}\right)^2}$$

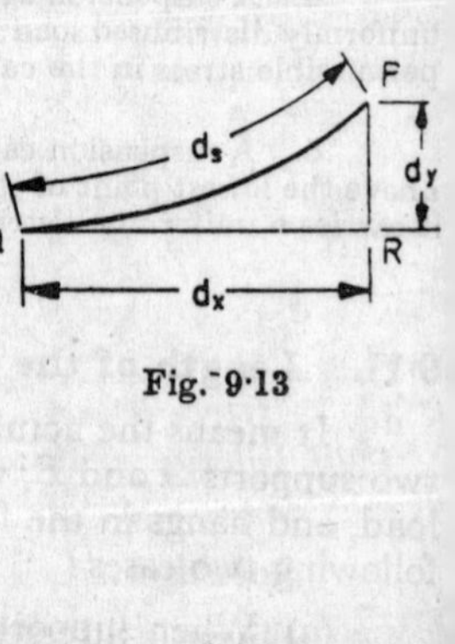

Substituting the value of $\frac{dy}{dx}$ from equa-

Fig. 9·13

tion (i) in the above equation,

$$ds=dx\sqrt{1+\left(\frac{wx}{H}\right)^2}$$

Now expanding the term inside the square root, by binomial theorem,

$$\sqrt{1+\left(\frac{wx}{H}\right)^2}=\sqrt{1+\frac{w^2x^2}{H^2}}=\left(1+\frac{w^2x^2}{H^2}\right)^{\frac{1}{2}}$$

$$=1+\frac{1}{2}\cdot\frac{w^2x^2}{H^2}+\cdots\cdots$$

$$\left(\text{Neglecting the higher powers of } \frac{w^2x^2}{H^2}\right)$$

$$\therefore \qquad ds=dx\left(1+\frac{1}{2}\cdot\frac{w^2x^2}{H^2}\right)$$

Now integrating the above equation between the limits $x=0$ and $x=\frac{l}{2}$,

$$s=\int_0^{\frac{l}{2}}\left(1+\frac{1}{2}\cdot\frac{w^2x^2}{H^2}\right)dx$$

$$=\left[x+\frac{w^2}{2H^2}\cdot\frac{x^3}{3}\right]_0^{\frac{l}{2}}=\frac{l}{2}+\frac{w^2}{2H^2}\cdot\frac{l^3}{24}$$

$$=\frac{l}{2}+\frac{w^2l^3}{48H^2}$$

A little consideration will show that since the limits of integration were from 0 to $l/2$ (taking C as origin) therefore the above equation gives the length of half of the cable.

$\therefore$ Total length of the cable,

$$L=l+\frac{w^2l^3}{24H^2}$$

Now substituting the value of $H=\frac{wl^2}{8y_C}$ in the above equation,

$$\therefore \qquad L=l+\frac{w^2l^3}{24}\times\frac{1}{\left(\frac{wl^2}{8y_C}\right)^2}$$

$$=l+\frac{8y_C^2}{3l}$$

Example 9·7. *A steel wire, of uniform section, is hung in the form of a parabola. Find the maximum horizontal span, if the central dip is 1/12th of the span and the stress in steel wire is not to exceed 1,200 kg/cm². Take density of steel as 7,800 kg/cm³.*

Solution.

Given. Stress in wire

$$f=1,200 \text{ kg/cm}^2$$

Density of steel $=7{,}800$ kg/m³

Let $l=$ Span of the parabola in metres.

$\therefore$ Central dip, $y_C = \dfrac{l}{12}$ m

Let $L=$ Length of the wire,

                 $A=$ Cross sectional area of the wire in cm²,

                 $H=$ Horizontal thrust in the wire, and

                 $T_{max}=$ Maximum tension in the wire.

Using the relation,

$$L = l + \frac{8y_C^2}{3l} \text{ with usual notations.}$$

$$= l + \frac{8\left(\dfrac{l}{12}\right)^2}{3l} = l + \frac{8l}{3 \times 144} = \frac{55l}{54} \text{ m}$$

We know that the total weight of the wire,

$$W = \frac{AL}{100^2} \times 7{,}800$$

$$= \frac{A}{10{,}000} \times \frac{55l}{54} \times 7{,}800 = 0{\cdot}7944 \ Al$$

Using the relation,

$$H = \frac{Wl}{8y_C} \text{ with usual notations.}$$

$$= \frac{Wl}{8 \times \dfrac{l}{12}} = \frac{3W}{2}$$

We also know that the vertical reaction,

$$R = \frac{wl}{2} = \frac{W}{2}$$

Now using the relation,

$$T_{max} = \sqrt{R^2 + H^2} \text{ with usual notations.}$$

$$= \sqrt{\left(\frac{W}{2}\right)^2 + \left(\frac{3W}{2}\right)^2}$$

$$= \sqrt{\frac{W^2}{4} + \frac{9W^2}{4}} = 1{\cdot}581 \ W$$

Now equating the maximum tension in the wire, to the maximum stress in it,

$$T_{max} = f. A$$

$$1{\cdot}581 \ W = 1{,}200 \times A$$

$$1{\cdot}581 \times 0{\cdot}7944 \ Al = 1{,}200 \times A$$

$$l = \frac{1{,}200}{1{\cdot}581 \times 0{\cdot}7944} = \textbf{955·5 m} \quad \textbf{Ans.}$$

9·13. Length of the cable, when supported at different levels

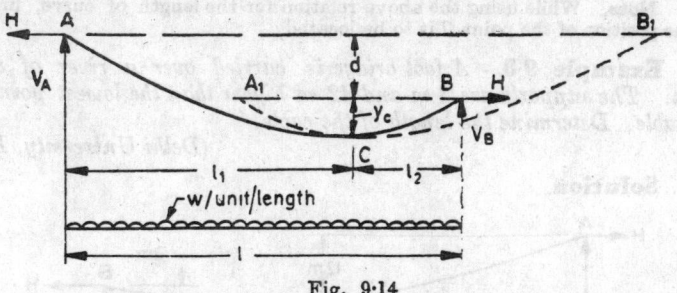

Fig. 9·14

Consider a cable ACB, supported at different levels A and B as shown in Fig. 9·14. Let C be the lowest point of the cable.

Let
l = Span of the cable,

y_C = Depth of the lowest point of the cable C from the support B,

d = Difference between the levels of the two supports,

l_1 = Horizontal length between A and C, and

l_2 = Horizontal length between C and B.

Let us imagine the portion CB of the cable to be extended to CB_1, such that the new support B_1 is at the same level as that of A. Similarly, imagine the portion AC of the cable to be cut short to A_1C, such that the new support A_1 is at the same level as that of B. From the geometry of the figure, we find that the cable ACB_1 has a span of $2l_1$ and a central dip of (y_C+d), whereas the cable A_1CB has a span of $2l_2$ and a central dip of y_C.

We have discussed in Art. 9·12 that the length of the cable,

$$L=l+\frac{8y_C^2}{3l}$$

∴ Length of the cable ACB_1

$$L_1=2l_1+\frac{8(y_C+d)^2}{3\times2l_1}=2l_1+\frac{8(y_C+d)^2}{6l_1} \qquad \ldots(i)$$

Similarly length of the cable A_1CB,

$$L_2=2l_2+\frac{8y_C^2}{3\times2l_2}=2l_2+\frac{8y_C^2}{6l_2} \qquad \ldots(ii)$$

Now the actual length of the cable ACB,

$$L=\frac{L_1+L_2}{2}=\frac{1}{2}\left(2l_1+\frac{8(y_C+d)^2}{6l_1}+2l_2+\frac{8y_C^2}{6l^2}\right)$$

$$=l_1+\frac{2(y_C+d)^2}{3l_1}+l_2+\frac{2y_C^2}{3l_2}$$

$$=l+\frac{2(y_C+d)^2}{3l_1}+\frac{2y_C^2}{3l_2} \qquad (\because \; l_1+l_2=l)$$

Note. While using the above relation for the length of curve, first of all the position of the point C is to be located.

Example 9·8. *A foot bridge is carried over a river of span 90 m. The supports are 3 m and 12 m higher than the lowest point of the cable. Determine the length of the cable.*

(*Delhi University, 1972*)

Solution.

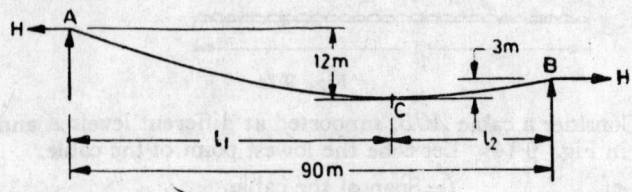

Fig. 9·15

Given. Span $l=90$ m

Depth of C from B, $y_C=3$ m

Difference of level between A and B,

$$d=12-3=9 \text{ m}$$

Let $L=$ Length of the cable ACB,

$l_1=$ Horizontal length of AC, and

$l_2=$ Horizontal length of CB.

Using the relation,

$$\frac{l_1}{l_2}=\sqrt{\frac{y_C+d}{y_C}} \text{ with usual notations.}$$

$$=\sqrt{\frac{3+9}{3}}=\sqrt{4}=2$$

$\therefore$ $l_1=2l_2$

We know that $l_1+l_2=90$ m

$\therefore$ $2l_2+l_2=90$ m   or $3l_2=90$

$\therefore$ $l_2=\frac{90}{3}=30$ m

and $l_1=90-30=60$ m

Now using the relation,

$$L=l+\frac{2(y_C+d)^2}{3l_1}+\frac{2y_C^2}{3l^2} \text{ with usual notations.}$$

$$=90+\frac{2(3+9)^2}{3\times 60}+\frac{2\times 3^2}{3\times 30} \text{ m}$$

$$=\textbf{91·8 m} \quad \textbf{Ans.}$$

Example 9·9. *A cable is suspended and loaded as shown in Fig. 9·16 below :*

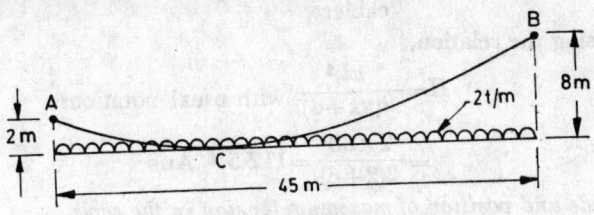

Fig. 9·16

(a) *Compute the length of the cable,*

(b) *Compute the horizontal component of the tension in the cable;* and

(c) *Determine the magnitude and position of the maximum tension occurring in the cable.*

Solution.

Given. Span, $l = 45$ m

Depth of C from A,

$$y_C = 2 \text{ m}$$

Difference of level between A and B,

$$d = 8 - 2 = 6 \text{ m}$$

Load on the cable,

$$w = 2\text{t/m}$$

Length of the cable

Let $\qquad L$ = Length of the cable ACB,

$\qquad\qquad l_1$ = Horizontal length of CB, and

$\qquad\qquad l_2$ = Horizontal length of AC.

Using the relation,

$$\frac{l_1}{l_2} = \sqrt{\frac{y_C + d}{y_C}} \text{ with usual notations.}$$

$$= \sqrt{\frac{2 + 6}{2}} = \sqrt{4} = 2$$

$$l_1 = 2l_2$$

We know that $l_1 + l_2 = 45$

$$\therefore \qquad 2l_2 + l_2 = 45 \qquad \text{or} \qquad 3l_2 = 45$$

$$\therefore \qquad l_2 = \frac{45}{3} = 15 \text{ m}$$

and $\qquad\qquad l_1 = 45 - 15 = 30 \text{ m}$

Now using the relation,

$$L = l + \frac{2(y_C + d)^2}{3l_1} + \frac{2y_C^2}{3l_2^2} \text{ with usual notations.}$$

$$= 45 + \frac{2(2 + 6)^2}{3 \times 30} + \frac{2 \times 2^2}{3 \times 15} = \textbf{46·6 m Ans.}$$

Horizontal component of the tension in the cable

Let H = Horizontal component of the tension in the cable.

Using the relation,

$$H = \frac{wl_1^2}{2(y_C + d)}. \text{ with usual notations.}$$

$$= \frac{2 \times 30^2}{2(2+6)} = 112 \cdot 5 \text{ t Ans.}$$

Magnitude and position of maximum tension in the cable

We know that the vertical reaction at the higher support B,

$$V_B = \frac{wl}{2} + \frac{H.d}{l} = \frac{2 \times 45}{2} + \frac{112 \cdot 5 \times 6}{45} = 60 \text{ t}$$

Now using the relation,

$$T_{max} = \sqrt{V_B^2 + H^2} \text{ with usual notations.}$$

$$= \sqrt{60^2 + 112 \cdot 5^2} = 127 \cdot 5 \text{ t}$$

The maximum tension in the cable will take place at the higher support and equal to 127·5, **t Ans.**

9·14.   Effect on the cable due to change in temperature.

Consider a cable ACB supported at A and B. Let C be the lowest point of the cable.

Let l = Span of the cable,

y_C = Central dip of the cable,

L = Length of the cable,

t = Rise in temperature, and

α = Coefficient of linear expansion for the cable material.

We know that the length of the cable must increase with the rise in temperature. Since the two supports A and B cannot undergo any displacement, therefore the downward movement of the point C will increase the central dip y_C of the cable.

We have already discussed in Art. 9·12 that the length of the cable,

$$L = l + \frac{8y_C^2}{3l}$$

Differentiating the above equation with respect to y_C,

$$\frac{dL}{dy_C} = \frac{16y_C}{3l}$$

∴ $$dL = \frac{16y_C}{3l} dy_C$$

or $$dy_C = \frac{3l}{16y_C}.dL \qquad\qquad ...(i)$$

As a result of rise in the temperature, increase in the length of cable,

CABLES AND SUSPENSION BRIDGES 351

$$dL = L.\alpha t = \alpha t \left(l + \frac{8y_C^2}{3l} \right) \quad \left(\because L = l + \frac{8y_C^2}{3l} \right)$$

$$= l.\alpha.t$$

$$\left(\text{neglecting } \frac{\alpha t \cdot 8y_C^2}{3l} \text{ as compared to } l.\alpha.t. \right)$$

Substituting this value of dL in equation (i),

$$dy_C = \frac{3l}{16y_C} \times l\alpha t = \frac{3l^2}{16y_C}\alpha t \qquad \ldots(ii)$$

We know that the horizontal thrust,

$$H = \frac{wl^2}{8y_C}$$

$$\therefore \qquad \propto \frac{1}{y_C}$$

or

$$\frac{dH}{H} = -\frac{dy_C}{y_C}$$

We also know that the stress in the cable,

$$f \propto H$$

or

$$\frac{df}{f} = \frac{dH}{H} = -\frac{dy_C}{y_C}$$

$$\therefore \qquad \frac{dH}{H} = -\frac{3l^2}{16y_C^2}\alpha t$$

Minus sign indicates that when the temperature rises the central dip increases, as a result of which the horizontal thrust decreases.

Example 9·10. *A cable, supported at 120 m apart at the same level, has a central dip of 10 m. Find the increase in dip due to rise of temperature of 20°C. Take $\alpha = 12 \times 10^{-6}$ per °C.*

Solution.

Given. Span, $l = 120$ m

Central dip, $y_C = 10$ m

Rise of temperature, $t = 20°C$

Coefficient of expansion,

$$\alpha = 12 \times 10^{-6} \text{ per } °C$$

Let $dy_C = $ Increase in the central dip.

Using the relation,

$$dy_C = -\frac{3l^2}{16y_C^2} \times \alpha t \text{ with usual notations.}$$

$$= \frac{3 \times 120}{16 \times 10} \times 12 \times 10^{-6} \times 20 \text{ m}$$

$$= 0.0648 \text{ m} = \mathbf{6·48 \text{ cm}} \quad \mathbf{Ans.}$$

Example 9·11. *A cable of span 120 m and central dip of 12 m is carrying a load of 2 t/m of horizontal length. Calculate the change in horizontal tension, when the temperature rises through 20°F. Take* $\alpha = 6 \times 10^{-6}$ *per °F.*

Solution.

Given. Span, $l = 120$ m

Central dip $y_C = 12$ m

Load, $w = 2$ t/m

Rise of temperature,

$$t = 20°F$$

Coefficient of linear expansion,

$$\alpha = 6 \times 10^{-6} \text{ per } °F$$

Let $H =$ Horizontal thrust on the cable before the increase of temperature.

$dH =$ Change in horizontal tension.

Using the relation,

$$H = \frac{wl^2}{8y_C} \text{ with usual notations.}$$

$$= \frac{2 \times 120^2}{8 \times 12} = 300 \text{ t}$$

Now using the relation,

$$\therefore \qquad \frac{dH}{H} = -\frac{3l^2}{16y_C{}^2} \cdot \alpha t \text{ with usual notations.}$$

$$= -\frac{3 \times 120^2}{16 \times 12^2} \times 6 \times 10^{-6} \times 20 = -\frac{9}{4} \times 10^{-3}$$

or $$H = \frac{9}{4} \times 10^{-3} \times H = \frac{9}{4} \times 10^{-3} \times 300 \text{ t}$$

$$= -0·675 \text{ t}$$

$$= 0·675 \text{ t (decrease) Ans.}$$

Exercise 9–2

1. A suspension cable of span 30 m has a central dip of 3 m. Find the length of the cable, if it carries a uniformly distributed load of 1·5 t/m.

[**Ans.** 30·8 m]

2. A wire is to be stretched between two pegs 50 m apart. Find necessary length of the line, if the central dip is 1/10th of the span.

(*Banaras Hindu University*)

[**Ans.** 51·33 m]

3. A suspension cable of 120 m span hangs between the points which are 4 m and 24 m above the lowest point of the cable. Find the length of the cable.

(*Kerala University*)

[**Ans.** 125·85 m]

4. A cable of span 50 m is suspended from two pegs 6 m and 1·25 m above the lowest point of the cable. Find the length of the cable, between the two pegs.

(*Utkal University*)

[**Ans.** 50·77 m]

9·15. Stiffening girders in the suspension bridges

We have already discussed that the cable is the main load bearing member in a suspension bridge. A little consideration will show that the curvature of the cable will go on distorting, as the load moves on the decking. In order to avoid the distortion of the cable and to make it to retain a parabolic shape, as the load moves from one support to another, two stiffening girders are suspended from the cables with the help of suspension rods. The stiffening girders may be three-hinged or two-hinged. The roadway is then provided on the stiffening girders.

The uniformly distributed dead load of the girders and roadway is transmitted by the hangers to the cables and is taken up entirely by the tension in the cables. A little consideration will show, that as the stiffening girders are supported, all along their lengths by the suspension rods, therefore the stiffening girders will have no S.F. or B.M. The live load on the bridge will be transmitted to the stiffening girders as point loads. The stiffening girders will transfer the live load to the cables as a uniformly distributed load. It is thus obvious, that the stiffening girders will be subjected to S.F. and B.M. throughout its length due to the live loads.

9·16. Suspension bridges with three-hinged stiffening girder

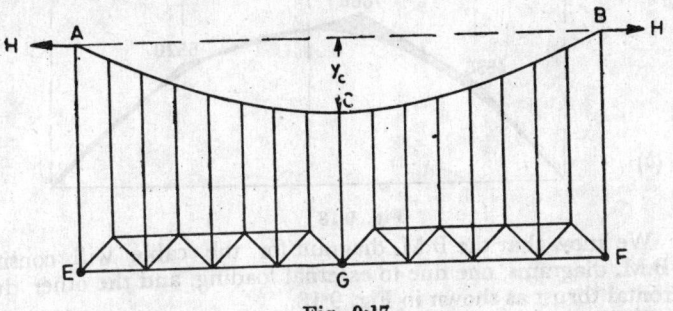

Fig. 9·17

A three-hinged stiffening girder, as the name indicates, is a girder hinged at three points i.e., at its two supports E and F as well as at its centre G as shown in Fig. 9·17. The dead load of the girder and roadway is transmitted by the suspension rods to the cables ; and is taken up entirely by the tension in the cables. The three-hinged stiffening girders, which are supported all along their lengths by the suspension rods, will have no S.F. or B.M. at any section because of uniformly distributed dead load. But the stiffening girder will be subjected to B.M. and S.F.

Example 9·12. *A cable of a suspension bridge has a span of 400 m over supports, which are at the same level and a sag of 40 m vertically from the line of support to the lowest point on the cable at mid-span. It is stiffened by a three-hinged girder with hinged supports at the two ends, and the third hinge at its mid-point. The girder carries 3 loads of 30 T, 50 T, 40 T acting at 70 m, 140 m, and 300 m respectively from the left hand end.*

(i) Draw the B.M. diagram for the girder giving values at salient points.

Solution.

Given. Span, $l = 400$ m
Central dip, $y_C = 40$ m

B.M. diagram

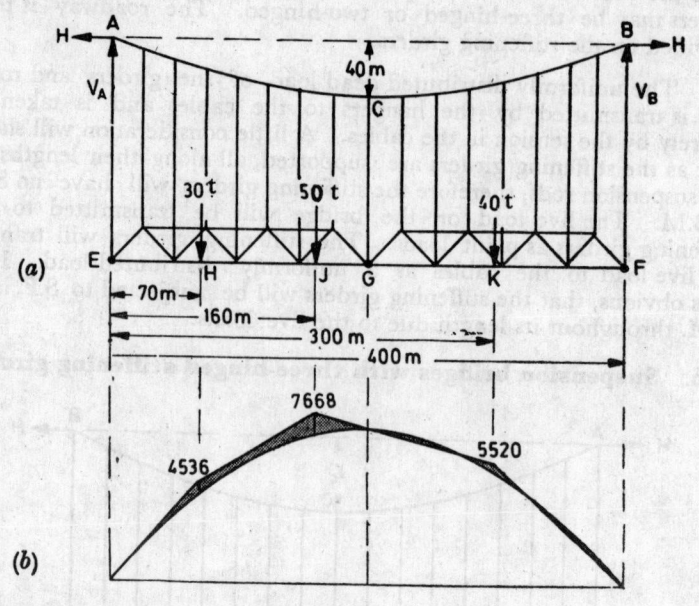

Fig. 9·18

We know that the B.M. diagram for the cable will consist of two B.M. diagrams, one due to external loading, and the other due to horizontal thrust as shown in Fig. 9·18.

Taking moments about A,

$$R_B \times 400 = 30 \times 70 + 50 \times 160 + 40 \times 300 = 22,100$$

$$\therefore \quad R_B = \frac{22,100}{400} = 55·2 \text{ t}$$

and $R_A = 30 + 50 + 40 - 55·2 = 64·8$ t'

∴ B.M. at H *i.e.* at a distance of 70 m from E,

$$\mu_{70} = R_A \times 70 = 64·8 \times 70 = 4,536 \text{ t-m}$$

Similarly $\mu_{160} = 64·8 \times 160 - 30 \times 90 = 7,668$ t-m

and $\mu_{300} = R_B \times 100 = 55·2 \times 100 = 5,520$ t-m

Now draw the B.M. diagram with the above values as shown in Fig. 9·18.

We know that the B.M. at C,

$$\mu_C = 55·2 \times 200 - 40 \times 100 = 7,040 \text{ t-m}$$

Using the relation,

$$H = \frac{\mu_C}{y_C} \text{ with usual notations.}$$

$$= \frac{7,040}{40} = 176 \text{ t}$$

Now draw the B.M. diagram for H as shown in Fig. 9·18.

We know that the depth of cable at a distance 70 m from A,

$$y = \frac{4y_C}{l^2}\, x(l-x) = \frac{4 \times 40}{400^2} \times 70(400-70)\ \text{m}$$

$$= \frac{4 \times 40 \times 70 \times 330}{400 \times 400} = 23 \cdot 1\ \text{m}$$

B.M. due to horizontal thrust at H i.e. 70 m from A
$$= H \times 23 \cdot 1 = 176 \times 23 \cdot 1 = 4{,}065\ \text{t-m}$$
and net B.M. at H i.e. 70 m from A,
$$M_{70} = 4{,}536 - 4{,}065 = +471\ \text{t-m}$$
Similarly depth of the cable at a distance of 160 m from A,

$$y = \frac{4y_C}{l^2}\, x(l-x) = \frac{4 \times 40}{400^2} \times 160(400-160)$$

$$= \frac{4 \times 40 \times 160 \times 240}{400 \times 400} = 38 \cdot 4\ \text{m}$$

$\therefore$ B.M. due to horizontal thrust at J i.e. at 160 m from A,
$$= H \times 38 \cdot 4 = 176 \times 38 \cdot 4 = 6758\ \text{t-m}$$
and net B.M. at H,
$$M_{160} = 7{,}668 - 6{,}758 = 910\ \text{t-m}$$
Similarly depth of the cable at a distance of 300 m from A,

$$y = \frac{4y_C}{l^2}\, x(l-x) = \frac{4 \times 40}{400^2} \times 300(400-300)$$

$$= \frac{4 \times 40 \times 300 \times 100}{400 \times 400} = 30\ \text{m}$$

$\therefore$ B.M. due to horizontal thrust at K i.e. at 300 m from A
$$= H \times 30 = 176 \times 30 = 5{,}280\ \text{t-m}$$
and net B.M. at K,
$$M_{300} = 5{,}520 = 5{,}280 = 240\ \text{t-m}$$

9·17. Influence lines for moving loads over the suspension bridges with three-hinged stiffening girders

We have already discussed in Art. 9·16, that a three-hinged stiffening girder will have no S.F. or B.M. due to uniformly distributed dead load. But the live load, rolling over the bridge, will be transmitted by the stiffening girders and suspension rods to the cables, as a uniformly distributed load. A little consideration will show that the above arrangement of transmitting the load will cause the cables to retain their parabolic shape, while the stiffening girders will have to resist the S.F. and B.M. due to live load.

9·18. Influence lines for a single concentrated load rolling over the suspension bridge with three-hinged stiffening girders

Consider a concentrated load W rolling over a suspension bridge ABC of span l and central rise y_C with three-hinged stiffening girders. We shall draw the influence lines for the following :

(a) Influence lines for horizontal thrust,

(b) Influence lines for shear force, and

(c) Influence lines for bending moment.

Influence lines for horizontal thrust

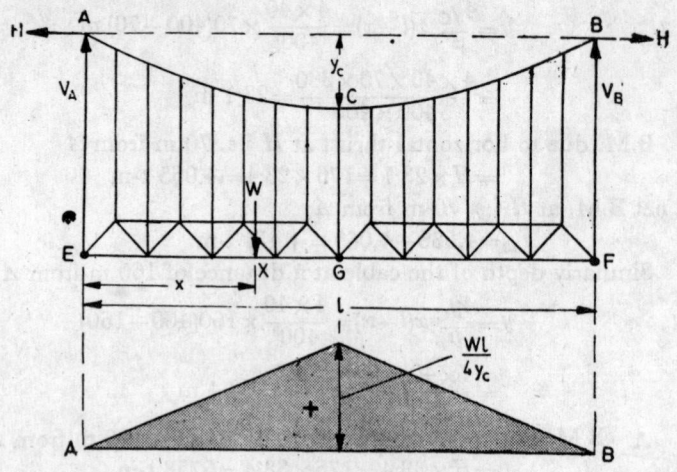

Fig. 9·19

Consider a section X at a distance x from A. When the load is on the section the vertical reaction at B,

$$R_B = \frac{Wx}{l} \quad \text{and} \quad R_A = \frac{W(l-x)}{l}$$

Now the B.M. at C due to loading on the bridge,

$$\mu_C = R_B \times \frac{l}{2} = \frac{Wx}{l} \times \frac{l}{2} = \frac{Wx}{2}$$

We know that the horizontal thrust,

$$H = \frac{\mu_C}{y_C} = \frac{Wx}{2} \times \frac{1}{y_C} = \frac{Wx}{2y_C}$$

From the above expression, we see that the horizontal thrust increases by a straight line law. Substituting the various values of x in the above equation, we find that when $x=0$, $H=0$ and when $x=\frac{l}{2}$, $H=\frac{Wl}{4y_C}$. We also see that when the load is in the portion CB of the bridge, the horizontal thrust,

$$H = \frac{\mu_C}{y_C} = \frac{R_A. \frac{l}{2}}{y_C} = \frac{\frac{W(l-x)}{l} \times \frac{l}{2}}{y_C} = \frac{W(l-x)}{2y_C}$$

Substituting the various values of x in the above equation, we find that when $x=\frac{l}{2}$, $H = \frac{W\left(l-\frac{l}{2}\right)}{2y_C} = \frac{Wl}{4y_C}$ and when $x=l$, $H=0$.

It is thus obvious that the horizontal thrust will be zero, when the load is at A. It will increase by a straight line law to $\frac{Wl}{4y_C}$

when the load is at C and will decrease by a straight line law to zero when the load is at B as shown in Fig. 9·19 (b).

Influence line for shear force

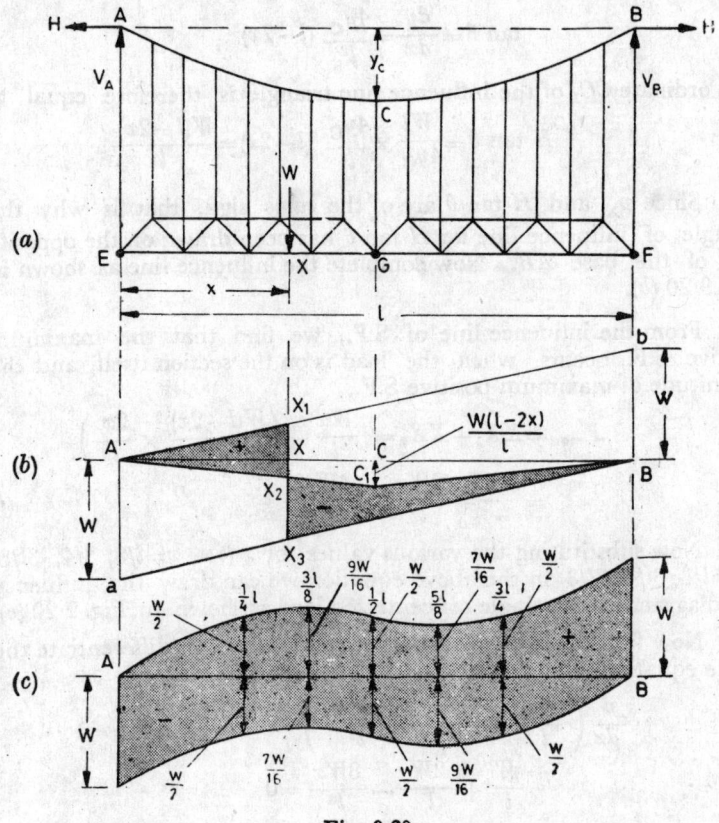

Fig. 9·20

Consider a section X, at a distance x from A. We know that the S.F. at X,

$$F_X = p_X + H \tan \theta$$

where p_X is the usual shear force as a freely supported beam, and $H \tan \theta$ is the vertical component of the tension in the cable at X. We have already discussed in Art. 3·3 that the influence lines for S.F. due to a single concentrated load will consist of two parallel straight lines Ab and aB as shown in Fig. 9·20(b). We have also discussed that the maximum positive and negative S.F. at a section occurs, when the load is on the section itself. The influence line for $H \tan \theta$ will be triangle having ordinate equal to $\dfrac{Wl}{4y_C} \tan \theta$ (because the value of θ is constant at a given section).

We know that the dip of the cable at a distance x from the support A,

$$y = \frac{4y_C}{l^2}\, x(l-x)$$

$\therefore$ $\tan\theta = \dfrac{dy}{dx} = \dfrac{4y_C}{l^2}\,(l-2x)$

The ordinate CC_1 of the influence line triangle is therefore equal to

$$H\tan\theta = \frac{Wl}{4y_C} \times \frac{4y_C}{l^2}\,(l-2x) = \frac{W(l-2x)}{l}$$

Since p_x and $H\tan\theta$ are of the *same* sign, that is why the triangle of influence line for $H\tan\theta$ has been drawn on the opposite side of the base AB. Now complete the influence line as shown in Fig. 9·20 (b).

From the influence line of S.F., we find that the maximum positive S.F. occurs, when the load is on the section itself, and the magnitude of maximum positive S.F.,

$$F_{max} = XX_1 + XX_2 = \frac{Wx}{l} + \left(\frac{W(l-2x)}{l} \times \frac{2x}{l} \right)$$

$$= \frac{Wx}{l} + \frac{2Wx}{l} - \frac{4Wx^2}{l^2} \qquad\qquad \ldots(i)$$

Now substituting the various values of x (*i.e.* $x = l/8$, $l/4$, $3l/8$, $l/2$, $5l/8$, $3l/4$, $7l/8$) in the above equation we can draw the influence line diagram for the single concentrated load as shown in Fig. 9·20 (c).

Now for the absolute maximum S.F., let us differentiate the above equation with respect to x and equate it to zero,

$$\frac{d}{dx}\left(\frac{Wx}{l} + \frac{2Wx}{l} - \frac{4Wx^2}{l^2} \right) = 0$$

$\therefore$ $\dfrac{W}{l} + \dfrac{2W}{l} - \dfrac{8Wx}{l^2} = 0$

or $\dfrac{8Wx}{l^2} = \dfrac{3W}{l}$

$$x = \frac{3l}{8}$$

Now substituting this value of x in equation, we get

$$F_{max\ max} = \left(\frac{W}{l} \cdot \frac{3l}{8} \right) + \left(\frac{2W}{l} \cdot \frac{3l}{8} \right) - \left(\frac{4W}{l^2} \cdot \frac{9l}{64} \right)$$

$$= \frac{9W}{16}$$

Now complete the influence line for the positive S.F. as shown in Fig. 9·20 (c).

Similarly complete the influence line for the negative S.F. as shown in Fig. 9·20 (c).

Influence lines for bending moment

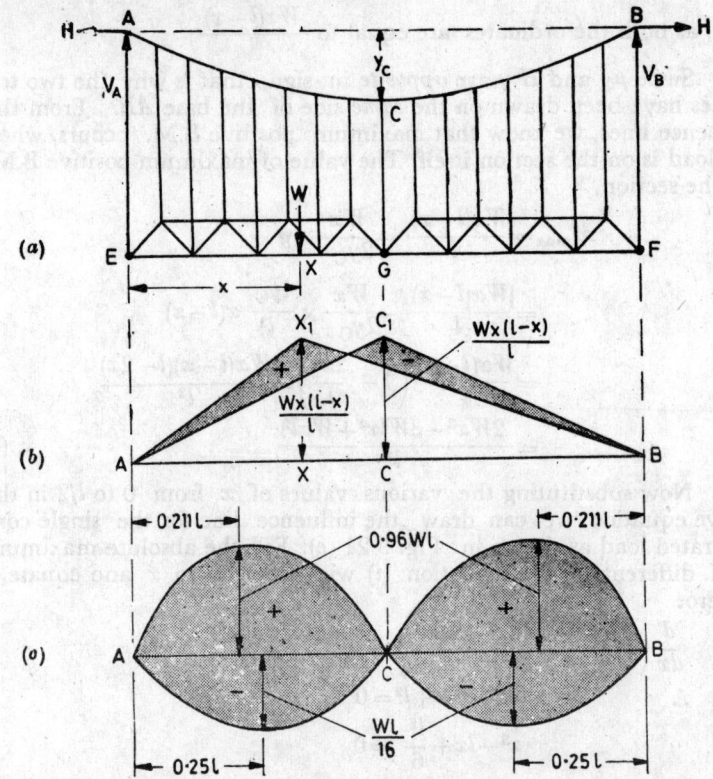

Fig. 9·21

Consider a section X, at a distance x from A. We know that the B.M. at x,

$$M_X = \mu_X - H \cdot y$$

where μ_X is the usual B.M. as a freely supported beam and $H.y$ is the moment due to horizontal thrust H. We also know that the influence line for μ_X will be a triangle with *ordinate XX_1 equal to $\dfrac{Wx(l-x)}{l}$. The influence line for $H.y$ will be a triangle with ordinate CC_1 equal to $\dfrac{Wl}{4y_C}$.

We know that the sag of the cable at a distance x from A,

$$y = \frac{4y_C}{l^2} \, x(l-x)$$

Substituting the value of y in the equation for the ordinate CC_1,

$$CC_1 = \frac{Wl}{4y_C} \cdot y = \frac{Wl}{4y_C} \times \frac{4y_C}{l^2} \, x(l-x)$$

$$= \frac{Wx(l-x)}{l}$$

*For details please refer to Art. 3·3.

It is thus obvious, that the ordinate XX_1 is equal to the ordinate CC_1, as both the ordinates are equal to $\frac{Wx(l-x)}{l}$.

Since μ_X and $H.y$ are *opposite* in sign, that is why the two triangles have been drawn on the *same* side of the base AB. From the influence lines, we know that maximum positive B.M. occurs, when the load is on the section itself. The value of maximum positive B.M. on the section,

$$M_{max} = \frac{Wx(l-x)}{l} - \frac{Wx}{2y_C} \times y$$

$$= \frac{Wx(l-x)}{l} - \frac{Wx}{2y_C} \times \frac{4y_C}{l^2}\, x(l-x)$$

$$= \frac{Wx(l-x)}{l}\left(1 - \frac{2x}{l}\right) = \frac{Wx(l-x)(l-2x)}{l^2}$$

$$= \frac{2Wx^3 - 3Wlx^2 + Wxl^2}{l^2} \qquad \ldots(i)$$

Now substituting the various values of x from 0 to $l/2$ in the above equation, we can draw the influence line for the single concentrated load as shown in Fig. 9·21 (c). For the absolute maximum B.M. differentiate the equation. (i) with respect to x and equate it to zero.

$$\frac{d}{dx}\left(\frac{2Wx^3 - 3Wlx^2 + Wxl^2}{l^2}\right) = 0$$

$$\therefore \qquad 6x^2 - 6lx + l^2 = 0$$

or $$\qquad x^2 - lx + \frac{l^2}{6} = 0$$

$$\therefore \qquad \left(x - \frac{l}{2}\right) = \frac{l^2}{12} \quad \left[\because \left(x - \frac{l}{2}\right)^2 = x^2 + \frac{l^2}{4} - lx\right]$$

$$= \frac{l}{2} \pm \frac{l}{2\sqrt{3}} = (0·5 \pm 0·289)l$$

$$= 0·211\, l \quad \text{or} \quad 0·789\, l$$

Substituting the value of x in equation (i), the absolute maximum B.M. at x,

$$M_{max\ max} = 0·096\ Wl$$

Now complete the influence line diagram for positive B.M. as shown in Fig. 9·21 (c).

Similarly, the maximum negative B.M. on the section occurs, when the load is at the centre of the span. The value of maximum negative B.M. on the section,

$$M_{max} = \frac{Wx(l-x)}{l} - \frac{Wx}{2} = \frac{Wx(l-2x)}{2l}$$

$$= \frac{Wlx - 2Wx^2}{2l} \qquad \ldots(ii)$$

Now substituting the various values of x from 0 to $l/2$ in the above equation, we can draw the influence line for the single concen-

rated load as shown in Fig. 9·21 (c). For the absolute maximum
B.M., differentiate the equation (ii) with respect to x and equate it
to zero.

$$\frac{d}{dx}\left(\frac{Wlx-2Wx^2}{2l}\right)=0$$

$$\therefore \qquad l-4x=0$$

or
$$x=\frac{l}{4}$$

Substitute this value of x in equation (ii), the absolute maxi‑
mum B.M. at x,

$$M_{max\ max}=\frac{Wl}{16}$$

Now complete the influence line diagram for the negative B.M.
as shown in Fig. 9·21 (c).

**9·19. Influence lines for a uniformly distributed load rolling
over the suspension bridge with three-hinged stiffening girders.**

All the influence lines for a uniformly distributed load are the
same, as those for a single concentrated load, except for the bending
moment which is discussed here.

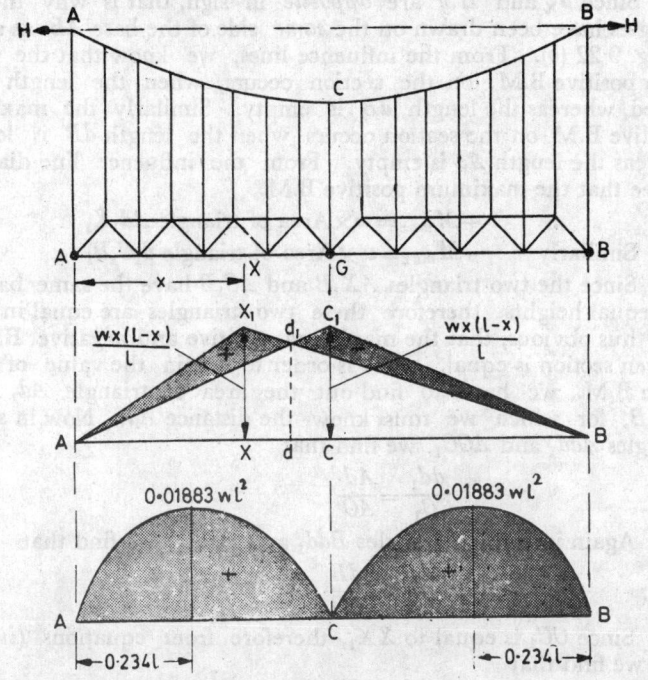

Fig. 9·22

Consider a uniformly distributed load of w per unit length rolling over a suspension bridge with three hinged stiffening girders.
Now consider a section X, at a distance x, from A. We know that
the B.M. at X, $M_X=\mu_x-H.y$

where μ_X is the usual B.M. as a freely supported beam and Hy is the moment due to horizontal thrust H. We also know that the influence line for μ_X will be a triangle with ordinate XX_1 equal to $\dfrac{wx(l-x)}{l}$. The influence line for $H.y$ will also be a triangle with ordinate CC_1 equal to $\dfrac{wl}{4y_C} . y$. We know that the sag of the cable at a distance x from A,

$$y = \frac{4y_C}{l^2} \, x(l-x)$$

Substituting the value of y in the equation for ordinate CC_1,

$$CC_1 = \frac{wl}{4y_C} . y = \frac{wl}{4y_C} \times \frac{4y_C}{l^2} \, x(l-x)$$
$$= \frac{wx(l-x)}{l}$$

It is thus obvious, that the crdinate XX_1 is equal to the ordinat CC_1, as both the ordinates are equal to $\dfrac{wx(l-x)}{l}$.

Since μ_X and $H.y$ are *opposite* in sign, that is why the two triangles have been drawn on the *same* side of the base AB as shown in Fig. 9·22 (b). From the influence lines, we know that the maxi mum positive B.M. on the section occurs, when the length Ad i loaded, whereas the length dB is empty. Similarly the maximum negative B.M. on the section occurs when the length dB is loaded whereas the length Ad is empty. From the influence line diagram we see that the maximum positive B.M.,

$$+M_{max} = w \times \text{Area of triangle } Ad_1X_1 \qquad ...(i$$
$$\text{Similarly} \qquad -M_{max} = w \times \text{Area of triangle } d_1C_1B_1 \qquad ...(ii$$

Since the two triangles AX_1B and AC_1B have the same base AB and equal heights, therefore these two triangles are equal in areas It is thus obvious, that the maximum positive and negative B.M. a a given section is equal. Now is order to obtain the value of maxi mum B.M., we have to find out the area of triangle $Ad_1 X_1$ o $d_1 C_1B$, for which we must know the distance Ad. Now in simila triangles Add_1 and ACC_1, we find that

$$\frac{dd_1}{CC_1} = \frac{Ad}{AC} \qquad ...(iii$$

Again in similar triangles Bdd_1 and BXX_1, we find that

$$\frac{dd_1}{XX_1} = \frac{dB}{XB} \qquad ...(iv$$

Since CC_1 is equal to XX_1, therefore from equations (iii) an (iv), we find that

$$\frac{Ad}{AC} = \frac{dB}{XB}$$
$$\therefore \qquad Ad.XB = AC.dB$$
$$Ad(l-x) = \frac{l}{2} \, (l - Ad)$$

$$Ad.l - Ad.x = \frac{l^2}{2} - Ad.\frac{l}{2}$$

$$Ad \times \frac{3l}{2} - Ad.x = \frac{l^2}{2}$$

$$Ad.3l - Ad.2x = l^2$$

$$Ad(3l - 2x) = l^2$$

or
$$Ad = \frac{l^2}{(3l - 2x)} \qquad \qquad ...(v)$$

Now substituting the value of Ad in equation (iii),

$$\frac{dd_1}{CC_1} = \frac{\frac{l^2}{(3l-2x)}}{AC} = \frac{\frac{l^2}{(3l-2x)}}{\frac{l}{2}} \qquad \left(\because AC = \frac{l}{2} \right)$$

$$= \frac{2l}{(3l-2x)}$$

or
$$dd_1 = \frac{2l}{(3l-2x)} \times CC_1$$

$$= \frac{2l}{(3l-2x)} \times \frac{x(l-x)}{l} \qquad \left[\because CC_1 = \frac{x(l-x)}{l} \right]$$

$$= \frac{2x(l-x)}{(3l-2x)}$$

From the geometry of the figure, we find that

Area of $\triangle\, Ad_1X_1 = $ Area of $\triangle AX_1B - $ Area of $\triangle Ad_1B$

$$= \left[\frac{l}{2} \times \frac{x(l-x)}{l} \right] - \left[\frac{l}{2} \times \frac{2x(l-x)}{(3l-2x)} \right]$$

$$= \frac{x(l-x)(l-2x)}{2(3l-2x)}$$

Thus the maximum positive as well as negative B.M. at a section is given by

$$M_{max} = \frac{wx(l-x)(l-2x)}{2(3l-2x)}$$

$$= \frac{w}{2} \left[\frac{l^2x - 3x^2l + 2x^3}{3l-2x} \right] \qquad ...(vi)$$

For the absolute maximum B.M. differentiating the equation (vi) with respect to x and equating the same to zero,

$$\frac{d}{dx} \left[\frac{w}{2} \left(\frac{l^2x - 3x^2l + 2x^3}{3l-2x} \right) \right] = 0$$

$$\frac{w}{2} \left[\frac{[(3l-2x)(l^2-6xl+bx^2)] - [l^2x-3x^2l+2x^3](-2)}{3l-2x^2} \right] = 0$$

$$(3l-2x)(l^2-6xl+6x^2) + 2(l^2x-3x^2+2x^3) = 0$$

$$3l^2 - 20xl^2 + 30x^2l - 12x^3 + 2l^2x - 6x^2l + 4x^3 = 0$$

$$3l^3 - 18xl^2 + 24x^2l - 8x^3 = 0$$

Solving this equation by trial and error, we get

$$x = 0.234\, l$$

Now substituting this value of x in equation (vi) the absolute maximum B.M. at X,

$$M_{max\ max} = 0.01883wl^2$$

Now complete the influence line diagram for B.M. as shown in Fig. 9·22(c).

Note. The influence line diagram for negative B.M. will be similar as that for positive B.M. but on its opposite side.

Example 9·13. *A suspension cable of 30 m span and 3 m dip is stiffened by a three-hinged girder. The dead load is 1 t/m. Determine the maximum tension in the cable and maximum B.M. in the girder due to concentrated load of 10 T crossing the girder, assuming that the whole dead load is carried by the cable without stressing the girder.*

(A.M.I.E., Summer, 1979)

Solution.

Span,	$l = 30$ m
Central dip	$y_C = 3$ m
Dead load	$= 1$ t/m
Rolling load,	$W = 10$ t

Maximum tension in the cable

Let T_{max} = Maximum tension in the cable.

We know that the maximum tension in the cable will take place, when the load of 10 T will be at the centre of the span

∴ Vertical reaction due to dead load,

$$R_1 = \frac{1 \times 30}{2} = 15 \text{ t}$$

and vertical reaction due to live load (*i.e.*, when 10 t is at the centre of the span),

$$R_2 = \frac{10}{2} = 5 \text{ t}$$

∴ Total vertical reaction,

$$V = 15 + 5 = 20 \text{ t}$$

Similarly, horizontal thrust due to dead load.

$$H_1 = \frac{wl^2}{8y_C} = \frac{1 \times 30^2}{8 \times 3} = 37.5 \text{ t}$$

and horizontal thrust due to live load,

$$H_2 = \frac{Wl}{4y_C} = \frac{10 \times 30}{4 \times 3} = 25 \text{ t}$$

∴ Total horizontal thrust,

$$H = 37.5 + 25 = 62.5 \text{ t}$$

Using the relation,

$$T_{max} = \sqrt{R^2 + H^2} \text{ with usual notations.}$$
$$= \sqrt{20^2 + 62.5^2} = \sqrt{4,306} \text{ t}$$
$$= 65.62 \text{ t Ans.}$$

Maximum positive B.M. in the girder

We know that the maximum positive B.M. occurs, when the

CABLES AND SUSPENSION BRIDGES 365

load is at a distance of $0.211\ l$ *i.e.*, $0.211 \times 30 = 6.33$ m from the support. At this instant

$$M_{max} = 0.096\ Wl = 0.096 \times 10 \times 30 \text{ t-m}$$

$$\textbf{28.8 t·m} \quad \textbf{Ans.}$$

Maximum negative B.M. in the girder

We also know that the maximum negative B.M. occurs, when the load is at a distance of $0.25\ l = 0.25 \times 30 = 7.5$ m from the support. At this moment

$$M_{max} = -\frac{Wl}{16} = -\frac{10 \times 30}{16} \text{ t-m}$$

$$= -187.5 \text{ t·m} \quad \textbf{Ans.}$$

9·20. Suspension bridges with two-hinged stiffening girders

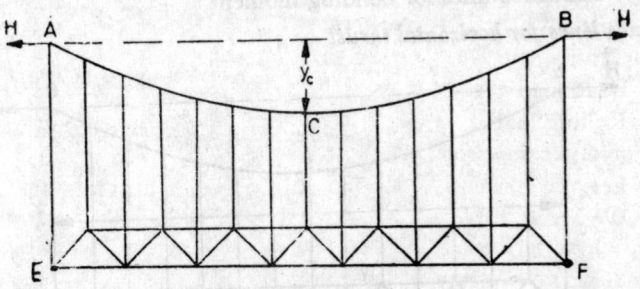

Fig. 9·23·

A two-hinged stiffening girder, as the name indicates, is a girder hinged at two points *i.e.* at its supports E and F as shown in Fig. 9·23. Like a three-hinged stiffening girder, the dead load of a two-hinged stiffening girder and roadways is transmitted by the suspension rods to the cables, and is taken up entirely by the tension in the cables.

A two-hinged stiffening girder is a statically indeterminate structure. But for simplicity, it is assumed that the girder is a rigid one and the load, irrespective of its position, is transmitted as a uniformly distributed load by the hangers to the cables, such that

$$W = w_e .\ l$$

where
$W =$ Total load on the bridge,

$w_e =$ Equivalent uniformly distributed load per horizontal unit length,

$l =$ Span of the cable.

Therefore corresponding horizontal thrust in the cable will be

$$H = \frac{w_e .\ l^2}{8y_C} = \frac{Wl}{8y_C} \qquad (\because\ W = w_e .\ l)$$

It is thus obvious, that the magnitude of horizontal thrust will be constant, since, it is independent of the load position.

9.21. Influence lines for a single concentrated load rolling over the suspension bridges with two-hinged stiffening girders.

We have a already discussed in Art. 9·20 that the dead load of the two-hinged stiffening girder and roadway is transmitted by the suspension rods to the cables. The live load is assumed to be transmitted as a uniformly distributed load by the hangers to the cables. Here we shall discuss the influence lines for the moving loads over the suspension bridges with two-hinged stiffening girders.

Consider a suspension bridge, of span l and a central dip y_0, with two-hinged stiffening girder. We shall draw the influence lines for the following :

(a) Influence lines for horizontal thrust,

(b) Influence lines for shear force, and

(c) Influence lines for bending moment.

Influence lines for horizontal thrust

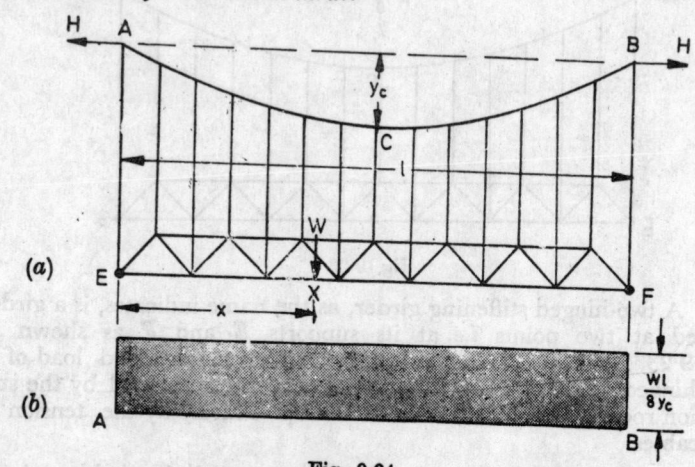

Fig. 9·24

Consider a section X, at a distance x from A. We know that in the case of two-hinged stiffening girder, it is assumed that the live load is assumed to be transmitted as a uniformly distributed load by the hangers to the cables. Therefore the equivalent uniformly distributed load per horizontal unit length,

$$w_e = \frac{W}{l}$$

We know that horizontal thrust in the cable, when it is loaded with a uniformly distributed load

$$H = \frac{w_e \cdot l^2}{8 \, y_C} = \frac{Wl}{8 y_C}$$

We see, from the above equation, that the magnitude of horizontal thrust is constant, since it is independent of the load position. It is thus obvious, that the influence line for the horizontal thrust will be horizontal line as shown in Fig. 9·24 (b).

Influence line for the shear force

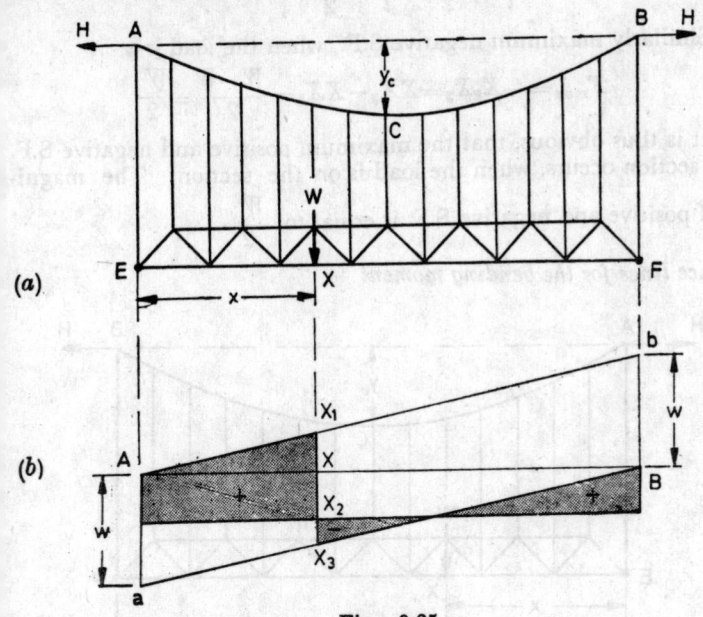

Fig. 9·25

Consider a section X, at a distance x from A. We know that the S.F. at x,

$$F_X = p_X + H \tan \theta$$

where p_X is the usual shear force as a freely supported beam and $H \tan \theta$ is the vertical component of the tension in the cable at X. We have already discussed in Art. 9·20 that the live load is assumed to be transmitted as a uniformly distributed load per horizontal unit length. Now we know that the influence lines for S.F. due to uniformly distributed load will consist of two parallel straight lines Ab and aB as shown in Fig. 9·25 (b). The influence lines for $H \tan \theta$ will be $\tan \theta$ times the influence lines for H. Since the influence line for H is a rectangle and the value of θ is constant at a given section, therefore the influence lines for $H \tan \theta$ will also be a rectangle.

We also know that the maximum positive and negative S.F. occurs, when the load is on the section itself.

Now $$F_X = p_X + H \tan \theta = p_X + \frac{Wl}{8y_C} \times \frac{4y_C}{l^2} (l - 2x)$$

$$= p_X + \frac{W}{2l} (l - 2x)$$

Now complete the influence line diagram as shown in Fig. 9·25 (b). From the geometry of the figure, we find that maximum positive S.F., when the load is at X,

$$F_{max} = +X_1 X_2 = X X_1 + X X_2 = \frac{Wx}{l} + \frac{W}{2l} (l - 2x)$$

$$F_{max} = \frac{Wx}{l} + \frac{W}{2} - \frac{Wx}{l} = \frac{W}{2}$$

Similarly maximum negative S.F., when the load is x

$$F_{max} = -X_2X_3 = XX_3 - XX_2 = \frac{W-W}{2} = \frac{W}{2}$$

It is thus obvious, that the maximum positive and negative S.F. on the section occurs, when the load is on the section. The magnitude of positive and negative S.F. is equal to $\frac{W}{2}$.

Influence lines for the bending moment

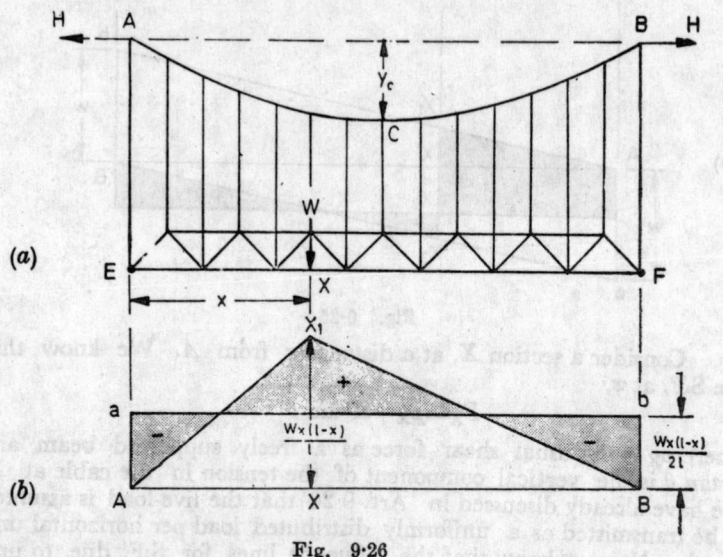

Fig. 9·26

Consider a section X, at a distance x from A. We know that B.M. at X,

$$M_X = \mu_X - H.y$$

where μ_X is the usual B.M. as a freely supported beam, and $H.y$ is the moment due to horizontal thrust H. We know that the influence line for μ_X will be a triangle having ordinate XX_1 equal to $\frac{Wx(l-x)}{l}$. The influence line for $H.y$ will be y times the influence line for H. Since the influence for H is a rectangle and the value of y is constant at a given section, therefore the influence lines for $H.y$ will also be a rectangle. The ordinate of the influence line for $H.y$

$$Aa = H.y = \frac{Wl}{8y_c} \times \frac{4y_c}{l^2} \cdot x(l-x) = \frac{Wx(l-x)}{2l}$$

Since μ_X and $H.y$ are *opposite* in sign, that is why the two figures have been drawn on the same side of the base AB. From the influence line, we find that the maximum positive B.M. occurs, when

CABLES AND SUSPENSION BRIDGES

369

the load is on the section itself. The value of maximum positive B.M. on the section,

$$M_{max} = \frac{Wx(l-x)}{l} - \frac{Wx(l-x)}{2l} = \frac{Wx(l-x)}{2l} \qquad ...(i)$$

and the maximum negative B.M. on the section occurs, when the load is on any one of the support. The value of maximum negative B.M. on the section,

$$M_{max} = -\frac{Wx(l-x)}{2l}$$

Now for the absolute maximum positive B.M. differentiate the equation (i) with respect to x and equate it to zero.

$$\frac{d}{dx}\left(\frac{Wx(l-x)}{2l}\right) = 0$$

$$\frac{d}{dx}\left(\frac{Wx}{2} - \frac{Wx^2}{2l}\right) = 0$$

$$\therefore \qquad \frac{W}{2} - \frac{Wx}{l} = 0$$

or

$$x = \frac{l}{2}$$

Substituting this value of x in equation (i) we get

$$M_{max\ max} = \frac{W \cdot \frac{l}{2}\left(l - \frac{l}{2}\right)}{2l} = \frac{Wl}{8}$$

Example 9·14. *A suspension cable of 60 m span and central dip of 6 m is strengthened by two-hinged stiffening girders. The dead load of stiffening girder is 2 t/m. The girder is also subjected to a live load of 3 t/m over the left half of the span. Find the maximum tension in the cable.*

Solution.

Given. Span $\qquad l = 60$ m

Central dip, $\qquad y_C = 6$ m

Dead load, $\qquad = 2$ t/m

Let $\qquad T_{max} =$ Maximum tension in the cable.

We know that the live load is assumed to be uniformly distributed over the entire span. Therefore total load on the cable,

$$w = \frac{\text{Total dead load} + \text{live load}}{\text{Span}}$$

$$= \frac{2 \times 60 + 3 \times 30}{60} = \frac{120 + 90}{60} = \frac{210}{60} = 3·5 \text{ t/m}$$

$\therefore$ Horizontal thrust, H

$$= \frac{wl^2}{8y_C} = \frac{3 \cdot 5 \times 60^2}{8 \times 6} = 262 \cdot 5 \text{ t}$$

and vertical reaction at supports,

$$R = \frac{210}{2} = 105 \text{ t}$$

Using the relation,

$$T_{max} = \sqrt{R^2 + H^2} \text{ with usual notations,}$$
$$= \sqrt{105^2 + 262 \cdot 5^2} \text{ t}$$
$$= 282 \cdot 8 \text{ t   Ans}.$$

Exercise 9·3

1. A suspension cable of 100 m span has 10 m dip. It is stiffened by a three-hinged girder, whose dead load is 2 t/m. Determine the maximum tension in the cable, if a point load of 50 T rolls over the girder. (*Madras University*)
[**Ans.** 395·3 t′]

2. A suspension cable of 40 m span and 4 m dip is stiffened by three-hinged girder. Find the values of maximum positive and negative B.M., when a concentrated load of 25 tonnes rolls over the girder. (*Bombay University*)
[**Ans.** 96 t ; −62·5 t′]

3. If in the above equation, the cable is stiffened by a two-hinged girder, find the value of maximum positive and negative B.M.
[**Ans.** 125 t′ ; −125 t]

HIGHLIGHTS

1. The equation of a cable is given by
$$y = k.x(l-x)$$
$$= \frac{4y_C}{l^2} x(l-x)$$

where $k = $A constant whose value is $\frac{4y_C}{l^2}$

$x = $Horizontal distance between the support and the section,

$l = $Span of the cable.

2. Horizontal thrust in a cable,
$$H = \frac{\mu_C}{y_C}$$

where $\mu_C = $Moment at the centre of the cable

$y_C = $Central dip of the cable.

3. Tension in the cable at support A,
$$T_A = \sqrt{R_A{}^3 + H^2}$$

Similarly $T_B = \sqrt{R_B{}^2 + H^2}$

where R_A and $R_B = $Vertical reaction on the supports at A and B,

$H = $Horizontal thrust in the cable.

4. When the cable is supported on a guide pulley, the tension in the suspension cable is equal to the tension in the anchor cable.

5. When the cable is supported on rollers, the horizontal components of the tension in the anchor cable and suspension cable are equal.

6. Length of cable,

$$L = l + \frac{8y_C^2}{3l} \quad \text{...(when supported at same level)}$$

$$= l + \frac{2(y_C + d)^2}{3l_1} + \frac{2y_C^2}{3l_2}$$

...(when supported at different levels)

where

$l =$ Span of the cable,

$y_C =$ Central dip of the cable,

$d =$ Difference in levels of the two supports,

$l_1 =$ Horizontal length between the higher support and the lowest point of the cable,

$l_2 =$ Horizontal length between the lower support and the lowest point of the cable.

7. If the temperature of a cable rises through $t°$,

$$dy_C = \frac{3l^2}{16y_C} \cdot \alpha t$$

or

$$\frac{dH}{H} = -\frac{3l^2}{16y_C} \cdot \alpha t$$

Where

$dy_C =$ Change in the central dip of the cable,

$l =$ Span of the cable,

$y_C =$ Central dip of the cable,

$dH =$ Change in horizontal thrust in the cable,

$H =$ Horizontal thrust in the cable.

8. In order to avoid the distortion of cable due to moving loads, it is stiffened by hanging either three-hinged or two-hinged stiffening girders from the cables.

9. When a point load W rolls over a three-hinged girder, the maximum positive B.M. $= 0.096 \, Wl$ and takes place when the load is at a distance of $0.211 \, l$ or $0.789 \, l$ from one of the supports. The maximum negative B.M. $= \dfrac{Wl}{16}$ takes place when the load is at a distance of $\dfrac{l}{4}$ or $\dfrac{3l}{4}$ from one of the supports.

10. When a point load W rolls over a two-hinged girder, the absolute maximum position B.M. $= \dfrac{Wl}{8}$ takes place, when the load is at the centre of the girder.

Do You Know ?

1. Derive an equation for the cable and show that it hangs in the form of a parabola.

2. Obtain a relation for the (1) horizontal thrust and (2) tension in the cable.

3. State the position of maximum and minimum tension in a cable supported at the (1) same level, and (2) different levels.

4. Explain the method of anchoring the cables. Differentiate clearly between the guide pulley support and roller support for suspension cables.

5. Derive an expression for the length of a cable.

6. What are stiffening girders ? Discuss their uses and types.

7. Describe an expression for the absolute maximum positive and negative B.M., when a point load rolls over a three-hinged stiffening girder supported by cables.

8. Show that the value of maximum positive and negative B.M. due to a rolling concentrated load on a two-hinged stiffening girder are same.

10

THREE—HINGED ARCHES

10·1. Introduction

An arch may be defined as a curved girder, having convexity upwards, and supported at its ends. It may be subjected to vertical, horizontal or even inclined loads. In the past, the arches had been the backbone of the important buildings. But in the modern building activity, the arches are becoming obsolete. Today the arches are being provided only for the architectural beauty in ultra modern buildings.

10·2. Theoretical arch or line of thrust

We have already discussed in the previous chapter, that a cable can support a given set of loads, by developing tension in its various segments. The shape of cable will correspond to the funicular polygon for the given system of loads. A little consideration will show, that the cable can be replaced by straight links, pin-jointed at the load points. The tension in each link will be the same, as in the corresponding segment of the cable.

Now, if the link work is inverted (the loads being downwards) as shown in Fig. 10·1, we can see that the stresses in each link will

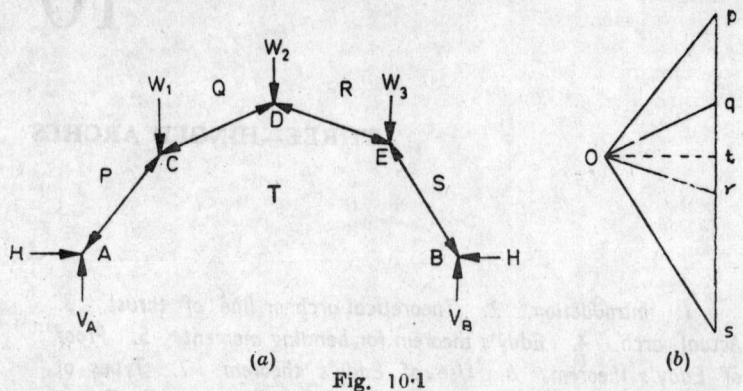

(a)   (b)

Fig. 10·1

now be compressive. Since the loads are acting vertically downwards, therefore we can see, by considering the equilibrium of forces at A, B, C,..., that the horizontal component of the thrust in each member is equal. It is thus obvious, that both the supports A and B must exert an equal horizontal inward reaction H to balance the thrusts in the links, besides exerting vertical reactions V_A and V_B. This pin-jointed linkwork $ACDEB$ can be used as an arrangement to support the given load system. These links are subjected to direct axial thrust only, and there is no bending moment or shear force anywhere. Such an arrangement will prove to be the most economical, as the links of a much smaller section are required to bear the induced thrusts ; as compared to a beam, which will require a larger section to bear the same bending moment and shear force caused by the given loads. Such a linkwork is known as a *linear arch* or *theoretical arch*.

The shape of linkwork and the forces in various links due to given loads may be obtained by drawing pq, qr and rs representing the vertical loads W_1, W_2 and W_3 to some suitable scale. Now select some suitable point o and join op, oq and or and os. Now, through the given support A, draw AC parallel to op. Similarly draw CD parallel to oq, DE parallel to or and EB parallel to os, thus completing the funicular polygon $ACDEB$. Now the lengths op, oq, or and os will give the required forces in the respective members of the link-work. A little consideration will show, that any member of funicular polygon can be drawn for a given system of loads and passing through the given supports A and B. This can be done by changing the position of the point O. Every funicular polygon will give different values of compression in the various links. It is thus obvious, that there can be only one funicular polygon, which will give the correct values of the forces in the links. Therefore a funicular polygon, which is encased in the corresponding arch is taken, and the values of forces in the various links are obtained. This funicular polygon is also known as *line of thrust.*

10·3. Actual arch

As a matter of fact, the position as well as the magnitude of loading over a structure goes on changing. Thus, it is not advisable to construct an arch according to its theoretical shape ; as it will be difficult to meet the varying load positions. Moreover, it is a difficult and troublesome task to construct a theoretical arch. Keeping all these factors, as well as the architectural beauty of an arch in view, its centre line is usually given a circular, parabolic or elliptical shape.

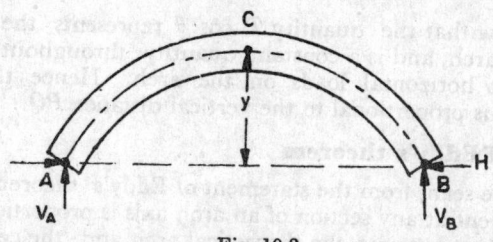

Fig. 10·2

The supports A and B of the arch are called *springings.* The centre line of the arch (shown by chain line) is called *axis* of the arch. The highest point on the arch axis C, is called *crown* of the arch and its height from the springings (y) is called *rise* of the arch as shown in Fig. 10·2.

10·4. Eddy's theorem for bending moment

It states. *"The bending moment, at any point of an arch axis, is proportional to the vertical intercept between the theoretical arch (i.e., line of thrust and the centre line of the actual arch (i.e., arch axis.)."*

10·5. Proof of Eddy's theorem

Consider an actual arch ACB as shown in Fig. 10·3. We have already discussed in Art. 10·2 that the funicular polygon, which represents the thrusts in the linkwork, consists of straight lines, intersecting the lines of action of the given loads. A little consideration will show, that when the arch is subjected to a uniformly distributed load,[*] the funicular polygon will become a continuous curve as shown by a dotted curve $A'B'C'$ in Fig. 10·3. This dotted curve will

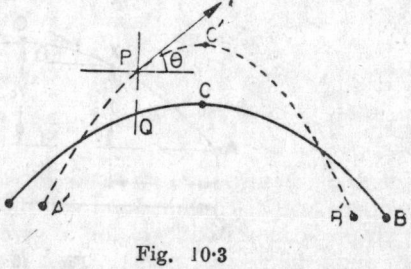

Fig. 10·3

also represent the line, along which the internal thrust acts.

---

[*]This can also be proved, when the arch is subjected to point loads.

Now take any point P on the line of thrust $A'C'B'$ and through this point P draw a vertical line meeting the actual arch ACB at Q. We know that the direction of thrust T, at the point P, will be along the tangent at this point as shown in Fig. 10·3. Let the direction of the thrust T make an angle θ with the horizontal.

This thrust T can be resolved into two components at P, such that its horizontal component,

$$T_H = t \cos \theta$$

and vertical component,

$$T_V = T \sin \theta$$

The bending moment* of this thrust at Q,

$$M = T \cos \theta \times \text{vertical distance } PQ.$$

We know that the quantity $T \cos \theta$ represents the horizontal thrust in the arch, and is a constant quantity throughout its length, if there are no horizontal loads on the arch. Hence the bending moment at Q is proportional to the vertical distance PQ.

10·6. Use of Eddy's theorem

We have seen, from the statement of Eddy's theorem, that the bending moment at any section of an arch axis is proportional to the vertical intercept between the theoretical arch and the centre line of the actual arch. A little consideration will show, that if the axis of the arch coincides with the theoretical arch (or line of thrust) the bending moment in the arch will be zero. It is thus obvious, that an ideal arch is one, whose axis coincides with the line of thrust for the given loads. We know that when the load is uniformly distributed, the bending moment diagram or the line of thrust is parabolic. Therefore a parabolic arch is best suited to support the uniformly distributed loads.

10·7. Types of three-hinged arches

A three-hinged arch may be either of the following two types, depending upon the geometry of its axis :

1. Parabolic arch, and   2. Circular arch.

10·8. Three-hinged parabolic arch

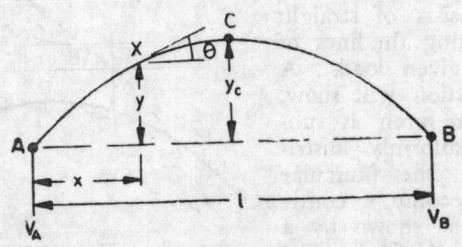

Fig. 10·4

A three-hinged arch, whose axis is parabolic, is known as a *three-hinged parabolic arch*. Consider a three-hinged parabolic arch

---

*The bending moment of the vertical components T_V will be zero.

ACB, having hinges at the supports A and B as well as at the crown C, as shown in Fig. 10·4. Now consider a point X, on the axis of the arch, at a distance x from A.

Let $l=$ Span AB of the arch,

       $y_C=$ Rise of the crown from the springings,

       $y=$ Rise of the point X from the springings, and

       $\theta=$ Angle, which the tangent at Q makes with the horizontal.

Now taking A as the origin, we know that the equation for the centre line of a parabolic arch is,

$$y=k.x(l-x) \qquad \qquad ...(i)$$

where k is a constant.

We know that when $x=l/2$, $y=y_C$. Therefore substituting these values of x and y in the equation (i), we get

$$y_C =k \cdot \frac{l}{2}\left(l-\frac{l}{2}\right)=\frac{kl^2}{4}$$

$$\therefore \qquad k=\frac{4y_C}{l^2}$$

Now substituting the value of k in equation (i) we get,

$$y=\frac{4y_C}{l^2} \cdot x(l-x) \qquad \qquad ...(ii)$$

This is the required equation for the rise y of an arch axis, from its springings, at a distance x from the support A or B.

Note. (1) The value of y, when $x=l/4$

$$y=\frac{4y_C}{l^2} \cdot \frac{l}{4}\left(l-\frac{l}{4}\right)=\frac{3y_C}{4}$$

(2) The slope of the angle θ may be found out by differentiating the equation (ii) with respect to x i.e.,

$$y=\frac{4y_C}{l^2} x(l-x)=\frac{4y_C}{l^2}(lx-x^2)$$

$$\therefore \qquad \frac{dy}{dx}=\tan \theta=\frac{4y_C}{l^2}(l-2x)$$

10·9. Three-hinged circular arch

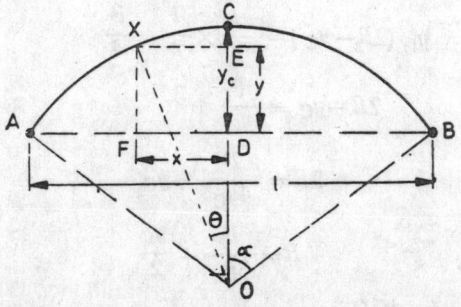

Fig. 10·5

A three-hinged arch, whose axis is circular, is known as a *three hinged circular arch*.

Consider a three-hinged circular arch ACB, having hinges at the supports A and B as well as at the crown C, and having O as centre as shown in Fig. 10·5. Join AB, and draw CD perpendicular to AB from the crown C. Now consider a point X, on the axis of the arch at a distance x from D.

Let
$l =$ Span AB of the arch,

$y_C =$ Rise of the crown from the springings,

$y =$ Rise of the point X, from the springings,

$2\alpha =$ Angle, which the two supports A and B make at the centre of the arch O,

$\theta =$ Angle, which OX makes with the centre line OC,

$R =$ Radius of the circular arch,

$x, y =$ Co-ordinates of the point X, with reference to the point D, which is middle of the span AB of the arch.

Since the arch is a circular one, therefore from the geometry of the figure, we find that

$$AO = OB = R$$

$$AD = DB = \frac{l}{2}$$

and
$$DC = y_C$$

From the given point X, draw XE perpendicular to OC. Then

$$OE = OD + DE = (R - y_C) + y$$

and
$$XE = x$$

Now in the right-angled triangle OEX, we know that

$$OX^2 = OE^2 + XE^2$$

$\therefore$
$$R^2 = [(R - y_C) + y]^2 + x^2$$

This is the required equation which gives the relation between x and y.

We also know that in a segment of a circle,

$$y_C (2R - y_C) = \frac{l}{2} \times \frac{l}{2} = \frac{l^4}{4}$$

$\therefore$
$$2R - y_C = \frac{l^2}{4y_C}$$

or
$$2R = \frac{l^2}{4y_C} + y_C$$

$\therefore$
$$R = \frac{l^2}{8y_C} + \frac{y_C}{2}$$

and in triangle AOD

$$\sin \alpha = \frac{BD}{AO} = \frac{l}{2} \times \frac{1}{R} = \frac{l}{2R}$$

and $$\cos \alpha = \frac{OD}{BO} = \frac{(R - y_C)}{R}$$

For a given section X,

$$x = OX \sin \theta = R \sin \theta$$

and $$y = OE - OD = R \cos \theta - R \cos \alpha$$

$$= R (\cos \theta - \cos \alpha)$$

10·10. Horizontal thrust in a three-hinged arch

The arches, having hinged supports, at their two ends and also having a third hinge, anywhere between the two ends, are known as *three-hinged arches*. The third hinge is, usually, placed at the crown of the arch. Since no bending moment can exist at the hinges, therefore the line of thrust, in a three-hinged arch, must pass through the three hinges. The reactions at the two ends A and B will have both vertical and horizontal components when an arch is subjected to vertical loads only. The horizontal components at the two supports will be equal and opposite. When the two ends of an arch are at the same level, the two vertical reactions R_A and R_B may be found out in the same way as in a simply supported beam.

Let $l =$ Span of the arch,

$y_C =$ Central rise of the arch, and

$H =$ Horizontal thrust on the arch.

A little consideration will show that the bending moment at the crown of the arch,

$$M_C = \mu_C - H . y_C$$

where $\mu_C =$ Beam moment at C due to loading (*i.e.* by considering the arch as a simply supported beam of span l).

$H . y_C =$ Moment due to horizontal thrust.

Since the arch is hinged at its crown, therefore the bending moment at the crown C will be zero.

∴ $$\mu_C - H . y_C = 0$$

or $$H . y_C = \mu_C$$

∴ $$H = \frac{\mu_C}{y_C}$$

This is the required equation for the horizontal thrust on an arch.

Example 10·1. *A three-hinged parabolic arch of span 40 m and rise 10 m is carrying a uniformly distributed load as shown in Fig. 10·6.*

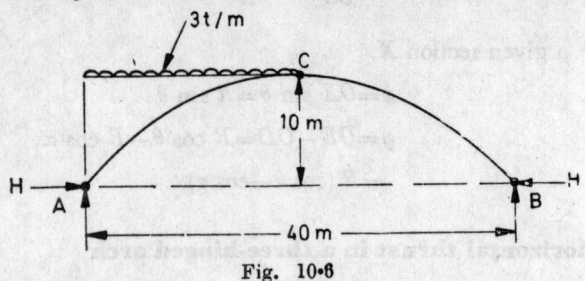

Fig. 10·6

Find the horizontal thrust at the springings.

Solution.

Given.   Span, $l = 40$ m

Central rise, $y_C = 10$ m

Let $H =$ Horizontal thrust at the springings,

                   $V_A =$ Vertical reaction at A, and

                   $V_B =$ Vertical reaction at B,

Taking moments about A,

$$V_B \times 40 = 3 \times 20 \times 10 = 600$$

∴ $$V_B = \frac{600}{40} = 15 \text{ t}$$

The beam moment at C due to external loading,

$$\mu_C = V_B \times 20 = 15 \times 20 = 300 \text{ t-m}$$

Using the relation,

$$H = \frac{\mu_C}{y_C} \text{ with usual notations.}$$

$$= \frac{300}{10} = 30 \text{ t Ans.}$$

Example 10·2. *A three-hinged circular arch of span 21 m has a rise of 4 m. The arch is loaded with a point load of 8 t at a horizontal distance of 6 m from the left support. Determine the horizontal thrust, two reactions and bending moment under the load.*

(*A.M.I.E. Summer, 1979*)

Solution.

Given.   Span, $l = 21$ m

Central rise, $y_C = 4$ m

Load, $W = 8$ t

Distance between the load and the left support

                   $= 6$ m

Let $\qquad$ V_A=Vertical reaction at A, and

V_B=Vertical reaction at B.

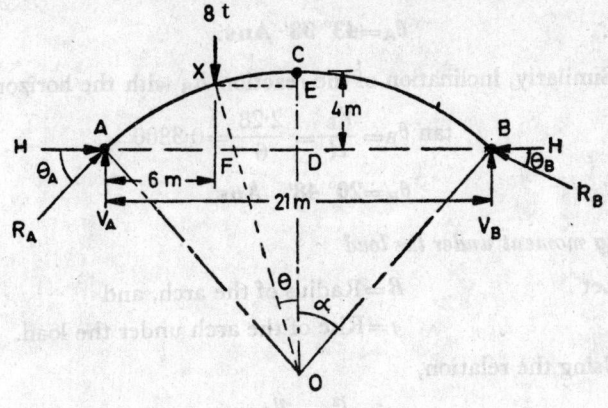

Fig. 10·7

Horizontal thrust

Let $\qquad$ H=Horizontal thrust.

Taking moment about A,

$$V_B \times 21 = 8 \times 6 = 48$$

$\therefore$, $\qquad$ $V_B = \dfrac{48}{21} = 2·28$ t

and $\qquad$ $V_A = 8 - 2·28 = 5·72$ t

Beam moment at C, due to external loading,

$$\mu_C = V_B \times \frac{21}{2} = \frac{48}{21} \times \frac{21}{2} = 24 \text{ t-m}$$

Using the relation,

$$H = \frac{\mu_C}{y_C} \text{ with usual notations}$$

$$= \frac{24}{4} = 6 \text{ t Ans.}$$

Reactions at the two supports

Let $\qquad$ R_A=Reaction at A, and

R_B=Reaction at B.

We know that the reaction at A,

$$R_A = \sqrt{V_A{}^2 + H^2} = \sqrt{5·72^2 + 6^2} \text{ 't'}$$

$$= 8·29 \text{ 't' Ans.}$$

Similarly, $\qquad$ $R_B = \sqrt{V_B{}^2 + H^2} = \sqrt{2·28^2 + 6^2}$ t

$$= 6·42 \text{ t Ans.}$$

Inclination of the reaction R_A with horizontal,

$$\tan \theta_A = \frac{V_A}{H} = \frac{5 \cdot 72}{6} = 0 \cdot 9533$$

$\therefore$ $\qquad\qquad\qquad \theta_A = \mathbf{43° \ 38'} \ \mathbf{Ans}.$

Similarly, inclination of the reaction R_B with the horizontal,

$$\tan \theta_B = \frac{V_B}{H} = \frac{2 \cdot 28}{6} = 0 \cdot 3800$$

$\therefore$ $\qquad\qquad\qquad \theta_B = \mathbf{20° \ 48'} \ \ \mathbf{Ans}.$

Bending moment under the load

Let $\qquad\qquad\qquad R =$ Radius of the arch, and

$\qquad\qquad\qquad\qquad y =$ Rise of the arch under the load.

Using the relation,

$$R = \frac{l^2}{8y_C} + \frac{y_C}{2} \ \text{with usual notations.}$$

$$= \frac{21^2}{8 \times 4} + \frac{4}{2} = 15 \cdot 8 \ \text{m}$$

From the geometry of the figure ODB, we find that

$$\sin \alpha = \frac{DB}{OB} = \frac{10 \cdot 5}{15 \cdot 8} = 0 \cdot 6645$$

$\therefore$ $\qquad\qquad\qquad \alpha = 41° \ 39'$

Similarly from the geometry of the figure OXE, we find that

$$\sin \theta = \frac{XE}{OX} = \frac{10 \cdot 5 - 6 \cdot 0}{15 \cdot 8} = \frac{4 \cdot 5}{15 \cdot 8} = 0 \cdot 2848$$

$\therefore$ $\qquad\qquad\qquad \theta = 16° \ 33'$

Now using the relation,

$\qquad\qquad y = R \ (\cos \theta - \cos \alpha) \ \text{with usual notations.}$

$\qquad\qquad\quad = 15 \cdot 8 \ (\cos 16° \ 33' - \cos 41° \ 39') \ \text{m}$

$\qquad\qquad\quad = 15 \cdot 8 \ (0 \cdot 9586 - 0 \cdot 7474) \ \text{m}$

$\qquad\qquad\quad = 15 \cdot 8 \times 0 \cdot 2112 = 3 \cdot 34 \ \text{m}$

$\therefore$ Bending moment at X,

$\qquad M_X = V_A \times 6 - H \times 3 \cdot 34 = 5 \cdot 72 \times 6 - 6 \times 3 \cdot 34 \ \text{t-m}$

$\qquad\qquad\quad = \mathbf{14 \cdot 28 \ t\text{-}m} \ \ \mathbf{Ans}.$

Example 10·3. *A three-hinged parabolic arch of span 20 metres and central rise of 5 metres carries a point load of 20 tonnes at 6 metres from the left hand support as shown in Fig. 10·8.*

(a) *Find the reaction at the supports A and B.*

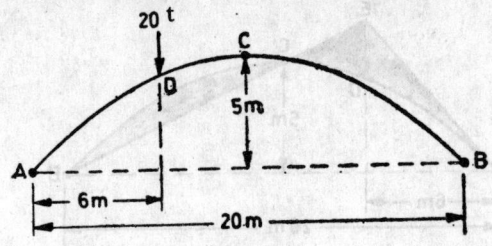

Fig. 10·8

(b) *Draw the bending moment diagram for the arch, and indicate the position of maximum bending moment.*

Solution.

Given. Span, $l = 20$ m

Central rise, $y_C = 5$ m

Horizontal distance between the load and the left support,

$$x = 6 \text{ m}$$

Reaction at the supports

Let V_A = Vertical reaction at A,

V_B = Vertical reaction at B,

H = Horizontal thrust at A and B.

Taking moments about A

$$V_B \times 20 = 20 \times 6 = 120$$

$$\therefore \qquad V_B = \frac{120}{20} = 6 \text{ 't'}$$

and

$$V_A = 20 - 6 = 14 \text{ 't'}$$

The beam moment at C, due to external loading,

$$\mu_C = V_B \times 10 = 6 \times 10 = 60 \text{ t-m}$$

Using the relation,

$$H = \frac{\mu_C}{y_C} \text{ with usual notations}$$

$$= \frac{60}{5} = 12 \text{ t}$$

We know that the reaction at A,

$$R_A = \sqrt{V_A^2 + H^2} = \sqrt{14^2 + 12^2} = \sqrt{340} \text{ t}$$

$$= 18·44 \text{ t} \quad \textbf{Ans.}$$

and the reaction at B,

$$R_B = \sqrt{V_B^2 + H^2} = \sqrt{6^2 + 12^2} = \sqrt{180} \text{ 't}$$

$$= 13·42 \text{ t} \quad \textbf{Ans.}$$

384

Position of maximum bending moment

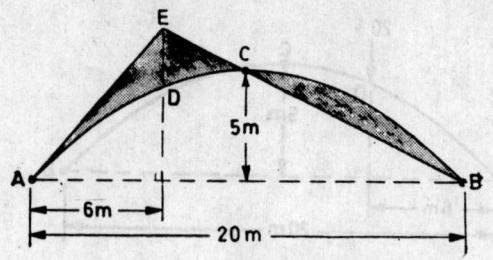

Fig. 10·9

First of all draw the bending moment diagram as discussed below :

1. Draw the arch ACB with the given span and rise.
2. Since the B.M. at A, B and C is zero, therefore join B and C and extend this line.
3. Draw a vertical line through D, meeting the line BC at E.
4. Join AE.

Now AEB is the required B.M. diagram. From the B.M. diagram, we see that maximum positive B.M. takes under the load.

Let $\qquad\qquad y=$ Rise of the arch at D.

Using the relation,

$$y=\frac{4y_C}{l^2}\,x(l-x)\text{ with usual notations.}$$

$$=\frac{4\times5}{20^2}\times6(20-6)=\frac{6\times14}{20}=4\cdot2\text{ m}$$

∴ Maximum positive B.M. at D,

$$M_{max}=V_A\times6-H_A\times4\cdot2=14\times6-12\times4\cdot2\text{ t-m}$$
$$=84-50\cdot4=\mathbf{33\cdot6\ t\text{-}m}\quad\mathbf{Ans.}$$

From the B.M. diagram, we also see that the maximum negative B.M. takes place in the section CB. Let the maximum negative B.M. take place at a distance of x from B. We know that the rise of arch at a distance x from B,

$$y=\frac{4y_C}{l^2}\,x(l-x)=\frac{4\times5}{20\times20}x(20-x)$$

$$=\frac{x}{20}\,(20-x)=x-\frac{x^2}{20}$$

∴ B.M. at X at a distance x from B,

$$M_X=V_B\times x-H.y=6x-12\left(x-\frac{x^2}{20}\right)$$

$$=6x-12x+\frac{3x^2}{5}=\frac{3x^2}{5}-6x$$

Now for maximum bending moment, let us differentiate the above equation with respect to x and equate it to zero.

THREE-HINGED ARCHES   385

i.e.,
$$\frac{dM_X}{dx}\left(\frac{3x^2}{5}-6x\right)=0$$

$$\frac{2\times 3x}{5}-6=0$$

or
$$x=5\text{ m}$$

∴ Rise of the arch at a distance of 5 m from B,

$$*y=\frac{4y_C}{l^2}\,x(l-x)=\frac{4\times5}{20\times20}\times5(20-5)\text{ m}$$

$$=\frac{5\times15}{20}=\frac{15}{4}\text{ m}$$

∴ Maximum negative B.M. at a distance of 5 m from B,

$$M_{max}=V_B\times x-H\,.y=6\times5-12\times\frac{15}{4}\text{ t-m}$$

$$=-15\text{ t·m}\quad\textbf{Ans.}$$

Example 10·4. *A parabolic three-pinned arch has a span of 20 m and central rise 4 m. It is loaded with a uniformly distributed load of 2 t/m for a length of 8 m from the left end support. Draw the B.M. diagram and find the position and magnitude of maximum B.M. over the arch.*

Solution.

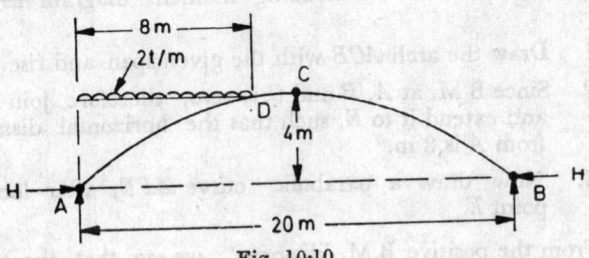

Fig. 10·10

Given. Span $\quad l=20$ m
Central rise, $\quad y_C=4$ m
Let $\quad H=$ Horizontal thrust,
$\quad V_A=$ Vertical reaction at A, and
$\quad V_B=$ Vertical reaction at B.

Taking moments about A
$$V_B\times20=2\times8\times4=64$$
∴
$$V_B=\frac{64}{20}=3\cdot2\text{ 't'}$$

*Rise of arch at a distance of $\frac{l}{4}$ from O,

$$y=\frac{3y_C}{4}=\frac{3\times5}{4}=\frac{15}{4}\text{ m}$$

and
$$V_A = (2 \times 8) - 3 \cdot 2 = 12 \cdot 8 \ t$$

Beam moment at C due to external loading,
$$\mu_C = V_B \times 10 = 3 \cdot 2 \times 10 = 32 \ \text{t-m}$$

Using the relation,

$$H = \frac{\mu_C}{y_C} \text{ with usual notations.}$$

$$= \frac{32}{4} = \mathbf{8 \ t \ Ans.}$$

Maximum positive bending moment

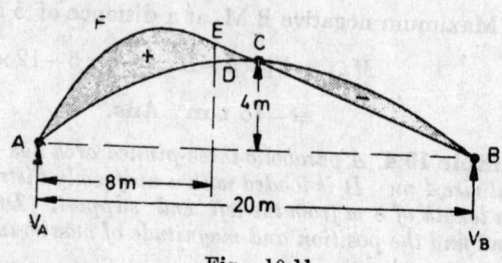

Fig. 10·11

First of all draw the bending moment diagram as discussed below :

1. Draw the arch ACB with the given span and rise.

2. Since B.M. at A, B and C is zero, therefore join B and C and extend it to E, such that the horizontal distance of E from A is 8 m.

3. Now draw a parabolic curve AFE, after locating the point E

From the positive B.M. diagram, we see that the maximum positive B.M. takes place in the section AD. Let the maximum positive B.M. take place at a distance x from A. We know that the rise of arch at distance x from A,

$$y = \frac{4y_C}{l^2} x(l-x) = \frac{4 \times 4}{20 \times 20} x(20-x)$$

$$= \frac{x}{25} (20-x) = \frac{4x}{5} - \frac{x^2}{25}$$

B.M. at X, at a distance x from A,

$$M_X = V_A . x - \frac{2x^2}{2} - H . y = 12 \cdot 8x - x^2 - 8\left(\frac{4x}{5} - \frac{x^2}{25}\right)$$

$$= 12 \cdot 8 \ x - x^2 - \frac{32x}{5} + \frac{8x^2}{25} = 6 \cdot 4x - \frac{17x^2}{25}$$

Now for maximum bending moment, let us differentiate the above equation with respect to x and equate it to zero.

∴.e.
$$\frac{dM_X}{dx}\left(6\cdot4x-\frac{17x^2}{25}\right)=0$$

$$6\cdot4-\frac{34x}{25}=0$$

or
$$x=\frac{6\cdot4\times25}{34}=4\cdot7\ \text{m}$$

∴ Rise of arch at a distance of $4\cdot7$ m from A,

$$y=\frac{4y_C}{l^2}\ x(l-x)=\frac{4\times4}{20\times20}\times4\cdot7(20-4\cdot7)=2\cdot88\ \text{m}$$

∴ Maximum positive B.M. at a distance of $4\cdot7$ m from A,

$$M_{max}=V_A\cdot x-\frac{2\times4\cdot7^2}{2}-H.y$$

$$=12\cdot8\times4\cdot7-4\cdot7^2-8\times2\cdot88\ \text{t-m}$$

$$=15\cdot0\ \text{t-m}\quad\textbf{Ans.}$$

Maximum negative bending moment

From the B.M. diagram, we also see that the maximum negative B.M. takes place in the section CB. Let the maximum negative
B.M. take place at a distance x from B. We know that the rise of
arch at a distance x from B,

$$y=\frac{4y_C}{l^2}\ x(l-x)=\frac{4\times4}{20\times20}x(20-x)$$

$$=\frac{x}{25}(20-x)=\frac{4x}{5}-\frac{x^2}{25}$$

∴ B.M. at x, at a distance x from B,

$$M_X=V_B.x-H.y=3\cdot2x-8\left(\frac{4x}{5}-\frac{x^2}{25}\right)$$

$$=3\cdot2x-6\cdot4x+\frac{8x^2}{25}=\frac{8x^2}{25}-3\cdot2x$$

Now for maximum bending moment, let us differentiate the
above equation with respect to x and equate it to zero.

i.e.
$$\frac{dM_X}{dx}\left(\frac{8x^2}{25}-3\cdot2x\right)=0$$

$$\frac{16x}{25}-3\cdot2=0$$

or
$$x=\frac{3\cdot2\times25}{16}=5\ \text{m}$$

∴ Rise of arch at a distance of 5 m from B (*i.e.* $l/4$),

$$y=\frac{3y_C}{4}=\frac{3\times4}{4}=3\ \text{m}$$

∴ Maximum negative B.M. at a distance of 5 m from B,

$$M_{max}=V_B\times x-H\times y=3\cdot2\times5-8\times3\ \text{t-m}$$

$$=-8\ \text{t-m}\quad\textbf{Ans.}$$

10·11. Three-hinged parabolic arch supported at different levels

Consider a three-hinged parabolic arch ACB, supported at different levels at A and B having hinges at the supports A and B as well as at the crown C as shown in Fig. 10·12.

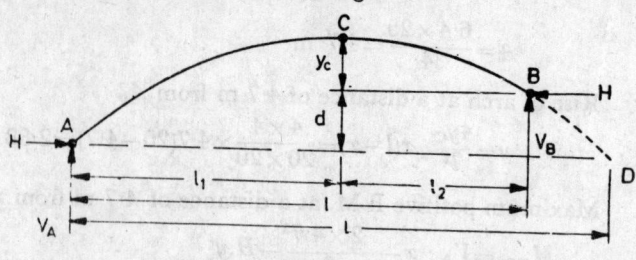

Fig. 10·12

Let $l =$ Span AB of the arch,

$y_C =$ Height of the crown C from the upper support B,

$d =$ Difference between the levels of the two supports,

$l_1 =$ Horizontal length between A and C, and

$l_2 =$ Horizontal length between C and B.

Let the arch ACB be extended to D, such that A and D are at the same level. Let the span of the imaginary arch ACD be L, such that L is equal to $2l_1$. A little consideration will show, that the rise of any point X on the arch axis at a distance x from the support A will be given by the relation,

$$y = \frac{4(y_C + d)}{L^2} x(L - x)$$

The reaction at the two ends A and B will have both vertical and horizontal components. When an arch is subjected to vertical loads only, the horizontal components at the two supports will be equal and opposite.

The vertical and horizontal components V_A, V_B and H may be found out by taking moments about any support; and then by taking moments about C and equating the same to zero.

Example 10·5. *A three-hinged parabolic arch of 60 m span is loaded as shown in Fig. 10·13.*

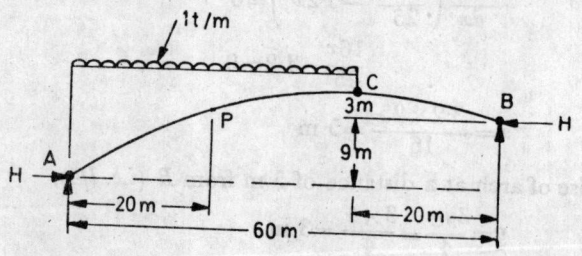

Fig. 10·13

Find the bending moment at a point P, 20 m from the left hand support.

(Allahabad University, 1970)

Solution.

Given. Span, $l = 60$ m

Horizontal distance between the crown and the left support A,

$$l_1 = 60 - 20 = 40 \text{ m}$$

Horizontal distance between the crown and the right support B,

$$l_2 = 20 \text{ m}$$

Height of the crown C from the higher support,

$$y_C = 3 \text{ m}$$

Difference between the levels of the two supports,

$$d = 9 \text{ m}$$

Let $L =$ Span of the imaginary arch ACD, such that A and D are at the same level.

$V_A =$ Vertical reaction at A, and

$V_B =$ Vertical reaction at B.

We know that the rise of arch, at any section X, at a distance x from A,

$$y = \frac{4(y_C + d)}{L^2} x(L - x) = \frac{4(3 + 9)x}{L^2} (L - x)$$

$$= \frac{48x}{L^2} (L - x) \qquad \qquad \ldots(i)$$

We also know that when $x = 60$, $y = 9$. Substituting these values of x and y in the above equation,

$$9 = \frac{48 \times 60}{L^2} (L - 60) = \frac{2,880}{L^2} (L - 60)$$

$$= \frac{2,880}{L} - \frac{1,72,800}{L^2}$$

$\therefore \qquad 9L^2 - 2,880L + 1,72,800 = 0$

or $L^2 - 320\,L + 19,200 = 0$

Solving the quadratic equation for L,

$$L = \frac{320 \pm \sqrt{320^2 - 4 \times 19,200}}{2}$$

$$= \frac{320 - 160}{2} = 80 \text{ m}$$

Substituting the value of L in equation (i),

$$y = \frac{48x}{80^2} (80 - x)$$

$\therefore$ Rise of arch at P *i.e.* at a distance of 20 m from A,

$$y = \frac{48 \times 20}{80^2} (80 - 20) = \frac{48 \times 20 \times 60}{80 \times 80} = 9 \text{ m}$$

Taking moments about A,

$$V_B \times 60 + H \times 9 = 1 \times 40 \times 20$$

$$H = \frac{800 - 60 V_B}{9} \qquad \qquad ...(ii)$$

Since the bending moment at the crown C is zero, therefore

$$3H = 20 V_B$$

Substituting the value of H in the above equation,

$$3 \left(\frac{800 - 60 V_B}{9} \right) = 20 V_B$$

$$\frac{800}{3} - 20 V_B = 20 V_B$$

$$\therefore \qquad 40\ V_B = \frac{800}{3}$$

or

$$V_B = \frac{800}{3} \times \frac{1}{40} = \frac{20}{3}\ t$$

$$\therefore \qquad V_A = 40 - \frac{20}{3} = \frac{100}{3}\ t$$

Substituting the value of V_B in equation (ii),

$$H = \frac{800 - 60 \times \dfrac{20}{3}}{3} = \frac{800 - 400}{9} = \frac{400}{9}\ t$$

$\therefore$ Bending moment at P,

$$M_P = V_A \times 20 - H \times 9 = \frac{100}{3} \times 20 - \frac{400}{9} \times 9\ \text{t-m}$$

$$= 266 \cdot 67\ \textbf{t-m}\quad \textbf{Ans}.$$

Example 10·6. *A parabolic arch has a span of 15 metres and is supported at different levels, such that the crown C is 9 metres from the left support A and 6 metres from the right support B. The right support is higher than left support by 2 metres and the crown is higher by 1·5 metre with respect to right support. The arch is hinged at the two supports and at the crown.*

Find the maximum B.M. in the arch at a section Q lying 4·5 metres from left support, when a point load W rolls over the span.

(Oxford University)

Solution.

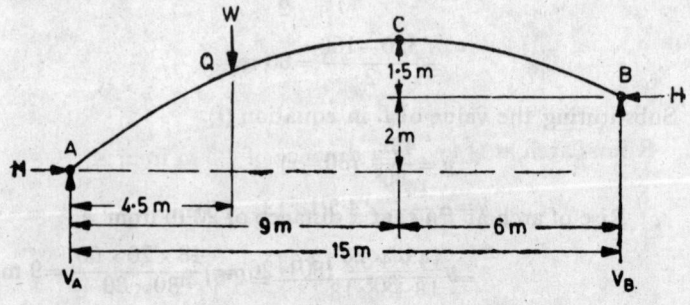

Fig. 10·14

Given. Span, $l = 15$ m

Horizontal distance between the crown and the left support A,
$$l_1 = 9 \text{ m}$$

Horizontal distance between the crown and the right support B,
$$l_2 = 6 \text{ m}$$

Difference between the levels of the two supports,
$$d = 2 \text{ m}$$

Height of the crown C from the right support B,
$$y_C = 1 \cdot 5 \text{ m}$$

Let $\quad\quad\quad L =$ Span of the imaginary arch ACD, such that A and D are at the same level.

$V_A =$ Vertical reaction at A, and

$V_B =$ Vertical reaction at B.

We know that the rise of arch at any section X, at a distance x from A,

$$y = \frac{4(y_C + d)}{L^2}\, x(L-x) = \frac{4(1 \cdot 5 + 2 \cdot 0)}{L^2}\, x(L-x)$$

$$= \frac{14}{L^2}\, x(L-x) \qquad\qquad\qquad \text{...(i)}$$

We also know that when $x = 15$, $y = 2$. Substituting these values of x and y in the above equation,

$$2 = \frac{14}{L^2} \times 15\,(L-15) = \frac{210}{L^2}\,(L-15)$$

$$= \frac{210}{L} - \frac{3{,}150}{L^2}$$

$\therefore \quad\quad\quad 2L^2 - 210\,L + 3{,}150 = 0$

Solving this quadratic equation for L,

$$L = \frac{210 \pm \sqrt{210^2 - 4 \times 2 \times 3{,}150}}{2 \times 2} \text{ m}$$

$$= \frac{210 - 137 \cdot 5}{2} = 18 \cdot 13 \text{ m}$$

Substituting this value of L in equation (i),

$$y = \frac{14}{18 \cdot 13^2}\, x(18 \cdot 13 - x)$$

$\therefore$ **Rise of arch at Q** i.e., at a distance of 4·5 m from A,

$$y = \frac{14}{18 \cdot 13^2} \times 4 \cdot 5 (18 \cdot 13 - 4 \cdot 5) \text{ m}$$

$$= \frac{14 \times 4 \cdot 5 \times 13 \cdot 63}{18 \cdot 13 \times 18 \cdot 13} = 2 \cdot 61 \text{ m}$$

A little consideration will show that the maximum B.M. in the arch at Q will take place, when the point load W will be at the section *i.e.*, at Q

Taking moments about B,

$$15 V_A = 2H + 10.5 W$$

$$H = \frac{15 V_A - 10.5 W}{2} \qquad \ldots(ii)$$

Since the bending moment at the crown C is zero, therefore

$$3.5 H + 4.5 W = 9 V_A$$

Substituting the value of H in the above equation,

$$3.5 \left(\frac{15 V_A - 10.5 W)}{2} \right) + 4.5 W = 9 V_A$$

$$26.25 V_A - 18.375 W + 4.5 W = 9 V_A$$

$$17.25 V_A = 13.875 W$$

$$\therefore \qquad V_A = \frac{13.875 W}{17.25} = 0.8 W$$

Substituting this value of V_A in equation (*iii*),

$$H = \frac{15 \times 0.8 W - 10.5 W}{2}$$

$$= \frac{1.5 W}{2} = 0.75 W$$

$\therefore$ Maximum bending moment at Q,

$$M_{max} = V_A \times 4.5 \times H \times 2.61$$

$$= 0.8 W \times 4.5 - 0.75 \times 2.61 \ W$$

$$= \mathbf{1.65 \ W \ \ Ans.}$$

Exercise 10-1

1. A three-hinged parabolic arch of 20 m span and 3 m rise is carrying a point load of 10 t at a section 7.5 m from the left support. Find the value of horizontal thrust and B.M. at a point 7.5 m from the right support.

[**Ans.** 12.5 T; 9.38 t-m]

2. A three-hinged parabolic arch of span l and central rise y_C carries a uniformly distributed load of w per unit length ever the left half of the span. Show that the maximum positive moment is equal to $\frac{wl^2}{64}$. (*Madras University*)

3. A three-hinged parabolic arch of a span 84 m has a central rise of 18 m. The arch is carrying a uniformly distributed load of 2 t/m over the 1/3 of the span from the left hand support. Calculate the bending moments at the quarter-span point. (*Jadavpur University*)

[**Ans.** 245 t-m ; −98 t-m]

4. A three-hinged circular arch of span 25 m and central rise 5 m is subjected to a point load of 10 t' at 7.5 m from the left support. Find the horizontal thrust and B.M. under the load. (*Calcutta University*)

[**Ans.** 7.5 T , 20.25 t-m]

5. A three-hinged parabolic arch of span 20 m has its crown 9 m high from the left hand support and 4 m higher than the right hand support. The crown of the arch is at a distance of 12 m from the left support and 8 m from the right support. Find the bending moment, at a section 4 m from the right hand support. (*Ranchi University*)

[**Ans.** 3.6 t-m]

10-12. Straining actions in a three hinged arch

We have already discussed in Art. 10-10 that the bending moment at each hinge, of a three-hinged arch, is zero. It is therefore essential, that the line of resultant thrust pass through the axis of an arch at the hinges.

Now consider a point X on the arch axis, where the line of resultant thrust does not coincide with the axis of the arch. There will be an eccentricity of the resultant thrust, from the arch axis, which will cause some bending moment on the arch axis at X. Now draw a normal on the arch axis at X and produce it to meet this line of action of the resultant thrust at K. Draw tangent to the arch axis at X, and through K draw a line parallel to the tangent at X. Let this line make an angle θ with the horizontal. Now resolve the resultant along tangent and normal at X. We see that there are following three straining actions on the point X :

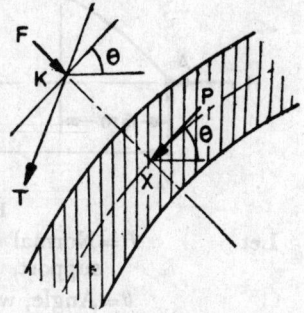

Fig. 10·15

(1) A normal thrust P (equal to the component of the resultant thrust T in the direction of the tangent X).

(2) A radial shear force F (equal to the component of the resultant thrust along the normal at X) ;

(3) A moment M (equal to the normal thrust $P \times$ the distance XK).

A little consideration will show that the normal thrust at any section X,

$$P_X = H \cos \theta + V \sin \theta \qquad \text{...(If } V \text{ is a net } upward \text{ force on the left of } X)$$

$$= H \cos \theta - V \sin \theta \qquad \text{...(If } V \text{ is a net } downward \text{ force on the left of } X)$$

and the radial shear force at X,

$$F_X = H \sin \theta - V \cos \theta \qquad \text{...(If } V \text{ is a net } upward \text{ force to the left of } X)$$

$$= H \sin \theta + V \cos \theta \qquad \text{.. (If } V \text{ is a net } downward \text{ force to the left of } X)$$

Example 10·7. *A three-hinged symmetrical parabolic arch has a span of 60 m between the ends, which are hinged supports, and a rise of 15 m up to its crown, where the third hinge is provided in the rib. It carries two concentrated loads, one 20 t on the left quarter point and the second 20 t on the right quarter point of span.*

Determine the normal thrust and shear force at the point 20 m from the left hand support.   (*Jaipur University, 1973*)

Solution.

Given. Span, $l = 60$ m

Central rise, $y_C = 15$ m
Load at D, $W_1 = 20$ t'
Load at E, $W_2 = 30$ t

Normal thrust at 20 m from the left hand support

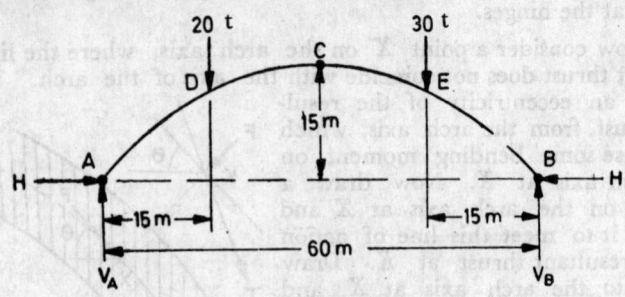

Fig. 10·16

Let $P =$ Normal thrust at 20 m from the left hand support,

             $\theta =$ Angle, which the tangent to the arch axis at 20 m from the left hand support, makes with the horizontal.

We know that the rise of the arch at D *i.e.*, at a distance of 15 m (*i.e. l/4*) from the left support,

$$y = \frac{3y_C}{4} = \frac{3 \times 15}{4} = \frac{45}{4} \text{ m}$$

Let $V_A =$ Vertical reaction at A,

             $V_B =$ Vertical reaction at B, and

             $H =$ Horizontal thrust at the springings.

Taking moments about A,

$$V_B \times 60 = 20 \times 15 + 30 \times 45 = 1,650$$

or $$V_B = \frac{1,650}{60} = \frac{165}{6} = 27 \cdot 5 \text{ t}$$

$$\therefore \quad V_A = (20 + 30) - 27 \cdot 5 = 22 \cdot 5 \text{ t}$$

Beam moment at C, due to external loadings,

$$\mu_C = V_B \times 30 - 30 \times 15 = 27 \cdot 5 \times 30 - 30 \times 15 \text{ t-m}$$
$$= 375 \text{ t-m}$$

Using the relation,

$$H = \frac{\mu_C}{y_C} \text{ with usual notations.}$$

$$= \frac{375}{15} = 25 \text{ t}$$

We know that the rise of arch axis at a distance x from A,

$$y = \frac{4y_C}{l^2} x(l - x) = \frac{4 \times 15}{60 \times 60} x(60 - x)$$

$$= \frac{x}{60} (60 - x) = x - \frac{x^2}{60}$$

Differentiating the above equation with respect to x,

i.e. $$\frac{dy}{dx} = \tan\theta = 1 - \frac{2x}{60} = 1 - \frac{x}{30}$$

Substituting the value of $x = 20$ m in the above equation,

$$\tan\theta = 1 - \frac{2 \times 20}{60} = \frac{1}{3}$$

$$\therefore \qquad \sin\theta = \frac{1}{\sqrt{10}} \qquad \text{and} \qquad \cos\theta = \frac{3}{\sqrt{10}}$$

Using the relation,

$P = H\cos\theta + V\sin\theta$ with usual notations,

$$= \frac{25 \times 3}{\sqrt{10}} + (22 \cdot 5 - 20) \times \frac{1}{\sqrt{10}} = \frac{77 \cdot 5}{\sqrt{10}} \quad \text{t}$$

$$= \textbf{24·6 t} \quad \textbf{Ans.}$$

Shear force at 20 m from the left hand support

Let $\qquad F =$ Shear force at 20 m from the left hand support.

Using the relation,

$F = H\sin\theta - V\cos\theta$ with usual notations.

$$= 25 \times \frac{1}{\sqrt{10}} - (22 \cdot 5 - 20) \times \frac{3}{\sqrt{10}} = \frac{17 \cdot 5}{\sqrt{10}} = -5 \cdot 55 \text{ t}$$

$$= \textbf{5·55 t (downward)} \quad \textbf{Ans.}$$

Example 10·8. *The equation of a three-hinged arch, with origin at its left support, is* $y = x - \dfrac{x^2}{40}$. *The span of the arch i 40 m.*

Find the normal thrust and radial shear force at a section X, 5 m from the left support, when the arch is carrying a uniformly distributed load of 3 t/m for the left half. (*U.P.S.C. Engg. Services, 1972*)

Solution.

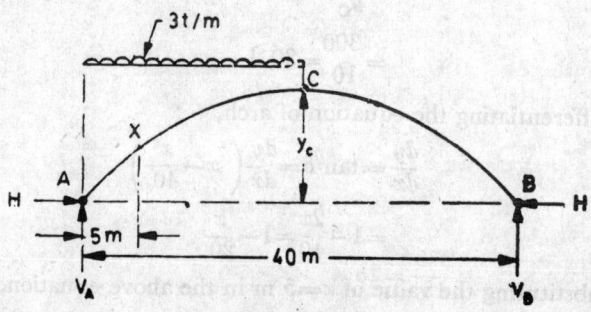

Fig. 10·17

Given. Equation of the arch,

$$y = x - \frac{x^2}{40} \qquad \qquad \qquad ...(i)$$

Span, $l = 40$ m

Distance between the section X and left support,

$x = 5$ m

Load, $w = 3$ t/m

Let $y_C =$ Central line of the arch,

$V_A =$ Vertical reaction at A,

$V_B =$ Vertical reaction at B, and

$H =$ Horizontal thrust.

Normal thrust at 5 m from the left hand support

Let $P =$ Normal thrust at 5 m from the left hand support,

$\theta =$ Angle, which the tangent to the arch axis at 5 m from the left hand support, makes with the horizontal.

Substituting the value of $x = 20$ in the equation of the arch,

$$y_C = 20 - \frac{20^2}{40} = 10 \text{ m} \qquad \left(\text{Given. } y = x - \frac{x^2}{40} \right)$$

Taking moments about A,

$$V_B \times 40 = 3 \times 20 \times 10 = 600$$

$\therefore$ $$V_B = \frac{600}{40} = 15 \text{ t}$$

and $$V_A = (3 \times 20) - 15 = 45 \text{ t}$$

Beam moment at C due to external loading,

$$\mu_C = V_B \times 40 = 15 \times 20 = 300 \text{ t-m}$$

Using the relation,

$$H = \frac{\mu_C}{y_C} \text{ with usual notations.}$$

$$= \frac{300}{10} = 30 \text{ t}$$

Differentiating the equation of arch,

i.e. $$\frac{dy}{dx} = \tan \theta = \frac{dy}{dx} \left(x - \frac{x^2}{40} \right)$$

$$= 1 - \frac{2x}{40} = 1 - \frac{x}{20}$$

Substituting the value of $x = 5$ m in the above equation,

$$\tan \theta = 1 - \frac{5}{20} = \frac{15}{20} = \frac{3}{4}$$

$\therefore$ $$\sin \theta = \frac{3}{5} \quad \text{and} \quad \cos \theta = \frac{4}{5}$$

Using the relation,

$$P = H \cos \theta + V \sin \theta \text{ with usual notations.}$$

$$= 30 \times \frac{4}{5} + (45 - 3 \times 5) \times \frac{3}{5} \text{ t}$$

$$= \textbf{42 t Ans.}$$

Radial shear force

Let $F =$ Shear force 5 m from the left hand support.

Using the relation,

$$F = H \sin \theta - V \cos \theta \text{ with usual notations.}$$

$$= 30 \times \frac{3}{5} - (45 - 3 \times 5) \times \frac{4}{5} \text{ t}$$

$$= 18 - 24 = -6 \text{ t}$$

$$= \textbf{6 t (downward) Ans.}$$

10·13. Effect of change in temperature on a three-hinged arch

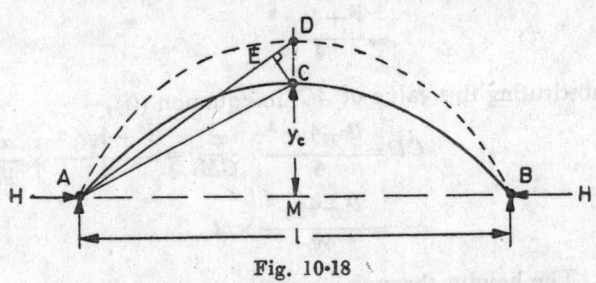

Fig. 10·18

Consider a three-hinged arch ACB of span l and central rise y_C as shown in Fig. 10·8. Let the arch be subjected to some rise of temperature.

Let $t =$ Rise of temperature,
                       $\alpha =$ Coefficient of linear expansion for the arch material.

We know that the length of the arch must increase, with rise in temperature. Since the end hinges A and B cannot undergo any displacement, therefore the crown C of the arch will rise upwards from C to D as shown in Fig. 10·18. Now the increase in the arch

$$= \text{Arc } AD - \text{Arc } AC$$

Considering the lengths of the arcs AD and AC equal to lengths of the chords AD and AC, therefore

Length of chord $AD =$ length of chord AC $(1 + \alpha t)$

$$= AC + AC. \alpha t$$

∴ Increase in the length of the arch

$$= \text{chord } AD - \text{chord } AC$$

$$= (AC + AC.\alpha t) - AC$$

$$= AC. \alpha t \qquad \qquad \ldots (i)$$

Now through C draw a perpendicular CE on AD. Considering the length AC equal to AE, the increase in the length of the arch,

$$ED = AD - AE = AC.\,\alpha t$$

Now the amount by which the crown C will rise

$$= CD = ED.\sec\angle EDC = AC.\alpha t\,\sec\angle EDC$$

Considering the angle ADC equal to angle ACM,

$$CD = AC.\,\alpha t\,\sec\angle ACM,$$

$$= AC.\alpha t.\,\frac{AC}{CM} \qquad \left(\sec.\ \angle ACM = \frac{AC}{CM}\right)$$

$$= \frac{AC^2\,\alpha t}{CM} \qquad\qquad\qquad \dots(ii)$$

Now, from the geometry of the right-angled triangle ACM, we know that

$$AC^2 = AM^2 + CM^2 = \left(\frac{l}{2}\right)^2 + y_C{}^2 = \frac{l^2}{4} + y_C{}^2$$

$$= \frac{l^2 + 4y_C{}^2}{4}$$

Substituting this value of AC^2 in equation (ii),

$$CD = \frac{l^2 + 4y_C{}^2}{4} \cdot \frac{\alpha t}{CM} = \frac{l^2 + 4y_C{}^2}{4} \times \frac{\alpha t}{y_C}$$

$$= \frac{l^2 + 4y_C{}^2}{4y_C} \times \alpha t$$

$\therefore$ The height, through which the crown will go up,

$$dy_C = \frac{l^2 + 4y_C{}^2}{4y_C} \times \alpha t$$

Example 10·9. *A three-hinged parabolic arch of span 30 m has a central rise of 5 m. Find the rise of the arch crown, if the temperature rises through 30°C. Take coefficient of linear expansion for the arch material as 12×10^{-6} per °C.*

Solution.

Span, $l = 30$ m

Central rise, $y_C = 5$ m

Rise of temperature, $t = 30°C$

Coefficient of linear expansion for the arch material,

$$\alpha = 12 \times 10^{-6}\ °C$$

Let $dy_C = $ Rise of the arch crown.

Using the relation,

$$dy_C = \frac{l^2 + 4y_C{}^2}{4y_C} \times \alpha t \text{ with usual notations.}$$

$$dy_C = \frac{30^2 + 4 \times 5^2}{4 \times 5} \times 12 \times 10^{-6} \times 30 \text{ m}$$

$$= 50 \times 12^{-6} \times 30 = 0.018 \text{ m}$$

$$= 1.8 \text{ cm} \quad \textbf{Ans.}$$

Exercise 10-2

1. A three-hinged parabolic arch of span 20 m and central rise 4 m is carrying a point load of 15 T at point P, 4 m from the left hand support. Find the values of normal thrust and radial shear force under the load.

(Banaras Hindu University)
[**Ans.** 11.95 t ; 7.57 t]

2. A parabolic arch rib of 20 m span and 4 m central rise is hinged at its springings and crown. Calculate the normal thrust and S.F. under a load of 4 T, when it is acting at a horizontal distance of 4 m from the left end support.

(Gujarat University)
[**Ans.** 3.18 t ; 2.02 t]

3. A three-hinged parabolic arch of span 36 m and central rise 8 m is carrying a uniformly distributed load of 4 t/m over the left hand half span. Find the horizontal thrust. Also determine the normal thrust and S.F. at a point 9 m from the left hand support.

(Panjab University)
[**Ans.** 40.5 t ; 44.325 t ; 0]

4. A three-hinged metal arch of span 20 m and central rise 4 m is subjected to a rise of temperature through 20°C. Find the vertical displacement of the arch crown. Take coefficient of linear expansion for the arch metal as 12×10^{-6} per °C.

(Mysore University)
[**Ans.** 0.69 cm]

10·14. Influence lines for moving loads over three-hinged arches

We shall discuss the following influence lines for a moving load over three-hinged arches, when, it rolls from one support to another.

(1) A single concentrated load and

(2) A uniformly distributed load.

10·15. Influence lines for a concentrated load moving over three-hinged circular arches

Consider a concentrated load W rolling over a three-hinged circular arch ACB, of span l and central rise y_C from A to B. Let us consider a section X, at a distance x from A. We shall draw the influence lines for the following :

(1) Influence line for the horizontal thrust,

(2) Influence lines for the bending moment,

(3) Influence lines for the normal thrust, and

(4) Influence lines for the radial shear force.

Influence lines for the horizontal thrust

Consider an instant when the load W is at any section X, at a distance x from A. At this instant the vertical reaction at B,

$$V_B = \frac{Wx}{l} \quad \text{and} \quad V_A = \frac{W(1-x)}{l}$$

Now B.M. at C due to loading on the arch,

$$\mu_C = V_B \times \frac{l}{2} = \frac{Wx}{l} \times \frac{l}{2} = \frac{Wx}{2}$$

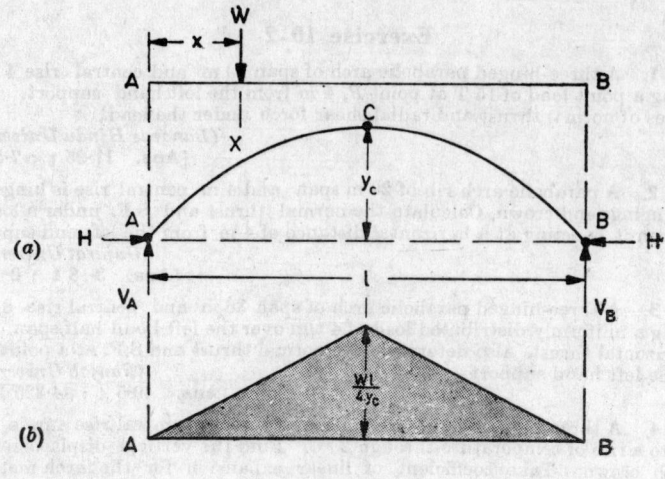

Fig. 10·19

We know that the horizontal thrust,

$$H = \frac{\mu_C}{y_C} = \frac{Wx}{2} \times \frac{1}{y_C} = \frac{Wx}{2y_C}$$

From the above expression, we see that the horizontal thrust increases by a straight line law. Substituting the various values of x in the above equation, we find that when $x=0$, $H=0$ and when $x=\frac{l}{2}$, $H=\frac{Wl}{4y_C}$.

We also see, that when the load is in the portion CB of the arch, the horizontal thrust,

$$H = \frac{\mu_C}{y_C} = \frac{V_A \cdot \frac{l}{2}}{y} = \frac{\frac{W(l-x)}{l} \times \frac{l}{2}}{y_C} = \frac{W(l-x)}{2y_C}$$

Substituting the various values of x in the above equation, we find that when

$$x = \frac{l}{2}, \quad H = \frac{W\left(l-\frac{l}{2}\right)}{2y_C} = \frac{Wl}{4y_C} \quad \text{and when } x=l, \ H=0.$$

It is thus obvious, that the horizontal thrust will be zero, when the load is at A. It will increase by a straight line law to $\frac{Wl}{4y_C}$, when the load is at the crown and will decrease by a straight line law to zero, when the load is at B as shown in Fig. 10·19 (b).

Influence lines for the bending moment

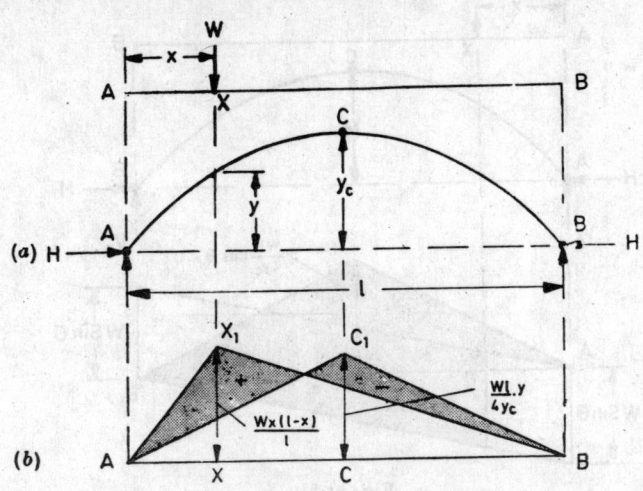

Fig. 10·20

Consider an instant, when the load W is at any section X at a distance x from A. Let the rise of the arch be y from the springings. We have discussed in Art. 10·10 that the bending moment at X,

$$M_x = \mu_x - H.y$$

where μ_x is the usual bending moment as a freely supported beam, and $H.y$ is the moment due to the horizontal thrust H.

We have seen in Art. 3·3 that the influence line for μ_x will be a triangle, having ordinate equal to $\dfrac{Wx\,(l-x)}{l}$. The influence line for $H.y$ will be y times the influence line for H^*. It is thus obvious, that the influence line for $H.y$ will also be a triangle having ordinate equal to $\dfrac{Wl}{4y_C} . y$ as shown in Fig. 10·20 (b). Since μ_x and $H.y$ are opposite in sign, that is why the two triangles have been drawn on the same side of the base AB. From the influence lines, we find that the maximum positive B.M. on the section occurs, when the load is on the section itself. The maximum negative B.M. on the section occurs, when the load is at the centre of the span.

Influence line for the normal thrust

Consider a section X, at a distance x from A. We have discussed in Art. 10·12 that the normal thrust at X,

$$P_x = H \cos \theta + V \sin \theta$$

*We know that the influence line for H will be zero at A, it will increase to $\dfrac{Wl}{4y_C}$ at C and then will decrease to zero at B.

where V is the ordinary vertical shear force, which can be found out by considering the arch as an ordinary beam.

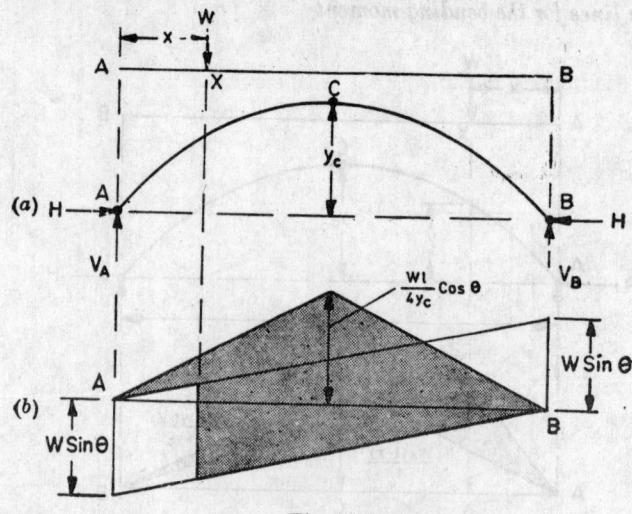

Fig. 11·21

Here we shall find out the magnitude of the normal thrust in the two stages, *i.e.* (*i*) when the load is in AX, and (*ii*) when the load is in XB. A little consideration will show that when the load is in AX, the normal thrust at X,

$$P_X = H \cos \theta - (W - V_A) \sin \theta$$
$$= H \cos \theta - V_B \sin \theta \qquad \qquad \ldots(i)$$

and when the load is in XB, the normal thrust at X,

$$P_X = H \cos \theta + V_A \sin \theta \qquad \qquad \ldots(ii)$$

We know that the influence lines for $V_B \sin \theta$ and $V_A \sin \theta$ will consist of two parallel lines having end ordinates equal to $W \sin \theta$. The influence line for $H \cos \theta$ will be $\cos \theta$ time the influence line for H^*. Since in the original equation (*i.e.* $P_X = H_A \cos \theta + V_B \sin \theta$) we have to add up the two influence lines, that is why the two influence lines are drawn on opposite sides of the base AB as shown in Fig. 10·21 (*b*).

Influence line for the radial shear force

Consider a section X, at a distance x from A. We have already discussed in Art. 10·12 that the radial shear force at X,

$$F_X = H \sin \theta - V \cos \theta$$

where V is the ordinary vertical shear force, which can be found out by considering the arch of an ordinary beam.

Here we shall find the magnitude of the radial shear force in the two stages, *i.e.* (*i*) when the load is in AX, and (*ii*) when the

*Please see foot note on page 401

load is in XB. A little consideration will show that when the load is in AX, the radial shear force at X,

$$F_X = H \sin \theta - (W - V_A) \cos \theta$$
$$= H \sin \theta + V_B \cos \theta \qquad \qquad ...(i)$$

and when the load is XB, the radial shear force at X,

$$F_X = H \sin \theta - V_A \cos \theta \qquad \qquad ...(ii)$$

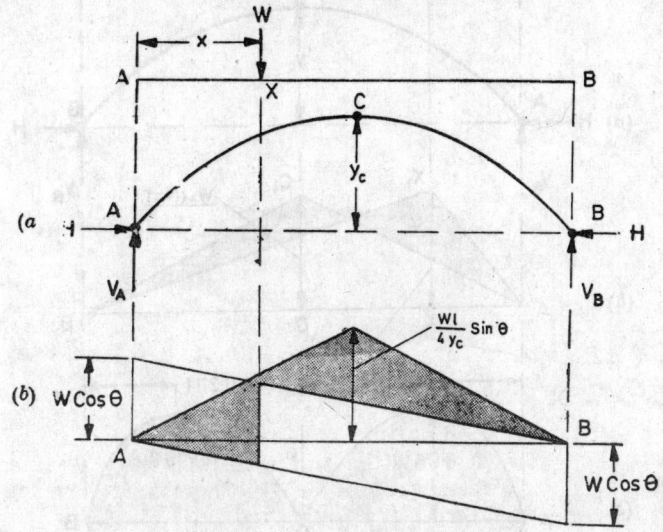

Fig. 10·22

We know that the influence lines for $V_B \cos \theta$ and $V_A \cos \theta$ will consist of two parallel lines having end ordinates equal to $W \cos \theta$. The influence line for $H \sin \theta$, will be $\sin \theta$ times the influence line for H^*. Since in the original equation (i.e. $F_X = H \sin \theta - V \cos \theta$), we have to subtract the two influence lines that is why the two influence lines are drawn on the same side of the base AB as shown in Fig. 10·22 (b).

10·16. Influence lines for a concentrated load moving over three-hinged parabolic arches

All the influence lines over a three-hinged parabolic arches are the same, as those of a three-hinged circular arch, except for the bending moment which is discussed here.

Consider a concentrated load W rolling over a three-hinged parabolic arch ACB of span l and central rise y_C from A to B. Let us consider a section X at a distance x from A, having a rise of y from the arch springings.

We know that B.M. at X,

$$M_X = \mu_X - H.y$$

where μ_X is the usual B.M. as a freely supported beam, and $H.y$ is the moment due to horizonal thrust H. We know that the influence line for μ_X will be a triangle, having *ordinate XX_1 equal to

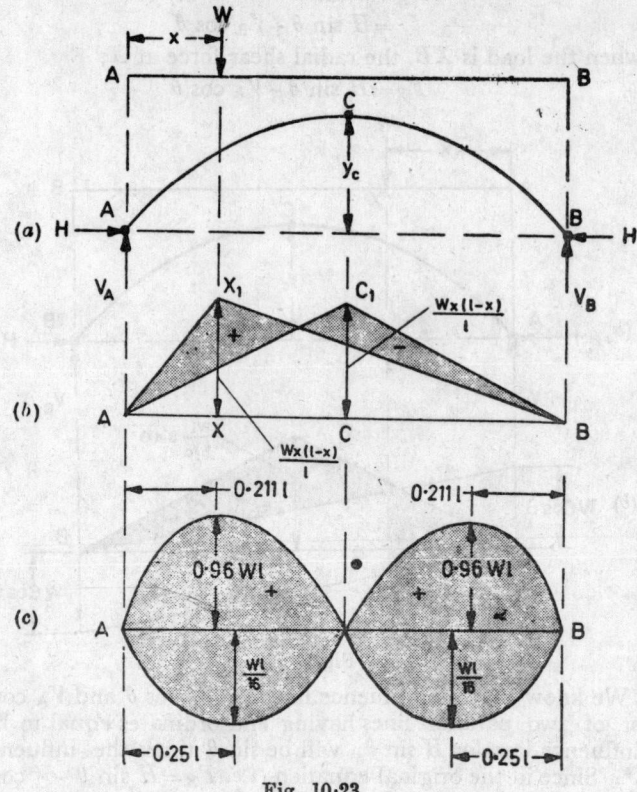

Fig. 10·23

$\dfrac{Wx(l-x)}{l}$. The influence line for $H.y$ will also be a triangle having

ordinate equal to $\dfrac{Wl}{4y_C} \cdot y$. But in the case of a parabolic arch, we

know that the rise of arch of a distance x from A,

$$y = \frac{4y_C}{l^2} \cdot x(l-x)$$

Substituting this value of y in the equation for ordinate CC_1, we get

$$CC_1 = \frac{Wl}{4y_C} \cdot y = \frac{Wl}{4y_C} \times \frac{4y_C}{l^2}\, x(l-x) = \frac{Wx(l-x)}{l}$$

It is thus obvious, that the ordinate XX_1 is equal to the ordinate CC_1, as both the ordinates are equal to $\dfrac{Wx(l-x)}{l}$.

*For details please refer to Art. 10·5.

Since μ_x and $H.y$ are opposite in sign, that is why the two triangles have been drawn on the same side of the base AB. From the influence lines, we find that the maximum positive B.M. on the section occurs, when the load is on the section itself. The value of maximum positive B.M. on the section

$$M_{max}=\frac{Wx(l-x)}{l}-\frac{Wx}{2y_C}\times y$$

$$=\frac{Wx(l-x)}{l}-\frac{Wx}{2y_C}\times\frac{4y_C}{l^2}\times x(l-x)$$

$$=\frac{Wx(l-x)}{l}\left(1-\frac{2x}{l}\right)=\frac{Wx(l-x)(l-2x)}{l^2}$$

$$=\frac{2Wx^3-3Wlx^2+Wxl^2}{l^2}\qquad\qquad\ldots(i)$$

Now substituting the various values of x from 0 to $\frac{l}{2}$ in the above equation, we can draw the influence line for the single concentrated load as shown in Fig. 10·23 (c). For the absolute maximum B.M., differentiate the equation (i) with respect to x and equate it to zero, i.e.

$$\frac{d}{dx}\left(\frac{2Wx^3-3Wlx^2+Wxl^2}{l^2}\right)=0$$

$$\therefore\qquad 6x^2-6lx+l^2=0$$

$$\text{or}\qquad x^2-lx+\frac{l^2}{6}=0$$

$$\left(x-\frac{l}{2}\right)^2=\frac{l^2}{12}\quad\left[\because\left(x-\frac{l}{2}\right)^2 x^2+\frac{l^2}{4}-lx\right]$$

$$\therefore\qquad x=\frac{l}{2}\pm\frac{l}{2\sqrt3}=(0\cdot5\pm0\cdot289)\,l$$

$$=0\cdot211\,l\text{ or }0\cdot789\,l$$

Substituting this value of x in equation (i), the absolute maximum B.M. at x,

$$M_{max\ max}=0\cdot096\ Wl$$

Now complete the influence line diagram for positive B.M. as shown in Fig. 10·23 (c).

Similarly, the maximum negative B.M. on the section occurs, when the load is at the centre of the span. The value of maximum negative B.M. on the section,

$$M_{max}=\frac{Wx(l-x)}{2}-\frac{Wx}{2}=\frac{Wx(l-x)}{2l}$$

$$=\frac{Wlx-2Wx^2}{2l}\qquad\qquad\ldots(ii)$$

Now substituting the various values of x from 0 to $\frac{l}{2}$ in the above equation, we can draw the influence line for the single concentrated load as shown in Fig. 10·23 (c). For the absolute maximum B.M. differentiate the equation (ii) with respect to x and equate it to zero, i.e.

$$\frac{d}{dx}\left(\frac{Wlx-2Wx^2}{2l}\right)=0$$

$$\therefore \qquad l-4x=0$$

or $\qquad\qquad x=\frac{l}{4}$

Substituting this value of x in equation (ii), the absolute maximum B.M. at x,

$$M_{max\ max}=-\frac{Wl}{16}=-0{\cdot}0625\ Wl$$

Now complete the influence line diagram for the negative B.M. as shown in Fig. 10·23 (c).

10·17. Influence lines for a uniformly distributed load moving over three-hinged parabolic arches

All the influence lines for a uniformly distributed load over three-hinged parabolic arch are the same as those of a concentrated load, except for the bending moment, which is discussed here :

Consider a uniformly distributed load of w per unit length rolling over a three-hinged parabolic arch from A to B. We know that the B.M. at any section X,

$$M_X=\mu_X-H.y$$

where μ_X is the usual B.M. as a freely supported beam, and $H.y$ is the moment due to horizontal thrust H. We know that the influence line for μ_X will be a triangle with ordinate XX_1 equal to $\dfrac{wx(l-x)}{l}$. The influence line for $H.y$ will also be a triangle with ordinate CC_1 equal to $\dfrac{wl}{4y_C}.y$. But in the case of a parabolic arch, we know that the rise of arch at a distance x from A,

$$y=\frac{4y_C}{l^2}\ x(l-x)$$

Substituting the value of y in the equation for ordinate CC_1,

$$CC_1=\frac{wl}{4y_C}\ .\ y=\frac{wl}{4y_C}\times\frac{4y_C}{l^2}\ x(l-x)$$

$$=\frac{wx\ (l-x)}{l}$$

It is thus obvious, that the ordinate XX_1 is equal to ordinate CC_1, as both the ordinates are equal to $\dfrac{wx\ (l-x)}{l}$.

Since μx and $H \cdot y$ are opposite in sign, that is why the two triangles have been drawn on the same side of the base AB as shown

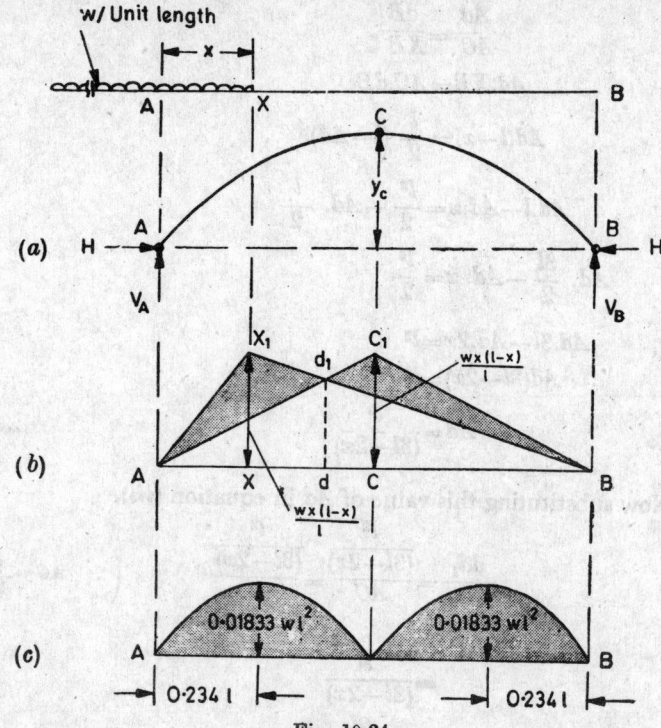

Fig. 10·24

in Fig. 10·24 (b). From the influence lines, we know that the maximum positive B.M. on the section occurs, when the length Ad is loaded, whereas the length dB is empty. Similarly, the maximum negative B.M. on the section occurs, when the length dB is loaded whereas the length Ad is empty. From the influence line diagram, we see that the maximum positive B.M.

$$+M_{max} = w \times \text{Area of triangle } Ad_1X_1 \qquad \ldots(i)$$

Similarly $\quad -M_{max} = w \times \text{Area of triangle } d_1C_1B \qquad \ldots(ii)$

Since the two triangles AX_1B and AC_1B have the same base AB and equal heights, therefore these two triangles are equal in areas. It is thus obvious, that the *maximum positive and negative B.M. at a given section is also equal.* Now in order to obtain the value of maximum B.M., we have to find out the area of triangle Ad_1X_1 or d_1C_1B, for which we must know the distance Ad. Now in similar triangles Add_1 and ACC_1, we find that

$$\frac{dd_1}{CC_1} = \frac{Ad}{AC} \qquad \ldots(iii)$$

Again in similar triangles Bdd_1, and BXX_1, we find that

$$\frac{dd_1}{XX_1} = \frac{dB}{XB} \qquad \ldots(iv)$$

Since CC_1 is equal to XX_1, therefore from equations (iii) and (iv) we find that

$$\frac{Ad}{AC} = \frac{dB}{XB}$$

∴ $Ad.XB = AC.dB$

$$Ad(l-x) = \frac{l}{2}\ (l - Ad)$$

or $$Ad.l - Ad.x = \frac{l^2}{2} - Ad.\ \frac{l}{2}$$

or $$Ad.\ \frac{3l}{2} - Ad.\ x = \frac{l^2}{2}$$

∴ $Ad.3l - Ad.2x = l^2$

or $Ad(3l - 2x) = l^2$

∴ $$Ad = \frac{l^2}{(3l - 2x)} \qquad\qquad\qquad ...(v)$$

Now substituting this value of Ad in equation (iii),

$$\frac{dd_1}{CC_1} = \frac{\dfrac{l^2}{(3l-2x)}}{AC} = \frac{\dfrac{l^2}{(3l-2x)}}{\dfrac{l}{2}} \qquad \left(\because\ AC = \frac{l}{2}\right)$$

$$= \frac{2l}{(3l - 2x)}$$

or $$dd_1 = \frac{2l}{(3l - 2x)} \times CC_1$$

$$= \frac{2l}{(3l-2x)} \times \frac{x(l-x)}{l} \qquad \left(\because\ CC_1 = \frac{x(l-x)}{l}\right)$$

$$= \frac{2x(l-x)}{(3l - 2x)}$$

From the geometry of the figure, we find that

Area of $\triangle\ Ad_1X_1 = $ Area of $\triangle\ AX_1B -$ Area of $\triangle\ Ad_1B$

$$= \left[\frac{l}{2} \times \frac{x(l-x)}{l}\right] - \left[\frac{l}{2} \times \frac{2x(l-x)}{3l-2x}\right]$$

$$= \frac{x(l-x)(l-2x)}{2(3l-2x)}$$

Thus the maximum positive as well as negative B.M. at a section is given by

$$M_{max} = \frac{wx(l-x)(l-2x)}{2(3l-2x)}$$

$$= \frac{w}{2}\left[\frac{l^2x - 3x^2l + 2x^3}{3l - 2x}\right] \qquad\qquad ...(vi)$$

For the absolute maximum B.M. differentiating the equation (vi) with respect to x and equating the same to zero,

$$\frac{d}{dx}\left[\frac{w}{2}\left(\frac{l^2x-3x^2l+2x^3}{3l-2x}\right)\right]=0$$

$$\frac{w}{2}\left[\frac{[(3l-2x)(l^2-6xl+6x^2)]-[(l^2x-3x^2l+2x^3)(-2)]}{3l-2x^2}\right]=0$$

or
$$(3l-2x)(l^3-6xl+6x^2)+2(l^2x-3x^2l+2x^3)=0$$

$$\therefore \quad 3l^3-20xl^2+30x^2l-12x^3+2l^2x-6x^2l+4x^3=0$$

or
$$3l^3-18xl^2+24x^2l-8x^3=0$$

Solving this equation by trial and error, we get

$$x=0.234\ l$$

Now substituting this value of x in equation (vi) the absolute maximum B.M. at X,

$$M_{max\ max}=0.01883\ wl^2$$

Now complete the influence line diagram for B.M. as shown in Fig. 10·24 (c).

Note. The influence line diagram for negative B.M. will be similar as that for positive B.M. but on its opposite side.

Example 10·10. *A parabolic arch, hinged at its springings and crown, has a span of 30 m and central rise 6 m. Determine the magnitude of maximum positive and negative bending moment at a section 10 m from left hand support, when a point load 9 tonnes rolls over the beam.* (*Nagpur University, 1978*)

Solution.

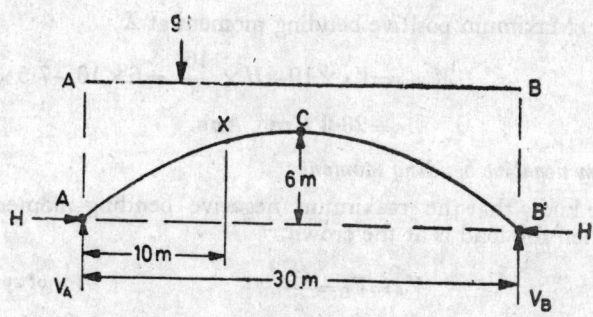

Fig. 10·25

Given. Span, $l=30$ m

Central rise, $y_C=6$ m

Load, $W=9$ t

Distance between the section and the left hand support,

$$x=10 \text{ m}$$

Maximum positive bending moment

Let $y=$ Rise of the arch at x,

$H =$ Horizontal thrust,

$V_A =$ Vertical reaction at A, and

$V_B =$ Vertical reaction at B.

Using the relation,

$$y = \frac{4y_C}{l^2} \, x(l-x) \text{ with usual notations.}$$

$$= \frac{4 \times 6}{30 \times 30} \, 10(30-10) = \frac{24 \times 10 \times 20}{30 \times 30} = \frac{16}{3} \text{ m}$$

..c know that the maximum positive bending moment takes place, when the load is on the section.

Taking moments about B,

$$V_A \times 30 = 9 \times 20$$

$$\therefore \qquad V_A = \frac{9 \times 20}{30} = 6 \text{ t}$$

and

$$V_B = 9 - 6 = 3 \text{ 't}$$

Beam moment at C, due to external loading

$$\mu_C = 3 \times 15 = 45 \text{ t-m}$$

Using the relation,

$$H = \frac{\mu_C}{y_C} \text{ with usual notations.}$$

$$= \frac{45}{6} = 7 \cdot 5 \text{ t}$$

$\therefore$ Maximum positive bending moment at X,

$$M_{max} = V_A \times 10 - H \times \frac{16}{3} = 6 \times 10 - 7 \cdot 5 \times \frac{16}{3} \text{ t-m}$$

$$= 20 \cdot 0 \text{ t-m} \quad \text{Ans.}$$

Maximum negative bending moment

We know that the maximum negative bending moment takes place, when the load is at the crown.

$$\therefore \qquad V_A = V_B = \frac{9}{2} \text{ t} \qquad\qquad (\because \text{ of symmetry})$$

and moment at C due to external loading,

$$\mu_C = \frac{9}{2} \times 15 = \frac{135}{2} = 67 \cdot 5 \text{ t-m}$$

Again using the relation,

$$H = \frac{\mu_C}{y_C} \text{ with usual notations,}$$

$$= \frac{67 \cdot 5}{6} = 11 \cdot 25 \text{ t}$$

∴ Maximum negative bending moment at X,

$$M_{max} = V_A \times 10 - H \times \frac{16}{3} = \frac{9}{2} \times 10 - 11 \cdot 25 \times \frac{16}{3} \text{ t-m}$$

$$= -15 \cdot 0 \text{ t-m} \quad \textbf{Ans.}$$

Example 10·11. *A three-hinged parabolic arch has a span of 25 m and central rise 4·5. Determine the maximum positive and negative bending moment at a section 8 m from the left hand support, if a uniformly distributed load of 5 t/m rolls over the arch. Also determine the absolute maximum bending moment.*

Solution.

Given. Span, $l = 25$ m

Central rise, $y_C = 4 \cdot 5$ m

Distance between the section and the left hand support,

$$x = 8 \text{ m}$$

Load, $w = 5$ t/m

Maximum positive bending moment

Let M_{max} = Maximum positive bending at the section.

Using the relation,

$$M_{max} = \frac{wx \, (l-x)(l-2x)}{2(3l-2x)} \text{ with usual notations.}$$

$$= \frac{5 \times 8(25-8)(25-16)}{2(75-16)} = \frac{5 \times 8 \times 17 \times 9}{2 \times 59} \text{ t-m}$$

$$= 51 \cdot 86 \text{ t-m} \quad \textbf{Ans.}$$

Maximum negative bending moment

We know that the magnitude of the maximum negative bending moment is equal to that of the maximum positive bending moment

$$M_{max} = -51 \cdot 86 \text{ t-m} \quad \textbf{Ans.}$$

Absolute maximum bending moment

Let $M_{max \; max}$ = Absolute maximum bending moment.

Using the relation,

$$M_{max \; max} = 0 \cdot 01883 \; wl^2 \text{ with usual notations.}$$

$$= 0 \cdot 01883 \times 5 \times 25^2 \text{ t-m}$$

$$= 58 \cdot 87 \text{ t-m} \quad \textbf{Ans.}$$

The absolute maximum bending moment will occur at a distance of $0 \cdot 234 \times 25 = 5 \cdot 85$ m from left hand support.

Exercise 10–3

1. A three-hinged parabolic arch of span 25 m has a central rise of 5 m. Determine the maximum positive and negative bending moment at a section X, 8 m from the left hand hinge, if a load of 15 T rolls over the arch.

(Jabalpur University)

[**Ans.** 29·38 t-m ; 21·6 t-m]

2. A three-hinged parabolic arch has a span of 20 m and central rise of 5 m. A point load of 10 T rolls over the arch from left to right. Find the maximum B.M. at a section 8 m from the left hand support. Also find the absolute maximum B.M., that will occur in the arch. *(Utkal University)*

[**Ans.** 27·39 t-m ; −18·65 t-m ; 28·8 t-m]

HIGHLIGHTS

1. Eddy's theorem states, *"The bending moment at any point of an arch axis is proportional to the vertical intercept between the theoretical arch (i.e. line of thrust) and the central line of the actual arch (i.e. arch axis)."*

2. In a parabolic arch rise of the arch axis,

$$y = \frac{4y_C}{l^2} \, x(l-x)$$

where
$y_C =$ central rise of the arch,
$l =$ Span of the arch,
$x =$ Horizontal distance between the support and the section.

3. In a circular arch

$$R = \frac{l^2}{8y_C} + \frac{y_C}{2}$$

and
$$y = R \, (\cos \theta - \cos \alpha)$$

where
$R =$ Radius of the arch,
$l =$ Span of the arch,
$y_C =$ Central rise of the arch,
$y =$ Rise of the arch axis at a section X,
$\theta =$ Angle, which OX makes with the central line OC,
$2\alpha =$ Angle, which the two supports A and B of the arch makes with the centre line OC.

4. Horizontal thrust in an arch,

$$H = \frac{\mu_C}{y_C}$$

where
$y_C =$ Moment at the crown due to external loading, and
$y_C =$ Central rise of the arch.

5. If the arch is supported at different levels, then extend the arch at the upper support to a point, whose level is the same as that of a lower support. The imaginary arch may, now, be analysed as usual.

6. The straining actions in a three-hinged arch are

(a) Normal thrust at any section X,

$$P_X = H \cos \theta + V \sin \theta \qquad \text{...(If } V \text{ is a net upward force on the left of } X)$$

$$= H \cos \theta - V \sin \theta \qquad \text{...(If } V \text{ is a net downward force on the left of } X)$$

(b) Radial shear force at X,

$$F_X = H \sin \theta - V \cos \theta \qquad \text{...(If } V \text{ is a net upward force to the left of } X)$$

$$= H \sin \theta + V \cos \theta \qquad \text{...(If } V \text{ is a net downward force to the left of } X)$$

7. When a three-hinged arch is subjected to a rise in temperature, then the vertical displacement of the arch crown,

$$dy_C = \frac{l^2 + 4y_C{}^2}{4y_C} \times \alpha t$$

where

l = Span of the arch,

y_C = Central rise of the arch,

α = Coefficient of linear expansion for the arch material, and

t = Change of temperature.

8. When a point load W rolls over an arch, then at any section X, (1) the maximum positive B.M. takes place when the load is at the section, (2) the maximum negative B.M. takes place when the load is at the crown.

9. When a point load W rolls over a parabolic three-hinged arch, the absolute maximum positive B.M. is $0.096\ Wl$ and absolute maximum negative B.M. is $0.0625\ Wl$.

10. When a uniformly distributed load of w per unit length rolls over a parabolic arch of span l, the maximum positive and negative B.M. at a section X, at a distance x from A,

$$M_{max} = \frac{wx(l-x)(l-2x)}{2(3l-2x)}$$

and the absolute maximum B.M. is $0.01883\ wl^2$.

Do You Know ?

1. Define clearly the terms 'theoretical arch' and 'actual arch'.

2. Enunciate Eddy's theorem and explain with the help of diagram, how it provides a convenient method for solving problems on three-hinged arches.

3. What do you understand by the term horizontal thrust ? Derive an equation for the same.

4. Define the terms normal thrust and radial shear force, as applied in three-hinged arches. Derive expressions for the same.

5. Derive an expression for the displacement of a three-hinged arch crown, subjected to change of temperature.

6. Draw influence lines for horizontal thrust B.M., normal thrust, and radial shear force, when a point load rolls over (a) circular three-hinged arch and (b) parabolic three-hinged arch.

7. Derive an expression for the maximum positive and negative B.M. at a section, when a uniformly distributed load rolls over a three-hinged arch. Also derive an expression for the absolute maximum B.M.

PART II

STATICALLY
INDETERMINATE STRUCTURES

PROPPED CANTILEVERS AND BEAMS

11

PROPPED CANTILEVERS AND BEAMS

11·1. Introduction

We have already discussed in chapter 7 that whenever a cantilever or a beam is loaded, it gets deflected. As a matter of fact, the amount, by which a cantilever or a beam may deflect, is so small that it is hardly detected by the residents. But sometimes, due to inaccurate design or bad workmanship, the deflection of the free end of a cantilever or centre of the beam is so much deflected that residents are always afraid of its falling down, and it effects their health. In order to set right the deflected cantilever or a beam ; or more precisely to avoid the deflection to some extent, it is propped up (*i.e.* supported by some vertical pole at the original level before deflection) at some suitable point.

11·2. Reaction* of a prop

Consider a cantilever beam *AB* fixed at *A* and propped at *B* as

---

*Very often the students commit the mistake of finding out the prop reaction by equating the clockwise moments (due to load on cantilever) to the anticlockwise moment (due to the prop reaction) about the fixed end ; as they would do in the case of a simply supported beam. This practice does not hold good in this case, as the net moment at the fixed end is not zero. There exists a fixing moment, which can not be determined unless the prop reaction is known.

shown in Fig. 11·1 (*a*). Let the cantilever be subjected to some load-ing, say uniformly distributed load.

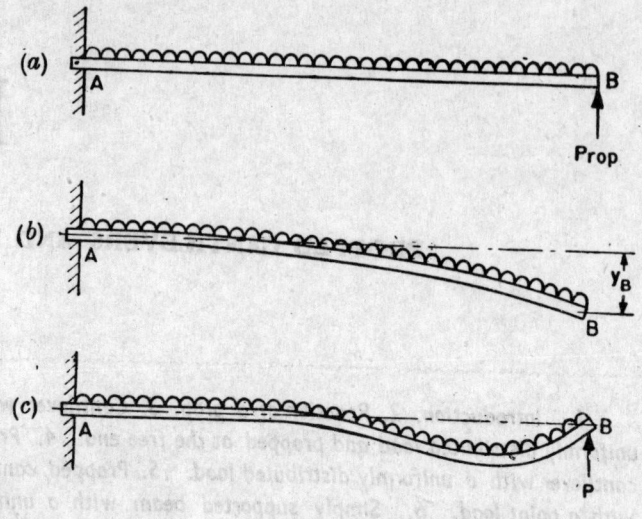

Fig. 11·1

It has been experimentally found, that this prop will be subjected to some reaction. This reaction can be obtained as discussed below :

(1) Imagine the prop to be removed and calculate the deflec-tion of the free end *B* as shown in Fig. 11·1 (*b*).

(2) Now imagine a prop to be introduced at *B*, which will exert an upward force *P* equal to the reaction of the prop. It will cause an upward deflection of *B* due to the prop reaction as shown in Fig. 11·1 (*c*).

(3) Now, by equating the downward deflection due to the load, and the upward deflection due to the prop reaction, the reaction of the prop may be found out.

11·3. Cantilever with a uniformly distributed load and propped at the free end.

Consider a cantilever *AB* of length *l* fixed at *A* and propped at *B*, and carrying a uniformly distributed load of *w* per unit length over the entire span, and propped at *B*, as shown in Fig. 11·2 (*a*). We know that the downward deflection of *B*, due to uniformly distributed load of *w* per unit length,

$$y_B = \frac{wl^4}{8EI} \qquad \qquad ...(i)$$

The effect of introducing the prop at *B* is to exert an upward reaction *P*. We also know that the upward deflection of the canti-lever due to the force *P*,

$$y_B = \frac{Pl^3}{3EI} \qquad \qquad ...(ii)$$

Since both the deflections are equal, therefore equating eqations (i) and (ii),

$$\frac{Pl^3}{3EI} = \frac{wl^4}{8EI}$$

$$\therefore \qquad P = \frac{3wl}{8} \qquad \qquad \qquad ...(iii)$$

Now let us analyse the propped cantilever for shear force, bending moment, slope and deflection at important sections of the cantilever.

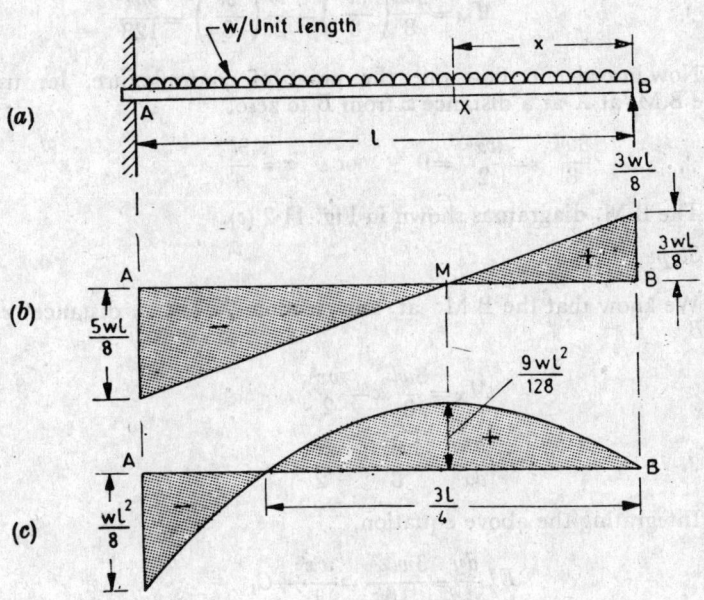

Fig. 11·2

(i) Shear force

We know that the shear force at B,

$$F_B = \frac{3wl}{8}$$

and

$$F_A = -\frac{5wl}{8}$$

Let M be the point at a distance x from B, where S.F. changes sign.

$$\therefore \qquad \frac{x}{l-x} = \frac{3}{8}$$

or

$$x = \frac{3l}{8}$$

Thus the S.F. is zero at a distance $3l/8$ from B. The S.F. diagram is shown in Fig. 11·2 (b).

(ii) Bending moment

We know that the bending moment at B,

$$M_B = 0$$

and

$$M_A = \frac{3wl}{8} . l - \frac{wl^2}{2} = -\frac{wl^2}{8}$$

We also know that the B.M. will be maximum at M, where S.F. changes sign

$$\therefore \qquad M_M = \frac{3wl}{8}\left(\frac{3l}{8}\right) - \frac{w}{2}\left(\frac{3l}{8}\right)^2 = \frac{9wl^2}{128}$$

Now in order to find out the point of contraflexure, let us equate B.M. at X at a distance x from B to zero.

$$\therefore \qquad \frac{3wl}{8} x - \frac{wx^2}{2} = 0 \qquad \text{or} \qquad x = \frac{3l}{4}$$

The B.M. diagram is shown in Fig. 11·2 (c).

(iii) Slope

We know that the B.M. at any section X, at a distance x from B,

$$M_X = \frac{3wl}{8} . x - \frac{wx^2}{2}$$

$$\therefore \qquad EI\frac{d^2y}{dx^2} = \frac{3wlx}{8} - \frac{wx^2}{2}$$

Integrating the above equation,

$$EI\frac{dy}{dx} = \frac{3wlx^2}{16} - \frac{wx^3}{6} + C_1$$

where C_1 is the constant of integration. We know that when $x = l$, $\frac{dy}{dx} = 0$. Therefore substituting these values is the above equation,

$$0 = \frac{3wl.l^2}{16} - \frac{wl^3}{6} + C_1$$

or

$$C_1 = -\frac{wl^3}{48}$$

$$\therefore \qquad EI\frac{dy}{dx} = \frac{3wlx^2}{16} - \frac{wx^3}{6} - \frac{wl^3}{48} \qquad \qquad ...(iv)$$

This is the required equation for *slope* at any section of the cantilever. Now for the slope at B, substituting $x = 0$ in the above equation,

$$EI.i_B = -\frac{wl^3}{48}$$

$$\therefore \qquad i_B = -\frac{wl^3}{48\,EI}$$

(Minus sign means that the tangent at B makes an angle with AB in the negative or anti-clockwise direction).

$$= \frac{wl^3}{48\,EI} \text{ radians} \qquad \qquad ...(v)$$

(iv) Deflection

Integrating the equation (vi) once again,

$$EI.y = \frac{3wlx^3}{48} - \frac{wx^4}{24} - \frac{wl^3x}{48} + C_2$$

where C_2 is the constant of integration. We know that when $x=l$, $y=0$. Therefore substituting the values in the above equation, we get $C_2=0$.

$$\therefore \qquad EI.y = \frac{3wlx^3}{48} - \frac{wx^4}{24} - \frac{wl^3x}{48} \qquad ...(vi)$$

This is the required equation for *deflection* at any section of the cantilever. Now for the deflection at the centre of the cantilever, substituting $x = \dfrac{l}{2}$,

$$EI.y_C = \frac{wl}{16}\left(\frac{l}{2}\right)^3 - \frac{w}{24}\left(\frac{l}{2}\right)^4 - \frac{wl^3}{48}\left(\frac{l}{2}\right) = -\frac{wl^4}{192}$$

$$\therefore \qquad y_C = -\frac{wl^4}{192\,EI}$$

(Minus sign means that the deflection is downwards)

$$= \frac{wl^4}{192\,EI} \qquad \qquad ...(vii)$$

(v) Maximum deflection

We know that the maximum deflection takes place, at a point where slope is zero. Therefore equating the equation (iv) to zero,

$$\frac{3wlx^2}{16} - \frac{wx^3}{6} - \frac{wl^3}{48} = 0$$

$$\therefore \qquad 9lx^2 - 8x^3 - l^3 = 0$$

Solving this equation by trial and error, we get $x = 0.422\,l$.

$$\therefore \qquad EI.y_{max} = \frac{wl}{16}(0.422\,l)^3 - \frac{w}{24}(0.422\,l)^4 - \frac{wl}{48}(0.422\,l)$$

$$= -0.005415\,wl^4$$

$$\therefore \qquad y_{max} = -\frac{0.005415\,wl^4}{EI}$$

(Minus sign means the deflection is downwards)

$$= \frac{0.005415\,wl^4}{EI} \qquad \qquad ...(viii)$$

Example 11·1. *A propped cantilever 10 m long has 15 cm wide and 40 cm deep cross-section. If the allowable bending stress and the deflection at the centre is 100 kg/cm² and 1·5 cm respectively, determine the safe uniformly distributed load, which the cantilever can carry. Take* $E = 1·2 \times 10^5$ *kg/cm².*   *sthan University 1977.*

Solution.

Given. Span $l = 10$ m $= 1,000$ cm

Width, $b = 15$ cm

Depth, $d = 40$ cm

∴   Moment of inertia,

$$I = \frac{bd^3}{12} = \frac{15 \times 40^3}{12} = 80,000 \text{ cm}^4$$

Allowable bending stress,

$$f = 100 \text{ kg/cm}^2$$

Allowable deflection at the centre,

$$y_C = 1.5 \text{ cm}$$

$$E = 1.2 \times 10^5 \text{ kg/cm}^2$$

Let $w = $ Uniformly distributed load over the cantilever.

We know that the maximum B.M. on a propped cantilever is at
its fixed end.

∴ $M = \frac{wl^2}{8} = \frac{w \times 1,000^2}{8} = 1,25,000 \ w \quad$ kg-cm

Using the relation,

$$\frac{M}{I} = \frac{f}{y} \text{ with usual notations.}$$

$$\frac{1,25,000 \ w}{80,000} = \frac{100}{20} \qquad\qquad \left(\because \ y = \frac{d}{2} \right)$$

or $w = \frac{100 \times 80,000}{20 \times 1,25,000} = 3.2$ kg-cm   ...(i)

Again using the relation,

$$y_C = \frac{wl^4}{192 \ EI} \text{ with usual notations.}$$

$$1.5 = \frac{w \times 1,000^4}{192 \times 1.2 \times 10^5 \times 8,000}$$

or $w = \frac{1.5 \times 192 \times 1.2 \times 10^5 \times 8,000}{1,000^4} = 2.76$ kg/cm

...(ii)

Thus the safe load over the cantilever is minimum of (i) and
(ii) i.e., 2.765 kg/cm = 276.5 kg/m.   **Ans.**

Example 11·2. *A beam AB 2 m long and carrying a uniformly
distributed load of 1 t/m is resting over a similar beam CD 1 m long as
shown in Fig. 11·3.*

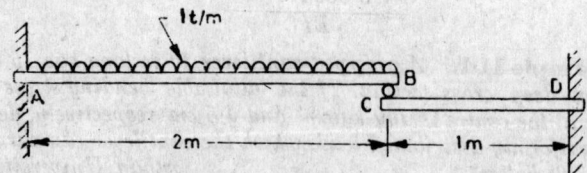

Fig. 11·3

Find the reaction at C.

Solution.

Given.   Span of beam AB,
$$l_{AB}=2 \text{ m}$$
Load, $w=1$ t/m
Span of beam CD, $l_{CD}=1$ m
Let $R_C=$ Reaction at C.

We know that downward deflection of the cantilever beam AB at B, due to the load of 1 t/m,
$$y=\frac{wl^4}{8EI}=\frac{1\times 2^4}{8EI}=\frac{2}{EI} \qquad \ldots(i)$$

and upward deflection of the cantilever AB at B due to the reaction at C
$$=\frac{R_C l^3}{3EI}=\frac{R_C \times 2^3}{3EI}=\frac{8R_C}{3EI}$$

$\therefore$ Net downward deflection of the cantilever AB at B,
$$y_B=\frac{2}{EI}-\frac{8R_C}{3EI} \qquad \ldots(ii)$$

We also know that the downward deflection of the beam CD at C due to the reaction R_C
$$=\frac{R_C l^3}{3EI}=\frac{R_C \times 1^3}{3EI}=\frac{R_C}{3EI} \qquad \ldots(iii)$$

Since both the deflections of B and C are equal, therefore equating (ii) and (iii),
$$\frac{2}{EI}-\frac{8R_C}{3EI}=\frac{R_C}{3EI}$$

$\therefore$ $R_C=\dfrac{2}{3}$ **t   Ans.**

11·4.   Propped cantilever with a uniformly distributed load

Sometimes, a cantilever is subjected to a uniformly distributed load over the entire span, or part of it. In some of the cases, the cantilever is propped at an intermediate point also. In such a case, the reaction of the prop is found out first by calculating the deflection of the cantilever at the point of prop, and then following the usual procedure as already discussed.

Example 11·3.   *A cantilever ABC is fixed at A, propped at C is loaded as shown in Fig. 11·4.*

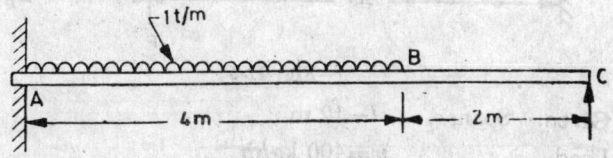

Fig. 11·4

Find the reaction at C.

Solution.

Given. Span, $l = 6$ m

Load, $w = 1$ t/m

Loaded length, $l_1 = 4$ m

Let $P =$ Reaction at the end C.

First of all, let us find out the deflection of cantilever at C due to load, but neglecting the prop.

Using the relation,

$$y_C = \frac{wl_1{}^4}{8EI} + \frac{wl_1{}^3}{6EI}(l - l_1) \text{ with usual notations.}$$

$$= \frac{1 \times 4^4}{8EI} + \frac{1 \times 4^3}{6EI}(6-4) = \frac{160}{3EI} \qquad ...(i)$$

Now let us find out the deflection of the cantilever at C due to the reaction on the prop.

Using the relation,

$$y_C = \frac{Wl^3}{3EI} \text{ with usual notations.}$$

$$= \frac{P \times 6^3}{3EI} = \frac{72P}{EI} \qquad ...(ii)$$

Equating (i) and (ii),

$$\frac{160}{3EI} = \frac{72P}{EI}$$

$$\therefore \qquad P = \frac{160}{3 \times 72} = 0.741 \text{ t Ans.}$$

Example 11·4. *A cantilever of 12 m span and carrying a uniformly distributed load of 400 kg/m is propped at a distance of 9 m from the free end at the same level. Determine the load on the prop.*

Ihi University, 1978

Solution.

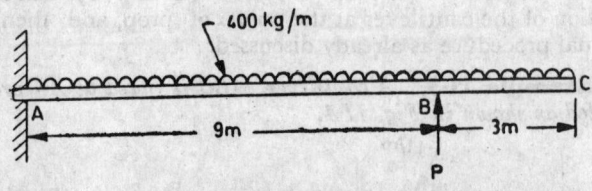

Fig. 11·5

Given. Span, $l = 12$ m

Load $w = 400$ kg/m

Distance between the prop and the free end,

$$l_1 = 9 \text{ m}$$

∴ Distance of prop from the free end,
$$x = 3 \text{ m}$$

Let $\quad\quad\quad\quad$ $P =$ Load on the prop.

First of all, let us find out the deflection of cantilever at B due to load, but neglecting the prop.

Using the relation,

$$EI.y = \frac{wl^3 x}{6} - \frac{wx^4}{24} - \frac{wl^4}{8} \text{ with usual notations.}$$

$$= \frac{400 \times 12^3 \times 3}{6} - \frac{400 \times 3^4}{24} - \frac{400 \times 12^4}{8}$$

$$= -6,92,550$$

$$y_B = -\frac{6,92,550}{EI} \text{ m} \quad \text{(Minus sign means that the deflection is downwards)}$$

$$= \frac{6,92,550}{EI} \text{ m} \quad\quad\quad ...(i)$$

Now let us find out the deflection of the cantilever at B due to relation of the prop.

Now using the relation,

$$y_B = \frac{Wl_1^3}{3EI} \text{ with usual notations.}$$

$$= \frac{P \times 9^3}{3EI} = \frac{243P}{EI} \quad\quad\quad ...(ii)$$

Equating (i) and (ii) wet get,

$$\frac{243P}{EI} = \frac{6,92,500}{EI}$$

$$P = \frac{6,92,500}{243} = 2,850 \text{ kg} \text{ Ans.}$$

Example 11·5. *A propped cantilever ABCD is loaded as shown in Fig. 11·6.*

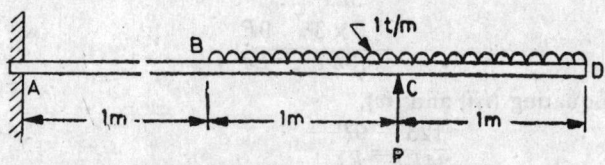

Fig. 11·6

Determine the prop reaction at C. $\quad\quad$ (*Kerala University 1973*)

Solution.

Given. Span, $\quad\quad$ $l = 3$ m

Load, $\quad\quad\quad\quad$ $w = 1$ t/m

Distance between the prop and the free end

$$x = 1 \text{ m}$$

Distance between the prop and the fixed end,
$$l_1 = 2 \text{ m}$$

Let $P = $ Prop reaction.

For the sake of simplicity let us assume that an upward and downward load of 1 t/m to be acting on AB.

Using the relation,

$$EI.y_C = \frac{wl^3 x}{6} - \frac{wx^4}{24} - \frac{wl^4}{8} \quad \text{with usual notations.}$$

$$= \frac{1 \times 3^3 \times 1}{6} - \frac{1 \times 1^4}{24} - \frac{1 \times 3^4}{8} = -\frac{16}{3}$$

(Due to downward load from A to D)

$$\therefore \qquad y_C = -\frac{16}{3EI} \qquad \text{(Minus sign means the deflection is downwards)}$$

$$= \frac{16}{3EI} \qquad\qquad\qquad ...(i)$$

Now using the relation,

$$y_C = -\left[\frac{wl_1^4}{8EI} + \frac{wl_1^2}{6EI}(l - l_1) \right] \quad \text{with usual notations.}$$

(Due to upward load from A to B)

$$= -\left[\frac{1 \times 1^4}{8EI} + \frac{1 \times 1^3}{6EI}(2 - 1) \right] \qquad \text{(Here } l = 2 \text{ m ; } l_1 = 1 \text{ m)}$$

$$= -\frac{5}{24EI} \qquad\qquad\qquad ...(ii)$$

$\therefore$ Net downward deflection of cantilever at C,

$$y_C = \frac{16}{3EI} - \frac{5}{24EI} = \frac{123}{24EI} \qquad ...(iii)$$

Now using the relation,

$$y_C = \frac{Pl^3}{3EI} \quad \text{with usual notations.}$$

$$= \frac{P \times 3^3}{3EI} = \frac{9P}{EI} \qquad\qquad\qquad ...(iv)$$

Equating (iii) and (iv),

$$\frac{123}{24EI} = \frac{9P}{EI}$$

$$P = \frac{123}{9 \times 24} = 0.57 \text{ t} \quad \textbf{Ans.}$$

11·5. Propped cantilever with a point load

Consider a cantilever AB of length l, fixed at A and propped at B, carrying a point load W at C at a distance l_1 from the fixed end A as shown in Fig. 11·7.

We know that the downward deflection of B due to load W at C,

$$y_C = \frac{Wl_1^3}{3EI} + \frac{Wl_1^2}{2EI}(l-l_1) \qquad \ldots(i)$$

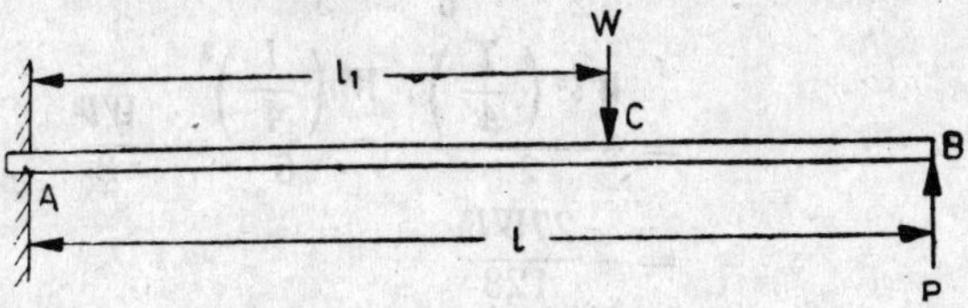

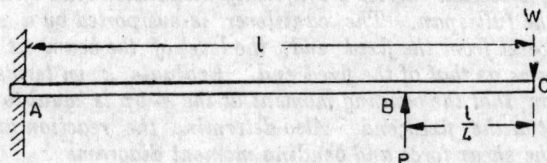

Fig. 11·7

and the upward deflection of B, due to the prop reaction P,

$$y_C = \frac{Pl^3}{3EI} \qquad \ldots(ii)$$

Equating (i) and (ii),

$$= \frac{Wl_1^3}{3EI} + \frac{Wl_1^2}{2EI}(l-l_1) = \frac{Pl^3}{3EI}$$

or

$$P = \frac{2Wl_1^3 + 3Wl_1^2(l-l_1)}{2l^3}$$

Cor. If $\quad l_1 = \dfrac{l}{2}$

Then $\qquad P = \dfrac{2W\left(\dfrac{l}{2}\right)^3 + 3W\left(\dfrac{l}{2}\right)^2 \times \dfrac{l}{2}}{2l^3} = \dfrac{5W}{16}$.

Example 11·6. *A cantilever of length l carries a point load W at its free end. It is propped at a distance of $l/4$ from the free end. Find out the prop reaction.*

Fig. 11·8

Solution.

Span $\qquad\qquad = l$

Load $\qquad\qquad = W$

Distance between the prop and the free end,

$$x = \frac{l}{4}$$

∴ Distance between the prop and the fixed end,

$$l_1 = \frac{3l}{4}$$

Let $\qquad\qquad P = $ Prop reaction.

First of all, let us find out the deflection of the cantilever at B due to the load, neglecting the prop.

Using the relation,

$$EI.y_B = \frac{Wl^2 x}{2} - \frac{Wx^3}{6} - \frac{Wl^3}{3} \quad \text{with usual notations.}$$

$$= \frac{Wl^2\left(\dfrac{l}{4}\right)}{2} - \frac{W\left(\dfrac{l}{4}\right)^3}{6} - \frac{Wl^2}{3}$$

$$= -\frac{27Wl^3}{128}$$

$$y_B = -\frac{27Wl^3}{128EI} \qquad \text{(Minus sign means that the slope is downwards)}$$

$$= \frac{27Wl^3}{128EI} \qquad \qquad \ldots(i)$$

Now let us find out the deflection of the cantilever B, due to the reaction P of the prop.

Now using the relation,

$$y_B = \frac{Pl_1^3}{3EI} = \frac{P\left(\dfrac{3l}{4}\right)^3}{3EI} = \frac{9Pl^3}{64EI} \qquad \ldots(ii)$$

Equating (i) and (ii) we get,

$$\frac{27Wl^3}{128EI} = \frac{9Pl^3}{64EI}$$

$$\therefore \qquad P = \frac{27W \times 64}{128 \times 9} = \frac{3W}{2} \quad \textbf{Ans.}$$

Example 11·7. *A horizontal cantilever beam of length l and of uniform cross-section carries a uniformly distributed load of w per unit length for the full span. The cantilever is supported by a rigid prop at a distance kl from the fixed end, the level of the beam at the prop being the same as that of the fixed end. Evaluate k in terms of l for the condition, that the bending moment at the prop is equal to the bending moment at the fixed end. Also determine the reaction at the prop and draw the shear force and bending moment diagrams.*

(Cambridge University,)

Solution.

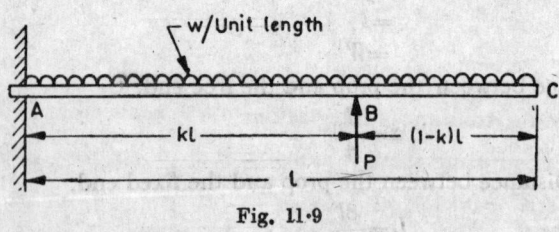

Fig. 11·9

Given.   Span $= l$

Load $= w$ per unit length

Distance of the prop from the fixed end
$$= kl$$

Let $\qquad\qquad P = \text{Prop reaction.}$

From the geometry of the cantilever, we find that the bending moment at the prop

$$= -\frac{w(1-k)^2 l^2}{2}$$

and bending moment at the fixed end A

$$= P \cdot kl - \frac{wl^2}{2}$$

Since both the bending moments are equal, therefore equating the same,

$$P \cdot kl - \frac{wl^2}{2} = -\frac{w(1-k)^2 l^2}{2}$$

$$\therefore \qquad P \cdot k = \frac{wl}{2} - \frac{wl(1-k)^2}{2}$$

$$= \frac{wl}{2} - \frac{wl(1+k^2-2k)}{2}$$

$$= \frac{wl}{2}(1-1-k^2+2k) = \frac{wlk}{2}(2-k)$$

or $\qquad\qquad P = \frac{wl}{2} \cdot (2-k) \qquad\qquad\qquad \ldots(i)$

First of all, let us find out the deflection of the cantilever at B due to load, but neglecting the prop.

We know that the bending moment at any section X, at a distance x from the fixed end,

$$M_X = -\frac{w(l-x)^2}{2}$$

$$\therefore \qquad EI\frac{d^2y}{dx^2} = -\frac{w(l-x)^2}{2} = -\frac{w(l^2+x^2-2lx)}{2}$$

Integrating the above equation,

$$EI\frac{dy}{dx} = -\frac{w}{2}\left(l^2x + \frac{x^3}{3} - \frac{2lx^2}{2}\right) + C_1$$

$$= -\frac{w}{2}\left(l^2x + \frac{x^3}{3} - lx^2\right) + C_1$$

where C_1 is the constant of integration. We know that when $x=0$, $\frac{dy}{dx}=0$. Substituting these values of x and $\frac{dy}{dx}$ in the above equation, we get $C_1 = 0$.

$$\therefore \qquad EI \cdot \frac{dy}{dx} = -\frac{w}{2}\left(l^2x + \frac{x^3}{3} - lx^2\right)$$

Integrating the above equation once again,

$$EI \cdot y = -\frac{w}{2}\left(\frac{l^2 x^2}{2} + \frac{x^4}{12} - \frac{lx^3}{3}\right) + C_2$$

where C_2 is the constant of integration. We know that when $x=0$, $y=0$. Substituting these values of x and y in the above equation, we get $C_2 = 0$.

$$EI.y = -\frac{w}{2}\left(\frac{l^2 x^2}{2} + \frac{x^4}{12} - \frac{lx^3}{3}\right)$$

Substituting $x=kl$ in the above equation,

$$EI.y_B = -\frac{w}{2}\left(\frac{l^2.k^2 l^2}{2} + \frac{k^4 l^4}{12} - \frac{l.k^3 l^3}{3}\right)$$

$$= -\frac{wl^4 k^2}{24}(6 + k^2 - 4k)$$

or

$$y_B = -\frac{wl^4 k^2}{24EI}(k^2 - 4k + 6)$$

$$\begin{bmatrix} \text{Minus sign means that the} \\ \text{deflection is downwards} \end{bmatrix}$$

$$= \frac{wl^4 k^2}{24EI}(k^2 - 4k + 6) \qquad \qquad ...(ii)$$

Similarly upward deflection of the cantilever due to the prop reaction,

$$y_B = \frac{P(kl)^3}{3EI} = \frac{Pk^3 l^3}{3EI}$$

Substituting the value of P from equation (i)

$$y_B = \frac{wl}{2}(2-k) \times \frac{k^3 l^3}{3EI}$$

$$= \frac{wl^4(2-k)k^3}{6EI} \qquad \qquad ...(iii)$$

Since the level of the beam, at the prop, is the same as that of the fixed end, therefore the net deflection at B is zero. Now equating (ii) and (iii),

$$\frac{wl^4 k^2}{24EI}(k^2 - 4k + 6) = \frac{wl^4(2-k)k^3}{6EI}$$

$$\therefore \qquad k^2 - 4k + 6 = 4k(2-k)$$

or $\qquad 5k^2 - 12k + 6 = 0$

Solving the above equation as a quadratic equation for k, we get

$$k = \frac{12 \pm \sqrt{144 - 4 \times 5 \times 6}}{2 \times 5} = \frac{12 - \sqrt{24}}{10}$$

$$= 0.71 \quad \textbf{Ans.}$$

Reaction at the prop

Substituting the value of k in equation (i) we get the reaction at the prop,

$$P=\frac{wl}{2}\ (2-k)=\ \frac{wl}{2}\ (2-0\cdot71)$$

$$=0\cdot645\ \mathbf{wl}\quad \mathbf{Ans.}$$

Shear force and bending moment diagrams

From the geometry of the cantilever, we find that the shear force at C,

$$F_C=0$$
$$F_B=-0\cdot29\ wl+0\cdot645\ wl=+0\cdot355\ wl$$
$$F_A=+0\cdot355\ wl-0\cdot71\ wl=-0\cdot355\ wl$$

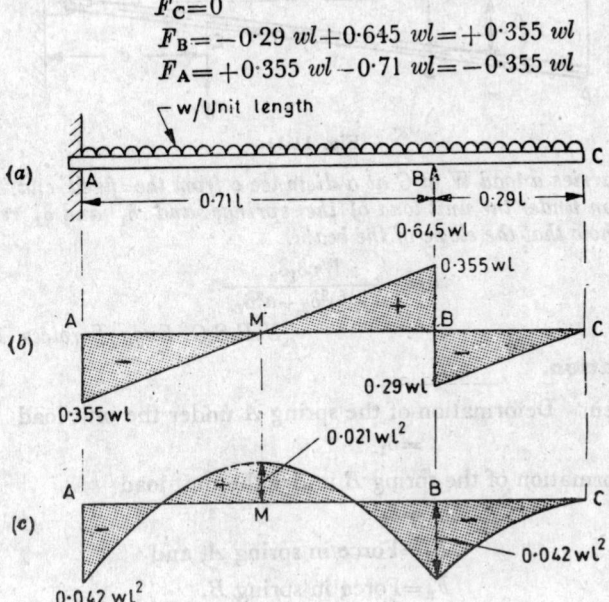

Fig. 11·10

Now draw the S.F. diagram as shown in Fig. 11·10 (b). From the geometry of the S.F. diagram, we find that the S.F. changes sign at M *i.e.* at the middle of AB *i.e.* at a distance of $0\cdot355\ l$ from A.

From the geometry of the cantilever, we also find that the bending moment at C

$$M_C=0$$

$$M_A=-\frac{w(0\cdot29l)^2}{2}=-0\cdot042\ wl^2$$

$$M_B=-\frac{wl^2}{2}+0\cdot645\ wl\times0\cdot71\ l=-0\cdot042\ wl^2$$

$$M_M=-\frac{w(0\cdot645\ l)^2}{2}+0\cdot645\ wl\times0\cdot355\ l$$

$$=+0\cdot021\ wl^2$$

Now draw the B.M. diagram as shown in Fig. 11·10 (c).

Example 11·8. *A rigid beam ABC is pinned to a wall at O and is supported by two springs A and B as shown in Fig. 11·11.*

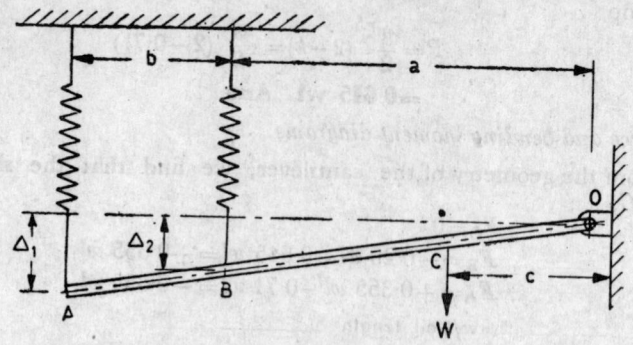

Fig. 11·11

It carries a load W at C at a distance c from the fixed end. The deformation under the unit load of the springs and δ_1 and δ_2 respectively. Show that the slope of the beam.

$$\theta = \frac{Wc\delta_1\delta_2}{(a+b)^2\delta_2 + a^2\delta_1}.$$

(U.P.S.C. Engg. Services, 1972)

Solution.

Given.   Deformation of the spring A under the unit load
$$= \delta_1$$

Deformation of the spring B under the unit load
$$= \delta_2$$

Let $F_1 =$ Force in spring A, and

$F_2 =$ Force in spring B.

∴   Deformation of spring A,

$$\triangle_1 = F_1\delta_1$$

or $F_1 = \dfrac{\triangle_1}{\delta_1}$ ...(i)

Similarly deformation of spring B,

$$\triangle_2 = F_2\delta_2$$

or $F_2 = \dfrac{\triangle_2}{\delta_2}$ ...(ii)

We know that the rigid beam ABC is hinged at O. Therefore equating the anticlockwise moments and the clockwise moments about O.

$$W.c = F_1(a+b) + F_2.a$$

$$= \frac{\triangle_1}{\delta_1}(a+b) + \frac{\triangle_2}{\delta_2}a \qquad ...(iii)$$

Now, from the geometry of the rigid beam, we find that

$$\frac{\triangle_1}{\triangle_2}=\frac{a+b}{a}$$

$$\therefore \qquad \triangle_1=\frac{a+b}{a}\cdot\triangle_2$$

Substituting the value of $\triangle_1$ in equation (iii),

$$W.c=\frac{(a+b)^2}{a}\frac{\triangle_2}{\delta_1}+\frac{\triangle_2}{\delta_2}a$$

$$=\triangle_2\left[\frac{(a+b)^2}{a\delta_1}+\frac{a}{\delta_2}\right]$$

$$=\triangle_2\left[\frac{\delta_2(a+b)^2+a^2\delta_1}{a\delta_1\delta_2}\right]$$

or $\qquad\qquad \triangle_2=\frac{W.c\times a\delta_1\delta_2}{(a+b)^2\delta_2+a^2\delta_1}$ ...(iv)

We know that slope of the beam,

$$\theta=\frac{\triangle_2}{a}=\frac{W.c.\;\delta_1\delta_2}{(a+b)^2\delta_2+a^2\delta_1}\quad\textbf{Ans.}$$

Example 11·9. *Fig. 11·12, given below, shows two cantilevers, the end of one being vertically above the other, and is connected to it by a spring AB.*

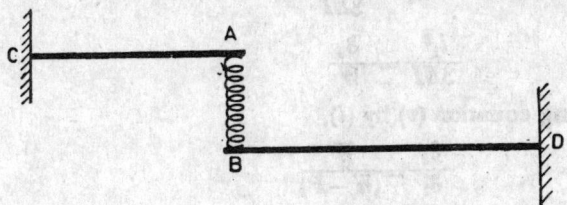

Fig. 11·12

Initially, the system is unstrained. A weight W placed at A causes a vertical deflection at A of δ_1 and a vertical deflection at B of δ_2. When the spring is removed, the weight W at A causes a deflection at A of δ_3. Find the extension in the spring when it is replaced and the weight W is transferred to B. (U.P.S.C. Engg. Services, 1973)

Solution.

Given. Weight at $\qquad\qquad A=W$

Deflection of A with spring $AB=\delta_1$

Deflection of B with spring $AB=\delta_2$

Deflection of A without spring$=\delta_3$

Let $\qquad\qquad l_1=$Length of cantilever AC,

$\qquad\qquad l_2=$Length of cantilever BD,

$P=$Force in the spring when the load is at A,

$T=$Force in the spring when the load is at B,

$\delta_4=$Deflection of A with load at B, and

$\delta_5=$Deflection of B with load at B.

We know that when the cantilever AC is loaded with W at A, the deflection of A,

$$\delta_1 = \frac{(W-P)l_1^3}{3EI} \qquad \qquad ...(i)$$

Similarly $\qquad \delta_2 = \frac{Pl_2^3}{3EI} \qquad \qquad ...(ii)$

or $\qquad \frac{l_2^3}{3EI} = \frac{\delta_2}{P} \qquad \qquad ....(iii)$

Therefore compression of the spring AB

$$= \delta_1 - \delta_2$$

and stiffness of the spring,

$$s = \frac{\text{Force}}{\text{Compression}} = \frac{P}{\delta_1 - \delta_2} \qquad \qquad ...(iv)$$

We also know that when the spring is removed and the cantilever AC is loaded with W at A, therefore deflection of A,

$$\delta_3 = \frac{Wl_1^3}{3EI} \qquad \qquad ...(v)$$

or $\qquad \frac{l_1^3}{3EI} = \frac{\delta_3}{W} \qquad \qquad ...(vi)$

Dividing equation (v) by (i),

$$\frac{\delta_3}{\delta_1} = \frac{W}{(W-P)}$$

$\therefore \qquad \frac{\delta_3 - \delta_1}{\delta_3} = \frac{P}{W} \qquad \qquad ...(vii)$

or $\qquad P = \frac{W(\delta_3 - \delta_1)}{\delta_3}$

Substituting this value of P in equation (iv),

$$s = \frac{\dfrac{W(\delta_3 - \delta_1)}{\delta_3}}{(\delta_1 - \delta_2)} = \frac{W(\delta_3 - \delta_1)}{\delta_3(\delta_1 - \delta_2)} \qquad \qquad ...(viii)$$

When the cantilever is loaded with W at B, the stiffness of the spring,

$$s = \frac{\text{Force}}{\text{Extension}} = \frac{T}{\delta_5 - \delta_4}$$

$\therefore \qquad \frac{T}{s} = \delta_5 - \delta_4 \qquad \qquad ...(ix)$

We know that when the cantilever BD is loaded with W at B, the deflection of B,

$$\delta_5 = \frac{(W-T)l_2^3}{3EI}$$

$$= \frac{(W-T)\delta_2}{P}$$

$$\left[\text{Substituting } \frac{l_2^2}{3EI} = \frac{\delta_2}{P} \text{ from eqn. } (iii)\right]$$

$$= \frac{(W-T)\ \delta_2 \times \delta_3}{W\ (\delta_3-\delta_1)}$$

[Substituting the value of P from eqn.(vii)]

Similarly $\qquad \delta_4 = \dfrac{Tl_1^3}{3EI} = \dfrac{T\delta_3}{W}$

$$\left[\text{Substituting } \frac{l_1^3}{3EI} = \frac{\delta_3}{W} \text{ from eqn. } (vi)\right]$$

The extension of the spring,

$$\delta_5 - \delta_4 = \frac{T}{s}$$

$$\left[\text{Substituting } \delta_5 - \delta_4 = \frac{T}{s} \text{ from eqn. } (ix)\right]$$

$$\delta_5 - \delta_4 = \frac{T \times \delta_2\ (\delta_1-\delta_2)}{W\ (\delta_3-\delta_1)}$$

[Substituting value of s from eqn. (x)]

We also know that the extension of the spring,

$$\delta_5 - \delta_4 = \frac{(W-T)\ \delta_2 \times \delta_3}{W\ (\delta_3-\delta_1)} - \frac{T\delta_3}{W}$$

$$= \frac{\delta_3}{W}\left[\frac{(W-T)\ \delta_2}{(\delta_3-\delta_1)} - T\right]$$

Equating both the values of $(\delta_5-\delta_4)$,

$$\frac{T \times \delta_3(\delta_1-\delta_2)}{W\ (\delta_3-\delta_1)} = \frac{\delta_3}{W}\left[\frac{(W-T)\ \delta_2}{(\delta_3-\delta_1)} - T\right]$$

$$\frac{T(\delta_1-\delta_2)}{(\delta_3-\delta_1)} = \frac{(W-T)\delta_2 - T(\delta_3-\delta_1)}{(\delta_3-\delta_1)}$$

or $\qquad\qquad T\delta_1 - T\delta_2 = W\delta_2 - T\delta_2 - T\delta_3 + T\delta_1$

$\therefore \qquad\qquad T\delta_3 = W\delta_2$

or $\qquad\qquad\qquad T = \dfrac{W\delta_2}{\delta_3}$

Substituting this value of T in equation (x) we get the extension of the spring,

$$\delta_5 - \delta_4 = \frac{\dfrac{W\delta_2}{\delta_2} \times \delta_3(\delta_1-\delta_2)}{W(\delta_3-\delta_1)}$$

$$= \frac{\delta_3(\delta_1-\delta_2)}{(\delta_3-\delta_1)} \quad \text{Ans.}$$

Exercise 11–1

1. A horizontal cantilever of length l supports a uniformly distributed load of w per unit length. If the cantilever is propped at a distance of $l/4$ from the free end, find the reaction of the prop.   (*Banaras Hindu University*)

$$\left[\; \text{Ans.}\quad \frac{19wl}{32}\;\right]$$

2. The free end of a cantilever of length l rests on the middle of a simply supported beam of the same span, and having the same section. Determine the reaction of the cantilever at its free end, if it is carrying a uniformly distributed load of w per unit length.   (*Nagpur University*)

$$\left[\; \text{Ans.}\quad \frac{6wl}{17}\;\right]$$

Hint. Net deflection of free end of the cantilever

$$= \frac{wl^2}{8EI} - \frac{Pl^3}{3EI} \qquad \qquad ...(i)$$

Deflection of the centre of the beam

$$= \frac{Pl^3}{48\,EI} \qquad \qquad ...(ii)$$

Equating (*i*) and (*ii*),

$$\frac{wl^2}{8EI} - \frac{Pl^3}{3EI} = \frac{Pl^3}{48EI}$$

or $$P = \frac{6wl}{17} \quad \text{Ans.}$$

3. A cantilever ABC of uniform section is fixed at A and propped at B. A point load W is applied at the free end C. Find the ratio of AB to BC, so that the reaction at B is $1.5\ W$.   (*Jodhpur University*)

[Ans.   3 : 1]

4. A cantilever AB 4 m long is fixed at A and propped at C at a distance 1 m from B. The cantilever carries a load, which varies gradually from zero at the free end to 6 t/m at the fixed end. Calculate the prop reaction.   (*Jadavpur University*)

[Ans.   3·91 t′]

11·6. Simply supported beam with a uniformly distributed load and propped at the centre

Consider a simply supported beam AB of length l and carrying a uniformly distributed load of w per unit length over the entire span, and propped at C as shown in Fig.11·13 (a). We know that the downward deflection of the beam at C due to u.d.l. of w per unit length,

$$y_C = \frac{5wl^4}{384EI} \qquad \qquad ...(i)$$

The effect of introducing the prop at B is to exert an upward reaction P. We know that the upward deflection of the beam due to the force P,

$$y_C = \frac{Pl^3}{48EI} \qquad \qquad ...(ii)$$

Equating (*i*) and (*ii*),

$$\frac{Pl^3}{48EI} = \frac{5wl^4}{384EI}$$

or
$$P = \frac{5wl}{8} = \frac{5W}{8} \qquad \text{(where } W = \text{Total load} = wl)$$

$$\therefore \quad \text{Reaction at } A, R_A = R_B = \frac{1}{2}\left(wl - \frac{5wl}{8}\right) = \frac{3wl}{16}$$

$$= \frac{3W}{16} \qquad \text{(where } W = \text{Total load} = wl)$$

Now let us analyse the propped beam for shear force, bending moment, slope and deflecticn at important section of the beam.

Shear force

We know that the shear force at A,

$$F_A = -\frac{3wl}{16}$$

$$F_C = -\frac{3wl}{16} + \frac{wl}{2} - \frac{5wl}{8} = -\frac{5wl}{16}$$

$$F_B = -\frac{5wl}{16} + \frac{wl}{2} = \frac{3wl}{16}$$

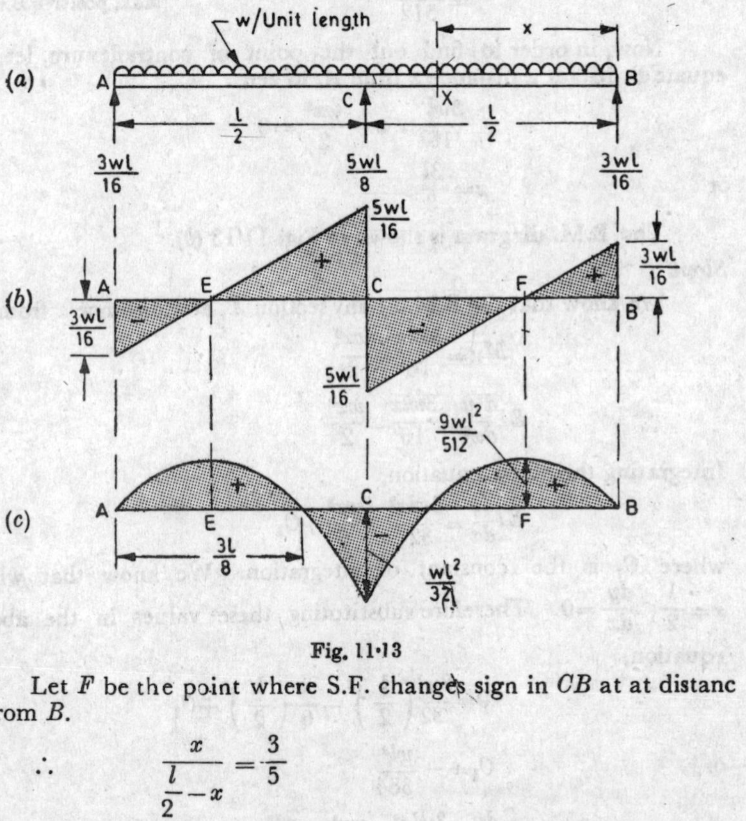

Fig. 11·13

Let F be the point where S.F. changes sign in CB at at distanc x from B.

$$\therefore \qquad \frac{x}{\frac{l}{2} - x} = \frac{3}{5}$$

or
$$x = \frac{3l}{16} \qquad \qquad \qquad \qquad \dots(iii)$$

Thus the S.F. is zero at a distance of $\dfrac{3l}{16}$ from B. The S.F. diagram is shown in Fig. 11·13 (b).

Bending moment

We know that the bending moment at A,

$$M_A = 0$$

$$M_C = \frac{3wl}{16} \times \frac{l}{2} - w \times \frac{l}{2} \times \frac{l}{4}$$

$$= -\frac{wl^2}{32} \qquad \qquad \text{(Max. negative B.M.)}$$

We also know that the B.M. will be maximum at F and E, where S.F. changes sign.

$$\therefore \qquad M_F = M_E = \frac{3wl}{16} \times \frac{3l}{16} - w \times \frac{3l}{16} \times \frac{3l}{32}$$

$$= \frac{9wl^2}{512} \qquad \qquad \text{(Max. positive B.M.)}$$

Now, in order to find out the point of contraflexure, let us equate B. M., at a distance x from A, to zero.

$$\frac{3wl}{16} \cdot x - \frac{wx^2}{2} = 0$$

or $\qquad\qquad\qquad x = \dfrac{3l}{8}$

The B.M. diagram is shown in Fig. 11·13 (b).

Slope

We know that the B.M. at any section X, at a distance x from B,

$$M_X = \frac{3wlx}{16} - \frac{wx^2}{2}$$

$$\therefore \qquad EI\frac{d^2y}{dx^2} = \frac{3wlx}{16} - \frac{wx^2}{2}$$

Integrating the above equation,

$$EI\frac{dy}{dx} = \frac{3wlx^2}{32} - \frac{wx^3}{6} + C_1$$

where C_1 is the constant of integration. We know that when $x = \dfrac{1}{2}$, $\dfrac{dy}{dx} = 0$. Therefore substituting these values in the above equation,

$$0 = \frac{3wl}{32}\left(\frac{l}{2}\right)^2 - \frac{w}{6}\left(\frac{l}{2}\right)^3 + C_1$$

or $\qquad\qquad\qquad C_1 = -\dfrac{wl^3}{384}$

$$\therefore \qquad EI\frac{dy}{dx} = \frac{3wlx^2}{32} - \frac{wx^3}{6} - \frac{wl^3}{384} \qquad \qquad ...(iv)$$

This is the required equation for *slope* at any section of the beam. Now for the slope at B, substituting $x = 0$ in the above equation,

$$EI.i_B = -\frac{wl^3}{384}$$

$$\therefore \qquad i_B = -\frac{wl^3}{384EI} \quad \text{(Minus sign means that the tangent at } B \text{ makes an angle with } AB \text{ in the negative or anticlockwise direction)}$$

$$= \frac{wl^3}{384EI} \text{ radians} \qquad \qquad ...(v)$$

By symmetry $i_A = \dfrac{wl^3}{384EI}$ radians

Deflection

Integrating the equation (*iv*) once again,

$$EI.y = \frac{3wlx^3}{96} - \frac{wx^4}{24} - \frac{wl^3x}{384} + C_2$$

where C_2 is the constant of integration. We know that when $x = 0$, $y = 0$. Therefore substituting these values in the above equation, we get $C_2 = 0$.

$$\therefore \qquad EI.y = \frac{3wlx^3}{96} - \frac{wx^4}{24} - \frac{wl^3x}{384} \qquad ...(vi)$$

This is the required equation for *deflection* at any section of the beam.

Maximum deflection

We know that the maximum deflection takes place, at a point, where slope is zero. Therefore equating the equation (*iv*) to zero,

i.e., $$\frac{3wlx^2}{32} - \frac{wx^3}{6} - \frac{wl^3}{384} = 0$$

$$\therefore \qquad 64x^3 - 36x + l^3 = 0$$

Solving the equation by trial and error, we get $x = 0.27l$

$$\therefore \quad EI.y_{max} = \frac{3wl}{96}(0.27l)^3 - \frac{w}{24}(0.27l)^4 - \frac{wl^3}{384}(0.27l)$$

$$= -0.0003062 \ wl^4$$

$$\therefore \qquad y_{max} = -\frac{0.0003062 \ wl^4}{EI} \quad \text{(Minus sign means that the deflection is downwards)}$$

$$= \frac{0.0003062 wl^4}{EI}$$

Example 11·10. *A uniform girder of length 8 m is subjected to total load of 20 tonnes, uniformly distributed over the entire length. The girder is freely supported at the ends. Calculate the deflection and bending moment at the mid-span.*

If a prop is introduced at the centre of the beam, so as to nullify the deflection already worked out, what would be the net bending moment at mid-point ?

Solution.

Given.   Span, $l = 8$ m

Total load, $W = 20$ t

$\therefore$ u.d.l. $= \dfrac{20}{8} = 2\cdot5$ t/m

Let $EI = $ Stiffness of the girder.

Deflection at the mid-span of the beam without prop

Let $y_C = $ Deflection at mid-span.

Using the relation,

$$y_C = \frac{5wl^4}{384EI} \text{ with usual notations.}$$

$$= \frac{5 \times 2\cdot5 \times 8^4}{384EI} = \frac{\mathbf{400}}{\mathbf{3EI}} \quad \textbf{Ans.}$$

Bending moment at the mid-span of the beam without prop

Let $M_1 = $ Bending at the mid-span.

Using the relation,

$$M = \frac{wl^2}{8} \text{ with usual notations.}$$

$\therefore$ $M_1 = \dfrac{2\cdot5 \times 8^2}{8} = \mathbf{20}$ **t-m**   **Ans.**

Bending moment at the mid-span of the beam with prop

Let $M_2 = $ Bending moment at the mid-span.

Now using the relation,

$$M = - \frac{wl^2}{32} \text{ with usual notations.}$$

$\therefore$ $M_2 = - \dfrac{2\cdot5 \times 8^2}{32} = \mathbf{-5\cdot0}$ **t-m**   **Ans.**

11·7. Sinking of the prop

In the previous articles, we have assumed that the prop in a cantilever or beam behaves like a rigid one *i.e.* it does not yield down due to the reaction of the beam.   But sometimes the prop sinks down, due to its elastic property and the reaction.   A sinking prop is called an *elastic prop* or *yielding prop*.

Let $\delta = $ Distance through which the prop has sunk down due to reaction,

$y_1 = $ Downward deflection of the beam, of the point of prop, and

$y_2 = $ Upward deflection of the beam due to the prop reaction.

A little consideration will show, that if the prop would not have sunk down, then $y_1 = y_2$.   But due to sinking of the prop

$$y_1 = y_2 + \delta$$

Now the prop reaction may be found out as usual.

Example 11·11. *A cantilever of length l is subjected to a point load W at its free end. The cantilever is also propped with an elastic* *prop at its free end. The prop sinks down in proportional to the load applied on it. Determine the value of proportionality k for sinking, when the reaction on the prop is half of the load W.*

Solution.

Given.   Span $= l$

Load at the free end $= W$

Reaction at the prop,

$$P = \frac{W}{2}$$

Constant of proportionality of sinking to the load

$$= k$$

$\therefore$ Sinking of the prop,

$$\delta = \frac{k.W}{2}$$

We know that the downward deflection of the cantilever due to load W at its free end,

$$y_1 = \frac{Wl^3}{3EI} \qquad \qquad ...(i)$$

and upward deflection of the cantilever due to prop reaction

$$y_2 = \frac{Pl^3}{3EI} = \frac{\frac{W}{2}(l)^3}{3EI} = \frac{Wl^3}{6EI} \qquad ...(ii)$$

Therefore sinking of the prop,

$$\delta = y_1 - y_2$$

$\therefore$

$$\frac{kW}{2} = \frac{Wl^3}{3EI} - \frac{Wl^3}{6EI} = \frac{Wl^3}{6EI}$$

or

$$k = \frac{l^3}{3EI} \quad \textbf{Ans.}$$

Example 11·12. *A simply supported beam of span l carries a uniformly distributed load of w per unit length. The beam was propped at the middle of the span. Find the amount, by which the prop should yield, in order to make all the three reactions equal.*

                                                                (Bihar University, 1978)

Solution.

Given   Span $= l$

Load $= w$ per unit length

$\therefore$ Total load $= wl$

and each reaction $= \dfrac{wl}{3}$

Let $\delta =$ Amount by which the prop should yield.

We know that the downward deflection of the beam, due to uniformly distributed load,

$$y_1 = \frac{5wl^4}{384EI}$$

and upward deflection due to the prop reaction,

$$y_2 = \frac{Pl^3}{48EI} = \frac{\frac{wl}{3} \times l^3}{48EI} = \frac{wl^4}{144EI}$$

Therefore yield of the prop

$$\delta = y_1 - y_2 = \frac{5wl^4}{384EI} - \frac{wl^4}{144EI} = \frac{7wl^4}{1,152EI} \quad \textbf{Ans.}$$

Exercise 11-2

1. A simply supported beam of length l is carrying a uniformly distributed load of w per unit length over its entire span. What upward load should be applied at the centre of the beam in order to neutralise the deflection ?

(*Utkal University*)

$$\left[\textbf{Ans.} \quad \frac{5wl}{8} \right]$$

2. A cantilever of length l is propped at its free end. The cantilever carries a uniformly distributed load of w per unit length over the entire span. If the prop sinks by δ, find the prop reaction.   (*Mysore University*)

$$\left[\textbf{Ans.} \quad \frac{3EI}{l^3}\left(\frac{wl^4}{8EI} - \delta \right) \right]$$

HIGHLIGHTS

1. The prop reaction is found out as discussed below :

   (1) Imagine the prop to be removed and calculate the deflection of the beam at the prop.

   (2) Now imagine the prop to be introduced, which will exert an upward force called the reaction of the prop. It will cause an upward deflection of the beam at the prop.

   (3) Now by equating the downward deflection due to the load and upward deflection due to the prop reaction, the reaction of the prop is found out.

2. When a cantilever AB with a *u.d.l.* is propped at the end B then

$$P = \frac{3wl}{8}$$

$$i_B = \frac{wl^3}{48\,EI} \text{ radians}$$

and

$$y_{max} = \frac{0.005415\ wl^4}{EI}$$

where

$$P = \text{Prop reaction,}$$

$i_B = $ Slope at the propped end B, and

$y_{max} = $ Maximum deflection of the cantilever. It will occur at a distance of $0.422\ l$ from the propped end.

3. When a simply supported beam AB with a *u.d.l.* is propped at centre then

$$P = \frac{5wl}{8}$$

$$i_B = i_A = \frac{wl^3}{384\,EI} \text{ radians}$$

$$y_{max} = \frac{0 \cdot 0003062\ wl^4}{EI}$$

where P = Prop reaction,

i_B = Slope at B, and·

y_{max} = Maximum deflection of the beam. It will occur at a distance of $0 \cdot 27\ l$ from A or B.

4. If a cantilever or a beam is supported on an elastic prop and it sinks down by δ due to the load on the prop, then

$$y_1 = y_2 + \delta$$

where y_1 = Downward deflection of the beam at the point of prop·

y_2 = Upward deflection of the beam due to prop reaction.

Do You Know ?

1 What do you understand by the term "prop" ? Discuss its importance.

2. Describe the procedure for finding out the prop reaction of a cantilever.

3. Derive an equation for the prop reaction in (a) a cantilever carrying a u.d.l. over the entire span and propped at the free end, and (b) a simply supported beam carrying a u.d.l. over the entire span and propped at the mid-span.

4. From first principles. derive a relation for the maximum deflection of a cantilever carrying a uniformly distributed load and propped at the free end.

5. Define 'sinking of a prop'. How does it differ from a rigid prop ?

6. Explain the procedure for finding out the reaction on an elastic prop.

12

FIXED BEAMS

12·1. Introduction

A beam, which is built in at its two supports, is called a constrained beam or a *fixed beam*. Since the beam is fixed at its two supports, therefore the slope of the elastic curve of the beam at its two ends, even after loading, will be zero. Thus, a fixed beam AB may be looked upon as a simply supported beam, subjected to end moments M_A and M_B, such that the slopes at two supports are zero. A little consideration will show, that this is only possible, if the direction of the restraining moments M_A and M_B are opposite to that of the bending moments under a given system of loading.

12·2. Advantages of fixed beams

A fixed beam has the following advantages, over a simply supported beam :

1. The beam is stiffer, stronger and more stable.
2. The slope at the two ends is zero.
3. The fixing moments are developed at the two ends, whose effect is to reduce the maximum bending moment at the centre of the beam.

4. The deflection of a beam, at its centre, is very much reduced.

12·3. Bending moment diagrams for fixed beams

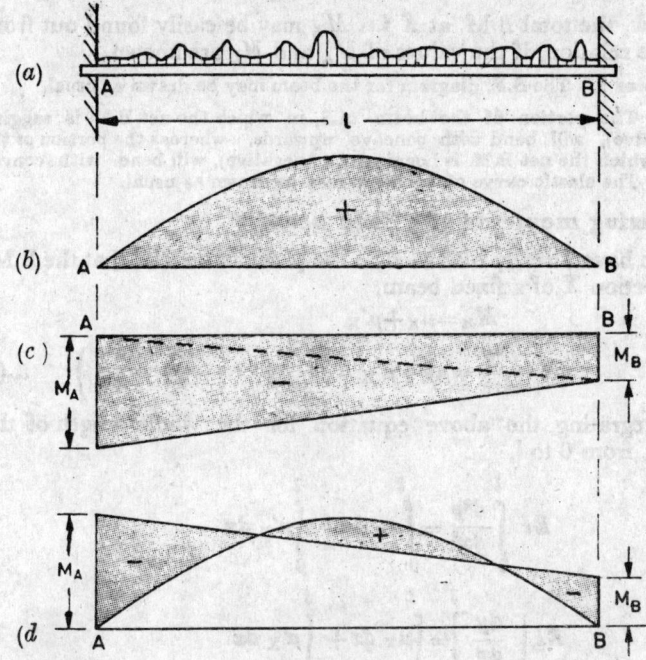

Fig. 12·1

Consider a fixed beam AB, of span l subjected to various types of loading as shown in Fig. 12·1 (a). Now let us analyse the beam into the following two categories :

1. A simply supported beam AB subjected to vertical loads and reactions.

2. A simply supported beam AB subjected to end moments.

The complete B.M. diagram may be drawn by superimposing the B.M. diagrams for the above two cases. We know that the beam AB, when treated as a simply supported beam carrying vertical loads and reactions, will be subjected to positive B.M. (*i.e.*, sagging) as shown in Fig. 12·1 (b). But the beam AB, when treated as a simply supported beam, having fixing moments M_A and M_B will be subjected to negative B.M. (*i.e.*, hogging) as shown in Fig. 12·1 (c). Since the direction of the above two moments are opposite to each other, therefore, their combining effect will be seen by drawing the two moments, on the same side of the base AB as shown in Fig. 12·1 (d).

Now consider any section X, at a distance x from A. Let the B.M. due to vertical loading be μ. The B.M. due to fixing moments M_A and M_B at X,

$$\mu'x = - \left[M_A + (M_B + M_A)\frac{x}{l} \right]$$

∴ **Total B.M. at X,**

$$M_X = \mu_X + \mu'_X = \mu_X - \left[M_A + (M_B - M_A)\frac{x}{l} \right]$$

Now, the total B.M. at X *i.e.* M_X may be easily found out from the above relation, if the values of M_A and M_B are known.

Notes 1. The S.F. diagram for the beam may be drawn as usual.

2. The portion of the beam AB, in which the net B.M. is sagging (*i.e.* positive), will bend with concave upwards, whereas the portion of the beam, in which the net B.M. is hogging (*i.e.* negative), will bend with convex upwards. The elastic curve of the beam may be drawn as usual.

12·4. Fixing moments of a fixed beam

We have already discussed in the previous article, that the B.M. at any section X of a fixed beam,

$$M_X = \mu_X + \mu'_X$$

∴ $$EI\frac{d^2y}{dx^2} = \mu_X + \mu'_X \qquad \left(\because \frac{M}{EI} = \frac{d^2y}{dx^2} \right) \quad ...(i)$$

Integrating the above equation for the whole length of the beam *i.e.* from 0 to l,

$$EI\int_0^l \frac{d^2y}{dx^2} = \int_0^l \mu_X \, dx + \int_0^l \mu'_X \, dx$$

$$EI\left[\frac{dy}{dx} \right]_0^l = \int_0^l \mu_X \, dx + \int_0^l \mu'_X \, dx$$

$$EI(i_B - i_A) = a + a' \qquad ...(ii)$$

where
i_B = Slope of the beam at B,

i_A = Slope of the beam at A,

a = Area of the μ diagram, and

a' = Area of the μ'-diagram

Since the slopes at A and B (*i.e.* i_A and i_B) are zero, therefore

$$a + a' = 0 \quad \text{or} \quad a = -a'$$

We know, that the shape of μ'-diagram is trapezoidal having end ordinates equal to M_A and M_B.

∴ Area μ'-diagram,

$$a' = \frac{l}{2}(M_A + M_B)$$

or $$\frac{l}{2}(M_A + M_B) = -a \qquad (\because a = -a')$$

∴ $$M_A + M_B = -\frac{2a}{l} \qquad ...(iii)$$

From equation (i) we know that

$$EI \frac{d^2y}{dx^2} = \mu_x + \mu' x$$

Multiplying the above equation by x, and integrating the same for the whole length of the beam i.e. from 0 to l.

$$EI \int_0^l x \frac{d^2y}{dx^2} = \int_0^l x.\mu_x \, dx + \int_0^l x.\mu' x \, dx$$

$$EI \left[x \frac{dy}{dx} - y \right]_0^l = a\bar{x} + a'\bar{x}'$$

or

$$EI [l(i_B - y_B) - 0(i_A - y_A)] = a\bar{x} + a'\bar{x}'$$

Since i_B and y_B are equal to zero, therefore

$$a\bar{x} + a'\bar{x}' = 0 \qquad \text{or} \qquad a\bar{x} = -a'\bar{x}'$$

where

$\bar{x} = $ Distance of centre of gravity of μ-diagram from A, and

$\bar{x}' = $ Distance of centre of gravity of μ'-diagram from A.

We know, that the shape of the μ'-diagram is trapezoidal, having end ordinates equal to M_A and M_B. Therefore splitting up the μ'-diagram into two triangles as shown in Fig. 12·1 (c).

$$a'\bar{x}' = \left(M_A \times \frac{l}{2} \times \frac{l}{3} \right) + \left(M_B \times \frac{l}{2} \times \frac{2l}{3} \right)$$

$$= (M_A + 2M_B) \frac{l^2}{6}$$

or

$$(M_A + 2M_B) \frac{l^2}{6} = -a\bar{x}$$

$$\therefore \quad M_A + 2M_B = -\frac{6a\bar{x}}{l^2} \qquad \qquad ...(iv)$$

Now subtracting equation (iii) from (iv),

$$M_B = -\frac{6a\bar{x}}{l^2} + \frac{2a}{l} = \frac{2a}{l^2}(-3\bar{x} + l)$$

and substituting the value of M_B in equation (iii),

$$M_A + \frac{2a}{l^2}(-3x + l) = -\frac{2a}{l}$$

or

$$M_A = -\frac{2a}{l} - \frac{2a}{l^2}(-3\bar{x} + l)$$

$$= -\frac{2a}{l^2}[l + (-3x + l)]$$

$$= -\frac{2a}{l^2}(2l - 3\bar{x})$$

These are the required equations for the fixing moments M_A and M_B of a fixed beam AB. Here we shall discuss the following standard cases for the fixing moments.

1. A fixed beam carrying a central point load.

2. A fixed beam carrying an eccentric point load.

3. A fixed beam carrying a uniformly distributed load.

4. A fixed beam carrying a gradually varying load from zero at one end to w per unit length at the other end.

12·5. Fixing moments of a fixed beam carrying a central point load

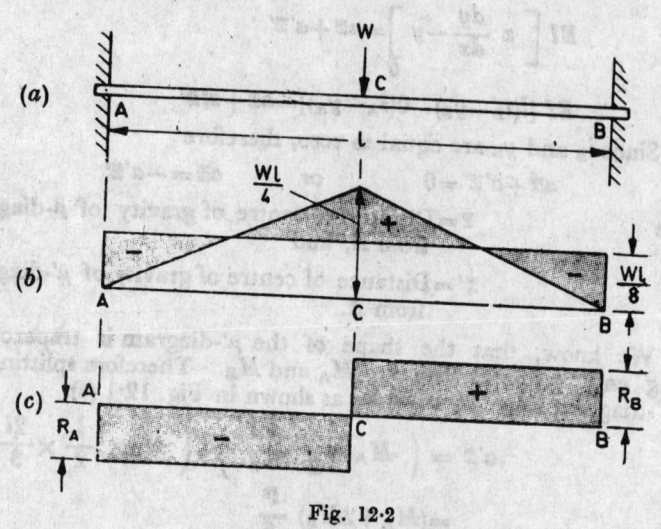

Fig. 12·2

Consider a beam AB of length l fixed at A and B, and carrying a central point load W as shown in Fig. 12·2 (a).

(i) B.M. Diagram

Let M_A=Fixing moment at A, and

M_B=Fixing moment at B.

Since the beam is symmetrical, therefore M_A and M_B will also be equal. Moreover the μ'-diagram will be a rectangle as shown in Fig. 12·2 (b). We know that μ-diagram will be a triangle with the central ordinate equal to $\dfrac{Wl}{4}$ as shown in Fig. 12·2 (b).

Now equating the areas of the two diagrams,

$$M_A.l = -\frac{1}{2}.l.\frac{Wl}{4} = -\frac{Wl^2}{8}$$

$$\therefore \quad M_A = -\frac{Wl}{8}. \quad \text{Similarly } M_B = -\frac{Wl}{8}$$

We know that the B.M. under the load, treating the beam as a simply supported

$$= \frac{Wl}{4}$$

Now complete the B.M. diagrams as shown in Fig. 12·2 (b).

S.F. diagram

Let $R_A =$ Reaction at A, and

$R_B =$ Reaction at B.

Equating the clockwise moments and anticlockwise moments about A,

$$R_B \times l + M_A = M_B + W \times \frac{l}{2}$$

$$R_B = \frac{W}{2}$$

Similarly $\qquad R_A = \frac{W}{2}$ \hfill (By symmetry)

Now complete the S.F. diagram as shown in Fig. 12·2 (c).

Deflection of the beam

From the geometry of the figure, we find that the *points of contraflexure will be at a distance of $l/4$ from both the ends of the beam.

We know that B.M. at any section X, at a distance x from A,

$$M_X = \mu x - \mu' x$$

or $\qquad EI \frac{d^2y}{dx^2} = \frac{Wx}{2} - \frac{Wl}{8}$ \hfill ...(i)

Integrating the above equation,

$$EI \frac{dy}{dx} = \frac{Wx^2}{4} - \frac{Wlx}{8} + C_1$$

where C_1 is the constant of integration. We know that when $x = 0$, $\frac{dy}{dx} = 0$. Therefore $C_1 = 0$.

or $\qquad EI \frac{dy}{dx} = \frac{Wx^2}{4} - \frac{Wlx}{8}$ \hfill ...(ii)

This is the required equation for the *slope* of the beam, at any section.

Integrating the equation (ii) once again,

$$EI.y = \frac{Wx^3}{12} - \frac{Wlx^2}{16} + C_2$$

---

*It is a point, where B.M. changes sign.

where C_2 is the constant of integration. We know that when $x=0$, $y=0$. Therefore $C_2=0$.

or
$$EI.y = \frac{Wx^3}{12} - \frac{Wlx^2}{16} \qquad \text{...(iii)}$$

This is the required equation for the *deflection* of the beam at any section.

We know that the maximum deflection occurs at the centre of the beam. Therefore substituting $x=l/2$ in the above equation,

$$EI.y_C = \frac{W}{12}\left(\frac{l}{2}\right)^2 - \frac{Wl}{16}\left(\frac{l}{2}\right)^2$$

$$= \frac{Wl^3}{96} - \frac{Wl^3}{64} = -\frac{Wl^3}{192}$$

or
$$y_C = -\frac{Wl^3}{192EI} \qquad \text{... (Minus sign means that the deflection is downwards)}$$

$$= \frac{Wl^3}{192EI}$$

Note. The term EI is known as flexural rigidity.

Example 12·1. *A fixed beam AB, 4 metres long, is carrying a central point load of 3 tonnes. Determine the fixing moments, and deflection of the beam under the load. Take flexural rigidity of the beam as 5×10^{10} kg-cm².*

Solution.

Given. Span, $\qquad l=4$ m

Load, $\qquad W=3$ t

Flexural rigidity, $EI=5 \times 10^{10}$ kg-cm²$=5 \times 10^6$ kg-m²

$$=5 \times 10^3 \text{ t-m}^2$$

Fixing moments

Let $\qquad M_A=$Fixing moment at A,

$\qquad M_B=$Fixing moment at B, and

$\qquad M_C=$Bending moment at C.

Using the relation,

$$M_C = -\frac{Wl}{8} \quad \text{with usual notations.}$$

$$= -\frac{3 \times 4}{8} = -\textbf{1·5 t-m} \quad \textbf{Ans.}$$

Similarly $\qquad M_A=M_C=-\textbf{1·5 t-m} \quad \textbf{Ans.}$

We know that the B.M. at the centre of the beam, treating it as a simply supported

$$= \frac{Wl}{4} = \frac{3 \times 4}{4} = 3·0 \text{ t-m}$$

Deflection of the beam under the load

Let $\qquad y_C=$Deflection of the beam under the load.

Using the relation,

$$y_C = \frac{Wl^3}{192EI} \quad \text{with usual notations.}$$

$$= \frac{3 \times 4^3}{192 \times 5 \times 10^3} = 0 \cdot 0002 \text{ m} = 0 \cdot 02 \text{ cm}$$

$$= 0 \cdot 2 \text{ mm} \quad \text{Ans.}$$

12·6. Fixing moments of a fixed beam carrying an eccentric point load

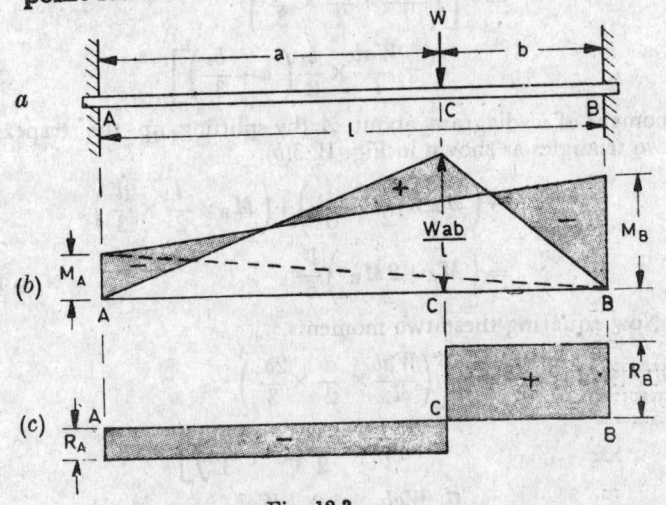

Fig. 12·3

Consider a beam AB fixed at A and B, and carrying an eccentric point load as shown in Fig. 12·3 (a).

Let l = Span of the beam,

         W = Load on the beam,

         a = Distance between the fixed end A, and the axis of the load,

         b = Distance between the fixed end B, and the axis of the load,

         M_A = Fixing moment at A, and

         M_B = Fixing moment at B.

B. M. diagram

Since the beam is not symmetrical, therefore M_A and M_B will also not be equal. Moreover the μ'-diagram will be a trapezium as shown in Fig. 12·3 (b).

We know that the μ-diagram will be a triangle, with ordinate equal to $\dfrac{Wab}{l}$ as shown in Fig. 12·3 (b).

Now equating the areas of the two diagrams,

$$(M_A + M_B)\frac{l}{2} = -\frac{Wab}{l} \cdot \frac{l}{2}$$

$$\therefore \qquad M_A + M_B = -\frac{Wab}{l} \qquad\qquad \ldots(i)$$

The moment of μ-diagram area about A (by splitting up the triangle into two right angled triangles

$$= -\left[\left(\frac{Wab}{l} \times \frac{a}{2} \times \frac{2a}{3}\right)\right.$$

$$\left. + \frac{Wab}{l} \times \frac{b}{2}\left(a + \frac{b}{3}\right)\right] \qquad\qquad \ldots(ii)$$

and moment of μ'-diagram about A (by splitting up the trapezium into two triangles as shown in Fig. 12·3(b).

$$= \left(M_A \times \frac{l}{2} \times \frac{l}{3}\right) + \left(M_B \times \frac{l}{2} \times \frac{2l}{3}\right)$$

$$= \left(M_A + 2M_B\right)\frac{l^2}{6} \qquad\qquad \ldots(iii)$$

Now equating these two moments,

$$\left(M_A + 2M_B\right)\frac{l^2}{6} = -\left[\left(\frac{Wab}{l} \times \frac{a}{2} \times \frac{2a}{3}\right)\right.$$

$$\left. + \frac{Wab}{l} \times \frac{b}{2}\left(a + \frac{b}{3}\right)\right]$$

$$= -\left[\left(\frac{Wab}{6l} \times 2a^2\right) + \frac{Wab}{2l}\left(ab + \frac{b^2}{3}\right)\right]$$

$$= -\left[\left(\frac{Wab}{6l} \times 2a^2\right) + \frac{Wab}{2l}\left(\frac{3ab + b^2}{3}\right)\right]$$

$$= -\left[\left(\frac{Wab}{6l} \times 2a^2\right) + \frac{Wab}{6l}\left(3ab + b^2\right)\right]$$

$$= -\frac{Wab}{6l}\left(2a^2 + 3ab + b^2\right)$$

$$= -\frac{Wab}{6l}\left[2(l-b)^2 + 3(l-b)b + b^2\right] \qquad (\because\ a+b=l)$$

$$= -\frac{Wab}{6l}\left[2(l^2 + b^2 - 2lb) + 3(lb - b^2) + b^2\right]$$

$$= -\frac{Wab}{6l}\left[2l^2 + 2b^2 - 4lb + 3lb - 3b^2 + b^2\right)$$

$$= -\frac{Wab}{6l}(2l^2 - lb) = -\frac{Wab}{6}(2l - b)$$

$$= -\frac{Wab}{6}\left[2(a+b) - b\right] \qquad (\because\ a+b=l)$$

$$= -\frac{Wab\,(2a+b)}{6}$$

$$\therefore\ M_A + 2M_B = -\frac{W\,b\,(2a+b)}{l^2} \qquad \ldots(iv)$$

Subtracting equation (i) from (iv),

$$M_B = -\frac{Wab(2a+b)}{l^2} + \frac{Wab}{l}$$

$$= -\frac{Wab}{l}\left(\frac{2a+b}{l}-1\right) = -\frac{Wab}{l}\left(\frac{2a+b-l}{l}\right)$$

$$= -\frac{Wab}{l^2}[2a+(l-a)-l]$$

$$= -\frac{Wa^2b}{l^2}$$

Substituting this value of M_B in equation (i),

$$M_A - \frac{Wa^2b}{l^2} = -\frac{Wab}{l}$$

$$\therefore\qquad M_A = -\frac{Wab}{l} + \frac{Wa^2b}{l^2} = -\frac{Wab}{l}\left(1+\frac{a}{l}\right)$$

$$= -\frac{Wab}{l}\left(\frac{l-a}{l}\right)$$

$$= -\frac{Wab^2}{l^2} \qquad (\because\ l-a=b)$$

Now complete the B. M. diagram as shown in Fig. 12·3 (b).

Deflection of the beam

We know, that the B.M. at any section X, at a distance x from A,

$$M_A = \mu_X - \mu'_X$$

or
$$EI\frac{d^2y}{dx^2} = \frac{Wb}{l}x - \left[M_A + (M_B - M_A)\frac{x}{l}\right]$$

$$= \frac{Wbx}{l} - \left[\frac{Wab^2}{l^2} + \left(\frac{Wa^2b}{l^2}-\frac{Wab^2}{l^2}\right)\frac{x}{l}\right]$$

$$= \frac{Wbx}{l} - \left[\frac{Wab^2}{l^2} + \frac{Wab\,(a-b)x}{l^3}\right]$$

$$= \frac{Wbx}{l} - \frac{Wab^2}{l^2} - \frac{Wab\,(a-b)x}{l^3}$$

Integrating the above equation,

$$EI\frac{dy}{dx} = \frac{Wbx^2}{2l} - \frac{Wab^2x}{l^2} - \frac{Wab\,(a-b)x^2}{2l^3} + C_1$$

where C_1 is the constant of integration. We know that when $x=0$, $\frac{dy}{dx}=0$. Therefore $C_1=0$.

or
$$EI \cdot \frac{dy}{dx} = \frac{Wbx^2}{2l} - \frac{Wab^2x}{l^2} - \frac{Wab\,(a-b)x^2}{2l^3}$$

$$= \frac{Wbx^2}{2l}\left(1 - \frac{a(a-b)}{l^2}\right) - \frac{Wab^2x}{l^2}$$

$$= \frac{Wbx^2}{2l}\left(\frac{l^2 - a^2 + ab}{l^2}\right) - \frac{Wab^2x}{l^2}$$

$$= \frac{Wbx^2}{2l^3}\left((a+b)^2 - a^2 + ab\right) - \frac{Wab^2x}{l^2}$$

$$= \frac{Wbx^2}{2l^3}\left(a^2 + b^2 + 2ab - a^2 + ab\right) - \frac{Wab^2x}{l^2}$$

$$= \frac{Wbx^2}{2l^3}\left(3ab + b^2\right) - \frac{Wab^2x}{l^2}$$

$$= \frac{Wb^2x^2\,(3a+b)}{2l^3} - \frac{Wab^2x}{l^2} \qquad \ldots(v)$$

Integrating the above equation once again,

$$EI \cdot y = \frac{Wb^2x^3\,(3a+b)}{6l^3} - \frac{Wab^2x^2}{2l^2} + C_2$$

where C_2 is the constant of integration. We know that when $x=0$, $y=0$. Therefore $C_2=0$.

or
$$EI \cdot y = \frac{Wb^2x^3\,(3a+b)}{6l^3} - \frac{Wab^2x^2}{2l^2}$$

$$= \frac{Wb^2x^2}{6l^3}\left[x(3a+b) - 3al\right] \qquad \ldots(vi)$$

We know that for maximum deflection, $\dfrac{dy}{dx}$ should be equal to zero. Therefore equating the equation (v) to zero,

$$\frac{Wb^2x^2\,(3a+b)}{2l^3} - \frac{Wab^2x}{l^2} = 0$$

$$\therefore \qquad x = \frac{2al}{(3a+b)}$$

Substituting this value of x in equation (vi),

$$EI \cdot y_{max} = \frac{Wb^2}{6l^3}\left(\frac{2al}{3a+b}\right)^2\left[\frac{2al}{(3a+b)}\,(3a+b) - 3al\right]$$

$$= \frac{Wb^2}{6l^3} \times \frac{4a^2l^2}{(3a+b)^2}\,(2al - 3al)$$

$$= -\frac{2}{3} \times \frac{Wa^3b^2}{(3a+b)^2}$$

$$\therefore \qquad y_{max} = -\frac{2}{3} \times \frac{Wa^3b^2}{(3a+b)^2 EI}$$

The deflection under the load may be found out by substituting $x=a$ in equation (vi).

$$EI.y = \frac{Wb^2a^2}{6l^3} [a(3a+b) - 3al]$$

$$= \frac{Wb^2a^2}{6l^3} [a(3a+b) - 3a(a+b)] \qquad [\because \quad l = a+b]$$

$$= -\frac{Wb^2a^2}{6l^3} [3a^2 + ab - 3a^2 - 3ab] = -\frac{Wa^3b^3}{3l^3}$$

$$y = -\frac{Wa^3b^3}{3l^3EI} \qquad \begin{array}{l} \text{(minus sign indicates that} \\ \text{the deflection is downwards)} \end{array}$$

$$= \frac{Wa^3b^3}{3l^3EI}$$

Now complete the S. F. diagram as shown in **Fig. 12·3** (c) as usual.

Example 12·2. *A fixed beam AB of 5 m span carries a point load of 2 tonnes at a distance of 2 m from A. Determine the values of fixing moments and the deflection under the load, if flexural rigidity the beam is* $1·0 \times 10^{10}$ *kg-cm.*^{2}

Solution.

Given.   Span, $l = 5$ m $= 500$ cm

Load, $W = 2$ t $= 2,000$ kg

Distance of load from A,

$$a = 2 \text{ m} = 200 \text{ cm}$$

$\therefore$ Distance of load from B,

$$b = 5 - 2 = 3 \text{ m} = 300 \text{ cm}$$

Flexural rigidity, $EI = 1·0 \times 10^{10}$ kg-cm^2

Fixing moments

Let $M_A = $Fixing moment at A, and

$M_B = $Fixing moment at B.

Using the relation,

$$M_A = -\frac{Wab^2}{l^2} \text{ with usual notations.}$$

$$= -\frac{2 \times 2 \times 3^2}{5^2} = -1·44 \text{ t-m} \quad \textbf{Ans.}$$

Now using the relation,

$$M_B = -\frac{Wa^2b}{l^2} \text{ with usual notations.}$$

$$= -\frac{2 \times 2^2 \times 3}{5^2} = -0·96 \text{ t-m Ans.}$$

Deflection under the load

Let $y = $Deflection under the load.

Using the relation,

$$y = \frac{W a^3 b^3}{3 l^3 E I} \text{ with usual notations.}$$

$$= \frac{2,000 \times 200^3 \times 300^3}{3 \times 500^3 \times 1 \cdot 0 \times 10^{10}} \text{ cm}$$

$$= \mathbf{0 \cdot 115 \text{ cm}} \quad \mathbf{Ans.}$$

Example 12·3. *A beam of span l is fixed at its both ends. It carries two concentrated loads of W each at a distance of l/3 from both the ends. Find out the fixing moments, and draw the bending moment diagram.*

Solution

Given.   Span $= l$

Load $= W$

Let M_A = Fixing moment at A, and

                    M_B = Fixing moment at B.

For the sake of convenience, let us first find out the fixing moments, separately due to loads at C and D, and then add up the

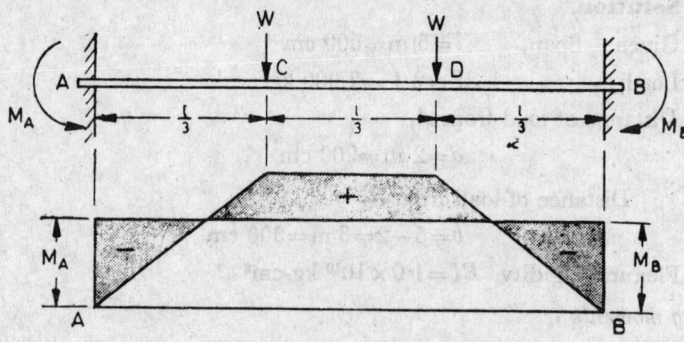

Fig. 12·4

moments. Since the beam and the loading is symmetrical, therefore both the fixing moments must be equal. Now consider the load W at C. From the geometry of the figure, we find that $a = \dfrac{l}{3}$ and $b = \dfrac{2l}{3}$

Using the relation,

$$M_A = -\frac{W a b^2}{l^2} \text{ with usual notations.}$$

$$M_{A_1} = -\frac{W \times \dfrac{l}{3} \times \left(\dfrac{2l}{3}\right)^2}{l^2} = -\frac{4Wl}{27} \qquad \text{...(}i\text{)}$$

Now consider the load W at D. From the geometry of the figure, we find that $a = \dfrac{2l}{3}$ and $b = \dfrac{l}{3}$

Again using the relation,

$$M_A = -\frac{Wab^2}{l^2} \text{ with usual notations.}$$

$$\therefore \quad M_{A2} = -\frac{W \times \frac{2l}{3} \times \left(\frac{l}{3}\right)^2}{l^2} = -\frac{2Wl}{27} \qquad ...(ii)$$

$\therefore$ Total fixing moment at A,

$$M_A = M_B = M_{A1} + M_{A2} = -\left(\frac{4Wl}{27} + \frac{2Wl}{27}\right)$$

$$= -\frac{6Wl}{27} = -\frac{2Wl}{9} \quad \textbf{Ans.}$$

We know that when the beam is considered as a simply supported, the reaction at A,

$$R_A = W$$

$\therefore$ Bending moment at C,

$$M_C = R_A \times \frac{l}{3} = W \times \frac{l}{3}$$

Now complete the bending moment diagram as shown in Fig. 12·4 (b).

12·7. Fixing moments of a fixed beam carrying a uniformly distributed load

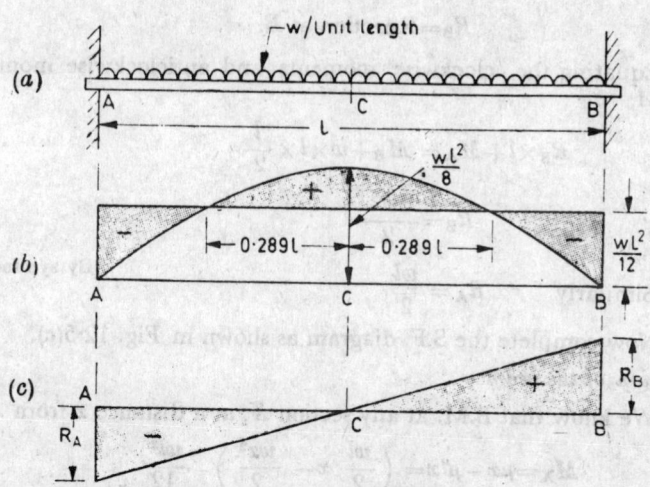

Fig. 12·5

Consider a beam AB of length l fixed at A and B, and carrying a uniformly distributed load w per unit length over the entire span as shown in Fig. 12·5 (a).

B.M. Diagram

Let M_A = Fixing moment at A, and

                       M_B = Fixing moment at B.

Since the beam is symmetrical, therefore M_A and M_B will also be equal. Moreover the μ'-diagram will be a rectangle, as shown in Fig. 12·5 (*b*). We know that the μ-diagram will be a parabola with the central ordinate equal to $\dfrac{wl^2}{8}$ as shown in Fig. 12·5(*b*).

Now equating the areas of the two diagrams,

$$M_A \cdot l = -\frac{2}{3} \cdot l \cdot \frac{wl^2}{8} = -\frac{wl^3}{12}$$

$$\therefore \qquad M_A = -\frac{wl^2}{12}$$

Similarly $M_B = -\dfrac{wl^2}{12}$

We know that the B.M. at the mid of the beam, treating it as a simply supported

$$= \frac{wl^2}{8}$$

Now complete the B.M. diagram as shown in Fig. 12·5(*b*).

S.F. diagram

Let R_A = Reaction at A, and

                       R_B = Reaction at B.

Equating the clockwise moments and anticlockwise moments about A,

$$R_B \times l + M_A = M_B + w \times l \times \frac{l}{2}$$

$$\therefore \qquad R_B = \frac{wl}{2}$$

Similarly $R_A = \dfrac{wl}{2}$ (By symmetry)

Now complete the S.F. diagram as shown in Fig. 12·5(*c*).

Deflection of the beam

We know that B.M. at any section X, at a distance x from A,

$$M_X = \mu x - \mu' x = \left(\frac{wl}{2} x - \frac{wx^2}{2}\right) - \frac{wl^2}{12}$$

The points of contraflexures may be found out by equating the above equation to zero, *i.e.*,

$$\frac{wl}{2} x - \frac{wx^2}{2} - \frac{wl^2}{12} = 0$$

or
$$lx - x^2 - \frac{l^2}{6} = 0$$

$$\therefore \qquad x^2 - lx + \frac{l^2}{6} = 0$$

Solving this quadratic equation for x,

$$x = \frac{l \pm \sqrt{l^2 - \frac{4l^2}{6}}}{2} = \frac{l}{2} \pm \frac{l}{2\sqrt{3}}$$

$$= 0.5\,l \pm 0.289\,l = 0.789\,l \text{ and } 0.211\,l$$

We know that the B.M. at X, at a distance x from A,

$$M_X = \mu_X - \mu'_x$$

$$\therefore \quad EI\,\frac{d^2y}{dx^2} = \left(\frac{wlx}{2} - \frac{wx^2}{2}\right) - \frac{wl^2}{12} \qquad \qquad \dots(i)$$

Integrating the above equation,

$$EI\,\frac{dy}{dx} = \frac{wlx^2}{4} - \frac{wx^3}{6} - \frac{wl^2x}{12} + C_1$$

where C_1 is the constant of integration. We know that when $x = 0$, $\frac{dy}{dx} = 0$. Therefore $C_1 = 0$.

or
$$EI\,\frac{dy}{dx} = \frac{wlx^2}{4} - \frac{wx^3}{6} - \frac{wl^2x}{12} \qquad \qquad \dots(ii)$$

Integrating the equation (ii) once again,

$$EI \cdot y = \frac{wlx^3}{12} - \frac{wx^4}{24} - \frac{wl^2x^2}{24} + C_2$$

where C_2 is the constant of integration. We know that when $x = 0$, $y = 0$. Therefore $C_2 = 0$.

or
$$EI \cdot y = \frac{wlx^3}{12} - \frac{wx^4}{24} - \frac{wl^2x^2}{24} \qquad \qquad \dots(iii)$$

We know that the maximum deflection occurs at the centre of the beam. Therefore substituting $x = \dfrac{l}{2}$ in the above equation,

$$EI \cdot y_C = \frac{wl}{48}\left(\frac{l}{2}\right)^3 - \frac{w}{24}\left(\frac{l}{2}\right)^4 - \frac{wl^2}{24}\left(\frac{l}{2}\right)^2$$

$$= \frac{wl^4}{96} - \frac{wl^4}{384} - \frac{wl^4}{96}$$

$$= -\frac{wl^4}{384}$$

or
$$y_C = -\frac{wl^4}{384\,EI} \qquad \qquad \text{(Minus sign means that the deflection is downwards)}$$

$$= \frac{wl^4}{384\,EI}$$

Example 12·4. *An encastre beam AB 2 m long is subjected to uniformly distributed load of 2 t/m over the entire length. Determine the values of maximum positive and negative bending moments. Also calculate the maximum deflection of the beam. Take flexural rigidity of the beam as $1·0 \times 10^{10}$ kg-cm².*

Solution.

Given.   Span $l=2$ m $=200$ cm

Load, $w=2$ t/m $=2,000$ kg/m $=20$ kg/cm

Flexural rigidity,

$$EI = 1·0 \times 10^{10} \text{ kg-cm}^2$$

Maximum negative bending moment

Let $M_A =$ Bending moment at A.

We know that the maximum negative bending moment on the beam is the fixing moment M_A or M_B.

Using the relation,

$$M_A = -\frac{wl^2}{12} \text{ with usual notations.}$$

$$= -\frac{2 \times 2^2}{12} = -0·67 \text{ t-m   Ans.}$$

Maximum positive bending moment

We know that the B.M. at the centre of the beam, treating it as a simply supported.

$$M_C = \frac{wl^2}{8} = \frac{2 \times 2^2}{8} = 1·0 \text{ t-m}$$

∴   Maximum positive bending moment,

$$= M_C - M_A = 1 - 0·67 \text{ t-m}$$

$$= 0·33 \text{ t-m   Ans.}$$

Maximum deflection of the beam

Let $y_C =$ Maximum deflection of the beam.

Using the relation,

$$y_C = \frac{wl^4}{384 \ EI} \text{ with usual notations.}$$

$$= \frac{20 \times 200^4}{384 \times 1·0 \times 10^{10}} = 0·008 \text{ cm}$$

$$= 0·08 \text{ mm   Ans.}$$

Example 12·5. *A fixed beam AB of span 6 m is carrying a uniformly distributed load of 4 t/m over the left half of the span. Find the fixing moments and support reactions.*

Solution.

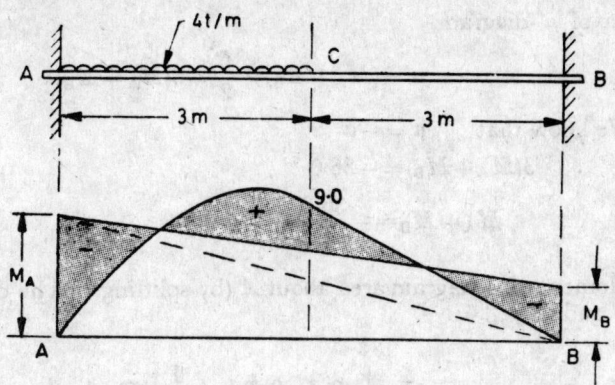

Fig. 12·6

Given. Span, $l = 6$ m

Load, $w = 4$ t/m

Let M_A = Fixing moment at A, and

M_B = Fixing moment at B.

Now consider the beam AB as a simply supported. Taking moments about A,

$$V_B \times 6 = 4 \times 3 \times 1·5 = 18$$

$$\therefore \qquad V_B = \frac{18}{6} = 3 \text{ T}$$

and $\qquad V_A = 3 \times 4 - 3 = 9 \text{ T}$

We know that μ-diagram will be parabolic from A to C and triangular from C to B as shown in Fig. 12·6 (b). The B.M. at C (treating the beam as a simply supported),

$$M_C = R_B \times 3 = 3 \times 3 = 9·0 \text{ t-m}$$

The B.M. at any section X in AC, at a distance x from A (treating the beam as a simply supported)

$$M_X = 9x - 4x \cdot \frac{x}{2} = 9x - 2x^2$$

$\therefore$ Area of μ-diagram from A to B,

$$a = \int_0^3 (9x - 2x^2)dx + \tfrac{1}{2} \times 9·0 \times 3$$

$$= \left[\frac{9x^2}{2} - \frac{2x^3}{3} \right]_0^3 + 13·5$$

$$= \frac{9 \times 3^2}{2} - \frac{2 \times 3^3}{3} + 13 \cdot 5 = 36 \cdot 0$$

and area of μ'-diagram,

$$a' = (M_A + M_B) \times \frac{6}{2} = 3(M_A + M_B)$$

We know that $\quad a' = -a$

$$\therefore \quad 3(M_A + M_B) = -36 \cdot 0$$

or $\qquad M_A + M_B = -\frac{36 \cdot 0}{3} = -12 \cdot 0$

Moment of μ-diagram area about A (by splitting up the diagram into AC and CB).

$$a\overline{x} = \int_0^3 (9x^2 - 2x^3)dx + \frac{1}{2} \times 9 \times 3 \times 4$$

$$= \left[\frac{9x^3}{3} - \frac{2x^4}{4} \right]_0^3 + 54$$

$$= \left[\frac{9 \times 3^3}{3} - \frac{2 \times 3^4}{4} \right] + 54 = 94 \cdot 5$$

and moment of μ'-diagram area about A (by splitting up the trapezium into two triangles as shown in Fig. 12·6 (a).

$$a'\overline{x} = \left(M_A \times \frac{6}{2} \times \frac{6}{3} \right) + M_B \times \frac{6}{2} \times \frac{2 \times 6}{3}$$

$$= 6M_A + 12M_B = 6(M_A + 2M_B)$$

We know that $\quad a'\overline{x}' = -a\overline{x}$

$$6(M_A + 2M_B) = -94 \cdot 5$$

$$\therefore \qquad M_A + 2M_B = -\frac{94 \cdot 5}{6} = -15 \cdot 75 \qquad \qquad \dots(ii)$$

Solving equations (i) and (ii) we get,

$$M_A = -8 \cdot 25 \text{ t-m} \quad \textbf{Ans.}$$

$$M_B = -3 \cdot 75 \text{ t-m} \quad \textbf{Ans.}$$

Now complete the B.M. diagram as shown in Fig. 12·6 (b).

Support reactions.

Let $\qquad\qquad R_A =$ Reaction at A, and

$\qquad\qquad\qquad R_B =$ Reaction at B.

Equating the clockwise moments and anticlockwise moments about A,

$$R_B \times 6 + 8 \cdot 25 = 4 \times 3 \times 1 \cdot 5 + 3 \cdot 75$$

$$\therefore \qquad R_B = 2{\cdot}25 \text{ t}$$

and $\qquad\qquad R_A = 4 \times 3 - 2{\cdot}25 = 9{\cdot}75 \text{ t}$

Example 12·6. *A beam AB of uniform section and 6 m span is built-in at the ends. A uniformly distributed load of 3 t/m runs over the left half of the span, and there is in addition a concentrated load of 4 tonnes at right quarter as shown in Fig. 12·7.*

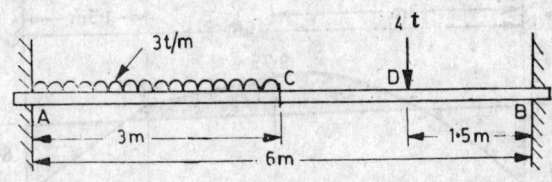

Fig. 12·7

Determine the fixing moments at the ends, and the reactions. Sketch neatly the bending moment and shearing force diagrams marking thereon sailent values. (U. (A.M.I.E. Summer 1977)

Solution.

Given. Span, $l = 6$ m

Load in AC, $w = 3$ t/m

Load at D, $W = 4$ t

Fixing moments at the ends

Let $\qquad\qquad M_A =$ Fixing moment at A, and

$\qquad\qquad M_B =$ Fixing moment at B.

Now consider the beam AB as a simply supported. Taking moments about A,

$$V_B \times 6 = 3 \times 3 \times 1{\cdot}5 + 4 \times 4{\cdot}5 = 31{\cdot}5$$

$$\therefore \qquad V_B = \frac{31{\cdot}5}{6} = 5{\cdot}25 \text{ t}$$

and $\qquad V_A = (3 \times 3 + 4) - 5{\cdot}25 = 7{\cdot}75 \text{ t}$

We know that the μ-diagram will be parabolic from A to C, trapezoidal from C to D and triangular from D to B as shown in Fig. 12·8 (*b*). The B.M. at D (treating the beam at a simply supported).

$$M_D = 5{\cdot}25 \times 1{\cdot}5 = 7{\cdot}875 \text{ t-m}$$

and $\qquad M_C = 5{\cdot}25 \times 3 - 4 \times 1{\cdot}5 = 9{\cdot}75 \text{ t-m}$

The B.M. at any section X in AC, at a distance x from A, (treating the beam as a simply supported),

$$M_X = 7.75x - 3x \frac{x}{2} = 7.75x - 1.5x^2$$

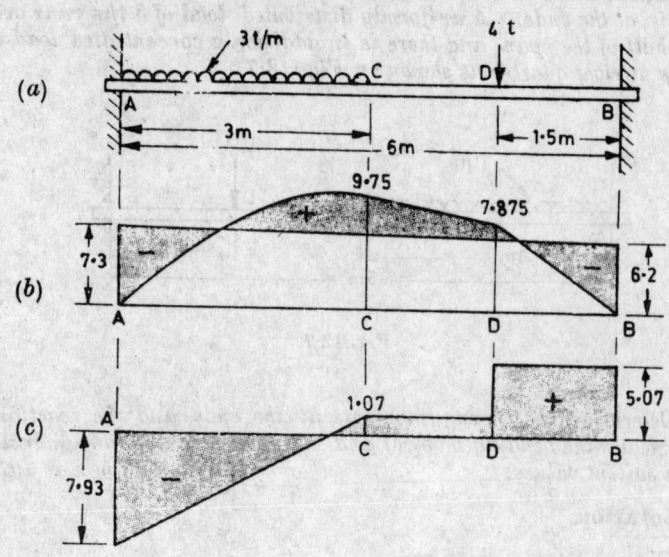

Fig. 12·8

Area of μ-diagram from A to B,

$$a = \int_0^3 (7.75x - 1.5x^2)\, dx + \frac{1}{2}(9.75 + 7.875) \times 1.5$$

$$+ \frac{1}{2} \times 7.875 \times 1.5$$

$$= \left[\frac{7.75x^2}{2} - \frac{1.5x^3}{3} \right]_0^3 + 19.175$$

$$= \frac{7.75 \times 3^2}{2} - \frac{1.5 \times 3^3}{3} + 19.175 = 40.5$$

and area of μ'-diagram,

$$a' = (M_A + M_B) \times \frac{6}{2} = 3(M_A + M_B)$$

We know that $a' = -a$

$\therefore \qquad 3(M_A + M_B) = -40.5 \qquad \qquad (\because \quad a = 40.5)$

or $$M_A + M_B = -13 \cdot 5 \qquad \qquad ...(i)$$

Moment of μ-diagram area about A (by splitting up the diagram into AC, CD and DB),

$$a\bar{x} = \int_0^3 (7 \cdot 75x^2 - 1 \cdot 5x^3)\, dx + \left(\frac{1}{2} \times 9 \cdot 75 \times 1 \cdot 5 \times 3 \cdot 5\right)$$

$$+ \left(\frac{1}{2} \times 7 \cdot 875 \times 1 \cdot 5 \times 4\right)$$

$$+ \left(\frac{1}{2} \times 7 \cdot 875 \times 1 \cdot 5 \times 5\right)$$

$$= \left[\frac{7 \cdot 75x^3}{3} - \frac{1 \cdot 5x^4}{4}\right] + 78 \cdot 75$$

$$= \frac{7 \cdot 75 \times 3^3}{3} - \frac{1 \cdot 5 \times 3^4}{4} + 78 \cdot 75 = 118 \cdot 2$$

and moment of μ'-diagram area about A (by splitting up the trapezium into two triangles).

$$a'\bar{x}' = \left(M_A \times \frac{6}{2} \times \frac{6}{3}\right) + \left(M_B \times \frac{6}{2} \times \frac{2 \times 6}{3}\right)$$

$$= 6M_A + 12M_B = 6(M_A + 2M_B)$$

We know that $a'\bar{x}' = -a\bar{x}$

$$\therefore \quad 6(M_A + 2M_B) = -118 \cdot 2 \qquad \qquad ...(ii)$$

or $$M_A + 2M_B = -\frac{118 \cdot 2}{6} = -19 \cdot 7$$

Solving equations (i) and (ii) we get,

$$M_A = -7 \cdot 3 \text{ t·m and } M_B = -6 \cdot 2 \text{ t·m}$$

Now complete the B.M. diagram as shown in Fig. 12·8 (b).

Shearing force diagram

Let $\qquad R_A = $ Reaction at A, and

$\qquad \qquad R_B = $ Reaction at B.

Equating the clockwise moments and anticlockwise moments about A,

$$R_B \times 6 + 7 \cdot 3 = 3 \times 3 \times 1 \cdot 5 + 4 \times 4 \cdot 5 + 6 \cdot 2$$

$$R_B = 5 \cdot 07 \text{ t}$$

and $$R_A = 3 \times 3 + 4 - 5 \cdot 07 = 7 \cdot 93 \text{ t}$$

Now complete the S.F. diagram as shown in Fig. 12·8 (c).

12·8. Fixing moments of a fixed beam carrying a gradually varying load from zero at one end to w per unit length at the other.

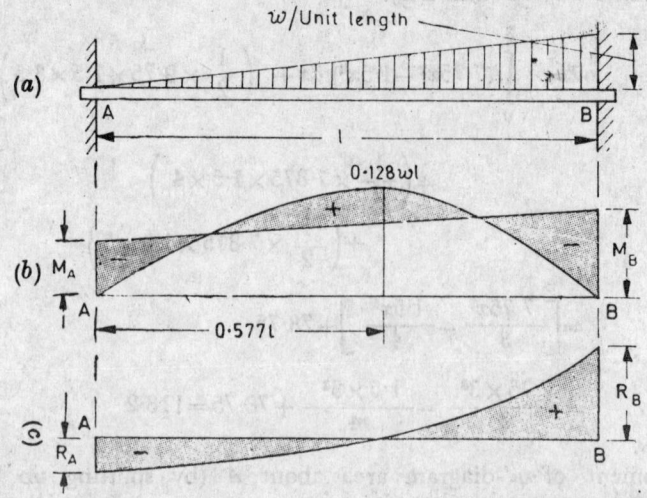

Fig. 12·9

Consider a beam AB of length l fixed at A and B, and carrying a gradually varying load from zero at A to w per unit length at B as shown in Fig. 12·9 (a).

Let $\qquad M_A =$ Fixing moment at A, and

$\qquad\qquad M_B =$ Fixing moment at B.

Now considering the beam AB as a simply supported, and taking moments about A,

$$V_B \times l = wl \times \frac{l}{2} \times \frac{2l}{3}$$

$$\therefore \qquad V_B = \frac{wl}{3}$$

and $\qquad V_A = \frac{wl}{2} - \frac{wl}{3} = \frac{wl}{6}$

We know that the μ-diagram will be parabolic from A to B. The B.M. at any section X, at a distance x from A (treating the beam as a simply supported),

$$M_X = \frac{wl}{6} \times x - \frac{wx}{l} \times \frac{x}{2} \times \frac{x}{3} = \frac{wlx}{6} - \frac{wx^3}{6l}$$

$\therefore$ Area of μ-diagram,

$$a = \int_0^l \left(\frac{wlx}{6} - \frac{wx^3}{6l} \right) dx$$

$$= \frac{w}{6} \int_0^l \left(lx - \frac{x^3}{l} \right) dx = \frac{w}{6} \left[\frac{lx^2}{2} - \frac{x^4}{4l} \right]_0^l$$

$$= \frac{w}{6} \left(\frac{l^3}{2} - \frac{l^3}{4} \right) = \frac{wl^3}{24}$$

and area of μ'-diagram,

$$a' = \frac{l}{2} (M_A + M_B)$$

We know that $a' = -a$

$$\therefore \qquad \frac{l}{2} (M_A + M_B) = -\frac{wl^3}{24}$$

or $$M_A + M_B = -\frac{wl^2}{12} \qquad \ldots(i)$$

Moment of μ-diagram area about A,

$$a\bar{x} = \int_0^l \left(\frac{wlx^2}{6} - \frac{wx^4}{6l} \right) dx$$

$$= \frac{w}{6} \int_0^l \left(lx^2 - \frac{x^4}{l} \right) dx = \frac{w}{6} \left[\frac{lx^3}{3} - \frac{x^5}{5l} \right]_0^l$$

$$= \frac{w}{6} \left(\frac{l^4}{3} - \frac{l^4}{5} \right) = \frac{wl^4}{45}$$

and moment of μ'-diagram about A (by splitting up the trapezium into two triangles).

$$a'\bar{x}' = M_A \times \frac{l}{2} \times \frac{l}{3} + M_B \times \frac{l}{2} \times \frac{2l}{3}$$

$$= M_A \times \frac{l^2}{6} + M_B \times \frac{l^2}{3}$$

$$= \frac{l^2}{6} \left(M_A + 2M_B \right)$$

We know that $a'\bar{x}' = -a\bar{x}$.

$$\therefore \qquad \frac{l^2}{6} \left(M_A + 2M_B \right) = -\frac{wl^4}{45}$$

or $$M_A + 2M_B = -\frac{2wl^2}{15} \qquad \ldots(ii)$$

Solving equations (*i*) and (*ii*),

$$M_A = -\frac{wl^2}{30} = -\frac{Wl}{15} \qquad \left(\because W = \frac{wl}{2}\right)$$

and

$$M_B = -\frac{wl^2}{20} = -\frac{Wl}{10} \qquad \left(\because W = \frac{wl}{2}\right)$$

Alternative method

Consider a strip of width dx at a distance x from the support A. We see that the lead at this section

$$=\frac{Wx}{l}$$

$\therefore$ Weight of the strip,

$$W = \frac{wx}{l} \cdot dx$$

We know that the fixing moment at A, due to the strip

$$= -\frac{Wab^2}{l^2} = -\frac{\dfrac{wx}{l}\, dx \cdot x(l-x)^2}{l^2}$$

$$= -\frac{wx^2(l-x)^2 dx}{l^3}$$

The total fixed end moment at A will be given by integrating the above equation from 0 to l,

$$M_A = \int_0^l -\frac{wx^2(l-x)^2 dx}{l^3}$$

$$= -\frac{w}{l^3}\int_0^l x^2(l^2 + x^2 - 2lx)dx$$

$$= -\frac{w}{l^3}\int_0^l (l^2x^2 + x^4 - 2lx^3)dx$$

$$= -\frac{w}{l^3}\left[\frac{l^2x^3}{3} + \frac{x^5}{5} - \frac{2lx^4}{4}\right]_0^l$$

$$= -\frac{w}{l^3}\left[\frac{l^5}{3} + \frac{l^5}{5} - \frac{l^5}{2}\right]$$

$$= -\frac{wl^2}{30} = -\frac{Wl}{15} \qquad \left(\because W = \frac{wl}{2}\right)$$

Similarly

$$M_B = \int_0^l -\frac{\dfrac{wx}{l}\, dx \cdot x^2(l-x)}{l^2}$$

$$M_B = -\frac{w}{l^3}\int_0^l x^3(l-x)dx$$

$$= -\frac{w}{l^3}\int_0^l (lx^3 - x^4)dx$$

$$= -\frac{w}{l^3}\left[\frac{lx^4}{4} - \frac{x^5}{5}\right]_0^l$$

$$= -\frac{w}{l^3}\left(\frac{l^5}{4} - \frac{l^5}{5}\right)$$

$$= -\frac{wl^2}{20} = -\frac{Wl}{10} \qquad \left(\because W = \frac{wl}{2}\right)$$

We know that the maximum bending moment equal to $0{\cdot}128\ Wl$ will take place at a distance of $0{\cdot}577\ l$ from A, treating the beam as a simply supported. Now complete the B.M. diagram as shown in Fig. 12·9(b).

S.F. diagram

Let
$$R_A = \text{Reaction at } A, \text{ and}$$
$$R_B = \text{Reaction at } B.$$

Equating the clockwise moments and anticlockwise moments about A,

$$R_B \times l + M_A = M_B + w \times \frac{l}{2} \times \frac{2l}{3}$$

$$= M_B + \frac{wl^2}{3}$$

$$R_B = \frac{M_B - M_A}{l} + \frac{wl}{3}$$

Similarly equating clockwise moments and anticlockwise moments about B,

$$R_A \times l + M_B = M_A + w \times \frac{l}{2} \times \frac{l}{3}$$

$$= M_A + \frac{wl^2}{6}$$

$$\therefore \qquad R_A = \frac{M_A - M_B}{l} + \frac{wl}{6}$$

Now complete the S.F. diagram as shown in Fig. 12·9 (c).

Example 12·7.   *A beam AB of span 5 m is built-in at its both ends. It carries a gradually varying load from zero at A to 4 t/m at B. Determine the fixed end moments and reactions at both ends of the beam.*

Solution

Given. Span AB, $l=5$ m

Load at B, $w=4$ t/m

Fixed end moments

Let M_A=Fixed end moment at A, and

M_B=Fixed end moment at B.

Using the relation

$$M_A=-\frac{wl^2}{30} \text{ with usual notations.}$$

$$=-\frac{4\times5^2}{30}=-3·33 \text{ t-m   Ans.}$$

Now using the relation,

$$M_B=-\frac{wl^2}{20} \text{ with usual notations}$$

$$=-\frac{4\times5^2}{20}=-5·0 \text{ t-m   Ans.}$$

Reactions

Let R_A=Reaction at A, and

R_B=Reaction at B.

Using the relation,

$$R_A=\frac{M_A-M_B}{l}+\frac{wl}{6} \text{ with usual notations.}$$

$$=\frac{3·33-5·0}{5}+\frac{4\times5}{6}=3·0 \text{ t   Ans.}$$

Now using the relation,

$$R_B=\frac{M_B-M_A}{l}+\frac{wl}{3} \text{ with usual notations.}$$

$$=\frac{5·0-3·33}{5}+\frac{4\times5}{3}=7·0 \text{ t   Ans.}$$

Exercise 12·1

1. A fixed beam of 2 m span is carrying a point load of 5 tonnes at its mid-point. Find the fixing moments and the deflection of beam at its mid-point. Take $EI=10·0\times10^{10}$ kg.-cm².

[**Ans.** $-1·2$ t-m ; $+1·2$ t-m ; $0·021$ cm]

2. A fixed beam AB of span 8 m carries a uniformly distributed load of 2 t/m from A to the mid-point of AB. Find the fixing moments at A and B.

(Bihar University)

[**Ans.** $-7·33$ t-m ; $-3·33$ t-m]

3. A built-in beam AB of span 6 m carries a concentrated load of 6 tonnes at 1·2 m and 12 tonnes at 4·5 m from A. Determine the fixing moments.

(*Roorkee University*)

[**Ans.** $-7·98$ t-m ; $-11·28$ t-m]

12·9. Fixing moments of a fixed beam due to sinking of a support

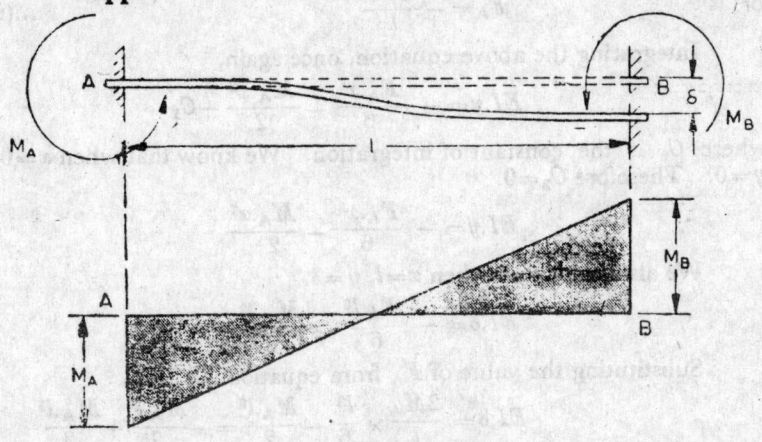

Fig. 12·10

Sometimes, one of the two supports sinks down, whereas the other remains at the same level. The effect of sinking of a support is to produce some fixing moment at the two supports. Consider a beam fixed at A and B without any load. Let the support B sink down from its original level as shown in Fig. 12·10 (*a*).

Let l = span of the fixed beam AB, and

                 δ = Amount, by which the support B has sunk down.

Taking A as the origin. We know that

$$EI \frac{d^4y}{dx^4} = 0 \qquad \text{(Since the beam is not loaded)}$$

Integrating the above equation,

$$EI \frac{d^3y}{dx^3} = -F_A \qquad (\because F_A \text{ is a negative})$$

where F_A is the shear force at A.

Integrating the above equation again,

$$EI \frac{d^2y}{dx^2} = -F_A x - M_A \qquad (\because M_A \text{ is negative})$$

where M_A is the bending moment at A.

Integrating the above equation once again,

$$EI \frac{dy}{dx} = -\frac{F_A \cdot x^2}{2} - M_A \cdot x + C_1$$

where C_1 is the constant of integration. We know that when $x = 0$, $\frac{dy}{dx} = 0$. Therefore $C_1 = 0$.

$$\therefore \qquad EI \frac{dy}{dx} = -\frac{F_A \cdot x^2}{2} - M_A \cdot x$$

We also know that when $x = l$, $\dfrac{dy}{dx} = 0$.

$$\therefore \quad 0 = -\frac{F_A \cdot l^2}{2} - M_A \cdot l$$

or

$$F_A = -\frac{2M_A}{l} \qquad \ldots(i)$$

Integrating the above equation, once again,

$$EI.y = -\frac{F_A \cdot x^3}{6} - \frac{M_A \cdot x^2}{2} + C_2$$

where C_2 is the constant of integration. We know that when $x = 0$, $y = 0$. Therefore $C_2 = 0$.

$$\therefore \quad EI.y = -\frac{F_A \cdot x^3}{6} - \frac{M_A \cdot x^2}{2}$$

We also know that when $x = l$, $y = \delta$.

$$\therefore \quad EI.\delta = -\frac{F_A \cdot l^3}{6} - \frac{M_A \cdot l^2}{2}$$

Substituting the value of F_A from equation (i),

$$EI.\delta = \frac{2M_A}{l} \times \frac{l^3}{6} - \frac{M_A \cdot l^2}{2} = -\frac{M_A \cdot l^2}{2} + \frac{M_A \cdot l^2}{3}$$

$$= -M_A \, l^2 \left(\tfrac{1}{2} - \tfrac{1}{3}\right) = -\frac{M_A \cdot l^2}{6}$$

$$\therefore \quad M_A = -\frac{6EI\delta}{l^2}$$

Substituting the value of M_A in equation (i),

$$F_A = +\frac{2}{l} \cdot \frac{6EI\delta}{l^2} = +\frac{12EI\delta}{l^3}$$

$$\therefore \quad M_B = F_A.l + M_A = \frac{12EI\delta}{l^3} \cdot l - \frac{6EI\delta}{l^2}$$

$$= +\frac{6EI\delta}{l^2}$$

The bending moment diagram is shown in Fig. 12·10 (b).

Note. If the support A sinks down by δ from its original level, then

$$M_A = +\frac{6EI\delta}{l^2}$$

and

$$M_B = -\frac{6EI\delta}{l^2}$$

Example 12·8. *A steel fixed beam AB of span 6 m is 6 cm wide and 10 cm deep. The support B sinks down by 6 mm. Find the fixing moments. Take $E = 2·0 \times 10^6$ kg/cm².*

Solution.

Given. Span, $l = 6$ m $= 600$ cm

Width of beam, $b = 6$ cm

Depth of beam, $d = 10$ cm

$\therefore$ Moment of inertia,

$$I = \frac{bd^3}{12} = \frac{6 \times 10^3}{12} = 500 \text{ cm}^4$$

Sinking of support B,

$$\delta = 6 \text{ mm} = 0.6 \text{ cm}$$

Young's modulus, $E = 2.0 \times 10^6 \text{ kg/cm}^2$

Let M_A = Fixing moment at A, and

        M_B = Fixing moment at B.

Using the relation,

$$M_A = - \frac{6EI\delta}{l^2} \text{ with usual notations.}$$

$$= - \frac{6 \times 2.0 \times 10^6 \times 500 \times 0.6}{600^2} \text{ kg-cm}$$

$$= -10,000 \text{ kg-cm}$$

$$= -100 \text{ kg-m Ans}.$$

Now using the relation,

$$M_B = \frac{6EI\delta}{l^2} \text{ with usual notations.}$$

$$= \frac{6 \times 2.0 \times 10^6 \times 500 \times 0.6}{600^2} = 10,000 \text{ kg-cm}$$

$$= 100 \text{ kg-m Ans}.$$

Example 12·9. *A built-up beam AB of 4 m span is carrying a uniformly distributed load of 1,500 kg/m. The support B sinks down by 1 cm. Find out the fixing moments. Take $E = 2.0 \times 10^6 \text{ kg/cm}^2$ and $I = 8,000 \text{ cm}^4$.*

Solution.

Given. Span, $l = 4 \text{ m} = 400 \text{ cm}$

Load $w = 1,500 \text{ kg/m} = 15 \text{ kg/cm}$

Sinking of support, $\delta = 1 \text{ cm}$

Young's modulus, $E = 2.0 \times 10^6 \text{ kg/cm}^2$

Moment of inertia, $I = 8,000 \text{ cm}^4$

Let M_A = Fixing moment in A, and

        M_B = Fixing moment at B.

Let us first find out the fixing moments separately, due to load and sinking, and then add up the moments.

Using the relation,

$$M_A = - \frac{wl^2}{12} \text{ with usual notations.}$$

$$M_{A1} = - \frac{15 \times 400^2}{12} = -2,00,000 \text{ kg-cm} = 2,000 \text{ kg-m}$$

Similarly, $M_{B1} = M_{A1} = -2,000 \text{ kg-m}$

Now using the relation,

$$M_A = - \frac{6EI\delta}{l^2} \text{ with usual notations.}$$

$$M_{A2} = - \frac{6 \times 2.0 \times 10^6 \times 8,000 \times 1}{400^2} = -6,00,000 \text{ kg-cm}$$

$$= -6,000 \text{ kg-m}$$

Similarly $M_{B2} = -M_{A2} = +6,000$ kg·m

$\therefore \qquad M_A = M_{A1} + M_{A2} = -2,000-6,000$ kg·m

$\qquad\qquad\qquad = -8,000$ kg·m **Ans**.

and $\qquad M_B = M_{B1} + M_{B2} = -2,000+6,000$ kg·m

$\qquad\qquad\qquad = 4,000$ kg·m **Ans**.

Exercise 12·2

1. A fixed beam AB is of span 4 m. The support B sinks down by 1 cm. Find the fixing moments, if the flexural rigidity of the beam is 4×10^8 kg·cm².

(Poona University)

[**Ans.** $M_A = -150$ kg·m ; $M_B = 150$ kg·m]

2. An encastre beam AB of span 8 m is carrying a uniformly distributed load of 2 t·m over the entire beam. The support B sinks down by 1·2 cm. Determine the fixing moments at A and B, if moment of inertia for the beam section is 9,875 cm⁴. Take $E = 2·0 \times 10^6$ kg/cm². *(Bombay University)*

[**Ans.** $-12·9$ t·m ; $+8·45$ t·m]

HIGHLIGHTS

1. A beam, which is built-in at its two supports, is called a fixed beam.

2. Following are the advantages of a fixed beam :

(a) The beam is stiffer, stronger and more stable.

(b) The slope at the two ends is zero.

(c) The fixing moments are developed at the two ends, whose effect is to reduce the maximum B.M. at the centre of the beam.

(d) The deflection of beam, at its centre, very much reduced.

3. The two areas of μ-diagram (*i.e.* B.M. diagram considering the beam as a simply supported beam) and μ'-diagram (*i.e.* B.M. diagram due to fixed end moments) are equal.

4. The two moments of μ'-diagram (*i.e.* $a\bar{x}$) and μ'-diagram (*i.e.* $a'\bar{x}'$) about the fixed ends A and B are equal.

5. The fixing moments in a fixed beam carrying a central point load,

$$M_A = M_B = -\frac{Wl}{8}$$

and the maximum deflection at the centre of the beam.

$$y_C = \frac{Wl^3}{192\,EI}$$

where $\qquad\qquad W =$ Point load on the beam,

$l =$ Span of the beam AB, and

$EI =$ Flexural rigidity of the beam.

6. The fixing moments in a fixed beam carrying a uniformly distributed load,

$$M_A = M_B = -\frac{wl^2}{12}$$

and maximum deflection at the centre of the beam,

$$y_C = \frac{wl^4}{384\,EI}$$

where
w = uniformly distributed load per unit length,

l = Span of the beam AB, and

EI = Flexural rigidity of the beam.

7. Fixing moments in a fixed beam carrying an eccentric point load

$$M_A = -\frac{Wab^2}{l^2} \quad \text{and} \quad M_B = -\frac{Wa^2b}{l^2}$$

and deflection under the load,

$$y = \frac{Wa^3l^3}{3l^3EI}$$

where
W = Point load on the beam AB,

l = Span of the beam AB,

a = Distance between the fixed end A and the axis of the load,

b = Distance between the fixed end B and the axis of the load, and

EI = Flexural rigidity of the beam.

8. Fixing moments in a fixed beam, when the support B sinks down from its original level,

$$M_A = -\frac{6EI\delta}{l^2} \quad \text{and} \quad M_B = \frac{6EI\delta}{l^2}$$

where
δ = Amount, by which the support B has sunk down,

l = Span of the beam AB, and

EI = Flexural rigidity of the beam.

Do You Know ?

1. What is your idea of an encastre beam and the arrangement of fixing it at its two supports ?

2. What is meant by an encastre beam ? Is there any advantage in using it ?

3. If a fixed beam AB carries a central load W, find out the value of maximum deflection.

4. Derive an expression for the maximum deflection of an encastre beam, carrying a uniformly distributed load of w per unit length.

5. A beam, built-in at its both ends, has a uniform flexural rigidity EI throughout its length l. It carries a single point load W, which is placed at a distance a from the left end. Calculate, from first principles, the fixed end moments developed at the two ends.

6. Derive an expression for the fixing moments, when one of the supports of a fixed beam sinks down by δ from its original position.

13

THEOREM OF THREE MOMENTS

13·1. Introduction

A beam, which is supported on more than two supports, is called a *continuous beam*. Such a beam, when loaded, will be deflected with convexity upwards, over the intermediate supports, and with concavity upwards over the mid of the spans. The intermediate supports of a continuous beam are always subjected to some bending moment. The end supports, if simply supported, will not be subjected to any bending moment. But the end supports, if fixed, will be subjected to fixing moments and the slope of the beam, at the fixed ends, will be zero.

13·2. Bending moment diagrams for continuous beams

The analysis of a continuous beam is similar to that of a fixed beam. The bending moment diagram for a continuous beam, under any system of loading, may be drawn in the following two stages :

1. By considering the beam as a series of discontinuous beams, from support to support, and drawing the usual μ-diagram due to vertical loads.

476

2. By superimposing the usual μ'-diagram, due to end moments, over μ-diagram.

13·3. Clapeyron's theorem of three moments

It states, *"If a beam has n supports, the end ones being fixed then the same number of equations required to determine the support moments may be obtained from the consecutive pairs of spans i.e. AB-BC, BC-CD, CD-DE and so on."*

13·4. Proof of Clapeyron's theorem of three moments

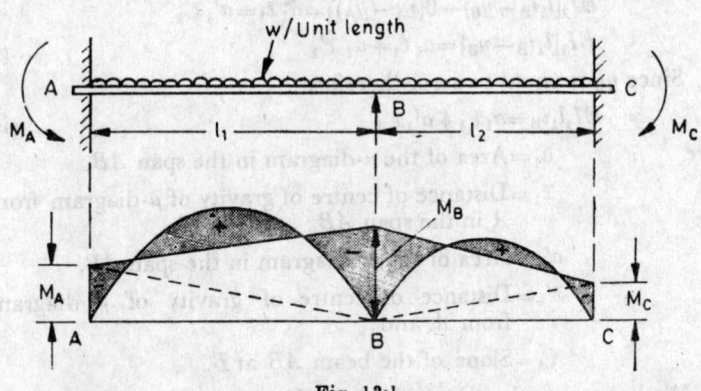

Fig. 13·1

Consider a continuous beam ABC, fixed at A and C, and supported at B as shown in Fig. 13·1.

Let $l_1 =$ Span AB of the beam,

$l_2 =$ Span BC of the beam,

$I_1 =$ Moment of inertia of the beam in span AB,

$I_2 =$ Moment of inertia of the beam in span BC,

$M_A =$ Support moment at A,

$M_B =$ Support moment at B, and

$M_C =$ Support moment at C,

$\mu_X =$ Bending moment at any section X, considering the beam between two supports as simply supported, and

$\mu'_X =$ Fixing moment at any section X, of the beam.

We know that, in the span AB, the B.M. at any section X at a distance x from A,

$$M_X = \mu_X + \mu'_X$$

$$\therefore \quad EI_1 \frac{d^2y}{dx^2} = \mu_X + \mu'_X \qquad \left(\because \frac{M}{EI} = \frac{d^2y}{dx^2} \right)$$

Multiplying the above equation by x, and integrating the same for the whole span AB *i.e.* from 0 to l.

$$EI_1 \int\limits_0^{l_1} \frac{x d^2 y}{dx^2} = \int\limits_0^{l_1} x.\mu_X dx + \int\limits_0^{l_1} x\,\mu'_X dx$$

$$EI_1 \left[x.\frac{dy}{dx} - y \right]_0^l = a_1 \bar{x}_1 + a'_1 \bar{x}'_1$$

or

$$EI_1[(l_1 i_B - y_B) - 0(i_A - y_A)] = a_1 \bar{x}_1 = a'_1 \bar{x}'_1$$

$$EI_1[l_1 i_B - y_B] = a_1 \bar{x}_1 + a_1' \bar{x}'_1 \qquad \qquad ...(i)$$

Since y_B is equal to zero, therefore

$$EI_1 l_1 i_B = a_1 \bar{x}_1 + a'_1 \bar{x}'_1 \qquad \qquad ...(ii)$$

where

$a_1 =$ Area of the μ-diagram in the span AB,

$\bar{x}_1 =$ Distance of centre of gravity of μ-diagram from A in the span AB,

$a'_1 =$ Area of the μ'-diagram in the span AB,

$\bar{x}'_1 =$ Distance of centre of gravity of μ'-diagram from A, and

$i_B =$ Slope of the beam AB at B.

We know that the shape of the μ'-diagram is trapezoidal, having end ordinates equal to M_A and M_B. Therefore splitting up the μ'-diagram into two triangles,

$$a'_1 \bar{x}'_1 = \left(M_A \times \frac{l_1}{2} \times \frac{l_1}{3} \right) + \left(M_B \times \frac{l_1}{2} \times \frac{l_1}{3} \right)$$

$$= (M_A + 2M_B) \frac{l_1^2}{6}$$

Substituting this value of $a'_1 \bar{x}'_1$ in equation (i),

$$EI_1 l_1 i_B = a_1 \bar{x}_1 + (M_A + 2M_B) \frac{l_1^2}{6}$$

or

$$EI_1 i_B = \frac{a_1 \bar{x}_1}{l_1} + (M_A + 2M_B) \frac{l_1}{6} \qquad \qquad ...(iii)$$

$\therefore$

$$E.i_B = \frac{a_1 \bar{x}_1}{I_1 l_1} + (M_A + 2M_B) \frac{l_1}{6 I_1}$$

Similarly, in the span BC, taking C as the origin and x positive to the left, we get

$$E i''_B = \frac{a_2 \bar{x}_2}{I_2 l_2} + (M_C + 2M_B) \frac{l_2}{6 I_2} \qquad \qquad ...(iv)$$

where

$a_2 =$ Area of the μ-diagram in the span BC

$\bar{x}'_2 =$ Distance of centre of gravity of μ-diagram from C in the span BC,

a_2 = Area of the μ'-diagram in the span BC,

$\bar{x}'_2$ = Distance of the centre of gravity of the μ'-diagram from C in the span BC,

i'_B = Slope of the beam BC at B.

Since i_B is equal to $-i'_B$, therefore $E.i_B$ is equal to $-E.i'_B$.

or $\dfrac{a_1 \bar{x}_1}{I_1 l_1} + (M_A + 2M_B)\dfrac{l_1}{6I_1} = -\left[\dfrac{a_2 \bar{x}_2}{I_2 l_2} + (M_C + 2M_B)\dfrac{l_2}{6I_2}\right]$

$\therefore (M_A + 2M_B)\dfrac{l_1}{I_1} + (M_C + 2M_B)\dfrac{l_2}{I_2} = -\dfrac{6a_1 \bar{x}_1}{I_1 l_1} - \dfrac{6a_2 \bar{x}_2}{I_2 l_2}$

$M_A\dfrac{l_1}{I_1} + 2M_B\dfrac{l_1}{I_1} + M_C\dfrac{l_2}{I_2} + 2M_B\dfrac{l_2}{I_2} = -\left(\dfrac{6a_1 \bar{x}_1}{I_1 l_1} + \dfrac{6a_2 \bar{x}_2}{I_2 l_2}\right)$

$\therefore M_A\left(\dfrac{l_1}{I_1}\right) + 2M_B\left(\dfrac{l_1}{I_1} + \dfrac{l_2}{I_2}\right) + M_C\left(\dfrac{l_2}{I_2}\right) = -\left(\dfrac{6a_1 \bar{x}_1}{I_1 l_1} + \dfrac{6a_2 \bar{x}_2}{I_2 l_2}\right)$

The three moment equation may also be found out by Mohr's second theorem (*i.e.*, area moment principle) which states, "*In any portion AB of a bent beam, the displacement of A from the tangent to the beam, at B is equal to the moment of the area of the M/EI diagram between A and B, taken about A.*"

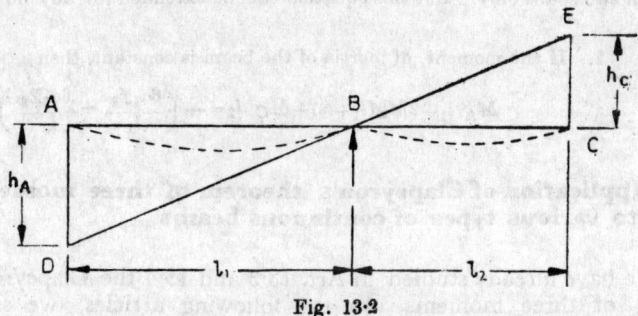

Fig. 13·2

Let the continuous beam ABC be deflected into elastic curve shown by dotted line. Draw tangent at B to the elastic curve, meeting the vertical line through A at D and the vertical line through C at E as shown in Fig. 13·2. Let the tangent DBE make an angle θ with ABC. From the geometry of the figure, we find that

$$\tan \theta = \frac{h_A}{l_1} = \frac{h_C}{l_2} \qquad \qquad \ldots(i)$$

Now as per Mohr's second theorem, we know that

$$h_A = \frac{1}{EI_1}\left[\left(a_1 \bar{x}_1 + M_A \times \frac{l_1}{2} \times \frac{l_1}{3}\right) + \left(M_B \times \frac{l_1}{2} \times \frac{2l_1}{3}\right)\right]$$

$$= \frac{1}{EI_1}\left[a_1\bar{x}_1 + \frac{M_A l_1^2}{6} + \frac{2M_B l_1^2}{6} \right]$$

$$= \frac{1}{6E}\left[\frac{6a_1\bar{x}_1}{I_1} + \frac{M_A l_1^2}{I_1} + \frac{2M_B l_1^2}{I_1} \right]$$

or $\qquad \dfrac{h_A}{l_1} = \dfrac{1}{6E}\left[\dfrac{6a_1\bar{x}_1}{I_1 l_1} + \dfrac{M_A l_1}{I_1} + \dfrac{2M_B l_1}{I_1} \right] \qquad$ ...(ii)

Similarly $\quad \dfrac{h_C}{l_2} = \dfrac{1}{6E}\left[\dfrac{6a_2\bar{x}_2}{I_2 l_2} + \dfrac{M_C l_2}{I_2} + \dfrac{2M_B l_2}{I_2} \right] \qquad$ ...(iii)

Equating the equations (ii) and (iii)

$$\frac{6a_1 x_1}{I_1 l_1} + \frac{M_A l_1}{I_1} + \frac{2M_B l_1}{I_1} = \frac{6a_2\bar{x}_2}{I_2 l_2} + \frac{M_C l_2}{I_2} + \frac{2M_B l_2}{I_2}$$

or $\qquad M_A\dfrac{l_1}{I_1} + 2M_B\left(\dfrac{l_1}{I_1} + \dfrac{l_2}{I_2} \right) + M_C\dfrac{l_2}{I_2} = -\left(\dfrac{6a_1\bar{x}_1}{I_1 l_1} + \dfrac{6a_2\bar{x}_2}{I_2 l_2} \right)$

This is the required Clapeyron's equation for three moments.

Note. 1. For the sake of simplicity, we have considered a continuous beam with two spans only. But this equation can be extended for any number of spans.

2. If the moment of inertia of the beam is constant, then

$$M_A\, l_1 + 2M_B(l_1 + l_2) + M_C\, l_2 = -\left(\frac{6a_1\bar{x}_1}{l_1} + \frac{6a_2\bar{x}_2}{l_2} \right)$$

13·5. Application of Clapeyron's theorem of three moments to various types of continuous beams

We have already studied in Art. 13·3 and 13·4 the Clapeyron's theorem of three moments. In the following articles, we shall discuss its application to the following types of continuous beams :

1. Continuous beams with simply supported ends,

2. Continuous beams with fixed end supports,

3. Continuous beams with the end span overhanging, and

4. Continuous beams with a sinking support.

13·6. Continuous beams with simply supported ends

Sometimes, a continuous beam is simply supported on its one or both the end supports. In such a case, the fixing moment on the simply supported end is zero.

Example 13·1. *A continuous beam ABC 10 m long rests on supports A, B and C at the same level and is loaded as shown in Fig. 13·3.*

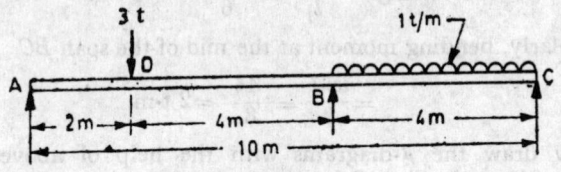

Fig. 13·3

Determine the moments over the beam, and draw the bending moment diagram. Also calculate the reactions at the supports and draw the shear force diagram.

Solution.

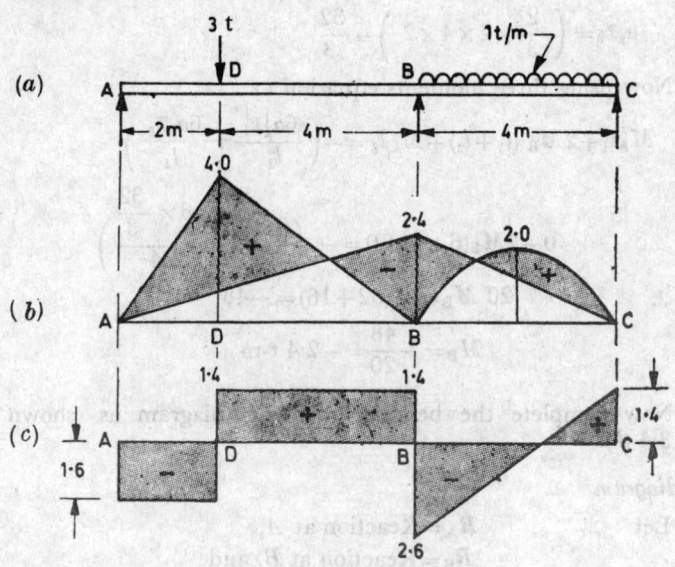

Fig. 13·4

Given. Span	$AB, l_1 = 6$ m
Span BC,	$l_2 = 4$ m
Load in AB	$W = 3$ t
Distance AD,	$a = 2$ m
Distance DB,	$b = 4$ m
Load in BC,	$w = 1$ t/m

Moments over the beam

Let *M_A = Fixing moment at A,

M_B = Fixing moment at B, and

*M_C = Fixing moment at C.

---

*Since the beam is simply supported at A and C, therefore the fixing moments M_A and M_C will be zero.

First of all, let us consider the beam AB as a simply supported beam. Therefore bending moment at D,

$$M_D = \frac{Wab}{l_1} = \frac{3 \times 2 \times 4}{6} = 4 \text{ t-m}$$

Similarly, bending moment at the mid of the span BC

$$= \frac{wl_2^2}{8} = \frac{4^2}{8} = 2 \text{ t-m}$$

Now draw the μ-diagrams with the help of above bending moments as shown in Fig. 13·4 (a).

From the geometry of the above bending moment diagrams, we find that

$$a_1\bar{x}_1 = \left[\left(\frac{1}{2} \times 2 \times 4 \times \frac{2 \times 2}{3} \right) + \left(\frac{1}{2} \times 4 \times 4 \right) \left(2 + \frac{4}{3} \right) \right] = 32$$

and

$$a_2\bar{x}_2 = \left(\frac{2}{3} \times 2 \times 4 \times 2 \right) = \frac{32}{3}$$

Now using three moments equation *i.e.*,

$$M_A l_1 + 2M_B (l_1 + l_2) + M_C l_2 = - \left(\frac{6a_1\bar{x}_1}{l_1} + \frac{6a_2\bar{x}_2}{l_2} \right)$$

$$0 + 2M_B(6 + 4) + 0 = - \left(\frac{6 \times 32}{6} + \frac{6 \times \frac{32}{3}}{4} \right)$$

$$\therefore \qquad 20\ M_B = -(32 + 16) = -48$$

or

$$M_B = -\frac{48}{20} = -2·4 \text{ t-m}$$

Now complete the bending moment diagram as shown in Fig. 13·4 (b).

S.F. diagram

Let R_A = Reaction at A,
                 R_B = Reaction at B, and
                 R_C = Reaction at C.

Taking moments about B,

$$R_A \times 6 - 3 \times 4 = -2·4 \qquad (\because \text{ B.M. at B is } -2·4 \text{ t-m})$$

$$\therefore \qquad R_A = \frac{-2·4 + 12·0}{6} = \frac{9·6}{6} = \mathbf{1·6\ t} \quad \textbf{Ans.}$$

Similarly $R_C \times 4 - 4 \times 2 = -2·4$

$$\therefore \qquad R_C = \frac{-2·4 + 8·0}{4} = \frac{5·6}{4} = \mathbf{1·4\ t\ Ans.}$$

$$\therefore \qquad R_B = (3 + 4) - (1·6 + 1·4) = \mathbf{4·0\ t} \quad \textbf{Ans.}$$

Now draw the S.F. diagram as shown in Fig. 13·4 (c).

Example 13·2. *A continuous beam ABCD, simply supported at A, B, C and D, is loaded as shown in Fig. 13·5.*

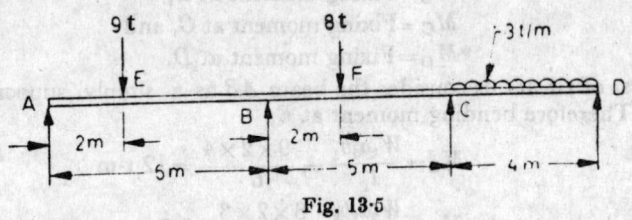

Fig. 13·5

Find the moments over the beam and draw the B.M. and S.F. diagrams. 〔*Kerala University, 1979*〕

Solution.

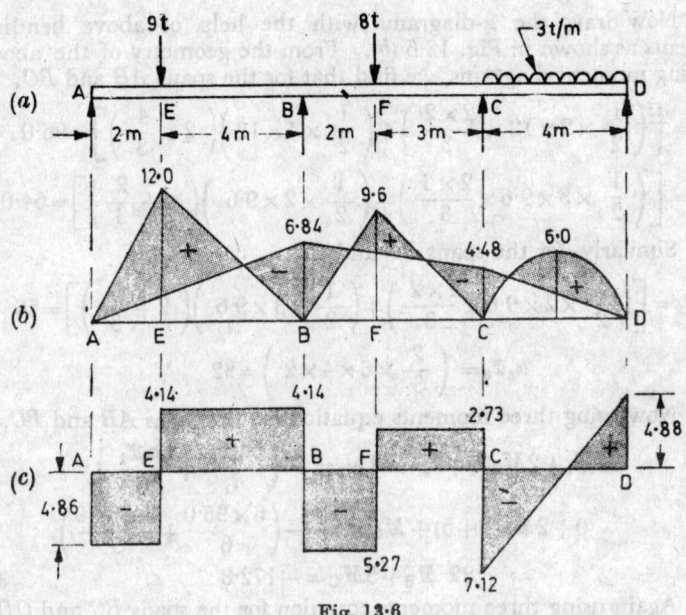

Fig. 13·6

Given. Span AB, $l_1 = 6$ m

Span BC, $\quad l_2 = 5$ m

Span CD, $\quad l_3 = 4$ m

Load in AB, $\quad W_1 = 9$ t

Distance AE $\quad a_1 = 2$ m

Distance EB, $\quad b_1 = 4$ m

Load in BC, $\quad W_2 = 8$ t

Distance BF, $\quad a_2 = 2$ m

Distance FC, $\quad b_2 = 3$ m

Load in CD, $\quad w = 3$ t/m

Moments over the beam

Let $\qquad$ *M_A=Fixing moment at A,

$\qquad\qquad\qquad$ M_B=Fixing moment at B,

$\qquad\qquad\qquad$ M_C =Fixing moment at C, and

$\qquad\qquad\qquad$ *M_D=Fixing moment at D.

First of all, let us consider the beam AB as a simply supported beam. Therefore bending moment at E,

$$M_E=\frac{W_1a_1b_1}{l_1}=\frac{9\times2\times4}{6}=12 \text{ t-m}$$

Similarly $\qquad$ $$M_C=\frac{W_2a_2b_2}{l_2}=\frac{8\times2\times3}{5}=9\cdot6 \text{ t-m}$$

and bending moment at the mid of the span CD

$$=\frac{wl_3{}^2}{8}=\frac{3\times4^2}{8}=6 \text{ t-m}$$

Now draw the μ-diagrams with the help of above bending moments as shown in Fig. 13·6 (b). From the geometry of the above bending moment diagrams, we find that for the spans AB and BC.

$$a_1\bar{x}_1=\left[\left(\frac{1}{2}\times2\times12\times\frac{2\times2}{3}\right)+\left(\frac{1}{2}\times4\times12\right)\left(2+\frac{4}{3}\right)\right]=96\cdot0$$

$$a_2\bar{x}_2=\left[\left(\frac{1}{2}\times3\times9\cdot6\times\frac{2\times3}{3}\right)+\left(\frac{1}{2}\times2\times9\cdot6\right)\left(3+\frac{2}{3}\right)\right]=64\cdot0$$

Similarly, for the spans BC and CD

$$**a_2\bar{x}_2=\left[\left(\frac{1}{2}\times2\times9\cdot6\times\frac{2\times2}{3}\right)+\left(\frac{1}{2}\times3\times9\cdot6\right)\left(2+\frac{3}{3}\right)\right]=56$$

and $\qquad\qquad$ $$a_3\bar{x}_3=\left(\frac{2}{3}\times6\times4\times2\right)=32$$

Now using three moments equation for the spans AB and BC,

$$M_Al_1+2M_B(l_1+l_2)+M_Cl_2=-\left(\frac{6a_1\bar{x}_1}{l_1}+\frac{6a_2\bar{x}_2}{l_2}\right)$$

$$0+2M_B(6+5)+M_C\times5=-\left(\frac{6\times96\cdot0}{6}+\frac{6\times64\cdot0}{5}\right)$$

$$22\ M_B+5M_C=-172\cdot8 \qquad\qquad\qquad ...(i)$$

Again using three moments equation for the spans BC and CD,

$$M_Bl_2+2M_C(l_2+l_3)+M_Dl_3=-\left(\frac{6a_2\bar{x}_2}{l_2}+\frac{6a_3\bar{x}_3}{l_3}\right)$$

$$M_B\times5+2M_C(5+4)+0=-\left(\frac{6\times56}{5}-\frac{6\times32}{4}\right)$$

$$5M_B+18M_C=-115\cdot2 \qquad\qquad\qquad ...(ii)$$

---

*Since the beam is simply supported at A and D, therefore the fixing moments M_A and M_D will be zero.

**The previous value of $a_2\bar{x}_2$ is with reference to the support C (being the end support of spans AB and BC). This value of $a_2\bar{x}_2$ is with reference to the support B (being the end support of the spans BC and CD).

Solving equations (*i*) and (*ii*) we get,

$$M_B = -6.84 \text{ t-m}$$
$$M_C = -4.48 \text{ t-m}$$

Now complete the bending moment diagram as shown in Fig. 13·6 (*b*).

Reactions at the supports

Let R_A = Reaction at A,

R_B = Reaction at B,

R_C = Reaction at C, and

R_D = Reaction at D.

Taking moments about B,

$$R_A \times 6 - 9 \times 4 = 6.84 \qquad (\because \text{ B.M. at } B \text{ is } -6.84 \text{ t-m})$$

$$R_A = \frac{-6.84 + 36}{6} = \frac{29.16}{6} = \textbf{4.86 t} \quad \textbf{Ans.}$$

Now taking moments about C,

$$R_D \times 4 - 12 \times 2 = -4.48 \qquad (\because \text{ B.M. at } C \text{ is } 4.48 \text{ t-m})$$

$$R_D = \frac{-4.48 + 24}{4} = \frac{19.52}{4} = \textbf{4.88 t} \quad \textbf{Ans.}$$

Again taking moments about C,

$$R_A \times 11 - 9 \times 9 + R_B \times 5 - 8 \times 3 = -4.48$$
$$4.86 \times 11 - 81 + 5R_B - 24 = -4.48$$

$$\therefore \qquad R_B = \frac{-4.48 - 53.46 + 81 + 24}{5} = \frac{47.06}{5} \text{ t}$$

$$= \textbf{9.41 t} \quad \textbf{Ans}.$$

$$\therefore \qquad R_C = (9 + 8 + 12) - (4.86 + 4.88 + 9.41) = 29 - 19.15 \text{ t}$$

$$= \textbf{9.85 t} \quad \textbf{Ans.}$$

Now draw the S.F. diagram as shown in Fig. 13·6 (*c*).

13·7. Continuous beams with fixed end supports

Sometimes, a continuous beam is fixed at its one or both ends. If the beam is fixed at the left end A, then an imaginary zero span is taken to the left of A and the three moments theorem is applied as usual. Similarly, if the beam is fixed at the right end, then an imaginary zero span is taken after the right end support, and the three moments theorem is applied as usual.

Note. The fixing moment at 0 *i.e.*, at the imaginary support of the zero span, is always equal to zero.

Example 13·3. *A continuous beam ABC of uniform section, with span AB as 8 m and BC as 6 m, is fixed at A and simply supported at B and C. The beam is carrying a uniformly distributed load of 1 tonne/m throughout its length. Find the moments along the beam, and the reactions at the supports. Also draw the B.M. diagram.*

Solution.

Given.   Span AB, $l_1 = 8$ m

Span BC, $l_2 = 6$ m

Load, $w = 1$ t/m

Moments along the beam

Since the beam is fixed at A, therefore assume a zero span to the left of A.

Let   *M_0 = Fixing moment at the left hand support of zero span,

               M_A = Fixing moment at A,

               M_B = Fixing moment at B, and

               *M_C = Fixing moment at C.

First of all, consider the beam AB as a simply supported beam. Therefore B.M. at the mid of the span AB

$$= \frac{wl_1^2}{8} = \frac{1 \times 8^2}{8} = 8 \text{ t-m}$$

Similarly, bending moment at the mid of the span BC

$$= \frac{wl_2^2}{8} = \frac{1 \times 6^2}{8} = 4.5 \text{ t-m}$$

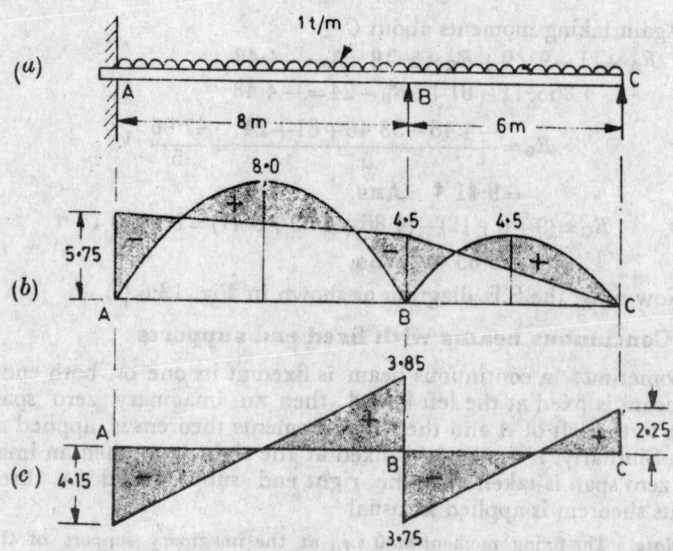

Fig. 13·7.

Now draw the μ-diagram with the help of above bending moments as shown in Fig. 13·7.

---

*Since there is a zero span on the left of A, therefore the fixing moment M_0 will be zero. Moreover as the beam is simply supported at C, therefore fixing moment M_C will also be zero.

From the geometry of the above bending moment diagrams, we find that for the spans OA and AB,

$$a_0\bar{x}_0 = 0$$

and

$$a_1\bar{x}_1 = \left(\frac{2}{3} \times 8 \times 8 \times 4\right) = \frac{512}{3}$$

Similarly, for the span AB and BC,

$$a_1\bar{x}_1 = \left(\frac{2}{3} \times 8 \times 8 \times 4\right) = \frac{512}{3}$$

and

$$a_2\bar{x}_2 = \left(\frac{2}{3} \times 4.5 \times 6 \times 3\right) = 54$$

Now using three moments equation for the spans OA and AB,

$$M_0 l_0 + 2M_A (0 + l_1) + M_B l_1 = -\left(\frac{6a_0\bar{x}_0}{l_0} + \frac{6a_1\bar{x}_1}{l_1}\right)$$

$$0 + 2M_A(0 + 8) + M_B \times 8 = -\left(0 + \frac{6 \times \dfrac{512}{3}}{8}\right)$$

$$16M_A + 8M_B = -128 \qquad \qquad ...(i)$$

Again using three moments equation for the spans AB and BC,

$$M_A l_1 + 2M_B (l_1 + l_2) + M_C l_2 = -\left(\frac{6a_1\bar{x}_1}{l_1} + \frac{6a_2\bar{x}_2}{l_2}\right)$$

$$M_A \times 8 + 2M_B(8 + 6) + 0 = -\left(\frac{6 \times \dfrac{512}{3}}{8} + \frac{6 \times 54}{6}\right)$$

$$8M_A + 28 M_B = -182 \qquad \qquad ...(ii)$$

Solving the equations (i) and (ii),

$$M_A = -5.75 \text{ t-m}$$
$$M_B = -4.5 \text{ t-m}$$

Now complete the bending moment diagram as shown in Fig. 13·7 (b).

Reactions at supports

Let
$$R_A = \text{Reaction at } A,$$
$$R_B = \text{Reaction at } B, \text{ and}$$
$$R_C = \text{Reaction at } C.$$

Taking moments about B,

$$R_C \times 6 - 6 \times 3 = -4.5 \qquad (\because \text{ B.M. at } B = -4.5 \text{ t-m})$$

$$\therefore \qquad R_C = \frac{-4.5 + 18}{6} = \frac{13.5}{6} \text{ t}$$

$$= 2.25 \text{ t' } \textbf{Ans.}$$

Now taking moments about A,

$$R_C \times 14 + R_B \times 8 - 14 \times 7 = -5.75 \qquad (\because \text{ B.M. at } A = -5.75 \text{ t-m})$$

$$2.25 \times 14 - 8 R_B - 98 = -5.75$$

$$\therefore \qquad R_B = \frac{-5.75 + 98 - 31.5}{8} = \frac{60.75}{8} \text{ t}$$

$$= 7.6 \text{ t } \textbf{Ans.}$$

$$\therefore \quad *R_A = 14 \cdot 0 - (2 \cdot 25 + 7 \cdot 6) = 14 \cdot 0 - 9 \cdot 85 \text{ t}$$
$$= 4 \cdot 15 \text{ 't'} \quad \textbf{Ans.}$$

Example 13·4. *Evaluate the bending moment and shear force diagrams of the beam shown in Fig. 13·8.*

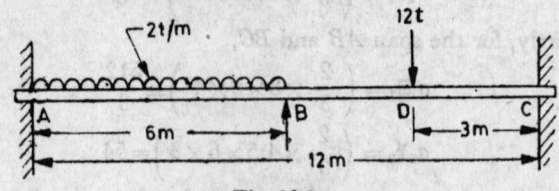

Fig. 13·8

What are the reactions at the supports ?

(U.P.S.C. Engg. Services, 1973)

Solution

Given. Span AB,	$l_1 = 6$ m
Span BC,	$l_2 = 6$ m
Load in AB,	$w = 2$ t/m
Load in BC,	$W = 12$ t

Support moments at A, B and C.

Since the beam is fixed at A and C, therefore assume a zero span to the left of A, and right of C.

Let $M_0 =$ Fixing moment at the imaginery support of zero span.
              $M_A =$ Fixing moment at A,
              $M_B =$ Fixing moment at B, and
              $M_C =$ Fixing moment at C.

First of all, consider the beam AB as a simply supported beam. Therefore bending moment at the mid of span AB

$$= \frac{wl_1^2}{8} = \frac{2 \times 6^2}{8} = 9 \cdot 0 \text{ t-m}$$

Similarly, bending moment at the mid of the span BC

$$= \frac{Wl}{4} = \frac{12 \times 6}{4} = 18 \cdot 0 \text{ t-m}$$

Now draw the μ-diagram with the help of above bending moments as shown in Fig. 13·9 (b).

From the geometry of bending moment diagrams, Fig. 13·9 (b), we find that for the spans OA and AB,

$$a_0 \bar{x}_0 = 0$$
$$a_1 \bar{x}_1 = \left(\frac{2}{3} \times 9 \cdot 0 \times 6 \times 3 \right) = 108$$

---

*The reaction at R_A may also be found out by taking moments about B, i.e.

$$R_A \times 8 - 8 \times 4 - M_A = -4 \cdot 5$$
$$8R_A - 32 - 5 \cdot 75 = -4 \cdot 5$$

$$\therefore \qquad R_A = \frac{-4 \cdot 5 + 32 + 5 \cdot 75}{8} = \frac{33 \cdot 25}{8} = 4 \cdot 15 \text{ t Ans.}$$

**This example is also solved on pages 517 and 559.

Similarly for the span AB and BC

$$a_1\bar{x}_1 = \left(\frac{2}{3} \times 9.0 \times 6 \times 3\right) = 108$$

$$a_2\bar{x}_2 = \left(\frac{1}{2} \times 18.0 \times 6 \times 3\right) = 162$$

and for span BC and CO

$$a_2\bar{x}_2 = \left(\frac{1}{2} \times 18.0 \times 6 \times 3\right) = 162$$

$$a_0\bar{x}_0 = 0$$

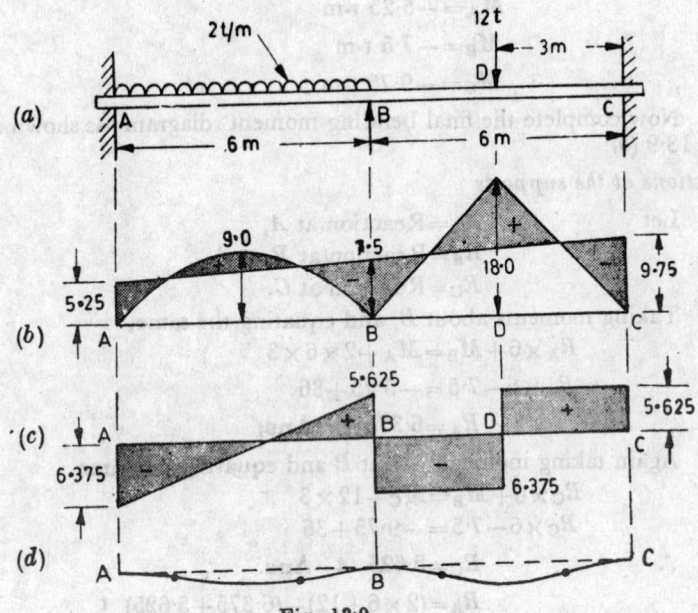

Fig. 13·9

Now using three moments equation for the spans OA and AB,

$$M_0 l_0 + 2M_A(0+l_1) + M_0 l_1 = -\left(\frac{6a_0\bar{x}_0}{l_0} + \frac{6a_1\bar{x}_1}{l_1}\right)$$

$$0 + 2M_A(0+6) + M_B \times 6 = -\left(0 + \frac{6 \times 108}{6}\right)$$

$$12M_A + 6M_B = -108$$

or $\qquad\qquad 2M_A + M_B = -18 \qquad\qquad\qquad ...(i)$

Again using three moments equation for the spans AB and BC,

$$M_A l_1 + 2M_B(l_1+l_2) + M_C l_2 = -\left(\frac{6a_1 x_1}{l_1} + \frac{6a_2 x_2}{l_2}\right)$$

$$M_A \times 6 + 2M_B(6+6) + M_C \times 6 = -\left(\frac{6 \times 108}{6} + \frac{6 \times 162}{6}\right)$$

$$6M_A + 24M_B + 6M_C = -270$$

$$M_A + 4M_B + M_C = -45 \qquad\qquad\qquad ...(ii)$$

Again using three moments equation for the spans BC and CO.

$$M_B l_2 + 2M_C(l_2 + l_0) + M_C l_0 = -\left(\frac{6a_2 \bar{x}_2}{l_2} + \frac{6a_0 \bar{x}_0}{l_0}\right)$$

$$M_B \times 6 + 2M_C(6 + 0) + 0 = -\left(\frac{6 \times 162}{6} + 0\right)$$

$$6M_B + 12M_C = -162$$

$$M_B + 2M_C = -27 \qquad\qquad \text{...(iii)}$$

Solving equations (i), (ii) and (iii),

$$M_A = -5.25 \text{ t-m}$$

$$M_B = -7.5 \text{ t-m}$$

$$M_C = -9.75 \text{ t-m}$$

Now complete the final bending moment diagram as shown in Fig. 13·9 (b).

Reactions at the supports

Let R_A = Reaction at A,

R_B = Reaction at B, and

R_C = Reaction at C.

Taking moments about B, and equating the same,

$$R_A \times 6 + M_B = M_A + 2 \times 6 \times 3$$

$$R_A \times 6 - 7.5 = -5.25 + 36$$

$$\therefore \qquad R_A = 6.375 \text{ t'} \quad \textbf{Ans.}$$

Again taking moments about B and equating the same,

$$R_C \times 6 + M_B = M_C + 12 \times 3$$

$$R_C \times 6 - 7.5 = -9.75 + 36$$

$$\therefore \qquad R_C = 5.625 \text{ t} \quad \textbf{Ans.}$$

and $$R_B = (2 \times 6 + 12) - (6.375 + 5.625) \text{ t}$$

$$= 12 \text{ t} \quad \textbf{Ans.}$$

Now draw the shear force diagram and elastic curve as shown in Fig. 13·9 (c) and 13·9 (d).

Exercise 13–1

1. A continuous beam $ABCD$ is simply supported over three spans, such that $AB = 8$ m, $BC = 12$ m and $CD = 5$ m. It carries uniformly distributed load of 4 t/m in span AB, 3 t/m in span BC and 6 t/m in span CD. Find the moments over the supports B and C.   [**Ans.** -35.9 t-m ; -31.0 t-m]

2. A simply supported beam ABC is continuous over two spans AB and BC of 6 m and 5 m respectively. The span AB is carrying a uniformly distributed load of 2 t/m and the span BC is carrying a point load of 5 t at a distance of 2 m from B. Find the support moment and the reactions.
[**Ans.** -7.1 t-m; 4·82 t ; 11·6 t ; 0·58 t)

3. A continuous beam $ABCD$ is simply supported over three spans of 5 m, 5 m and 4 m respectively. The first two spans are carrying a uniformly distributed load of 4 t/m, whereas the last span is carrying a uniformly distributed load of 5 t/m. Find the support moments at B and C.
[**Ans.** -10.37 t-m ; -8.52 t-m]

13·8. Continuous beams with end span overhanging

Sometimes, a continuous beam is overhanging, at its one or both ends. In such a case, the overhanging part of the beam behaves like a cantilever. The fixing moment on the end support may be found out by the cantilever action of the overhanging part of the beam.

Example 13·5. *A beam ABCD 9 m long is simply supported at A, B and C, such that the span AB is 3 m, span BC is 4·5 m and the overhung CD is 1·5 m. It carries a uniformly distributed load of 1·5 t/m in span AB and a point load of 1 tonne at the free end D. The moment of inertia of the beam in span AB is I and that in the span BC is 2I. Draw the B.M. and S.F. diagrams for the beam.*

*Solution

Given. Span AB, $l_1=4$ m

Span BC, $l_2=4·5$ m

Span CD, $l_3=1·5$ m

M.I. of span AB $=I$

M.I. of span BC, $=2I$

Load in AB, $w=1·5$ t/m

Load at D, $W=1$ t

Bending moment diagram

Let **M_A=Fixing moment at A,

M_B=Fixing moment at B,

M_C=Fixing moment at C.

First of all, consider the beam AB as a simply supported beam. Therefore B.M. at the mid of span AB

$$=\frac{wl_1^2}{8}=\frac{1·53^2}{8}=1·69 \text{ t-m}$$

From the geometry of the figure, we find that the fixing moment at C,

$$M_C=-1·0\times1·5=-1·5 \text{ t-m}$$

Now draw the μ-diagram, with the help of above bending moments as shown in Fig. 13·10 (b).

---

*This example is also solved on page **566**

**Since the beam is simply supported at A, therefore the fixing moment M_A will be zero.

From the geometry of the above bending moment diagram, we find that for the spans AB and BC

$$a_1 \bar{x}_1 = \frac{2}{3} \times 1.69 \times 3 \times 1.5 = 5.07$$

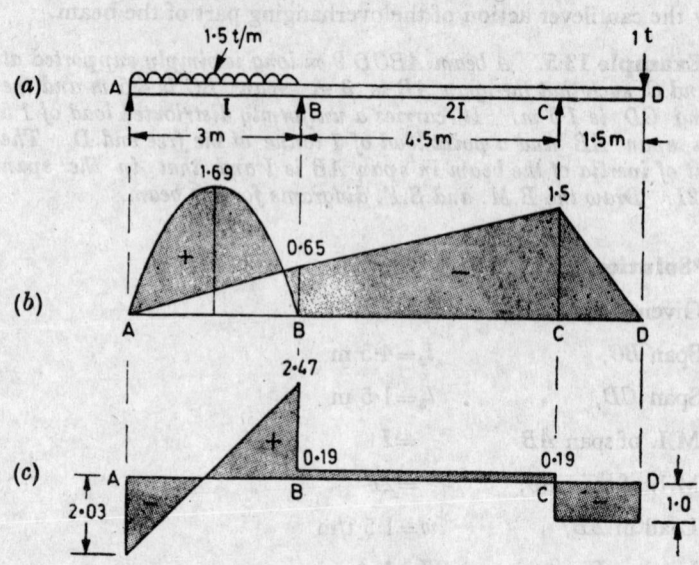

Fig. 13·10

Now using three moments equation for the spans AB and BC,

$$M_A \frac{l_1}{I_1} + 2 M_B \left(\frac{l_1}{I_1} + \frac{l_2}{I_2} \right) + M_C \frac{l_2}{I_2} = - \left[\frac{6 a_1 \bar{x}_1}{I_1 l_1} + \frac{6 a_2 \bar{x}_2}{I_2 l_2} \right]$$

$$0 + 2 M_B \left(\frac{3}{I} + \frac{4 \cdot 5}{2I} \right) - 1 \cdot 5 \times \frac{4 \cdot 5}{2I} = - \frac{6 \times 5 \cdot 07}{I \times 3}$$

$$\frac{10 \cdot 5 \, M_B}{I} - \frac{6 \cdot 75}{2I} = - \frac{10 \cdot 14}{I}$$

or
$$10 \cdot 5 M_B - 3 \cdot 375 = - 10 \cdot 14$$

∴
$$M_B = \frac{-10 \cdot 14 + 3 \cdot 375}{10 \cdot 5} = \frac{-6 \cdot 765}{10 \cdot 5} \text{ t-m}$$

$$= - 0 \cdot 65 \text{ t-m}$$

Now complete the B.M. diagram as shown in Fig. 13·10 (b).

Shear force diagram

Let
$$R_A = \text{Reaction at } A,$$
$$R_B = \text{Reaction at } B, \text{ and}$$
$$R_C = \text{Reaction at } C.$$

Taking moments at B,

$$R_A \times 3 - 1.5 \times 3 \times 1.5 = -0.65 \qquad (\because \text{ B.M. at } B \text{ is } -0.65 \text{ t-m})$$

$$\therefore \qquad R_A = \frac{-0.65 + 6.75}{3} = \frac{6.1}{3} = 2.03 \text{ t}$$

Again taking moments about B,

$$R_C \times 4.5 - 1 \times 6 = -0.65$$

$$\therefore \qquad R_C = \frac{-0.65 + 6}{4.5} = \frac{5.35}{4.5} = 1.19 \text{ t}$$

and $\qquad R_B = (3 \times 1.5 + 1) - (2.03 + 1.19) = 2.28 \text{ t}$

Now complete the S.F. diagram as shown in Fig. 13·10 (c).

Example 13·6. *A continuous beam, ABCD, is pinned at A and simply supported at B and C, these points being at the same level CD is an overhang. AB=3·0 m, BC=3·6 m and CD=1·8 m. It carries a point load of 10 tonnes at the mid point of BC and a uniformly distributed load of 1·6 tonnes/metre run from A to D as shown in Fig. 13·11.*

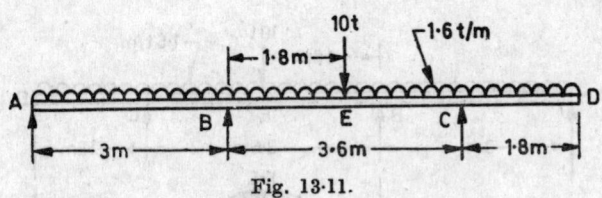

Fig. 13·11.

Find the bending moment at B, using the theorem of three moments and draw the bending moment and shear force diagrams for the continuous beam.   *(A.M.I.E., Summer, 1974)*

Solution.

Given. Span $AB, l_1 = 3$ m

Span BC, $l_2 = 3.6$ m

Span CD, $l_3 = 1.8$ m

Load in AD, $w = 1.6$ t/m

Load at E $W = 10$ t

Bending moment diagram

Let   *M_A = Fixing moment at A,

M_B = Fixing moment at B, and

M_C = Fixing moment at C.

---

* Since the beam is simply supported at A, therefore the fixing moment M_A will be zero.

First of all, consider the beam AB as a simply supported beam. Therefore B. M. at the mid of span AB

$$= \frac{wl_1^2}{8} = \frac{1 \cdot 6 \times 3^3}{8} = 1 \cdot 8 \text{ t-m}$$

Similarly B. M. at the mid of the span BC due to point load,

$$M_1 = \frac{wl}{4} = \frac{10 \times 3 \cdot 6}{4} = 9 \cdot 0 \text{ t-m} \qquad \ldots(i)$$

and B. M. due to uniformly distributed load,

$$M_2 = \frac{wl^2}{8} = \frac{1 \cdot 6 \times 3 \cdot 6^2}{2} = 2 \cdot 6 \text{ t-m} \qquad \ldots(ii)$$

∴ Total B. M. at the mid of the span. $9 \cdot 0 + 2 \cdot 6 = 11 \cdot 6$ t-m

From the geometry of the figure, we find that the fixing moment at C,

$$M_C = -\frac{wl_3^2}{2} = -\frac{1 \cdot 6 \times 1 \cdot 8^2}{2} = -2 \cdot 6 \text{ t m}$$

Now draw the μ-diagram with the help of the above bending moment as shown in Fig. 13·12 (b).

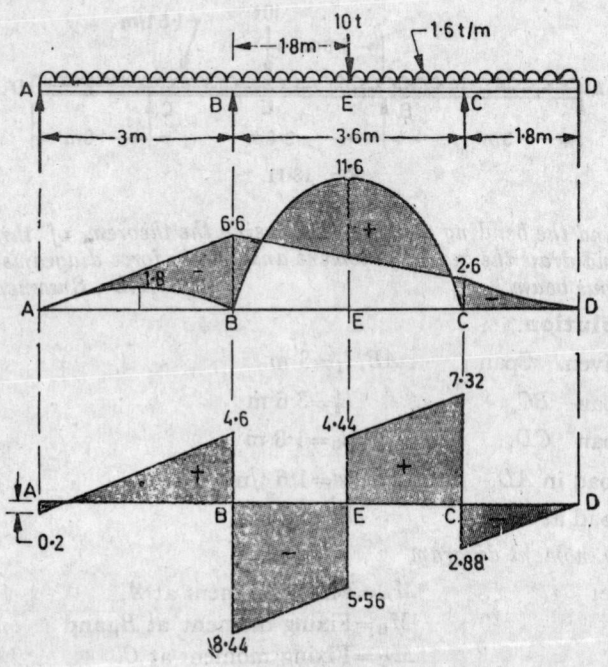

Fig. 13·12.

From the geometry of the bending moment diagram, we find that for the spans AB and BC,

$$a_1 \bar{x}_1 = \left(\frac{2}{3} \times 3 \times 1 \cdot 8 \times 1 \cdot 5 \right) = 5 \cdot 4$$

$$*a_2 \bar{x}_2 = \left(\frac{1}{2} \times 3 \cdot 6 \times 9 \cdot 0 \times 1 \cdot 8 \right)$$

$$+ \left(\frac{2}{3} \times 3 \cdot 6 \times 2 \cdot 6 \times 1 \cdot 8 \right) = 40 \cdot 39$$

Now using three moments equation for the spans AB and AC
i.e.,

$$M_A.\ l_1 + 2M_B(l_1+l_2) + M_C.\ l_2 = -\left(\frac{6\ a_1 \bar{x}_1}{l_1} + \frac{6\ a^2 x^2}{l_2} \right)$$

$$0 + 2M_B\ (3 \cdot 0 + 3 \cdot 6) + 2 \cdot 6 \times 3 \cdot 6 = -\left(\frac{6 \times 5 \cdot 4}{3} + \frac{6 \times 40 \cdot 39}{3 \cdot 6} \right)$$

$$13 \cdot 2\ M_B + 9 \cdot 36 = -\left(10 \cdot 8 + 67 \cdot 17 \right)$$

or $$13 \cdot 2\ M_B = -87 \cdot 33$$

$$\therefore \qquad M_B = -\frac{87 \cdot 33}{13 \cdot 2} = -6 \cdot 6 \text{ t-m}$$

Now complete the B. M. diagram as shown in Fig. 13·12. (b).

Shear force diagram

Let $\qquad R_A$ = Reaction at A,

$\qquad R_B$ = Reaction at B, and

$\qquad R_C$ = Reaction at C

Taking moments about B,

$$R_A \times 3 - 1 \cdot 6 \times 3 \times 1 \cdot 5 = -6 \cdot 6$$

$$R_A = \frac{-6 \cdot 6 + 7 \cdot 2}{3} = +0 \cdot 2 \text{ t}$$

Similarly $\quad R_C \times 3 \cdot 6 - 5 \cdot 4 \times 1 \cdot 6 \times 2 \cdot 7 - 10 \times 1 \cdot 8 = -6 \cdot 6$

or $$R_C = \frac{-6 \cdot 6 + 41 \cdot 33}{3 \cdot 6} = 10 \cdot 2 \text{ t}$$

and $\qquad R_B = (10 + 1 \cdot 6 \times 8 \cdot 4) - (0 \cdot 2 + 10 \cdot 2) = 1 \cdot 04 \text{ t}$

Now draw the S. F. diagram as shown in Fig. 13·12.) (c).

---

* The B. M. diagram consists of a triangle of ordinate 9·0 (due to point load) and parabola of ordinate 2·6 (due to uniformly distributed load 1).

Example 13·7. *A beam ABCDE has a built-in support at A and roller supports at B, C and D, DE being an overhung. AB=7 m, BC=5 m, CD=4 m and DE=1·5 m. The values of moment of inertia of the section over each of these lengths are 3I, 2I, I and I respectively. The beam carries a point load of 10 t at a point 3 m from A, a uniformly distributed load of 4·5 t/m over whole of BC and concentrated load of 9 t in CD 1·5 m from C and another point load of 3 t at E, the top of overhung as shown in Fig. 13·13.*

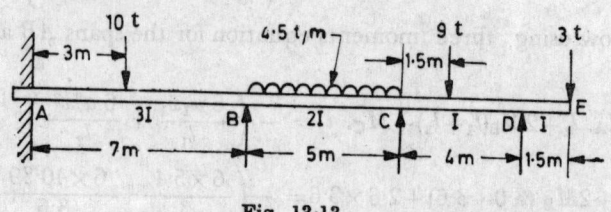

Fig. 13·13

Determine (i) the moments developed over each of the support A, B, C and D, and (ii) draw the B.M. diagram for the entire beam, stating values at salient points.

*Solution

Given.   Span AB, $l_1=7$ m   and   M.I.$=3I$

Span BC, $l_2=5$ m   and   M.I.$=2I$

Span CD, $l_3=4$ m   and   M.I.$=I$

Span DE, $l_4=1·5$ m   and   M.I.$=I$

Load in AB, $W_1=10$ t

Load in BC, $w_2=4·5$ t/m

Load in BC. $W_3=9$ t

Since the beam is fixed at A, therefore assume a zero span to the left of A.

Let   ****M_0**=Fixing moment at left hand support of zero span,

M_A=Fixing moment at A,

M_B=Fixing moment at B,

M_C=Fixing moment at C, and

M_D=Fixing moment at D.

First of all, let us consider the beam AB as a simply supported beam.   Therefore bending moment under the 10 t load

$$=\frac{Wab}{l}=\frac{10\times3\times4}{7}=\frac{120}{7}=17·14 \text{ t-m}$$

*This example is also solved on page 571

**Since there is a zero span on the left of A, therefore the fixing moment M_0 will be zero.

Similarly, B.M. at the mid of the span BC,

$$\frac{w_2 l_2^2}{8} = \frac{4 \cdot 5 \times 5^2}{8} = \frac{112 \cdot 5}{8} = 14 \cdot 06 \text{ t-m}$$

and B.M. under the 9 T load in span CD,

$$= \frac{9 \times 1 \cdot 5 \times 2 \cdot 5}{4} = \frac{33 \cdot 75}{4} = 8 \cdot 44 \text{ t-m}$$

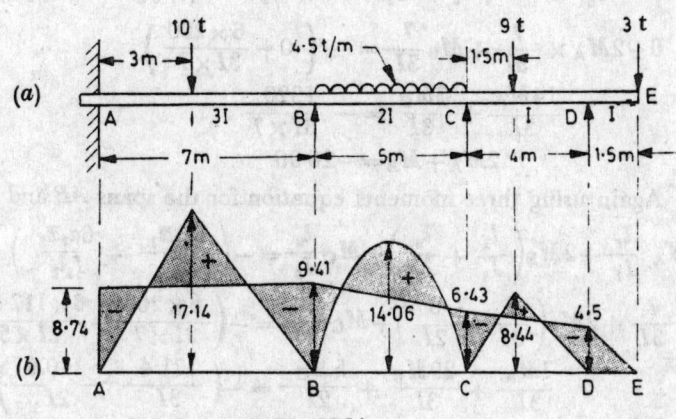

Fig. 13·14

From the geometry of the figure, we find that the fixing moment at D,

$$M_D = -3 \times 1 \cdot 5 = -4 \cdot 5 \text{ t·m}$$

Now draw μ-diagram with the help of above bending moments as shown in Fig. 13·14. (b).

From the geometry of the above bending moment diagram, we find that for the span OA nnd AB,

$$a_0 \bar{x}_0 = 0$$

and
$$a_1 \bar{x}_1 = \left[\left(\frac{1}{2} \times 4 \times 17 \cdot 14 \times \frac{2 \times 4}{3} \right) \right.$$
$$\left. + \left(\frac{1}{2} \times 3 \times 17 \cdot 14 \right) \left(4 + \frac{3}{3} \right) \right] = 220$$

Similarly for the spans AB and BC,

$$a_1 \bar{x}_1 = \left[\left(\frac{1}{2} \times 3 \times 17 \cdot 14 \times \frac{3 \times 2}{3} \right) \right.$$
$$\left. + \left(\frac{1}{2} \times 4 \times 17 \cdot 14 \right) \left(3 + \frac{4}{3} \right) \right] = 200$$

and
$$a_2 \bar{x}_2 = \frac{2}{3} \times 5 \times 14 \cdot 06 \times 2 \cdot 5 = 117 \cdot 2$$

Similarly for the spans BC and CD,

$$a_2 \bar{x}_2 = \frac{2}{3} \times 5 \times 14 \cdot 06 \times 2 \cdot 5 = 117 \cdot 2$$

and $\qquad a_3\bar{x}_3 = \left[\left(\dfrac{1}{2}\times 2\cdot 5\times 8\cdot 44\times\dfrac{2\times 2\cdot 5}{3}\right)\right.$

$\qquad\qquad\qquad\left.+\left(\dfrac{1}{2}\times 1\cdot 5\times 8\cdot 44\right)\left(2\cdot 5+\dfrac{1\cdot 5}{3}\right)\right] = 36\cdot 6$

Now using three moments equation for the spans OA and AB,

$$M_0\dfrac{l_0}{I_0}+2M_A\left(\dfrac{l_0}{I_0}+\dfrac{l_1}{I_1}\right)+M_B\dfrac{l_1}{I_1}=-\left(\dfrac{6a_0\bar{x}_0}{I_0 l_0}+\dfrac{6a_1\bar{x}_1}{I_1 l_1}\right)$$

$$0+2M_A\times\dfrac{7}{3I}+M_B\dfrac{7}{3I}=-\left(0+\dfrac{6\times 220}{3I\times 7}\right)$$

$$\dfrac{14M_A}{3I}+\dfrac{7M_B}{3I}=-\dfrac{1320}{3I\times 7}$$

or $\qquad\qquad 2M_A+M_B=-26\cdot 96 \qquad\qquad\qquad \text{...}(i)$

Again using three moments equation for the spans AB and BC,

$$M_A\dfrac{l_1}{I_1}+2M_B\left(\dfrac{l_1}{I_1}+\dfrac{l_2}{I_2}\right)+M_C\dfrac{l_2}{I_2}=-\left(\dfrac{6a_1\bar{x}_1}{I_1 l_1}+\dfrac{6a_2\bar{x}_2}{I_2 l_2}\right)$$

$$M_A\dfrac{7}{3I}+2M_B\left(\dfrac{7}{3I}+\dfrac{5}{2I}\right)+M_C\dfrac{5}{2I}=-\left(\dfrac{6\times 200}{3I\times 7}+\dfrac{6\times 117\cdot 2}{2I\times 5}\right)$$

$$\dfrac{7M_A}{3I}+\dfrac{29M_B}{3I}+\dfrac{5M_C}{2I}=-\left(\dfrac{171\cdot 4}{3I}+\dfrac{140.6}{2I}\right)$$

$$14M_A+58M_B+15M_C=-764\cdot 6 \qquad\qquad \text{...}(ii)$$

Again using three moments equation for the spans BC and CD,

$$M_B\dfrac{l_2}{I_2}+2M_C\left(\dfrac{l_2}{I_2}+\dfrac{l_3}{I_3}\right)+M_D\dfrac{l_3}{I_3}=-\dfrac{6a_2\bar{x}_2}{I_2 l_2}+\left(\dfrac{6a_3\bar{x}_3}{I_3 l_3}\right)$$

$$M_B\dfrac{5}{2I}+2M_C\left(\dfrac{5}{2I}+\dfrac{4}{I}\right)-4\cdot 5\dfrac{4}{I}=-\left(\dfrac{6\times 117\cdot 2}{2I\times 5}+\dfrac{6\times 36\cdot 6}{I\times 4}\right)$$

$$\dfrac{5M_B}{2I}+\dfrac{26M_C}{2I}-\dfrac{18}{I}=-\left(\dfrac{140.6}{2I}+\dfrac{54\cdot 9}{I}\right)$$

$$5M_B+26M_C=-214\cdot 4 \qquad\qquad\qquad \text{...}(iii)$$

Now solving equations (i), (ii) and (iii), we get

$$M_A=-8\cdot 78 \text{ t-m Ans.}$$
$$M_B=-9\cdot 41 \text{ t-m Ans.}$$
$$M_C=-6\cdot 43 \text{ t-m Ans.}$$

Now complete the B.M. diagram as shown in Fig. $13\cdot 14'$ (b).

13·9. Continuous beams with a sinking support

Sometimes, one of the supports, of a continuous beam sinks down due to loading, with respect to other supports, which remain at the same level. The sinking of support, effects the fixing moments of the supports.

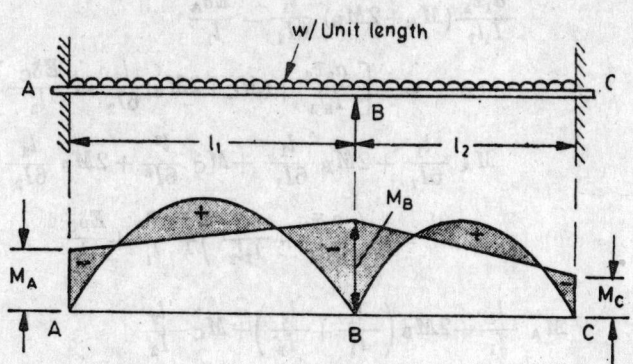

Fig. 13·15

Now consider a continuous beam ABC fixed at A and C and supported at B as shown in Fig. $13·15$ Let the support B sink down from its original position.

Let $\quad \delta_A =$ Height of support A from B,

and $\qquad \delta_C =$ Height of support C from B.

We have already discussed in Art. $13·4$ that in the span AB,

$$EI_1[l_1 i_B - y_B] = a_1 \bar{x}_1 + a_1' \bar{x}_1'$$

or
$$EI_1[l_1 i_B - \delta_A] = a_1 \bar{x}_1 + a_1' \bar{x}_1'$$

We have also discussed that

$$a_1' \bar{x}_1' = (M_A + 2M_B) \frac{l_1^2}{6} \cdot$$

$$\therefore \quad EI_1[l_1 i_B + \delta_A] = a_1 \bar{x}_1 + (M_A + 2M_B) \frac{l_1^3}{6}$$

$$EI_1 l_1 \, i_B + EI_1 \delta_A = a_1 \bar{x}_1 + (M_A + 2M_B) \frac{l_1^2}{6}$$

$$Ei_B + \frac{E\delta_A}{l_1} = \frac{a_1 \bar{x}_1}{I_1 l_1} + (M_A + 2M_B) \frac{l_1}{6I_1}$$

or
$$Ei_B = \frac{a_1 \bar{x}_1}{I_1 l_1} + (M_A + 2M_B) \frac{l_1}{3I_1} - \frac{E\delta_A}{l_1}$$

Similarly, in the span BC taking C as the origin and x positive to the left.

$$Ei'_B = \frac{a_2 x_2}{I_2 l_2} + (M_C + 2M_B) \frac{l_2}{6I_2} - \frac{E\delta_C}{l_2}$$

Since i_B is equal to $-i'_B$, therefore Ei_B is equal to $-Ei'_B$.

or
$$\frac{a_1\bar{x}_1}{I_1l_1}(M_A+2M_B)\frac{l_1}{6I_1} - \frac{E\delta_A}{l_1}$$

$$=-\left[\frac{a_2\bar{x}_2}{I_2l_2}+(M_C+2M_B)\frac{l_2}{6I_2} - \frac{E\delta_C}{l_2}\right]$$

$$M_A\frac{l_1}{6I_1}+2M_B\frac{l_1}{6I_1}+M_C\frac{l^2}{6I^2}+2M_B\frac{l_2}{6I_2}$$

$$=-\left(\frac{a_1\bar{x}_1}{I_1l_1}+\frac{a_2\bar{x}_2}{I_2l_2}\right)+\frac{E\delta_A}{l_1}+\frac{E\delta_C}{l_2}$$

$$\therefore \quad M_A\frac{l_1}{I_1}+2M_B\left(\frac{l_1}{I_1}+\frac{l_2}{I_2}\right)+M_C\frac{l_2}{I_2}$$

$$=-\left(\frac{6a_1\bar{x}_1}{I_1l_1}+\frac{6a_2\bar{x}_2}{I_2l_2}\right)+\frac{6E\delta_A}{l_1}+\frac{6E\delta_C}{l_2}$$

This equation may also be found out by Mohr's second theorem as discussed below:

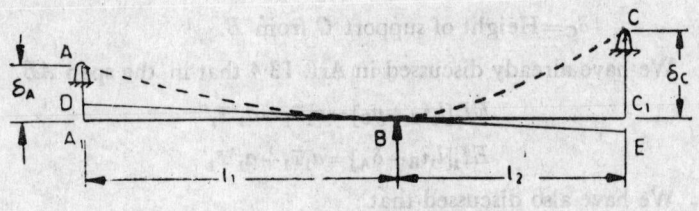

Fig. 13·16

Let the continuous beam be deflected into elastic curve ABC as shown by dotted line. At B draw a horizontal line A_1BC_1 and tangent to the elastic curve meeting the vertical line through A at D and vertical line through C at E as shown in Fig. 13·16 Let the tangent DBE make an angle θ with the horizontal line A_1BC_1. From the geometry of the figure, we find that

$$\tan\theta=\frac{A_1D}{l_1}=\frac{C_1E}{l_2}$$

We also find that

$$A_1D=\delta_A-AD$$
and
$$C_1E=CE-\delta_C$$
$$\therefore \quad \frac{\delta_A}{l_1}-\frac{AD}{l}=\frac{CE}{l_2}-\frac{\delta_C}{l_2} \quad \dots(i)$$

Now as per Mohr's second theorem (as we have discussed in Art. 13·4) we know that

$$AD = \frac{1}{6E}\left[\frac{6a_1\bar{x}_1}{I_1} + \frac{M_A l_1^2}{I_1} + \frac{2M_B l_1^2}{I_1}\right]$$

$$\therefore \quad \frac{AD}{l_1} = \frac{1}{6E}\left[\frac{6a_1\bar{x}_1}{I_1 l_1} + \frac{M_A l_1}{I_1} + \frac{2M_B l_1}{I_1}\right]$$

Similarly $\quad \dfrac{CE}{l_2} = \dfrac{1}{6E}\left[\dfrac{6a_2\bar{x}_2}{I_2 l_2} + \dfrac{M_C l_2}{I_2} + \dfrac{2M_B l_2}{I_2}\right]$

Now substituting these values in equation (*i*), and equating the same,

$$\frac{\delta_A}{l_1} - \frac{1}{6E}\left[\frac{6a_1\bar{x}_1}{I_1 l_1} + \frac{M_A l_1}{I_1} + \frac{2M_B l_1}{I_1}\right]$$

$$= \frac{1}{6E}\left[\frac{6a_2\bar{x}_2}{I_2 l_2} + \frac{M_C l_2}{I_2} + \frac{2M_B l_2}{I_2}\right] - \frac{\delta_C}{l_2}$$

$$\therefore \quad \frac{6E\delta_A}{l_1} - \frac{6a_1\bar{x}_1}{I_1 l_1} - \frac{M_A l_1}{I_1} - \frac{2M_B l_1}{I_1}$$

$$= \frac{6a_2\bar{x}_2}{I_2 l_2} + \frac{M_C l_2}{I_2} + \frac{2M_B l_2}{I_2} - \frac{6E\delta_C}{l_2}$$

or $\quad M_A\left(\dfrac{l_1}{I_1}\right) + 2M_B\left(\dfrac{l_1}{I_1} + \dfrac{l_2}{I_2}\right)M_C\left(\dfrac{l_2}{I_2}\right)$

$$= -\left(\frac{6a_1\bar{x}_1}{I_1 l_1} + \frac{6a_2\bar{x}_2}{I_2 l_2}\right) + \frac{6E\delta_A}{l_1} + \frac{6E\delta_C}{l_2}$$

This is the required Clapeyron's equation for three moments.

Note 1. If the moment of inertia of the beam is constant then

$$M_A\, l_1 + 2M_B(l_1 + l_2) + M_C\, l_2 = -\left(\frac{6a_1\bar{x}_1}{l_1} + \frac{6a_2\bar{x}_2}{l_2}\right) + \frac{6EI\delta_A}{l_1} + \frac{6EI\delta_C}{l_2}$$

2. The above formula has been derived by taking δ_A and δ_C as positive. But, while solving the numericals on sinking supports, care should always be taken to use the proper sign. The following guide rules should be kept in mind for the purpose :

I.   The three moments equation always reter to two adjacent spans. The reference level should always be taken as that of the common support.

2.   The sign of δ for the left and right support, should then be used by comparing its height, with the central support (positive for higher and negative for lower), *e.g.*

Consider a continuous beam *ABCD* in which let the support *B* sink down by an amount equal to δ. The three supports, namely, *A, C* and *D* will remain at the same level. Now, while using three moments equation for the spans *AB* and *BC*, the values of δ_A and δ_C will be positive (because both the supports *A* and *B* are at higher level than that of *B*). But while using three moments equation for the spans *BC* and *CD*, the value of δ_B will be negative (because the support *B* is at a lower level than that of *C*) and the value of δ_D will be zero (because the support *D* is at the same level as that of *C*).

Example 13·8 *A continuous beam ABC, shown in Fig. 13·17, carries a uniformly distributed load of 5 tonnes/m on AB and BC. The support B sinks by 5 mm below A and C, and the values of EI is constant throughout the beam.*

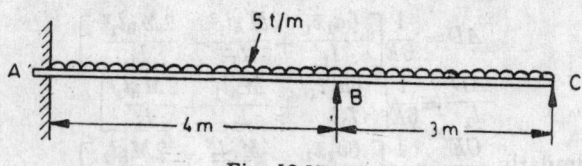

Fig. 13·17

Taking E=2,000 tonnes/cm² and I=33,200 cm⁴, find the bending moment at supports A and B, and draw the bending moment diagram.

*Solution

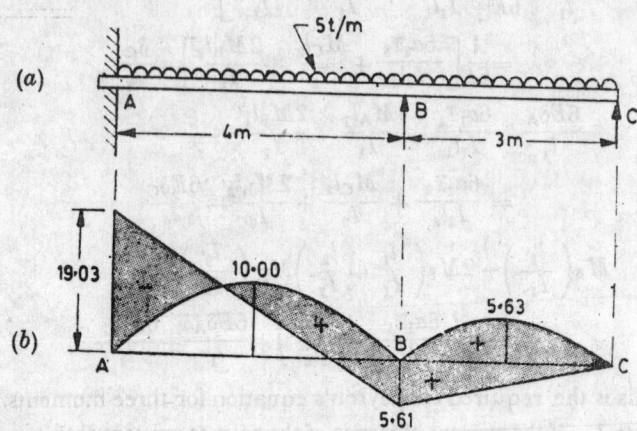

Fig. 13·18

Given. Span *AB*, $l_1=4$ m
Span, *BC*, $l_2=3$ m
Load, $w=5$ t/m

Sinking of support $B=5$ mm $=\dfrac{5}{1000}$ m

∴ $\delta_A=\delta_C=5$ mm $=\dfrac{5}{1000}$ m

Young's modulus $E=2,000$ t/cm²$=2\times10^7$ t/m²

Moment of inertia, $I=33,200$ cm⁴$=\dfrac{3\cdot32}{10^4}$m⁴

Since the beam is fixed at *A*, therefore assume a zero span to the left of *A*.

$M_A=$Fixing moment at *A*,
$M_B=$Fixing moment at *B*,
**$M_C=$Fixing moment at *C*.

*This example is also solved on page 575

†**Since the beam is simply supported at *C*, therefore the fixing moment $\bar M_c$ will be zero.

First of all, consider the beam AB as a simply supported. Therefore B.M. at the mid of the span AB

$$=\frac{wl_1^2}{8}=\frac{5\times 4^2}{8}=10 \text{ t-m}$$

Similarly B.M. at the mid of the span BC

$$=\frac{wl_2^2}{8}=\frac{5\times 3^2}{8}=\frac{45}{8}=5\cdot 63 \text{ t-m}$$

Now draw μ-diagram, with the help of above bending moments as shown in Fig. $13\cdot 18$ (b).

From the geometry of the above bending moment diagram, we find that for the spans OA and AB,

$$a_0\bar{x}_0=0$$

and

$$a_1\bar{x}_1=\frac{2}{3}\times 10\times 4\times 2=\frac{160}{3}$$

Similarly, for the spans AB and BC,

$$a_1\bar{x}_1=\frac{2}{3}\times 10\times 4\times 2=\frac{160}{3}$$

and

$$a_2\bar{x}_2=\frac{2}{3}\times 5\cdot 63\times 3\times 1\cdot 5=\frac{135}{8}$$

Now using three moments equation for the spans OA and AB,

$$M_0l_0+2M_A(0+l_1)+M_Bl_1=-\left[\frac{6a_0\bar{x}_0}{l_0}+\frac{6a_1\bar{x}_1}{l_1}\right]$$
$$+\frac{6E\delta_0 I}{l_1}+\frac{6E\delta_B I}{l_2}$$

$$0+2M_A(0+4)+M_B\times 4=-\left[0+\frac{6\times\dfrac{160}{3}}{4}\right]$$
$$+0-\frac{6\times 2\times 10^7\times\dfrac{5}{1000}\times\dfrac{3\cdot 32}{10^4}}{4}$$

$$8M_A+4M_B=-80-49\cdot 8$$

$$\therefore\qquad 2M_A+M_B=-32\cdot 45 \qquad\qquad\dots(i)$$

Now using three moments equation for the spans AB or BC,

$$M_Al_1+2M_B(l_1+l_2)+M_Cl_2=-\left[\frac{6a_1\bar{x}_1}{l_1}+\frac{6a_2\bar{x}_2}{l_2}\right]$$
$$+\frac{6EI\delta_A}{l_1}+\frac{6EI\delta_C}{l_2}$$

$$M_A\times 4+2M_B(4+3)+0=-\left[\frac{6\times\dfrac{160}{3}}{4}+\frac{6\times\dfrac{135}{8}}{3}\right]$$

$$+ \frac{6 \times 2 \times 10^7 \times \frac{3 \cdot 32}{10^4} \times \frac{5}{1000}}{4}$$

$$+ \frac{6 \times 2 \times 10^7 \times \frac{3 \cdot 32}{10^4} \times \frac{5}{1000}}{3}$$

$$4M_A + 14M_B = -80 - 33 \cdot 75 + 49 \cdot 8 + 66 \cdot 4$$

$$\therefore \qquad 4M_A + 14M_B = 2 \cdot 45 \qquad \qquad \dots (ii)$$

Solving equations (i) and (ii) we get,

$$M_A = -19 \cdot 03 \text{ t-m} \quad \text{Ans.}$$

$$M_B = 5 \cdot 61 \text{ t-m} \quad \text{Ans.}$$

Now complete the B.M. diagram as shown in Fig. 13·18 (b).

Example 13·9 A continuous beam is built-in at A and is carried over rollers at B and C as shown in Fig. 13·19 AB=BC=12 m.

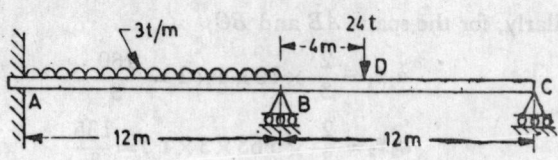

Fig. 13·19

It carries a uniformly distributed load of 3 t/m over AB and a point load of 24 tonnes over BC, 4 m from the support B, which sinks 3 cm. The values of E and I are $2 \cdot 0 \times 10^6$ kg/cm² and 2×10^5 cm⁴ respectively and uniform throughout.

Calculate the support moments and draw B.M. diagram and S.F. diagram, giving critical values.

*Solution

Given.   Span AB, $l_1 = 12$ m

Load in AB, $w = 3$ t/m

Span   BC, $l_2 = 12$ m

Load in BC, $W = 24$ t

Sinking of support B, $\delta = 3$ cm $= 0 \cdot 03$ m

Young's modulus, $E = 2 \cdot 0 \times 10^6$ kg/cm² $= 2 \cdot 0 \times 10^7$ t/m²

Moment of inertia, $I = 2 \times 10^5$ cm⁴ $= 2 \times 10^{-3}$ m⁴

Support moments at A, B and C

Let $M_A =$ support moment at A,

                   $M_B =$ support moment at B, and

              ** $M_C =$ support moment at C.

---

*This example is also solved on pages 520 , 578

**Since the beam is simply supported at C, threfore the support moment M_C will be zero.

First of all, consider the beam AB as a simply supported beam. Therefore B.M. at the mid of span AB

$$= \frac{wl_1^2}{8} = \frac{3 \times 12^2}{8} = 54 \cdot 0 \text{ t-m}$$

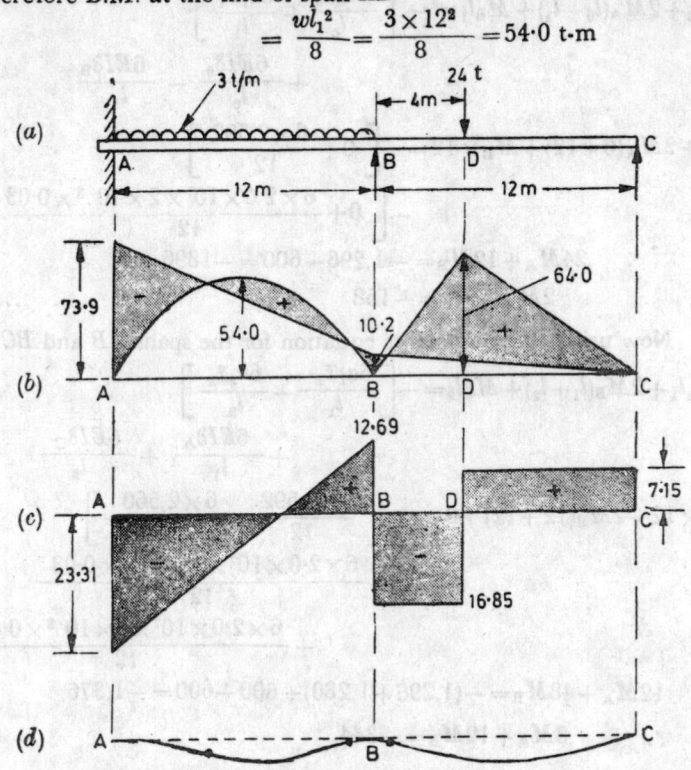

Fig. 13·20

Similarly B.M. under the 24 tonnes load

$$= \frac{Wab}{l_2} = \frac{24 \times 4 \times 8}{12} = 64 \cdot 0 \text{ t-m}$$

Now draw μ-diagram with the help of above bending moments as shown in Fig. 13·20(b). From the geometry of the above B.M. diagram, we find that for the spans $0A$ and AB,

$$a_0 \bar{x}_0 = 0$$

and

$$a_1 \bar{x}_1 = \frac{2}{3} \times 54 \times 12 \times 6 = 2,592$$

Similarly for the spans AB and BC

$$a_1 \bar{x}_1 = \frac{2}{3} \times 54 \times 12 \times 6 = 2,592$$

and

$$a_2 \bar{x}_2 = \left[\left(\frac{1}{2} \times 64 \times 8 \times \frac{2 \times 8}{3} \right) \right.$$
$$\left. + \left(\frac{1}{2} \times 64 \times 4 \right) \left(8 + \frac{4}{3} \right) \right]$$
$$= 2,560$$

Now using three moments equation for the spans $0A$ and AB,

$$M_0 l_0 + 2M_A(l_0 + l_1) + M_B l_1 = -\left[\frac{6a_0 \bar{x}_0}{l_0} + \frac{6a_1 \bar{x}_1}{l_1}\right]$$

$$+ \frac{6EI\delta_0}{l_0} - \frac{6EI\delta_B}{l_1}$$

$$0 + 2M_A(0+12) + M_B \times 12 = -\left[0 + \frac{6 \times 2{,}592}{12}\right]$$

$$-\left[0 + \frac{6 \times 2 \cdot 0 \times 10^7 \times 2 \times 10^{-3} \times 0 \cdot 03}{12}\right]$$

$$24M_A + 12M_B = -1{,}296 - 600 = -1896$$

$$2M_A + M_B = -158 \qquad \qquad \qquad \ldots(i)$$

Now using three moments equation for the spans AB and BC,

$$M_A l_1 + 2M_B(l_1 + l_2) + M_C l_2 = -\left[\frac{6a_1 \bar{x}_1}{l_1} + \frac{6a_2 \bar{x}_2}{l_2}\right]$$

$$+ \frac{6EI\delta_A}{l_1} + \frac{6EI\delta_C}{l_2}$$

$$M_A \times 12 + 2M_B(12+12) + 0 = -\left[\frac{6 \times 2{,}592}{12} + \frac{6 \times 2{,}560}{12}\right]$$

$$+ \frac{6 \times 2 \cdot 0 \times 10^7 \times 2 \times 10^{-3} \times 0 \cdot 03}{12}$$

$$+ \frac{6 \times 2 \cdot 0 \times 10^7 \times 2 \times 10^{-3} \times 0 \cdot 03}{12}$$

$$12M_A + 48M_B = -(1{,}296 + 1{,}280) + 600 + 600 = -1{,}376$$

$$\therefore \qquad 3M_A + 12M_B = -344 \qquad \qquad \ldots(ii)$$

Solving equations (i) and (ii) we get,

$$M_A = -73 \cdot 9 \text{ t-m} \quad \textbf{Ans.}$$
$$M_B = -10 \cdot 2 \text{ t-m} \quad \textbf{Ans.}$$
$$M_C = 0 \quad \textbf{Ans.}$$

Now complete the B.M. diagram as shown in Fig. 13·20 (b)

Shear force diagram

Let R_A = Reaction at A,
        R_B = Reaction at B, and
        R_C = Reaction at C.

Taking moments about B,

$$-10 \cdot 2 = R_C \times 12 - 24 \times 4 \qquad (\because \text{ B.M. at } B = -10 \cdot 2 \text{ t-m})$$

$$\therefore \qquad R_C = \frac{-10 \cdot 2 + 96}{12} = 7 \cdot 15 \text{ t}$$

Now taking moments about A,

$$-73 \cdot 9 = 7 \cdot 15 \times 24 + R_B \times 12 - 24 \times 16 - 3 \times 12 \times 6$$
$$= 171 \cdot 6 + 12R_B - 384 - 216$$
$$= 12R_B - 428 \cdot 4$$

$$\therefore \qquad R_B = \frac{-73 \cdot 9 + 428 \cdot 4}{12} = 29 \cdot 54 \ \text{t}$$

and $\qquad R_A = (3 \times 12 + 24) - (7 \cdot 15 - 29 \cdot 54) = 23 \cdot 31 \ \text{t}$

Now complete the S.F. diagram as shown in Fig. 13·20 (c). Also draw the elastic curve as shown in Fig. 13·20 (d).

13·10. Continuous beams subjected to a couple

Sometimes, a continuous beam is subjected to a couple in one (or more) of the spans. Such a beam is also analysed in the similar manner. The couple will cause negative B. M. in one part and positive B. M. in the other part of the span. While taking moment of the B. M. diagram, due care should be taken for the positive and negative bending moments.

Example 13·10. *A continuous beam ABC of constant moment of inertia carries a central load of 10,000 kg in span AB, and a central clockwise moment of 30,000 kgm in span BC as shown in Fig.*

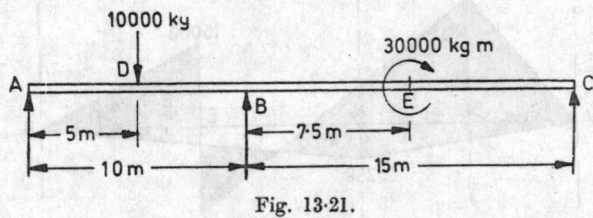

Fig. 13·21.

Find the support moment using the theorem of three moments, and plot the shear force and bending moment diagrams.

(A.M.I.E., Winter, 1974)

Solution.

Given. Span *AB*, $\qquad l_1 = 10 \ \text{m}$

Load in *AB*, $\qquad W = 10{,}000 \ \text{kg.}$

Span *BC*, $\qquad l_2 = 15 \ \text{m}$

Moment in *BC*, $\qquad M = 30{,}000 \ \text{kg}$

Bending moment diagram.

Let $\qquad \qquad *M_A =$ Support moment at *A*,

$\qquad M_B =$ Support moment at *B*, and

$\qquad *M_C =$ Support moment at *C*.

First of all, let us consider the beam *AB* and *CD* as a simply supported beam. Therefore bending moment at *D*,

$$M_D = \frac{W l_1}{4} = \frac{4000 \times 10}{4} = 10{,}000 \ \text{kg-m}$$

* Since the beam is simply supported at *A* and *C*, therefore support moment at M_A and M_C will be zero.

Similarly moment just on the right side of E

$$= \frac{\mu}{2} = \frac{30,000}{2} = 15,000 \text{ kg-m}$$

and moment just on the left side of E

$$= -\frac{\mu}{2} = -\frac{30,000}{2} = -15,000 \text{ kg-m}$$

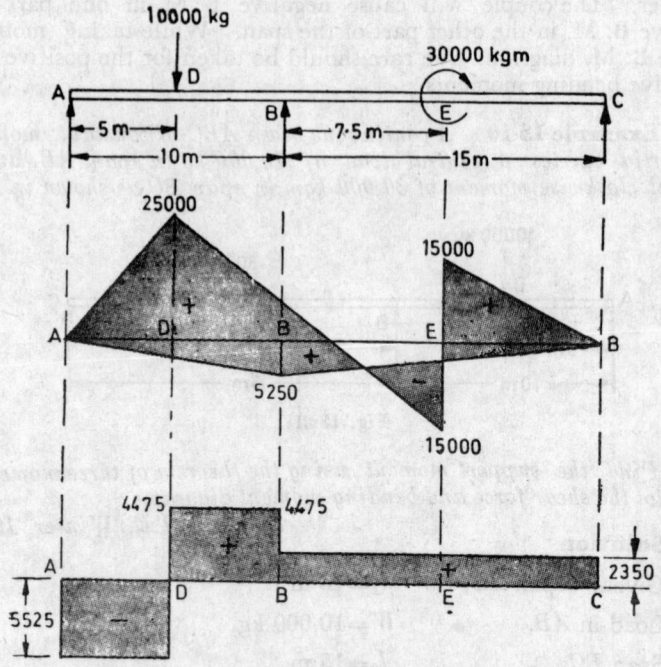

Fig. 13·22.

Now draw the μ-diagram with the help of above bending moments as shown in Fig. 13·22 (b). From the geometry of the bending moment diagrams, we find that for spans AB and BC,

$$a_1\bar{x}_1 = \left[\left(\frac{1}{2} \times 5 \times 25,000 \times \frac{5 \times 2}{2}\right) + \left(\frac{1}{2} \times 5 \times 25,000\right)\left(5 + \frac{5}{3}\right)\right]$$

$$= 6,25,000 \text{ kg-m}$$

$$a_2\bar{x}_2 = \left[\left(\frac{1}{2} \times 7{\cdot}5 \times 15,000 \times 5\right) - \left(\frac{1}{2} \times 7{\cdot}5 \times 15,000 \times 10\right)\right]$$

$$= -2,81,250 \text{ kg-m}$$

Now using three moments equation,

$$M_A l_1 + 2M_B(l_1+l_2) + M_C l_2 = -\left(\frac{6a_1\bar{x}_1}{l_1} + \frac{6a_2\bar{x}_2}{l_2}\right)$$

$$0 + 2M_B(10+15) + 0 = -\left(\frac{6\times6,25,000}{10} - \frac{6\times2,81,250}{15}\right)$$

$$50\ M_B = 2,62,500$$

$$M_B = 5250 \text{ kg-m}$$

Now plot the bending moment diagram as shown in Fig. 13·22 (b).

Shear force diagram.

Let
$$R_A = \text{Reaction at } A,$$
$$R_B = \text{Reaction at } B, \text{ and}$$
$$R_C = \text{Reaction at } C.$$

Taking moments about B,

$$R_A \times 10 - 10,000 \times 5 = 5250$$

$$\therefore \qquad R_A = \frac{5250+20,000}{10} = 5525 \text{ kg}$$

Similarly $\quad R_C \times 15 - 30,000 = 5250$

$$\therefore \qquad R_C = \frac{5250+30,000}{15} = 2,350 \text{ kg}$$

and $$R_B = 10,000 - (5525 + 2,350)$$
$$= 2,125 \text{ kg}$$

Now draw the shear force diagram as shown in Fig 13·22 (c).

Exercise 13-2

1. A continuous beam $ABCD$ is fixed at A and simply supported at B and C, the beam CD is overhanging. The span $AB=6$ m, $BC=5$ m and overhanging $CD=2·5$ m. The moment of inertia of the span BC is $2I$ and that of AB and CD is I. The beam is carrying a uniformly distributed load of 2 t/m over the span AB, a point load of 5 T in BC at a distance of 3 m from B and a point load of 3 T at the free end.

Determine the fixing moments at A, B and C, and draw the B.M. diagram. (*Ranchi University*)

[Ans. —8·11 t-m ; —1·79 t-m ; —20·0 t-m]

2. A beam $ABCD$ is continuous over three spans $AB=8$ m, $BD=4$ m and $CD=8$ m. The beam AB and BC is subjected to a uniformly distributed load of 1·5 t/m, whereas there is a central point load of 4 t in CD. The moment of inertia of AB and CD is $2I$ and that of BC is I. The ends A and D are fixed. During loading, the support B sinks down by 1 cm. Find the fixed end moments. Take $E=2·0 \times 10^3$ t/m² and $I=16,000$ cm⁴.

(Bombay University)

[**Ans.** —16·53 t-m ; —30·66 t-m ; —77·33 t-m ; —21·33 t-m]

3. A continuous beam $ABCD$ 20 m long is supported at B and C and fixed at A and D. The spans AB, BC and CD are 6 m, 8 m and 6 m respectively. The span AB carries a uniformly distributed load of 1 t/m, the span BC carries a central point load of 10 t and the span CD carries a point load of 5 t at a distance of 3 m from C. During loading, the support B sinks by 1 cm. Find the fixed end moments and draw the B.M. diagram. Take $E=2·0 \times 10^3$ t/cm² and $I=30,000$ cm⁴. The moment of inertia of the spans AB and CD is I and that of $BC=2I$.

(Allahabad University)

[**Ans.** —5·74 t-m ; —2·57 t-m ; —9·2 t-m ; —2·07 t-m]

HIGHLIGHTS

1. A beam, which is supported on more than two supports, is called a continuous beam. The B.M. diagram for a continuous beam is drawn in the following two stages :

(a) By considering the beam as a series of discontinuous beams, from support to support, and drawing the usual μ-diagram due to vertical loads.

(b) By superimposing the usual μ'-diagram due to end moments over μ-diagram.

2. Clapeyron's theorem of three moments states : *"If a beam has n supports, the end ones being fixed, then the same number of equations required to determine the support moments, may be obtained from the consecutive pairs of spans i.e. AB-BC, BC-CD, CD-DE and so on."*

3. When a continuous beam is simply supported on its one, or both the end supports, the fixing end moments on the simply supported end is zero.

4. When a beam is fixed at its one or both ends, then an imaginary zero span is taken and the three moments theorem is applied as usual.

5. When a beam is overhanging at its one or both ends, the overhanging beam behaves like a cantilever. The fixing moment on the end support is found out by the cantilever action of the overhanging part of the beam.

6. When one of the supports of a continuous beam sinks down, it affects the fixing moments on the supports.

Do You Know ?

1. What is a continuous beam ? Explain the way in which a continuous beam gets deflected as a result of loading.

2. Explain the theorem of three moments.

3. Prove the Clapeyron's theorem of three moments.

4. How will you apply the theorem of three moments to a fixed beam ?

5. Explain the effect on a continuous beam, when one of the intermediate supports sinks down.

14

SLOPE DEFLECTION METHOD

14·1. Introduction

This method was first proposed by Prof. G.A. Maney in 1915, as a general method for the analysis of indeterminate beams and frames. This method had been widely used, till Prof. Hardy Cross introduced moment distribution method in 1930, which is very simple to understand. and has scientific and systematic approach. Today, the slope deflection method is not encouraged in design offices, because of universal use of moment distribution method.

In this method, if the slopes at the ends, and the relative displacement of the ends are known, the moment at the ends can be found in terms of slopes, deflections, stiffness and length of the members.

14·2. Assumptions in the slope deflection method

The slope deflection method is based on the following simplified assumptions :

1. All the joints of the frame are rigid, *i.e.*, the angles between the members at the joints do not change, when the frame is loaded.

2. Whenever the beams or frames are deflected, the rigid joints are considered to rotate as a whole, *i.e.* the angles

512

> between the tangents to the various branches of the elastic curve meeting at a joint, remain the same as those in the original structure.

3. Distortions, due to axial and shear stresses, being very small, are neglected.

14·3. Sign Conventions

Though there are different sign conventions used in different books, yet the following sign conventions will be used in this book, which are widely used and internationally recognised.

1. *Moments.* All the clockwise moments at the ends of a member will be taken as positive ; whereas all the anticlockwise moments at the ends will be taken as negative.

2. *Rotation.* Whenever a joint rotates, all the angles of rotation, whose tangent on the elastic curve rotates in the clockwise direction from the joint will be taken as positive ; whereas those whose tangent rotates in the anticlockwise direction will be taken as negative.

3. *Sinking.* Whenever a right support sinks down with respect to the left support, the moments on both the supports will be negative, whereas if the left support sinks down the moments on both the supports will be positive. A little consideration will show that this sign convention is derived from the first convention.

Note : These sign conventions are used only for this method.

14·4. Slope deflection equations

The general slope deflection equations give the support moments of a beam in terms of slopes, deflection, stiffness and lengths of the beams. In the succeeding pages, we shall derive the slope deflection equations for the following two cases :

(1) When the supports are at the same level, and

(2) When one of the support is at a lower level.

14·5. Slope deflection equations when the supports are at the same level

Consider an intermediate span AB of a continuous beam, subjected to some loading as shown in Fig. 14·1 (a).

Let $l =$ Length of the span AB,

         $E =$ Young's modulus for the beam material,

         $I =$ Moment of inertia of the beam section,

     $M_{AB} =$ Support moment at A, and

     $M_{BA} =$ Support moment at B.

We know that due to external loading, the beam will deflect, as a result of which the ends A and B of the beam will rotate as shown in Fig 14·1 (b).

Let
$$i_A = \text{Slope at } A, \text{ and}$$
$$i_B = \text{Slope at } B.$$

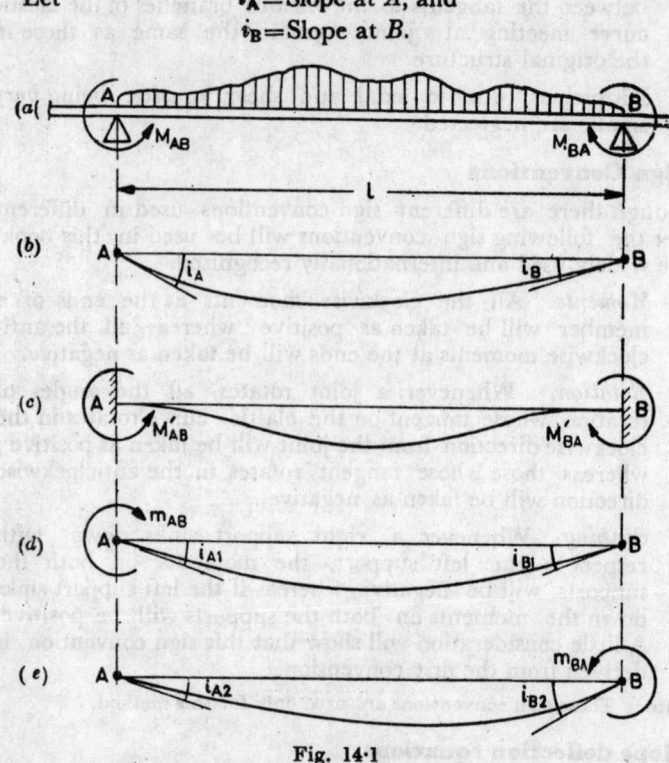

Fig. 14·1

First of all, let us apply end moments M'_{AB} and M'_{BA} at the ends A and B respectively, of such magnitude, that the slopes i_A and i_B due to external loading are reduced to zero as shown in Fig. 14·1 (c). A little consideration will show that the applied moments M'_{AB} and M'_{BA} has action similar to that of fixed end moments. These moments may be easily found out from the standard fixed beam formulae, for a given system of loading.

Now consider the additional moments m_{AB} and m_{BA} at the ends A and B, whose net effect is to cause slopes equal to i_A and i_B. Let the moment m_{AB} cause slopes equal to i_{A1} and i_{B1} at the ends A and B respectively as shown in Fig. 14·1 (d). Similarly let the moment m_{BA} cause slopes equal to i_{A2} and i_{B2} at the ends A and B respectively as shown in Fig. 14·1 (e). From the geometry of the figure, we find that

$$i_A = +(0 + i_{A1} + i_{A2}) \qquad (\because i_A \text{ is positive})$$
$$= i_{A1} + i_{A2} \qquad\qquad\qquad\qquad ...(i)$$

and
$$i_B = -(0 + i_{B2} + i_{B1}) \qquad (\because i_B \text{ is negative})$$
$$= -(i_{B2} + i_{B1}) \qquad\qquad\qquad ...(ii)$$

By super imposing the moments, we also find that
$$M_{AB} = -(-M'_{AB} + m_{AB})$$
$$= -m_{AB} + M'_{AB} \qquad\qquad\qquad ...(iii)$$

and $\qquad M_{BA} = -m_{BA} + M'_{BA}$ $\qquad\qquad ...(iv)$

We know that

$$i_{A1} = \frac{m_{AB}\,l}{3EI} \qquad\qquad i_{B1} = \frac{m_{AB}.l}{6EI}$$

$$i_{A2} = \frac{m_{BA}.l}{6EI} \qquad\qquad i_{B2} = \frac{m_{BA}.l}{3EI}$$

Now substituting the above values in equations (i) and (ii),

$$i_A = i_{A1} + i_{A2} = \frac{m_{AB}.l}{3EI} + \frac{m_{BA}.l}{6EI}$$

$$= \frac{l}{6EI}\,(2m_{AB} + m_{BA}) \qquad\qquad ...(v)$$

and $\qquad i_B = -(i_{B2} + i_{B1}) = -\left[\frac{m_{BA}.l}{3EI} + \frac{m_{AB}.l}{6EI}\right]$

$$= -\frac{l}{6EI}\,(2m_{BA} + m_{AB}) \qquad\qquad ...(vi)$$

Now solving equations (iii) and (iv) for m_{AB} and m_{BA} we get,

$$m_{AB} = -\frac{2EI}{l}\,(2i_A + i_B)$$

and $\qquad m_{BA} = -\frac{2EI}{l}\,(2i_B + i_A)$

Now substituting the above values of m_{AB} and m_{BA} in equations (iii) and (iv),

$$M_{AB} = \frac{2EI}{l}\,(2i_A + i_B) + M'_{AB} \qquad\qquad ...(vii)$$

and $\qquad M_{BA} = \frac{2EI}{l}\,(2i_B + i_A) + M'_{BA} \qquad\qquad ...(viii)$

The above four equations *i.e.* (v), (vi), (vii) and $(viii)$ are the general slope deflection equations, when the supports are at the same level.

14·6. Slope deflection equations when one of the supports is at a lower level.

We have already discussed in Art. 12·9 that whenever one of the supports of a beam is at a lower level as compared to the other it will cause a moment at both the ends. We know that the magnitude of this moment is $\dfrac{6EI\delta}{l^2}$.

Now consider an intermediate span AB of a continuous beam having its support B at a lower level δ as compared to the support A. We know that the moment at A,

$$M_A = M_B = -\frac{6EI\delta}{l^2}$$

Adding these additional moments to the support moments obtained in the previous article, we get the support moment at A,

$$M_{AB} = \frac{2EI}{l}(2i_A + i_B) + M'_{AB} - \frac{6EI\delta}{l^2}$$

$$= \frac{2EI}{l}\left(2i_A + i_B - \frac{3\delta}{l}\right) + M'_{AB} \qquad ...(i)$$

and

$$M_{BA} = \frac{2EI}{l}(2i_B + i_A) + M'_{BA} - \frac{6EI\delta}{l^2}$$

$$= \frac{2EI}{l}\left(2i_B + i_A - \frac{3\delta}{l}\right) + M'_{BA}$$

The above two equations are the general slope deflection equations, when one of the supports is at a lower level.

14·7. Application of slope deflection equations

We have already derived the general equations for slope deflection in the previous article. Now we shall study the application of these equations for the following types of beams and frames :

1. Continuous beams,
2. Simple frames, and
3. Portal frames.

14·8. Continuous beams

A continuous beam is essentially a statically indeterminate structure, which must satisfy the geometrical conditions and statical equilibrium. In the slope deflection method, the geometrical conditions are satisfied by the slope deflection equations. In order to satisfy the statical equilibrium, the algebraic sum of the moments acting at a joint must be equal to zero. The procedure, for the solution of a problem, on a continuous beam, is given below :

1. First of all, treat each span as a fixed beam, and calculate the fixed end moments.
2. Now express all the end moments in terms of fixed end moments and joint rotations by using slope deflection equations.
3. Then write down the equilibrium equations for the individual joints.
4. Solve the end rotations for all supports.
5. Now substitute the values of end rotations, back into the slope deflection equations and find out the end moments.
6. Determine all the reactions, and draw shear force and bending moment diagrams.

Example 14·1. *Evaluate the bending moment and shear force diagrams of the beam shown in Fig. 14·2.*

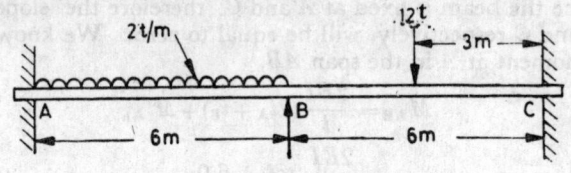

Fig. 14·2

What are the reactions at the supports ?

(U.P.S.C., Engg. Services, 1973)

*Solution.

Given. Span AB, $l_1 = 6$ m
Span BC, $\qquad l_2 = 6$ m
Load in AB, $\qquad w = 2$ t/m
Load in BC $\qquad W = 12$ t

Support moments at A, B and C

Let $\qquad M_{AB} = $ Moment at A,

$M_{BA} = $ Moment at B in span AB,

$M_{BC} = $ Moment at B in span BC, and

$M_{CB} = $ Support Moment at C.

(i) Fixed end moments

First of all, let us assume the beam ABC to be made up of fixed beams AB and BC. From the geometry of the figure, we find that the fixed end moment at A in the span AB.

$$M'_{AB} = -\frac{wl^2}{12} \qquad \text{(Minus sign due to anticlockwise)}$$

$$= -\frac{2 \times 6^2}{12} = -6·0 \text{ t-m}$$

and $\qquad M'_{BA} = +\frac{wl^2}{12} \qquad \text{(Plus sign due to clockwise)}$

$$= \frac{2 \times 6^2}{12} = 6·0 \text{ t-m}$$

Similarly fixed end moment at B in the span BC,

$$M'_{BC} = -\frac{Wl}{8} \qquad \text{(Minus sign due to anticlockwise)}$$

$$= -\frac{12 \times 6}{8} = -9·0 \text{ t-m}$$

and $\qquad M'_{CB} = +\frac{Wl}{8} \qquad \text{(Plus sign due to clockwise)}$

$$= \frac{12 \times 6}{8} = +9·0 \text{ t-m}$$

_____

*This example is also solved on pages 488 and 559.

(ii) Slope deflection equations

Since the beam is fixed at A and C, therefore the slopes i_A and i_C, at A and C respectively, will be equal to zero. We know that the support moment at A in the span AB,

$$M_{AB} = \frac{2EI}{l}(2i_A + i_B) + M'_{Ab}$$

$$= \frac{2EI}{6} \cdot i_B - 6 \cdot 0 \qquad (\because\ i_A = 0)$$

$$= \frac{EI}{3} \times i_B - 6 \cdot 0 \qquad\qquad ...(i)$$

and

$$M_{BA} = \frac{2EI}{l}(2i_B + i_A) + M'_{BA}$$

$$= \frac{2EI}{6} \times 2i_B + 6 \cdot 0 \qquad (\because\ i_A = 0)$$

$$= \frac{2EI}{3} \times i_B + 6 \cdot 0 \qquad\qquad ...(ii)$$

Similarly support moment at B in the span BC,

$$M_{BC} = \frac{2EI}{l}(2i_B + i_C) + M'_{BC}$$

$$= \frac{2EI}{6} \times 2i_B - 9 \cdot 0 \qquad (\because\ i_C = 0)$$

$$= \frac{2EI}{3} \times i_B - 9 \cdot 0 \qquad\qquad ...(iii)$$

and

$$M_{CB} = \frac{2EI}{l}(2i_C + i_B) + M'_{CB}$$

$$= \frac{2EI}{6} \times i_B + 9 \cdot 0 \qquad (\because\ i_C = 0)$$

$$= \frac{EI}{3} \times i_B + 9 \cdot 0 \qquad\qquad ..(iv)$$

(iii) Equilibrium equations

Since the joint B is in equilibrium, therefore equating $M_{BA} + M_{BC}$ equal to zero,

$$\frac{2EI}{3} \times i_B + 6 \cdot 0 + \frac{2EI}{3} \times i_B - 9 \cdot 0 = 0$$

$$\frac{4EI}{3} \times i_B = 3 \cdot 0$$

$$\therefore \qquad\qquad i_B = \frac{9}{4EI}$$

(iv) Final moments.

Substituting the value of i_B in equations (i) to (iv),

$$M_{AB} = \frac{EI}{3} \times \frac{9}{4EI} - 6 \cdot 0 = -5 \cdot 25\ \text{t·m} \quad \textbf{Ans.}$$

$$M_{BA} = \frac{2EI}{3} \times \frac{9}{4EI} + 6 \cdot 0 = +7 \cdot 5\ \text{t·m} \quad \textbf{Ans.}$$

$$M_{BC} = \frac{2EI}{3} \times \frac{9}{4EI} - 9.0 = -\mathbf{7.5 \ t \cdot m} \quad \textbf{Ans}.$$

and

$$M_{CB} = \frac{EI}{3} \times \frac{9}{4EI} + 9.0 = +\mathbf{9.75 \ t \cdot m} \quad \textbf{Ans}$$

We know that the B.M. at the mid of the span AB

$$= \frac{wl^4}{8} = \frac{2 \times 6^2}{8} = 9.0 \ t \cdot m$$

Load B.M. under the load D

$$= \frac{Wl}{4} = \frac{12 \times 6}{4} = 18.0 \ t \cdot m$$

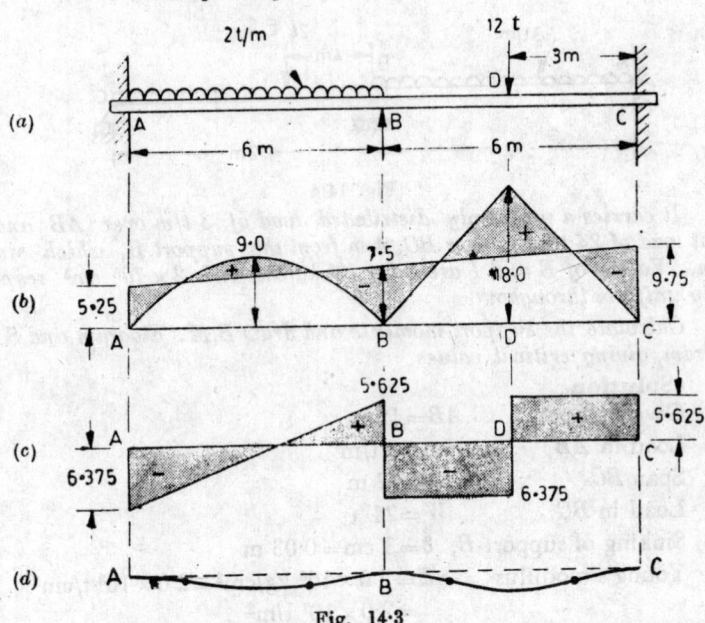

Fig. 14.3

Now complete the final bending diagram* as shown in Fig. 14.3 (b).

Reactions at the supports

Let R_A = Reaction at A,

R_B = Reaction at B, and

R_C = Reaction at C

Taking moments about B, and equating the same,

$$\bar{R}_A \times 6 + M_B = M_A + 2 \times 6 \times 3$$

$$R_A \times 6 - 7.5 = -5.25 + 36$$

$$\therefore \qquad R_A = \mathbf{6.375 \ t} \quad \textbf{Ans.}$$

*The most important point, while drawing the B.M. diagram, is that the moment $M_{BA} = +7.5$ t-m and $M_{CB} = +9.75$ t-m. But in the B.M. diagram these moments have been taken as -7.5 t-m and -9.75 t-m respectively. The simple reason for the same, is that these moments have been taken as negative as they tend to bend the beam with convexity upwards, and thus are hogging moments.

Again taking moments about B, and equating the same,

$$R_C \times 6 + M_B = M_C + 12 \times 3$$
$$R_C \times 6 - 7.5 = -9.75 + 36$$
$$R_C = 6.625 \text{ t} \quad \textbf{Ans}.$$

and
$$R_B = (2 \times 6 + 12) - (6.375 + 5.625) \text{ t}$$
$$= 12 \text{ 't' } \quad \textbf{Ans}.$$

Now draw the shear force diagram and elastic curve as shown in Fig. 14·3 (c) and (d).

Example 14·2. *A continuous beam is built-in at A and is carried over rollers at B and C as shown in Fig. 14·4. AB=BC=12 m.*

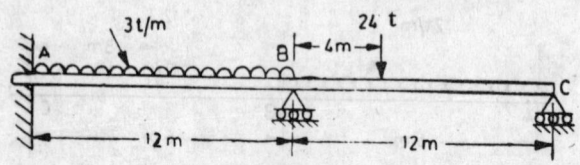

Fig. 14·4

It carries a uniformly distributed load of 3 t/m over AB and a point load of 24 tonnes over BC, 4 m from the support B, which sinks 3 cm. Values of E and I are $2 \cdot 0 \times 10^6$ kg/cm² and 2×10^5 cm⁴ respectively uniform throughout.

Calculate the support moments and draw B.M. diagram and S.F. diagram, giving critical values.

*Solution.

Given. Span, $AB = 12$ m

Load in AB, $w = 3$ t/m

Span BC $= 12$ m

Load in BC, $W = 24$ t

Sinking of support B, $\delta = 3$ cm $= 0.03$ m

Young's modulus $E = 2 \cdot 0 \times 10^6$ kg/cm² $= 2 \cdot 0 \times 10^3$ t/cm²
$$= 2 \cdot 0 \times 10^7 \text{ t/m}^2$$

Moment of inertia, I $= 2 \times 10^5$ cm⁵ $= 2 \times 10^{-3}$ m²

Support moments at A, B, and C

Let M_{AB} = Support moment at A,

                M_{BA} = Support moment at B in span AB,

                M_{BC} = Support moment at B in span BC,

                M_{CB} = Support moment at C.

(i) Fixed end moments

First of all, let us assume the beam ABC to be made up of fixed beams AB and BC. From the geometry of the figure, we find that the fixed end moment at A in the span AB,

$$M'_{AB} = -\frac{wl^2}{12} = -\frac{3 \times 12^2}{12} = -36 \cdot 0 \text{ t-m}$$

and
$$M'_{BA} = +\frac{wl^2}{12} = +\frac{3 \times 12^2}{12} = +36 \cdot 0 \text{ t-m}$$

---

*This example is also solved on pages 495 and 578

Similarly fixed end moments at B in the span BC,

$$M'_{BC} = -\frac{Wab^2}{l^2} = -\frac{24 \times 4 \times 8^2}{12^2} = -42.67 \text{ t-m}$$

and

$$M'_{CB} = +\frac{Wab^2}{l} = +\frac{24 \times 8 \times 4^2}{12^2} = +21.33 \text{ t-m}$$

(*ii*) **Slope deflection equations**

Since the beam is fixed at A, therefore the slope i_A will be equal to zero. We know that the support moment at A in the span AB,

$$M_{AB} = \frac{2EI}{l}\left(2i_A + i_B - \frac{3\delta}{l}\right) + M'_{AB}$$

$$= \frac{2EI}{12}\left(i_B - \frac{3 \times 0.03}{12}\right) - 36.0 \qquad (\because \; i_A = 0)$$

$$= \frac{EI}{6}\left(i_B - \frac{3}{400}\right) - 36.0 \qquad \qquad \ldots(i)$$

and

$$M_{BA} = \frac{2EI}{l}\left(2i_B + i_A - \frac{3\delta}{l}\right) + M'_{BA}$$

$$= \frac{2EI}{12}\left(2i_B - \frac{3 \times 0.03}{12}\right) + 36.0 \qquad (\because \; i_A = 0)$$

$$= \frac{EI}{6}\left(2i_B - \frac{3}{400}\right) + 36.0 \qquad \qquad \ldots(ii)$$

Similarly support moment at B in the span BC,

$$M_{BC} = \frac{2EI}{l}\left(2\,i_B + i_C + \frac{3\delta}{l}\right) + M'_{BC}$$

(In this case δ is plus)

$$= \frac{2EI}{12}\left(2i_B + i_C + \frac{3 \times 0.03}{12}\right) - 42.67$$

$$= \frac{EI}{6}\left(2i_B + i_C + \frac{3}{400}\right) - 42.67 \qquad \ldots(iii)$$

and

$$M_{CB} = \frac{2EI}{l}\left(2i_C + i_B + \frac{3\delta}{l}\right) + 21.33$$

$$= \frac{2EI}{12}\left(2i_C + i_B + \frac{3 \times 0.03}{12}\right) + 21.33$$

$$= \frac{EI}{6}\left(2i_C + i_B + \frac{3}{400}\right) + 21.33 \qquad \ldots(iv)$$

(*iii*) **Equilibrium equations**

Since the beam is simply supported at C, therefore the support moment at C will be zero. Now equating the equation (*iv*) to zero,

$$\frac{EI}{6}\left(2i_C + i_B + \frac{3}{400}\right) + 21.33 = 0$$

or

$$\frac{EI}{6}\left(2i_C + i_B + \frac{3}{400}\right) = -21.33$$

$$EI\left(2i_C + i_B + \frac{3}{400}\right) = -128.0$$

$$2.0 \times 10^7 \times 2 \times 10^{-3}\left(2i_C + i_B + \frac{3}{400}\right) = -128$$

$$80,000 \ i_C + 40,000 \ i_B + 300 = -128$$
$$80,000 \ i_C + 40,000 \ i_B = -428 \qquad \qquad ...(v)$$

Since the joint B is in equilibrium, therefore equating $M_{BA} + M_{BC}$ equal to zero,

$$\frac{EI}{6}\left(2i_B - \frac{3}{400}\right) + 36 + \frac{EI}{6}\left(2i_B + i_C + \frac{3}{400}\right) - 42 \cdot 67 = 0$$

$$\frac{EI}{6}\left(2i_B - \frac{3}{400}\right) + \frac{EI}{6}\left(2i_B + i_C + \frac{3}{400}\right) = 6 \cdot 67$$

$$EI\left(2i_B - \frac{3}{400}\right) + EI\left(2i_B + i_C + \frac{3}{400}\right) = 40$$

$$EI(4i_B + i_C) = 40$$
$$2 \cdot 0 \times 10^7 \times 2 \times 10^{-3} \ (4i_B + i_C) = 40$$
$$1,60,000 \ i_B + 40,000 i_C = 40$$

or $\qquad\qquad 3,20,000 \ i_B + 80,000 \ i_C = 80 \qquad\qquad$ (Multiplying by 2)

Subtracting equation (v) from the above equation,

$$2,80,000 \ i_B = 508$$

$$\therefore \qquad\qquad i_B = \frac{508}{2,80,000} = \frac{127}{70,000}$$

Substituting this value of i_B in equation (v),

$$80,000 \ i_C + 40,000 \times \frac{127}{70.000} = -428$$

$$\therefore \qquad\qquad i_C = -\frac{3504}{7 \times 80,000}$$

(iv) Final moments

Substituting the values of i_B and i_C in equation (i) to (iv),

$$M_{AB} = \frac{2 \cdot 0 \times 10^7 \times 2 \times 10^{-3}}{6}\left(\frac{127}{70,000} - \frac{3}{400}\right) - 36 \cdot 0$$

$$= 73 \cdot 9 \ \textbf{t} \cdot \textbf{m} \quad \textbf{Ans}.$$

$$M_{BA} = \frac{2 \cdot 0 \times 10^7 \times 2 \times 10^{-3}}{6}\left(2 \times \frac{127}{70,000} - \frac{3}{400}\right) + 36 \cdot 0$$

$$= +10 \cdot 2 \ \textbf{t-m} \quad \textbf{Ans}.$$

$$M_{BC} = \frac{2 \cdot 0 \times 10^7 \times 2 \times 10^{-3}}{6}\left(2 \times \frac{127}{70,000} - \frac{3,504}{7 \times 80,000} + \frac{3}{400}\right)$$
$$\qquad\qquad\qquad -42 \cdot 67 \ \text{t-m}$$

$$= -10 \cdot 2 \ \textbf{t-m} \quad \textbf{Ans}.$$

Bending moment diagram

We know that B.M. at the mid of the span AB

$$= \frac{wl^2}{8} = \frac{3 \times 12^2}{8} = 54 \cdot 0 \ \text{t-m}$$

Similarly B.M. under the 24 tonnes load

$$= \frac{Wab}{l} = \frac{24 \times 4 \times 8}{12} = 64 \cdot 0 \ \text{t-m}$$

Now complete the B.M. diagram as shown in Fig. 14·5 (b)

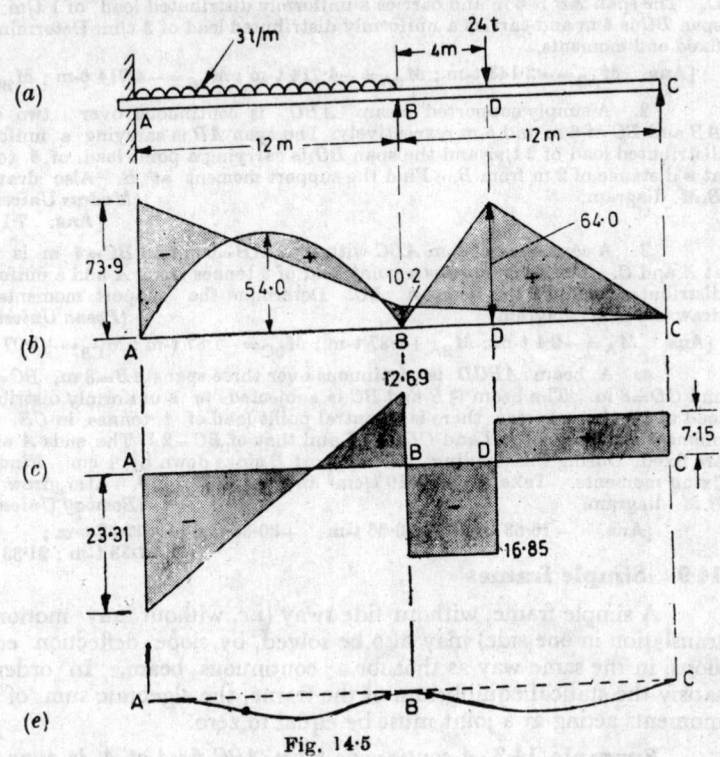

Fig. 14·5

Shear force diagram

Let R_A = Reaction at A,
              R_B = Reaction at B, and
              R_C = Reaction at C.

Taking moments about B,

$$-10·2 = R_C \times 12 - 24 \times 4 \qquad \text{(B.M. at } B = -10·2 \text{ t-m)}$$

∴ $$R_C = \frac{-10·2 + 96·0}{12} = 7·15 \text{ t}$$

Now taking moments about A,

$$-73·9 = 7·15 \times 24 + R_B \times 12 - 24 \times 16 - 3 \times 12 \times 6$$
$$= 171·6 + 12 R_B - 384 - 216$$
$$R_B = \frac{-73·9 + 428·4}{12} = 29·54 \text{ t}$$

and $$R_A = (3 \times 12 + 24) - (7·15 + 29·54) = 23·31 \text{ t}$$

Now complete the S.F. diagram as shown in Fig. 14·5 (c).

The elastic curve *i.e.* deflected shape of the centre line of the beam is shown in Fig. 14·5 (d).

EXERCISE 14–1

1. A continuous beam ABC is fixed at A and simply supported at B and C. The span AB is 6 m and carries a uniformly distributed load of 1 t/m. The span BC is 4 m and carries a uniformly distributed load of 3 t/m. Determine the fixed end moments.

[Ans. $M_{AB} = -2.143$ t-m ; $M_{BA} = +4.714$ t-m ; $M_{BC} = -4.714$ t-m ; $M_{BC} = 0$]

2. A simply supported beam ABC is continuous over two spans AB and BC of 6 m and 5 m respectively. The span AB is carrying a uniformly distributed load of 2 t/m and the span BC is carrying a point load of 5 tonnes at a distance of 2 m from B. Find the support moment at B. Also draw the B.M. diagram.   (Madras University)

[Ans.   7.1 t-m]

3. A continuous beam ABC with span $AB = 6$ m and $BC = 4$ m is fixed at A and C. The beam carries a point load of 3 tonnes from A and a uniformly distributed load of 2 t/m in span BC. Determine the support moments and draw the B.M. diagram.   (Poona University)

[Ans. $M_A = -2.4$ t-m ; $M_{BA} + 1.87$ t-m ; $M_{BC} = -1.87$ t-m ; $M_{CB} = +3.07$ t-m]

4. A beam $ABCD$ is continuous over three spans $AB = 8$ m, $BC = 4$ m and $CD = 8$ m. The beam AB and BC is subjected to a uniformly distributed load of 1.5 t/m, whereas there is a central point load of 4 tonnes in CB. The moment of inertia of AB and CD is 2 I and that of $BC = 2$ I. The ends A and D are fixed. During the loading the support B sinks down by 1 cm. Find the fixing moments. Take $E = 2.0 \times 10$ t/cm^2 and $I = 16,000$ cm^4. Also draw the B.M. diagram.   (Bombay University)

[Ans. -16.53 t-m ; -30.66 t-m; $+30.66$ t-m ; -77.33 t-m ;
                                          $+77.33$ t-m ; 21.33 t-m]

14.9. Simple frames

A simple frame, without side sway (i.e. without any motion of translation in one side) may also be solved, by slope deflection equations, in the same way as that for a continuous beam. In order to satisfy the statical equilibrium of the frame, the algebraic sum of the moments acting at a joint must be equal to zero.

Example 14.3. *A continuous beam ABC fixed at A, is supported on an elastic column BD and is loaded as shown in Fig. 14.6.*

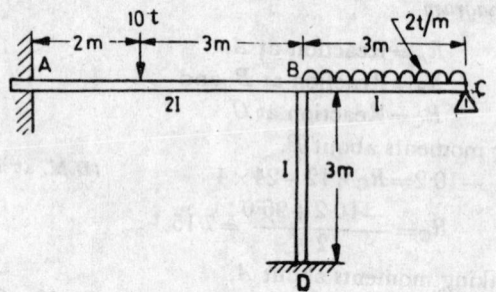

Fig. 14.6

Analyse the frame and draw the B.M. diagram. Take moment of inertia of AB as 2 I and that of BC and CD as I.

*Solution.

Given.   Span AB, $= 5$ m

Load in AB, $W = 10$ t

---

*This example is also solved on page 584

Moment of inertia of $AB=2I$

Span BC, $=3$ m

Load in BC, $w=2$ t/m

Moment of inertia of $BC=I$

Span BD, $=3$ m

Moment of inertia of $BD=I$

Moments at A, B, C and D

Let $M_{AB}=$ Moment at A,

$M_{BA}=$ Moment at B in span AE,

$M_{BC}=$ Moment at B in span BC,

$M_{CB}=$ Moment at C in span CB,

$M_{BD}=$ Moment at B in span BD, and

$M_{DB}=$ Moment at D in span DB.

(i) Fixed end moments.

First of all, let us assume the frame to be made up of fixed beams AB, BC and BD. From the grometry of the figure, we find that the fixed end moment at A in the span AB,

$$M'_{AB}=-\frac{Wab^2}{l^2}=-\frac{10\times2\times3^2}{5^2}=-7\cdot2 \text{ t-m}$$

and $$M'_{BA}=+\frac{Wa^2b}{l^2}=+\frac{10\times2^2\times3}{5^2}=+4\cdot8 \text{ t-m}$$

Similarly fixed end moment at B in span BC,

$$M'_{BC}=-\frac{wl^2}{12}=-\frac{2\times3^2}{12}=-1\cdot5 \text{ t-m}$$

and $$M'_{CB}=+\frac{wl^2}{12}=+\frac{2\times3^2}{15}=+1\cdot5 \text{ t-m}$$

Since the member BD of the frame is not carrying any load, therefore fixed end moment at B in span BD,

$$M'_{BD}=M'_{DB}=0$$

(ii) Slope deflection equations

Since the beam is fixed at A and D, therefore the slopes i_A and i_D will be equal to zero. We know that the support moment at A in the span AB,

$$M_{AB}=\frac{2E.I_{AB}}{l}(2i_A+i_B)+M'_{AB}$$

$$=\frac{2E\times2I}{5}(0+i_B)-7\cdot2 \qquad\qquad (\because\ i_A=0)$$

$$M_{AB} = \frac{4EI \cdot i_B}{5} - 7 \cdot 2 \qquad \qquad \ldots(i)$$

and

$$M_{BA} = \frac{2E \cdot I_{BA}}{l} (2\,i_B + i_A) + M'_{BA}$$

$$= \frac{2E \times 2I}{5} (2i_B + 0) + 4 \cdot 8 \qquad (\because \; i_A = 0)$$

$$= \frac{8EI \cdot i_B}{5} + 4 \cdot 8 \qquad \qquad \ldots(ii)$$

Similarly moment at B in span BC,

$$M_{BC} = \frac{2E \cdot I_{BC}}{l} (2i_B + i_C) + M'_{BC} = \frac{2EI}{3} (2i_B + i_C) - 1 \cdot 5 \quad \ldots(iii)$$

and

$$M_{CB} = \frac{2E \cdot I_{CB}}{l} (2i_C + i_B) + M'_{CB}$$

$$= \frac{2EI}{3} (2i_C + i_B) + 1 \cdot 5 \qquad \qquad \ldots(iv)$$

and moment at B in span BD,

$$M_{BD} = \frac{2EI_{BD}}{l} (2i_B + i_D) + M'_{BD}$$

$$= \frac{2EI}{3} (2i_B + 0) + 0 \qquad (\because \; i_D \text{ and } M'_{BD} = 0)$$

$$= \frac{4EI \cdot i_B}{3} \qquad \qquad \ldots(v)$$

$$M_{DB} = \frac{2EI}{l} (2i_D + i_B) + M'_{DB}$$

$$= \frac{2EI}{3} (0 + i_B) + 0 \qquad (\because \; i_D \text{ and } M'_{DB} = 0)$$

$$= \frac{2EI \cdot i_B}{3} \qquad \qquad \ldots(vi)$$

(iii) Equilibrium equations

Since the beam is simply supported at C, therefore the moment at C will be zero. Now equating the equation (iii) to zero,

$$= \frac{2EI}{3} (2i_C + i_B) + 1 \cdot 5 = 0$$

$$\frac{4EI.i_C}{3} + \frac{2EI.i_B}{3} - 1 \cdot 5$$

$$4i_C + 2i_B = -\frac{4 \cdot 5}{EI} \qquad \left(\text{Multiplying both sides by } \frac{3}{EI} \right)$$

$$20i_C + 10i_B = \frac{22 \cdot 5}{EI} \qquad \qquad \ldots(vii)$$

(Multiplying both sides by 5)

Since the joint B is in equilibrium, therefore equating $M_{BA} + M_{BC} + M_{BD}$ equal to zero,

$$\frac{8EI.i_B}{5} + 4.8 + \frac{2EI}{3}(2i_B + i_C) - 1.5 + \frac{4EI.i_B}{3} = 0$$

$$\frac{8EI.i_B}{5} + \frac{4EI.i_B}{3} + \frac{2EI.i_C}{3} + \frac{4EI.i_B}{3} = -3.3$$

$$\frac{64EI.i_B}{15} + \frac{2EI.i_C}{3} = -3.3 \qquad \qquad ...(viii)$$

$$64i_B + 10i_C = -49.5 \qquad \left(\text{Multiplying both sides by } \frac{15}{EI}\right)$$

or $\qquad\qquad 128\ i_B + 20i_C = -99 \qquad$ (Multiplying both sides by 2)

Subtracting equation (vii) from equation $(viii)$,

$$118i_B = -76.5$$

$$i_B = -0.648$$

Substituting this value of i_B in equation (vii),

$$i_C = -0.801$$

(iv) Final moments

Substituting the values of $EI.i_B$ and $EI.i_C$ in equations (i) to (vi),

$$M_{AB} = \frac{4}{5}(-0.648) - 7.2 = -7.72 \text{ t-m}$$

$$M_{BA} = \frac{8}{5}(-0.648) + 4.8 = 3.76 \text{ t-m}$$

$$M_{BC} = \frac{4}{3}(-0.648) + \frac{2}{3}(-0.801) - 1.5 = -2.9 \text{ t-m}$$

$$M_{CB} = 0$$

$$M_{BD} = \frac{4}{3}(-0.648) = -0.86 \text{ t-m}$$

$$M_{DB} = \frac{2}{3}(-0.648) = -0.43 \text{ t-m}$$

B.M. diagram

From the geometry of the figure, we find that the B.M. under the 10 tonnes load

$$= \frac{Wab}{l} = \frac{10 \times 2 \times 3}{5} = 12.0 \text{ t-m}$$

Similarly B.M. at the mid of the span BC

$$= \frac{wl^2}{8} = \frac{2 \times 3^2}{8} = 2.25 \text{ t-m}$$

Now complete the B.M. diagram as shown in Fig. 14·7

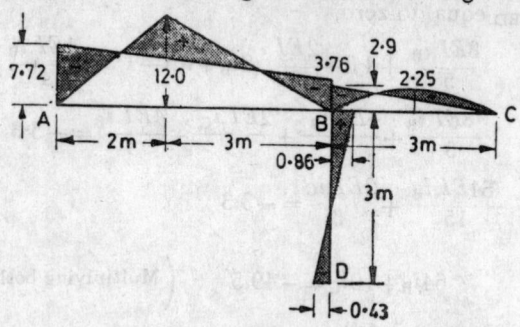

Fig. 14·7

Example 14·4. *A beam ABC supported on a column BD is loaded as shown in Fig. 14·8*

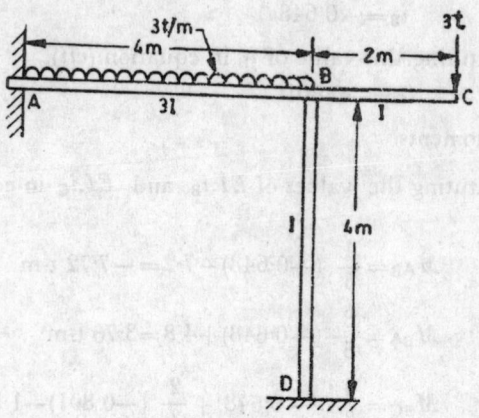

Fig. 14·8

Analyse the frame, by slope deflection method, and draw the bending moment diagram.

(*Nagpur University, 1973*)

Solution.

Given.   Span AB $= 4\ m$

Load in AB, $w = 3\ t/m$

Span BC $= 2\ m$

Load at C $= 3\ t$

Span BD $= 4\ m$

Moment of intertia of $AB = 3I$

Moment of ihertia of $BC = I$

Moment of inertia of $BD = I$

Let $M_{AB} =$ Moment at A,

$M_{BA} =$ Moment at B in span BA,

M_{BC}=Moment at B in span BC,

M_{BD}=Moment at B in span BD, and

M_{DB}=Moment at D in span DB.

(i) Fixed end moments

First of all, let us assume the frame to be made up of fixed beams AB and BD and a cantilever BC. From the geometry of the figure, we find that the fixed end moment at A in the span AB,

$$M'_{AB}=-\frac{wl^2}{12}=-\frac{3\times 4^2}{12}=-4\cdot 0 \text{ t-m}$$

$$M'_{BA}=+\frac{wl^2}{12}=+\frac{3\times 4^2}{12}=+4\cdot 0 \text{ t-m}$$

$$M'_{BC}=-3\times 2=-6 \text{ t-m} \qquad (\because BC \text{ is a cantilever})$$

Since the member BD of the frame is not carrying any load, therefore the fixed end moment at B in span BD,

$$M'_{BD}=M'_{DB}=0$$

(ii) Slope deflection equations

Since the beam is fixed at A and D, therefore the slopes i_A and i_D will be equal to zero. We know that the moment at A,

$$M_{AB}=\frac{2E\times I_{AB}}{l}(2i_A+i_B)+M'_{AB}$$

$$=\frac{2E\times 3I}{4}(0+i_B)-4\cdot 0 \qquad (\because i_A=0)$$

$$=\frac{3EI}{2}\times i_B-4\cdot 0 \qquad \qquad \dots(i)$$

and $\qquad M_{BA}=\frac{2E\times i_{BA}}{l}(2i_B+i_A)+M'_{BA}$

$$=\frac{2E\times 3I}{4}(2i_B+0)+4\cdot 0 \qquad (\because i_A=0)$$

$$=3EI.i_B+4\cdot 0 \qquad \qquad \dots(ii)$$

Similarly moment at B in span BD,

$$M_{BD}=\frac{2E\times I_{BD}}{l}(2i_B+i_D)+M'_{BD}$$

$$=\frac{2EI}{4}(2i_B+0)+0 \qquad (\because i_D \text{ and } M'_{BD}=0)$$

$$=EI.i_B \qquad \qquad \dots(iii)$$

and $\qquad M_{DB}=\frac{2E\times I_{DB}}{l}(2i_D+i_B)+M'_{DB}$

$$=\frac{2EI}{4}(0+i_B)+0 \qquad (\because i_D \text{ and } M'_{DB}=0)$$

$$=\frac{EI}{2}\times i_B \qquad \qquad \dots(iv)$$

(iii) Equilibrium equations

Since the joint B is in equilibrium, therefore equating

$$M_{BA} + M_{BC} + M_{BD} = 0$$
$$3EI.i_B + 4\cdot0 - 6\cdot0 + EI.i_B = 0$$
$$4EI.i_B = 2$$
$$\therefore \qquad i_B = \frac{1}{2EI} \qquad\qquad\qquad ...(v)$$

(iv) Final moments

Substituting the values of i_B in equations (i) to (iv),

$$M_{AB} = \frac{3EI}{2} \times \frac{1}{2EI} - 4\cdot0 = -3\cdot25 \text{ t-m}$$

$$M_{BA} = 3EI \times \frac{1}{2EI} + 4\cdot0 = +5\cdot5 \text{ t-m}$$

$$M_{BD} = EI \times \frac{1}{2EI} = +0\cdot5 \text{ t-m}$$

$$M_{DB} = \frac{EI}{2} \times \frac{1}{2EI} = +0\cdot25 \text{ t-m}$$

From the geometry of the figure, we find that the B.M. at the centre of the span AB, treating it as a simply supported beam,

$$= \frac{wl^2}{8} = \frac{3 \times 4^2}{8} = 6\cdot0 \text{ t-m}$$

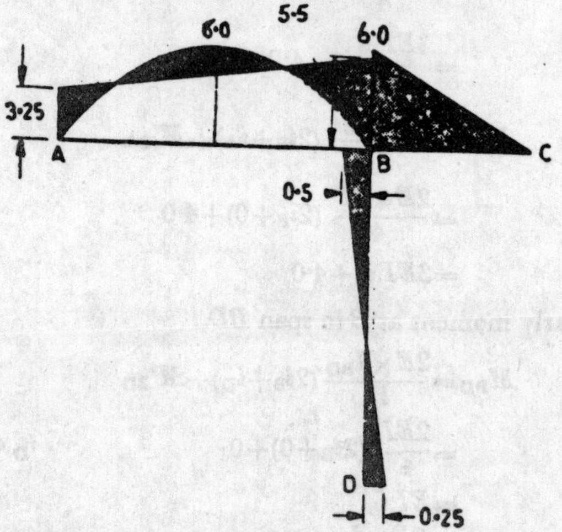

Fig. 14·9

Now complete the B.M. diagram as shown in Fig. 14·9.

14·10. Portal frames

A simple portal frame consists of a beam resting over two columns (A portal frame may also consist of a beam resting over more than two supports or having more than one storey). The junction of the beams and columns behaves like a rigid joint. In general, the portal frames are of the following two types :

    (1) Symmetrical portal frames, and

    (2) Unsymmetrical portal frames.

14·11. Symmetrical portal frames

    A symmetrical portal frame is that, in which both the columns are of the same length, having similar end conditions (*i.e.*, either hinged or fixed) moment of inertia, modulus of electricity as well as subjected to symmetrical loading. A little consideration will show, that the joints of such a portal frame will not be subjected to any translation on side sway.

    All the single bay and single storey symmetrical portal frames are analysed by opening them out, and treating them exactly like a three span continuous beam.

    Example 14·5. *A portal frame ABCD shown in Fig. 14·10 is loaded with a uniformly distributed load of 2 t/m on the horizontal member. The moment of inertia of member AB = that of CD = I and that of BC = 3I.*

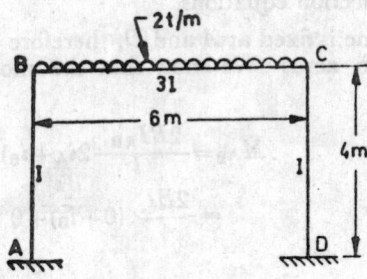

Fig. 14·10.

    Find the support reactions and bending moment in the frame by slope-deflection method and draw the bending moment diagram.

(A.M.I.E., Summer 1974)

Solution

Given. Length of $AB = 4$ m

Length of BC = 6m

Length of CD = 4m

Load on BC, $w = 2t/m$

M. I. of AB, $I_{AB} = I_{CD} = I$

M. I. of BC, $I_{BC} = 2I$

Support reactions

Let M_{AB} = Moment at A,

$$M_{BA} = \text{Moment at } B \text{ in span } BA,$$
$$M_{BC} = \text{Moment at } B \text{ in span } BC,$$
$$M_{CB} = \text{Moment at } C \text{ in span } CB,$$
$$M_{CD} = \text{Moment } C \text{ in span } CD, \text{ and}$$
$$M_{DC} = \text{Moment at } D.$$

(i) Fixed end moments

First of all, let us assume the frame to be made up of fixed beams AB, BC and CD. From the geometry of the figure, we find that the fixed end moment at A,

$$M'_{AB} = M_{BA} = 0$$

($\because$ AB is not carrying any load)

$$M'_{BC} = -\frac{wl^2}{12} = -\frac{2 \times 6^2}{12} = -6 \text{ t-m}$$

$$M'_{CB} = +\frac{wl^2}{12} = +\frac{2 \times 6^2}{12} = +6 \text{ t-m}$$

$$M'_{CD} = M'_{DC} = 0$$

($\because$ CD is not carrying any load)

(ii) Slope deflection equations.

Since the frame is fixed at A and D, therefore the slopes i_A and i_D will be equal to zero. We know that the moment at A in the span AB,

$$M_{AB} = \frac{2EI_{AB}}{l} 2i_A + i_B) + M'_{AB}$$

$$= \frac{2EI}{4} (0 + i_B) + 0$$

($\because$ i_A and $M'_{AB} = 0$)

$$= \frac{E.I\, i_B}{2} \qquad \qquad \ldots(i)$$

and $$M_{BA} = \frac{2E.I_{BA}}{l} (2i_B + i_A) + M'_{BA}$$

$$= \frac{2EI}{4} (2i_B + 0) + 0 + 6.0 \qquad (\because \ i_A \text{ and } M'_{BA} = 0)$$

$$= EI.\, i_B \qquad \qquad \ldots(ii)$$

Similarly moment at B in span BC,

$$M_{BC} = \frac{2E.I_{BC}}{l} (2i_B + i_C) + M'_{BC}$$

$$= \frac{2E \times 3I}{6} (2i_B + i_C) - 6.0$$

$$=EI(2i_B+i_C)-6\cdot0 \qquad ...(iii)$$

and

$$M_{CB}=\frac{2E.I_{CB}}{l}(2i_C+i_B)+M'_{CB}$$

$$=\frac{2E.3I}{6}(2i_C+i_B)+6\cdot0$$

$$=EI(2i_C+i_B)+6\cdot0 \qquad ...(iv)$$

and moment at C in span CD,

$$M_{CD}=\frac{2E.I_{CD}}{l}(2i_C+i_D)+M'_{CD}$$

$$=\frac{2EI}{4}(2i_C+0)+0 \qquad (\because i_D \text{ and } M_{CD}=0)$$

$$=EI.i_C \qquad ...(v)$$

and

$$M_{DC}=\frac{2EI_{DC}}{l}(2i_D+i_C)+M'_{DC}$$

$$=\frac{2EI}{4}(0+i_C)+0 \qquad (\because i_D \text{ and } M_{DC}=0.)$$

$$=\frac{EI.i_C}{2} \qquad ...(vi)$$

(iii) Equilibrium equations

Since the joints B and C are in equilibrium, therefore first equating $M_{BA}+M_{BC}$ equal to zero,

$$EI.i_B+EI(2i_B+i_C)-6\cdot0=0$$

$$EI.i_B+2EI.i_B+EI.i_C-6\cdot0=0$$

$$3EI.i_B+EI.i_C=6\cdot0 \qquad ...(vii)$$

Now equating $M_{CB}+M_{CD}$ equal to zero,

$$EI.(2i_C+i_B)+6\cdot0+EI.i_C=0$$

$$2EI.i_C+EI.i_B+6\cdot0+EI.i_C=0$$

$$3EI.i_C+EI.i_B+6\cdot0=0 \qquad ...(viii)$$

We know that by symmetry, $i_B=-i_C$, therefore substituting these values in equations (vii) and (viii),

$$2EI.i_B=6\cdot0$$

$$\therefore \qquad EI.i_B=\frac{6\cdot0}{2}=3\cdot0$$

and $\quad\quad 2EI.\ i_C = -6\cdot0$

$\therefore\quad\quad EI.\ i_C = -\dfrac{6\cdot0}{2} = -3\cdot0$

(iv) Final moments

Substituting the above values in equations (i) to (vi),

$$M_{AB} = \frac{EI.\ i_B}{2} = \frac{3\cdot0}{2} = 1\cdot5 \text{ t-m}$$

$$M_{BA} = EI.\ i_B = 3\cdot0 \text{ t-m}$$

$$M_{BC} = EI\ (2i_B + i_C) - 6\cdot0$$

$$= 2EI.\ i_B + EI.\ i_C - 6\cdot0$$

$$= 6\cdot0 - 3\cdot0 - 6\cdot0 = -3\cdot0 \text{ t-m}$$

$$M_{CB} = EI\ (2i_C + i_B) + 6\cdot0$$

$$= 2EI.\ i_C + EI.\ i_B + 6\cdot0$$

$$= -6\cdot0 + 3\cdot0 + 6\cdot0 = +3\cdot0 \text{ t-m}$$

$$M_{CD} = EI.\ i_C = -3\cdot0 \text{ t-m}$$

$$M_{DC} = \frac{EI.\ i_C}{2} = \frac{-3\cdot0}{2} = -1\cdot5 \text{ t-m}$$

Vertical reaction at A,

$$V_A = V_D = \frac{2 \times 6}{2} = 6\cdot0 \text{ t}$$

Horizontal reaction at A,

$$H_A = \frac{1\cdot5 + 3\cdot0}{4} = 1\cdot125 \text{ t } (\rightarrow)$$

Similarly $\quad H_D = 1\cdot125 \text{ t } (\leftarrow)$

Bending moment diagram

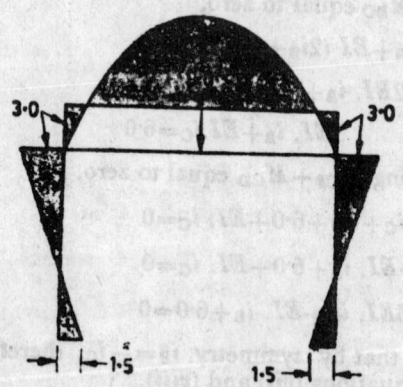

Fig. 14·11.

The bending moment at the mid of the span BC, by consider ing it as a simply supported beam

$$= \frac{wl^2}{8} = \frac{2 \times 6^2}{8} = 9\cdot0 \text{ t-m}$$

Now complete the final bending moment diagram as shown in Fig. 14·11.

14 12. Unsymmetrical portal frames

An unsymmetrical portal frame is that in which either the columns are not symmetrical, or the frame is not symmetrically loaded. A little consideration will show that the unsymmetrical portal frame will be subjected to some horizontal movement, known as sway, to one side or the other. As a result of this, the rigid joints between the columns and beams will have a motion of translation. In such a case, the sway of the columns will become additional unknown quantities. It is thus obvious, that some additional equations are required for analysing the frame. These additional equations of equilibrium are obtained from the consideration of shear force exerted on the columns by the horizontal external forces. This force is obtained by algebraically adding the moments at the ends of a column, divided by its length. These equations are known as *shear equations*.

Note : 1. If the frame is assumed to sway to the right side, the value of δ is taken as negative, and if on the left side the value of δ is taken as positive.

2. Only vertical columns of a frame will be assumed to have some sway ; whereas the horizontal beam over the columns will be assumed just move with the columns. A little consideration will show that the horizontal beam will not be subjected to any sway of its own, as the two ends do not move up or down but remain at the same level.

Example 14·6. *A portal frame shown in Fig. 14·12 is fixed at A and hinged at D. It is subjected to a horizontal load of 10 tonnes at B,*

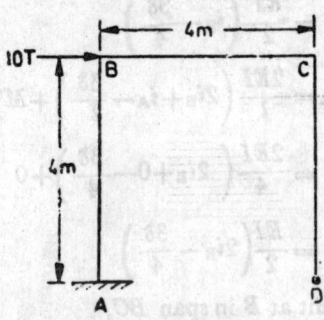

Fig. 14·12

Analyse the frame, by slope deflection method, and draw the bending moment diagram. Also sketch the deflected shape of the frame. Take flexural rigidity for all the members to be constant.

Solution.

Given. Length of $AB = BC = CD = 4$ m

Load at B $= 10$ t

Flexural rigidity of AB,

$$EI_{AB}=EI_{BC}=EI_{CD}$$

Let

$M_{AB}=$ Moment at A,

$M_{BA}=$ Moment at B in span BA.

$M_{BC}=$ Moment at B in span BC,

$M_{CB}=$ Moment at C in span CB,

$M_{CD}=$ Moment at C in span CD, and

$M_{DC}=$ Moment at D.

(i) Fixed end moments

Since the horizontal load of 10 tonnes is acting at the joint B, and there is no other load on any span, therefore the fixed end moments at all the joints will be zero *i.e.*, $M'_{AB}=M'_{BA}=M'_{BC}=M_{CB}=M'_{CD}=M'_{DC}=0.$

(ii) Slope deflection equations

Let us assume that the frame will sway to the right side by an amount equal to δ. Since the frame is fixed at A, therefore the slope at A will be equal to zero. We know that the moment at A,

$$M_{AB}=\frac{2EI}{l}\left(2i_A+i_B-\frac{3\delta}{l}\right)+M'_{AB}$$

$$=\frac{2EI}{4}\left(0+i_B-\frac{3\delta}{4}\right)+0 \quad [\because i_A \text{ and } M'_{AB}=0]$$

$$=\frac{EI}{2}\left(i_B-\frac{3\delta}{4}\right) \qquad ...(i)$$

and

$$M_{BA}=\frac{2EI}{l}\left(2i_B+i_A-\frac{3\delta}{l}\right)+M'_{BA}$$

$$=\frac{2EI}{4}\left(2i_B+0-\frac{3\delta}{4}\right)+0 \quad (\because i_A \text{ and } M'_{BA}=0)$$

$$=\frac{EI}{2}\left(2i_B-\frac{3\delta}{4}\right) \qquad ...(ii)$$

Similarly moment at B in span BC,

$$M_{BC}=\frac{2EI}{l}\left(2i_B+i_C-\frac{3\delta}{l}\right)+M'_{BC}$$

$$=-\frac{2EI}{4}\left(2i_B+i_C-0\right)+0$$

$$(\because \delta \text{ for } BC \text{ and } M'_{BC}=0)$$

$$=\frac{EI}{2}\left(2i_B+i_C\right) \qquad ...(iii)$$

and
$$M_{CB} = \frac{2EI}{l}\left(2i_C + i_B - \frac{3\delta}{l}\right) + M'_{CB}$$

$$= \frac{2EI}{4}\left(2i_C + i_B - 0\right) + 0$$

$$\text{(}\because \ \delta \text{ for } CB \text{ and } M'_{CB} = 0\text{)}$$

$$= \frac{EI}{2}\left(2i_C + i_B\right) \qquad \qquad ...(iv)$$

and moment at C in span CD,
$$M_{CD} = \frac{2EI}{l}\left(2i_C + i_D - \frac{3\delta}{l}\right) + M'_{CD}$$

$$= \frac{2EI}{4}\left(2i_C + i_D - \frac{3\delta}{4}\right) + 0 \qquad \text{(}\because \ M'_{CD} = 0\text{)}$$

$$= \frac{EI}{2}\left(2i_C + i_D - \frac{3\delta}{4}\right) \qquad \qquad ...(v)$$

and
$$M_{DC} = \frac{2EI}{l}\left(2i_D + i_C - \frac{3\delta}{l}\right) + M'_{DC}$$

$$= \frac{2EI}{4}\left(2i_D + i_C - \frac{3\delta}{4}\right) + 0 \qquad \text{(}\because \ M'_{DC} = 0\text{)}$$

$$= \frac{EI}{2}\left(2i_D + i_C - \frac{3\delta}{4}\right) \qquad \qquad ...(vi)$$

(iii) Equilibrium equations

Since the frame is hinged at D, therefore M_{DC} will be equal to zero. Now equating the equation (vi) to zero,

$$\frac{EI}{2}\left(2i_D + i_C - \frac{3\delta}{4}\right) = 0$$

or
$$2i_D + i_C - \frac{3\delta}{4} = 0 \qquad \left(\text{Multiplying with side } \frac{2}{EI}\right)$$

$$3\delta = 8i_D + 4i_C \qquad \qquad ...(vii)$$

Moreover, since the joints B and C are in equilibrium, therefore first equating $M_{BA} + M_{BC}$ equal to zero,

$$\frac{EI}{2}\left(2i_B - \frac{3\delta}{4}\right) + \frac{EI}{2}\left(2i_B + i_C\right) = 0$$

or
$$2i_B - \frac{3\delta}{4} + 2i_B + i_C = 0 \qquad \left(\text{Multiplying both sides by } \frac{2}{EI}\right)$$

$$4i_B + i_C - \frac{3\delta}{4} = 0$$

$$3\delta = 16i_B + 4i_C \qquad \qquad ...(viii)$$

Equating the values of 3δ from equations (vii) and $(viii)$,

$$8i_B + 4i_C = 16i_B + 4i_C$$

$$\therefore \qquad \qquad i_D = 2i_B \qquad \qquad ...(ix)$$

Now equating $M_{CB} + M_{CD}$ equal to zero,

$$\frac{EI}{2}(2i_C + i_B) + \frac{EI}{2}\left(2i_C + i_D - \frac{3\delta}{4}\right) = 0$$

$$2i_C + i_B + 2i_C + i_D - \frac{3\delta}{4} = 0 \quad \left(\text{Multiplying both sides by } \frac{2}{EI}\right)$$

$$i_B + 4i_C + i_D - \frac{3\delta}{4} = 0$$

or $\qquad\qquad 4i_B + 16i_C + 4i_D - 3\delta = 0$

Now substituting the values of i_D and 3δ from equations (ix) and $(viii)$ in the above equation,

$$4i_B + 16i_C + 4 \times 2i_B - 16i_B - 4i_C = 0$$

$$12i_C - 4i_B = 0$$

or $\qquad\qquad i_C = \dfrac{i_B}{3}$ $\qquad\qquad\qquad$ ...(x)

Now substituting the value of i_C in equation $(viii)$,

$$3\delta = 16i_B + \frac{4 \times i_B}{3} = \frac{52i_B}{3}$$

or $\qquad\qquad \delta = \dfrac{52i_B}{9}$ $\qquad\qquad\qquad$ ...(xi)

(iv) Shear equations

Since the frame is subjected to a horizontal load of 10 tonnes, therefore sum of the horizontal reactions at A and D must be equal to 10 tonnes, i.e.

$$H_A + H_B = 10 \text{ T}$$

or $\qquad\qquad \dfrac{M_{BA} + M_{AB}}{4} + \dfrac{M_{CD} + M_{DC}}{4} = 10$

$$\frac{\frac{EI}{2}\left(2i_B - \frac{3\delta}{4}\right) + \frac{EI}{2}\left(i_B - \frac{3\delta}{4}\right)}{4} + \frac{\frac{EI}{2}\left(2i_C + i_D - \frac{3\delta}{4}\right) + 0}{4} = 10$$

$$2i_B - \frac{3\delta}{4} + i_B - \frac{3\delta}{4} + 2i_C + i_D - \frac{3\delta}{4} = \frac{80}{EI}$$

$$3i_B + 2i_C + i_D - \frac{9\delta}{4} = \frac{80}{EI}$$

Substituting the value of i_C, i_D and δ from equations (x), (ix) and (xi) respectively in the above equation,

$$3i_B + 2 \times \frac{i_B}{3} + 2i_B - \frac{9}{4} \times \frac{52i_B}{9} = \frac{80}{EI}$$

$$-\frac{22i_B}{3} = \frac{80}{EI}$$

or
$$i_B = -\frac{80}{EI} \times \frac{3}{22} = -\frac{120}{11EI}$$

$$\therefore \quad i_C = \frac{i_B}{3} = -\frac{120}{11EI} \times \frac{1}{3} = -\frac{40}{11EI}$$

$$i_D = 2i_B = 2\left(-\frac{120}{11EI}\right) = -\frac{240}{11EI}$$

and
$$\delta = \frac{52i_B}{9} = \frac{52}{9}\left(-\frac{120}{11EI}\right) = -\frac{2080}{33EI}$$

(v) **Final equations**

Now substituting the various values in the slope deflection equations,

$$M_{AB} = \frac{EI}{2}\left(i_B - \frac{3\delta}{4}\right) = \frac{EI}{2}\left(-\frac{120}{11EI} - \frac{3}{4} \times -\frac{2080}{33EI}\right) \text{ t-m}$$

$$= \frac{EI}{2} \times \frac{400}{11EI} = \frac{200}{11} = 18 \cdot 18 \text{ t-m}$$

$$M_{BA} = \frac{EI}{2}\left(2i_B - \frac{3\delta}{4}\right)$$

$$= \frac{EI}{2}\left(2 \times -\frac{120}{11EI} - \frac{3}{4} \times -\frac{2080}{33EI}\right) \text{ t-m}$$

$$= \frac{EI}{2} \times \frac{280}{11EI} = \frac{140}{11} = 12 \cdot 73 \text{ t-m}$$

$$M_{BC} = \frac{EI}{2}(2i_B + i_C) = \frac{EI}{2}\left(2 \times -\frac{120}{11EI} - \frac{40}{11EI}\right) \text{ t-m}$$

$$= \frac{EI}{2} \times -\frac{280}{11EI} = -\frac{140}{11} = -12 \cdot 73 \text{ t-m}$$

$$M_{CB} = \frac{EI}{2}(2i_C + i_B) = \frac{EI}{2}\left(2 \times -\frac{40}{11EI} - \frac{120}{11EI}\right) \text{ t-m}$$

$$= \frac{EI}{2} \times -\frac{200}{11EI} = -\frac{100}{11} = -9 \cdot 09 \text{ t-m}$$

$$M_{CD} = \frac{EI}{2}\left(2i_C + i_D - \frac{3\delta}{4}\right)$$

$$= \frac{EI}{2}\left(2 \times -\frac{40}{11EI} - \frac{240}{11EI} - \frac{3}{4} \times -\frac{2080}{33EI}\right) \text{ t-m}$$

$$= \frac{EI}{2} \times \frac{200}{11EI} = \frac{100}{11} = 9 \cdot 09 \text{ t-m}$$

$$M_{DC} = 0$$

∴   Horizontal reaction at A.

$$H_A = \frac{18 \cdot 18 + 12 \cdot 73}{4} = \frac{30 \cdot 91}{4} = 7 \cdot 73 \text{ t} \leftarrow$$

and horizontal reaction at D,

$$H_D = \frac{9 \cdot 09}{4} = 2 \cdot 27 \text{ t} \leftarrow$$

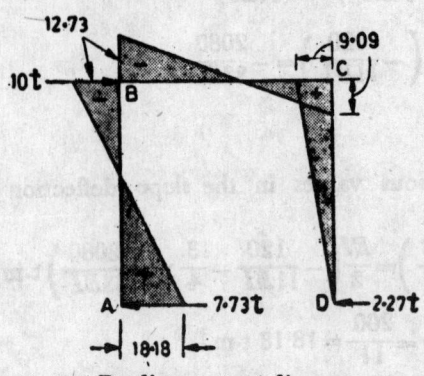 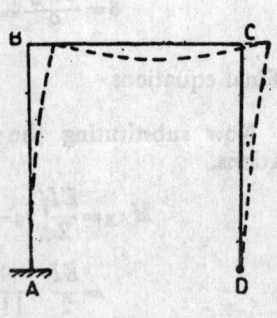

(a) Bending moment diagram   (b) Deflection shape of the frame

Fig. 14·13

Now draw the B.M. diagram and deflected shape of the frame as shown in Fig 14·13 (a) and (b).

Example 14·7. *A portal frame ABCD fixed at A and hinged at D carries a point load as shown in Fig. 14·14.*

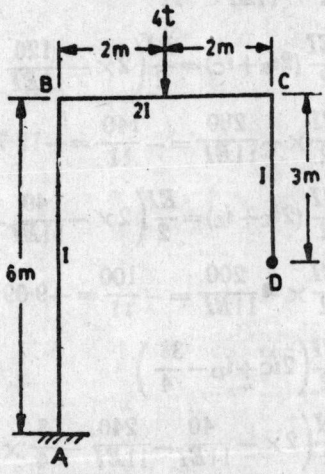

Fig. 14·14

Analyse the frame, by slope deflection method, and draw the bend-ing moment diagram.

Solution.

Given. Length of $AB = 6$ m

Length of BC $= 2 + 2 = 4$ m

Length of CD $= 3$ m

Load on BC, $W = 4$ t

Moment of inertia for AB,

$$I_{AB} = I_{CD} = I$$

and moment of inertia for BC

$$I_{BC} = 2I$$

Let $M_{AB} =$ Moment at A,

         $M_{BA} =$ Moment at B in span BA,

         $M_{BC} =$ Moment at B in span BC,

         $M_{CB} =$ Moment at C in span CB,

         $M_{CD} =$ Moment at C in span CD, and

         $M_{DC} =$ Moment at D.

(i) Fixed end moments

First of all, let us assume the frame to be made up of fixed beams AB, BC and CD. From the geometry of the figure, we find that the fixed end moment at A,

$$M'_{AB} = M'_{BA} = 0 \qquad (\because \quad AB \text{ is not carrying any load})$$

$$M'_{BC} = -\frac{Wl}{8} = -\frac{4 \times 4}{8} = -2 \cdot 0 \text{ t-m}$$

$$M'_{BC} = +\frac{Wl}{8} = +\frac{4 \times 4}{8} = +2 \cdot 0 \text{ t-m}$$

$$M'_{CD} = M'_{DC} = 0 \qquad (\because \quad CD \text{ is not carrying any load})$$

(ii) Slope deflection equations

Let us assume that the frame will sway to the left side by an amount equal to δ. Since the frame is fixed at A, therefore the slope at A will be equal to zero. We know that the moment at A,

$$M_{AB} = \frac{2E.I_{AB}}{l}\left(2i_A + i_B + \frac{3\delta}{l}\right) + M'_{AB}$$

$$= \frac{2E \times I}{6}\left(0 + i_B + \frac{3\delta}{6}\right) + 0 \qquad (\because i_A \text{ and } M'_{AB} = 0)$$

$$= \frac{EI}{3}\left(i_B + \frac{\delta}{2}\right) \qquad \qquad \ldots(i)$$

and $$M_{BA} = \frac{2EI.BA}{l}\left(2i_B + i_A + \frac{3\delta}{l}\right) + M'_{BA}$$

$$= \frac{2E \times I}{6}\left(2i_B + 0 + \frac{3\delta}{6}\right) + 0 \qquad (\because \quad i_A \text{ and } M'_{BA} = 0)$$

$$= \frac{EI}{3}\left(2i_B + \frac{\delta}{2}\right) \qquad \qquad \ldots(ii)$$

Similarly moment at B in span BC,

$$M_{BC} = \frac{2E.I_{BC}}{l}\left(2i_B + i_C + \frac{3\delta}{l}\right) + M'_{BC}$$

$$= \frac{2E \times 2I}{4}(2i_B + i_C + 0) - 2\cdot0 \qquad (\because \ \delta \text{ for } BC = 0)$$

$$= EI(2i_B + i_C) - 2\cdot0 \qquad\qquad\qquad \ldots(iii)$$

and $$M_{CB} = \frac{2E.I_{CB}}{l}\left(2i_C + i_B + \frac{3\delta}{l}\right) + M'_{CB}$$

$$= \frac{2E \times 2I}{4}(2i_C + i_B + 0) + 2\cdot0 \qquad (\because \ \delta \text{ for } CB = 0)$$

$$= EI(2i_C + i_B) + 2\cdot0 \qquad\qquad\qquad \ldots(iv)$$

and moment at C in span CD,

$$M_{CD} = \frac{2E.I_{CD}}{l}\left(2i_C + i_D + \frac{3\delta}{l}\right) + M'_{CD}$$

$$= \frac{2E \times I}{3}\left(2i_C + i_D + \frac{3\delta}{3}\right) + 0 \qquad (M'_{CD} = 0)$$

$$= \frac{2EI}{3}(2i_C + i_D + \delta) \qquad\qquad\qquad \ldots(v)$$

and $$M_{DC} = \frac{2E.I_{DC}}{l}\left(2i_D + i_C + \frac{3\delta}{l}\right) + M'_{DC}$$

$$= \frac{2E \times I}{3}\left(2i_D + i_C + \frac{3\delta}{3}\right) + 0 \qquad (\because \ M'_{DC} = 0)$$

$$= \frac{2EI}{3}(2i_D + i_C + \delta) \qquad\qquad\qquad \ldots(vi)$$

(iii) Equilibrium equations

Since the frame is hinged at D, therefore M_{DC} will be equal to zero. Now equating the equation (vi) to zero,

$$\frac{2EI}{3}(2i_D + i_C + \delta) = 0$$

$$2i_D + i_C + \delta = 0 \qquad \left(\textbf{Multiplying both sides by } \frac{3}{2EI}\right)$$

or $$2i_D = -i_C - \delta \qquad\qquad\qquad \ldots(vii)$$

Moreover, since the joints B and C are in equilibrium, therefore first equating $M_{BA} + M_{BC}$ equal to zero,

$$\frac{EI}{3}\left(2i_B + \frac{\delta}{2}\right) + EI(2i_B + i_C) - 2\cdot0 = 0$$

$$2i_B + \frac{\delta}{2} + 3(2i_B + i_C) - \frac{6}{EI} = 0$$

$$\left(\textbf{Multiplying both sides by } \frac{3}{EI}\right)$$

$$2i_B + \frac{\delta}{2} + 6i_B + 3i_C = \frac{6}{EI}$$

$$\therefore \qquad 8i_B + 3i_C + \frac{\delta}{2} = \frac{6}{EI} \qquad \ldots(viii)$$

Now equating $M_{CB} + M_{CD}$ equal to zero,

$$EI\,(2i_C + i_B) + 2 \cdot 0 + \frac{2EI}{3}\,(2i_C + i_D + \delta) = 0$$

$$3(2i_C + i_B) + \frac{6}{EI} + 2(2i_C + i_D + \delta) = 0$$

$$\left(\text{Multiplying both sides by } \frac{3}{EI}\right)$$

$$6i_C + 3i_B + 4i_C + 2i_D + 2\delta = -\frac{6}{EI}$$

$$3i_B + 10i_C + 2i_D + 2\delta = -\frac{6}{EI}$$

Substituting the value of $2i_D$ from equation (vii) in the above equation,

$$3i_B + 10i_C - i_C - \delta + 2\delta = -\frac{6}{EI}$$

$$3i_B + 9i_B + \delta = -\frac{6}{EI}$$

or $\qquad i_B + 3i_C + \dfrac{\delta}{3} = -\dfrac{2}{EI}$ $\qquad \ldots(ix)$

Now subtracting equation (ix) from equation $(viii)$,

$$7i_B + \frac{\delta}{6} = \frac{8}{EI}$$

or $\qquad i_B = \dfrac{8}{7EI} - \dfrac{\delta}{42}$ $\qquad \ldots(x)$

Substituting this value of i_B in equation (ix),

$$\frac{8}{7EI} - \frac{\delta}{42} + 3i_C + \frac{\delta}{3} = -\frac{2}{EI}$$

$$\therefore \qquad 3i_C = -\frac{2}{EI} - \frac{8}{7EI} - \frac{\delta}{3} + \frac{\delta}{42}$$

$$= -\frac{22}{7EI} - \frac{13\delta}{42}$$

or $\qquad i_C = -\dfrac{22}{21EI} - \dfrac{13\delta}{126}$ $\qquad \ldots(xi)$

Substituting this value of i_C in equation (vii),

$$2i_D = \frac{22}{21EI} + \frac{13\delta}{126} - \delta = \frac{22}{21EI} - \frac{113\delta}{126}$$

$$\therefore \qquad i_D = \frac{11}{21EI} - \frac{113\delta}{252} \qquad \qquad \ldots (xii)$$

(iv) Shear equations

Since the frame is subjected to vertical load only, therefore the horizontal reactions at A and D must be equal and opposite i.e.,

$$H_A + H_D = 0$$

or $$\frac{M_{BA} + M_{AB}}{6} + \frac{M_{CD} + M_{DC}}{3} = 0$$

$$\frac{\frac{EI}{3}\left(2i_B + \frac{\delta}{2}\right) + \frac{EI}{3}\left(i_B + \frac{\delta}{2}\right)}{6} + \frac{\frac{2EI}{3}(2i_C + i_D + \delta) + 0}{3} = 0$$

$$\frac{EI}{3}\left(2i_B + \frac{\delta}{2}\right) + \frac{EI}{3}\left(i_B + \frac{\delta}{2}\right) + \frac{4EI}{3}(2i_C + i_D + \delta) = 0$$

(Multiplying both sides by 6)

$$\left(2i_B + \frac{\delta}{2}\right) + \left(i_B + \frac{\delta}{2}\right) + 4(2i_C + i_D + \delta) = 0$$

$$\left(\text{Multiplying both sides by } \frac{3}{EI}\right)$$

$$2i_B + \frac{\delta}{2} + i_B + \frac{\delta}{2} + 8i_C + 4i_D + 4\delta = 0$$

or $$3i_B + 8i_C + 4i_D + 5\delta = 0$$

Substituting the values of i_B, i_C and i_D from equations (x), (xi), (xii) respectively in the above equation,

$$3\left(\frac{8}{7EI} - \frac{\delta}{42}\right) + 8\left(-\frac{22}{21EI} - \frac{13\delta}{126}\right) + 4\left(\frac{11}{21EI} - \frac{113\delta}{252}\right) + 5\delta = 0$$

$$\frac{24}{7EI} - \frac{\delta}{14} - \frac{176}{21EI} - \frac{104\delta}{126} + \frac{44}{21EI} - \frac{113\delta l}{63} + 5\delta = 0$$

$$\therefore \qquad \delta = \frac{120}{97EI} \qquad \qquad \ldots(xiii)$$

Now substituting the values of δ in equations (x), (xi) and (xii),

$$i_B = \frac{8}{7EI} - \frac{1}{42} \times \frac{120}{97EI} = \frac{756}{679EI}$$

$$i_C = -\frac{22}{21EI} - \frac{13}{126} \times \frac{120}{97EI} = -\frac{2,394}{2,037EI}$$

and $$i_D = \frac{11}{21EI} - \frac{113}{252} \times \frac{120}{97EI} = -\frac{3}{97}$$

(v) Final equations

Now substituting the various values in the slope deflection equations,

$$M_{AB} = \frac{EI}{3}\left(i_B + \frac{\delta}{2}\right) = \frac{EI}{3}\left(\frac{756}{679EI} + \frac{120}{97EI}\right) \text{ t-m}$$

$$= 0.577 \text{ t-m}$$

$$M_{BA} = \frac{EI}{3}\left(2i_B + \frac{\delta}{2}\right) = \frac{EI}{3}\left(2 \times \frac{756}{679EI} + \frac{120}{97EI}\right) \text{ t-m}$$

$$= 0 \cdot 948 \text{ t-m}$$

$$M_{BC} = EI(2i_B - i_C) - 2 \cdot 0 = EI\left(2 \times \frac{756}{679EI} - \frac{2,394}{2,037EI}\right) - 2 \cdot 0 \text{ t-m}$$

$$= -0 \cdot 948 \text{ t-m}$$

$$M_{CB} = EI(2i_C + i_B) + 2 \cdot 0$$

$$= EI\left(2 \times -\frac{2,394}{2,037EI} + \frac{756}{679EI}\right) + 2 \cdot 0 \text{ t-m}$$

$$= 0 \cdot 763 \text{ t-m}$$

$$M_{CD} = \frac{2EI}{3}(2i_D + i_C + \delta)$$

$$= \frac{2EI}{3}\left(2 \times -\frac{3}{97} - \frac{2,394}{2,037EI} + \frac{120}{97EI}\right) \text{ t-m}$$

$$= -0 \cdot 763 \text{ t-m}$$

$$M_{DC} = 0$$

We know that the *B.M.* under the load, treating the beam *BC* as a simply supported,

$$M = \frac{Wl}{4} = \frac{4 \times 4}{4} = 4 \cdot 0 \text{ t-m}$$

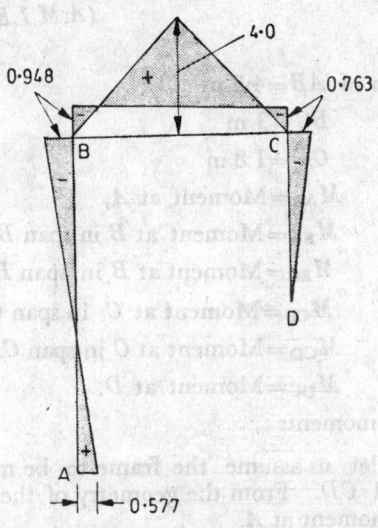

Fig. 14·15

Now draw the B.M. diagram as shown in Fig. 14·15.

Example 14·8. *A stiff-jointed frame ABCD is of constant section throughout and has the following dimensions : AB=4·5 m, BC=3·0 m and CD=1·8 m. It carries concentrated loads of 9 tonnes at each of the third points of the horizontal beam as shown in Fig. 14·16.*

Calculate the bending moments at the joints by slope deflection method, if the feet of the stanctions AB and CD are encastre, and if a

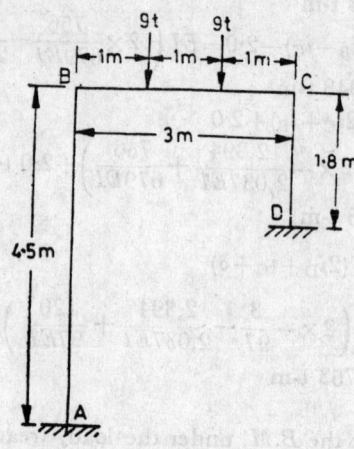

Fig·14·16

simple support is provided at B (not shown in the figure) which prevents horizontal displacements but allows rotation of the joint.

(*A.M.I.E., Winter, 1974*)

Solution

Given. Length of $AB = 4 \cdot 5$ m

Length of $\quad BC = 3$ m

Length of $\quad CD = 1 \cdot 8$ m

Let $\quad M_{AB} =$ Moment at A,

$\qquad M_{BA} =$ Moment at B in span BA,

$\qquad M_{BC} =$ Moment at B in span BC,

$\qquad M_{CB} =$ Moment at C in span CB,

$\qquad M_{CD} =$ Moment at C in span CD, and

$\qquad M_{DC} =$ Moment at D.

(*i*) Fixed end moments

First of all, let us assume the frame to be made up of fixed beams AB, BC and CD. From the geometry of the figure, we find that the fixed end moment at A,

$$M'_{AB} = M'_{BA} = 0 \qquad (\because \; AB \text{ is not carrying any load})$$

$$M'_{BC} = -\left(\frac{9 \times 1 \times 2^2}{3^2} + \frac{9 \times 1^2 \times 2}{3^2}\right) = -6 \cdot 0 \text{ t-m}$$

$$M'_{CB} = +\frac{9 \times 1^2 \times 2}{3^2} + \frac{9 \times 1 \times 2^2}{3^2} = +6 \cdot 0 \text{ t-m}$$

$$M'_{CD} = M'_{DC} = 0 \qquad (\because \; CD \text{ is not carrying any load})$$

(ii) Slope deflection equations

Since a simple support is provided at B, which prevents horizontal displacement, but allows the rotation of the joint, therefore the frame will not sway. Moreover, as the frame is fixed at A and D, therefore the slopes at A and D will be equal to zero. We know that the moment at A,

$$M_{AB} = \frac{2EI}{l}\ (2i_A + i_B) + M'_{AB}$$

$$= \frac{2EI}{4 \cdot 5}\ (0 + i_B) + 0 \qquad (\because \ i_A \text{ and } M'_{AB} = 0)$$

$$= \frac{4EI \cdot i_B}{9} \qquad\qquad \ldots(i)$$

and

$$M_{BA} = \frac{2EI}{l}\ (2i_B + i_A) + M'_{BA}$$

$$= \frac{2EI}{4 \cdot 5}\ (2i_B + 0) + 0 \qquad (\because \ i_A \text{ and } M'_{BA} = 0)$$

$$= \frac{8EI \cdot i_B}{9} \qquad\qquad \ldots(ii)$$

Similiarly moment at B in span BC,

$$M_{BC} = \frac{2EI}{l}\ (2i_B + i_C) + M'_{BC}$$

$$= \frac{2EI}{3}\ (2i_B + i_C) - 6 \cdot 0 \qquad\qquad \ldots(iii)$$

and

$$M_{CB} = \frac{2EI}{l}\ (2i_C + i_B) + M'_{CB}$$

$$= \frac{2EI}{3}\ (2i_C + i_B) + 6 \cdot 0 \qquad\qquad \ldots(iv)$$

and moment at C in span CD,

$$M_{CD} = \frac{2EI}{l}\ (2i_C + i_B) + M'_{CD}$$

$$= \frac{2EI}{1 \cdot 8}\ (2i_C + 0) + 0 \qquad (\because \ i_D \text{ and } M'_{CD} = 0)$$

$$= \frac{20EI \cdot i_C}{9} \qquad\qquad \ldots(v)$$

and

$$M_{DC} = \frac{2EI}{l}\ (2i_D + i_C) + M'_{DC}$$

$$= \frac{2EI}{1 \cdot 0}\ (0 + i_C) + 0 \qquad (\because \ i_D \text{ and } M'_{DC} = 0)$$

$$= \frac{10\,EI.\,i_C}{9} \qquad \ldots(vi)$$

(iii) Equilibrium equations

Since the joints B and C are in equilibrium, therefore first equating $M_{BA} + M_{BC}$ equal to zero,

$$\frac{8EI.\,i_B}{9} + \frac{2EI}{3}\,(2i_B + i_C) - 6\cdot0 = 0$$

$$8i_B + 6(2i_B + i_C) - \frac{54}{EI} = 0 \quad \left(\text{Multiplying both sides by} \frac{9}{EI}\right)$$

$$8i_B + 12i_B + 6i_C = \frac{54}{EI}$$

$$20i_B + 6i_C = \frac{54}{EI}$$

$$30i_B + 9i_C = \frac{81}{EI} \qquad \ldots(viii)$$

Now equating $M_{CB} + M_{CD}$ equal to zero,

$$\frac{2EI}{3}\,(2i_C + i_B) + 6\cdot0 + \frac{20EI.\,i_C}{9} = 0$$

$$6\,(2i_C + i_B) + \frac{54}{EI} + 20i_C = 0 \quad \left(\text{Multiplying both sides by} \frac{9}{EI}\right)$$

$$12i_C + 6i_B + 20i_C = -\frac{54}{EI}$$

$$6i_B + 32i_C = -\frac{54}{EI}$$

$$30i_B + 160i_C = -\frac{270}{EI} \qquad \ldots(vii)$$

Subtracting equation (vii) from $(viii)$,

$$151i_C = -\frac{351}{EI}$$

or

$$i_C = \frac{351}{151\,EI} \qquad \ldots(ix)$$

Now substituting the value of i_C in equation $(viii)$,

$$30i_B + 9 - \left(\frac{351}{151\,EI}\right) = \frac{81}{EI}.$$

$$10i_B - \frac{1,053}{151\,EI} = \frac{27}{EI} \qquad \text{(Dividing both sides by 3)}$$

$$10i_B = \frac{27}{EI} + \frac{1053}{151\,EI} = \frac{5,130}{151\,EI}$$

$$i_B = \frac{5,130}{1,510\,EI} \qquad \qquad ...(x)$$

(iv) Final equations

Now substituting the values of i_B and i_C in the slope deflection equations,

$$M_{AB} = \frac{4\,EI.\,i_B}{9} = \frac{4\,EI}{9} \times \frac{5,130}{1,510\,EI} = \mathbf{1 \cdot 51\ t\text{-}m}\quad \mathbf{Ans.}$$

$$M_{BA} = \frac{8EI.\,i_B}{9} = \frac{8\,EI}{9} \times \frac{5,130}{1,510\,EI} = \mathbf{3 \cdot 02\ t\text{-}m}\quad \mathbf{Ans.}$$

$$M_{BC} = \frac{2\,EI}{3}\,(2i_B + i_C) - 6 \cdot 0\ \text{t-m}$$

$$= \frac{2\,EI}{3}\left(2 \times \frac{5,130}{1,510\,EI} - \frac{351}{151\,EI}\right) - 6 \cdot 0\ \text{t-m}$$

$$= -\mathbf{3 \cdot 02\ t\text{-}m}\quad \mathbf{Ans.}$$

$$M_{CB} = \frac{2\,EI}{3}\,(2i_C + i_B) + 6 \cdot 0\ \text{t-m}$$

$$= \frac{2\,EI}{3}\left[2\left(-\frac{351}{151\,EI}\right) + \frac{5,13}{1,510\,EI}\right] + 6 \cdot 0\ \text{t-m}$$

$$= \mathbf{5 \cdot 17\ t\text{-}m}\quad \mathbf{Ans.}$$

$$M_{CD} = \frac{20\,EI.\,i_C}{9} = \frac{20\,EI}{9}\left(-\frac{351}{151EI}\right)\ \text{t-m}$$

$$= -\mathbf{5 \cdot 17\ t\text{-}m}\quad \mathbf{Ans.}$$

$$M_{DC} = \frac{10\,EI.\,i_C}{9} = \frac{10\,EI}{9}\left(-\frac{351}{151\,EI}\right)\ \text{t-m}$$

$$= -\mathbf{2 \cdot 58\ t\text{-}m}\quad \mathbf{Ans.}$$

Exercise 14-2

1. A portal frame $ABCD$ of span 6 m and height 4 m has its vertical members fixed into the ground and horizontal member carries a uniformly distributed load of 2 t/m. The moment of inertia of the vertical members is I and that of the horizontal members is $2I$. Determine the moments at A, B, C, and D. Also draw the B.M. diagram. (*Calcutta University*)

[**Ans.** 1·5 t-m ; 3·0 t-m ; 3·0 t-m ; 1·5 t-m]

2. A portal frame $ABCD$ is hinged at A and D. The member $AB = 2$ m, $BC = 2$m and $CD = 3$ m. There is a central point load of 8 tonnes on BC. Determine the moments at B and C, if the flexural rigidity is constant for all the members. Also draw the B.M. diagram. (*Ranchi University*)

[**Ans.** 1·26 t-m ; 0·94 t-m]

HIGHLIGHTS

1. The slope deflection method is based on the following assumptions :

(a) All the joints of the frame are rigid *i.e.*, the angles between the members at the joints do not change, when the frame is loaded.

(b) Whenever the beams or frames are deflected, the rigid joints are considered to rotate as a whole, *i.e.*, the angles between the tangents to the various branches of the elastic curve meeting at a joint, remain the same as those in the original structures.

(c) Distortions, due to axial and shear stresses being very small, are neglected.

2. The general slope deflection equations are :

$$M_{AB} = \frac{2EI}{l}\left(2i_A + i_B - \frac{3\delta}{l}\right) + M'_{AB}$$

and

$$M_{BA} = \frac{2EI}{l}\left(2i_B + i_A - \frac{3\delta}{l}\right) + M'_{BA}$$

where M_{AB} = Final moment at A,

i_A = Slope at A,

M'_{AB} = Fixed end moment at A,

δ = Amount, by which the beam or frame, is deflected

M_{BA}, i_B, M'_{BA} = Corresponding values at B.

3. The procedure for solving a continuous beam by slope deflection method is as follows.

(a) First of all, treat each span as a fixed beam and calculate the fixed end moments.

(b) Now express all the end moments in terms of fixed end moments and joint rotations, by using slope deflection equations.

(c) Solve end rotations for all the supports or joints.

(d) Now substitute the values of end rotations back into the slope deflection equations, and find out the end moments.

4. The simple frames are analysed in the same way as that for a continuous beam.

5. A symmetrical portal frame is that, in which both the columns are of the same length, having similar end conditions (*i.e.*, either hinged or fixed) moments of inertia and modulus of elasticity as well as subjected to symmetrical loading. All the symmetrical portal frames are analysed in the same way as that for a continuous beam.

6. In case of unsymmetrical portal frames, additional equations, known as shear equations, are required for their analysis. The shear equations are obtained from the consideration–of shear force exerted on the columns by the horizontal external forces.

Do You Know ?

1. Define the slope deflection method and explain why its use in not encouraged in the modern design offices these days.

2. State the assumptions made in the slope deflection method.

3. Derive the slope deflection equations when

(*i*) the supports of a continuous are at the same level, and

(*ii*) the supports of a continuous beam are not at the same level.

4. Explain the procedure for analysing a statical indeterminate structure by slope deflection method.

5. How will you analyse a symmetrical portal frame by slope deflection method ?

6. What are shear equations ? Explain their uses. How will you write them ?

15

MOMENT DISTRIBUTION METHOD

15·1. Introduction

The method of moment distribution, which was first introduced by Prof. Hardy Cross in 1930, is widely used for the analysis of indeterminate structures. In this method, all the members of a structure are first assumed to be fixed in position and direction, and fixed end moments due to external loads are obtained. Now all the hinged joints are released, by applying an equal and opposite moment, and their effects are evaluated on the opposite joints. The unbalanced moment at a joint, is distributed in the two spans in the ratio of their distribution factors. This process is continued, till we reach the required degree of precision. Before entering into the details of the method, the following basic definitions should be clearly understood.

15·2. Sign conventions.

Though different types of sign conventions are adopted by different authors in their books, yet the following sign conventions, which are widely used and internationally recognised, will be used in this book.

1. All the *clockwise moments* at the ends are taken as *positive*.

2. All the *anticlockwise moments* at the ends are taken as *negative*.

15·3. Carry over factor

We have already discussed in Art. 15·1 that the moments are applied on all the end joints of a structure, whose effects are evaluated on the other joints. The ratio of moment produced at a joint to the moment applied at the other joint, without displacing it, is called *carry over factor*. Now we shall find out the value of carry over factor in the following two cases of beams :

1. When the beam is fixed at one end, and simply supported at the other.

2. When the beam is simply supported at both the ends.

15·4. Carry over factor for a beam fixed at one end, and simply supported at the other

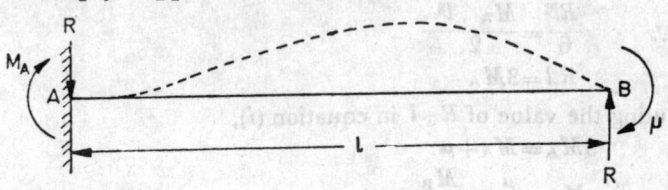

Fig. 15·1

Consider a beam AB fixed at A, and simply supported at B. Let a clockwise moment be applied at the support B of the beam as shown in Fig. 15·1.

Let $\quad l$ = Span of the beam.

$\qquad \mu$ = Clockwise moment applied at B (*i.e.* M_B), and

$\qquad M_A$ = Fixing moment at A.

Since the beam is not subjected to any external loading, therefore the two reactions must be equal and opposite as shown in Fig. 15·1.

Taking moments about A,

$$R.l = M_A + \mu \qquad \qquad ...(i)$$

Now consider any section X, at a distance x from A. We know that the moment at X,

$$M_X = M_A - R \,.\, x$$

or $\qquad EI\ \dfrac{d^2y}{dx^2}=M_A-R\ .\ x \qquad\qquad \left(\because\ M=EI\dfrac{d^2y}{dx^2}\right)$

Integrating the above equation,

$$EI\frac{dy}{dx}=M_A\ .\ x-\frac{Rx^2}{2}+C_1$$

where C_1 is the constant of integration. We know that when $x=0$, $\dfrac{dy}{dx}=0$. Therefore $C_1=0$.

or $\qquad EI\ \dfrac{dy}{dx}=M_A\ .\ x-\dfrac{Rx^2}{2} \qquad\qquad\qquad\qquad …(ii)$

Integrating the above equation once again,

$$EI\ .\ y=\frac{M_A\ .\ x^2}{2}-\frac{Rx^3}{6}+C_2$$

where C_2 is the constant of integration. We know that when $x=0$, $y=0$. Therefore $C_2=0$

or $\qquad EI\ .\ y=\dfrac{M_A\ .\ x^2}{2}-\dfrac{Rx^3}{6} \qquad\qquad\qquad\qquad …(iii)$

We also know that when $x=l$, $y=0$. Therefore substituting these values in equation (iii),

$$0=\frac{M_A.l^2}{2}-\frac{R\ .\ l^3}{6}$$

$\therefore \qquad \dfrac{Rl^3}{6}=\dfrac{M_A\ .\ l^2}{2}$

or $\qquad R.l=3M_A$

substituting the value of $R\ .\ l$ in equation (i),

$$3M_A=M_A+\mu$$

or $\qquad M_A=\dfrac{\mu}{2}=\dfrac{M_B}{2} \qquad\qquad\qquad\qquad\qquad …(iv)$

$\therefore \qquad \dfrac{M_A}{M_B}=\dfrac{1}{2}$

It is thus obvious, that carry over factor is *one-half* in this case. We see from equation (ii) that

$$EI\ \frac{dy}{dx}=M_A\ .\ x-\frac{Rx^2}{2}$$

Now for slope at B, substituting $x=l$ in the above equation,

$$EI.i_B=M_A\ .\ l-\frac{Rl^2}{2}$$

$$=M_A.\ l-\frac{3}{2}\ M_A\ .\ l \qquad\qquad (\because\ R.l=3M_A)$$

$$=-\frac{M_A\ .\ l}{2}=-\frac{\mu l}{4} \qquad\qquad \left(\because\ M_A=\frac{\mu}{2}\right)$$

$$i_B = \frac{\mu l}{4EI}$$

or

$$\mu = \frac{4EI \cdot i_B}{l}$$

15·5. Carry over factor for a beam simply supported at both ends

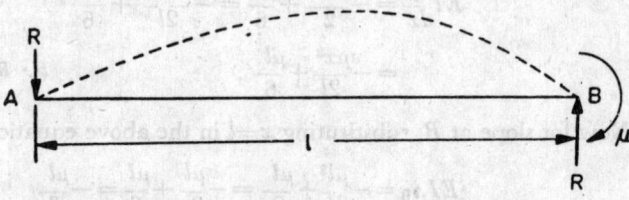

Fig. 15·2

Consider a beam AB simply supported at A and B. Let a clockwise moment be applied at the support B of the beam is shown in Fig. 15·2.

Let $\qquad$ $l =$ Span of the beam, and

$\qquad\qquad\qquad$ $\mu =$ Clockwise moment at B.

Since the beam is simply supported at A, therefore there will be no fixing moment at A. Moreover, as the beam is not subjected to any external loading, therefore the two reactions must be equal and opposite as shown in Fig. 15·2.

Taking moments about A,

$$R \cdot l = \mu \qquad\qquad\qquad ...(i)$$

Now consider any section X, at a distance x from A. We know that the moment at X,

$$M_X = R \cdot x$$

or

$$EI \frac{d^2y}{dx^2} = -R \cdot x \qquad \left(\because M = EI \frac{d^2y}{dx^2} \right)$$

Integrating the above equation,

$$EI \frac{dy}{dx} = -\frac{Rx^2}{2} + C_1 \qquad\qquad ...(ii)$$

where C_1 is the constant of integration. Integrating the above equation once again,

$$EI.y = -\frac{Rx^3}{6} + C_1 x + C_2$$

where C_2 is the constant of integration. We know that when $x = 0$, $y = 0$. Therefore $C_2 = 0$.

or

$$EI \cdot y = -\frac{Rx^3}{6} + C_1 x \qquad\qquad ...(iii)$$

We also know that when $x = l$, $y = 0$. Therefore substituting these values in the above equation,

$$0 = -\frac{Rl^3}{6} + C_1\, l$$

$$\therefore \qquad C_1 = \frac{Rl^2}{6} = \frac{\mu l}{6} \qquad\qquad (\because\ R.l=\mu)$$

Substituting this value of C_1 in equation (ii),

$$EI\frac{dy}{dx} = -\frac{Rx^2}{2} + \frac{\mu l}{6} = -\frac{Rlx^2}{2l} + \frac{\mu l}{6}$$

$$= -\frac{\mu x^2}{2l} + \frac{\mu l}{6} \qquad\qquad (\because\ R.l=\mu)$$

Now for slope at B, substituting $x=l$ in the above equation,

$$EI.i_{\text{B}} = -\frac{\mu l^2}{2l} + \frac{\mu l}{6} = -\frac{\mu l}{2} + \frac{\mu l}{6} = -\frac{\mu l}{3}$$

or

$$i_{\text{B}} = \frac{\mu l}{3EI}$$

$$\therefore \qquad \mu = \frac{3EI.i_{\text{B}}}{l}$$

15 6. Stiffness factor

It is the moment required to rotate the end, while acting on it, through a unit angle, without translation of the far end. We have seen in Art. 15·4 that the moment on a beam having one end fixed and the other freely supported,

$$\mu = \frac{4EI.i_{\text{B}}}{l}$$

$\therefore$ The stiffness factor for such a beam (substituting $i_{\text{B}}=1$)

$$k_1 = \frac{4EI}{l}$$

Similarly, we have seen in Art. 15·5 that the moment on a beam having simply supported ends,

$$\mu = \frac{3EI.i_{\text{B}}}{l}$$

$\therefore$ The stiffness factor for such a beam (substituting $i_{\text{B}}=1$)

$$k_2 = \frac{3EI}{l}$$

15·7. Distribution factors

Sometimes a moment is applied on a structural joint, to produce rotation, without the translation of its members. This moment is distributed among all the connecting members of the joint, in the proportion of their stiffness.

Consider four members OA, OB, OC and OD meeting at A. Let the member OA and OC be fixed at A and C, whereas the members.

OB and OD be hinged at B and D. Let the joint O be subjected to a moment μ as shown in Fig. 15·3.

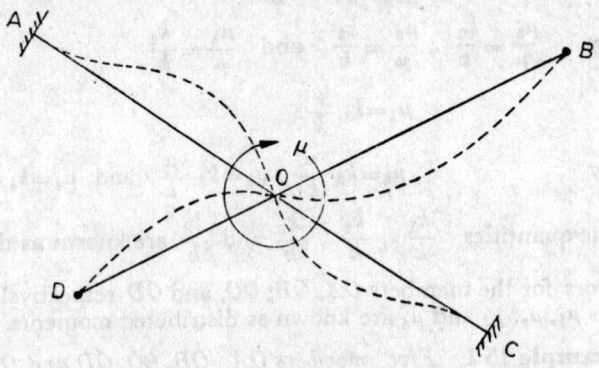

Fig. 15·3

Let $\qquad l_1 =$ Length of the member OA,

$\qquad\qquad I_1 =$ Moment of inertia of the member OA,

$\qquad\qquad E_1 =$ Modulus of elasticity of the member OA,

$\qquad l_2, I_2, E_2 =$ Corresponding values of the member OB,

$\qquad l_3, I_3, E_3 =$ Corresponding values of the member OC,

and $\qquad l_4, I_4, E_4 =$ Corresponding values of the member OD.

A little consideration will show that, as a result of the moment μ, each member gets rotated through an equal angle. Let this angle, through which each member is rotated, be θ.

We know that the stiffness of member OA,

$$k_1 = \frac{4E_1 I_1}{l_1} \qquad\qquad \text{...($\because$ End A is fixed)}$$

Similarly $$k_2 = \frac{3E_2 I_2}{l_2} \qquad\qquad \text{...($\because$ End B is hinged)}$$

and $$k_3 = \frac{4E_3 I_3}{l_3} \qquad\qquad \text{...($\because$ End C is fixed)}$$

and $$k_4 = \frac{3E_4 I_4}{l_4} \qquad\qquad \text{...($\because$ End D is hinged)}$$

Now total stiffness of all the members,

$$k = k_1 + k_2 + k_3 + k_4$$

and the total moment applied at the joint,

$$\mu = k\theta$$

$\therefore$ Moment on the member OA,

$$\mu_1 = k_1 \theta$$

Similarly $\mu_2 = k_2\theta$; $\mu_3 = k_3\theta$; and $\mu_4 = k_4\theta$

$$\therefore \qquad \frac{\mu_1}{\mu} = \frac{k_1\theta}{k\theta} = \frac{k_1}{k}.$$

Similarly $\quad \dfrac{\mu_2}{\mu} = \dfrac{k_2}{k}$; $\dfrac{\mu_3}{\mu} = \dfrac{k_3}{k}$ and $\dfrac{\mu_4}{\mu} = \dfrac{k_4}{k}.$

$$\therefore \qquad \mu_1 = k_1 \, \frac{\mu}{k}$$

Similarly $\qquad \mu_2 = k_2 \, \dfrac{\mu}{k}$; $\mu_3 = k_3 \, \dfrac{\mu}{k}$ and $\mu_4 = k_4 \, \dfrac{\mu}{k}.$

The quantities $\dfrac{k_1}{k}$, $\dfrac{k_2}{k}$, $\dfrac{k_3}{k}$ and $\dfrac{k_4}{k}$ are known as distribution factors for the members OA, OB, OC, and OD respectively. The moments μ_1, μ_2, μ_3 and μ_4 are known as distributed moments.

Example 15·1. *Five members OA, OB, OC, OD and OE meeting at O, are hinged at A and C and fixed at B, D and E. The lengths of OA, OB, OC, OD and DE are 3 m, 4 m, 2 m, 3 m, and 5 m, and their moments of inertia are 400 cm⁴, 300 cm⁴, 200 cm⁴, 300 cm⁴ and 250 cm⁴ respectively. Determine the distribution factors for the members and the distributed moments, when a moment of 4,000 kg·m is applied at O.*

Solution.

Given.   Moment = 4,000 kg-m

The moment of 4,000 kg-m applied at the joint O, will be distributed among the members as obtained from the following chart

$$\text{for fixed} = \frac{4\,EI}{l} \; ; \quad \text{for hinged} = \frac{3\,EI}{l}.$$

Member	Length m	M.I. cm⁴	Stiffness	Distribution factor	Distributed moments kg-m
OA	3	400	$\dfrac{E \times 400}{3} = 400E$	$\dfrac{400\,E}{1600\,E} = \dfrac{1}{4}$	1,000
OB	4	300	$\dfrac{4E \times 300}{4} = 300E$	$\dfrac{300\,E}{1600\,E} = \dfrac{3}{16}$	750
OC	2	200	$\dfrac{3E \times 200}{2} = 300E$	$\dfrac{300\,E}{1600\,E} = \dfrac{3}{16}$	750
OD	3	300	$\dfrac{4E \times 300}{3} = 400E$	$\dfrac{400\,E}{1600\,E} = \dfrac{1}{4}$	1,000
OA	5	250	$\dfrac{4E \times 250}{5} = 200E$	$\dfrac{200\,E}{1600\,E} = \dfrac{1}{8}$	500
			$\Sigma\ 1600\ E.$		

∴   Distribution factors for OA, OB, OC, OD, and OE are

$$\frac{1}{4}, \ \frac{3}{16}, \ \frac{3}{16}, \ \frac{1}{4} \text{ and } \frac{1}{8} \text{ respectively } \textbf{Ans.}$$

and distributed moments for OA, OB, OC, OD and OE are 1,000, 750, 750, 1,000, 500 kg·m respectively.   **Ans.**

15·8. Application of moment distribution method to various types of continuous beams

In the previous articles, we have studied the principles of the moment distribution method. First of all, all the supports are assumed to be clamped and fixing moments are found out for all the spans. The unbalanced moment at a support is distributed among the two spans in the ratio of their stiffness factors and their effects are evaluated on the opposite joints. This process is continued, till we reach the required degree of precision. In the proceeding articles, we shall discuss its application to the following types of beams.

1. Beams with fixed end supports,
2. Beams with simply supported ends,
3. Beams with the end span overhanging, and
4. Beams with a sinking support.

15·9. Beams with fixed end supports

Sometimes, a continuous beam is fixed at its one or both ends. In such a case, the fixed end moments are obtained and the unbalanced moment is distributed in two spans in the ratio of their distribution factors.

Example 15·2. *Evaluate the bending moment and shear force diagrams of the beam shown in Fig. 15·4.*

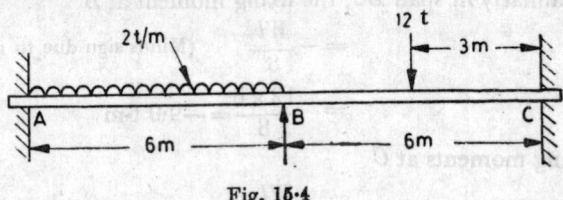

Fig. 15·4

What are the reactions at the supports ?

(*U.P.S.C. Engg. Services, 1973*)

*Solution.

Given. Span $AB = 6$ m
Span $BC = 6$ m
Load in $AB = 2$ t/m
Load in $BC = 12$ t

First of all, let us consider the continuous beam ABC to be split into two fixed beams AB and BC as shown in Fig. 15·5 (*b*) and (*c*):

---

*This example is also solved on pages 488 and 517

Now in span AB, the fixing moment at A

$$= -\frac{wl^2}{12} \qquad \text{(Minus sign due to anticlockwise)}$$

$$= -\frac{2 \times 6^2}{12} = -6 \cdot 0 \text{ t-m}$$

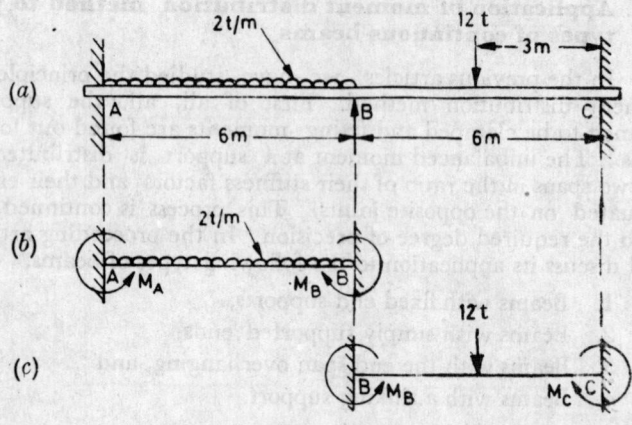

Fig. 15·5

and fixing moment at B

$$= +\frac{wl^2}{12} \qquad \text{(Plus sign due to clockwise)}$$

$$= \frac{2 \times 6^2}{12} = 6 \cdot 0 \text{ t-m}$$

Similarly in span BC, the fixing moment at B

$$= -\frac{Wl}{8} \qquad \text{(Minus sign due to anticlockwise)}$$

$$= -\frac{12 \times 6}{8} = -9 \cdot 0 \text{ t-m}$$

and fixing moments at C

$$= +\frac{Wl}{8} \qquad \text{(Plus sign due to clockwise)}$$

$$= \frac{12 \times 6}{8} = 9 \cdot 0 \text{ t-m}$$

Now let us find out the distribution factors at B. From the geometry of the figure, we find that the stiffness factor for BA,

$$k_B = \frac{4EI}{l} = \frac{4EI}{6} = \frac{2EI}{3}$$

$$(\because \text{ the beam is fixed at } A)$$

Similarly $\qquad k_B = \frac{4EI}{l} = \frac{4EI}{6} = \frac{2EI}{3}$

$$(\because \text{ the beam is fixed at } C)$$

Therefore distribution factors for BA and BC will be $\frac{1}{2}$ and $\frac{1}{2}$.

Now draw the beam and fill up the distribution factors [and initial moments as shown in Fig. 15·6 (a).

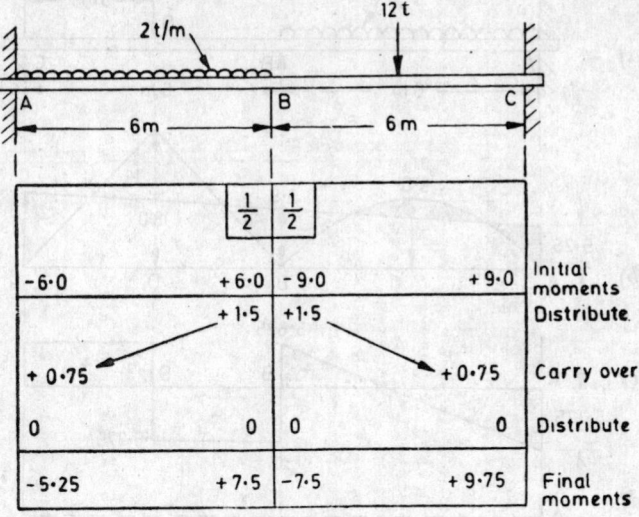

Fig. 15·6

We obtained the moment at B in span AB is $+6·0$ and that in span BC is $-9·0$. Thus there is an unbalanced moment equal to $-9·0+6·0=-3·0$ t-m. Now distribute this unbalanced moment (equal to $-3·0$ t-m) into the spans AB and BC in the ratio of their distribution factors i.e. $+1·5$ and $+1·5$. Now *carry over the effects of these distributed moments at A and D equal to $\frac{1}{2}\times 1·5=+0·75$. Then distribute the unbalanced moment at B (In this case, there is no carry over moment from from A or D at B. So the distribution of moment at B is zero). Now find out the final moments at A, B and C in the spans AB and BC by algebraically adding the respective values.

Now calculate the bending moment in the spans AB and BC, by considering them as simply supported beams.

We know that the bending moment at the centre of the span AB treating it as a simply supported

$$=\frac{wl^2}{8}=\frac{2\times 6^2}{8}=9·0 \text{ t-m}$$

Similarly bending moment under the load in the span BC

$$=\frac{Wl}{4}=\frac{12\times 6}{4}=18·0 \text{ t-m}$$

*As per Art. 15·t the carry over factor is $\frac{1}{2}$.

Now complete the final bending moment *diagram as shown in Fig. 15·7 (b).

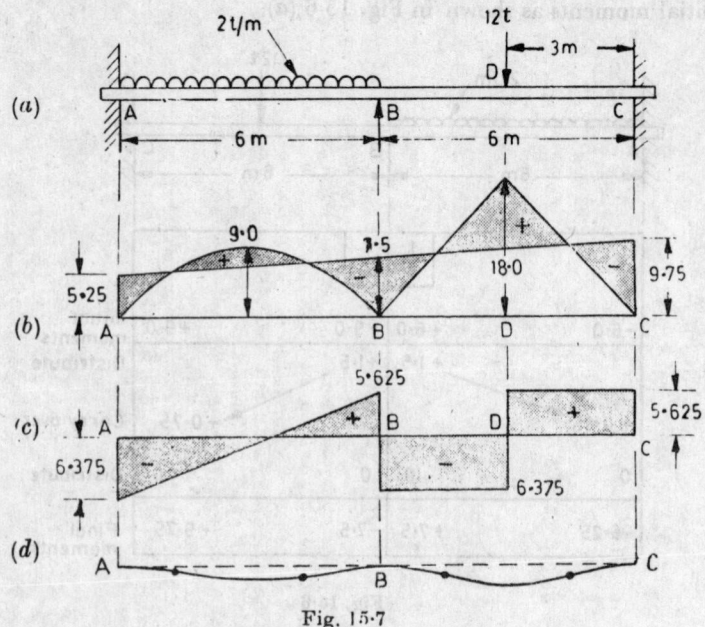

Fig. 15·7

Let R_A = Reaction at A,
          R_B = Reaction at B, and
          R_C = Reaction at C.

Taking moments at B, and equating the same,

$$R_A \times 6 + M_B = M_A + 2 \times 6 \times 3$$
$$R_A \times 6 - 7 \cdot 5 = 5 \cdot 25 + 36$$
$$\therefore \quad R_A = \mathbf{6 \cdot 375 \ t \ Ans.}$$

Again taking moments about B, and equating the same,

$$R_C \times 6 + M_B = M_C + 12 \times 3$$
$$R_C \times 6 - 7 \cdot 5 = -9 \cdot 75 + 36$$
$$\therefore \quad R_C = \mathbf{6 \cdot 625 \ t \ Ans.}$$

and
$$R_B = (2 \times 6 + 12) - (6 \cdot 375 + 5 \cdot 625) \ t$$
$$= \mathbf{12 \ t \ Ans.}$$

Now complete the shear force diagram and eastic over as shown in Fig. 15·7 (c) and (d).

---

*Though the moment at B in the span AB is positive and moment at C in the span BC is *positive* in the table, yet these moments are taken as *negative* in the B.M. diagram. The reason for the same is that at these points the moments tend to bend the beam with convexity upwards as shown in Fig. 15·5 (b) and (c).

15·10. Beams with simply supported ends

Sometimes, a continuous beam is simply supported over one or both of its ends. We know that the fixing moment on a simply supported end is *zero*. Therefore, in such a case, the simply supported ends are released by applying equal and opposite moments, and their effects are carried over on the opposite joints. It may also be noted that no moment is carried over from the opposite joint to the simply supported end.

Example 15·3. *A continuous beam ABC 10 m long rests on supports A, B and C at the same level and is loaded as shown in Fig. 15·8.*

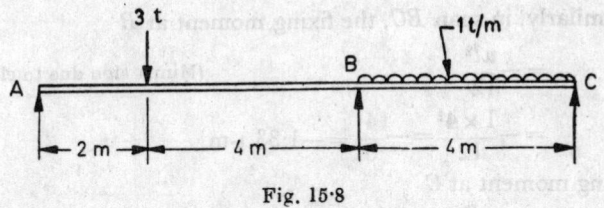

Fig. 15·8

Determine the moments over the beam and draw the bending moment diagram.

Solution.

Given. Span $AB = 6$ m

Span BC $= 4$ m

Load in AB, $W = 3$ t

Load in BC, $w = 1$ t/m

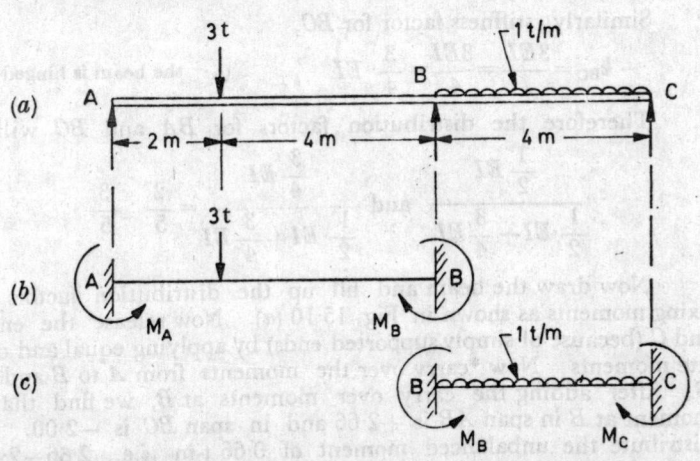

Fig. 15·9

First of all, assume the continuous beam ABC to be split up into two fixed beams AB and BC as shown in Fig. 15·9 (b) and (c).

In span AB, the fixing moment at A

$$= -\frac{Wab^2}{l^2} \qquad \text{(Minus sign due to anticlockwise)}$$

$$= -\frac{3 \times 2 \times 4^2}{6^2} = -\frac{8}{3} = -2 \cdot 67 \text{ t-m}$$

and fixing moment at B

$$= +\frac{Wa^2b}{l^2} \qquad \text{(Plus sign due to clockwise)}$$

$$= +\frac{3 \times 2^2 \times 4}{6^2} = +\frac{4}{3} = +1 \cdot 33 \text{ t-m}$$

Similarly, in span BC, the fixing moment at B

$$= -\frac{wl^2}{12} \qquad \text{(Minus sign due to clockwise)}$$

$$= -\frac{1 \times 4^2}{12} = -\frac{4}{3} = -1 \cdot 33 \text{ t-m}$$

and fixing moment at C

$$= +\frac{wl^2}{12} \qquad \text{(Plus sign due to clockwise)}$$

$$= +\frac{1 \times 4^2}{12} = +\frac{4}{3} = +1 \cdot 33 \text{ t-m}$$

Now let us find out the distribution factors at B. From the geometry of the figure, we find that the stiffness factor for BA,

$$k_{BA} = \frac{3EI}{l} = \frac{3EI}{6} = \frac{1}{2} EI \qquad (\because \text{ the beam is hinged at } A)$$

Similarly, stiffness factor for BC,

$$k_{BC} = \frac{3EI}{l} = \frac{3EI}{4} = \frac{3}{4} EI \qquad (\because \text{ the beam is hinged at } C)$$

Therefore the distribution factors for BA and BC will be

$$\frac{\frac{1}{2} EI}{\frac{1}{2} EI + \frac{3}{4} EI} \quad \text{and} \quad \frac{\frac{3}{4} EI}{\frac{1}{2} EI + \frac{3}{4} EI} = \frac{2}{5} : \frac{3}{5}$$

Now draw the beam and fill up the distribution factors and fixing moments as shown in Fig. 15·10 (a). Now release the ends A and C (because of simply supported ends) by applying equal and opposite moments Now *carry over the moments from A to B and C to B. After adding the carry over moments at B, we find that the moment at B in span AB is $+2 \cdot 66$ and in span BC is $-2 \cdot 00$. Now distribute the unbalanced moment of $0 \cdot 66$ t-m (i.e.. $2 \cdot 66 - 2 \cdot 00 = 0 \cdot 66$ t-m) into the two spans of AC in the ratio of their distribution factors, i.e., $-0 \cdot 26$ and $-0 \cdot 40$. Now find out the final moments by

---

*As per Art. 15·4, the carry over factor is $\frac{1}{2}$.

adding all the values. Now calculate the bending moments in the spans *AB* and *BC*, by considering them as simply supported beams *i.e.*

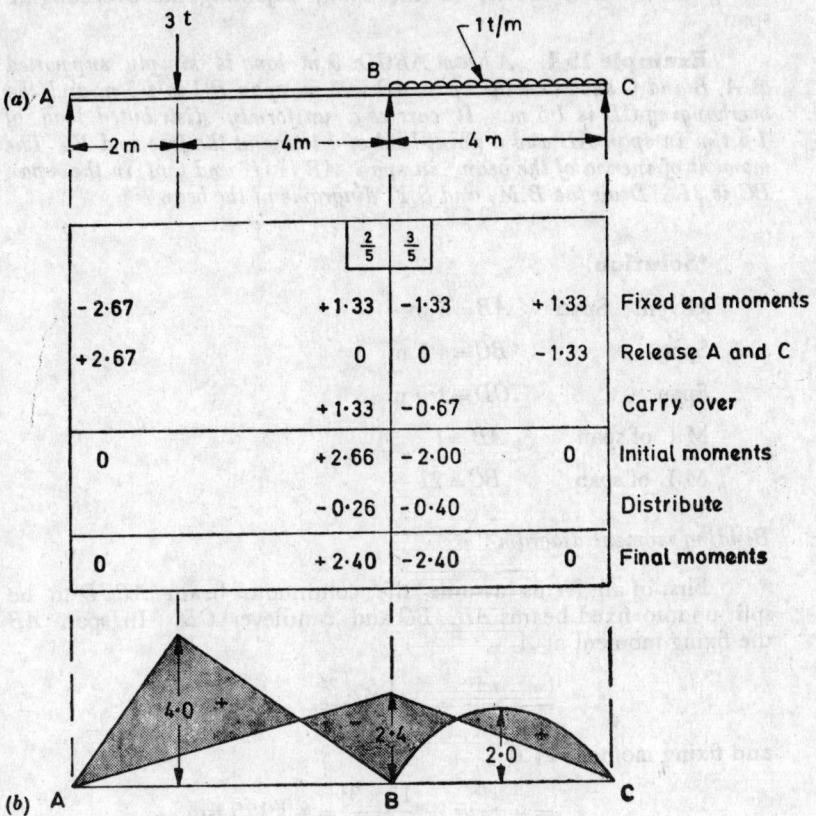

Fig. 15·10

B.M. under 3 tonnes load in the span *AB*

$$= \frac{Wab}{l} = \frac{3 \times 2 \times 4}{6} = 4 \text{ t-m}$$

Similarly B.M. at the mid of the span *BC* due to uniformly distributed load

$$= \frac{wl^2}{8} = \frac{1 \times 4^2}{8} = 2 \text{ t-m}$$

Now complete the final B.M. diagram as shown in Fig. 15·10 (*b*).

15·11. Beams with end span overhanging

Sometimes, a beam is overhanging at its one or both the end supports. In such a case, the bending moment at the support near the overhanging end will be due to the load over the overhanging portion and will remain constant, irrespective of the moments on the other supports. It is thus obvious, that the distribution factors over the support having one span overhanging will be 1 and 0. Moreover this support is considered as a simply supported for the purpose of calculating distribution factors in the span, adjoining the overhanging span.

Example 15·4. *A beam ABCD 9 m long is simply supported at A, B and C such that the span AB is 3 m, span BC is 4·5 m and the overhanging CD is 1·5 m. It carries a uniformly distributed load of 1·5 t/m in span AB and a point load of 1 tonne at the free end B. The moment of inertia of the beam in span AB is I and that in the span BC is 2I. Draw the B.M. and S.F. diagrams of the beam.*

*Solution.

Given.　Span　　$AB = 3$ m

Span　　　　　$BC = 4·5$ m

Span　　　　　$CD = 1·5$ m

M.I. of span　$AB = I$

M.I. of span　$BC = 2I$

Bending moment diagram

First of all, let us assume the continuous beam $ABCD$ to be split up into fixed beams AB, BC and cantilever CD. In span AB the fixing moment at A

$$= -\frac{wl^2}{12} = -\frac{1·5 \times 3^2}{12} = -1·125 \text{ t-m}$$

and fixing moment at B

$$= +\frac{wl^2}{12} = +\frac{15 \times 3^2}{12} = +1·125 \text{ t-m}$$

Since the span BC is not carrying any load, therefore fixing moment at B and C will be zero. The moment at C, for the cantilever CD, will be $1 \times 1·5 = -1·5$ t-m.

Now let us find out the distribution factors at B and C. From the geometry of the figure, we find that the stiffness factor for BA,

$$k_{BA} = \frac{3EI}{l} = \frac{3EI}{3} = 1\,EI \quad (\because \text{ the end } A \text{ is simply supported})$$

Similarly $k_{BC} = \dfrac{3EI}{l} = \dfrac{3E \times 2I}{4·5} = \dfrac{6EI}{4·5} = \dfrac{4}{3}\,EI$

$$(\because \text{ beam is overhanging beyond } C)$$

Therefore the distribution factor for BA and BC will be

$$\frac{1\,EI}{1EI+\frac{4}{3}EI}\quad\text{and}\quad\frac{\frac{4}{3}\,EI}{1\,EI+\frac{4}{3}\,EI}=\frac{3}{7}\quad\text{and}\quad\frac{4}{7}$$

We know that distribution factors for CA and CD will be 1 and 0, because the beam is overhanging at C.

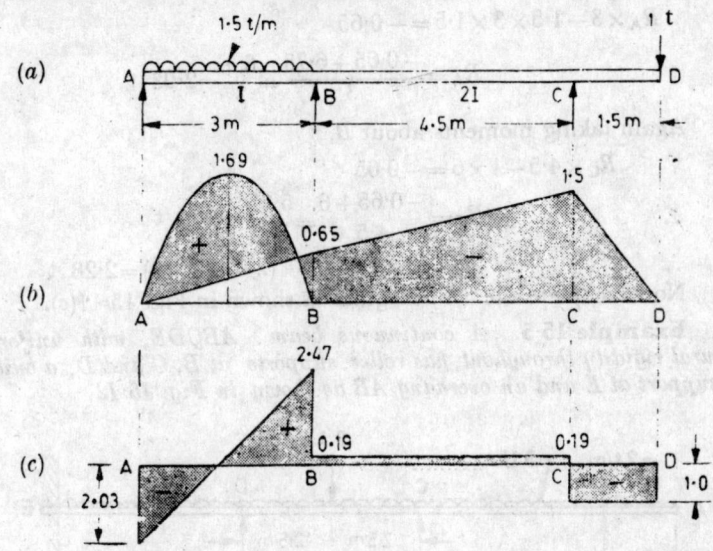

Fig. 15·11

Now prepare the following table.

A	B		C	D	
	$\frac{3}{7}$	$\frac{4}{7}$	1	0	
$-1\cdot125$	$+1\cdot125$	0	0	$1\cdot5$	Fixed end moments
$+1\cdot125$	0	0	$+1\cdot5$	0	Release A and balance C
	$+0\cdot562$	$+0\cdot75$	0	0	Carry over
0	$+1\cdot687$	$+0\cdot75$	$+1\cdot5$	$-1\cdot5$	Initial moments
	$-1\cdot037$	$-1\cdot40$			Distribute
0	$+0\cdot65$	$-0\cdot65$	$+1\cdot5$	$-1\cdot5$	Final moments

Bending moment in the middle of span AB, by considering it as a simply supported beam

$$=\frac{wl^2}{8}=\frac{1\cdot5\times3^2}{8}=1\cdot69\ \text{t·m}$$

Now complete the final B.M. diagram as shown in Fig. 15·11(b).

Shear force diagram

Let R_A=Reaction at A,
                      R_B=Reaction at B, and
                      R_C=Reaction at C

Taking moments at B,

$$R_A \times 3 - 1.5 \times 3 \times 1.5 = -0.65$$

or $$R_A = \frac{-0.65 + 6.75}{3} = \frac{6.1}{3} = 2.03 \text{ t}$$

Again taking moments about B,

$$R_C \times 4.5 - 1 \times 6 = -0.65$$

∴ $$R_C = \frac{-0.65 + 6}{4.5} = \frac{5.35}{4.5} = 1.19 \text{ t}$$

∴ $$R_B = (3 \times 1.5 + 1) - (2.03 + 1.19) = 2.28 \text{ t}$$

Now complete the S.F. diagram as shown in Fig. 15·11(c).

Example 15·5. *A continuous beam ABCDE, with uniform flexural rigidity throughout, has roller supports at B, C and D, a built-in support at E and an overhang AB as shown in Fig. 15·12.*

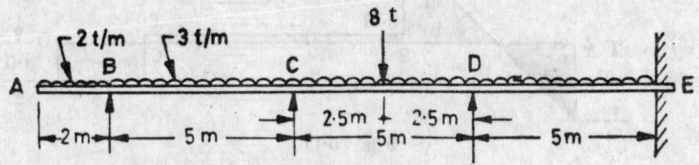

Fig. 15·12

It carries a uniformly distributed load of intensity of 2 t/m on AB and another of intensity of 3 t/m over BCDE. In addition to it, a point load of 8 tonnes is placed midway between C and D. The span lengths are AB=2 m, BC=CD=DE=5 m.

Obtain the support moments by the moment distribution method and sketch the B.M. diagram giving values at salient points.

                                                    (*Oxford University*)

Solution.

Given. Span *AB*   =2 m
Span   *BC=CD=DE* =5 m
Load in *AB*, w=2 t/m
Load in *BC=CD=DE* =3 t/m
Point load in *CD* =8 t

Support moments

First of all, let us assume the continuous beam *ABCDE* to be split up into cantilever *AB* and fixed beams *BC*, *CD* and *DE*.

The B.M. at *B* for the cantilever *AB*

$$=2 \times 2 \times 1 = 4 \text{ t·m}$$

In span BC, fixing moment at B

$$= -\frac{wl^2}{12} = -\frac{3 \times 5^2}{12} = \frac{75}{12} \text{ t-m}$$

$$= -6 \cdot 25 \text{ t-m}$$

and fixing moment at B

$$= +\frac{wl^2}{12} = +\frac{3 \times 5^2}{12} = +6 \cdot 25 \text{ t-m}$$

In span CD, the fixing moment at C,

$$= -\left[\frac{wl^2}{12} + \frac{Wab^2}{l^2}\right]$$

$$= -\left[\frac{3 \times 5^2}{12} + \frac{8 \times 2 \cdot 5 \times 2 \cdot 5^2}{5^2}\right] \text{ t-m}$$

$$= -11 \cdot 25 \text{ t-m}$$

and fixing moment at D

$$= +\frac{wl^2}{12} + \frac{Wa^2b}{l^2} = \frac{3 \times 5^2}{12} + \frac{8 \times 2 \cdot 5 \times 2 \cdot 5^2}{5^2} \text{ t-m}$$

$$= +11 \cdot 25 \text{ t-m}$$

In span DE, the fixing moment at D

$$= -\frac{wl^2}{12} = -\frac{3 \times 5^2}{12} = -6 \cdot 25 \text{ t-m}$$

and fixing moment at $E = +\frac{wl^2}{12} = \frac{3 \times 5^2}{12} = +6 \cdot 25 \text{ t-m}$

Now let us find out the distribution factors at B, C and D. From the geometry of the figure, we find that the distribution factors BA and BC will be 0 and 1.

Stiffness factor for CB,

$$k_{CB} = \frac{3EI}{l} = \frac{3EI}{5} = \frac{3}{5} EI \qquad (\because \text{ the beam is over-hanging beyond } B)$$

and

$$k_{CD} = \frac{4EI}{l} = \frac{4}{5} EI \quad (\because \text{ the beam is continuous at } D)$$

Therefore the distribution factors for CB and CD will be

$$\frac{\frac{3}{5} \cdot EI}{\frac{3}{5} EI + \frac{4}{5} EI} \quad \text{and} \quad \frac{\frac{4}{5} EI}{\frac{3}{5} EI \text{ and } \frac{4}{5} EI} = \frac{3}{7} \text{ and } \frac{4}{7}$$

Similarly stiffness factor for DC,

$$k_{DC} = \frac{4EI}{l} = \frac{4}{5} EI \qquad (\because \text{ the beam is continuous at } C)$$

and

$$k_{DE} = \frac{4EI}{l} = \frac{4}{5} EI \qquad (\because \text{ the beam is fixed at } E)$$

Distribution factors for DC and DE will be

$$\frac{\frac{4}{5} EI}{\frac{4}{5} EI + \frac{4}{5} EI} \quad \text{and} \quad \frac{\frac{4}{5} EI}{\frac{4}{5} EI + \frac{4}{5} EI} = \frac{1}{2} \text{ and } \frac{1}{2}$$

Now prepare the following table

A	B		C		D		E	
	0	1	$\frac{3}{7}$	$\frac{4}{7}$	$\frac{1}{2}$	$\frac{1}{2}$		
0	+4·00	−6·25	+6·25	−11·25	+11·25	−6·25	+6·25	Fixed end moments
		+2·25						Balance B
			+1·13					Carry over
0	+4·00	−4·00	+7·38	−11·25	+11·25	−6·25	+6·25	Initial moments
			+1·66	+ 2·21	− 2·50	−2·50		Distribute
				− 1·25	+ 1·10		−1·25	Carry over
			+0·54	+ 0·71	− 0·55	−0·55		Distribute
				− 0·28	+ 0·36		−0·28	Carry over
			+0·12	+ 0·16	− 0·18	−0·18		Distribute
				− 0·09	+ 0·08		−0·09	Carry over
			+0·04	+ 0·05	− 0·04	−0·04		Distribute
0	+4·00	−4·00	+9·74	− 9·74	+ 9·52	−9·52	+4·63	Final moments

Bending moment diagram

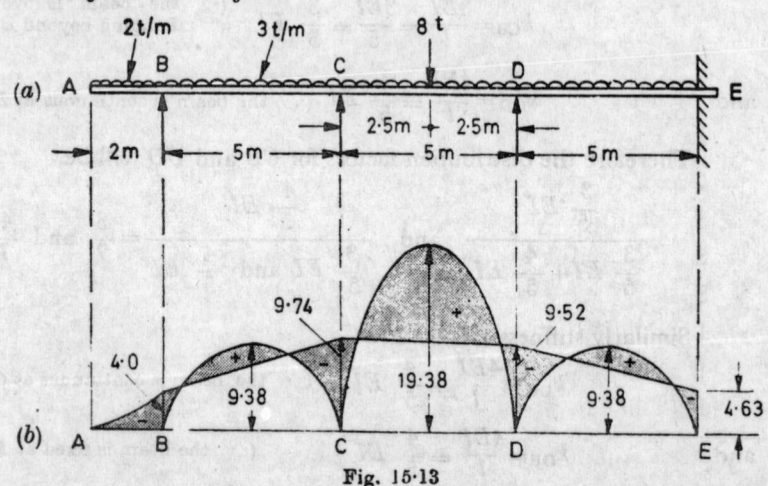

Fig. 15·13

The bending moment in the middle of span BC, by considering it as a simply supported beam

$$= \frac{wl^2}{8} = \frac{3 \times 5^2}{8} = 9 \cdot 38 \text{ t-m}$$

Similarly B.M. in the middle of span CD, by considering it as a simply supported beam

$$= \frac{wl^2}{8} + \frac{Wab}{l} = \frac{3 \times 5^2}{8} + \frac{8 \times 2 \cdot 5 \times 2 \cdot 5}{5} = 19 \cdot 38 \text{ t-m}$$

and B.M. in the middle of span DE, by considering it as simply supported beam

$$= \frac{wl^2}{8} = \frac{3 \times 5^2}{8} = 9 \cdot 38 \text{ t-m}$$

Now complete the B.M. diagram as shown in Fig. 15·13 (b)

Example 15·6. *A beam ABCDE has a built-in support A and roller supports at B, C and D, DE being an overhung. AB=7 m, BC=5 m, CD=4 m and DE=1·5 m. The values of I, the moment of inertia of the section, over each of these lengths are 3I, 2I, I and I respectively.*

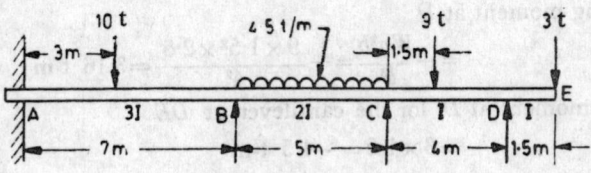

Fig. 15·14

The beam carries a point load of 10 tonnes at a point 3 m from A, a uniformly distributed load of 4·5 tonnes/m over whole of BC and a concentrated load of 9 tonnes at CD 1·5 m from C and another point load of 3 tonnes at E. the top of overhang as shown in Fig. 15·14.

Determine (i) the moments developed over each of the supports A, B, C and D, and (ii) draw B.M. diagram for the entire beam, stating values at salient points.

*Solution

Given   Span $AB = 7$ m and M.I. $= 3I$

Span $BC = 5$ m and M.I. $= 2I$

Span $CD = 4$ m and M.I. $= I$

Span $DE = 1·5$ m and M.I. $= I$

Load in AB, $W = 10$ t

Load in BC, $w = 4·5$ t/m

Load in CD, $W = 9$ t

---

*This example is also solved on page 496

Moments developed over each of the support

First of all, let us assume the continuous beam *ABCDE* to be split up into fixd beams *AB*, *BC*, *CD* and cantilever *DE*.

In span *AB*, fixing moment at *A*,

$$= -\frac{Wab^2}{l^2} = -\frac{10 \times 3 \times 4^2}{7^2} = -9 \cdot 78 \text{ t-m}$$

and fixing moment at *B*,

$$= + \frac{Wa^2b}{l^2} = \frac{10 \times 3^2 \times 4}{7^2} = 7 \cdot 35 \text{ t-m}$$

In span *BC*, fixing moment at *B*,

$$= -\frac{wl^2}{12} = -\frac{4 \cdot 5 \times 5^2}{12} = -9 \cdot 38 \text{ t-m}$$

and fixing moment at *C*

$$= + \frac{wl^2}{12} = \frac{4 \cdot 5 \times 5^2}{12} = 9 \cdot 38 \text{ t-m}$$

In span *CD*, fixing moment at *C*

$$= -\frac{Wab^2}{l^2} = -\frac{9 \times 1 \cdot 5 \times 2 \cdot 5^2}{4^2} = -5 \cdot 26 \text{ t-m}$$

and fixing moment at *D*

$$= + \frac{Wa^2b}{l^2} = -\frac{9 \times 1 \cdot 5^2 \times 2 \cdot 5}{4^2} = 3 \cdot 16 \text{ t-m}$$

and the moment at *D*, for the cantilever at *DE*

$$= 3 \times 1 \cdot 5 = -4 \cdot 5 \text{ t-m}$$

Now let us find out the distribution factors at *B*, *C*, and *D*. From the geometry of the figure, we find that the stiffness factor for *BA*,

$$k_{BA} = \frac{4EI}{l} = \frac{4E \times 3I}{7} = \frac{12}{7} EI \qquad (\because \text{ the beam is fixed at } A)$$

and

$$k_{BC} = \frac{4EI}{l} = \frac{4E \times 2I}{5} = \frac{8}{5} EI \quad (\because \text{ the beam is continuous at } C)$$

Therefore distribution factors for *BA* and *BC* will be

$$\frac{\frac{12}{7} EI}{\frac{12}{7} EI + \frac{8}{5} EI} \text{ and } \frac{\frac{8}{5} EI}{\frac{12}{7} EI + \frac{8}{5} EI} = \frac{15}{29} \text{ and } \frac{14}{29}$$

Similarly stiffness factors for *CB*,

$$k_{CB} = \frac{4EI}{l} = \frac{4E \times 2I}{5} = \frac{8}{5} EI \qquad [\because \text{ the beam is continuous at } B]$$

and

$$k_{CD} = \frac{3EI}{l} = \frac{3EI}{4} = \frac{3}{4} EI \qquad [\because \text{ the beam is overhanging beyond } D]$$

Therefore distribution factors for *CB* and *CD* will be

$$\frac{\frac{8}{5}\,EI}{\frac{8}{5}\,EI+\frac{3}{4}\,EI}\quad\text{and}\quad\frac{\frac{3}{4}\,EI}{\frac{8}{5}\,EI+\frac{3}{4}\,EI}=\frac{32}{47}\ \text{and}\ \frac{15}{47}$$

We know that the distribution factors for DE and DE will be 1 and 0, because the beam is overhanging at D.

Now prepare the following table :

A		B		C			D	E	
	$\dfrac{15}{29}$	$\dfrac{14}{29}$		$\dfrac{32}{47}$	$\dfrac{15}{47}$		1	0	
−9·78	+7·35	−9·38	+9·88	−5·26			+3·16	−4·5	Fixed end moments
							+1·34		Balance D
				+0·67					Carry over
−9·78	+7·35	−9·38	+9·38	−4·59			+4·5	−4·5	Initial moments
	+1·05	+0·98	−3·26	−1·53					Distribute
+0·53		−1·63	+0·49						Carry over
	+0·84	+0·79	−0·33	−0·16					Distribute
+0·42		−0·17	+0·40						Carry over
	+0·09	+0·08	−0·27	−0·13					Distribute
+0·05		−0·13	+0·04						Carry over
	+0·07	+0·06	−0·03	−0·01					Distribute
+0·04		−0·02	+0·03						Carry over
	+0·01	+0·01	−0·02	−0·01					Distribute
−8·74	+9·41	−9·41	+6·43	−6·43			+4·5	−4·5	Final moments

Bending moment diagram

B.M. under the 10-tonne load

$$=\frac{Wab}{l}=\frac{10\times3\times4}{7}=17\cdot14\ \text{t-m}$$

Similarly, B.M. at the mid of the span BC

$$=\frac{wl^2}{8}=\frac{4\cdot5\times5^2}{8}=14\cdot06 \text{ t-m}$$

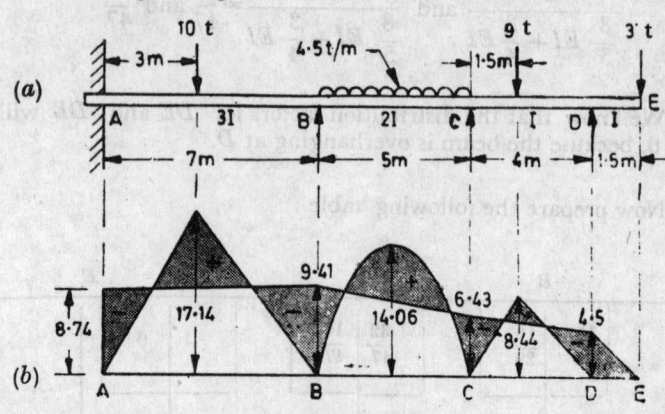

Fig. 15·15

and B.M. under the 9-tonne load in span CD

$$=\frac{Wab}{l}=\frac{9\times1\cdot5\times2\cdot5}{4}=8\cdot44 \text{ t-m}$$

Now complete the B.M. diagram as shown in Fig. 15·15 (b).

15·12.   Beams with a sinking support

Sometimes one of the supports of a continuous beam sinks down, with respect to others, as a result of loading. As a result of sinking, some moments are caused on the two supports, in addition to the moments due to loading.

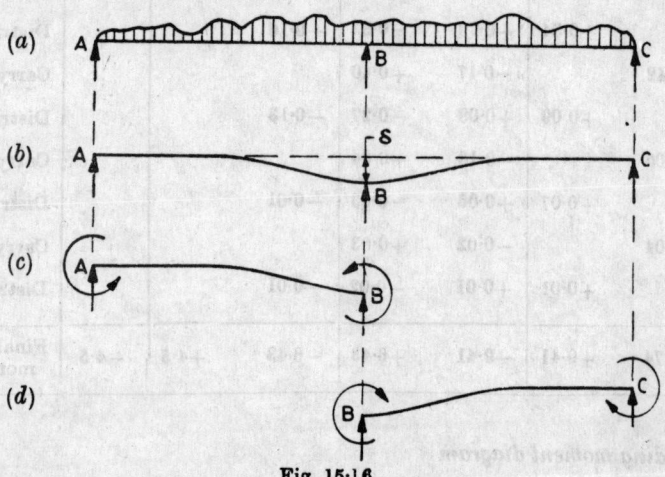

Fig. 15·16

We have already discussed in Art. 12·9 that the moment due to sinking

$$=\frac{6EI\delta}{l^2}$$

Consider a beam *ABC* simply supported at *A*, *B* and *C* subjec ted to any loading, As a result of this loading, let the support *B* sink down by an amount equal to δ as shown in Fig. 15·16 (*b*). Now assume the beam *ABC* to be split up into two beams *AB* and *BC*. A little consideration will show, that the beam *AB* will be subjected to *anticlockwise* moment at *A* and *B*, and the beam *BC* will be subjected to *clockwise* moment at *B* and *C* as shown in Fig. 15·16 (*c*) and (*d*). It is thus obvious, that if the right hand support sinks down with respect to the left hand, the moment caused on both

supports will be $-\dfrac{6EI\delta}{l^2}$ in both the supports. Similarly, if the

left hand support sinks down with respect to the right hand support,

the moments caused on both the supports will be $+\dfrac{6EI\delta}{l^2}$.

Example 15·7. *A continuous beam ABC shown in Fig. 15·17 carries a uniformly distributed load of 5 tonnes/m on AB and BC. The support B sinks by 5 mm below A and C and the value of EI is constant throughout the beam.*

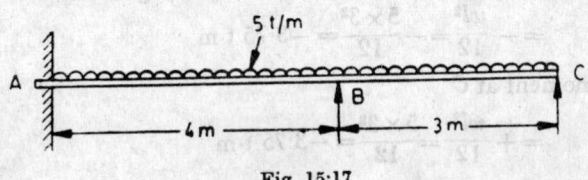

5 t/m

A ⌒⌒⌒⌒⌒⌒⌒⌒⌒⌒⌒ B ⌒⌒⌒⌒⌒⌒⌒⌒⌒⌒⌒ C

|← 4 m →| |← 3 m →|

Fig. 15·17

Taking E=2,000 tonnes/cm² and I=33,200 cm⁴, find the bending moment at supports A and B and draw the bending moment diagram.

(Madras University , 1974)

*Solution

Given. Span   *AB*=4 m

Span   *BC*=3 m

Load,   *w*=5 t/m

Sinking of support *B*=5 mm=$\dfrac{5}{1000}$ m

Young's modulus, $E=2,000$ t/cm²$=2\times10^7$ t/m²

Moment of inertia, $I=33,200$ cm⁴$=\dfrac{3\cdot32}{10^4}$ m⁴

Bending moment at supports

First of all let us assume the continuous beam *ABC* to be split up into fixed beams *AB* and *BC*. In span *AB* fixing moment at *A* due to loading—

$$=-\frac{wl^2}{12}=-\frac{5\times4^2}{12}=-6\cdot67\text{ t-m}$$

---

*This example is also solved on page 502

and fixing moment at B

$$= + \frac{wl^2}{12} = \frac{5 \times 4^2}{12} = 6 \cdot 67 \text{ t-m}$$

Now in span AB moment A due to sinking of support B

$$= - \frac{6EI\delta}{l^2} \qquad \text{(Minus sign due to right support sinking)}$$

$$= - \frac{6 \times 2 \times 10^7 \times \dfrac{3 \cdot 32}{10^4} \times \dfrac{5}{1000}}{4^2} = -12 \cdot 45 \text{ t-m}$$

Similarly moment at B due to sinking of support B

$$= -12 \cdot 45 \text{ t-m}$$

$\therefore$ Total moment at A

$$= -6 \cdot 67 - 12 \cdot 45 = -19 \cdot 12 \text{ t-m}$$

and total moment at B

$$= +6 \cdot 67 - 12 \cdot 45 = -5 \cdot 78 \text{ t-m}$$

In span BC fixing moment at B due to loading

$$= - \frac{wl^2}{12} = - \frac{5 \times 3^2}{12} = -3 \cdot 75 \text{ t-m}$$

and fixing moment at C

$$= + \frac{wl^2}{12} = \frac{5 \times 3^2}{12} = -3 \cdot 75 \text{ t-m}$$

Now in span BC moment at B due to sinking of support B

$$= + \frac{6EI\delta}{l^2} \qquad \text{(Plus sign due to left support sinking)}$$

$$= \frac{6 \times 2 \times 10^7 \times \dfrac{3 \cdot 32}{10^4} \times \dfrac{5}{1000}}{3^2} = 22 \cdot 13 \text{ t-m}$$

Similarly moment at C due to sinking of support B

$$= 22 \cdot 13 \text{ t-m}$$

Total moment at

$$B = -3 \cdot 75 + 22 \cdot 13 = 18 \cdot 38 \text{ t-m}$$

and total moment at

$$C = +3 \cdot 75 + 22 \cdot 13 = 25 \cdot 88 \text{ t-m}$$

Now let, us find out the distribution factors at B. From the geometry of the figure, we find that the stiffness factor for BA,

$$k_{BA} = \frac{4EI}{l} = \frac{4EI}{4} = 1 \ EI \qquad (\because \text{ the beam is fixed at } A)$$

or

$$k_{BC} = \frac{3EI}{l} = \frac{3EI}{3} = 1 EI \ (\because \text{ the beam is simply supported at } C)$$

Therefore the distribution factors for BA and BC will be $1/2$ and $1/2$.

Now prepare the following table :

A	B		C	
	$\frac{1}{2}$	$\frac{1}{2}$		
−19·12	−5·78	+18·38	+25·88	Fixed end moment
			−25·88	Release C
		−12·94		Carry over
−19·12	−5·78	+5·44	0	Initial moments
	+0·17	+0·17		Distribute
+0·09				Carry over
0	0	0	0	Distribute
−19·03	−5·61	+5·61	0	Final moments

Bending moment diagram

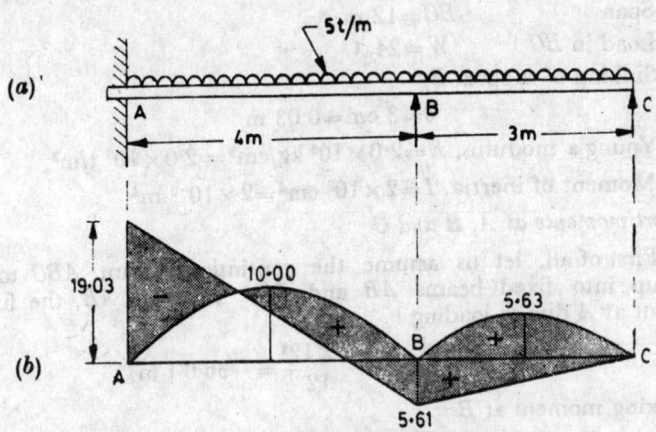

Fig. 15·18

Bending moment at the mid of the span *AB*, treating it as a simply supported beam,

$$= \frac{wl^2}{8} = \frac{5 \times 4^2}{8} = 10 \text{ t-m}$$

Similarly B.M. at the mid of the span *BC*

$$= \frac{wl^2}{8} = \frac{5 \times 3^2}{8} = 5·63 \text{ t-m}$$

Now complete the final B.M. diagram as shown in Fig. 15·18(*b*).

Example 15·8. *A continuous beam is built-in at A and is carried over rollers at B and C as shown in Fig. 15·19.*

$$AB = BC = 12 \ m.$$

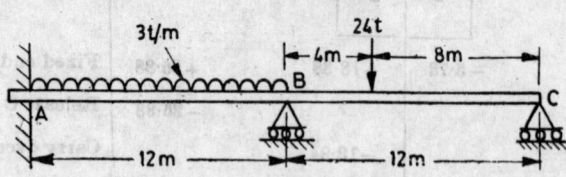

Fig. 15·19

It carries a *uniformly distributed load of 3 t/m over AB and a point load of 24 tonnes over BC, 4 m from the support B, which sinks 3 cm. Values of E and I are $2·0 \times 10^6$ kg/cm² and 2×10^5 cm⁴ respectively uniform throughout.*

Calculate the support moments and draw B.M. diagram and S.F. diagram, giving critical values.   1975)

*Solution.

Given. Span, $AB = 12$ m
Load in *AB*, $w = 3$ t/m
Span $BC = 12$ m
Load in *BC* $W = 24$ t
Sinking of support *B*,

$$\delta = 3 \ cm = 0·03 \ m$$

Young's modulus, $E = 2·0 \times 10^6$ kg/cm² $= 2·0 \times 10^7$ t/m²

Moment of inertia, $I = 2 \times 10^5$ cm⁴ $= 2 \times 10^{-3}$ m⁴

Support moments at A, B and C

First of all, let us assume the continuous beam *ABC* to be split up into fixed beams *AB* and *BC*. In span *AB*, the fixing moment at *A* due to loading

$$= -\frac{wl^2}{12} = -\frac{3 \times 12^2}{12} = -36·0 \text{ t-m}$$

and fixing moment at *B*

$$= +\frac{wl^2}{12} = +\frac{3 \times 12^2}{12} = +36·0 \text{ t-m}$$

Now in span *AB*, the moment at *A* due to sinking of support *B*

$$= -\frac{6EI\delta}{l^2} \qquad \left(\begin{array}{l}\text{Minus sign due to right}\\ \text{support sinking}\end{array}\right)$$

$$= -\frac{6 \times 2·0 \times 10^7 \times 2 \times 10^{-3} \times 0·03}{12^2} = -50·0 \text{ t-m}$$

_____

*This example is also solved on pages 504 and 520·

Similarly moment at B due to sinking of support B

$$= -\frac{6 \times 2 \cdot 0 \times 10^7 \times 2 \times 10^{-3} \times 0 \cdot 03}{12^2} = -50 \cdot 0 \text{ t-m}$$

$\therefore$ Total moment at A

$$= -36 \cdot 0 - 50 \cdot 0 = -86 \cdot 0 \text{ t-m}$$

and total moment at B

$$= +36 \cdot 0 - 50 \cdot 0 = -14 \cdot 0 \text{ t-m}$$

Now in span BC, the fixing moment at B due to loading

$$= -\frac{Wab^2}{l^2} = -\frac{24 \times 4 \times 8^2}{12^2} = -42 \cdot 67 \text{ t-m}$$

and fixing moment at B

$$= +\frac{Wa^2b}{l^2} = \frac{24 \times 4^2 \times 8}{12^2} = +21 \cdot 33 \text{ t-m}$$

Now in span BC, the moment at B, due to sinking of support B

$$= +\frac{6EI\delta}{l^2} \qquad \text{(Plus sign due to left support sinking)}$$

$$= +\frac{6 \times 2 \cdot 0 \times 10^7 \times 2 \times 10^{-3} \times 0 \cdot 03}{12^2} = +50 \cdot 0 \text{ t-m}$$

Similarly moment at C due to sinking of support B,

$$= +\frac{6 \times 2 \cdot 0 \times 10^7 \times 2 \times 10^{-3} \times 0 \cdot 03}{12^2} = +50 \cdot 0 \text{ t-m}$$

$\therefore$ Total moment at B

$$= -42 \cdot 67 + 50 \cdot 0 = +7 \cdot 33 \text{ t-m}$$

and total moment at C,

$$= +21 \cdot 33 + 50 \cdot 0 = +71 \cdot 33 \text{ t-m}$$

Now let us find out the distribution factors at B. From the geometry of the figure, we find that the stiffness factor for BA,

$$k_{BA} = \frac{4EI}{l} = \frac{4EI}{12} = \frac{EI}{3} \qquad (\because \text{ the beam is fixed at } A)$$

and $$k_{BC} = \frac{3EI}{l} = \frac{3EI}{12} = \frac{EI}{4} \qquad \cdots \left(\begin{array}{l} \because \text{ the beam is simply} \\ \text{supported at } C \end{array} \right)$$

Therefore the distribution factors for BA and BC will be

$$\frac{\dfrac{EI}{3}}{\dfrac{EI}{3} + \dfrac{EI}{4}} \text{ and } \frac{\dfrac{EI}{4}}{\dfrac{EI}{3} + \dfrac{EI}{4}} = \frac{4}{7} \text{ and } \frac{3}{7}$$

Now prepare the following table.

A				C	
	$\frac{4}{7}$	$\frac{3}{7}$			
−86·00	−14·00	+7·33	+71·33		Fixed end moment
			−71·33		Release C
		−35·67			Carry over
−86·00	−14·00	−28·34	0		Initial moments
	+24·20	+18·14			Distribute
+12·10					Carry over
0	0	0	0		Distribute
−73·9	+10·2	−10·2	0		Final moments

Bending moment diagram

From the geometry of the figure, we find that the B.M. at the mid of the span AB

$$= \frac{wl^2}{8} = \frac{3 \times 12^2}{8} = 54\cdot0 \text{ t-m}$$

Similarly B.M. under the 24 tonnes load

$$= \frac{Wab}{l} = \frac{24 \times 4 \times 8}{12} = 64\cdot0 \text{ t-m}$$

Now complete the B.M. diagram as shown in Fig. 15·20 (*b*)

Shear force diagram

Let R_A = Reaction at A,

R_B = Reaction at B, and

R_C = Reaction at C.

Taking moments about B,

$$-10\cdot2 = R_C \times 12 - 24 \times 4$$

$$(\because \text{ B.M. at } B = -10\cdot2 \text{ t-m})$$

$$\therefore \qquad R_C = \frac{-10\cdot2 + 96\cdot0}{12} = 7\cdot15 \text{ t}$$

Now taking moments about A,

$$-73 \cdot 9 = 7 \cdot 15 \times 24 + R_B \times 12 - 24 \times 16 - 3 \times 12 \times 6$$

$$(\because \text{ B.M. at } A = -73 \cdot 9 \text{ t-m})$$

$$= 171 \cdot 6 + 12 R_B - 384 - 216$$

$$= 12 R_B - 428 \cdot 4$$

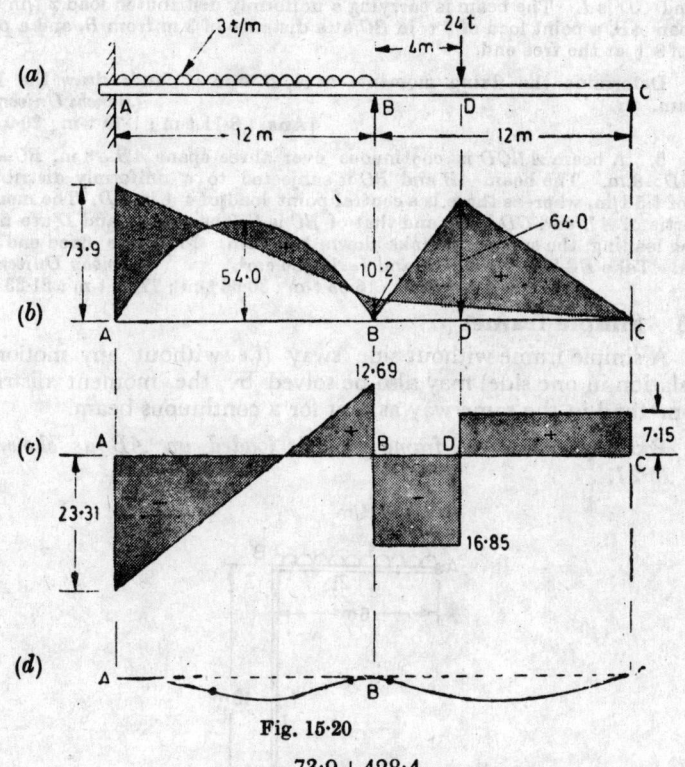

Fig. 15·20

$$\therefore \qquad R_B = \frac{-73 \cdot 9 + 428 \cdot 4}{12} = 29 \cdot 54 \text{ t}$$

and

$$R_A = (3 \times 12 + 24) - (7 \cdot 15 - 29 \cdot 54) = 23 \cdot 31 \text{ t}$$

Now complete the S.F. diagram as shown in Fig. 15·20 (c).

The elastic curve *i.e.* deflected shape of the centre line of the beam is shown in Fig. 15·20 (d).

Exercise 15–1

1. A continuous beam ABC is fixed at A and is simply supported at and C. The span AB is 6 m and carries a uniformly distributed load of 1 t/m. The span BC is 4 m and carries a uniformly distributed load of 3 t/m. Determine the fixed end moments. ($M_A = 2 \cdot 143$ t-m ; $M_B = 4 \cdot 714$ t-m ; $M_C = 0$)

2. A continuous beam $ABCD$ is simply supported over three spans, such that $AB = 8$ m, $BC = 12$ m and $CD = 5$ m. It carries uniformly distributed load of 4 t/m in span AB, 3 t/m in span BC and 6 t/m in span CD. Find the moments over the supports B and C. [Ans. 35·9 t-m ; 31·0 t-m]

3. A simply supported beam ABC is continuous over two spans AB and BC of 6 m and 5 m respectively. The span AB is carrying a uniformly distributed load of 2 t/m and the span BC is carrying a point load of 5 T at a distance of 2 m from B. Find the support moment at B.   *(Madras University)*

[**Ans.** 7·1 t-m]

4. A continuous beam $ABCD$ is fixed at A and simply supported at B and C, the beam CD is overhanging. The spans $AB=6$ m, $BC=5$ m and overhanging $CD=2\cdot5$ m. The moment of inertia of the span BC is $2I$ and that of AB and CD is I. The beam is carrying a uniformly distributed load 2 t/m over the span AB, a point load of 5 t in BC at a distance of 3 m from B, and a point load of 8 t at the free end.

Determine the fixing moments at A, B and C and draw the B.M. diagram.   *(Ranchi University)*

[**Ans.** 8·11 t-m ; 1·79 t-m ; 20·0 t-m]

5. A beam $ABCD$ is continuous over three spans $AB=8$ m, $BC=4$ m and $CD=8$ m. The beam AB and BC is subjected to a uniformly distributed load of 1·5 t/m, whereas there is a central point load of 4 t in CD. The moment of inertia of AB and CD is $2I$ and that of BC is I. The ends A and D are fixed. During loading, the support B sinks down by 1 cm. Find the fixed end moments. Take $E=2\cdot0\times10^3$ t/cm^2 and $I=16,000$ cm^4.   *(Bombay University)*

[**Ans.** 16·53 t-m ; 30·66 t-m ; 77·33 t-m ; 21·33 t-m]

15·13. Simple frames

A simple frame without side sway (*i.e.* without any motion of translation in one side) may also be solved by the moment distribution method in the same way as that for a continuous beam.

Example 15·9. *A frame ABC is loaded on AB as shown in Fig. 15·21.*

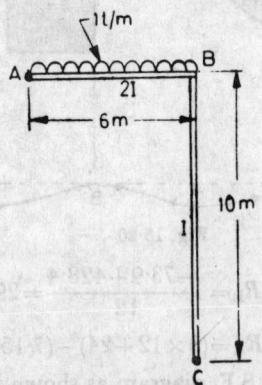

Fig. 15·21

Analyse the frame, with the help of moment distribution method, and draw the bending moment diagram. Take moment of inertia of BC and AB as I and $2I$ respectively.

Solution.

Given.   Span AB $=6$ m

Span BC $=10$ m

Load on AB, $w=1$ t/m

Moment of inertia of BC $=I$

Moment of inertia of AB $=2I$

First of all, let us assume the frame ABC to be split up into fixed beams AB and BC. In span AB, the fixing moment at A due to loading

$$= -\frac{wl^2}{12} = -\frac{1 \times 6^2}{12} = -3.0 \text{ t-m}$$

and fixing moment at B,

$$= +\frac{wl^2}{12} = +\frac{1 \times 6^2}{12} = +3.0 \text{ t-m}$$

Since the beam BC is not loaded, therefore fixing moments at B and C in span BC are zero.

Now, let us find out the distribution factors at B. From the geometry of the figure, we find that the stiffness factor for BA,

$$k_{BA} = \frac{3EI}{l} = \frac{3E \times 2I}{6} = 1EI \qquad (\because \text{ the beam is hinged at } A)$$

and $\quad k_{BC} = \frac{3EI}{l} = \frac{3EI}{10} = \frac{3EI}{10} \qquad (\because \text{ the beam is hinged at } C)$

Therefore the distribution factors for BA and BC will be

$$\frac{1EI}{1EI + \frac{3EI}{10}} \quad \text{and} \quad \frac{\frac{3EI}{10}}{1EI + \frac{3EI}{10}} = \frac{10}{13} \text{ and } \frac{3}{13}.$$

Now prepare the following table :

A		B		C	
	$\frac{10}{13}$	$\frac{3}{13}$			
-3.0	$+3.0$	0		0	Fixed end moments
$+3.0$					Release A
	$+1.5$				Carry over
0	$+4.5$	0		0	Initial moments
	$+3.46$	-1.04			Distribute
0	$+1.04$	-1.04		0	Final moments

From the geometry of the figure we find that the B.M. at the mid of the span AB treating it as a simply supported beam

$$\frac{wl^2}{8} = \frac{1 \times 6^2}{8} = 4.5 \text{ t-m}$$

Now complete the B.M. diagram as shown in Fig. 15·22.

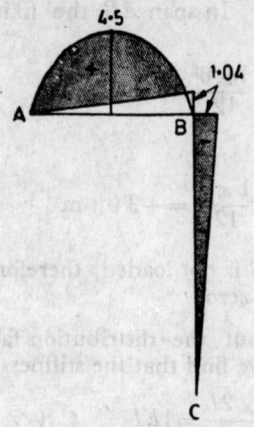

Fig. 15·22

Example 15·10. *A continuous beam ABC, fixed at A, is supported on an elastic column BD and is loaded as shown in Fig. 15 23.*

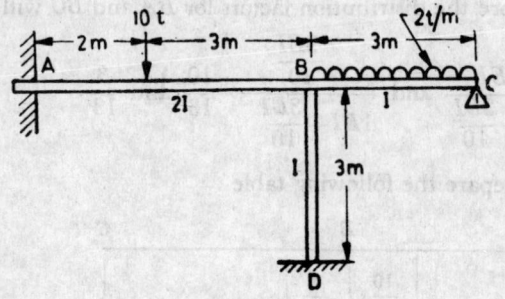

Fig. 15·23

Analyse the frame and draw the B.M. diagram. Take moment of inertia of AB as 2I and that of BC and CD as I

A.M.I.E.. Summer. 1978

*Solution.

Given. Span $AB = 5$ m
Load in *AB* $W = 10$ t
Moment of inertia of $AB = 2I$
Span $BC = 3$ m
Load in *BC*, $w = 2$ t/m
Moment of inertia of $BC = I$
Span $BD = 3$ m
Moment of inertia of $BD = I$

Support moments at A, B, C and D

First of all, let us assume the continuous beam *ABC* to be split up into fixed beams *AB* and *BC*. In span *AB*, the fixing moment at *A* due to loading,

---
*This example is also solved on page **524**

$$-\frac{Wab^2}{l^2} = -\frac{10 \times 2 \times 3^2}{5^2} = -7 \cdot 2 \text{ t-m}$$

and fixing moment at B

$$= +\frac{Wa^2b}{l^2} = \frac{10 \times 2^2 \times 3}{5^2} = +4 \cdot 8 \text{ t-m}$$

Similarly in span BC, the fixing moment at B due to loading

$$= -\frac{wl^2}{12} = -\frac{2 \times 3^2}{12} = -1 \cdot 5 \text{ t-m}$$

and fixing moment at C

$$= +\frac{wl^2}{12} = +\frac{2 \times 3^2}{12} = +1 \cdot 5 \text{ t-m}$$

Now, let us find out the distribution factors at B. From the geometry of the figure, we find that the stiffness factor for BA,

$$k_{BA} = \frac{4EI}{l} = \frac{4E \times 2I}{5} = \frac{8EI}{5} \qquad (\because \text{ the beam is fixed at } A)$$

$$k_{BC} = \frac{3EI}{l} = \frac{3EI}{3} = EI \qquad (\because \text{ the beam is simply supported at } C)$$

and $$k_{BD} = \frac{4EI}{l} = \frac{4EI}{3} \qquad (\because \text{ the beam is fixed at } D)$$

Therefore distribution factors for BA, BC and BD will be

$$\frac{\frac{8EI}{5}}{\frac{8EI}{5} + EI + \frac{4EI}{3}}, \quad \frac{EI}{\frac{8EI}{5} + EI + \frac{4EI}{3}} \text{ and } \frac{\frac{4EI}{3}}{\frac{8EI}{5} + EI + \frac{4EI}{3}}$$

$$= 0 \cdot 406, \ 0 \cdot 254 \text{ and } 0 \cdot 340.$$

Now prepare the following table.

A	B		C		
	BA 0.406	BD 0.340	BC 0.254		
−7.20	+4.80	0.00	−1.50	+1.50	Fixed end moments
				−1.50	Release C
			−0.75		Carry over
−7.20	+4.80	0.00	−2.25	0	Initial moments
	−1.04	−0.86	−0.90		Distribute
−0.52	0	0	0	0	Carry over
−7 72	+3.76	−0.86	−2.90	0	Final moments

Bending moment diagram

From the geometry of the figure, we find that the B.M. under the 10 tonnes load

$$= \frac{Wab}{l} = \frac{10 \times 2 \times 3}{5} = 12 \cdot 0 \text{ t-m}$$

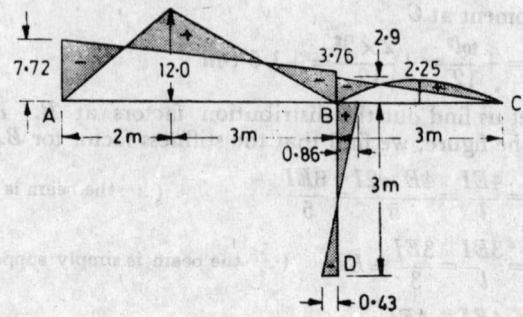

Fig. 15·24

Similarly B.M. at the mid of the span *BC*

$$= \frac{wl^2}{8} = \frac{2 \times 3^2}{8} = 2 \cdot 25 \text{ t-m}$$

Now complete the B.M. diagram as shown in Fig. 15·24.

15·14. Portal frames

A simple portal frame consists of a beam, resting over two columns. The junction of the beam and column behaves like a rigid joint. In general, the portal frames are of the following two types :

(a) Symmetrical portal frames, and

(b) Unsymmetrical portal frames.

15·15. Symmetrical portal frames

A symmetrical portal frame is that, in which both the columns are of the same length, having similar end conditions (*i.e.*, either hinged or fixed), moments of inertia and modulus of elasticity as well as subjected to symmetrical loading. A little consideration will show, that the joints of such a portal frame will not be subjected to any translation or side sway.

All the single bay and single storey symmetrical portal frames may be analysed by opening them out, and treating them exactly like a three-span continuous beam.

Example 15·11. *A portal frame ABCD shown in Fig. 15·25 is loaded with a uniformly distributed load of 2,000 kg/m on the horizontal member. The moment of inertia of member AB=that oj CD=I and that of member BC=3I.*

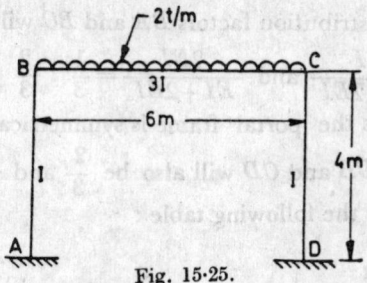

Fig. 15·25.

Find the support reactions and bending moment in the frame by moment distribution method and draw the bending moment diagram.

(A.M.I.E. Winter, 1974)

Solution

Given. Length of $AB = 4$ m

Length of $BC = 6$ m

Length of $CD = 4$ m

Length on BC, $w = 2$ t/m

M. I. of AB and $CD = I$

M. I. of $BC = 3I$

Since the portal frame is symmetrical, therefore it will be analysed by assuming it to be a continuous beam $ABCD$.

Support reactions

First of all let us assume the portal frame $ABCD$ to be split up into fixed beams, AB, BC and CD. Since the beams AB and CD are not subjected to any load, therefore fixing moments in the spans AB and CD at A, B, C and D are zero.

In span BC, the fixing moment at B

$$= -\frac{wl^2}{12} = -\frac{2 \times 6^2}{12} = -6·0 \text{ t-m}$$

$$= -6,000 \text{ kg-m}$$

Similarly fixing moment at C

$$= +\frac{wl^2}{12} = +\frac{2 \times 6^2}{12} = +6·0 \text{ t-m}$$

Now, let us find out the distribution factors at B and C. From the geometry of the figure, we find that the stiffness factor for BA,

$$k_{BA} = \frac{4\,EI}{l} = \frac{4\,EI}{4} = EI$$

($\because$ the beam is fixed at A)

and
$$k_{BC} = \frac{4\,EI}{l}$$

$$= \frac{4E \times 3I}{6} = 2EI \quad (\because \text{ the beam is continuous at } C)$$

Therefore distribution factors BA and BC will be

$$\frac{EI}{EI+2EI} \text{ and } \frac{2EI}{EI+2EI} = \frac{1}{3} : \frac{2}{3}$$

Moreover, as the portal frame is symmetrical therefore distribution factors for CB and CD will also be $\frac{2}{3}$ and $\frac{1}{3}$

Now prepare the following table :

A	B			C		D	
	$\frac{1}{3}$	$\frac{2}{3}$		$\frac{2}{3}$	$\frac{1}{3}$		
0	0	−6·0	+6·0	0		0	Fixed end initial moments
	+2·0	+4·0	−4·0	−2·0			
+1·0		−2·0	+2·0		−1·0		
	+0·67	+1·33	−1·33	−0·67			
+0·34		−0·66	+0·66		−0·34		
	+0·22	+0·44	−0·44	−0·22			
+0·11		−0·22	+0·22		−0·11		
	+0·07	+0·15	−0·15	−0·07			
+0·03		−0·8	+0·08		−0·03		
	+0·03	+0·05	−0·05	−0·03			
+0·02		−0·03	+0·03		−0·02		
	+0·03	+0·02	−0·02	−0·01			
+1·5	+3·0	−3·0	+3·0	−3·0	+1·5		Final moments

Vertical reaction at A,
$$V_A = V_D = \frac{2 \times 6}{2} = 6 \text{ t}$$

Horizontal reaction at A,

$$H_A = \frac{1·5 \times 30}{4} = 1·125 \text{ t} (\rightarrow)$$

Similarly $H_D = 1·125$ t $(\leftarrow)$

Bending moment

First of all draw the free body diagram as shown in Fig. 15·26 (a).

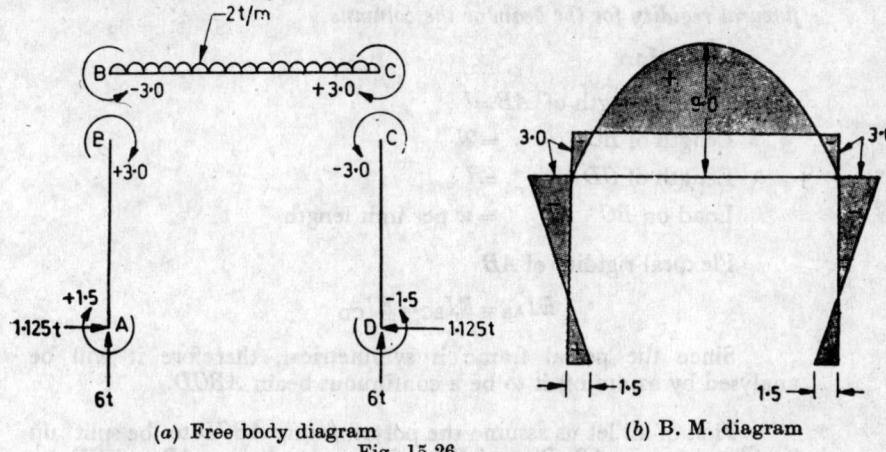

(a) Free body diagram   (b) B. M. diagram

Fig. 15·26.

The bending moment at the mid of the span BC, by considering it as a simply supported beam

$$= \frac{wl^2}{8} = \frac{2 \times 6^2}{8} = 9 \cdot 0 \text{ t-m}$$

Now complete the final bending moment diagram as shown in Fig. 15·26 (b).

Example 15·12. *A rectangular portal frame of uniform flexural rigidity EI carries a uniformly distributed load as shown in Fig. 15·27.*

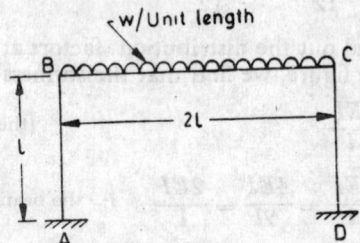

Fig. 15·27

---

*The most important point, while drawing the B.M. diagram, is that in the table, the B.M. at B in the span BA is $+3\cdot0$ t-m, whereas in the B.M. diagram it has been taken as $-3\cdot0$ t-m. Similarly the B.M. at D in the span CD is $-1\cdot55$ t-m, whereas in the B.M. diagram it has been taken as $+1\cdot5$ t-m. The reason for the same is that the B.M. at A will tend to bend the beam AB with convexity on the right side, whereas the B.M. at B will tend to bend the beam AB with convexity on the left side. That is why, if the B.M. at A is taken as positive, the B.M. at B will be taken as negative. Similarly the B.M. at C in the span CD is taken as negative, and B.M. at D is taken as positive.

Draw the bending moment diagram and sketch the deflected curve. Hence or otherwise indicate with reason, whether you will provide greater flexural regidity for the beam or the columns.

*Solution.

Given. Length of $AB = l$

Length of BC $= 2l$

Length of CD $= l$

Load on BC $= w$ per unit length

Flexural rigidity of AB

$$EI_{AB} = EI_{BC} = EI_{CD}$$

Since the portal frame is symmetrical, therefore it will be analysed by assuming it to be a continuous beam $ABCD$.

First of all let us assume the portal frame $ABCD$ to be split up into fixed beams AB, BC and CD. Since the beams AB and CD are not subjected to any load, therefore fixing moments in spans AB and CD at A, B, C, and D are zero.

In span BC, the fixing moment at B,

$$= -\frac{wl^2}{12} = -\frac{w(2l)^2}{12} = -0.33\ wl^2$$

Similarly fixing moment at C

$$= +\frac{wl^2}{12} = +\frac{w(2l)^2}{12} = +0.33\ wl^2$$

Now let us find out the distribution factors at B and C. From the geometry of the figure, we find that the stiffness factor for BA.

$$k_{BA} = \frac{4EI}{l} \qquad\qquad \text{(the beam is fixed at } A)$$

and $k_{BC} = \dfrac{4EI}{l} = \dfrac{4EI}{2l} = \dfrac{2EI}{l}$ ($\because$ the beam is continuous at C)

Therefore distribution factor for BA and BC will be

$$\frac{\dfrac{4EI}{l}}{\dfrac{4EI}{l} + \dfrac{2EI}{l}} \text{ and } \frac{\dfrac{2EI}{l}}{\dfrac{4EI}{l} + \dfrac{2EI}{l}} = \frac{2}{3} \text{ and } \frac{1}{3}$$

Since the portal frame is symmetrical, therefore the distribution factors for CB and CD will also be $\dfrac{1}{3}$ and $\dfrac{2}{3}$

' *This example is also solved on pages 638.

Now prepare the following table.

A		B		C		D	
	$\frac{2}{3}$	$\frac{1}{3}$		$\frac{1}{3}$		$\frac{2}{3}$	
0	0	−0·33	+0·33	0		0	Fixed end initial moments
	+0·22	+0·11	−0·11	−0·22			Distribute
+0·11		−0·06	+0·06		−0·11		Carry over
	+0·04	+0·02	−0·02	−0·04			Distribute
+0·02		−0·01	+0·01		−0·02		Carry over
	+0·01	0	0	−0·01			Distribute
+0·13	+0·27	−0·27	+0·27	−0·27	−0·13		Final moments

We know that the B.M. at the centre of the beam BC, treating it as a simply supported beam,

$$M = \frac{w_1 2l)^2}{8} = \frac{wl^2}{2} = 0 \cdot 5 \; wl^2$$

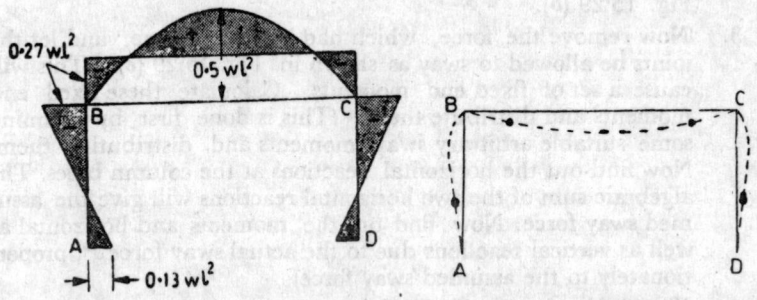

(a) Bending moment diagram.   (b) Deflected curve.

Fig. 15·28

Now draw the B.M. diagram and the deflected curve as shown in Fig. 15·28 (a) and (b).

Since the B.M. is more at the ends B and C of the column, therefore greater flexural rigidity is to be provided for the columns.

Ans.

15·16. Unsymmetrical portal frames

An unsymmetrical portal frame is that in which either the columns are not symmetrical, or the frame is not symmetrically loaded. A little consideration will show that an unsymmetrical portal frame will be subjected to some horizontal movement known, as sway, to one side or the other. As a result of this, rigid joints between the columns and beam will have a motion of translation. The portal frame which are subjected to side sway, are analysed as discussed below :

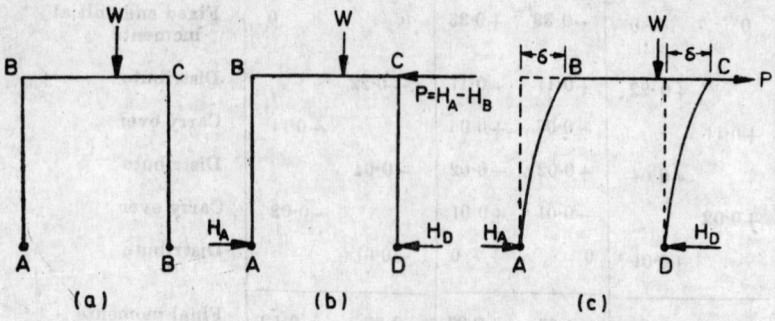

Fig. 15·29

1. *Ignoring the side sway of the frame, calculate the usual fixed end moments due to external loading and distribute them.

2. Now calculate horizontal as well as vertical reactions at the column bases. The algebraic sum of the two horizontal reactions will give the magnitude and direction of the sway force, (P), which had held the frame against side sway as shown in Fig. 15·29 (b).

3. Now remove the force, which had held the frame, and let the joints be allowed to sway as shown in Fig. 15·29 (c). This will cause a set of fixed end moments. Calculate these fixed end moments and distribute them. (This is done first by assuming some suitable arbitrary sway moments and distributing them. Now find out the horizontal reactions at the column bases. The algebraic sum of the two horizontal reactions will give the assumed sway force. Now find out the moments and horizontal as well as vertical reactions due to the actual sway force P, proportionately to the assumed sway force).

4. The final moment at each joint may now be obtained by adding algebraically the moments obtained in paras 1 and 3 above.

5. The horizontal as well as vertical reactions at the column bases may also be obtained by adding algebraically the reactions obtained in paras 2 and 3 above.

15·17. Ratio of sway moments at the joints of column heads and beam

We have already discussed in Art. 15·16 that an unsymmetrical portal frame is always subjected to some sway moments on all the

_____

*Strictly speaking, the effect of side sway is not ignored. But a force P is assumed to act at one of the joints between the columns and the beam, which will hold the frame against sway.

fixed joints. A little consideration will show that magnitude of sway moments depends upon the end conditions of the portal frame. Now we shall discuss the ratios of sway moments at the joints of column heads and beam, under the following end conditions.

(a) When both the ends are hinged,

(b) When both the ends are fixed, and

(c) When one end is fixed and the other hinged.

15·18. Ratio of sway moments at the joints of column heads and beam, when both the ends are hinged.

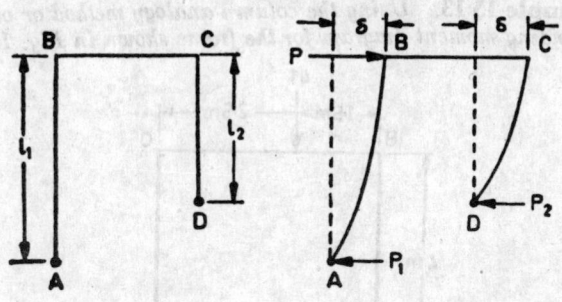

(a) Before sway   (b) After sway

Fig. 15·30

Consider an unsymmetrical portal frame $ABCD$ having stiff joints B and C, and hinged ends A and D as shown in Fig. 15·30 (a).

Let l_1 = Length of the column AB,

           I_1 = Moment of inertia of column AB,

           l_2, I_2 = Corresponding values for the column CD.

Now consider a sway force P to act at B, and as a result of this force, let the frame deflect through a small distance δ as shown in Fig. 15·30 (b).

Let P_1 = Horizontal reaction at A, and

           P_2 = Horizontal reaction at D.

Now consider the column AB of the frame, which will behave like a cantilever fixed at B and carrying a point load P_1 at A. We know that the deflection of a cantilever with point load at the free end,

$$\delta = \frac{P_1 l_1^3}{3EI_1} \qquad \ldots(i)$$

Similarly, consider the column CD of the frame, which will also behave like a cantilever fixed at C and carrying point load P_2 at D. Now deflection of the CD,

$$\delta = \frac{P_2 l_2^3}{3EI_2} \qquad \ldots(ii)$$

Since the deflection of AB and CD is equal, therefore equating (i) and (ii),

$$\frac{P_1 l_1^3}{3EI_1} = \frac{P_2 l_2^3}{3EI_2}$$

or

$$\frac{P_1 l_1}{P_2 l_2} = \frac{I_1}{I_2} \times \frac{l_2^2}{l_1^2}$$

Since $P_1 l_1$ and $P_2 l_2$ are the moments at B and C in the members BA and CD, respectively, therefore

$$\frac{M_{\mathrm{BA}}}{M_{\mathrm{CD}}} = \frac{I_1}{I_2} \times \frac{l_2^2}{l_1^2}$$

Example 15·13. *Using the column analogy method or otherwise, plot the bending moment diagram for the frame shown in Fig. 15·31.*

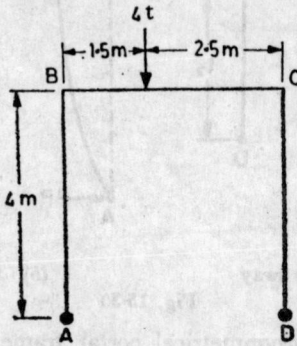

Fig. 15·31

Also draw the deflected shape of the frame.

*Solution.

Length of $AB=CD=4$ m
Length of $BC=4$ m
Load on BC, $W=4$ t

Since the portal frame is unsymmetrical, therefore it will be analysed first by assuming it as a continuous beam and then applying a sway correction.

First of all, let us assume the portal frame $ABCD$ to be split up into fixed beams AB, BC and CD. Since the beams AB and CD are not subjected to any load, therefore fixing moments in spans AB and CD at A, B, C and D are zero.

In span BC, the fixing moment at B,

$$= -\frac{Wab^2}{l^2} = -\frac{4 \times 1·5 \times 2·5^2}{4^2} = -2·34 \text{ t-m}$$

Similarly fixing moment at C,

$$= +\frac{Wa^2 b}{l^2} = \frac{4 \times 1·5^2 \times 2·5}{4^2} = +1·40 \text{ t-m}$$

---

*This example is also solved on page 642

Now let us find out the distribution factors at B and C. From the geometry of the figure, we find that the stiffness factor for BA,

$$k_{BA} = \frac{3EI}{l} = \frac{3EI}{4} \qquad (\because \text{ the beam is hinged at } A)$$

and

$$k_{BC} = \frac{4EI}{l} = \frac{4EI}{4} = EI \ (\because \text{ the beam is continuous at } C)$$

Therefore distribution factors for BA and BC will be

$$\frac{\frac{3EI}{4}}{\frac{3EI}{4} + EI} \text{ and } \frac{EI}{\frac{3EI}{4} + EI} = \frac{3}{4} \text{ and } \frac{4}{7}.$$

Similarly stiffness factor for CB,

$$k_{CB} = \frac{4EI}{l} = \frac{4EI}{4} = EI \qquad (\because \text{ the beam is continuous at } B)$$

and

$$k_{CD} = \frac{3EI}{l} = \frac{3EI}{4} \qquad (\because \text{ the beam is linged at } D)$$

Therefore distribution factors for CB and CD will be

$$\frac{EI}{EI + \frac{3EI}{4}} \text{ and } \frac{\frac{3EI}{4}}{EI + \frac{3EI}{4}} = \frac{4}{7} \text{ and } \frac{3}{7}$$

Now complete the following table :

A	B		C		D	
	$\frac{3}{7}$	$\frac{4}{7}$	$\frac{4}{7}$	$\frac{3}{7}$		
0	0	−2·34	+1·40	0	0	Initial moments
	+1·01	+1·33	−0·80	−0·60		Distribute
0		−0·40	+0·67		0	Carry over
	+0·18	+0·22	−0·38	−0·29		Distribute
0		−0·19	+0·11		0	Carry over
	+0·08	+0·11	−0·06	−0·05		Distribute
	−0·03	+0·05		0		Carry over
	+0·01	+0·02	−0·03	−0·02		Distribute
0	+1·28	−1·28	+0·96	−0·96	0	Final moments

$\therefore$ Horizontal reaction at A

$$= \frac{1\cdot 28}{4} = 0\cdot 32 \text{ t} \rightarrow$$

and horizontal reaction at B

$$= -\frac{0\cdot 96}{4} = 0\cdot 24 \text{ t} \leftarrow$$

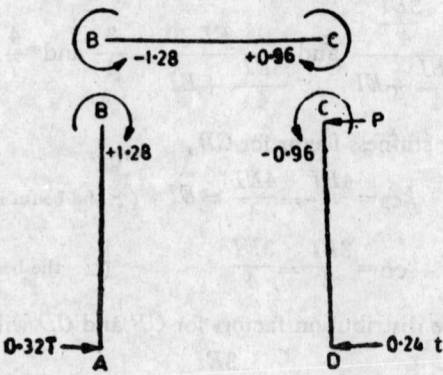

Fig. 15·32

From the free body diagram, as shown in Fig. 15·32, we find that the unbalanced horizontal force at C,

$$P = 0\cdot 32 - 0\cdot 24 = 0\cdot 08 \text{ t} \leftarrow$$

We have worked out the moments and the reactions, under the assumption, that a horizontal force $P = 0\cdot 08$ t $\leftarrow$ is acting at C to prevent the side sway. Actually there is no such force acting at C. We know that if a force of $0\cdot 08$ t $\rightarrow$ is applied at C, it will neutralise the effect of assumed force $P = 0\cdot 08$ t $\rightarrow$ at C. A little consideration will show that if a force of $0\cdot 08$ t $\rightarrow$ is applied at C, it will cause some sway moments at the joints B and C of the portal frame.

Let M_{BA} = Moment at B in member BA, and

               M_{CD} = Moment at C in member CD.

We know that $\dfrac{M_{BA}}{M_{CD}} = \dfrac{I \times l_2^2}{I \times l_1^2} = \dfrac{I \times 4^2}{I \times 4^2} = \dfrac{1}{1}$

Now assume* some arbitrary moments at B and C (*i.e.* M_{BA} and M_{CD}) in the ratio of $1 : 1$ as already calculated and distribute the same at other joints also. Let us assume the sway moments M_{BA} and M_{CD} as $2\cdot 0$ t-m and $2\cdot 0$ t-m respectively.

*As a matter of fact, there is no hard and fast rule for assuming the sway moments. But in order to have the same relative accuracy in the second set of calculations, we must assume the moments of the same order, as in the first set of calculations.

Now complete the following table :

A	B			C		D	
	$\frac{3}{7}$	$\frac{4}{7}$		$\frac{4}{7}$	$\frac{3}{7}$		
0	+2·0	0		0	+2·0	0	Assumed moments
	−0·86	−1·14		−1·14	−0·86		Distribute
0		−0·57		−0·57		0	Carry over
	+0·24	+0·33		+0·33	+0·24		Distribute
0		+0·16		+0·16		0	Carry over
	−0·07	−0·09		−0·09	−0·07		Distribute
		−0·05		−0·05			Carry over
	+0·02	+0·03		+0·03	+0·02		Distribute
0	+1·33	−1·33		−1·33	+133	0	Final moments

∴ Horizontal reaction at A

$$=+\frac{1\cdot33}{4}=0\cdot33 \text{ t} \rightarrow$$

and horizontal reaction at D

$$=+\frac{1\cdot33}{4}=0\cdot33 \text{ t} \rightarrow$$

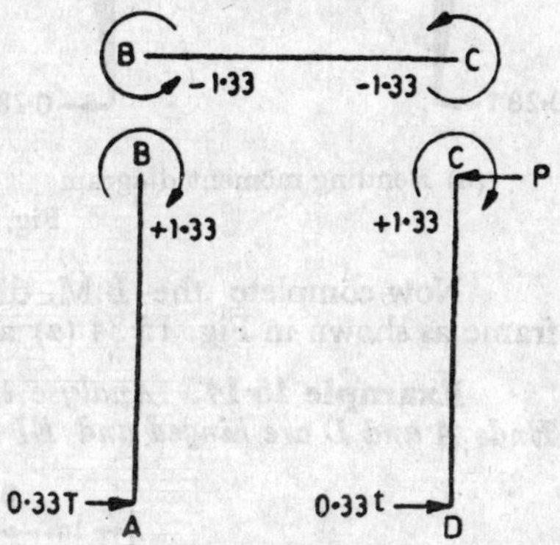

From the free body diagram as shown in Fig. 15·33, we find that the unbalanced horizontal force at C,

$$P=0\cdot33+0\cdot33=0\cdot66 \text{ t} \leftarrow$$

But the actual unbalanced horizontal force is $0\cdot08$ t←.

Fig. 15·33

Therefore the actual sway moments in the members due to a sway force of $0\cdot08$ t → may be found out by proportion.

Now complete the following table :

	A	B	C		D
Assumed sway = 0·66 T←	0	+1·33	−1·33	−1·33	+1·33 0
Actual sway = 0·08 T→	0	−0·16	+0·16	+0·16	−0·16 0
Non-sway	0	+1·28	−1·28	+0·96	−0·96 0
Final moments	0	+1·12	−1·12	+1·12	−1·12 0

Now horizontal reaction at A,

$$H_A = \frac{1 \cdot 12}{4} = 0 \cdot 28 \ \text{t} \rightarrow$$

and horizontal reaction at D,

$$H_D = -\frac{1 \cdot 12}{4} = 0 \cdot 28 \ \text{t} \leftarrow$$

Now bending moment under the 4 tonnes load on BC, considering it as a simply supported beam,

$$M = \frac{Wab}{l} = \frac{4 \times 1 \cdot 5 \times 2 \cdot 5}{4} = 3 \cdot 75 \ \text{t-m}$$

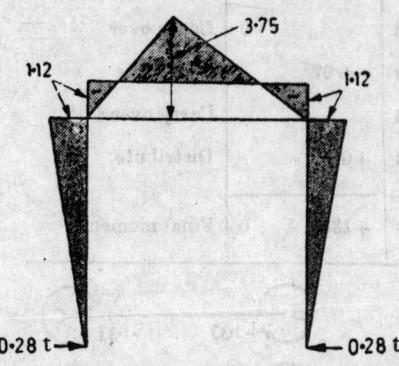
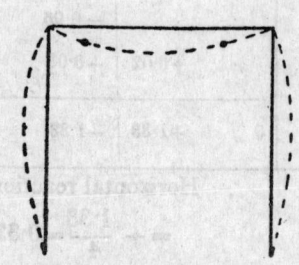

(a) Bending moment diagram   (b) Deflected shape

Fig. 15·34

Now complete the B.M. diagram and deflected shape of the frame as shown in Fig. 15·34 (a) and (b).

Example 15·14. *Analyse the portal frame shown in Fig. 15·35. Ends A and D are hinged and EI is constant.*

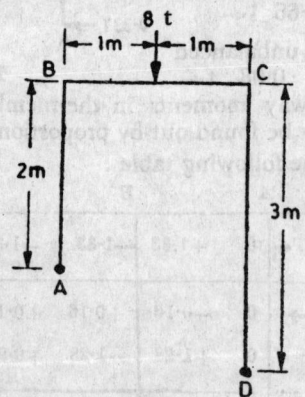

Fig. 15·35

Also draw the B.M. diagram.

Solution.

Given. Length of AB,

$$l_1 = 2 \text{ m}$$

Length of BC $= 2$ m

Length of CD, $l_2 = 3$ m

Load on BC, $W = 8$ ^{t}

Since the portal frame is unsymmetrical, therefore it will be analysed first by assuming it as a continuous beam, and then applying a sway correction.

First of all, let us assume the portal frame $ABCD$ to be split up into fixed beams AB, BC and CD. Since the beams AB and CD are not subjected to any load, therefore fixing moments in span AB and CD, at A, B, C and D are zero.

In span BC, fixing moment at B

$$= -\frac{Wl}{8} = -\frac{8 \times 2}{8} = -2 \text{ t·m}$$

Similarly fixing moment at C

$$= +\frac{Wl}{8} = +\frac{8 \times 2}{8} = +2 \text{ t·m}$$

Now let us find out the distribution factors at B and C. From the geometry of the figure, we find that the stiffness factor for BA,

$$k_{BA} = \frac{3EI}{l} = \frac{3EI}{2} = \frac{3}{2} EI \quad (\because \text{ the beam is hinged at } A)$$

and

$$k_{BC} = \frac{4EI}{l} = \frac{4EI}{2} = 2 EI$$

$$(\because \text{ the beam is continuous at } C)$$

Therefore the distribution factors for BA and BC will be

$$\frac{\frac{3}{2} EI}{\frac{3}{2} EI + 2 EI} \quad \text{and} \quad \frac{2 EI}{\frac{3}{2} EI + 2 EI} = \frac{3}{7} \text{ and } \frac{4}{7}$$

Similarly stiffness factors for CB,

$$k_{CB} = \frac{4EI}{l} = \frac{4EI}{2} = 2 EI$$

$$(\because \text{ the beam is continuous at } B)$$

and

$$k_{CD} = \frac{3EI}{l} = \frac{3EI}{3} = 1 EI \quad (\because \text{ the beam is hinged at } A)$$

Therefore the distribution factors for CB and CB will be

$$\frac{2EI}{2EI + 1EI} \quad \text{and} \quad \frac{1EI}{2EI + 1EI} = \frac{2}{3} \text{ and } \frac{1}{3}.$$

Now complete the following table.

A		B		C	D	
	$\frac{3}{7}$	$\frac{4}{7}$		$\frac{2}{3}$	$\frac{1}{3}$	
0	0	−2·00	+2·00	0	0	Fixed end initial moments
	+0·86	+1·14	−1·33	−0·67		Distribute
		−0·67	−0·57			Carry over
	+0·29	+0·38	−0·38	−0·19		Distribute
		−0·19	+0·19			Carry over
	+0·08	+0·11	−0·13	−0·06		Distribute
		−0·07	+0·06			Carry over
	+0·03	+0·04	−0·04	−0·02		Distribute
o	+1·26	−1·26	+0·94	−0·94	0	Final moments

∴ Horizontal reaction at A,

$$H_A = + \frac{1·26}{2} = 0·63 \text{ t} \rightarrow$$

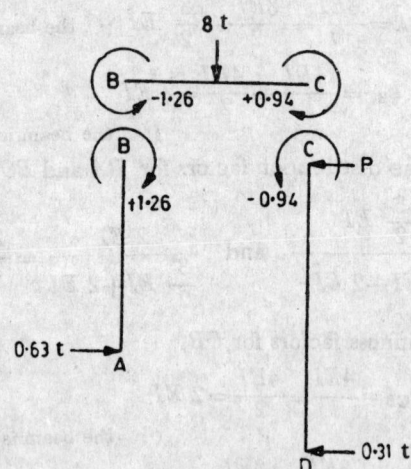

Fig. 15·36

and horizontal reaction at D,

$$H_D = - \frac{0·94}{3} = 0·31 \text{ t} \leftarrow$$

From the free body diagram, as shown in Fig. 15·34, we find that the unbalanced horizontal force at C,

$$P = 0.63 - 0.31 = 0.32 \text{ t} \leftarrow$$

We have worked out the moments and the reactions, under the assumption that a horizontal force $P = 0.32$ t $\leftarrow$ is acting at C to prevent the side sway. Actually there is no such force acting at C. We know that if a force of 0.32 t $\rightarrow$ is applied at C it will neutralise the effect of assumed force $P = 0.32$ t $\leftarrow$ at C. A little consideration will show, that if a force of 0.32 t $\rightarrow$ is applied at C, it will cause some sway moments at the joints B and C of the portal frame.

Let $\quad\quad\quad M_{BA}$ = Moment at B in member BA and

$\quad\quad\quad\quad\quad\quad M_{CD}$ = Moment at C in member CD

We know that $\dfrac{M_{BA}}{M_{CD}} = \dfrac{I_1 \times l_2{}^2}{I_2 \times l_1{}^2} = \dfrac{I \times 3^2}{I \times 2^2} = \dfrac{9}{4}$.

Now assume some arbitrary moments at B and C (i.e., M_{BA} and M_{CD}) in the ratio of $9 : 4$ as already calculated and distribute the same at other joints also. Let us assume the sway moments M_{BA} and M_{CD} as 9 t-m and 4 t-m respectively.

Now complete the following table :

A	B		C		D	
	$\dfrac{3}{7}$	$\dfrac{4}{7}$	2	$\dfrac{1}{3}$		
0	+9·00	0	0	+4·00	0	Assumed moments
	−3·86	−5·14	−2·67	−1·33		Distribute
		−1·33	−2·57			Carry over
	+0·57	+0·76	+1·71	+0·86		Distribute
		+0·85	+0·38			Carry over
	−0·37	−0·48	−0·25	−0·13		Distribute
		−0·13	−0·24			Carry over
	+0·06	+0·07	+0·16	+0·08		Distribute
0	+5·40	−5·40	−3·48	+3·48	0	Final moments

∴ Horizontal reaction at A,

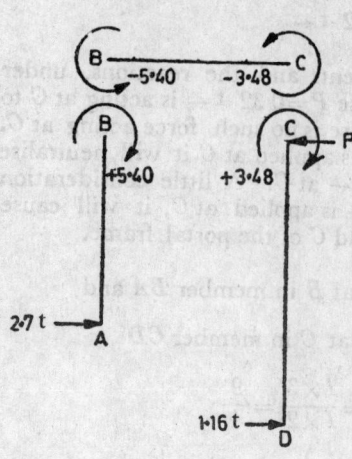

Fig. 15·37

$$H_A = \frac{5·40}{2} = 2·7 \text{ t} \rightarrow$$

and horizontal reaction at D,

$$H_D = \frac{3·48}{3} = 1·16 \text{ t} \rightarrow$$

From the free body diagram as shown in Fig. 15·37, we find that the unbalanced horizontal force at C,

$$P = 2·7 + 1·16$$
$$= 3·86 \text{ t} \leftarrow$$

But the actual unbalanced horizontal force is $0·32$ t ←. Therefore the actual sway moments in the members due to a sway force of $0·32$ t → may be found out by proportion.

Now complete the following table :

	A	B		C		D
Assumed sway = 3·86 t ←	0	+5·40	−5·40	−3·48	+3·48	0
Actual sway = 0·32 t →	0	−0·44	+0·44	+0·29	−0·29	0
Non-sway	0	+1·26	−1·26	+0·94	−0·94	0
Final moments	0	+0·82	−0·82	+1·23	−1·23	0

∴ Horizontal reaction at A,

$$H_A = \frac{0·82}{2} = 0·41 \text{ t} \rightarrow$$

and horizontal reaction at D,

$$H_D = \frac{-1·23}{3} = -0·41 \text{ t} = 0·41 \text{ t} \leftarrow$$

Now B.M. under the 8 t load on BC, by considering it as a simply supported beam,

$$M = \frac{Wl}{4} = \frac{8 \times 2}{4} = 4·0 \text{ t-m}$$

Now complete the B.M. diagram as shown in **Fig. 15·38.**

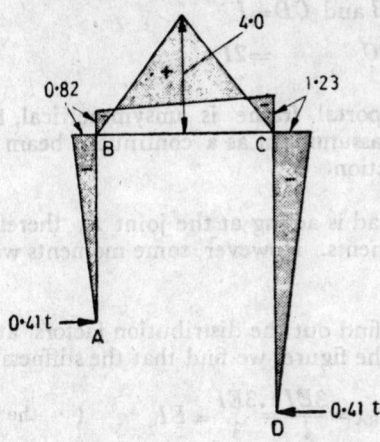

Fig. 15·38

Example 15·15. *A portal frame shown in Fig. 15·39 is to resist a horizontal load of 5 tonnes applied at B, the moment of inertia of the column section is I and the moment of inertia of the beam is 2I.*

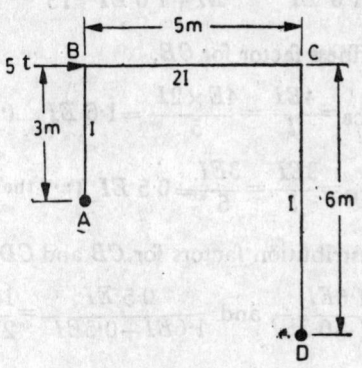

Fig. 15·39

Determine by moment distribution, or otherwise, the bending moment developed at B and C and sketch the bending moment diagram for the entire span.

Solution.

Given. Length of $AB = 3$ m

Length of BC $= 5$ m

Length of CD $\qquad = 6$ m

Horizontal load at $B = 5$ t

M.I. for AB and $CD = I$

M.I. for BC $\qquad = 2I$

Since the portal frame is unsymmetrical, therefore it will be analysed first by assuming it as a continuous beam and then applying a sway correction.

Since the load is acting at the joint B, therefore there will be no fixed end moments. However, some moments will be induced due to side sway.

Now let us find out the distribution factors at B and C. From the geometry of the figure, we find that the stiffness factor for BA,

$$k_{BA} = \frac{3EI}{l} = \frac{3EI}{3} = EI \qquad (\because \text{ the beam is hinged at } A)$$

and
$$k_{BC} = \frac{4EI}{l} = \frac{4E \times 2I}{5} = 1 \cdot 6 \, EI \qquad (\because \text{ the beam is continuous at } C)$$

Therefore distribution factors for BA and BC will be

$$\frac{EI}{EI + 1 \cdot 6 \, EI} \text{ and } \frac{1 \cdot 6 \, EI}{EI + 1 \cdot 6 \, EI} = \frac{5}{13} \text{ and } \frac{8}{13}$$

Similarly stiffness factor for CB,

$$k_{CB} = \frac{4EI}{l} = \frac{4E \times 2I}{5} = 1 \cdot 6 \, EI \qquad (\because \text{ the beam is continuous at } C)$$

and
$$k_{CD} = \frac{3EI}{l} = \frac{3EI}{6} = 0 \cdot 5 \, EI \qquad (\because \text{ the beam is hinged at } D)$$

Therefore distribution factors for CB and CD will be

$$\frac{1 \cdot 6EI}{1 \cdot 6EI + 0 \cdot 5EI} \text{ and } \frac{0 \cdot 5 \, EI}{1 \cdot 6EI + 0 \cdot 5EI} = \frac{16}{21} \text{ and } \frac{5}{21}$$

Let $\qquad M_{BA} = $ Moment at B in member BA, and

$M_{CD} = $ Moment at C in member CD.

We know that $\dfrac{M_{BA}}{M_{CD}} = \dfrac{I_1 \times l_2^2}{I_2 \times l_1^2} = \dfrac{I \times 6^2}{I \times 3^2} = \dfrac{4}{1}$.

Now assume some arbitrary moments at B and C (i.e., M_{BA} and M_{CD}) in the ratio of $4 : 1$ as already calculated, and distribute the same. Let us assume the sway moments M_{BA} and M_{CD} as $4 \cdot 0$ t-m and $1 \cdot 0$ t-m respectively.

Now complete the following table :

A	B			C		D	
	$\frac{5}{13}$	$\frac{8}{13}$		$\frac{16}{21}$	$\frac{5}{21}$		
0	+4·00	0		0	+1·00	0	Assumed moments
	−1·54	−2·46		−0·76	−0·24		Distribute
		−0·38		−1·23			Carry over
	+0·15	+0·03		+0·94	+0·29		Distribute
		+0·47		+0·12			Carry over
	−0·18	−0·29		−0·09	−0·03		Distribute
0	+2·43	−2·43		−1·02	+1·02	0	Final moment

Horizontal reaction at A

$$=\frac{2\cdot 43}{3}=0\cdot 81 \text{ t} \leftarrow$$

and horizontal reaction at D

$$=\frac{1\cdot 02}{6}=0\cdot 17 \text{ t} \leftarrow$$

Therefore the sway force causing the assumed moments

$$=0\cdot 81+0\cdot 17=0\cdot 98 \text{ t} \leftarrow$$

But the actual sway force is $5\cdot 0$ tonnes. Therefore the actual sway moments in the members due to a sway force of $5\cdot 0$ tonnes may be found out by proportion.

Now complete the following table :

	A		B		C		D
							→
Assumed sway −0·98 t	0		+2·43	−2·43	−1·02	+1·02	0
Actual sway −5 t	0		+12·4	−12·4	−5·2	+5·2	0

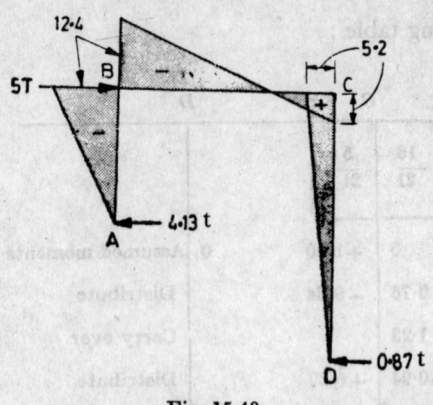

∴ Horizontal reaction at A,

$$H_A = \frac{12 \cdot 4}{3} \text{ t}$$

$$= 4 \cdot 13 \text{ t} \leftarrow$$

and horizontal reaction at D,

$$H_D = \frac{5 \cdot 2}{6} \text{ t}$$

$$= 0 \cdot 87 \text{ t} \leftarrow$$

Fig. 15·40

Now complete the B.M. diagram as shown in Fig. 15·40.

15·19. Ratio of sway moments at the joints of column heads and beam, when both the ends are fixed.

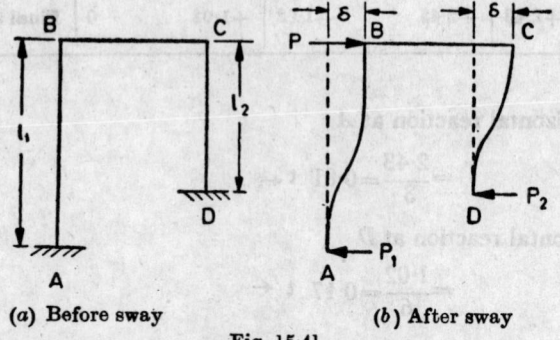

(a) Before sway   (b) After sway

Fig. 15·41

Consider an unsymmetrical portal frame $ABCD$ having stiff joints B and C, and fixed ends A and D as shown in Fig. 15·41 (a).

Let l_1=Length of the column AB,

I_1=Moment of inertia of column AB,

l_2, I_2=Corresponding values for the column CD.

Now consider a sway force P to act at B, and as a result of this force, let the frame deflect through a small distnce δ as shown in Fig. 15·41 (b).

Let P_1=Horizontal reaction at A, and

P_2=Horizontal reaction at D.

Now consider the column AB of the frame, which will behave like a beam fixed at A and B, and deflected through a small distance δ. We have already discussed in Art. 12·9 that moment at joint B, when one of the joints A of the fixed beam AC deflects through δ.

$$M_{BA} = \frac{6EI_1\delta}{l_1{}^2} \qquad \qquad ...(i)$$

Similarly consider the column CD of the frame, which will also behave like a beam fixed at C and D, and deflected through a small distance δ. Therefore moments at the joint C,

$$M_{CD} = \frac{6EI_2\delta}{l_2^2} \qquad \qquad ...(ii)$$

Now dividing the equation (i) by (ii),

$$\frac{M_{BA}}{M_{CD}} = \frac{\dfrac{6EI_1\delta}{l_1^2}}{\dfrac{6EI_2\delta}{l_2^2}} = \frac{I_1}{I_2} \times \frac{l_2^2}{l_1^2}.$$

Example 15·16. *A portal frame ABCD fixed at A and D is loaded as shown in Fig. 15·42.*

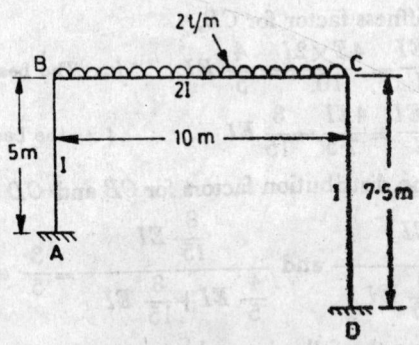

Fig. 15·42

The moment of inertia of members AB and CD is I and that of member BC is 2I. Draw the B.M. diagram for the frame.

(U.P.S.C., Engg. Services, 1975)

Solution.
Given. Length of $AB = 5$ m
Length of BC $\qquad = 10$ m
Length of CD $\qquad = 7\cdot5$ m
Load on BC, $\qquad w = 2$ t/m
M.I. of AB and CD $= I$
M.I. of BC $\qquad = 2I$

Since the portal frame is unsymmetrical, therefore it will be analysed, first by assuming it as a continuous beam, and then applying a sway correction.

First of all, let us assume the portal frame $ABCD$ to be split up into fixed beams AB, BC and CD. Since the beams AB and CD are not subjected to any load, therefore the fixing moments in span AB and CD at A, B, C and D are zero. In span BC, fixing moment at B.

$$= \frac{-wl^2}{12} = -\frac{2 \times 10^2}{12} = -16\cdot66 \text{ t-m}$$

Similarly fixing moment at C

$$= +\frac{wl^2}{12} = \frac{2 \times 10^2}{12} = +16\cdot66 \text{ t-m}$$

Now let us find out the distribution factors at B and C. From the geometry of the figure, we find that the stiffness factor for BA,

$$k_{BA} = \frac{4EI}{l} = \frac{4EI}{5} = \frac{4}{5} EI \qquad (\because \text{ the beam is fixed at } A)$$

and

$$k_{BC} = \frac{4EI}{l} = \frac{4E \times 2I}{10} = \frac{4}{5} EI$$

$$(\because \text{ the beam is continuous at } C)$$

Therefore the distribution factors for BA and BC will be

$$\frac{\frac{4}{5} EI}{\frac{4}{5} EI + \frac{4}{5} EI} \text{ and } \frac{\frac{4}{5} EI}{\frac{4}{5} EI + \frac{4}{5} EI} = \frac{1}{2} \text{ and } \frac{1}{2}$$

Similarly stiffness factor for CB,

$$k_{CB} = \frac{4EI}{l} = \frac{4E \times 2I}{10} = \frac{4}{5} EI \qquad (\because \text{ The beam is continuous at } B)$$

and

$$k_{CD} = \frac{4EI}{l} = \frac{4EI}{7 \cdot 5} = \frac{8}{15} EI \qquad (\because \text{ the beam is fixed at } D)$$

Therefore, the distribution factors for CB and CD will be

$$\frac{\frac{4}{5} EI}{\frac{4}{5} EI + \frac{8}{15} EI} \text{ and } \frac{\frac{8}{15} EI}{\frac{4}{5} EI + \frac{8}{15} EI} = \frac{3}{5} \text{ and } \frac{2}{5}$$

Now complete the following table

A		B		C	D	
	$\frac{1}{2}$	$\frac{1}{2}$		$\frac{3}{5}$	$\frac{2}{5}$	
0	0	−16·66	+16·66	0	0	Initial moments
	+ 8·33	+ 8·33	−10·00	−6·66		Distribute
+4·16		− 5·00	+ 4·16		−3·33	Carry over
	+ 2·50	+ 2·50	− 2·5	−1·66		Distribute
+1·25		− 1·25	+ 1·25		−0·83	Carry over
	+ 0·625	+ 0·625	− 0·75	−0·50		Distribute
+0·31		− 0·38	+ 0·31		−0·05	Carry over
	+ 0·19	+ 0·19	− 0·20	−0·11		Distribute
+0·10		− 0·10	+ 0·10		−0·05	Carry over
	+ 0·05	+ 0·05	− 0·06	−0·04		Distribute
+5·82	+11·695	−11·695	+ 8·97	−8·97	−4·46	Final moments

∴ Horizontal reaction at A,

$$H_A = \frac{5 \cdot 82 + 11 \cdot 695}{5} = 3 \cdot 5 \text{ t} \rightarrow$$

and horizontal reaction at D,

$$H_D = \frac{-8 \cdot 97 - 4 \cdot 46}{7 \cdot 5} = 1 \cdot 8 \text{ t} \leftarrow$$

From the free body diagram, as shown in Fig. 15·41, we find that the unbalanced horizontal force at C,

$$P = 3 \cdot 5 - 1 \cdot 8 = 1 \cdot 7 \text{ t} \leftarrow$$

We have worked out the moments and the reactions, under the assumption that a horizontal force, $P = 1 \cdot 7$ t ← is acting at C, to prevent the side sway. Actually there is no such force acting at C. We know that if a force of $1 \cdot 7$ t → is applied at C, it will neutralise the effect of assumed force $P = 1 \cdot 7$ t ← at C. A little consideration will show that if a force of $1 \cdot 7$ t → is applied at C, it will cause some sway moments at the joints B and C of the portal frame.

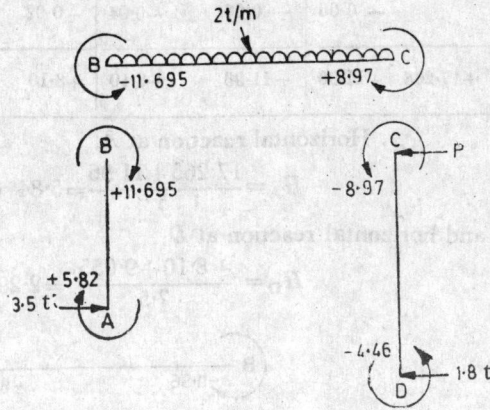

Fig. 15·43

Let
M_{BA} = Moment at

B in member BA

and M_{CD} = Moment at C in member CD.

We know that $\dfrac{M_{BA}}{M_{CD}} = \dfrac{I_1 \times l_2{}^2}{I_2 \times l_1{}^2} = \dfrac{I \times 5^2}{I \times 7 \cdot 5^2} = \dfrac{4}{9}$

Now assume some arbitrary moments at B and C (i.e. M_{AB} and M_{CD}) in the ratio of 9 : 4 as already calculated, and distribute the same at their joints also. Let us assume the sway moments M_{BA} and M_{CD} as 22·5 t-m and 10 t-m respectively.

Now complete the following table :

A	B			O		D	
	$\frac{1}{2}$	$\frac{1}{2}$		$\frac{3}{5}$	$\frac{2}{5}$		
+22·5	+22·5	0		0	+10·00	+10·00	Assumed moments
	—11·25	—11·25		—6·00	—4·00		Distribute
— 5·625		— 3·00		—5·625		—2·00	Carry over
	+ 1·50	+ 1·50		+3·375	+2·25		Distribute
+ 0·75		+ 1·68		+0·75		+1·125	Carry over
	— 0·84	— 0·84		—0·45	—0·30		Distribute
— 0·42		— 0·22		—0·42		—0·15	Carry over
	+ 0·11	+ 0·11		+0·25	+0·17		Distribute
+ 0·06		+ 0·12		+0·06		+0·08	Carry over
	— 0·06	— 0·06		—0·04	—0·02		Distribute
+17·265	+11·96	—11·96		—8·10	+8·10	+9·055	Final moments

$\therefore$ Horizontal reaction at A,

$$H_A = \frac{17·265 + 11·96}{5} = 5·84 \; \iota \rightarrow$$

and horizontal reaction at D,

$$H_D = \frac{+8·10 + 9·055}{7·5} = 2·29 \; t \rightarrow$$

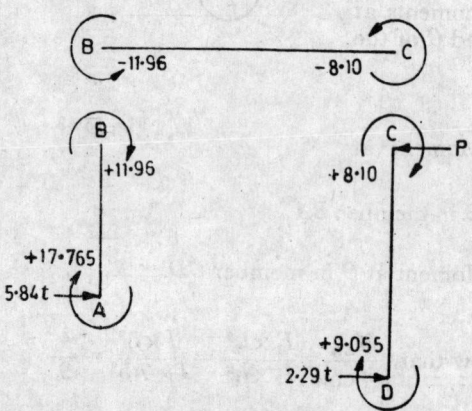

Fig. 15·44

From the free body diagram, as shown in Fig. 15·44, we find that the unbalanced horizontal force at C,

$$P = 5·84 + 2·29 = 8·13 \; t \leftarrow$$

But the actual unbalanced horizontal force is $1\cdot7$ t→. Therefore the actual sway moments in the members due to a sway force of $1\cdot7$ t → may be found out by proportion.

Now complete the following table :

	A		B		C		D
Assumed sway $=8\cdot13$ t ←	$+17\cdot265$	$+11\cdot96$	$-11\cdot96$	$-8\cdot10$	$+8\cdot10$		$+9\cdot055$
Actual sway $=1\cdot7$ t →	$-3\cdot72$	$+2\cdot50$	$+2\cdot50$	$+1\cdot69$	$-1\cdot69$		$-1\cdot89$
Non-sway	$+5\cdot82$	$+11\cdot695$	$-11\cdot695$	$+8\cdot97$	$-8\cdot97$		$-4\cdot46$
Final moments	$+2\cdot10$	$+9\cdot195$	$-9\cdot195$	$+10\cdot66$	$-10\cdot66$		$-6\cdot35$

∴ Horizontal reaction at A,

$$H_A = \frac{+2\cdot10 + 9\cdot195}{5} = 2\cdot26 \text{ t} \rightarrow$$

and horizontal reaction at D,

$$H_D = \frac{-10\cdot66 - 6\cdot35}{7\cdot5} = 2\cdot26 \text{ t} \leftarrow$$

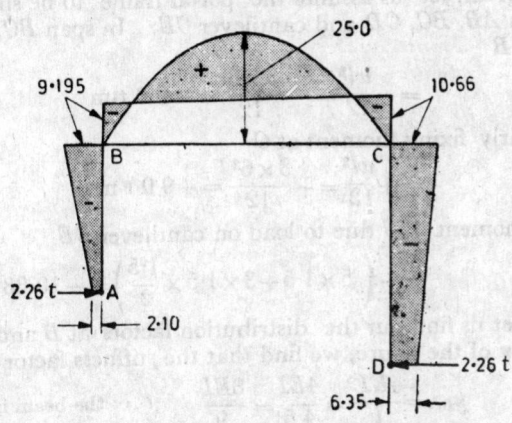

Fig. 15·45

Now B.M. at the centre of BC, considering it as a simply supported beam

$$= \frac{wl^2}{8} = \frac{2 \times 10^2}{8} = 25\cdot0 \text{ t-m}$$

Now complete B.M. diagram as shown in Fig. 15·45.

Example 15·17. *A portal frame shown in Fig. 15·46 is subjected to a loading as shown.*

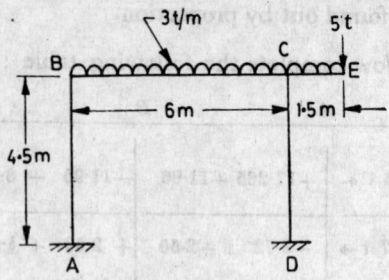

Fig. 15·46

Analyse the frame and draw the bending moment diagram. The flexural rigidity for all the members is constant.

Solution.

Length of $AB=CD=4·5$ m
Length of $BC=6$ m
Length of $CE=1·5$ m
Load on BE, $w=3$ t/m
Load at $E=W=5$ t

Since the portal frame is unsymmetrical, therefore, it will be analysed first by assuming it without sway, and then applying a sway correction.

First of all, let us assume the portal frame to be split up into fixed beams AB, BC, CD and cantilever CE. In span BC, the fixing moment at B

$$=-\frac{wl^2}{12}=-\frac{3\times6^2}{12}=-9·0 \text{ t-m}$$

Similarly fixing moment at C

$$=+\frac{wl^2}{12}=+\frac{3\times6^2}{12}=+9·0 \text{ t-m}$$

The moment at C due to load on cantilever CE

$$=-\left(5\times1·5+3\times1·5\times\frac{1·5}{2}\right)=-10·88 \text{ t-m}$$

Now let us find out the distribution factors at B and C. From the geometry of the figure, we find that the stiffness factor for BA,

$$k_{BA}=\frac{4EI}{l}=\frac{4EI}{4·5}=\frac{8EI}{9} \quad (\because \text{ the beam is fixed at } A)$$

and $$k_{BC}=\frac{4EI}{l}=\frac{4EI}{6}=\frac{2EI}{3} \quad (\because \text{ the beam is continuous at } G)$$

Therefore distribution factors for BA and BC will be

$$\frac{\frac{8EI}{9}}{\frac{8EI}{9}+\frac{2EI}{3}} \quad \text{and} \quad \frac{\frac{2EI}{3}}{\frac{8EI}{9}+\frac{2EI}{3}}=\frac{4}{7} \text{ and } \frac{3}{7}.$$

Similarly stiffness factor for CB,

$$k_{CB} = \frac{4EI}{l} = \frac{4EI}{6} = \frac{2EI}{3}$$

$(\because$ the beam is continuous at $B)$

and

$$k_{CD} = \frac{4EI}{l} = \frac{4EI}{4\cdot5} = \frac{8EI}{9}$$

$(\because$ the beam is fixed at $D)$

Therefore distribution factors for CB and CD will be

$$\frac{\dfrac{2EI}{3}}{\dfrac{2EI}{3}+\dfrac{8EI}{9}} \quad \text{and} \quad \frac{\dfrac{8EI}{9}}{\dfrac{2EI}{3}+\dfrac{8EI}{9}} = \frac{3}{7} \text{ and } \frac{4}{7}.$$

Now complete the following table :

A	B		O			D	
	$\frac{4}{7}$	$\frac{3}{7}$	$\frac{3}{7}$	CE	$\frac{4}{7}$		
0	0	−9·0	+9·0	−10·88	0		0	Initial moments
	+5·14	+3·86	+0·80	—	+1·08			Distribute
+2·57		+0·40	+1·93				+0·54	Carry over
	−0·23	−0·17	−0·83		−1·10			Distribute
−0·12		−0·42	−0·09				−0·55	Carry over
	+0·24	+0·18	+0·04	—	+0·05			Distribute
+0·12		+0·02	+0·09				+0·02	Carry over
	−0·01	−0·01	−0·04	—	−0·05			Distribute
+2·57	+5·14	−5·14	+10·90	−10·88	−0·02		−0·01	Final moments

$\therefore$ Horizontal reaction at A,

$$H_A = \frac{+2\cdot57+5\cdot14}{4\cdot5} = 1\cdot71 \text{ t} \rightarrow$$

and horizontal reaction at D,

$$H_D = \frac{-0\cdot01-0\cdot02}{4\cdot5} = 0\cdot01 \text{ t} \leftarrow$$

From the free body diagram as shown in Fig. 15·47, we find that the unbalanced horizontal force at B

$$P = 1\cdot71 - 0\cdot01 = 17 \text{ t} \leftarrow$$

We have worked out the moments and the reactions, under the assumption that a horizontal force, $P = 1\cdot7$ t $\leftarrow$ is acting at B to prevent the side sway. Actually there is no such force acting at B. We know that if a force of $1\cdot7$ t $\rightarrow$ is applied at B, it will neutralise the effect of assumed force $P = 1\cdot7$ t $\leftarrow$ at B. A little consideration will show that if a force of $1\cdot7$ t $\rightarrow$ is applied at B, it will cause some sway moments at the joints B and C of the portal frame.

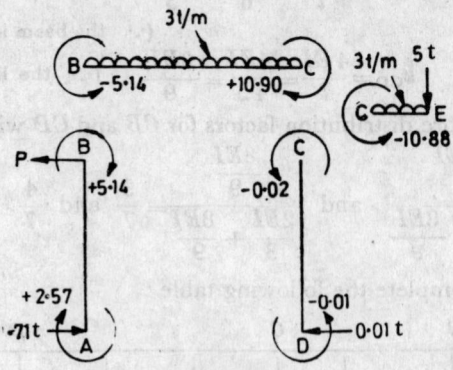

Fig. 15·47

Let $\qquad\qquad M_{BA}=$ Moment at B in member BA, and

$\qquad\qquad\qquad M_{CD}=$ Moment at C in member CD

We know that $\dfrac{M_{BA}}{M_{CD}} = \dfrac{I_1 \times l_2{}^2}{I_2 \times l_1{}^2} = \dfrac{I \times 4\cdot5^2}{I \times 4\cdot5^2} = 1$.

Now assume some arbitrary moments at B and C (*i.e.*, M_{BA} and M_{CD}) in the ratio of $1:1$ as already calculated, and distribute the same at their joints also. Let us assume the sway moments M_{BA} and M_{CD} as $10\cdot0$ t-m and $10\cdot0$ t-m respectively.

Now complete the following table.

A		B			C	D		
	$\dfrac{4}{7}$	$\dfrac{3}{7}$	$\dfrac{3}{7}$	CE	$\dfrac{4}{7}$		
$+10\cdot0$	$+10\cdot0$				$+10\cdot0$	$+10\cdot0$	Assumed moments
	$-5\cdot72$	$-4\cdot28$	$-4\cdot28$	—	$-5\cdot72$		Distribute
$-2\cdot86$		$-2\cdot14$	$-2\cdot14$			$-2\cdot86$	Carry over
	$+1\cdot22$	$+0\cdot92$	$+0\cdot92$	—	$+1\cdot22$		Distribute
$+0\cdot61$		$+0\cdot46$	$+0\cdot46$			$+0\cdot61$	Carry over
	$-0\cdot26$	$-0\cdot20$	$-0\cdot20$	—	$-0\cdot26$		Distribute
$-0\cdot13$		$+0\cdot10$	$-0\cdot10$			$-0\cdot13$	Carry over
	$+0\cdot06$	$+0\cdot04$	$+0\cdot04$	—	$+0\cdot06$		Distribute
$+7\cdot62$	$+5\cdot30$	$-5\cdot30$	$-5\cdot30$		$+5\cdot30$	$+7\cdot62$	Final moments

$\therefore$ Horizontal reaction at A,

$$H_A = \frac{7\cdot62 + 5\cdot30}{4\cdot5} = 2\cdot87 \text{ t} \rightarrow$$

and horizontal reaction at D,

$$H_D = \frac{+5 \cdot 30 + 7 \cdot 62}{4 \cdot 5} = 2 \cdot 87 \ t \rightarrow$$

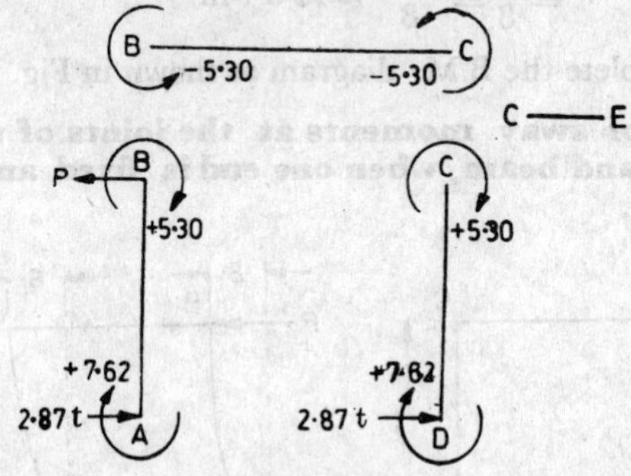

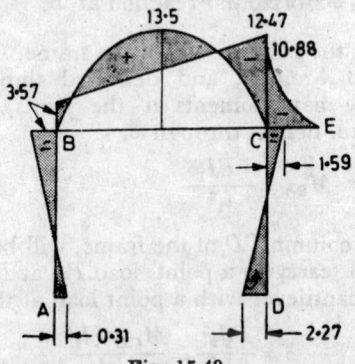

Fig. 15·48

From the free body diagram, as shown in Fig. 15·48, we find that the unbalanced horizontal force at B,

$$P = 2 \cdot 87 + 2 \cdot 87 = 5 \cdot 74 \ t \leftarrow$$

But the actual unbalanced horizontal force is $1 \cdot 7 \ t \rightarrow$. Therefore the actual sway moments in the members due to a sway force of $1 \cdot 7 \ t \rightarrow$ may be found out by proportion.

Now complete the following table :

	A		B		O	D	
Assumed away =5·74 t ←	+7·62	+5·30	−5·30	−5·30	CE —	+5·30	+7·62
Actual sway =1·7 t →	−2·26	−1·57	+1·57	+1·57		−1·57	−2·26
Non-sway moments	+2·57	+5·14	−5·14	+10·90	−10·88	−0·02	−0·01
Final moments	+0·31	+3·57	−3·57	+12·47	−10·88	−1·59	−2·27

Fig. 15·49

We know that the B.M. at the centre of BC, considering it as a simply supported beam

$$= \frac{wl^2}{8} = \frac{3 \times 6^2}{8} = 13 \cdot 5 \text{ t-m}$$

Now complete the B.M. diagram as shown in Fig. 15·49.

15·20. Ratio of sway moments at the joints of the column heads and beam, when one end is fixed and the other hinged.

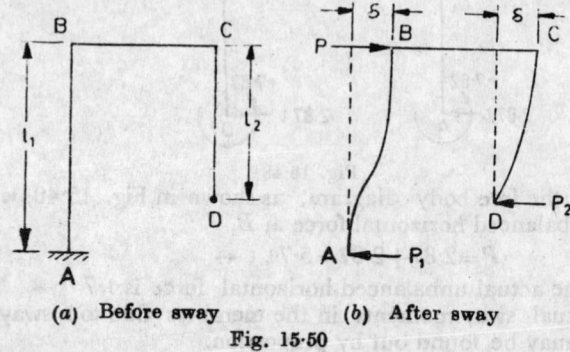

(a) **Before sway**   (b) **After sway**

Fig. 15·50

Consider an unsymmetrical portal frame $ABCD$ having stiff joints B and C and fixed end A, and hinged end D as shown in Fig. 5·50 (a).

Let $l_1 = $ Length of the column AB,

$I_1 = $ Moment of inertia of column AB, and

$l_2, I_2 = $ Corresponding values for the column CD.

Now consider a sway force P to act at B, as a result of this force, let the frame deflect through a small distance δ as shown in Fig. 15·50 (b).

Let $P_1 = $ Horizontal reaction at A, and

$P_2 = $ Horizontal reaction at B.

Now consider the column AB of the frame, which will behave like a beam fixed at A and B, and deflected through a small distance δ. We know that moments at the joints, when one of the joints of a fixed beam deflect through δ,

$$M_{BA} = \frac{6EI_1\delta}{l_1{}^2} \qquad\qquad ...(i)$$

Similarly the column CD of the frame, will behave like a cantilever fixed at C and carrying a point load P_2 at D. We know that the deflection of a cantilever with a point load at the free end,

$$\delta = \frac{P_2 l_2{}^3}{3EI} = \frac{M_{CD} \times l_2{}^2}{3EI_2} \qquad ...(M_{CD} = P_2 l_2)$$

or
$$M_{CD} = \frac{3EI_2 \, \delta}{l_2{}^2} \qquad (ii)$$

Now dividing the equation (i) by (ii),

$$\frac{M_{BA}}{M_{CD}} = \frac{\dfrac{6EI_1 \, \delta}{l_1{}^2}}{\dfrac{3EI_2 \, \delta}{l_2{}^2}} = \frac{2I_1}{I_2} \times \frac{l_2{}^2}{l_1{}^2}$$

Example 15·17. *Fig. 15·51 shows a portal frame ABCD fixed at A and hinged at D.*

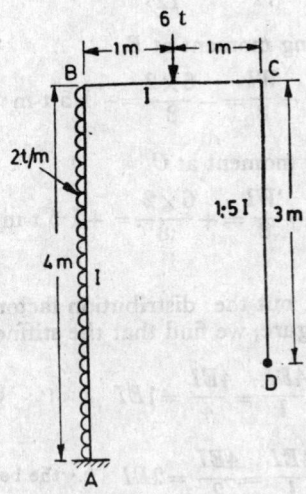

Fig. 15·51

Analyse the frame and draw the B.M. diagram

(*London University*)

Solution

Given.   Length of $AB = 4$ m

Length of BC $= 2$ m

Length of CD $= 3$ m

Load on AB, $w = 2$ t/m

Load on BC, $W = 6$ t

M.I. of AB and BC $= I$

M.I. of CD $= 1·5I$

Since the portal frame is unsymmetrical, therefore, it will be analysed first by assuming it as a continuous beam, and then applying a sway correction.

First of all, let us assume the portal frame $ABCD$ to be split up into fixed beams AB, BC and CD. Since the beam CD is not subjected to any load, therefore, the fixing moments in span CD at C and D are zero.

In span AB, fixing moment at A

$$= -\frac{wl^2}{12} = -\frac{2 \times 4^2}{12} = -2 \cdot 67 \text{ t-m}$$

Similarly fixing moment at B

$$= +\frac{wl^2}{12} = +\frac{2 \times 4^2}{12} = +2 \cdot 67 \text{ t-m}$$

In span BC, fixing moment at B

$$= -\frac{Wl}{8} = -\frac{6 \times 2}{8} = -1 \cdot 5 \text{ t-m}$$

Similarly, fixing moment at C

$$= +\frac{Wl}{8} = +\frac{6 \times 2}{8} = +1 \cdot 5 \text{ t-m}$$

Now, let us find out the distribution factors at B and C. From the geometry of the figure, we find that the stiffness factor for BA,

$$k_{BA} = \frac{4EI}{l} = \frac{4EI}{4} = 1EI \quad \dots (\because \text{ the beam is fixed at } A)$$

and $$k_{BC} = \frac{4EI}{l} = \frac{4EI}{2} = 2EI \quad (\because \text{ the beam is continuous at } C)$$

Therefore the distribution factors for BA and BC will be

$$\frac{1EI}{1EI + 2EI} \quad \text{and} \quad \frac{2EI}{1EI + 2EI} = \frac{1}{3} \quad \text{and} \quad \frac{2}{3}.$$

Similarly stiffness factor for CB,

$$k_{CB} = \frac{4EI}{l} = \frac{4EI}{2} = 2EI$$

$$\dots(\because \text{ the beam is continuous at } B$$

and $$k_{CD} = \frac{3EI}{l} = \frac{3E \times 1 \cdot 5 \, I}{3} = 1 \cdot 5 \, EI$$

$$\dots(\because \text{ the beam is hinged at } D)$$

$\therefore$ Therefore distribution factors for CB and CD will be

$$\frac{2EI}{2EI + 1 \cdot 5EI} \quad \text{and} \quad \frac{1 \cdot 5EI}{2EI + 1 \cdot 5EI} = \frac{4}{7} \quad \text{and} \quad \frac{3}{7}.$$

Now complete the following table.

A		B		C		D	
	$\frac{1}{3}$	$\frac{2}{3}$		$\frac{4}{7}$	$\frac{3}{7}$		
$-2\cdot67$	$+2\cdot67$	$-1\cdot5$	$+1\cdot5$	0	0		Initial moment
	$-0\cdot39$	$-0\cdot78$	$-0\cdot86$	$-0\cdot64$	0		Distribute
$-0\cdot19$		$-0\cdot43$	$-0\cdot39$		0		Carry over
	$+0\cdot14$	$+0\cdot29$	$+0\cdot2$	$+0\cdot17$			Distribute
$+0\cdot07$		$+0\cdot11$	$+0\cdot15$		0		Carry over
	$-0\cdot04$	$-0\cdot07$	$-0\cdot09$	$-0\cdot06$			Distribute
$-0\cdot02$		$-0\cdot05$	$-0\cdot04$		0		Carry over
	$+0\cdot02$	$+0\cdot03$	$+0\cdot02$	$+0\cdot02$			Distribute
$-2\cdot81$	$+2\cdot40$	$-2\cdot40$	$+0\cdot51$	$-0\cdot51$	0		Final moments

For horizontal reaction at A, equate the moments about B.

$$H_A \times 4 = -(2 \times 4 \times 2) + 2\cdot4 - 2\cdot81 = -16\cdot41$$

$$\therefore \quad H_A = -\frac{16\cdot41}{4} \ t$$

$$= 4\cdot1 \ t \leftarrow$$

and horizontal reaction at D,

$$H_D = -\frac{0\cdot51}{3} \ t$$

$$= -0\cdot17 \ t \leftarrow$$

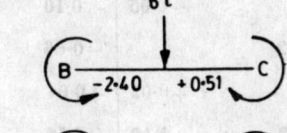

From the free body diagram, as shown in Fig. 15·52, we find that the unbalanced horizontal force at C,

$$P = (2 \times 4) - 4\cdot1 - 0\cdot17$$

$$= 3\cdot73 \ t \leftarrow$$

We have worked out the moments and the reactions, under the assumption that a horizontal force, $P = 3\cdot73$ t $\leftarrow$ is acting at C to prevent the side sway. Actually there is no such force acting at

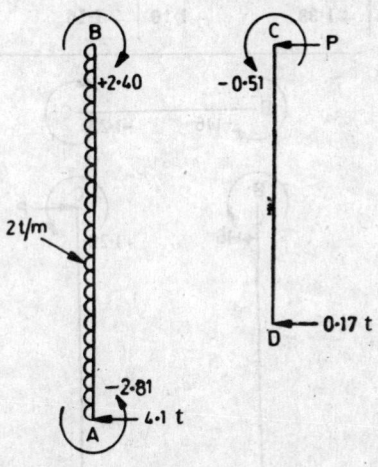

Fig. 15·52

C. We know that if a force of $3\cdot73$ t $\rightarrow$ is applied at C, it will neutralise the effect of assumed force $P = 3\cdot73$ t $\leftarrow$ at C. A little consideration will show, that if a force of $3\cdot73$ t $\rightarrow$ is applied at C, it will cause some sway moments at the joints B and C of the portal frame.

Let M_{BA} = Moment at B in member BA, and

and M_{CD} = Moment at C in member CD.

We know that $\dfrac{M_{BA}}{M_{CD}} = \dfrac{2I_1 \times l_2{}^2}{I_2 l_1{}^2} = \dfrac{2I \times 3^2}{1 \cdot 5 I \times 4^3} = \dfrac{3}{4}.$

Now assume some arbitrary moments at B and C (*i.e.*, M_{BA} and M_{CD}) in the ratio of $3 : 4$ as already calculated, and distribute the same at their joints also. Let us assume the sway moments M_{BA} and M_{CD} at $1 \cdot 5$ t-m and $2 \cdot 0$ t-m.

Now complete the following table :

A	B			C		D	
	$\dfrac{1}{3}$	$\dfrac{2}{3}$		$\dfrac{4}{7}$	$\dfrac{3}{7}$		
+1·5	+1·5	0	0		+2·0	0	Assumed moments
	−0·5	−1·0		−1 14	−0·86		Distribute
−0·25		−0·57		−0·50		0	Carry over
	+0·19	+0·38		+0·29	+0·2		Distribute
+0·10		+0·15		+0·19		0	Carry over
	−0·05	−0·10		−0·11	−0·08		Distribute
−0·02		−0·06		−0·05		0	Carry over
	+0·02	+0·04		+0·03	+0·02		Distribute
+1·33	+1·16	−1·16		−1·29	+1·29	0	Final moments

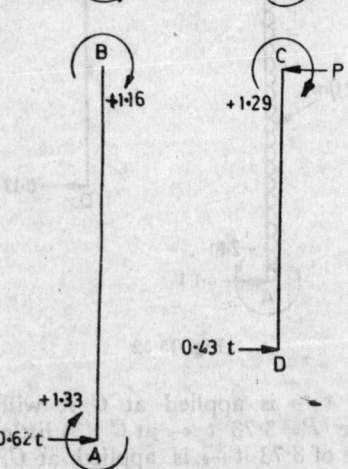

Fig. 15·53

$\therefore$ Horizontal reaction at A,

$$H_A = \frac{+1 \cdot 33 + 1 \cdot 16}{4} = 0 \cdot 62 \quad t \rightarrow$$

and horizontal reaction at D,

$$H_D = \frac{+1 \cdot 29}{3} = 0 \cdot 43 \quad t \rightarrow$$

From the free body diagram as shown in Fig. 15·53, we find that the unbalanced vertical figure at C,

$$P = 0 \cdot 62 + 0 \cdot 43 = 1 \cdot 05 \quad t \leftarrow$$

But the actual unbalanced horizontal force is $3 \cdot 73$ t $\rightarrow$. Therefore the actual sway moments in the members due to a sway force of $3 \cdot 73$ t $\rightarrow$ may be found out by proportion.

Now complete the following table :

	A		B		C		D
Assumed sway=1·05 T←	+1·33	+1·16	−1·16	−1·29	+1·29		0
Actual sway=3·73 T→	−4·71	−4·11	+4·11	+4·58	−4·58		0
Non-sway	−2 81	+2·40	−2·40	+0·51	−0·51		0
Final moments	−7·52	−1·71	+1·71	+5·09	−5·09		0

For horizontal reaction at A, equate the moments about B.

$$H_A \times 4 = (2 \times 4 \times 2) - 7 \cdot 52 - 1 \cdot 71 = -25 \cdot 23$$

$$\therefore H_A = \frac{-25 \cdot 23}{4} = 6 \cdot 31 \text{ t} \leftarrow$$

and horizontal reaction at D,

$$H_D = -\frac{5 \cdot 09}{3} = 1 \cdot 69 \text{ t} \leftarrow$$

Now the B.M. at the centre of AB, considering it as a simply supported beam

$$= \frac{wl^2}{8} = \frac{2 \times 4^2}{4} = 4 \text{ t-m}$$

and B.M. under the load of BC, considering it as a simply supported beam

$$= \frac{Wl}{4} = \frac{6 \times 2}{4} = 3 \text{ t-m}$$

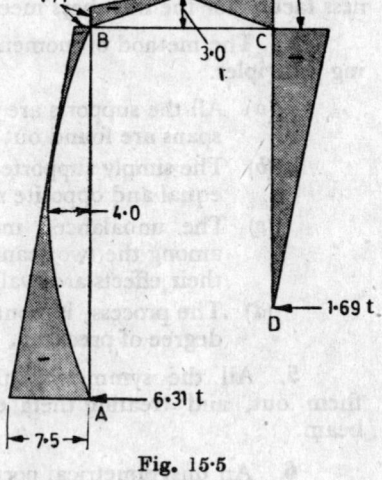

Fig. 15·5

Now complete the B.M. diagram as shown in Fig. 15·54.

Exercise 15·2

1. A portal frame $ABCD$ of span 6 m and height 4 m has its vertical members **fixed** into the ground and the horizontal member carries a uniformly distributed load of 2 t/m. The moment of inertia of the vertical members is I and that of horizontal member is $2I$. Determine the moments at A, B, C and D. Also draw the B.M. diagram. *(Calcutta University)*

[**Ans.** 1·5 t-m ; 3·0 t-m ; 3·0 t-m ; 1·5 t-m]

2. A portal frame $ABCD$ of span 4 m is subjected to a uniformly distributed load of 1 t/m over the member BC. The members AB and CD of lengths 4 m and 2 m are hinged at A and D. Determine the members at A, B, C and D and draw the B.M. diagram. *(Nagpur University)*

[**Ans.** 0 ; 1·04 t-m ; 0·52 t-m ; 0]

3. A portal frame $ABCD$ is fixed at A and hinged at B. The members AB and CD are of 4 m length. Find the moments at A, B, C and D, if a horizontal force of 10 tonnes is acting at B. *(London University)*

[**Ans.** 18·18 t-m ; 12·73 t-m ; 9·09 t-m ; 0]

HIGHLIGHTS

1. The stiffness factor for a beam fixed at one end, and simply supported at the other is $\dfrac{4EI}{l}$.

2. The stiffness factor for a beam simply supported at both ends is $\dfrac{3EI}{l}$.

where E = Young's modulus for the beam material,
           I = Moment of inertia of the beam section, and
           l = Length of the beam.

3. The distribution factors, at a joint, are in the ratio of stiffness factors for the members meeting at the joint.

4 The method of moment distribution is based on the following principles.

     (a) All the supports are assumed to be clamped and fixing spans are found out for all the spans.

     (b) The simply supported ends are released, by applying an equal and opposite moment on the end.

     (c) The unbalanced moment, at a support, is distributed among the two spans in the ratio of their stiffness factors, their effects are evaluated on the opposite joints.

     (d) The process is continued, till we reach the required degree of precision.

5. All the symmetrical portal frames are analysed by opening them out, and treating them exactly like a three-span continuous beam.

6. An unsymmetrical portal frame is subjected to side sway and is analysed as discussed below :

     (a) Ignoring the side sway, calculate the usual fixed end moments.

     (b) Now calculate the horizontal as well as vertical reactions at the column bases. The algebraic sum of the two horizontal reactions will give the magnitude and direction of the side sway force, which had held the frame against side sway.

     (c) Now remove the sway force. This will cause a set of fixed end moments and distribute them (First by assuming some arbitrary moments and then by proportion).

     (d) The final moment at each joint is given by the algebraic sum of the moment given above.

7. The ratio of sway moments at the joints of column heads and beam (a) when both the ends are hinged

$$\frac{M_{BA}}{M_{CD}} = \frac{I_1}{I_2} \times \frac{l_2^2}{l_1^2}$$

(b) when both the ends are fixed

$$\frac{M_{BA}}{M_{CD}} = \frac{I_1}{I_2} \times \frac{l_2^2}{l_1^2}$$

(c) when end A is fixed and C is hinged,

$$\frac{M_{BA}}{M_{CD}} = \frac{2I_1}{I_2} \times \frac{l_2^2}{l_1^2}$$

Do You Know ?

1. Define the term 'carry over factor'. Derive a relation for the stiffness factor for a beam simply supported at its both ends

2. What do you understand by the term 'distribution factor' ? Discuss its importance in the method of moment distribution.

3. 'Explain the procedure for finding out the fixed end moments in :

   (a) beams with fixed end supports.
   (b) beams with simply supported ends.
   (c) beams with end span overhanging.
   (d) beams with a sinking support.

4. What is a portal frame ? How will you distinguish between a symmetrical portal frame, and an unsymmetrical portal frame ?

5. Describe the meaning of 'sway force'. How it is obtained in an unsymmetrical frame ?

6. Explain the procedure for finding out the moments and reactions in an unsymmetrical frame.

7. From first principles, derive the ratio of sway moments at the joints of column heads and beam, when

   (a) both the ends are hinged,
   (b) both the ends are fixed and
   (c) one of the ends is fixed and the other hinged.

16

COLUMN ANALOGY METHOD

1. Introduction. 2. Sign conventions. 3. Theory of column analogy. 4. Application of column analogy method. 5. Fixed beams. 6. Portal frames. 7. Symmetrical Portal frames. 8. Portal frame with hinged legs. 9. Unsymmetrical portal frames.

16·1. Introduction.

The column analogy method was proposed by Prof. Hardy Cross and is most useful for the analysis of beams with fixed supports or rigid frames up to third degree of redundancy. These members may be of uniform moment of inertia throughout their lengths, or it may be varying. This method is based on a mathematical similarity (*i.e.*, analogy) between the stresses developed on a column section subjected to eccentric load and the moments imposed on a member due to fixity of its supports.

16·2. Sign Conventions.

Though there are different sign conventions used in different books, yet the following sign conventions will be used in this book, which are widely used and internationally recognised.

1. *Moment*. A moment which tends to bend the beam with a curvature having concavity at the top will be taken as positive. But a moment which tends to bend the beam with curvature having convexity at the top will be taken as negative.

2. A downward load on the column will be taken as positive, whereas an upward load on the column will be taken as negative.

3. The axes of reference will be taken $x-x$ and $y-y$ axes respectively as usual. The value of x will be taken as positive, if the reference point is on the right, but negative if it is on the left of the axis of reference. Similarly the value of y will be taken as positive if the reference point is above, but negative if it is below the axis of reference. A little consideration will show that the above mentioned values are the general values in mathematics.

16·3. Theory of column analogy.

We have already discussed in Art. 11·2 that in a fixed beam, AB, the slope at both the ends, even after loading, will be zero. We have also discussed that a fixed beam AB may be looked upon as a simply supported beam subjected to end moments M_A and M_B, of such magnitude that the slopes at A and B are reduced to zero. A little consideration will show that, this is only possible, if the direction of the fixing or restraining moments M_A and M_B is opposite to that of the bending moment, under a given system of loading.

We know, that since the slope at A and B is equal to zero, therefore as per *Mohr's theorem I, the area of B.M. diagram between A and B should also be equal to zero. Or in other words, the area of sagging B.M. diagram (considering the beam as a simply supported) should be equal to the area of hogging B.M. diagram (considering the end moments).

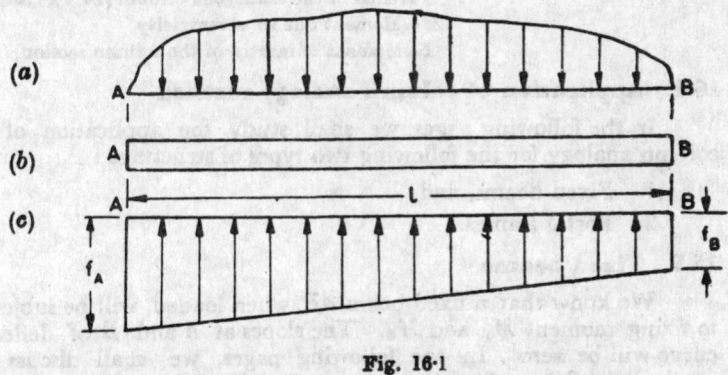

Fig. 16·1

Now let us imagine a column section of uniform unit width having length equal to the span of the column and of a very small height as shown in Fig. 16·1 (b). A little consideration will show that if the imaginary column is subjected to a vertical loading equal to static B.M. (i.e. M_s) as shown in Fig. 16·1(a), the base of the column will be subjected to a compressive stress, which will have a linear variation from f_A to f_B at the ends A and B as shown in Fig. 16·1 (c). Since the compressive stress in the column corresponds

---

*It states, "The change of slope between any two points on an elastic curve is equal to the area of B.M. diagram between these points divided by EI."

to the moment, opposite in nature to the applied moment, it will represent the indeterminate moment (*i.e.* M_i). It is thus obvious that $M_A = f_A$ and $M_B = f_B$. The values of M_A and M_B may be easily obtained by the following considerations.

1. The total pressure at the base of the column is equal to the total pressure at the top of the column.

2. The moment of total pressure at the base of the column, about any support A or B, is equal to the moment of the total pressure at the top of the column about the same support. A little consideration will show that since the two pressures and their moments are equal, therefore their centres of gravity must lie in the same vertical line.

3. The net moment at any point on a beam will be given by the relation, $M = M_s - M_i$.

Note : 1. If the beam is symmetrically loaded, the indeterminate moment at A,

$$Mi_A = Mi_B = \frac{P}{A}$$

2. If the beam is not symmetrically loaded, the indeterminate moment at A,

$$Mi_A = f_A = \frac{P}{A} \pm \frac{M}{I}$$

where
$P = $ Total load or total area of M_s diagram
$A = $ Area of the analogous column (i.e. $l \times 1 = l$)
$M = $ Moment due to eccentricity.
$I = $ Moment of inertia of the column section.

16·4. Application of column analogy method

In the following pages, we shall study the application of the column analogy for the following two types of structures :

1. Fixed beams, and
2. Portal frames.

16·5. Fixed beams

We know that a fixed beam AB, when loaded, will be subjected to fixing moment M_A and M_B. The slopes at A and B of deflected curve will be zero. In the following pages, we shall discuss the application of the column analogy method for various types of loading.

(i) A fixed beam carrying a central point load

Consider a beam AB fixed at A and B of span l, and carrying a central point load W at C as shown in Fig. 16·2 (a).

We know that the B.M. under the load considering it as a simply supported,

$$M = \frac{Wl}{4}$$

Now draw an analogous column AB of length l with a width equal to $1/EI$. Let this beam be subjected to pressure equal to the

B.M. diagram. such that pressure at the centre of the column be $\dfrac{Wl}{4}$ as shown in Fig. 16·2 (b). From the geometry of the figure, we find that the total pressure on the column

$$P = \frac{Ml}{2EI} = \frac{Wl}{4} \cdot \frac{l}{2EI} \cdot \frac{Wl^2}{8EI}$$

Area of column, $\quad A = l \times \dfrac{1}{EI} = \dfrac{l}{EI}$

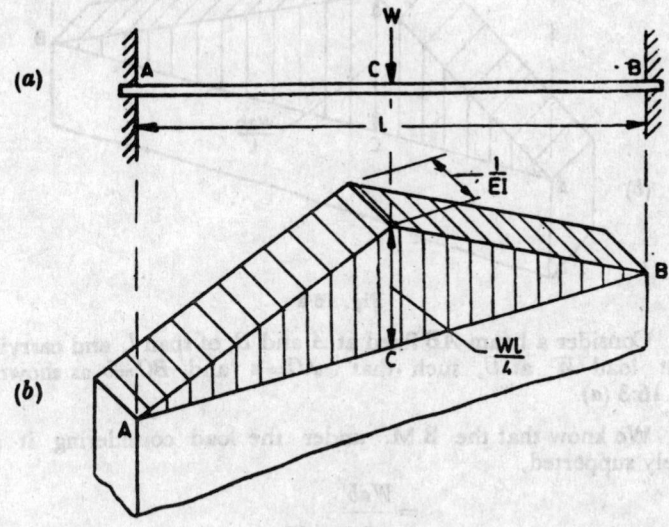

Fig. 16·2

Since the *c.g.* of the pressure diagram coincides with the centre of the beam, therefore there is no eccentricity of the loading. It is thus obvious, that the indeterminate moment at *A*,

$$\therefore \quad M_{iA} = f_A = \frac{P}{A} = \frac{\dfrac{Wl^2}{8EI}}{\dfrac{l}{EI}} = \frac{Wl}{8}$$

From the geometry of the figure, we also find that the static moment at *A*,

$$M_{SA} = 0$$

∴ Now moment at *A*,

$$M_A = M_{SA} - M_{iA} = 0 - \frac{Wl}{8}$$

$$= -\frac{Wl}{8} \qquad \text{(Same as before)}$$

Similarly, $\quad M_B = -\dfrac{Wl}{8} \qquad$ (By symmetry)

(ii) A fixed beam carrying an eccentric point load

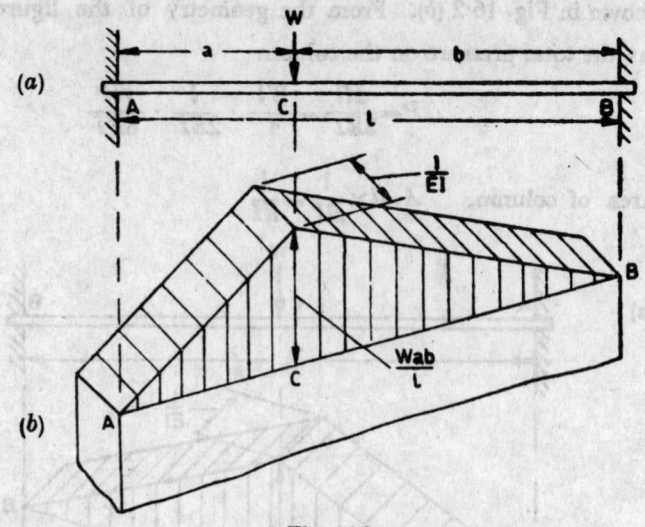

Fig. 16·3

Consider a beam AB fixed at A and B, of span l, and carrying a point load W at C, such that $AC=a$ and $BC=b$ as shown in Fig. 16·3 (a).

We know that the B.M. under the load considering it as a simply supported,

$$= \frac{Wab}{l}$$

Now draw an analogous column AB of length l, with a width equal to $1/EI$. Let this beam be subjected to the pressure equal to the B.M. diagram, such that pressure at C be $\dfrac{Wab}{l}$ as shown in Fig. 16·3 (b). From the geometry of the figure, we find that the total pressure on the column,

$$P = \frac{Ml}{2EI} = \frac{Wab}{l} \cdot \frac{l}{2EI} = \frac{Wab}{2EI}$$

Area of column, $A = l \times \dfrac{1}{EI} = \dfrac{l}{EI}$

and moment of inertia of the analogous column section about $y-y$ axis,

$$I_{YY} = \frac{\dfrac{1}{EI} \times l^3}{12} = \frac{l^3}{12EI}$$

Let $\bar{x}$ be the distance between the $c.g.$ of the pressure and the fixed end A.

$$\therefore \quad \bar{x} = \frac{a_1 x_1 + a_2 x_2}{a_1 + a_2}$$

$$= \frac{\left[\frac{1}{2} \times M \times a \times \frac{2a}{3}\right] + \left[\frac{1}{2} \times M \times b \left(a + \frac{b}{3}\right)\right]}{\left(\frac{1}{2} \times M \times a\right) + \left(\frac{1}{2} \times M \times b\right)}$$

$$= \frac{\frac{2a^2}{3} + ab + \frac{b^2}{3}}{a + b} = \frac{2a^2 + 3ab + b^2}{3(a+b)}$$

$$= \frac{(a^2 + b^2 + 2ab) + a^2 + ab}{3(a+b)} = \frac{(a+b)^2 + a(a+b)}{3(a+b)}$$

$$= \frac{l^2 + al}{3l} = \frac{l+a}{3} \qquad [(\because \quad a+b=l)]$$

and eccentricity,

$$e = \frac{l}{2} - \frac{l+a}{3} = \frac{3l - 2l - 2a}{6} = \frac{l - 2a}{6}$$

$$= \frac{a+b-2a}{6} = \frac{b-a}{6} \qquad (\because \quad l = a+b)$$

$\therefore$ Moment,

$$M' = P.e = \frac{Wab}{2EI} \times \frac{b-a}{6}$$

We know that the indeterminate moment at A,

$$M_{iA} = \frac{P}{A} + \frac{M'.x}{I} = \frac{\dfrac{Wab}{2EI}}{\dfrac{l}{EI}} + \frac{\left(\dfrac{Wab}{2EI} \times \dfrac{b-a}{6}\right) \times \dfrac{l}{2}}{\dfrac{l^3}{12EI}} \qquad \left(\because \quad x = \frac{l}{2}\right)$$

$$= \frac{Wab}{2l} + \frac{Wab\,(b-a)}{2l^2} = \frac{Wab}{2l^2}\,(l+b-a)$$

$$= \frac{Wab}{2l^2}\,(a+b+b-a) \qquad (\because \quad l = a+b)$$

$$= \frac{Wab^2}{l^2}$$

From the geometry of the figure, we also find that the static moment at A,

$$Ms_A = 0$$

$\therefore$ Net moment at A,

$$M_A = M_sA = M_{iA} = 0 - \frac{Wab^2}{l^2}$$

$$= -\frac{Wab^2}{l^2} \qquad \ldots\text{(Same as before)}$$

Similarly,

$$M_B = -\frac{Wa^2 b}{l^2} \qquad \ldots\text{(Substituting } b \text{ for } a \text{ and } vice\ versa)$$

(iii) *A fixed beam carrying a uniformly distributed load*

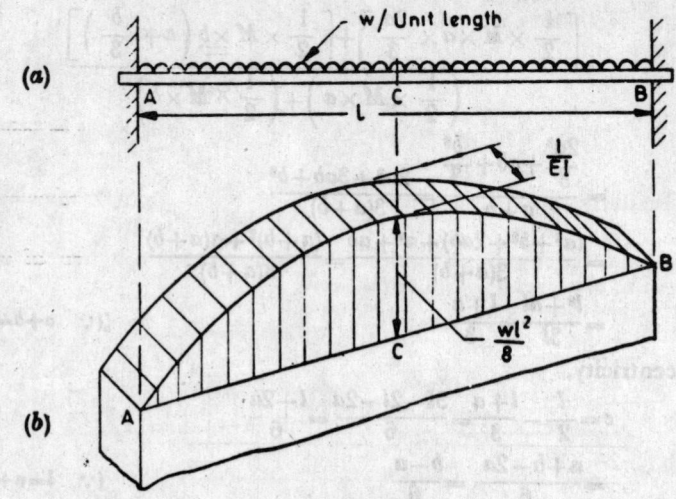

Fig. 16·4

Consider a beam AB fixed at A and B, and of span l, carrying a uniformly distributed load of w per unit length as shown in Fig. 16·4 (a).

We know that the B.M. at the centre of the beam, considering it as a simply supported,

$$M = \frac{wl^2}{8}$$

Now draw an analogous column AB of length l, with a width equal to $1/EI$. Let this beam be subjected to the pressure equal to the B.M. diagram, such that the pressure at the centre of the beam be $\frac{wl^2}{8}$ as shown in Fig. 16·4 (b). From the geometry of the figure, we find that the total pressure on the column,

$$P = \frac{2}{3} \times M \times l \times \frac{1}{EI} = \frac{2}{3} \times \frac{wl^2}{8} \cdot l \times \frac{1}{EI} = \frac{wl^3}{12EI}$$

area of column,

$$A = l \times \frac{1}{EI} = \frac{l}{EI}$$

Since the *c.g.* of the pressure diagram coincides with the centre of the beam, therefore there is no eccentricity of the loading. It is thus obvious, that the indeterminate moment at A,

$$M_{iA} = \frac{P}{A} = \frac{\dfrac{wl^3}{12EI}}{\dfrac{l}{EI}} = \frac{wl^2}{12}$$

From the geometry of the figure, we also find that the static moment at A,

$$M_{SA} = 0$$

∴ Net moment at A,

$$M_A = M_{SA} - M_{iA} = 0 - \frac{wl^3}{12}$$

$$= -\frac{wl^3}{12} \qquad \text{(same as before)}$$

Similarly $\qquad M_B = -\frac{wl^2}{12} \qquad$ (By symmetry)

(iv) A fixed beam carrying a gradually varying load from zero at one end to w per unit length at the other.

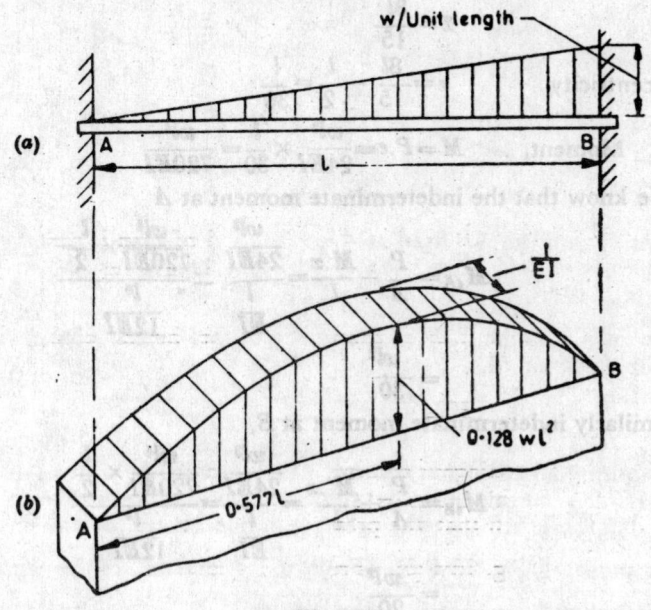

Fig. 16·5

Consider a beam AB fixed at A and B and of span l, carrying gradually varying load from zero at one end to w per unit length at the other as shown in Fig. 16·5 (a).

We know that the maximum B.M. on the beam occurs at a distance of $0.577\,l$, considering it as a simply supported beam,

$$M_{max} = 0.128\, wl^2$$

Now draw an analogous column AB of length l, with a width equal to $1/EI$. Let this beam be subjected to the pressure equal to the B.M. diagram, such that the pressure at a distance of $0.577\,l$

from A be $0.128\ wl^2$ as shown in Fig. 16·5 (b). From the geometry of the figure, we find that pressure on the column,

$$P = \frac{wl^3}{24EI}$$

Area of column, $\quad A = l \times \dfrac{1}{EI} = \dfrac{l}{EI}$

and moment of inertia of the analogous column sector above $y \cdot y$ axis, as,

$$I_{YY} = \frac{\dfrac{1}{EI} \times l^3}{12} = \frac{l^3}{12EI}$$

From the geometry of the figure, we also find that the $c.g.$ of the pressure and the fixed end A,

$$\bar{x} = \frac{8l}{15}$$

and eccentricity $\quad e = \dfrac{8l}{15} - \dfrac{l}{2} = \dfrac{l}{30}$

$\therefore$ Moment, $\quad M = P.e = \dfrac{wl^3}{24EI} \times \dfrac{l}{30} = \dfrac{wl^4}{720EI}$

We know that the indeterminate moment at A

$$M_{iA} = \frac{P}{A} - \frac{M.x}{I} = \frac{\dfrac{wl^3}{24EI}}{\dfrac{l}{EI}} - \frac{\dfrac{wl^4}{720EI} \times \dfrac{l}{2}}{\dfrac{l^3}{12EI}}$$

$$= \frac{wl^2}{30}$$

Similarly indeterminate moment at B,

$$M_{iB} = \frac{P}{A} + \frac{M'.x}{I} = \frac{\dfrac{wl^3}{24EI}}{\dfrac{l}{EI}} = \frac{\dfrac{wl^4}{720EI} \times \dfrac{l}{2}}{\dfrac{l^3}{12EI}}$$

$$= \frac{wl^2}{20}.$$

From the geometry of the figure, we also find that the static moment at A,

$$M_{SA} = 0 \qquad \text{Similarly} \quad M_{SB} = 0$$

$\therefore$ Net moment at A,

$$M_A = M_{SA} - M_{iA} = 0 - \frac{wl^2}{30}$$

$$= -\frac{wl^2}{30} \qquad \qquad \text{(same as before)}$$

and $\qquad M_B = M_{SB} - M_{iB} = 0 - \dfrac{wl^2}{20}$

$$= -\frac{wl^2}{20} \qquad \qquad \text{(same as before)}$$

Example 16·1. *A beam AB of span l, fixed at its both ends, is carrying a central point load W. Determine the fixed end moments, if the moment of inertia for the central half of the span is 2I and that for the both the quarters of the span is I.*

Solution.

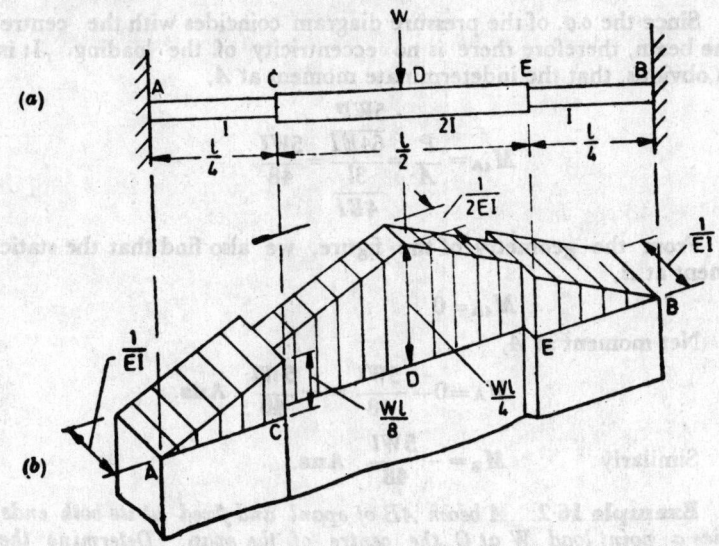

Fig. 16·6

Given   Span $= l$
Load at $D = W$
M.I. for $CE = 2I$
M.I. for $AC = EB = I$

We know that the bending moment under the load, considering it as a simply supported beam,

$$M = \frac{Wl}{4}$$

Now draw the analogous column $ACDEB$ of width equal to $\frac{1}{EI}$ for AC and EB and $\frac{1}{2EI}$ for CE. Let the beam be subjected to pressure equal to the B.M. diagram, such that the pressure at D be equal to $\frac{Wl}{4}$ and that at C be equal to $\frac{Wl}{8}$ as shown in Fig. 16·6(b).

Area of analogous column section,

$$A = \frac{l}{4} \times \frac{1}{EI} + \frac{l}{2} \times \frac{1}{2EI} + \frac{l}{4} + \frac{1}{EI} = \frac{3l}{4EI}$$

and total pressure on the column,

$$P = 2\left[\left(\frac{1}{2} \times \frac{l}{4} \times \frac{Wl}{8} \times \frac{1}{EI}\right) + \frac{1}{2}\left(\frac{Wl}{4} + \frac{Wl}{8}\right) \times \frac{l}{4} \times \frac{1}{2EI}\right]$$

$$= 2\left[\frac{Wl^2}{64EI} + \frac{3Wl^2}{128EI}\right] = \frac{5Wl^2}{64EI}$$

Since the *c.g.* of the pressure diagram coincides with the centre of the beam, therefore there is no eccentricity of the loading. It is thus obvious, that the indeterminate moment at A,

$$M_{iA} = \frac{P}{A} = \frac{\dfrac{5Wl^2}{64EI}}{\dfrac{3l}{4EI}} = \frac{5Wl}{48}$$

From the geometry of the figure, we also find that the static moment at A,

$$M_{iA} = 0$$

Net moment at A,

$$M_A = 0 - \frac{5Wl}{48} = -\frac{5Wl}{48} \quad \textbf{Ans.}$$

Similarly $M_B = -\dfrac{5Wl}{48}$ **Ans.**

Example 16·2. *A beam AB of span l and fixed at its both ends carries a point load W at C, the centre of the span. Determine the fixed end moments, if the moment of inertia for AC is 2I and that for BC = I*

Solution.

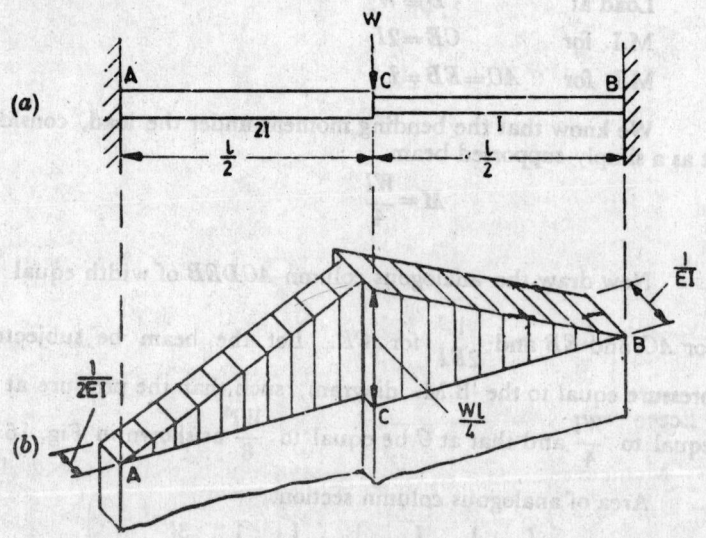

Fig. 16·7.

Given.   Span $= l$
Load at C $= W$
M.I. for AC $= 2I$
M.I. for CB $= I$

We know that the bending moment under the load, considering it as a simply supported beam,

$$M = \frac{Wl}{4}$$

Now draw the analogous column ACB of width equal to $1/2EI$ for AC and $1/EI$ for CB. Let the beam be subjected to pressure equal to the B.M. diagram, such that the pressure at C be equal to $\frac{Wl}{4}$ as shown in Fig. 16·7 (b).

Area of the analogous column section,

$$A = \frac{l}{2} \times \frac{1}{2EI} + \frac{l}{2} \times \frac{1}{EI} = \frac{3l}{4EI}$$

and total pressure on the column,

$$P = \left(\frac{1}{2} \times \frac{l}{2} \times \frac{Wl}{4} \times \frac{1}{2EI} \right) + \left(\frac{1}{2} \times \frac{l}{2} \times \frac{Wl}{4} \times \frac{1}{EI} \right)$$

$$= \frac{Wl^2}{32EI} + \frac{Wl^2}{16EI} = \frac{3Wl^2}{32EI}$$

Let $x =$ Distance of the c.g. of the analogous column section from A.

$h =$ Distance of the c.g. of the pressure from the fixed end A.

$$\therefore \quad \frac{3l}{4EI} \times x = \left(\frac{l}{2} \times \frac{1}{2EI} \times \frac{l}{4} \right) + \left(\frac{l}{2} \times \frac{1}{EI} \times \frac{3l}{4} \right) = \frac{7l^2}{16EI}$$

or $$x = \frac{7l^2}{16EI} \times \frac{4EI}{3l} = \frac{7l}{12} \qquad \text{...(i)}$$

and $$\frac{3Wl^2}{32EI} \times h = \frac{Wl^2}{32EI} \left(\frac{2}{3} \times \frac{l}{2} \right) + \frac{Wl^2}{16EI} \left(\frac{l}{2} + \frac{1}{3} \times \frac{l}{2} \right)$$

$$= \frac{Wl^3}{96EI} + \frac{Wl^3}{24EI} = \frac{5Wl^3}{96EI}$$

or $$h = \frac{5Wl^3}{96EI} \times \frac{32EI}{3Wl^2} = \frac{5l}{9} \qquad \text{...(ii)}$$

$\therefore$ Eccentricity, $\quad e = \frac{7l}{12} - \frac{5l}{9} = \frac{l}{36} \qquad \text{...(iii)}$

Moment of inertia of the analogous column section about AA,

$$I_{AA} = \frac{1}{3} \times \frac{1}{2EI} \left(\frac{l}{2} \right)^3 \times \left[\frac{1}{12} \times \frac{1}{EI} \times \left(\frac{l}{2} \right)^3 \right.$$
$$\left. + \frac{l}{2EI} \left(\frac{3l}{4} \right)^2 \right]$$

$$I_{AA} = \frac{l^3}{48EI} + \frac{l^3}{96EI} + \frac{9l^3}{96EI} = \frac{30l^3}{96EI} = \frac{5l^3}{16EI}$$

Therefore moment of inertia of the analogous column section about $y-y$ axis,

$$I_{YY} = I_{AA} - Ax^2 = \frac{5l^3}{16EI} - \frac{3l}{4EI} \times \left(\frac{7l}{12}\right)^2$$

$$= \frac{5l^3}{16EI} - \frac{147l^3}{576EI} = \frac{33l^3}{576EI} = \frac{11l^3}{192EI}$$

∴ Moment, $M' = P.e = \dfrac{3Wl^2}{32EI} \times \dfrac{l}{36} = \dfrac{Wl^3}{384EI}$

We know that the indeterminate moment at A,

$$M_{iA} = \frac{P}{A} + \frac{M'.x}{I} = \frac{\dfrac{3Wl^2}{32EI}}{\dfrac{3l}{4EI}} + \frac{\dfrac{Wl^3}{384EI} \times \dfrac{7l}{12}}{\dfrac{11l^3}{192EI}} \qquad \left(\because \quad x = \frac{7l}{12}\right)$$

$$= \frac{Wl}{8} + \frac{7Wl}{264} = \frac{40Wl}{264} = \frac{5Wl}{33}$$

Similarly indeterminate moment at B

$$M_{iB} = \frac{P}{A} - \frac{M'.x}{I} = \frac{\dfrac{3Wl^2}{32EI}}{\dfrac{3l}{4EI}} - \frac{\dfrac{Wl^3}{384EI} \times \dfrac{5l}{12}}{\dfrac{11l^3}{192EI}} \qquad \left(\because \quad x = \frac{5l}{12}\right)$$

$$= \frac{Wl}{8} - \frac{5Wl}{264} = \frac{28Wl}{264} = \frac{7Wl}{66}$$

From the geometry of the figure, we also find that the static moment at A and B is equal to zero.

∴ Net moment at A,

$$M_A = 0 - \frac{5Wl}{33} = -\frac{5Wl}{33} \quad \textbf{Ans.}$$

and $$M_B = 0 - \frac{7Wl}{66} = -\frac{7Wl}{66} \quad \textbf{Ans.}$$

Exercise 16·1

1. A fixed beam AB of 5 m span carries a point load of 2 tonnes at a distance of 2 m from A. Determine, by column analogy method, the values of fixing moments. (*Kerala University*)

[**Ans**. —1·44 t-m ; —0·96 t-m]

2. A beam ACB fixed at A and B is subjected to a load of 3 t at C. The lengths AC and CB are 2 m and 4 m respectively. Obtain from first principles, the fixing moment at A and B, if the moments of inertia for AC and CB are I and $2I$. (*Jiwaji University*)

[**Ans**. —2·27 t-m ; —1·62 t-m]

3. A fixed beam ACB of span 7·5 m is made up of $AC = 5$ m, $CB = 2·5$ m with moment of inertia for AC as twice that for CB. A point load of 20 tonnes acts at a distance of 3 m A. Find the fixing moments at A and B. (*Andhra University*)

[**Ans**. —23·23 t-m ; —10·7 t-m]

16·6. Portal frames

A simple portal frame consists of a beam resting over two columns (A portal frame may also consist of a beam resting over more than two supports or having more than one storey). The junction of the beams and colum is behaves like a rigid joint.

In general, the portal frames are of the following two types :

(1) Symmetrical portal frames, and
(2) Unsymmetrical portal frames.

16·7. Symmetrical portal frames

A symmetrical portal frame is that in which both the columns are of the same length, having similar end conditions (*i.e.* either hinged or fixed) moment of inertia, modulus of elasticity as well as subjected to symmetrical loading.

All the single bay and single storey symmetrical portal frames may be analysed as discussed below :

1. First of all, determine the total pressure (P) on the column as usual.

2. Then draw the analogous column and determine its area (A) as usual.

3. Now determine the *c.g.* of the analogous column and calculate I_{XX} and I_{YY}.

4. Then determine the moment due to the total pressure about the *c.g.* of the analogous column in X-X and Y-Y directions (*e.g.*, $M_x = P.\bar{x}$, where $\bar{x}$ is the distance between the *c.g.* of the pressure and the *c.g.* of the analogous column).

The indeterminate moment at any point may then be found out as the fibre stress, and by the relation,

$$M_i = f = \frac{P}{A} + \frac{M_X.y}{I_{XX}} + \frac{M_Y.x}{I_{YY}}$$

Notes : 1. The length of every member of the analogous column will be equal to the length of the corresponding member of the frame, and its width will be equal to $1/EI$.

2. Since the width of the column $1/EI$ is extremely small as compared to its length, therefore the values with power of EI (*i.e.* $I/(EI)^2$ or $1/(EI^3)$ and are generally neglected in calculations.

16·8. Portal frames with hinged legs

Sometimes, one or both the legs of a portal frame are hinged. Since a hinge does not offer any resistance to rotation, therefore the flexural rigidity *i.e.*, EI at the hinge is zero. It results in the following.

(1) As the width of the analogous column is $1/EI$, therefore its area corresponding to the hinge is infinite.

(2) If one of the legs of a portal frame is fixed and the other hinged, both the axes (*i.e.*, $x-x$ and $y-y$) will pass

through the hinge. Moreover, as the area of the analogous column is infinite, therefore it will be assumed to be concentrated at the hinge.

The indeterminate moment or the fibre stress may then be found out by the relation,

$$M_i = f = \frac{P}{\infty} + \frac{M_X \cdot y}{I_{XX}} + \frac{M_Y \cdot x}{I_{YY}}$$

$$= \frac{M_X \cdot y}{I_{XX}} + \frac{M_Y \cdot x}{I_{YY}}.$$

(3) If both the legs of the portal frame are hinged, the *c.g.* of analogous column section lies mid-way between the hinges. As the area A and moment of inertia about $y-y$ axis becomes infinite, the neutral axis of the column section will pass through the hinges. The indeterminate moment or the fibre stress may then be found out by the relation,

$$M_i = \frac{P}{\infty} + \frac{M_X \cdot y}{I_{XX}} + \frac{M_Y \cdot x}{\infty}$$

$$= \frac{M_X \cdot y}{I_{XX}}$$

Example 16·3. *A rectangular portal frame of uniform flexural rigidity EI carries a uniformly distributed load as shown in Fig. 16·8.*

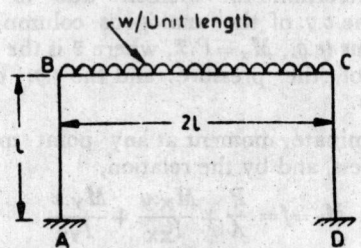

Fig. 16·8

Draw the bending moment diagram and sketch the deflected curve. Hence, or otherwise indicate with reason, whether you will provide greater flexural rigidity for the beam or the columns.

·Solution.

Given. Length of $AB = l$

Length of $\qquad BC = 2l$

Length of $\qquad CD = l$

Load on $\qquad BC = w$ per unit length

Flexural rigidity for AB,

$$EI_{AB} = EI_{BC} = EI_{CD}$$

——————————————————————————

·This example is also solved on pages 590

We know that the B.M. at the centre of the beam BC, considering it as a simply supported,

$$M = \frac{w(2l)^2}{8} = \frac{wl^2}{2}$$

Now draw an analogous column frame $ABCD$ with width equal to $1/EI$. Let the beam BC be subjected to pressure equal to the B.M. diagram, such that the pressure at the centre of the beam be $\frac{wl^2}{2}$ as shown in Fig. 16·9 (a).

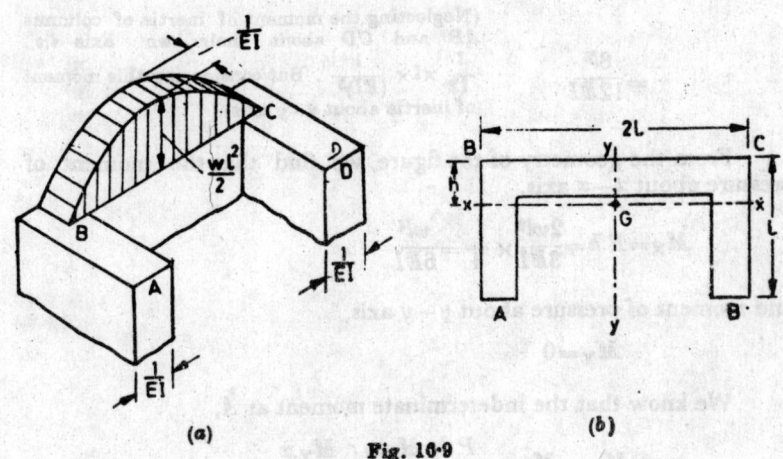

(a)

Fig. 16·9

(b)

Area of column, $A = \frac{1}{EI} \ (l + 2l + l) = \frac{4l}{EI}$

Total pressure on the column,

$$P = \frac{2}{3} \times M \times 2l \times \frac{1}{EI} = \frac{2}{3} \times \frac{wl^2}{2} \times 2l \times \frac{1}{EI} = \frac{2wl^3}{3EI}$$

Let $h = $ Distance between the $c.g.$ of the analogous column section and the extreme edge of the beam BC.

$$\therefore \quad h = \frac{\left(l \times \frac{1}{EI} \times \frac{l}{2} \right) + \left(2l \times \frac{1}{EI} \times \frac{1}{2EI} \right) + \left(l \times \frac{1}{EI} \times \frac{l}{2} \right)}{\frac{1}{EI} \ (l + 2l + l)}$$

$$= \frac{l}{4} \qquad\qquad \left(\text{Neglecting } \frac{1}{(EI)^2} \right)$$

Moment of inertia of the analogous column section about BC,

$$I_{BC} = 2 \left(\frac{1}{3} \times \frac{1}{EI} \times l^3 \right) \qquad \left(\text{Neglecting moment of inertia} \right.$$

$$= \frac{2l^3}{3EI} \qquad\qquad \left. \text{of } BC = \frac{1}{3} \times 2l \times \frac{1}{(EI)^3} \right)$$

and moment of inertia of the analogous column section about its $c.g.$, and parallel to $x-x$ axis,

$$I_{XX} = I_{BC} - Ah^2 = \frac{2l^3}{3EI} - \frac{4l}{EI} \times \left(\frac{l}{4}\right)^2$$

$$= \frac{5l^3}{12EI}$$

and

$$I_{YY} = \frac{1}{12} \times \frac{1}{EI} \times (2l)^3 + 0 + \left(2 \times l \times \frac{1}{EI}\right) \times l^2$$

$$= \frac{8l^3}{12EI}$$

(Neglecting the moment of inertia of columns AB and CD about their own axis i.e., $\frac{1}{12} \times l \times \frac{1}{(EI)^3}$. But considering this moment of inertia about $y-y$ axis).

From the geometry of the figure, we find that the moment of pressure about $x-x$ axis,

$$M_X = P.h = \frac{2wl^3}{3EI} \times \frac{l}{4} = \frac{wl^4}{6EI}$$

and moment of pressure about $y-y$ axis,

$$M_Y = 0$$

We know that the indeterminate moment at A,

$$M_{iA} = M_{iD} = \frac{P}{A} + \frac{M_X.y}{I_{XX}} + \frac{M_Y.x}{I_{YY}}$$

$$= \frac{\frac{2wl^3}{3EI}}{\frac{4l}{EI}} + \frac{\frac{wl^4}{6EI} \times \left(-\frac{3l}{4}\right)}{\frac{5l^3}{12EI}} + 0$$

$$\left(\because M_Y = 0 \text{ and } y = -\frac{3l}{4}\right)$$

$$= -\frac{2wl^3}{15}$$

Similarly

$$M_{iB} = M_{iC} = \frac{P}{A} + \frac{M_X.y}{I_{XX}} + \frac{M_Y.x}{I_{YY}}$$

$$= \frac{\frac{2wl^3}{3EI}}{\frac{4l}{EI}} + \frac{\frac{wl^4}{6EI} \times \frac{l}{4}}{\frac{5l^3}{12EI}} + 0 \quad \left(\because M_Y = 0 \text{ and } y = \frac{l}{4}\right)$$

$$= +\frac{4wl^3}{15}$$

From the geometry of the figure, we also find that the static moment at A, B, C and D is equal to zero.

$\therefore$ Net moment at A,

$$M_A = M_D = 0 - \left(-\frac{2wl^2}{15}\right) = +\frac{2wl^2}{15} = +0.13\ wl^2$$

Similarly

$$M_B = M_C = 0 - \frac{4wl^2}{15} = -\frac{4wl^2}{15} = -0.27\ wl^2$$

We know that the B.M. at the centre of the beam BC treating it as a simply supported beam,

$$M = \frac{wl^2}{2} = 0.5\ wl^2$$

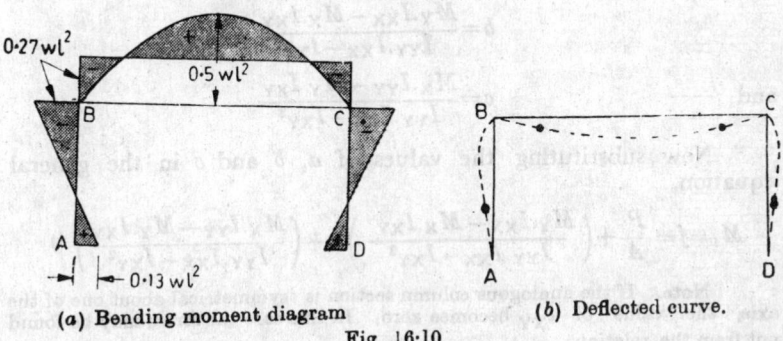

(a) Bending moment diagram   (b) Deflected curve.

Fig. 16·10

Now draw the B.M. diagram and the deflected curve as shown in Fig. 16·10 (a) and (b).

Since the B.M. is more at the ends B and C of the columns, therefore greater flexural rigidity is to be provided for the columns.

Ans.

16·9. Unsymmetrical portal frames

An unsymmetrical portal frame is that, in which either the columns are not symmetrical, or the frame is not symmetrically loaded. In such cases, the pressure of bending moment diagram will be eccentric about x-x axis and y-y axis, provided it is not hinged at both the ends.

We know that the stress at any point, having co-ordinates as x and y.

$$f = a + bx + cy$$

where a, b and c are constants.

We also know that the total force on the section,

$$P = \int f.dA = \int (a + bx + cy)dA$$
$$= \int a.dA + \int bx.dA + \int cy.dA$$

642   COLUMN ANALOGY METHOD

In case, the axes of reference pass through the centre of gravity of the section, then $\int x.dA$ and $\int y.dA$ are equal to zero. Then

$$P=\int a.dA=a.A \qquad \text{(where A is the total area)}$$

or
$$a=\frac{P}{A} \qquad (i)$$

and the moment of the total force about x-x axis,

$$
\begin{aligned}
M_X&=\int(f.dA)y=\int(a+bx+cy)y.dA\\
&=a\int y\,dA+b\int xydA+c\int y^2dA\\
&=0+bI_{XY}+cI_{XX} \qquad \ldots(ii)
\end{aligned}
$$

Similarly
$$
\begin{aligned}
M_Y&=\int (fdA)x=\int(a+bx+cy)x.dA\\
&=a\int xdA+b\int x^2dA+c\int xydA\\
&=0+bI_{YY}+cI_{XY} \qquad \ldots(iii)
\end{aligned}
$$

Solving equations (ii) and (iii), we get

$$b=\frac{M_Y.I_{XX}-M_X.I_{XY}}{I_{YY}.I_{XX}-I_{XY}^2}$$

and
$$c=\frac{M_X.I_{YY}-M_Y.I_{XY}}{I_{YY}.I_{XX}-I_{XY}^2}$$

Now substituting the values of a, b and c in the general equation,

$$M_i=f=\frac{P}{A}+\left(\frac{M_Y.I_{XX}-M_X.I_{XY}}{I_{YY}.I_{XX}-I_{XY}^2}\right)x+\left(\frac{M_X.I_{YY}-M_Y.I_{XY}}{I_{YY}.I_{XX}-I_{XY}^2}\right)y$$

Note. If the analogous column section is symmetrical about one of the axis, the value of I_{XY} becomes zero. In this case the stress may be found out from the relation.

$$f=\frac{P}{A}+\frac{M_Y.x}{I_{YY}}+\frac{M_X.y}{I_{XX}}$$

Example 16·4. *Using column analogy method, or otherwise, plot the bending moment diagram for the frame shown in Fig. 16·11.*

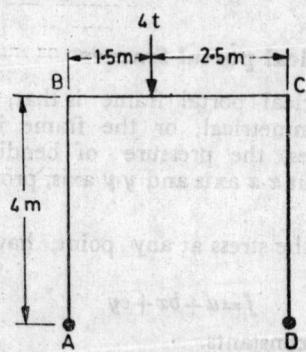

Fig. 16·11

Also draw the deflected shape of the frame.

*Solution.

Length of $AB=CD=4$ m
Length of $BC=4$ m
Load on BC, $W=4$ t

We know that the B.M. under the load, considering the beam BC as a simply supported,

$$M=\frac{Wab}{l}=\frac{4\times1\cdot5\times2\cdot5}{4}=3\cdot75 \text{ t-m}$$

Now draw the analogous column frame $ABCD$ with width equal to $1/EI$. Let the beam BC be subjected to pressure equal to the B.M. diagram, such that the pressure under the load be $3\cdot75$ t-m as shown in Fig. $16\cdot12$ (a).

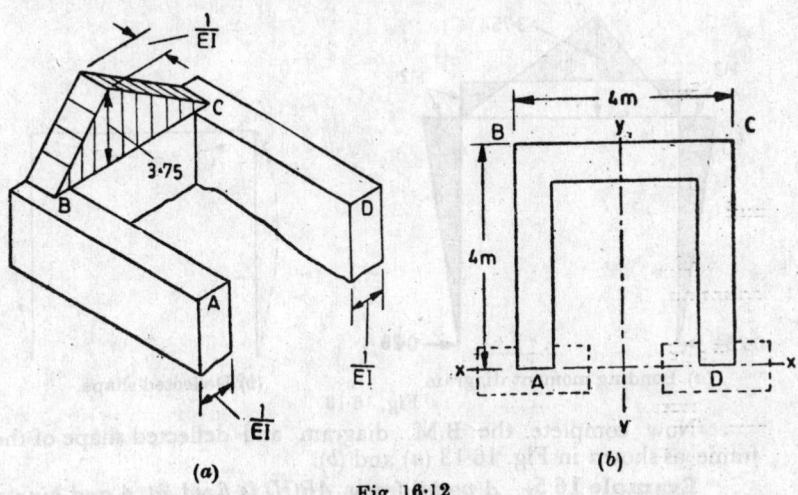

(a)   (b)

Fig. 16·12

Total pressure on the column,

$$P=\frac{1}{EI}\left(\frac{1}{2}\times3\cdot75\times4\right)=\frac{7\cdot5}{EI}$$

Since the ends A and B of the frame are hinged, therefore the analogous column will have infinite area at the hinges. The remaining area of the legs AB and CD as well as that of beam BC becomes negligible. Moreover the x-x axis will pass through the hinges and y-y axis will pass through the axis of symmetry as shown in Fig. $16\cdot12$ (b).

The moment of inertia of the analogous column section about x-x axis,

$$I_{\text{XX}}=I_{\text{AD}}=2\left(\frac{1}{3}\times\frac{1}{EI}\times4^{3}\right)+0+\left(\frac{1}{EI}\times4\times4^{2}\right)$$

(Neglecting the moment of inertia of the beam BC about its own axis $i.e.$ $\frac{1}{12}\times4\times\frac{1}{(EI)^{3}}$. But considering this moment of inertia about x-x axis).

$$=\frac{128}{3EI}+\frac{64}{EI}=\frac{320}{3EI}.$$

*This example is also solved on page 594

From the geometry of the figure, we find that the moment of pressure about x-x axis,

$$M_X = \frac{7.5}{EI} \times 4 = \frac{20}{EI}$$

We know that the indeterminate moment at B,

$$M_{iB} = M_{iC} = \frac{M_X \cdot y}{I_{XX}} = \frac{\dfrac{30}{EI} \times 4}{\dfrac{320}{3EI}} = 1.12 \text{ t-m}$$

From the geometry of the figure, we also find that the static moment at A, B, C and is equal to zero.

∴   Net moment at B, C $M_B = M_C = 0 - 1.12 = -1.12$ t-m

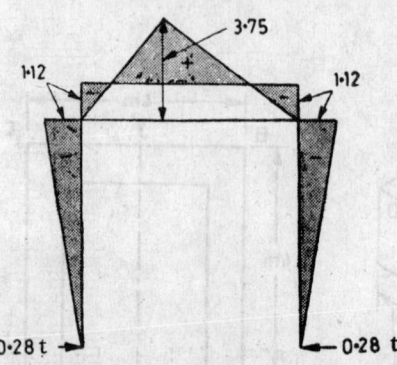
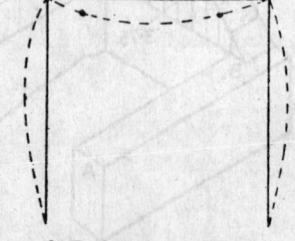

(a) Bending moment diagram   (b) Deflected shape

Fig. 16·13

Now complete the B.M. diagram and deflected shape of the frame as shown in Fig. 16·13 (a) and (b).

Example 16·5. *A portal frame ABCD is fixed at A and hinged at D. The frame is loaded as shown below :*

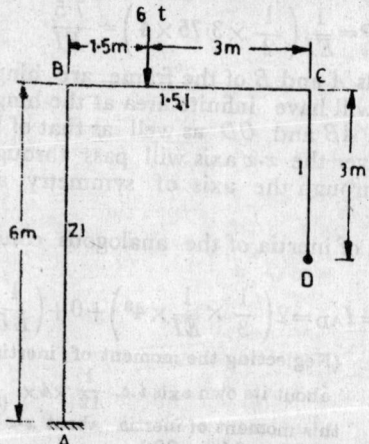

Fig. 16·14

Determine the moment at A, B, C and D by column analogy method and draw the bending moment diagram. (Cambridge University)

Solution

Given. Length of $AB = 6$ m

Length of $\quad BC = 1\cdot5 + 3 = 4\cdot5$ m

Length of $\quad CB = 3$ m

Load on $\quad BC = 6$ ᵗ

M.I. of $\quad AB = 2I$

M.I. of $\quad BC = 1\cdot5I$

M.I. of $\quad CD = I$

We know that the B.M. under the load, considering the beam BC as a simply supported,

$$M = \frac{Wab}{l} = \frac{6 \times 1\cdot5 \times 3}{4\cdot5} = 6\cdot0 \text{ t-m}$$

Now draw the analogous column frame $ABCD$ with width equal to $1/2EI$ for AB, $1/1\cdot5EI$ for BC and $1/EI$ for CD. Let the beam BC be subjected to pressure equal to the B.M. diagram, such that the pressure under the load be $6\cdot0$ t-m as shown in Fig. $16\cdot15$ (*a*).

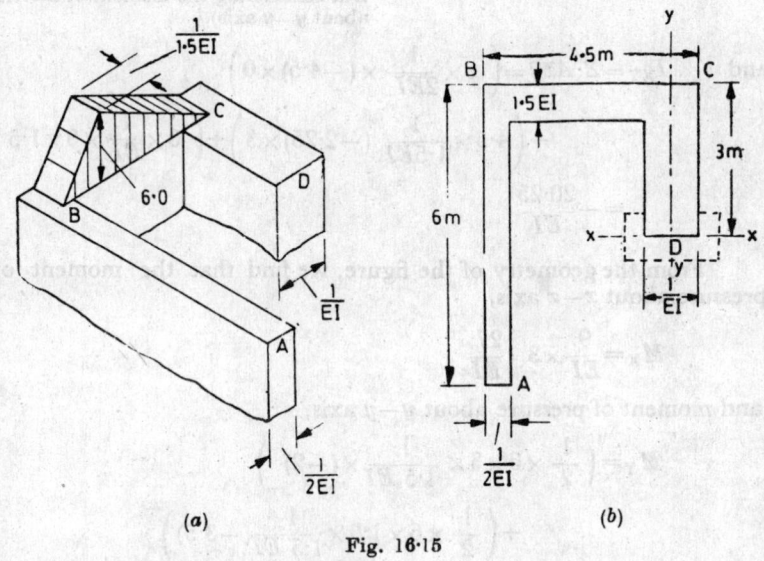

(*a*)　　　　　　　　　　(*b*)

Fig. 16·15

Total pressure on the column,

$$P = \frac{1}{1\cdot5EI}\left(\frac{1}{2} \times 4\cdot5 \times 6\right) = \frac{9}{EI}$$

Since the end D of the frame is hinged, therefore the analogous column will have an infinite area at the hinge D. The remaining area of the legs AB and CD as well as that of beam BC becomes negligible. Moreover the $x-x$ and $y-y$ axes will pass through the hinge D.

The moment of inertia of the analogous column section about $x-x$ axis,

$$I_{XX}=\left(\frac{1}{12}\times\frac{1}{2EI}\times6^3\right)+\left(\frac{1}{3}\times\frac{1}{EI}\times3^3\right)+0+\left(\frac{1}{1\cdot5EI}\times4\cdot5\times3^2\right)$$

$$=\frac{9}{EI}+\frac{9}{EI}+\frac{27}{EI}$$

$$=\frac{45}{EI}$$

(Neglecting the moment of inertia of the beam BC about its own axis, *i.e.* $\frac{1}{12}\times4\cdot5\times\frac{1}{(1\cdot5EI)^2}$. But considering this moment of inertia about $x-x$ axis)

Similarly

$$I_{YY}=\left(\frac{1}{3}\times\frac{1}{1\cdot5EI}\times4\cdot5^3\right)+0+\left(6\times\frac{1}{2EI}\times4\cdot5^2\right)+0$$

$$=\frac{20\cdot25}{EI}+\frac{60\cdot75}{EI}$$

$$=\frac{81}{EI}$$

(Neglecting the moment of inertia of leg CD i.e., $\frac{1}{12}\times3\times\frac{1}{(EI)^3}$. Also neglecting the moment of inertia of leg AB about its own axis i.e., $\frac{1}{12}\times6\times\frac{1}{(EI)^3}$. But considering the moment of inertia about $y-y$ axis).

and $$I_{XY}=\Sigma\ A\bar{x}\bar{y}=\left(6\times\frac{1}{2EI}\times(-4\cdot5)\times0\right)$$

$$+\left(4\cdot5\times\frac{1}{1\cdot5EI}(-2\cdot25)\times3\right)+\left(3\times\frac{1}{EI}\times0\times1\cdot5\right)$$

$$=-\frac{20\cdot25}{EI}$$

From the geometry of the figure, we find that the moment of pressure about $x-x$ axis,

$$M_X=\frac{9}{EI}\times3=\frac{27}{EI}$$

and moment of pressure about $y-y$ axis,

$$M_Y=\left(\frac{1}{2}\times6\times3\times\frac{1}{1\cdot5EI}\times(-2)\right)$$

$$+\left(\frac{1}{2}\times6\times1\cdot5\times\frac{1}{1\cdot5EI}(-3\cdot5)\right)$$

$$=-\frac{12}{EI}-\frac{10\cdot5}{EI}=-\frac{22\cdot5}{EI}$$

We know that

$$\frac{M_Y\cdot I_{XX}-M_X\cdot I_{XY}}{I_{YY}\cdot I_{XX}-I_{XY}{}^2}=\frac{\left(-\frac{22\cdot5}{EI}\times\frac{45}{EI}\right)-\left(\frac{27}{EI}\times-\frac{20\cdot25}{EI}\right)}{\frac{81}{EI}\times\frac{45}{EI}-\left(-\frac{20\cdot25}{EI}\right)^2}$$

$$=-0\cdot144$$

and $\dfrac{M_X.I_{YY}-M_Y.I_{XY}}{I_{YY}.I_{XX}-I_{XY}{}^2}$ $\dfrac{\left(\dfrac{27}{EI}\times\dfrac{81}{EI}\right)-\left(-\dfrac{22\cdot5}{EI}\times-\dfrac{20\cdot25}{EI}\right)}{\dfrac{81}{EI}\times\dfrac{45}{EI}-\left(-\dfrac{20\cdot25}{EI}\right)^2}$

$$=+0\cdot535$$

We also know that the stress (*i.e.*, indeterminate moment) at a point,

$$f=\frac{P}{\infty}+(-0\cdot144)\,x+(0\cdot535)\,y$$

$$=0+(-0\cdot144)\,x+(0\cdot535)\,y$$

∴ Indeterminate moment at A,

$$M_{iA}=(-0\cdot144)\times(-4\cdot5)+(0\cdot535)\times(-3\cdot0)\text{ t-m}$$

$$=-0\cdot96\text{ t-m}$$

Similarly,

$$M_{iB}=(-0\cdot144)\times(-4\cdot5)$$
$$+(0\cdot535)\times(3\cdot0)\text{ t-m}$$
$$=+2\cdot25\text{ t-m}$$

$$M_{iC}=(-0\cdot144)\times0$$
$$+(0\cdot535)\times3\cdot0\text{ t-m}$$
$$=+1\cdot60\text{ t-m}$$

and $\quad M_{iD}=0$

From the geometry of the figure, we also find that the static moment at A, B, C and D is zero.

∴ Net moment at A,

$$M_A=0-(-0\cdot96)$$
$$=+0\cdot96\text{t-m}$$
$$M_B=0-2\cdot25=-2\cdot25\text{ t-m}$$
$$M_C=0-1\cdot60=-1\cdot60\text{ t-m}$$
$$M_C=0$$

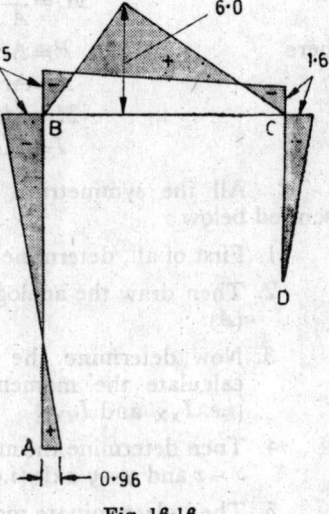

Fig. 16·16

Now complete the B.M. diagram as shown in Fig. 16·16.

Exercise 16·2

1. A portal frame $ABCD$ fixed at A and D is carrying a uniformly distributed load of 6 t/m over BC. The lengths of AB and CD is 2 m and that for BC is 4 m. Determine the moments at A, B, C and D, if the moment of inertia is constant throughout. *(Andhra University)*
[**Ans.** 3·9 t-m ; 6·39 t-m ; 6·39 t-m ; 3·9 t-m]

2. A portal frame $ABCD$ is fixed at A and D and has rigid joints at B and C. The columns AB and CD are 3 m and 2 m long respectively. The beam BC is 2 m long and is carrying uniformly distributed load of 6 t/m, Determine the moments and draw the B.M. diagram. *(Bombay University)*
[**Ans.** 0·88 t-m ; 1·60 t-m ; 1·19 t·m ; 0·49 t-m]

3. A portal frame $ABCD$ of span 4 m is subjected to a uniformly distributed load of 1 t/m over the member BC. The members AB and CD of the length 4 m and 2 m are hinged at A and D. Determine the moments at A B, C, and D and also draw the B.M. diagram.   (*Rajasthan University*)

[**Ans.** 0, 1·04 t-m ; 0·52 t-m ; 0]

HIGHLIGHTS

1. The analogous column is considered to be subjected to pressure, equal to the B.M. diagram. As a result of this the base of the column will be subjected to compressive stress. The values of M_A and M_B is then found out by the following considerations.

1. The total pressure at the base of the column is equal to the pressure at the top of the column.

2. The moment of total pressure at the base of the column about any support, is equal to the moment of the total pressure at the top of the column about the same support.

3. The indeterminate moment at a point,

$$M_i = \frac{P}{A} \pm \frac{M}{I}$$

where P = Area of total M_i diagram
                          A = Area of the analogous column
                          M = Moment due to eccentricity
                          I = Moment of inertia of the column section.

2. All the symmetrical portal frames may be analysed as discussed below :

1. First of all, determine the total pressure (P) on the column.

2. Then draw the analogous column and determine its area (A).

3. Now determine the c.g. of the analogous column and calculate the moment of inertia about $x-x$ and $y-y$ axis (*i.e.* I_{XX} and I_{YY}).

4. Then determine the moment due to the total pressure about $x-x$ and $y-y$ axis (*i.e.*, M_X and M_Y).

5. The indeterminate moment is then given by the relation,

$$M_i = \frac{P}{A} \pm \frac{M_X \cdot y}{I_{XX}} \pm \frac{M_Y \cdot x}{I_{YY}}$$

where M_X = Moment due to eccentricity about x-x axis.
                          I_{XX} = Moment of inertia about $x-x$ axis.
                          M_Y, I_{YY} = Corresponding values about $y-y$ axis.

3. A portal frame has zero flexural rigidity at its hinged ends, and results in the following :

1. The area at the hinged end of the analogous column is infinite.

2. If one of the legs of a portal frame is fixed and the other hinged, both the axes will pass through the hinge.

Moreover the area of the analogous column is assumed to be concentrated at the hinge.

3. If both the legs of a portal frame are hinged, the *c.g.* of the analogous column section lies mid-way between the hinges. The indeterminate moment at point is given by

$$M_i = \frac{M_X \cdot y}{I_{XX}}$$

4. The indeterminate moment at a point in an unsymmetrical portal frame is given by,

$$M_i = \frac{P}{A} + \left(\frac{M_Y \cdot I_{XX} - M_X \cdot I_{YY}}{I_{YY} \cdot I_{XX} - I_{XY}{}^2} \right) x + \left(\frac{M_X \cdot I_{YY} - M_Y \cdot I_{XY}}{I_{YY} \cdot I_{XX} - I_{XY}{}^2} \right) y$$

where I_{XY} = sum of the moments of the total area of the analogous column section about *c.g.*

Do You Know ?

1. What is column analogy method ? How does it differ from moment distribution method.

2. Explain the theory of column analogy.

3. Obtain a relation for the fixing moment of a fixed beam, when it is subjected to an eccentric point load.

4. Give the values of fixing moment, when a fixed beam of length l carries a gradually varying load from zero at one end to w per unit length at the other.

5. Explain the important point, which is adopted, at the hinged end of a portal frame. Explain your point.

6. Give the procedure adopted for an unsymmetrical portal frame.

17

TWO-HINGED ARCHES

17·1. Introduction

A two-hinged arch, as the name indicates, is an arch hinged on
its two supports only. Like a three-hinged arch, the reactions at the
supports of a two-hinged arch, will consist of a vertical component
V_A and V_B as well as horizontal component H. The vertical components of the support reactions may be easily found out by equating
the clockwise and anticlockwise moments of all the forces, on the
arch about any one hinge.

As a matter of fact, a two-hinged arch is a statically indeterminate structure, since the horizontal thrust can not be found out by
the equations of statical equilibrium.

17·2. Horizontal thrust in two-hinged arches

We have already discussed in Art. 17·1 that a two-hinged arch
is an indeterminate structure. Therefore the horizontal thrust in a
two-hinged arch is found out on the following assumptions.

(1) Before loading, the arch is free from stresses.

(2) The supports, A and B, are rigidly hinged.

(3) After loading, the horizontal span AB of the arch remains unchanged.

The horizontal thrust in a two-hinged arch may be found out either by the strain energy method or by the flexural deformation.

17·3. Horizontal thrust by strain energy method

Consider an arch ACB, hinged, at its two supports A and B,

Let $l =$ Span of the arch,

$y_C =$ Central rise of the arch, and

$H =$ Horizontal thrust in the arch,

We know that the bending moment at any section X of the arch,

$$M = \mu_X - H.y \qquad \qquad ...(i)$$

where μ_X is the B.M. at X, by considering the arch as a simply supported beam of span l, and $H.y$ is the B M. at X due to the horizontal thrust M.

Now consider a small element of the arch. We also know that the total strain energy stored in the element,

$$\frac{dU}{dH} = \frac{M^2 ds}{2EI}$$

and the total strain energy in the arch,

$$U = \int\limits_0^l \frac{M^2 ds}{2EI}$$

$$= \int\limits_0^l \frac{(\mu_X - H.y)^2 ds}{2EI} \qquad (\because \quad M = \mu_X - H.y)$$

$$= \int\limits_0^l \frac{2(\mu_X - H.y)(-y)ds}{2EI}$$

$$= - \int\limits_0^l \frac{\mu_X.y\,ds}{EI} + H\int\limits_0^l \frac{y^2 ds}{EI}$$

Since, no strain energy can be stored in a two-hinged arch, therefore equating the above equation to zero,

$$H \int\limits_0^l \frac{y^2 ds}{EI} - \int\limits_0^l \frac{\mu_X.y\, ds}{EI} = 0$$

$$\therefore \qquad H \int\limits_0^l \frac{y^2 ds}{EI} = \int\limits_0^l \frac{\mu_X.y ds}{EI}$$

or
$$H = \dfrac{\displaystyle\int_0^l \dfrac{\mu x . y ds}{El}}{\displaystyle\int_0^l \dfrac{y^2 . ds}{EI}}$$

17·4. Horizontal thrust by flexural deformation

Consider an arch rib ACB subjected to some bending moment as shown in Fig. 17·1.

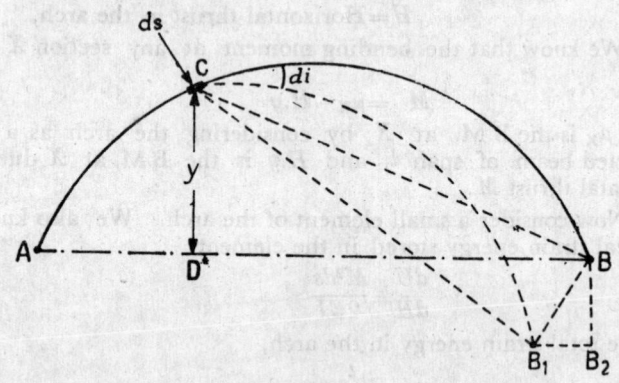

Fig. 17·1

Now consider a small element ds of the arch at C. Let this arch element ds be subjected to bending moment M. As a result of the B M., let the arch rib be bent through a small angle di and the support B occupy a new position B_1 as shown in Fig. 17·1. Through B draw a vertical BB_2 and through B_1 draw B_1B_2 a perpendicular to it. Sin~e the angle di, through which portion of arch CB has rotated is very small, therefore length of the chord CB may be taken to be equal to CB_1.

Now, a little consideration will show that the vertical displacement of B is BB_2, whereas the horizontal displacement of B is B_1B_2. We see that in triangle BB_1B_2

$$B_1B_2 = BB_1 \cos \angle BB_1B_2 = CB . di . \cos \angle BB_1B_2$$
$$(\because \ BB_1 = CB.di)$$
$$= di.CB \cos \angle BCD \quad (\because \ \angle BB_1B_2 = \angle BCD)$$
$$= CD.di \quad (CB \cos \angle BCD = CD)$$
$$= y.di \quad (CD = y = \text{rise at } D)$$

We know that in a curved rib,

$$\frac{M}{I} = \frac{E}{R} = \frac{E}{\dfrac{ds}{di}} = \frac{E.di}{ds}$$

$$\therefore \qquad\qquad di = \frac{M.ds}{EI}$$

Substituting the value of di in the equation for horizontal displacement.

$$B_1B_2 = y.di = y.\frac{Mds}{EI} = \frac{Myds}{EI}$$

Therefore the total horizontal displacement of B due to B.M. on the arch rib from A to B

$$= \sum \frac{My.ds}{EI}$$

Since the arch is rigidly hinged at the support B, therefore, the support B cannot undergo any horizontal displacement,

$$\therefore \qquad \sum \frac{My.ds}{EI} = 0$$

or

$$\sum \frac{(\mu_x - H.y)yds}{EI} = 0 \qquad (\because M = \mu_x - H.y)$$

$$\int_0^l \frac{\mu_x.yds}{EI} - H\int_0^l \frac{y^2ds}{EI} = 0$$

or

$$H\int_0^l \frac{y^2ds}{EI} = \int_0^l \frac{\mu_x.yds}{EI}$$

$$H = \frac{\displaystyle\int_0^l \frac{\mu_x.y.ds}{EI}}{\displaystyle\int_0^l \frac{y^2ds}{EI}}$$

The quantity EI is called flexural rigidity of the arch. Sometimes the moment of inertia is not constant throughout the arch. If the value of moment of inertia at the supports is I, then the moment of inertia at the crown I_0 is generally made to have a relation with the moment of inertia at the supports, such that

$$I = I_0 \sec \theta$$

where θ is the angle, which the tangent at a section makes with the horizontal. Now taking $ds = dx \cos \theta$ and substituting the values of I and dx in equation (i),

$$H = \frac{\displaystyle\int_0^l \frac{\mu_x.y.dx \sec \theta}{EI_0 \sec \theta}}{\displaystyle\int_0^l \frac{y^2dx \sec \theta}{EI_0 \sec \theta}} = \frac{\displaystyle\int_0^l \frac{\mu_x.y.dx}{EI}}{\displaystyle\int_0^l \frac{y^2 dx}{EI}}$$

$$H = \cfrac{\displaystyle\int_0^l \mu_X . y \, dx}{\displaystyle\int_0^l y^2 \, dx}$$

A little consideration will show that the values of part $\int Mx \, . \, y \, . \, dx$ of the equation depends upon the property and loading of the arch, whereas the value of part $\int y^2 \, dx$ of the equation depends upon the property of the arch alone.

17·5. Types of two-hinged arches

The two-hinged arches, like three-hinged arches, may also be of the following two types, depending upen the geometry of their axes :

   1.   Parabolic arches, and

   2.   Circular arches.

As a matter of fact, the determination of horizontal thrust is an important step in analysing a two-hinged arch. Here we shall derive an expression for the horizontal thrust in the above two types of two-hinged arches.

17·6. Horizontal thrust in a two-hinged parabolic arch carrying a concentrated load

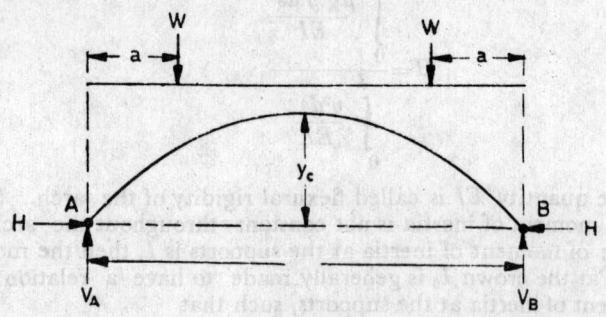

Fig. 17·2

A two-hinged arch, having parabolic axis, is known as a *two-hinged parabolic arch*. Consider a two-hinged parabolic arch AB, supported at A and B, and carrying a point load W at a distance a from the support A as shown in Fig. 17·2.

Let $l =$ Span AB of the arch, and

                            $y_C =$ Rise of theorem from the support.

For simplification, let us assume another load W to act at a distance a from the support B as shown in Fig. 17·2. A little consideration will show, that as a result of this assumed load the horizontal thrust at A and B will be twice the horizontal thrust due to the given load only. Now, the vertical reactions V_A and V_B will be both equal to W. Consider any section X at a distance x from A. We find that the beam moment, when $x = 0$ to $x = a$

$$\mu_{X1} = W \cdot x \qquad\qquad \ldots(i)$$

and the beam moment, when $x=a$ to $x=l/2$

$$\mu_{X2} = Wx - W(x-a) = Wa \qquad\qquad \ldots(ii)$$

We know that

$$\int_0^l \mu_X \cdot y \cdot dx = 2\int_0^{l/2} \mu_X \cdot y \cdot dx$$

$$= 2\int_0^a \mu_{X1} \cdot y \, dx + 2\int_0^{l/2} \mu_{X2} \cdot y \, dx$$

$$= 2\left[\int_0^a \mu_{X1}\, y \cdot dx + \int_a^{l/2} \mu_{X2} \cdot y.dx\right]$$

$$= 2\left[\int_0^a Wx \cdot \frac{4y_C}{l^2}x(l-x)dx\right.$$

$$\left. + \int_0^{l/2} Wa \cdot \frac{4y_C}{l^2}x(l-x)dx\right]$$

$$= 2\left[\frac{4Wy_C}{l^2}\int_0^a (lx^2-x^3)dx + \frac{4Wy_Ca}{l^2}\int_a^{l/2}(lx-x^2)dx\right]$$

$$= \frac{8Wy_C}{l^2}\left[\left(\frac{lx^3}{3}-\frac{x^4}{4}\right)_0^a + a\left(\frac{lx^2}{2}-\frac{x^3}{3}\right)_a^{l/2}\right]$$

$$= \frac{8Wy_C}{l^2}\left\{\frac{la^2}{3}-\frac{a^4}{4}+a\left\{\frac{l}{2}\left[\left(\frac{l}{2}\right)^2-a^2\right]-\frac{1}{3}\left[\left(\frac{l}{2}\right)^3-a^3\right]\right\}\right\}$$

$$= \frac{8Wy_Ca}{l^2}\left[\frac{la^2}{3}-\frac{a^3}{4}+\frac{l^3}{8}-\frac{la^2}{2}+\frac{a^3}{3}\right]$$

$$= \frac{8Wy_Ca}{24l^2}[8la^2-6a^3+3l^3-12la^2-l^3+8a^3]$$

$$= \frac{Wy_Ca}{3l^2}(2l^3-4la^2+2a^3) = \frac{2Wy_Ca}{3l^2}(l^3-2la^2+a^3)$$

$$= \frac{2Wy_Ca}{3l^2}(l^3-la^2-la^2+a^3) = \frac{2Wy_Ca}{3l^2}\left[l(l^2-a^2)-a^2(l-a)\right]$$

$$= \frac{2Wy_Ca}{3l^2}(l-a)\left[l(l+a)-a^2\right] = \frac{2Wy_Ca}{3l^2}(l-a)(l^2+la-a^2) \quad \ldots(iii)$$

Now $\quad\displaystyle\int_0^l y^2 dx = 2\int_0^{l/2} y^2 dx = 2\int_0^{l/2}\left(\frac{4y_C}{l^2}x(l-x)\right)^2 dx$

or $\qquad\qquad = 2\left[\frac{16 y_C^2}{l^4}\int_0^{l/2} x^2(l^2+x^2-2lx)dx\right]$

$\qquad\qquad = \frac{32 y_C^2}{l^4}\int_0^{l/2}(l^2 x^2 + x^4 - 2lx^3)dx$

$\qquad\qquad = \frac{32 y_C^2}{l^4}\left[\frac{l^2 x^3}{3} + \frac{x^5}{5} - \frac{lx^4}{2}\right]_0^{l/2}$

$\qquad\qquad = \frac{32 y_C^2}{l^4}\left[\frac{l^2}{3}\left(\frac{l}{2}\right)^3 + \frac{1}{5}\left(\frac{l}{2}\right)^5 - \frac{l}{2}\left(\frac{l}{2}\right)^4\right]$

$\qquad\qquad = \frac{32 y_C^2}{l^4}\left[\frac{l^5}{24} + \frac{l^5}{5\times 32} - \frac{l^5}{32}\right]$

$\qquad\qquad = \frac{8}{15}y_C^2 l \qquad\qquad\qquad\qquad\qquad\qquad \dots(iv)$

Therefore twice the horizontal thrust in the arch,

$$2H = \frac{\displaystyle\int_0^l \mu_x.y.dx}{\displaystyle\int_0^l y^2.dx} = \frac{\dfrac{2W y_C a}{3l^2}(l-a)(l^2+la-a^2)}{\dfrac{8}{15}y_C^2 l}$$

$$\qquad\qquad = \frac{5}{4}\times\frac{Wa}{y_C l^3}(l-a)(l^2+al-a^2)$$

or $\qquad\qquad H = \frac{5}{8}\times\frac{Wa}{y_C l^3}(l-a)(l^2+al-a^2)$

Note : 1. If the load is at a distance x from A, the horizontal thrust,

$$H = \frac{5}{8}\times\frac{Wx}{y_C l^3}(l-x)(l^2+lx-x^2)$$

2. If $\quad x=kl$, then

$$H = \frac{5}{8}\times\frac{Wkl}{y_C l^3}(l-kl)(l^2+kl-k^2 l^2)$$

$$\qquad\quad = \frac{5}{8}\times\frac{Wl}{y_C}(k^4-2k^3+k)$$

Example 17·1. *A two-hinged parabolic arch of span 36 m and central rise 6 m, has the moment of inertia varying as the secant of slope of rib axis. Find the horizontal thrust on the arch, if a point load of 10 T acts at a distance of 9 m from the left end support.*

(*A.M.I.E., Winter, 1981*)

Solution.

Given. Span, $l = 36$ m

Central rise, $y_C = 6$ m

Point load, $W = 10$ t

Horizontal distance between the load and the left end support,

$$x = 9 \text{ m}$$

Let $H =$ Horizontal thrust on the arch.

Using the relation,

$$H = \frac{5}{8} \times \frac{Wx}{y_C l^3}(l-x)(l^2 + lx - x^2) \text{ with usual notations.}$$

$$= \frac{5}{8} \times \frac{10 \times 9}{6 \times 36^3}(36-9)(36^2 + 36 \times 9 - 9^2) \text{ t}$$

$$= \frac{5 \times 10 \times 9 \times 27 \times 1{,}449}{8 \times 6 \times 36 \times 36 \times 36} = \frac{25{,}650}{3{,}072} \text{ t}$$

$$= \mathbf{8 \cdot 35 \text{ t}} \quad \mathbf{Ans.}$$

Example 17·2. *A two-hinged parabolic arch of span 12 m and central rise 2·4 m has secant variation for the moment of inertia of the rib and is loaded as shown in Fig. 17·3.*

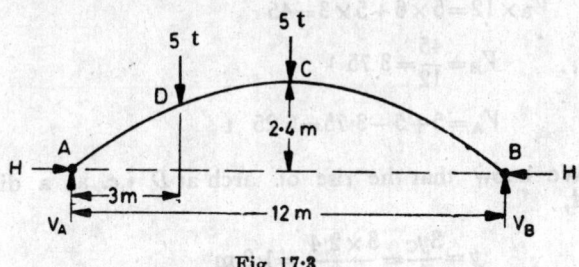

Fig. 17·3

Find the horizontal thrust on the arch and bending moment at D.

Solution.

Given. Span, $l = 12$ m

Central rise, $y_C = 2·4$ m

Horizontal thrust

Let $H =$ Horizontal thrust on the arch.

For simplification, let us find out the values of horizontal thrust on the arch, due to the two loads separately, and then the total horizontal thrust will be given by the sum of the two horizontal thrusts.

Let $H_1 =$ Horizontal thrust due to the load at D, and

$H_2 =$ Horizontal thrust due to the load C.

Using the relation,

$$H = \frac{5}{8} \times \frac{Wx}{y_C l^3} \ (l-x)(l^2+lx-x^2) \text{ with usual notations.}$$

$$H_1 = \frac{5}{8} \times \frac{5 \times 3}{2 \cdot 4 \times 12^3} \ (12-3)(12^2+12 \times 3-3^2) \text{ t}$$

$$= \frac{5 \times 5 \times 3 \times 9 \times 171}{8 \times 2 \cdot 4 \times 12 \times 12 \times 12} = \frac{7,125}{2,048} = 3 \cdot 48 \text{ t}$$

Similarly $\quad H_2 = \frac{5}{8} \times \frac{5 \times 6}{2 \cdot 4 \times 12^3} \ (12-6)(12^2+12 \times 6 \times 6^2) \text{ t}$

$$= \frac{5 \times 5 \times 6 \times 6 \times 180}{8 \times 2 \cdot 4 \times 12 \times 12 \times 12} = 4 \cdot 88 \text{ t}$$

∴ $H = H_1 + H_2 = 3 \cdot 48 + 4 \cdot 88 = \textbf{8} \cdot \textbf{36} \text{ t} \quad \textbf{Ans.}$

Bending moment at D

Let V_A = Vertical reaction at A, and

$\quad\quad V_B$ = Vertical reaction at B.

Taking moments about A,

$$V_B \times 12 = 5 \times 6 + 5 \times 3 = 45$$

∴ $V_B = \frac{45}{12} = 3 \cdot 75 \text{ t}$

and $V_A = 5 + 5 - 3 \cdot 75 = 6 \cdot 25 \text{ t}$

We know that the rise of arch at D *i.e.*, at a distance $l/4$ from A,

$$y = \frac{3y_C}{4} = \frac{3 \times 2 \cdot 4}{4} = 1 \cdot 8 \text{ m}$$

∴ B.M. at D,
$$M_D = V_A \times 3 - H \times 1 \cdot 8 = 6 \cdot 25 \times 3 - 8 \cdot 36 \times 1 \cdot 8 \text{ t-m}$$
$$= 3 \cdot 7 \text{ t-m} \quad \textbf{Ans.}$$

17·7. Horizontal thrust in a two-hinged circular arch carrying a concentrated load

A two-hinged arch, having circular axis, is known as a *two-hinged circular arch*. Consider a two-hinged circular arch AB supported at A and B, and carrying a point load W at a distance a from the support A as shown in Fig. 17·4.

Let l = Span AB of the arch,

$\quad\quad\quad y_C$ = Rise of the crown from the supports, and

$\quad\quad\quad R$ = Radius of the arch.

For simplification, let us assume another load W to act at a distance a from the support B as shown in Fig. 17·4. A little considera-

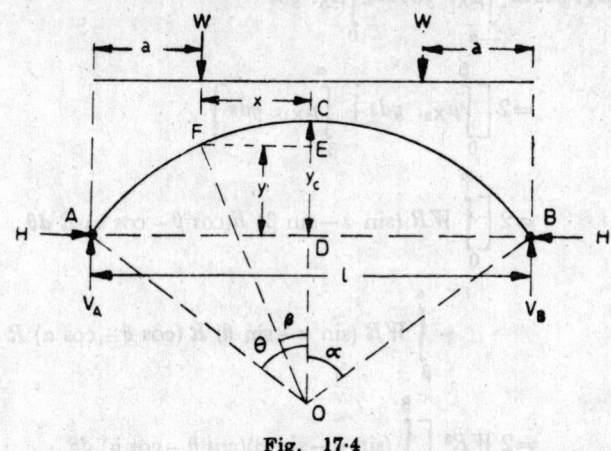

Fig. 17·4

tion will show, that as a result of this assumed load, the horizontal thrust at A and B will be twice the horizontal thrust due to the given load only, and the vertical reactions V_A and V_B will be both equal to W. Now let us consider any section X, at a distance x from C. We find that the beam moment when $x=0$ to $x=a$

$$\mu_{X1}=W.x \qquad\qquad ...(i)$$

and the beam moment, when $x=a$ to $x=\dfrac{l}{2}$

$$\mu_{X2}=W.a-W(x-a)=W.a \qquad\qquad ...(ii)$$

Taking OC as the axis of reference, let the angle subtended at the centre, by the arch $BC=\alpha$ and angle subtended by the arc $FC=\beta$ as shown in Fig. 17·4.

From the geometry of the circular arch, we find

$$x=R(\sin\alpha-\sin\theta)$$
$$a=R(\sin\alpha\cdot\sin\beta)$$
$$y=R(\cos\theta-\cos\alpha)$$

and $$ds=R\,d\theta$$

where ds is the length of a small element of the arch and $d\theta$ is the angle subtended by ds at the centre of the arch.

Substituting these values in equations (i) and (ii),

$$\mu_{X1}=W.x=W.R(\sin\alpha-\sin\theta)$$

and $$\mu_{X2}=W.a=WR(\sin\alpha-\sin\beta)$$

We know that

$$\int_0^l \mu_{\text{X}} . \; yds = \int_{-\alpha}^{+\alpha} \mu_{\text{X}} . \; yds = 2\int_0^\alpha \mu_{\text{X}} . \; yds$$

$$= 2\left[\int_0^\beta \mu_{\text{X}_2} . \; yds + \int_\beta^\alpha \mu_{\text{X}_1} . \; yds\right]$$

$$= 2\left[\int_0^\beta WR\,(\sin\,\alpha - \sin\,\beta)\; R(\cos\,\theta - \cos\,\alpha)\; R\; d\theta \right.$$

$$\left. + \int_\beta^\alpha WR\,(\sin\,\alpha - \sin\,\theta)\; R\,(\cos\,\theta - \cos\,\alpha)\; R\; d\theta\right]$$

$$= 2\,WR^3\left[\int_0^\beta (\sin\,\alpha - \sin\,\beta)(\cos\,\theta - \cos\,\alpha)\; d\theta \right.$$

$$\left. + \int_\beta^\alpha (\sin\,\alpha - \sin\,\theta)\,(\cos\,\theta - \cos\,\alpha)\; d\theta\right]$$

on simplification we get,

$$\int_0^l \mu_{\text{X}} . \; y\; ds = 2WR^3\left[\; \sin^2\,\alpha - \sin^2\,\beta - 2\cos\,\alpha\,(\cos\,\alpha - \cos\,\beta \right.$$

$$\left. + \alpha\,\sin\,\alpha - \beta\,\sin\,\beta)\;\right]$$

Now $$\int_0^l y^2\; ds = \int_{-\alpha}^{+\alpha} R^2\,(\cos\,\theta - \cos\,\alpha)^2\; R\; d\theta = 2\int_0^\alpha R^2\,(\cos\,\theta - \cos\,\alpha)^2\; Rd\theta$$

$$= 2\,R^3 \int_0^\alpha (\cos^2\,\theta + \cos^2\,\alpha - 2\cos\,\theta\,\cos\,\alpha)\; d\theta$$

$$= R^3 \int_0^\alpha (2\cos^2\,\theta + 2\cos^2\,\alpha - 4\cos\,\theta\,\cos\,\alpha)\; d\theta$$

$$= R^3 \int_0^\alpha \left[(1 + \cos\,2\theta) + 2\cos^2\,\alpha - 4\cos\,\theta\,\cos\,\alpha\right]\; d\theta$$

$$= R^3 \left[\theta + \frac{\sin 2\theta}{2} + 2 \cos^2 \alpha \theta - 4 \sin \theta \cos \theta \right]_0^\alpha$$

on simplification we get,

$$\int_0^l y^2 \, ds = R^3 (4\alpha \cos^2 \alpha + 2\alpha - 3 \sin 2\alpha)$$

Therefore twice the horizontal thrust in the arch,

$$2H = \frac{\displaystyle\int_0^l \mu_X . y \, ds}{\displaystyle\int_0^l y^2 \, ds}$$

$$= \frac{2 \, WR^3 [\sin^2 \alpha - \sin^2 \beta - 2 \cos \alpha \, (\cos \alpha - \cos \beta + \alpha \sin \alpha - \beta \sin \beta)]}{R^3 \, (4\alpha \cos^2 \alpha + 2\alpha - 3 \sin 2\alpha)}$$

$$\therefore \quad H = \frac{W \, [\sin^2 \alpha - \sin^2 \beta - 2 \cos \alpha \, (\cos \alpha - \cos \beta + \alpha \sin \alpha - \beta \sin \beta)]}{(4\alpha \cos^2 \alpha + 2\alpha - 3 \sin 2\alpha)}$$

Cor. 1. If the arch is a semicircular, then substituting $\alpha = \frac{\pi}{2}$ in the above equation, we get

$$H = \frac{W \cos^2 \beta}{\pi}$$

2. If the load is acting and the crown of the semicircular arch,

$$H = \frac{W}{\pi}$$

Example 17·3. *A two-hinged semicircular arch of the span 20 m is loaded as shown in Fig. 17·5.*

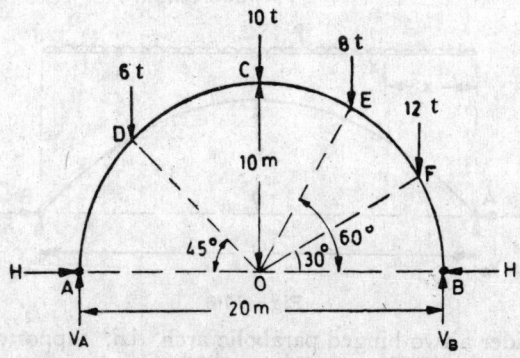

Fig. 17·5

Find the horizontal thrust on the arch.

Solution.

Given. Span, $l = 20$ m
Central rise. $y_C = 10$ m
Let $H_1 =$ Horizontal thrust due to load at D,
                     $H_2 =$ Horizontal thrust due to load at C,
                     $H_3 =$ Horizontal thrust due to load at E, and
                     $H_4 =$ Horizontal thrust due to load at F.

Using the relation,

$$H = \frac{W \cos^2 \beta *}{\pi} \text{with usual notations.}$$

$\therefore \qquad H_1 = \dfrac{6 \cos^2 45°}{\pi} = \dfrac{6\left(\dfrac{1}{\sqrt{2}}\right)^2}{\pi} = \dfrac{3}{\pi} \qquad \qquad \dots(i)$

Similarly $\qquad H_2 = \dfrac{10 \cos^2 90°}{\pi} = \dfrac{10}{\pi} \qquad \qquad \dots(ii)$

and $\qquad H_3 = \dfrac{8 \cos^2 30°}{\pi} = \dfrac{8\left(\dfrac{\sqrt{3}}{2}\right)^2}{\pi} = \dfrac{6}{\pi} \qquad \dots(iii)$

and $\qquad H_4 = \dfrac{12 \cos^2 60°}{\pi} = \dfrac{12\left(\dfrac{1}{2}\right)^2}{\pi} = \dfrac{3}{\pi} \qquad \dots(iv)$

Total horizontal thrust

$$H = H_1 + H_2 + H_3 + H_4$$
$$= \frac{3}{\pi} + \frac{10}{\pi} + \frac{6}{\pi} + \frac{3}{\pi} = \frac{22}{\pi} = 7 \text{ t} \quad \textbf{Ans.}$$

17·8. Horizontal thrust in a two-hinged parabolic arch carrying a uniformly distributed load over the entire span

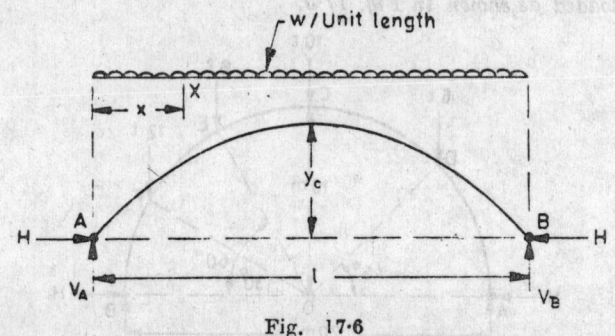

Fig. 17·6

Consider a two-hinged parabolic arch AB, supported at A and B, and carrying a uniformly distributed load.

---

*It may be remembered that β is the angle, which the line OD makes with the line OC.

Let $\qquad$ $l=$Span AB of the arch

$\qquad$ $y_C=$Rise of the crown from the supports, and

$\qquad$ $w=$Uniformly distributed load per horizontal unit length.

We know that the vertical reactions at the two supports,

$$V_A=V_B=\frac{wl}{2}$$

Now consider any section X, at a distance x from A, the beam moment at X,

$$\mu_X=\frac{wl}{2}\cdot x-\frac{wx^2}{2}=\frac{wx}{2}(l-x)$$

We know that

$$\int_0^l \mu_X.y\,dx=\int_0^l\frac{wx}{2}(l-x).\frac{4y_C}{l^2}x(l-x)\,dx$$

$$=\frac{2wy_C}{l^2}\int_0^l x^2(l-x^2)\,dx$$

$$=\frac{2wy_C}{l^2}\int_0^l l^2x^2+x^4-2lx^3\,dx$$

$$=\frac{2wy_C}{l^2}\left[\frac{l^2x^3}{3}+\frac{x^5}{5}-\frac{lx^4}{2}\right]_0^l$$

$$=\frac{2wy_C}{l^2}\left[\frac{l^5}{3}+\frac{l^5}{5}-\frac{l^5}{2}\right]$$

$$=\frac{wy_Cl^3}{15} \qquad\qquad (i)$$

Now $\quad\displaystyle\int_0^l y^2\,dx=\int_0^l\left[\frac{4y_C}{l^2}x(l-x)\right]^2 dx=\int_0^l\frac{16y_C^2}{l^4}x^2(l-x)^2\,dx$

$$=\frac{16y_C^2}{l^4}\int_0^l l^2x^2+x^4-2lx^3\,dx$$

$$=\frac{16y_C^2}{l^4}\left[\frac{l^2x^3}{3}+\frac{x^5}{5}-\frac{lx^4}{2}\right]_0^l$$

$$= \frac{16y_C^2}{l^4}\left(\frac{l^5}{3} + \frac{l^5}{5} - \frac{l^5}{2}\right) = \frac{16y_C^2 l}{30}$$

$$= \frac{8y_C^2 l}{15} \qquad\qquad\qquad ...(ii)$$

$\therefore$ Horizontal thrust in the arch,

$$H = \frac{\displaystyle\int_0^l \mu_x \cdot y \cdot dx}{\displaystyle\int_0^l y^2 \cdot dx} = \frac{\dfrac{w \cdot y_C \cdot l^3}{15}}{\dfrac{8y_C^2 l}{15}}$$

$$= \frac{wl^2}{8y_C}$$

Example 17·4. *A two-hinged parabolic arch of span 20 m has central rise 4 m. The arch is loaded with a triangular load of intensity zero at left abutment and 5 t/m at right abutment, as shown in*

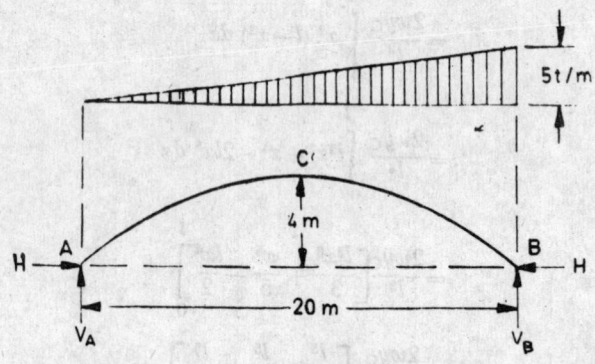

Fig. 17·7

Fig. 17·7. Determine the horizontal thrust on the arch.

(*A.M.I.E.*, *Summer, 1978*)

Solution.

Given.   Span, $l = 20$ m

Central rise, $y_C = 4$ m

Let $H =$ Horizontal thrust on the arch.

For simplification, let us assume another triangular load of intensity 5 t/m at left abutment, and zero at right abutment to be acting over the arch. We know that as a result of this assumed load*, the horizontal thrust at A and B will be twice the thrust due to given load only.

---

*Now the arch will be subjected to a uniformly distribute load of 5 t/m over the entire span.

Now using the relation,

$$2H = \frac{wl^2}{8y_C} \text{ with usual notations.}$$

$$= \frac{5 \times 20^2}{8 \times 4} = \frac{125}{2} \text{ t}$$

$$\therefore \quad H = \frac{125}{2 \times 2} = 31 \cdot 25 \text{ t} \quad \textbf{Ans.}$$

Example 17·5. *A parabolic arch, hinged at its springings, of span 36 m and rise 8 m is loaded as shown in Fig. 17·8.*

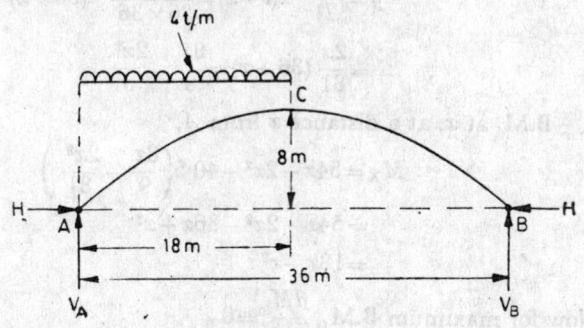

Fig. 17·8

Determine the values of horizontal thrust as well as maximum positive and negative B.M. over the arch. (...M.I.E. Summer, 1979)

Solution.

Given.   Span, $l = 36$ m
Central rise, $y_C = 8$ m
Load, $w = 4$ t/m

Horizontal thrust

Let $H =$ Horizontal thrust over the arch.

For simplification, let us assume another uniformly distributed load of 4 t/m to be acting over the right half of the arch. We know that as a result of this assumed load, the horizontal thrust at A and B will be twice the horizontal thrust to the given load only.

Using the relation,

$$2H = \frac{wl^2}{8y_C} \text{ with usual notations.}$$

$$= \frac{4 \times 36^2}{8 \times 8} = 81 \text{ t}$$

$$\therefore \quad H = \frac{81}{2} = 40 \cdot 5 \text{ t} \quad \textbf{Ans}.$$

Maximum positive B.M.

Let $V_A =$ Vertical reaction at A, and
$V_B =$ Vertical reaction at B.

Taking moment about A,

$$V_B \times 36 = 4 \times 18 \times 9 = 648$$

or

$$V_B = \frac{648}{36} = 18 \text{ t}$$

$$\therefore \quad V_A = 4 \times 18 - 18 = 54 \text{ t}$$

We know that the maximum positive B.M. takes place in the section AC. Let the maximum positive B.M. take place at a distance x from A. We know that the rise of arch at a distance x from A,

$$y = \frac{4y_C}{l^2} x(l-x) = \frac{4 \times 8}{36 \times 36} x(36-x)$$

$$= \frac{2x}{81}(36-x) = \frac{8x}{9} - \frac{2x^2}{81}$$

$\therefore$ B.M. at x, at a distance x from A,

$$M_X = 54x - 2x^2 - 40 \cdot 5 \left(\frac{8x}{9} - \frac{2x^2}{81} \right)$$

$$= 54x - 2x^2 - 36x + x^2$$

$$= 18x - x^2$$

Now for maximum B.M., $\frac{dM}{dx} = 0$

i.e.,

$$\frac{dM}{dx}(18x - x^2) = 0$$

$$18 - 2x = 0$$

or

$$x = \frac{18}{2} = 9 \text{ m}$$

We know that the rise of arch at a distance of 9 m from A i.e. $l/4$ from A,

$$y = \frac{3y_C}{4} = \frac{3 \times 8}{4} = 6 \cdot 0 \text{ m}$$

$\therefore$ Maximum positive B.M. at a distance of 9 m from A,

$$M_{max} = 54 \times 9 - \frac{4 \times 9^2}{2} - 40 \cdot 5 \times 6 \text{ t·m}$$

$$= 81 \text{ t·m} \quad \textbf{Ans.}$$

Maximum negative B.M.

We know that the maximum negative B.M. takes place in the section CB. Let the maximum negative B.M. take place at a distance x from B. We know that the rise of arch at a distance x from B,

$$y = \frac{4y_C}{l^2} x(l-x) = \frac{4 \times 8}{36 \times 36} x(36-x)$$

$$= \frac{2x}{81}(36-x) = \frac{8x}{9} - \frac{2x^2}{81}$$

$$\therefore \quad \text{B.M. at a distance } x \text{ from } B,$$

$$M_X = 18x - 40 \cdot 5\left(\frac{8x}{9} - \frac{2x^2}{81}\right) = 18x - 36x + x^2$$

$$= x^2 - 18x$$

Now for maximum B.M., $\dfrac{dM}{dx} = 0$

i.e.,
$$\frac{dM}{dx}(x^2 - 18x) = 0$$

$$2x - 18 = 0$$

or
$$x = \frac{18}{2} = 9 \text{ m}$$

We know that the rise of the arch at a distance 9 m from B, *i.e.* $l/4$ from B,

$$y = \frac{3y_C}{4} = \frac{3 \times 8}{4} = 6 \cdot 0 \text{ m}$$

$\therefore$ Maximum negative B.M. at a distance of 9 m from B,
$$M_{max} = V_B.x - Hy = 18 \times 9 - 40 \cdot 5 \times 6 \text{ t-m}$$
$$= -81 \text{ t-m} \quad \textbf{Ans.}$$

Exercise 17-1

1. A two-hinged parabolic arch of span 30 m and central rise 5 m is carrying a point load of 10 t at a distance of 10 m from the left support. Determine (*i*) horizontal thrust and (*ii*) B.M. under the load.
[**Ans.** 10·19 t ; 21·43 t-m]

2. A parabolic arch of span 100 m and central rise 25 m is hinged at its ends. The arch is carrying a uniformly distributed horizontal load of 1 t/m over the left half of the span. Find the values of horizontal thrust, and maximum B.M. over the arch. (*Bihar University*)
[**Ans.** 25 t ; +156·25 t-m ; −156·25 t-m]

3. A two-hinged parabolic arch of span l has central rise of one-fourth of the span. Show that the horizontal thrust on the arch due to a point load of W at its quarter span is $\dfrac{5W}{8}$. Take $I = I_0 \sec \theta$. (*Mysore University*)

4. A two-hinged parabolic arch of span 30 m and central rise 6 m is carrying a uniformly distributed load of 2 t/m for a length of 12 m from the left support towards the centre. Find the horizontal thrust and the reaction at the hinges. (*Karnatak University*)
[**Ans.** 12·98 t ; 23·17 t ; 13·85 t]

17·9. Effect of change in temperature in a two-hinged arch

Consider a two-hinged arch AB, supported at A and B, of span l and central rise y_C. Let the arch be subjected to some rise of temperature.

Let t = Rise of temperature, and
α = Coefficient of linear expansion for the arch material.

We know that the length of arch will tend to increase with the rise in temperature. Since the arch is rigidly hinged at A and B, therefore the horizontal span AB must remain the same. A little consideration will show that the hinges will exert a horizontal thrust on the arch to prevent the ends from moving out. The horizontal expansion prevented by the hinges

$$= l\alpha t \qquad \qquad ...(i)$$

If H_t is the horizontal thrust, due to rise in temperature, exerted by the hinges, the B.M. on an element of the arch at a height y from the support,

$$M = H_t \cdot y \qquad \qquad ...(ii)$$

We have already discussed in Art. 17·4 that the horizontal increase in span due to bonding moment

$$= \int_0^l \frac{My.ds}{EI} \qquad \qquad ...(iii)$$

Now equating (i) and (iii),

$$l\alpha t = \int_0^l \frac{My.ds}{EI}$$

$$= \int_0^l \frac{H_t.y.y.ds}{EI} \qquad (\because \quad M = H_t.y)$$

$$= \int_0^l \frac{H_t.y^2 ds}{EI}$$

$$= \int_0^l \frac{H_t.y^2 dx}{EI_0} \qquad \begin{array}{l}\text{(Substituting } I = I_0 \sec \theta, \text{ and}\\ ds = dx \cos \theta)\end{array}$$

or

$$H_t = \frac{l\alpha t}{\displaystyle\int_0^l \frac{y^2 dx}{EI_0}} = \frac{lxt}{\dfrac{1}{EI_0}\displaystyle\int_0^l y^2 dx} = \frac{EI_0 l\alpha t}{\displaystyle\int_0^l y^2 dx}$$

Cor 1. If the given arch is a parabolic arch, then substituting the value of $\int_0^l y^2 dx$ in the above equation, we get

$$H_t = \frac{EI_0 \, l \, \alpha \, t}{\dfrac{8}{15} \, y_C^2 \, l} = \frac{15 \, EI_0 \, \alpha \, t}{8 y_C^2}.$$

Cor. 2. Sometimes, as a result of horizontal thrust, one of the arch abutments yield by an amount δ. In such a case, the horizontal thrust.

$$H_t = \frac{l \, \alpha \, t - \delta}{\dfrac{1}{EI_0}\displaystyle\int_0^l y^2 dx} = \frac{EI_0(l \, \alpha \, t \, \delta)}{\displaystyle\int_0^l y^2 dx}$$

and if the arch is a parabolic one, then

$$H_t = \frac{EI_0(l \, \alpha \, t - \delta)}{\frac{8}{15} \, y_C^2.l} = \frac{15 \, EI_0(l \, \alpha \, l - \delta)}{8y_C^2.l}$$

Example 17·6. *A two-hinged parabolic arch has a span 30 m and a rise of 7·5 m. The moment of inertia of arch section is proportional to sec θ, where θ is the slope of the arch axis at any point with the horizontal. Calculate the horizontal thrust caused in the arch due to a rise of temperature by 25°F. The value of $E = 2·0 \times 10^6$ kg/cm² and coefficient of expansion $= 6 \times 10^{-6}$ per degree F. The moment of inertia at crown is 125×10^4 cm⁴.*

Solution.

Given. Span, $l = 30$ m $= 3,000$ cm

Central rise, $y_C = 7·5$ m $= 750$ cm

Rise of temperature, $t = 25°F$

Modulus of elasticity, $E = 2·0 \times 10^6$ kg/cm²

Coefficient of linear expansion,

$$\alpha = 6 \times 10^{-6} \text{ per } °F$$

Moment of inertia at the crown,

$$I_0 = 125 \times 10^4 \text{ cm}^4$$

Let $H_t =$ Horizontal thrust due to rise of temperature.

Using the relation,

$$H_t = \frac{15EI_0 \, \alpha \, t}{8y_C^2} \text{ with usual notations.}$$

$$= \frac{15 \times 2·0 \times 10^6 \times 125 \times 10^4 \times 6 \times 10^{-6} \times 25}{8 \times 750 \times 750} \text{ kg}$$

$$= \mathbf{1,250 \text{ kg}} \quad \mathbf{Ans}.$$

Example 17·7. *A two-hinged circular arch of span 30 m and central rise 5 m has 90 cm deep rib. Determine the maximum bending stress in the arch, when it is subjected to a rise of temperature of 20°C. Take the value of $E = 2·1 \times 10^6$ kg/cm² and $\alpha = 12 \times 10^{-6}$ per °C for the arch material.*

Solution.

Given. Span, $l = 30$ m $= 3,000$ cm

Central rise $y_C = 5$ m $= 500$ cm

Depth of the rib, $d = 90$ cm

Rise of temperature,

$$t = 20°C$$

Modulus of elasticity,

$$E = 2·1 \times 10^6 \text{ kg/cm}^2$$

Coefficient of linear expansion,

$$\alpha = 12 \times 10^{-6} \text{ per } °C$$

Let $f =$ Maximum bending stress in arch,

$R =$ Radius of the arch,

H_t = Horizontal thrust on the arch,

I = Moment of inertia of the arch rib.

2α = Angle, which the two supports A and B make at the centre of the arch.

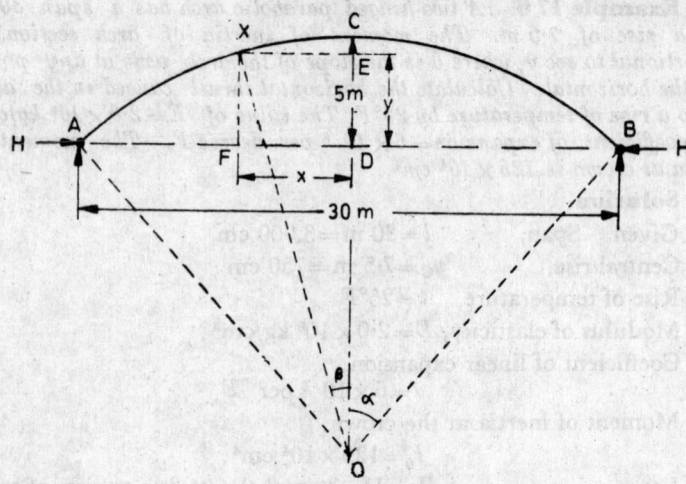

Fig. 17·9

We know that the radius of the arch

$$R = \frac{l^2}{8y_C} + \frac{y_C}{2} = \frac{30^2}{8 \times 5} + \frac{5}{2} = 25 \text{ m}$$

$$= 2,500 \text{ cm}$$

Now consider a point X on the arch axis, at a distance x from D.

Let β = Angle, which OX makes with the centre line OC

x, y = Coordinates of the point X, with reference to the point D, which is the middle of the span AB of the arch.

From the given data, we find that

$$l\alpha t = 3,000 \times 12 \times 10^{-6} \times 20 = 72 \times 10^{-2} \qquad \text{...}(i)$$

From the geometry of the arch, we see that

$$\sin \alpha = \frac{l}{2R} = \frac{30}{2 \times 25} = 0.6$$

or $\alpha = 36°52' = 0.6435$ radian

$$2\alpha = 1.2870 \text{ radians}$$

$\therefore$ $\sin 2\alpha = \sin 73°44' = 0.96$

and $\cos 2\alpha = \cos 73°44' = 0.2801$

Now consider a small element δs of the arch, at an angle β with the centre line OC, subtending an angle $\delta\beta$ at the centre E. We know that

$$\delta s = R.\delta\beta$$

From the geometry of the figure, we find that

$$y = R \cos \beta - R \cos \alpha = R (\cos \beta - \cos \alpha)$$

$$\therefore \int y^2 ds = \int_{-\alpha}^{+\alpha} R^2 (\cos \beta - \cos \alpha)^2 \, R d\beta$$

$$= \int_{0}^{\alpha} 2R^3 (\cos \beta - \cos \alpha)^2 \, d\beta$$

$$= 2R^3 \int_{0}^{\alpha} (\cos^2 \beta + \cos^2 \alpha - 2 \cos \alpha \cos \beta) \, d\beta$$

$$= R^3 \int_{0}^{\alpha} [(1 + \cos 2\beta) + 2 \cos^2 \alpha - 4 \cos \alpha \cos \beta] d\beta$$

$$= R^3 \left[\beta + \frac{\sin 2\beta}{2} + 2\beta \cos^2 \alpha - 4 \cos \alpha \sin \beta \right]_{0}^{\alpha}$$

$$= R^3 \left[\alpha + \frac{\sin 2\alpha}{2} + 2\alpha \cos^2 \alpha - 4 \cos \alpha \sin \alpha \right]$$

$$= R^3 \left[\alpha - \frac{3}{2} \sin 2\alpha + \alpha(1 + \cos 2\alpha) \right]$$

$$= R^3 [2\alpha - 1\cdot 5 \sin 2\alpha + \alpha \cos 2\alpha]$$

$$= R^3 [1\cdot 2870 - 1\cdot 5 \times 0\cdot 96 + 0\cdot 6435 \times 0\cdot 2801]$$

$$= 0\cdot 0273 \, R^3 \qquad \qquad \ldots(ii)$$

Now using the relation,

$$H_t = \cfrac{l\alpha t}{\cfrac{1}{EI} \displaystyle\int_{0}^{l} y^2 \, ds} \text{ with usual notations.}$$

$$= \cfrac{72 \times 10^{-2}}{\cfrac{1}{2.1 \times 10^6 I} \times 0\cdot 0273 \times 2,500^3} = 0\cdot 003545 \, I$$

Moment at the crown,

$$M_C = H_t \cdot y_C = 0\cdot 003545 \, I \times 500 = 1\cdot 7725 \, I$$

We know that the modulus of rib section,

$$Z = \frac{I}{\dfrac{d}{2}} = \frac{I}{\dfrac{90}{2}} = \frac{I}{45}$$

Using the relation,

$$f = \frac{M_C}{Z} \text{ with usual notations.}$$

$$\therefore \qquad f = \frac{1 \cdot 7725\, I}{\dfrac{I}{45}} = 79 \cdot 8 \text{ kg/cm}^2 \quad \textbf{Ans}.$$

17·10. Straining actions in a two-hinged arch

Like a three-hinged arch, the axis of a two-hinged arch is also subjected to the following three straining actions :

(1) Moment of the thrust,

(2) Normal thrust, and

(3) Radial shear force.

The normal thrust at any section x,

$$P_X = H \cos \theta + V \sin \theta \qquad \text{...(If } V \text{ is a net } upward \text{ force on the left of } X)$$

$$= H \cos \theta - V \sin \theta \qquad \text{...(If } V \text{ is a net } downward \text{ force on the left of } X)$$

and the radial shear force at x,

$$F_X = H \sin \theta - V \cos \theta \qquad \text{... (If } V \text{ is a net } upward \text{ force to the left of } X)$$

$$= H \sin \theta + V \cos \theta \qquad \text{...(If } V \text{ is a net } downward \text{ force to the left of } X)$$

Example 17·8. *A two-hinged parabolic arch of 30 m span and 5 m central rise has a varying second moment of area, which is proportional to the secant of the slope of its neutral axis. It carries a point load of 15 t at a distance of 10 m from the left end. Determine the horizontal thrust, B.M., normal thrust and radial shear force under the load.*

(A.M.I.E., Summer, 1981)

Solution.

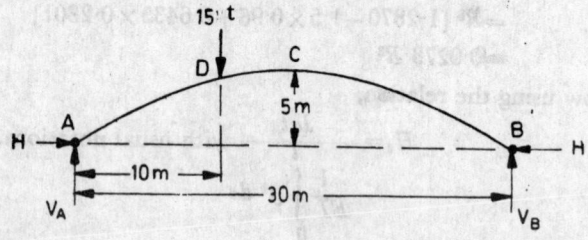

Fig. 17·10

Given. Span, $\qquad l = 30$ m

Central rise, $\qquad y_C = 5$ m

Load, $\qquad\qquad W = 15$ t

Distance between the section and the left end,

$$x = 10 \text{ m}$$

Horizontal thrust

Let $\qquad\qquad H = $ Horizontal thrust on the arch.

Using the relation,

$$H = \frac{5}{8} \times \frac{Wx}{y_C l^3}\, (l-x)(l^2 + lx - x^2) \qquad \text{with usual notations.}$$

$$H = \frac{5}{8} \times \frac{15 \times 10}{5 \times 30^3}(30-10)(30^2+30 \times 10-10^2) \text{ t}$$

$$= \frac{5 \times 15 \times 10 \times 20 \times 1100}{8 \times 5 \times 30 \times 30 \times 30} = \frac{275}{18} \text{ t}$$

$$= 15 \cdot 28 \text{ t} \quad \textbf{Ans.}$$

Bending moment at 10 m from A

Let $\qquad V_A$ = Vertical reaction at A, and

$\qquad\qquad\qquad V_B$ = Vertical reaction at B.

Taking moments about B,

$$V_B \times 30 = 15 \times 20 = 300$$

$$\therefore \qquad\qquad V_A = \frac{300}{30} = 10 \text{ t}$$

We know that rise of arch at D *i.e.* at a distance 10 m from A,

$$y = \frac{4y_C}{l} x(l-x) = \frac{4 \times 5 \times 10}{30 \times 30}(30-10) = \frac{40}{9}\text{m}$$

$\therefore$ B.M. at D, $M_D = V_A.x - Hy = 10 \times 10 - 15 \cdot 28 \times \dfrac{40}{9}$ t-m

$$= 32 \cdot 1 \text{ t-m} \quad \textbf{Ans.}$$

Normal thrust at 10 m from A

Let $\qquad P$ = Normal thrust at 10 m from A,

$\qquad\qquad\qquad \theta$ = Angle, which the tangent to the arch axis at 10 m from the left hand support, makes with the horizontal.

We know that the rise of arch axis, at a distance x from A,

$$y = \frac{4y_C}{l^2} x(l-x) = \frac{4 \times 5}{30 \times 30} x (30-x)$$

$$= \frac{x}{45}(30-x) = \frac{2x}{3} - \frac{x^2}{45}$$

Differentiating the above equation with respect to x,

$$\frac{dy}{dx} = \tan \theta = \frac{2}{3} - \frac{2x}{45}$$

Substituting the value of $x = 10$ m in above equation,

$$\tan \theta = \frac{2}{3} - \frac{2 \times 10}{45} = \frac{2}{9} = 0 \cdot 2222$$

$$\therefore \qquad\qquad \theta = 12° \ 31'$$

$$\therefore \qquad \sin\ \theta = \sin\ 12°\ 31' = 0.2167$$

and $\qquad\qquad \cos\ \theta = \cos\ 12°\ 31' = 0.9762$

Using the relation,

$$P = H\cos\theta + V\sin\theta \text{ with usual notations.}$$
$$= 15.28 \times 0.9762 + 10 \times 0.2167\ t$$
$$= \mathbf{17.077\ t\quad Ans.}$$

Shear force at 10 m from A

Let $\qquad\qquad F = $ Shear force at 10 m from A.

Using the relation,

$$F = H\sin\theta - V\cos\theta \text{ with usual notations.}$$
$$= 15.28 \times 0.2167 - 10 \times 0.9762 = -6.45\ t$$
$$= \mathbf{6.45\ t\ (downward)\ Ans.}$$

Example 17·9. *Find the horizontal thrust, B.M., normal thrust and radial shear at 10 m from the left abutment in the case of the two-hinged parabolic arch shown in Fig. 17·11.* $I_X \propto I_0 \sec \alpha$.

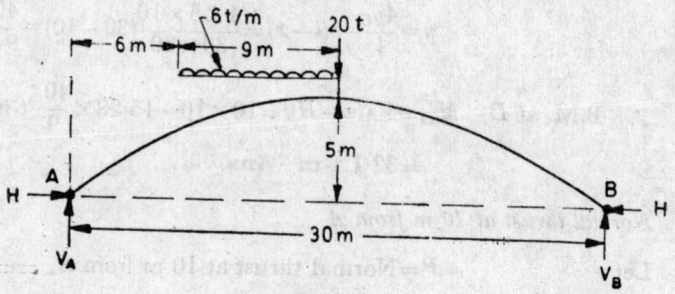

Fig. 17·11

Solution.

Given. Span, $\qquad l = 30$ m

Central rise, $\qquad y_C = 5$ m

Horizontal thrust on the arch

Let $\qquad\qquad H = $ Horizontal thrust on the arch,

$V_A = $ Vertical reaction at A, and

$V_B = $ Vertical reaction at B.

We know that rise of arch at a distance x from A,

$$y = \frac{4y_C}{l^2}x\ (l-x) = \frac{4\times 5}{30\times 30}\ x(30-x)$$
$$= \frac{x}{45}\ (30-x) \qquad\qquad ..(i)$$

Taking moments about A,

$$V_B \times 30 = 6 \times 9 \times 10.5 + 20 \times 15 = 867$$

or $\qquad\qquad V_B = \frac{867}{30} = 28.9\ t$

$$\therefore \quad V_A = 9 \times 6 + 20 - 28 \cdot 9 = 45 \cdot 1 \text{ t}$$

Let us split up the arch into the three sections AD, DC and CB. We see that the value of beam moment

(i) between A and D, with origin at A
$$= V_A \cdot x = 45 \cdot 1 \ x \qquad \qquad \dots(ii)$$

(ii) between D and C, with origin at A
$$= V_A \cdot x - \frac{6(x-6)^2}{2} = 45 \cdot 1 \ x - 3(x-6)^2$$
$$= 45 \cdot 1 \ x - 3(x^2 + 36 - 12x)$$
$$= 45 \cdot 1 \ x - 3x^2 - 108 + 36 \ x$$
$$= 81 \cdot 1 \ x - 3x^2 - 108 \qquad \qquad \dots(iii)$$

and (iii) between C and B with origin at B
$$= V_B \cdot x = 28 \cdot 9 \ x$$

Now, the total value of $\displaystyle\int_0^l \mu_X \cdot y \cdot dx$ may be obtained first by

finding out the values of $\displaystyle\int_0^l \mu_X \cdot y \cdot dx$ for the arch in the above three

sections individually, and then by algebraically adding these values.

Considering the section AD of the arch,

$$\int_A^D \mu_X \cdot y \cdot dx = \int_A^D 45 \cdot 1 \ x \times \frac{x}{45}(30-x) \ dx$$

[Substituting value of y from equation (i)]

$$= \frac{45 \cdot 1}{45} \int_0^6 (30x^2 - x^3) dx = \frac{45 \cdot 1}{45} \left[\frac{30x^3}{4} - \frac{x^4}{4} \right]_0^6$$

$$= \frac{45 \cdot 1}{45} \left[10 \times 6^3 - \frac{6^4}{4} \right] = \frac{45 \cdot 1 \times 1836}{45}$$

$$= 1,840 \qquad \qquad \dots(v)$$

Similarly in section DC of the arch,

$$\int_D^C \mu_X \cdot y \, dx = \int_6^{15} (81 \cdot 1 \ x - 3x^2 - 108) \frac{x}{45}(30-x) dx$$

$$= \frac{1}{45} \int_6^{15} (81 \cdot 1 x^2 - 3x^3 - 108x)(30-x) dx$$

$$= \frac{1}{45} \int_6^{15} (2,433 \ x^2 - 90x^3 - 3,240 \ x - 81 \cdot 1 \ x^3$$

$$+ 3x^4 + 108x^2) dx$$

$$= \frac{1}{45} \int_{6}^{15} (3x^4 - 171 \cdot 1x^3 + 2{,}541x^2 - 3{,}240x)dx$$

$$= \frac{1}{45} \left[\frac{3x^5}{5} - \frac{171 \cdot 1x^4}{4} + \frac{2{,}541x^3}{3} - \frac{3{,}240x^2}{2} \right]_{6}^{15}$$

$$= \frac{1}{45} \left[\frac{3}{5}(15^5 - 6^5) - \frac{171 \cdot 4}{4}(15^4 - 6^4] \right.$$

$$\left. + 847(15^3 - 6^3) - 1{,}620(15^2 - 6^2) \right]$$

$$= \frac{1}{45} \left[4{,}51{,}000 - 21{,}21{,}000 + 26{,}77{,}00 - 3{,}06{,}200 \right]$$

$$= \frac{7{,}00{,}800}{45} = 15{,}580 \qquad \qquad ...(vi)$$

Similarly in section CB of the arch

$$\int_{B}^{C} \mu_x \cdot y \cdot dx = \int_{0}^{15} 28 \cdot 9x \times \frac{x}{45}(30-x) \ dx = \frac{28 \cdot 9}{45} \int_{0}^{15} (30x^2 - x^3)dx$$

$$= \frac{28 \cdot 9}{45} \left[\frac{30x^3}{3} - \frac{x^4}{4} \right]_{0}^{15} = \frac{28 \cdot 9}{45} \left[10 \times 15^3 - \frac{15^4}{4} \right]$$

$$= \frac{28 \cdot 9 \times 21{,}090}{45} = 13{,}300 \qquad \qquad ...(vii)$$

$$\therefore \quad \int_{0}^{l} \mu_x \cdot y \cdot dx = 1{,}840 + 15{,}580 + 13{,}300 = 30{,}720 \quad ...(viii)$$

From the given data, we also find that

$$\int_{0}^{l} y^2 dx^* = \int_{0}^{l} \frac{x^2}{45 \times 45} (30-x)^2 \ dx$$

$$= \frac{1}{2{,}025} \int_{0}^{l} x^2(900 + x^2 - 60x)dx$$

----

*The value of $\int_{0}^{l} y^2 dx$ of a parabolic arch may also be found out directly from the relation

$$\frac{8}{15} \ y_2{}_C{}^l = \frac{8}{15} \times 5 \times 5 \times 30 = 400$$

$$= \frac{1}{2,025} \int_0^l (900x^2 + x^4 - 60x^3)\, dx$$

$$= \frac{1}{2,025} \left[\frac{900x^3}{3} + \frac{x^5}{5} - \frac{60x^4}{4} \right]_0^{30}$$

$$= \frac{1}{2,025} \left[300 \times 30^3 + \frac{30^5}{5} - \frac{60 \times 30^4}{4} \right]$$

$$= \frac{30^3}{2,025} [300 + 180 - 450] = 400 \qquad ...(ix)$$

Now using the relation,

$$H = \frac{\displaystyle\int_0^l \mu \cdot_x \cdot y \cdot dx}{\displaystyle\int_0^l y^2 \cdot dx} \quad \text{with usual notations.}$$

$$= \frac{30,720}{400} = 76 \cdot 8 \text{ t} \quad \textbf{Ans.}$$

B.M. at 10 m from A

We know that rise of arch at 10 m from A,

$$y = \frac{4y_C}{l^2} x(l-x) = \frac{4 \times 5}{30 \times 30} \times 10(30-10) \text{ m}$$

$$= \frac{4 \times 5 \times 10 \times 20}{30 \times 30} = \frac{40}{9} \text{ m}$$

$\therefore$ B.M. at a point 10 m from A,

$$M = 45 \cdot 1 \times 10 - 76 \cdot 8 \times \frac{40}{9} - 6 \times 4 \times 2 \text{ t-m}$$

$$= \textbf{62 t-m Ans.}$$

Normal thrust at 10 m from the left abutment

Let P = Normal thrust at 10 m from the left abutment,

θ = Angle, which the tangent to the arch axis at 10 m from the left hand support makes with the horizontal.

We know that the rise of arch axis at a distance x from A.

$$y = \frac{4y_C}{l^2} x(l-x) = \frac{4 \times 5}{30 \times 30} x(30-x)$$

$$= \frac{x}{45}(30-x) = \frac{2x}{3} - \frac{x^2}{45}$$

Differentiating the above equation with respect to x

$$\frac{dy}{dx} = \tan \theta = \frac{2}{3} - \frac{2x}{45}$$

Substituting the value of $x = 10$ m in the above equation.

$$\tan \theta = \frac{2}{3} - \frac{2 \times 10}{45} = \frac{2}{9} = 0.2222$$

$\therefore \qquad \theta = 12° \ 31$

$\therefore \qquad \sin \theta = \sin 12°31' = 0.2167$

and $\qquad \cos \theta = \cos 12°31' = 0.9762$

Using the relation,

$\qquad P = H \cos \theta + V \sin \theta$ with usual notations.

$\qquad = 76.8 \times 0.9762 + (45.1 - 24) \ 0.2167$ t

$\qquad = \mathbf{79 \ 55 \ t} \qquad \mathbf{Ans}.$

Radial shear at 10 m from the left abutment

Let $\qquad F =$ Radial shear at 10 m from the left abutment.

Using the relation,

$\qquad F = H \sin \theta - V \cos \theta$ with usual notations.

$\qquad = 76.8 \times 0.2167 - (45.1 - 24) \ 0.9762 = 4.05$ t

$\qquad = \mathbf{4.05 \ t \ (downwards)} \qquad \mathbf{Ans}.$

17·11. Influence lines for moving loads over two-hinged arches

Consider a concentrated load W rolling from A to B over a two-hinged arch AB of span l and central rise y_C.

Here we shall discuss the following influence lines for a single concentrated load, when it rolls from one support to another.

1. Influence lines for horizontal thrust,
2. Influence lines for bending moment,
3. Influence lines for normal thrust, and
4. Influence lines for radial shear force.

Influence lines for the horizontal thrust

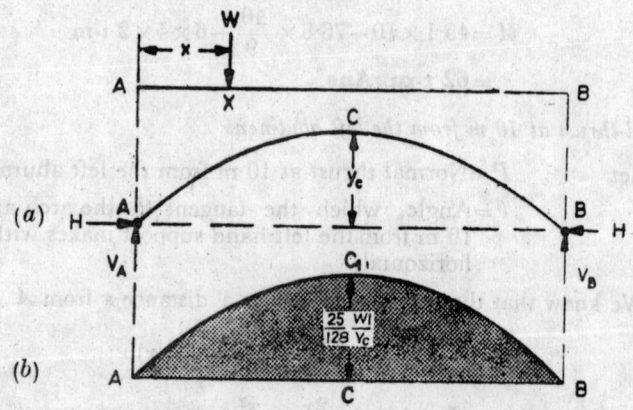

Fig. 17·12

Consider a section X, at a distance x from A. We have already discussed in Art. 17·6 that the horizontal thrust in a two-hinged arch carrying a concentrated load,

$$H = \frac{5}{8} \times \frac{Wx}{y_C l^3}(l-x)(l^2 + lx - x^2)$$

We know that the horizontal thrust is always maximum, when the load is at the crown C of the arch, therefore substituting $x = l/2$ in the above equation,

$$H = \frac{5}{8} \cdot \frac{W \cdot \dfrac{l}{2}}{y_C l^3}\left(l - \frac{l}{2}\right)\left[l^2 + l \cdot \frac{l}{2} - \left(\frac{l}{2}\right)^2\right]$$

$$= \frac{25}{128} \times \frac{Wl}{y_C}$$

Now complete the influence line as shown in Fig. 17·12(b).

Influence lines for the bending moment

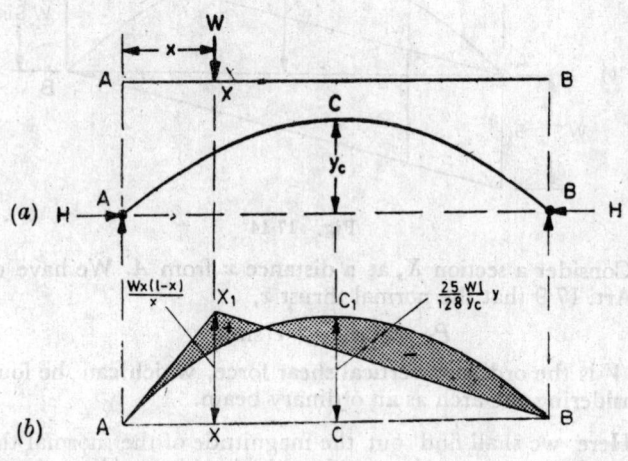

Fig. 17·13

Consider a section X at a distance x from A. We know that B.M. at X,

$$M_X = \mu_X - H.y$$

where μ_X is the usual B.M. as a freely supported beam, and $H.y$ is the moment due to the horizontal thrust H.

We also know that the influence line for μ_X will be a triangle having ordinate XX_1 equal to $\dfrac{Wx(l-x)}{l}$. The influence line for $H.y$ will be y times the influence line for H. It is thus obvious, that the influence line for $H.y$ will be a parabola having central ordinate CC_1

equal to $\dfrac{25}{128}\cdot\dfrac{Wl}{y_C}$. y as shown in Fig. 17·13 (b). Since μ_X and $H.y$ are opposite in sign, that is why the two diagrams have been drawn on the same side of the base AB. From the influence line diagram, we find that the maximum positive B.M. on the section occurs, when the load is on the section itself. The maximum negative B.M. on the section occurs, when the load is at the crown of the arch.

Influence line for the normal thrust

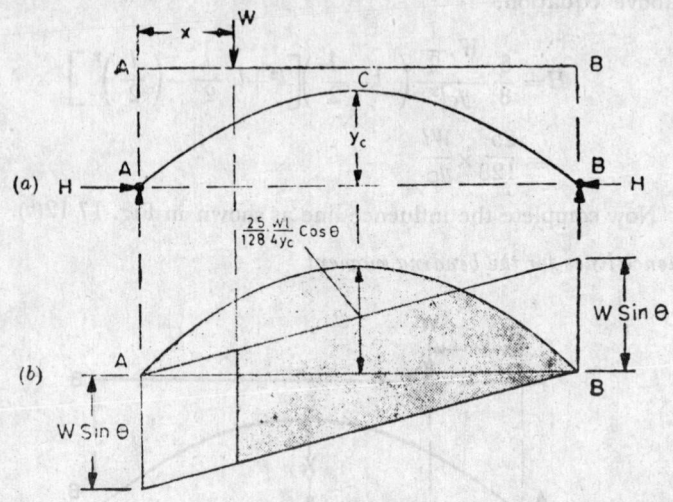

Fig. 17·14

Consider a section X, at a distance x from A. We have discussed in Art. 17·9 that the normal thrust x,

$$P_X = H\cos\theta + V\sin\theta$$

where V is the ordinary vertical shear force, which can be found out by considering the arch as an ordinary beam.

Here we shall find out the magnitude of the normal thrust in the following two stages *i.e.* (i) when the load is in AX, and (ii) when the load is in XB. A little consideration will show that when the load is in AX, the normal thrust at x,

$$P_X = H\cos\theta - (W - V_A)\sin\theta = H\cos\theta - V_B\sin\theta \qquad …(i)$$

and when the load is in XB, the normal thrust at X,

$$P_X = H\cos\theta + V_A\sin\theta \qquad …(ii)$$

We know that the influence lines for $V_B\sin\theta$ and $V_A\sin\theta$ will consist of two parallel lines having end ordinates equal to $W\sin\theta$. The influence line for $H\cos\theta$ will be $\cos\theta$ times the influence line for H. Since in the original equation (*i.e.* $P_X = H\cos\theta + V\sin\theta$) we have to add up the two influence lines, that is why the two influence lines are drawn on the opposite sides of the base AB as shown in Fig. 17·14 (b).

Influence lines for the radial shear force

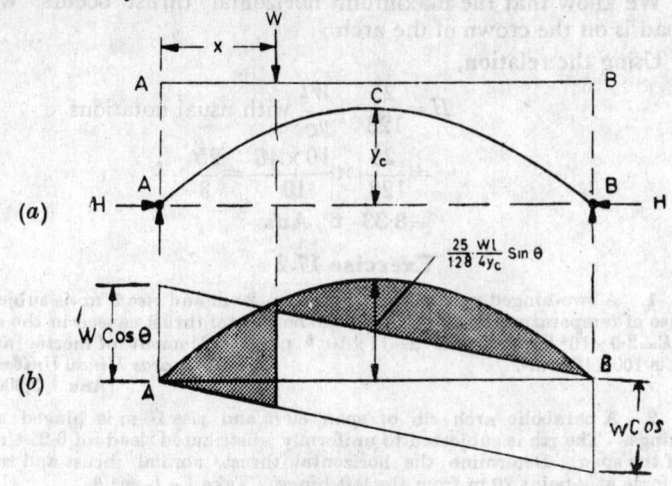

Fig. 17·15

Consider a section X, at a distance x from A. We have already discussed in Art. 17·9 that the radial S.F. at X,

$$F_X = H \sin \theta - V \cos \theta$$

where V is the ordinary vertical shear force, which can be found out by considering the arch as an ordinary beam.

Here we shall find the magnitude of the radial shear force in the following two stages, *i.e.*, (*i*) when the load is in AX, and (*ii*) when load is in XB. A little consideration will show that when the load is in AX, the radial shear force at X,

$$F_X = H \sin \theta + (W - V_A) \cos \theta = H \sin \theta + V_B \cos \theta$$

and when the load is in XB, the radial shear force at X,

$$F_X = H \sin \theta - V_A \cos \theta$$

We know that the influence lines for $V_B \cos \theta$ and $V_A \cos \theta$ will consist of two parallel lines having end ordinates equal to $W \cos \theta$. The influence line for $H \sin \theta$ will be equal to $\sin \theta$ times the influence line for H. Since in the general equation (*i.e.* $F_X = H \sin \theta - V \cos \theta$) we have to subtract the two influence lines, that is why the two influence lines are drawn on the same side of the base AB as shown in Fig. 17·15 (*b*).

Example 17·10. *A parabolic arch, hinged at its two abutments, of span 48 m has a central rise of 10 m. A point load of 5 T rolls over the arch from one abutment to another. Determine the magnitude of maximum horizontal thrust.*

Solution.

Given. Span, $l = 48$ m

Central rise, $y_C = 10$ m

Load, $W = 10$ T

Let $H =$ Horizontal thrust on the arch.

We know that the maximum horizontal thrust occurs, when the load is on the crown of the arch.

Using the relation,

$$H = \frac{25}{128} \cdot \frac{Wl}{y_C} \text{ with usual notations.}$$

$$= \frac{25}{128} \times \frac{10 \times 48}{10} = \frac{25}{3} \text{ t}$$

$$= 8 \cdot 33 \text{ t   Ans.}$$

Exercise 17·2

1. A two-hinged parabolic arch of span 20 m and rise 5 m is subjected to a rise of temperature by 20°C. Find the horizontal thrust caused in the arch. Take $E = 2 \cdot 0 \times 10^6$ kg/cm² and $\alpha = 11 \times 10^{-6}$ per °C. Moment of inertia at the crown is 100×10^4 cm⁴.   (*Banaras Hindu University*)

[**Ans.** 3,300 kg]

2. A parabolic arch rib of span 80 m and rise 10 m is hinged at its springings. The rib is subjected to uniformly distributed load of 0 25 t/m for half of the span. Determine the horizontal thrust, normal thrust and radial shear force at a point 20 m from the left hinge. Take $I = I_0 \sec \theta$.

(*Punjab University*)

[**Ans** 10 t ; 10·31 t ; 0]

3. A two-hinged parabolic arch of span 30 m and rise 5 m carries a point load of 15 t at a distance 10 m from the left hinge. Determine the normal thrust and radial shear. Take $I = I_0 \sec \theta$.   (*Calcutta University*)

[**Ans.** 17·08 t ; 6·45 t]

HIGHLIGHTS

1. Horizontal thrust on a two-hinged arch

$$H = \frac{\displaystyle\int_0^l \mu_x \cdot y \cdot dx}{\displaystyle\int_0^l y^2 \cdot dx}$$

where $\mu_x =$ B.M. at any point x, considering the arch as a simply supported beam

$y =$ Rise of the arch from the springings at x.

2. In the case of a parabolic two-hinged arch carrying a point load W,

$$H = \frac{5}{8} \times \frac{Wx}{y_C l^3} (l-x)(l^2 + lx - x^2)$$

$$= \frac{5}{8} \times \frac{Wl}{y_C} (k^4 - 2k^3 + k)$$

where $x =$ Horizontal distance between the load, and the end of the arch,

$l =$ Span of the arch,

$y_C =$ Central rise of the arch, and

$x = kl$

3. In the case of a semi-circular two-hinged arch, carrying a point load W,

$$H = \frac{W \cos^2 \beta}{\pi}$$

where β = Angle, subtended at the centre of the arch, by an arc between the load and centre of the arch.

4. In the case of a parabolic two-hinged arch, carrying a uniformly distributed load of w/unit horizontal length,

$$H = \frac{wl^2}{8y_C}$$

where l = Span of the arch, and
y_C = Central rise of the arch.

5. The horizontal thrust caused by the rise of temperature in a two-hinged arch,

$$H = \frac{\acute{E}I_0 \cdot l\alpha t}{\displaystyle\int_0^l y^2 \, dx} \qquad \text{...(general relation)}$$

$$= \frac{15EI_0\alpha t}{8y_C^2} \qquad \text{...(If the arch is parabolic)}$$

where E = Modulus of elasticity of the arch material,
I_0 = Moment of inertia of the arch at the crown,
l = Span of the arch,
α = Coefficient of linear expansion for the arch material,
t = Rise of temperature,
y = Rise of the arch axis at a distance x from the hinge,
y_C = Central rise of the arch.

6. The straining actions in a two-hinged arch are

(*a*) Normal thrust at any section X,

$$P_X = H \cos \theta + V \sin \theta \qquad \text{...(If V is a net upward force on the left of X)}$$

$$= H \cos \theta - V \sin \theta \qquad \text{...(If V is a net downward force on the left of X)}$$

(*b*) radial shear force at X,

$$F_X = H \sin \theta - V \cos \theta \qquad \text{...(If V is a net upward force to the left of X)}$$

$$= H \sin \theta + V \cos \theta \qquad \text{...(If V is a net downward force to the left of X)}$$

7. When a point load W rolls over a two-hinged parabolic arch, maximum horizontal thrust takes place, when the load is at the crown of the arch.

8. When a point load W rolls over a two-hinged parabolic arch, maximum positive B.M. on the section occurs, when the load is on the section. The maximum negative B.M. on the section occurs, when the load is at the crown of the arch.

Do You Know ?

1. Derive an equation for the horizontal thrust in a two-hinged arch.

2. A symmetrical two-hinged parabolic arch rib has its both supports l distance apart at the same level, the rise up to the crown being r. The second moment of area of the section at any point is $I_0 \sec \theta$, where I_0 is the corresponding value at the crown and θ is the inclination of the tangent at the point with the horizontal.

Calculate, from first principles, the horizontal thrust developed at the supports due to a unit load placed on the arch at a distance kl, measured horizontally from the left support.

3. Show that the horizontal thrust on a semi-circular two-hinged arch, when subjected to a load W at its crown is $\dfrac{W}{\pi}$.

4. From first principles, derive an expression for the horizontal thrust, when a parabolic two-hinged arch of span l and central y_C is subjected to a uniformly distributed load of w per unit horizontal length.

5. Obtain an expression for the horizontal thrust, caused by the rise of temperature of a two-hinged arch.

6. What are the straining actions in a two-hinged arch ? Obtain expressions for them.

7. Draw influence lines for the horizontal thrust and B.M., when a point load rolls over the arch from left to right.

18

FORCES IN REDUNDANT FRAMES

---

1. Introduction. 2. Redundant frame. 3. Castigliano's first theorem. 4. Proof of Castigliano's first theorem. 5. Maxwell's method for the forces in redundant frames. 6. Forces due to error in the length of a member in a redundant frame. 7. Frames with two or more redundant members. 8. Trussed beams.

18·1. Introduction

We have already discussed in the chapter on "Forces in perfect frames" in *Strength of Materials*, the various methods of finding out the forces in members of a perfect frame. Both the analytical methods (*i.e.* method of joints and method of sections) are based on the principles of statics. That is why, a perfect frame is also called a statically determinate structure.

18·2. Redundant frame

We have already discussed in chapter 8 that the number of members and number of joints, in a perfect frame may be expressed by the relation :

$$n = 2j - 3$$

where n = No. of members, and
               j = No. of joints.

But a frame, which does not satisfy the above equation is known as an imperfect frame. The number of members is an imperfect frame may be more or less than $(2j-3)$. An imperfect frame, in which the number of members are less than $(2j-3)$, is known as a deficient frame ; whereas an imperfect member, in which the number of members are more than $(2j--3)$ is known as a *redundant frame*. In this chapter, we shall discuss the methods of finding out the forces in redundant frames only.

18·3. Castigliano's first theorem

It states, "*If an elastic structure is subjected to a system of loads in equilibrium, produce a total strain energy U in the structure, the partial differential coefficient of the total strain energy U with respect to any load gives the displacement of the load in its own line of action.*"

Mathematically

$$y_1 = \frac{dU}{dW_1} \; ; \quad y_2 = \frac{dU}{dW_2} \; ; \quad y_3 = \frac{dU}{dW_3}$$

where $y_1, y_2, y_3 \dots \dots$ = Vertical deflections of the joints 1, 2, 3 …… of the structure

$W_1, W_2, W_3 \dots \dots$ = Loads on the joints, 1, 2, 3 …… of the structure

18·4. Proof of Castigliano's first theorem

Consider a frame supported at A and D and carrying a load at E as shown in Fig. 18·1.

Let $\quad W \dots \dots$ = Load on the frame,

$F_1, \; F_2, \; F_3 \dots \dots$ = Forces in the members 1, 2, 3,... of the frame as a result of external loads,

$l_1, \; l_2, \; l_3, \dots \dots$ = Lengths of the members 1, 2, 3...of the frame

$A_1, \; A_2, \; A_3, \dots \dots$ = Cross sectional areas of the members 1, 2, 3...

$y_1, \; y_2, \; y_3, \dots \dots$ = Deflection of the joints 1, 2, 3,……

Now assume all the external loads on the frame to be removed and a unit vertical load to act at any joint whose deflection is requir-

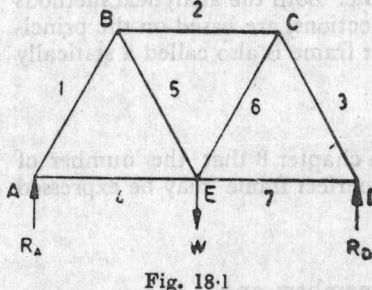

Fig. 18·1

ed to be found out. As a result of this unit vertical load, let v_1, v_2, v_3 …… be the forces in the members 1, 2, 3,…… of the frame. Now assume a vertical load W_1 to act at the same joint, where the unit load was acting then the forces in the members due to the load W will be v_1W, v_2W, v_3W …… A little consideration will show, that the total forces P_1, P_2, P_3 ……in the members due to the loads $W_1, W_2,$ W_3 ……and W will be $(F_1+v_1W), (F_2+v_2W), (F_3+v_3W)$ ……

We know that the strain energy stored in a body

$$U = \frac{P^2 l}{2AE}$$

The total strain energy stored in a frame.

$$U = \sum \frac{P^2 l}{2AE} = \frac{1}{2E} \sum \frac{P^2 l}{A}$$

$$= \frac{1}{2E} \sum \frac{(F_1 + v_1 W)^2}{A_1}$$

Differentiating the above equation with respect to W_1,

$$\frac{dU}{dW_1} = \frac{1}{2E} \sum \frac{2(F_1 + v_1 W)v_1 l}{A_1}$$

$$= \frac{1}{E} \sum \frac{(F_1 + v_1 W)v_1 l}{A_1}$$

$$= \frac{1}{E} \sum \frac{P_1 v_1 l}{A_1} \qquad (\because \; P_1 = F_1 + v_1 W)$$

$$= \frac{\sum pvl}{E} \qquad \left(\because \; p = \frac{p}{A} \right)$$

We have already discussed that the deflection,

$$y_1 = \frac{\sum pvl}{E}$$

$$\therefore \quad \frac{dU}{dW_1} = y_1$$

similarly, it can be proved that

$$y_2 = \frac{dU}{dW_2} \text{ and } y_3 = \frac{dU}{dW_3} \text{ and so on.}$$

Note. This theorem is also applicable to the system of moments and the resulting angular deformations. Thus if $M_1, M_2, M_3 \ldots$ are the moments acting on a body, the angular deformations, $i_1 = \frac{dU}{dM_1}$, $i_2 = \frac{dU}{dM}$ and so on.

18·5. Maxwell's method for the forces in redundant frames

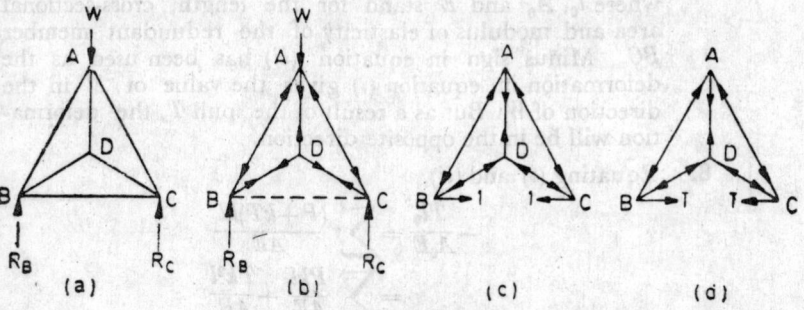

Fig. 18·2

Though there are many methods for finding out the force in the members of a redundant frame, yet the Maxwell's method is important from the subject point of view. According to this method, the forces in the various members of a redundant frame may be found out as discussed below :

1. Select any member of the redundant frame (say BC) to be a redundant member. Now imagine this redundant mem-

ber BC to be removed, so that the remaining frame becomes a perfect one, as shown in Fig. 18·2 (a) and (b).

2.   Now find out the forces P_1, P_2, P_3 (i.e., pulls or pushes) in the various members of the perfect frame, due to the external loads.

3.   Now apply a unit pull in the redundant member, BC, which was assumed to be removed as shown in Fig. 18·2 (c) and find out the forces k_1, k_2, k_3 (i.e., pulls or pushes) in the various members of the frame, due to the unit pull.

4.   Now, if T is the actual pull in the redundant member BC, then the actual force in the members will be (P_1+k_1T), (P_2+k_2T), $(P_3+k_3T)\ldots\ldots$

5.   The total strain energy in the frame (excluding the redundant member),

$$U = \sum \frac{(P+kT)^2 l}{2AE}$$

6.   As per Castigliano's first theorem, we know that the partial differential coefficient of the above equation with respect to T, will give us the deformation of the member BC in the direction of T

$$\delta l = \sum \frac{2(P+kT)kl}{2AE} = \sum \frac{(P+kT)kl}{AE} \qquad \ldots(i)$$

7.   As a result of the pull T, in the redundant member BC, we can also find out the deformation of the member BC i.e.

$$\delta l = -\frac{Tl_0}{A_0 E} \qquad \ldots(ii)$$

where l_0, A_0 and E stand for the length, cross-sectional area and modulus of elasticity of the redundant member BC. Minus sign in equation (ii) has been used as the deformation in equation (i) gives the value of δl in the direction of W. But as a result of the pull T, the deformation will be in the opposite direction.

8.   Equating (i) and (ii),

$$-\frac{Tl_0}{A_0 E} = \sum \frac{(P+kT)kl}{AE}$$

$$= \sum \frac{Pkl}{AE} + \frac{Tk^2 l}{AE}$$

$$\therefore \quad \sum \frac{Tk^2 l}{AE} + \frac{Tl_0}{A_0 E} = -\sum \frac{Pkl}{AE}$$

$$T\left[\sum\left(\frac{k^2 l}{AE}\right) + \frac{l_0}{A_0 E}\right] = -\sum \frac{Pkl}{AE}$$

or

$$T = -\frac{\displaystyle\sum \frac{Pkl}{AE}}{\displaystyle\sum\left(\frac{k^2 l}{AE}\right) + \frac{l_0}{A_0 E}}$$

Once the value of T has been obtained, the actual force in the remaining members *i.e.* (P_1+k_1T), (P_2+k_2T), (P_3+k_3T)......may be easily found out.

Note. The selection of the redundant member should be such that after its removal, the frame does not get distorted.

Example 18·1. *Three rods AC, DC and BC are hinged at A, D and B and connected at C, carry a load of 2,000 kg as shown in Fig. 18·3.*

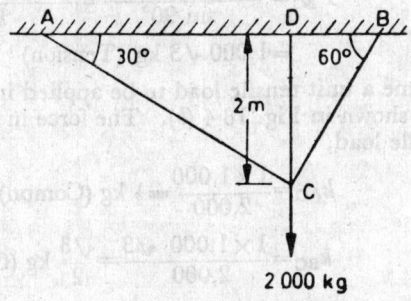

Fig. 18·3

Find the forces in the various members of the frame, if the cross-sectional areas of AC and BC is 8 cm², and that of CD is 4 cm²

Solution.

Given. Load at $C=2,000$ kg

Areas of AC and $BC=8$ cm²

Area of CD $A_0=4$ cm⁴   [considering the member CD to be a redundant member]

Length of CD $l_0=2$ m

Let $T=$Force in the redundant member CD.

From the geometry of the figure, we find that the length of member $AC=4$ m and that of BC is $\dfrac{4}{\sqrt{3}}$ m.

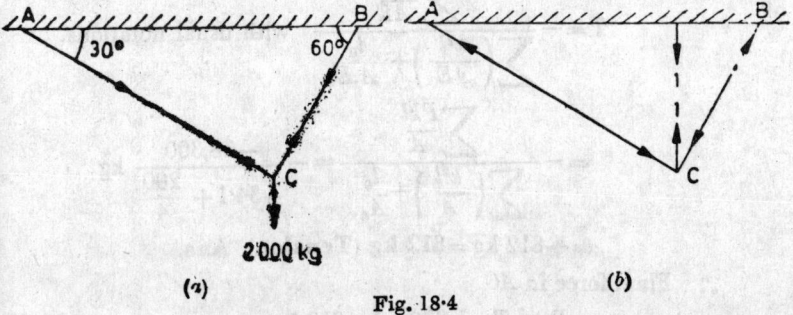

Fig. 18·4

Now assume the redundant member CD to be removed as shown in Fig. 18·4(a).

Using Lami's equation at CD,

$$\frac{2,000}{\sin 90°} = \frac{P_{AC}}{\sin 150°} = \frac{P_{BC}}{\sin 120°}$$

$$\therefore \qquad P_{AC} = \frac{2,000 \times \sin 150°}{\sin 90°} = \frac{2,000 \times 0.5}{1}$$

$$= 1,000 \text{ kg (Tension)}$$

and

$$P_{BC} = \frac{2,000 \times \sin 120°}{\sin 90°} = \frac{2,000 \times \frac{\sqrt{3}}{2}}{1}$$

$$= 1,000 \sqrt{3} \text{ kg (Tension)}$$

Now, assume a unit tensile load to be applied in the redundant member CD, as shown in Fig. 18·4 (b). The force in member AC due to this unit tensile load,

$$k_{AC} = \frac{1 \times 1,000}{2,000} = \tfrac{1}{2} \text{ kg (Compn)}$$

and

$$k_{BC} = \frac{1 \times 1,000 \sqrt{3}}{2,000} = \frac{\sqrt{3}}{2} \text{ kg (Compn)}$$

Now complete the following table :

Tension +ve ; Compn —ve

Member	l cm	A cm^2	P kg	k kg	$\frac{Pkl}{A}$	$\frac{k^2l}{A}$	Final force $P+kT$
AC	400	8	$+1,000$	$-\tfrac{1}{2}$	$-25,000$	$+12·5$	$+\ \ 594$ kg
BC	$\frac{400}{\sqrt{3}}$	8	$+1,000\sqrt{3}$	$-\frac{\sqrt{3}}{2}$	$-43,300$	$+21·6$	$+1,029$ kg

$$\Sigma-68,300 \qquad \Sigma+34·1$$

Using the relation,

$$T = -\frac{\sum \dfrac{Pkl}{AE}}{\sum \left(\dfrac{k^2l}{AE}\right) + \dfrac{l_0}{A.E}} \quad \text{with usual notations.}$$

$$= -\frac{\sum \dfrac{Pkl}{A}}{\sum \left(\dfrac{k^2l}{A}\right) + \dfrac{l_0}{A_0}} = -\frac{-68,300}{+34·1 + \dfrac{200}{4}} \text{ kg}$$

$$= +812 \text{ kg} = 812 \text{ kg (Tension) Ans.}$$

$\therefore$ Final force in AC

$$= P + kT = 1,000 - \tfrac{1}{2} \times 812 \text{ kg}$$

$$= 594 \text{ kg (Tension) Ans.}$$

Final force in BC

$$=P+kT=1,000\sqrt{3}-\frac{\sqrt{3}}{2}\times 812 \text{ kg}$$

$$=1,029 \text{ kg (Tension) Ans.}$$

Now complete the last column of the table.

Example 18·2. *A statically indeterminate frame shown in Fig. 18·5, carries a load of 10 tonnes at A.*

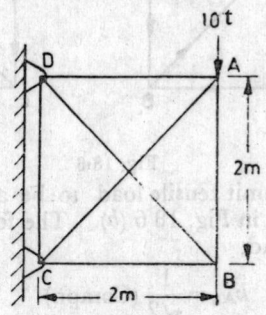

Fig. 18·5

Find the forces in all the members. The cross-sectional area and E are the same for all the members. 'Mysore University, 1978)

Solution.

Given.   Load at A $=10$ t

Lengths of member
     AB, BC, CD and $DA=2$ m

Cross-sectional area of
     AB, BC, CD, DA and $BD=A$

Length of member $BD=2\sqrt{2}$ m

Cross-sectional area of AC, $A_0=A$ [Considering the member AC to be a redundant member]

Length of member AC, $l_0=2\sqrt{2}$ m

Let $T=$ Force in redundant member AC

Now assume the redundant member AC to be removed as shown in Fig. 18·6 (a). Taking moments about D, we find that horizontal reaction at C is equal to 10 t. Therefore vertical and horizontal reactions at C are also equal to 10 t as shown in Fig. 18·6 (b). From the figure, we find that the force in member AB,

$$P_{AB}=10 \text{ t (Compn)}$$
$$P_{BC}=10 \text{ t (Compn)}$$
$$P_{BD}=10\sqrt{2} \text{ t (Tension)}$$

$$P_{CD}=0$$

and
$$P_{DA}=0.$$

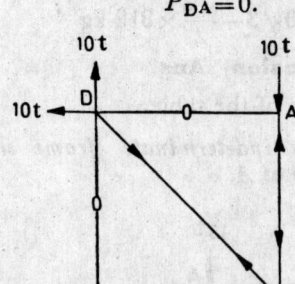

Fig. 18·6

Now assume a unit tensile load to be applied in the redundant member AC as shown in Fig. 18 6 (b). The force in member AB, due to this unit tensile load,

$$k_{AB}=\frac{1}{\sqrt{2}} \text{ (Compn)}$$

$$k_{BC}=\frac{1}{\sqrt{2}} \text{ (Compn)}$$

$$k_{BD}=1 \quad \text{(Tension)}$$

$$k_{CD}=\frac{1}{\sqrt{2}} \text{ (Compn)}$$

$$k_{DA}=\frac{1}{\sqrt{2}} \text{ (Compn)}$$

Now complete the following table.

Tension $+ve$; Comp. $-ve$

Member	l	A	P	k	Pkl	k^2l	Final force $P+kT$
AB	2·0	A	$-10\cdot0$	$-\dfrac{1}{\sqrt{2}}$	14·14	$+1$	-5 t
BC	2·0	A	$-10\ 0$	$-\dfrac{1}{\sqrt{2}}$	14·14	$+1$	-5 t
BD	$2\sqrt{2}$	A	$+10\sqrt{2}$	$+1$	40·0	$+2\cdot828$	$+7\cdot07$ t
CD	2·0	A	0	$-\dfrac{1}{\sqrt{2}}$	0	$+1$	-5 t
DA	2·0	A	0	$-\dfrac{1}{\sqrt{2}}$	0	$+1$	-5 t
				Σ	68·28	6 828	

Using the relation,

$$T = -\frac{\sum \dfrac{Pkl}{AE}}{\sum \left(\dfrac{k^2l}{AE}\right) + \dfrac{l_0}{A_0E}} \quad \text{with usual notations.}$$

$$= -\frac{\Sigma .Pkl}{\Sigma(k^2l) + l} = -\frac{68 \cdot 28}{6 \cdot 828 + 2\sqrt{2}} \text{ t}$$

$$= -7 \cdot 07 \text{ t}$$

$$= 7 \cdot 07 \text{ t (Compn) Ans.}$$

$\therefore$ Final force in $AB = P + kT = -10 \cdot 0 - \dfrac{1}{\sqrt{2}} \times 7 \cdot 07 = -15$ t

Final force in $BC = P + kT = -10 \cdot 0 - \dfrac{1}{\sqrt{2}} \times 7 \cdot 07 = -15$ t

Final force in $BD = P + kT = +10\sqrt{2} - 7 \cdot 07 = +7 \cdot 07$ t

Final force in $CD = P + kT = 0 - \dfrac{1}{\sqrt{2}} \times 7 \cdot 07 = -5$ t

and Final force in $DA = P + kT = 0 - \dfrac{1}{\sqrt{2}} \times 7 \cdot 07 = -5$ t

Now complete the last column of the table.

Example 18·3. *Determine the forces in the members AC and BE of a pin-jointed truss shown in Fig. 18·7.*

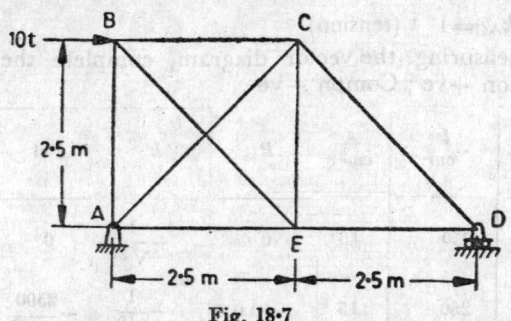

Fig. 18·7

Assume cross-sectional area of each member to be 15 cm².

University, 1979

Solution.

Given. Cross-sectional area of each member,

$$A = 15 \text{ cm}^2$$

Consider BE to be a redundant member, and assume it to be removed.

Force in member BE

Let T = Force in the redundant member BE

Now assume a unit tensile load to be applied in place of the redundant member BE as shown in Fig. 18·9.

In order to determine the forces in the various members of the truss, let us find out the reaction and draw the space and vector diagrams as shown in Fig. 18·8 (a) and (b).

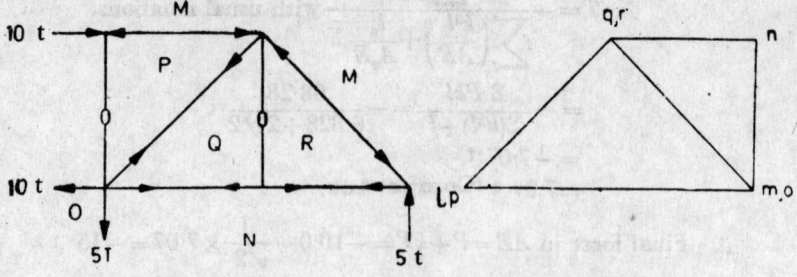

(a) Space diagram   (b) Vector diagram

Fig. 18·8.

Now the force in member AB due to this unit tensile load,

$$k_{AB} = \frac{1}{\sqrt{2}} \text{ t (Compn.)}$$

$$k_{BC} = \frac{1}{\sqrt{2}} \text{ t (Compn.)}$$

$$k_{CE} = \frac{1}{\sqrt{2}} \text{ t (Compn.)}$$

$$k_{EA} = \frac{1}{\sqrt{2}} \text{ t (Compn.)}$$

Fig. 18·9

$$k_{AC} = 1 \text{ t (tension)}$$

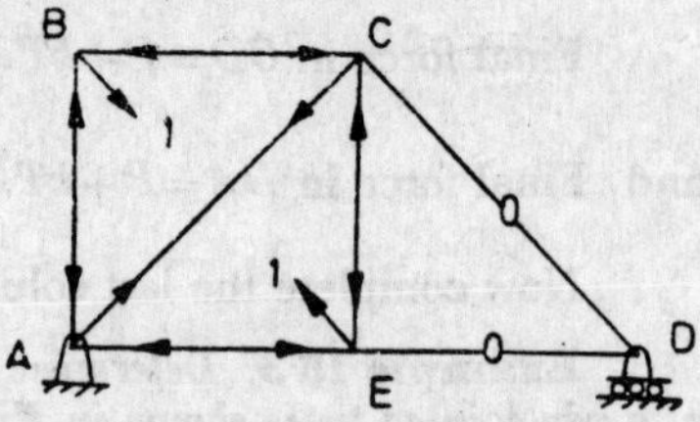

After measuring the vector diagram, complete the following table. Tension +ve ; Compn −ve

Member	l cm	A cm^2	P	k	Pkl	$k^2 l$
AB	250	15	0	$-\dfrac{1}{\sqrt{2}}$	0	$+125$
BC	250	15	$-10\cdot0$	$-\dfrac{1}{\sqrt{2}}$	$+\dfrac{2500}{\sqrt{2}}$	$+125$
CD	$250\sqrt{2}$	15	$-5\sqrt{2}$	0	0	0
DE	250	15	$+5\cdot0$	0	0	0
EA	250	15	$+5\cdot0$	$-\dfrac{1}{\sqrt{2}}$	$-\dfrac{1250}{\sqrt{2}}$	$+125$
CE	250	15	0	$-\dfrac{1}{\sqrt{2}}$	0	$+125$
AC	$250\sqrt{2}$	15	$+5\sqrt{2}$	$+1\cdot0$	$+2500$	$250\sqrt{2}$
				Σ	$+3,380$	$853\cdot5$

Using the relation,

$$T = -\frac{\displaystyle\sum \frac{Pkl}{AE}}{\displaystyle\sum \left(\frac{k^2 l}{AE}\right) + \frac{l_0}{A_0 E}} \quad \text{with usual notations.}$$

$$= -\frac{\Sigma Pkl}{\Sigma k^2 l + l_0} = -\frac{3,380}{853 \cdot 5 + 250\sqrt{2}} = -2 \cdot 8 \text{ t}$$

$$= 2 \cdot 8 \text{ t (Compn) Ans.}$$

Force in member AC

We know that the force in member AC

$$= P + kT = 5\sqrt{2} - 1 \times 2 \cdot 8 \text{ t}$$

$$= 4 \cdot 27 \text{ t (Tension) Ans.}$$

Example 18·4. *Find the forces in the various members of the truss given in Fig. 18·10.*

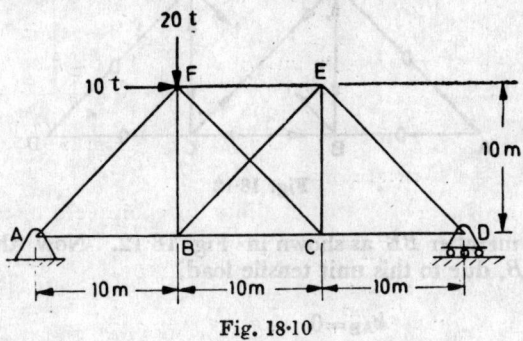

Fig. 18·10

The ratio of length to the cross-sectional area for all the members is the same. The frame is pinned at A and rests on rollers at D.

(*Oxford University,*)

Solution.

Given. $\dfrac{l}{A}$ is constant for all the members.

Consider BE to be a redundant member, and assume it to be removed.

Let $\qquad T =$ Force in the redundant member BE.

In order to determine the forces in the various members of the truss, let us find out the reactions and draw the space and vector diagrams as shown in Fig. 18·11 (a) and (b).

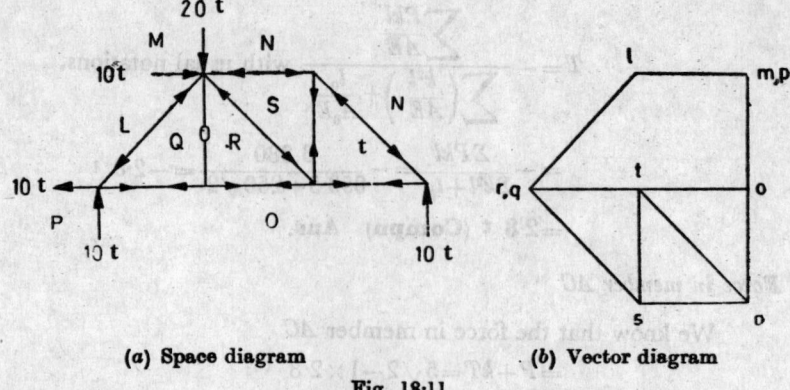

(a) Space diagram (b) Vector diagram

Fig. 18·11

Now assume a unit tensile load to be applied in place of the

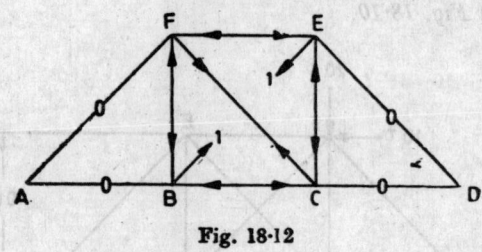

Fig. 18·12

redundant member BE as shown in Fig. 18·12. Now the force in member AB, due to this unit tensile load,

$$k_{AB}=0$$

$$k_{BC}=\frac{1}{\sqrt{2}} \text{ 't' (Compn)}$$

$$k_{CD}=0$$

$$k_{DE}=0$$

$$k_{EF}=\frac{1}{\sqrt{2}} \text{ t' (Compn)}$$

$$k_{FA}=0$$

$$k_{BF}=\frac{1}{\sqrt{2}} \text{ (Compn)}$$

$$k_{CF}=1 \text{ (Tension)}$$

$$k_{CE}=\frac{1}{\sqrt{2}} \text{ (Compn)}$$

After measuring the vector diagram complete the following table* :

Tension +ve ; Compn. —ve

Member	P	k	Pk	k	Final force P+kt
AB	+20·0	0	0	0	+20·0
BC	+20·0	$-\dfrac{1}{\sqrt{2}}$	$-\dfrac{20}{\sqrt{2}}$	$+\frac{1}{2}$	+10·00
CD	+10·0	0	0	0	+10·0
DE	−10√2	0	0	0	−14·14
EE	−10·0	$-\dfrac{1}{\sqrt{2}}$	$+\dfrac{10}{\sqrt{2}}$	$+\frac{1}{2}$	−20·00
FA	− 0√2	0	0	0	−14·14
BF	0	$-\dfrac{1}{\sqrt{2}}$	0	$+\frac{1}{2}$	−10·00
CF	−10√2	+1	−10√2	+1·0	−7·07
CE	+10·0	$-\dfrac{1}{\sqrt{2}}$	$-\dfrac{10}{\sqrt{2}}$	$+\frac{1}{2}$	+0·00
		Σ	$-\dfrac{40}{\sqrt{2}}$	+3·0	

Using the relation,

$$T = -\frac{\displaystyle\sum \frac{Pkl}{AE}}{\displaystyle\sum \left(\frac{k^2 l}{AE}\right) + \frac{l_0}{A_0 E}} \quad \text{with usual notations.}$$

$$= -\frac{\Sigma Pk}{\Sigma k^2 + 1} = -\frac{-\dfrac{40}{\sqrt{2}}}{4} \; t'$$

$$= 7{\cdot}07 \; \text{t (Tension)} \quad \textbf{Ans.}$$

Now complete the last column of the table :

Exercise 18-1

1. Three rods AC, DC and BC are hinged at A, D and carry a load of 1 tonne as shown in Fig. 18·13.

─────────

*There is no necessity of providing the columns for l and A in the table, as the value of l/A is constant for all the members.

Determine the forces in the members AC, BC and CD, if their cross-sectional areas of all the members is the same.

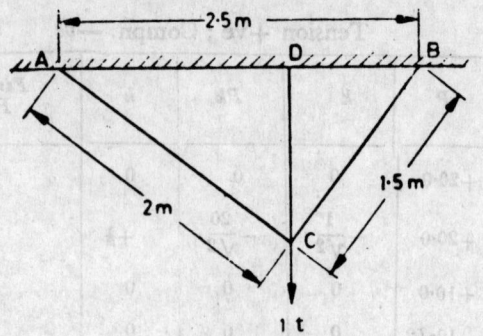

Fig. 18·13

[**Ans.** +0·25 t ; +0·83 t ; +0·583 t]

2. A steel frame consists of similar bars of 75 cm long as shown in Fig. 18·14.

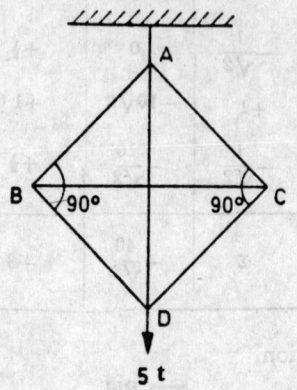

Fig. 18·14

Find the forces in members AB, AC, BD, CD, BC and AD, when a load of 5 tonnes is suspended from the joint D.

[**Ans.** +1·04 t ; +1·04 t ; +1·04 t ; + 1·04 t ; —1·0 t ; +3·54 t]

3. A steel framework shown in Fig. 18·15 is made up bars of same cross sectional areas.

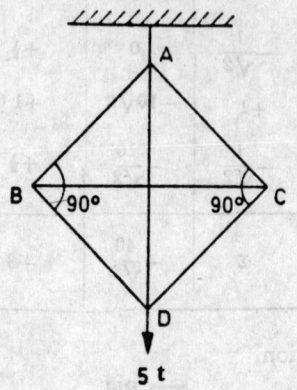

Fig. 18·15

Determine the forces in all the members of the frame, when it is subjected to vertical loads of 10 tonnes each at A and D.

[**Ans.** $AB, CD = -7\cdot07$ t ; $BC = -3\cdot96$ t ; $AF, DE = +7\cdot07$ t
$BF, CE = -3\cdot96$ t ; $BE, CF = -1\cdot46$ t ; $FE = +6\cdot03$ t]

18·6. Forces due to errors in the length of a member in a redundant frame

Sometimes, at the time of assembling a frame in a workshop, the length of some member is found to be slightly shorter or longer than the correct length of the member. In such cases, if the error in the length is very small, the member is forced in position. A little consideration will show, that if the frame is perfect one, no force is induced in the members of the frame. But if the frame is a redundant one, it will induce forces in all the members of the frame.

Now consider a redundant frame $ABCD$, in which the member AB is slightly *shorter than the correct length* as shown in Fig. 18·16.

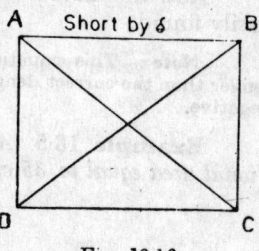

Fig. 18·16

In order to force the member AB in position, the joints A and B have to be pulled together by applying forces at them. After the member is fitted in the frame, the external forces on the joints A and B are released. As soon as the external forces are released, the two joints will tend to move outwards and will cause tension in the member AB. The forces in the various members of the frame due to this error in the length of the member AB may be found out as discussed below :

1. Assume the member AB to be removed and apply a unit pull at A and B, along the centre line of the member AB. Now find out the forces $k_1, k_2, k_3...$(*i.e.* pulls or pushes) in the various members of the frame.

2. Now, if T is the actual pull in the member AB, then the actual forces in the members will be $k_1T, k_2T, k_3T...$

3. The total strain energy in the frame (excluding the member AB),

$$= \sum \frac{(kT)^2 l}{2AE}$$

4. As per Castigliano's first theorem, we know that the partial differential coefficient of the above equation with respect to T will give us the deformation of the joints A and B in the direction of T.

$$\delta l_1 = \sum \frac{2kT.kl}{2AE} = T \sum \frac{k^2 l}{AE} \qquad ...(i)$$

5. As a result of the pull T, the deformation of the member AB,

$$\delta l_2 = \frac{Tl_0}{A_0 E}$$

where l_0, A_0 and E stand for the length, cross-sectional area and modulus of elasticity of the member AB.

6.   **Now we know that**

Total error=Deformation of joints A and B

+Deformation of the member AB.

$$\delta = \delta l_1 + \delta l_2 = T \sum \frac{k^2 l}{AE} + \frac{T l_0}{A_0 E}$$

$$= T \left[\sum \left(\frac{k^2 l}{AE} \right) + \frac{l_0}{A_0 E} \right]$$

$$\therefore \qquad T = \frac{\delta}{\Sigma \left(\dfrac{k^2 l}{AE} \right) + \dfrac{l_0}{A_0 E}}$$

Now the actual forces in the members $k_1 T$, $k_2 T$, $k_3 T$...may be easily found.

Note : This equation also holds good if the member AB is slightly longer than the correct length. In this case the value of δ should be taken as negative.

Example 18·5. *A triangular frame shown in Fig. 18·17 has sectional area equal to 35 cm² for all the members.*

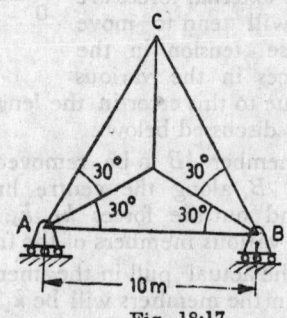

Fig. 18·17

The member AB was found to be 1 mm shorter than its correct length at the time of assembling. Find the forces in the members, if the member AB is forced in position. Take Young's Modulus as $2·1 \times 10^3$ t/cm².

Solution.

Given.   Area of members $A = 35$ cm²

Error in length, $\delta = 1 \text{mm} = 0·1$ cm

Young's modulus $E = 2·1 \times 10^3$ t/cm²

Let $T =$ Force in the member AB due to lack of fit.

Now assume the member AB to be removed and apply a unit tensile force at the joints in the direction of the member AB as shown in Fig. 18·18.

In order to determine the forces in the various members of the frame, let us draw the space and vector diagrams as shown in Fig. 18·18 (a) and (b).

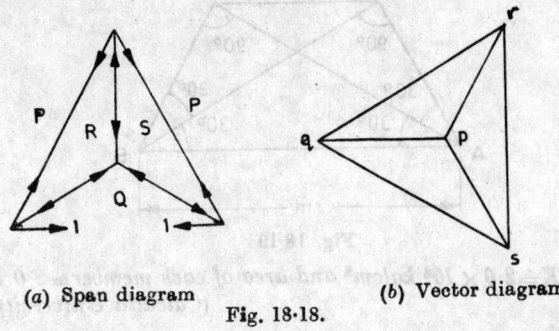

(a) Span diagram   (b) Vector diagram
Fig. 18·18.

Now measuring the various sides of the vector diagram complete the following table :

Member	l cm	k	$k^2 l$	Actual force (kT) in force
AC	1000	+1	1000	+0·9
BC	1000	+1	1000	+0·9
AD	576·7	$-\sqrt{3}$	1730	−1·56
CD	576·7	$-\sqrt{3}$	1730	−1·56
BD	576·7	$-\sqrt{3}$	1730	−1·56
		Σ	7,190	

Using the relation,

$$T = \frac{\delta}{\Sigma\left(\dfrac{k^2 l}{AE}\right) + \dfrac{l_0}{A_0 E_0}} \quad \text{with usual notations.}$$

$$= \frac{0·1}{\dfrac{7,190}{35 \times 2·1 \times 10^3} + \dfrac{1000}{35 \times 2·1 \times 10^3}} = 0·9 \text{ t}$$

Now complete the last column of the table.

Example 18·6. *The diagonal AC of the framework, shown in Fig. 18·19, was found to be 0·625 mm longer than its correct length, at the time of assembling.*

The member AC was forced in position.   Find the forces induced in the members of the frame.

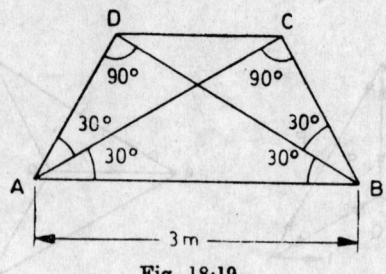

Fig. 18·19

Take E=2·0 × 10⁶ kg/cm² and area of each member = 2·0 cm.
                                                      (Calcutta University,·

Solution.

Given.   Error in length AC

$$\delta = -0·625 \text{ mm} = -0·0625 \text{ cm}$$

(Minus sign due to longer of the member AC)

Young's modulus, $E = 2·0 × 10⁶ \text{ kg/cm}²$

Dia. of members $= 2 \text{ cm}$

Area, $A = \dfrac{\pi}{4} × 2² = \pi \text{ cm}²$

Let $T = $Force in member AC due to lack of fit.

Now assume the member AC to be removed and apply a unit tensile force at the joints in the direction of member AC as shown in Fig. 18·20.

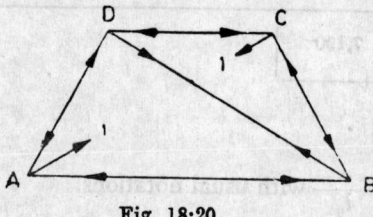

Fig. 18·20

Resolving the forces vertically at A,

$$P_{AD} \cos 30° = 1 \cos 60°$$

$$\therefore \quad P_{AD} = \frac{1 × 0·5}{\frac{\sqrt{3}}{2}} = \frac{1}{\sqrt{3}} \text{ (Compn.)}$$

Now resolving the forces horizontally at A,

$$P_{AB} = 1 \cos 30° - P_{AD} \cos 60°$$

$$= 1 × \frac{\sqrt{3}}{2} - \frac{1}{\sqrt{3}} × \frac{1}{2} = \frac{1}{\sqrt{3}} \text{ (Compn)}$$

Resolving the forces vertically at C,

$$P_{BC} \cos 30° = 1 \cos 60°$$

$$\therefore \quad P_{BC} = \frac{1 × 0·5}{\frac{\sqrt{3}}{2}} = \frac{1}{\sqrt{3}} \text{ (Compn)}$$

Now resolving the forces horizontally,
$$P_{CD} = 1 \cos 30° + P_{AD} \cos 60°$$

$$= 1 \times \frac{\sqrt{3}}{2} + \frac{1}{\sqrt{3}} \times \frac{1}{2} = \frac{2}{\sqrt{3}} \text{ (Compn)}$$

Now resolving the forces vertically at B,
$$P_{BD} \cos 60° = P_{BC} \cos 30°$$

$$\therefore \qquad P_{BD} = \frac{P_{BC} \cos 30°}{\cos 60°} = \frac{\frac{1}{\sqrt{3}} \times \frac{\sqrt{3}}{2}}{\frac{1}{2}} = 1 \text{ (Tension)}$$

Now prepare the following table :

Tension $+ve$;   Compn $-ve$

Member	l cm	A cm^2	k	$k^2 l$	Actual force (kT) kg
AB	300	π	$-\dfrac{1}{\sqrt{3}}$	100	$+246\cdot5$
BC	150	π	$-\dfrac{1}{\sqrt{3}}$	50	$+246\cdot5$
CD	150	π	$-\dfrac{2}{\sqrt{3}}$	200	$+493\cdot0$
AD	150	π	$-\dfrac{1}{\sqrt{3}}$	50	$+246\cdot5$
BD	260	π	$+1\cdot0$	260	$-427\cdot0$

Σ 660

Using the relation,

$$T = \frac{\delta}{\sum \left(\dfrac{k^2 l}{AE} \right) + \dfrac{l_0}{A_0 E}} \quad \text{with usual notations.}$$

$$= \frac{-0\cdot0625}{\dfrac{660}{\pi \times 2\cdot0 \times 10^6} + \dfrac{200}{\pi \times 2\cdot0 \times 10^6}} = -427 \text{ kg}$$

$$= 427 \text{ kg (Compn)   Ans.}$$

Now complete the last column of the table.

18·7. Frames with two or more redundant members

The procedure of finding out the forces in the various members of a frame, with two or more redundant members, is similar to that for the frames with one redundant member as discussed below :

1. Select any two members of the redundant frame (say *DF* and *DH*) to be redundant members. Now imagine these

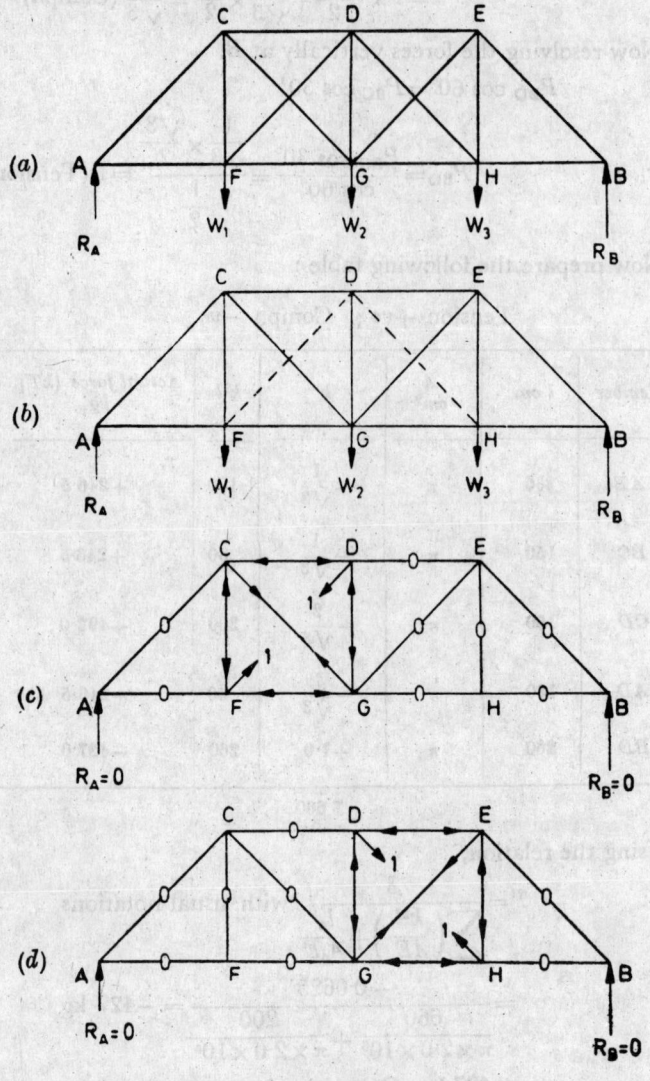

Fig. 18·21

redundant members to be removed, so that the remaining frame becomes a perfect one as shown in Fig. 18·21 (a) and (b).

2. Now find out the forces $P_1, P_2, P_3 \ldots$ (i.e., pulls or pushes in the various members of the perfect frame due to external loads.

3. Now apply a unit pull in the redundant member DF, which was assumed to be removed as shown in Fig. 18·2 (c)l and find out the forces k_1, k_2, k_3......(*i.e.* pulls or pushes) in the various members of the frame, due to the unit pull.

4. Now, if T is the actual pull in the redundant member DF, then the force in the members will be

$$(P_1 + k_1 T),\ (P_2 + k_2 T),\ (P_3 + k_3 T)......$$

5. Similarly, apply a unit pull in the redundant member DH, which was assumed to be removed as shown in Fig. 18·21 (d) and find out the forces k_1', k_2', k_3'......(*i.e.*, pulls or pushes) in the various members of the frame, due to the unit pull.

6. Now, if T' is the actual pull in the redundant member DH then the actual forces in the members will be

$$(P_1 + k_1 T + k_1' T'),\ (P_2 + k_2 T + k_2' T'),\ (P_3 + k_3 T + k_3 T')......$$

7. The total strain energy in the frame (excluding the redundant members),

$$U = \sum \frac{(P + kT + k'T')^2}{2AE}$$

8. As per Castigliano's first theorem, we know that the partial differential coefficient of the above equation with respect to T and T' will give us the deformation of the members DF and DH in the directions of T and T'

∴
$$\delta l = \sum \frac{2(P + kT + k'T')kl}{2AE}$$

$$= \sum \frac{(P + kT + k'T')kl}{AE}$$

$$= \sum \frac{Pkl}{AE} + T \sum \frac{k^2 l}{AE} + T' \sum \frac{kk'l}{AE} \qquad ...(i)$$

9. As a result of the pull T in the redundant member DE, we can also find out the deformation of the member DF, *i.e.*,

$$\delta l = - \frac{Tl}{AE} \qquad ...(ii)$$

where l, A and E stand for the length, cross-section area and modulus of elasticity of the redundant member DF. Minus sign in equation (*ii*) has been used as the deformation in equation (*i*) gives the value of δl in the direction of W. But as a result of the pull T the deformation will be in the opposite direction.

10. Equating (*i*) and (*ii*)

$$\sum \frac{Pkl}{AE} + T \sum \frac{k^2 l}{AE} + T' \sum \frac{kkl}{AE} = - \frac{Tl}{AE} \qquad ...(iii)$$

11. Similarly by considering the redundant member DH, we get

$$\sum \frac{Pk'l}{AE} + T' \sum \frac{k^2 l}{AE} + T \sum \frac{kk'l}{AE} = - \frac{T'l}{AE}$$

From the above equations (*iii*) and (*iv*), we can find out the values of T and T'. Now the actual force *i.e.*,

$$(P_1 + k_1 T + k_1' T'),\ (P_2 + k_2 T + k_2' T')\ldots\ldots$$

in the remaining members may be easily found out.

Note. The selection of redundant members, unless mentioned otherwise should be such that after their removal, the frame does not get distorted.

Example 18·7. *Find the forces in the members FD and DH of the frame shown in Fig. 18·22.*

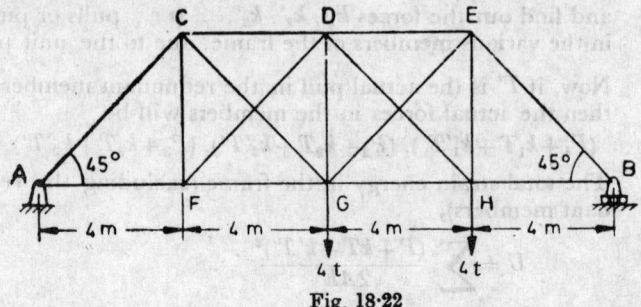

Fig. 18·22

The ratio of length to the cross-sectional area for all the members is the same.

Solution.

Given. $\dfrac{l}{A}$ is constant for all the members.

Consider the members FD and DH to be redundant members and assume them to be removed.

Let $T =$ Force in the redundant member FD, and

$T' =$ Force in the redundant member DH.

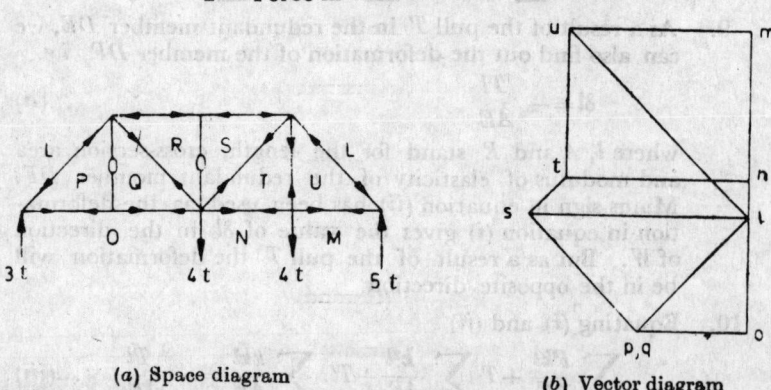

(a) Space diagram   (b) Vector diagram

Fig. 18·23

In order to determine the forces in various members of the truss, find out the reactions and draw the vector diagrams as shown in Fig. 18·23 (*b*).

Now assume a unit tensile load to be applied in place of the redundant member *FD* as shown in Fig. 18·24. Now the force in member *CD*,

$$k_{CD}=\frac{1}{\sqrt{2}} \text{ (Compn)}$$

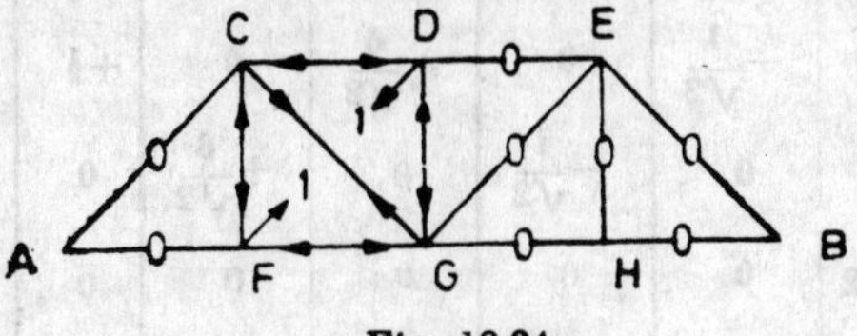

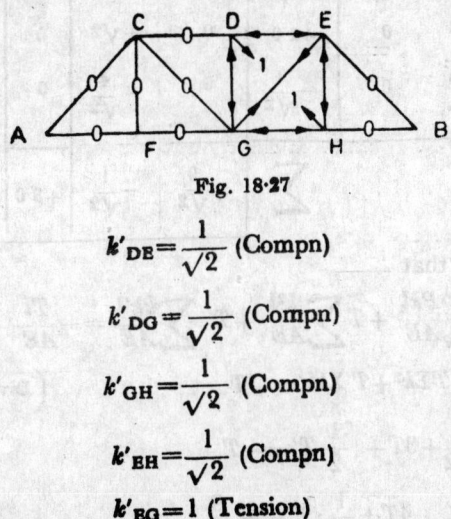

Fig. 18·24

$$k_{DG}=\frac{1}{\sqrt{2}} \text{ (Compn)}$$

$$k_{FG}=\frac{1}{\sqrt{2}} \text{ (Compn)}$$

$$k_{CF}=\frac{1}{\sqrt{2}} \text{ (Compn)}$$

$$k_{CG}=1 \text{ (Tension)}$$

Similarly, now assume a unit tensile load to be applied in place of the redundant member *DH* as shown in Fig. 18·25. Now the force in member *DE*,

Fig. 18·27

$$k'_{DE}=\frac{1}{\sqrt{2}} \text{ (Compn)}$$

$$k'_{DG}=\frac{1}{\sqrt{2}} \text{ (Compn)}$$

$$k'_{GH}=\frac{1}{\sqrt{2}} \text{ (Compn)}$$

$$k'_{EH}=\frac{1}{\sqrt{2}} \text{ (Compn)}$$

$$k'_{EG}=1 \text{ (Tension)}$$

After measuring the vector diagram, complete the following table.*

*There is no need of providing the columns for *l* and *A*, in the table as the value of *l/A* is constant for all the members.

Tension +ve ; Compn. −ve

Member	P	k	k'	Pk	Pk'	k^2	$k^{2\prime}$	kk'
AC	$-3\sqrt{2}$	0	0	0	0	0	0	0
CD	$-6\cdot0$	$-\dfrac{1}{\sqrt{2}}$	0	$+\dfrac{6}{\sqrt{2}}$	0	$+\tfrac{1}{2}$	0	0
DE	$-6\cdot0$	0	$-\dfrac{1}{\sqrt{2}}$	0	$+\dfrac{6}{\sqrt{2}}$	0	$+\tfrac{1}{2}$	0
EB	$-5\sqrt{2}$	0	0	0	0	0	0	0
AF	$+3\cdot0$	0	0	0	0	0	0	0
FG	$+3\cdot0$	$-\dfrac{1}{\sqrt{2}}$	0	$-\dfrac{3}{\sqrt{2}}$	0	$+\tfrac{1}{2}$	0	0
GH	$+5\cdot0$	0	$-\dfrac{1}{\sqrt{2}}$	0	$-\dfrac{5}{\sqrt{2}}$	0	$+\tfrac{1}{2}$	0
HB	$+5\cdot0$	0	0	0	0	0	0	0
CF	0	$-\dfrac{1}{\sqrt{2}}$	0	0	0	$+\tfrac{1}{2}$	0	0
CG	$+3\sqrt{2}$	$+1\cdot0$	0	$+3\sqrt{2}$	0	$+1\cdot0$	0	0
DG	0	$-\dfrac{1}{\sqrt{2}}$	$-\dfrac{1}{\sqrt{2}}$	0	0	$+\tfrac{1}{2}$	$+\tfrac{1}{2}$	$+\tfrac{1}{2}$
EG	$+\sqrt{2}$	0	$+1\cdot0$	0	$+\sqrt{2}$	0	$+1\cdot0$	0
EH	$+4\cdot0$	0	$-\dfrac{1}{\sqrt{2}}$	0	$-\dfrac{4}{\sqrt{2}}$	0	$+\tfrac{1}{2}$	0
$\sum$				$+\dfrac{9}{\sqrt{2}}$	$-\dfrac{1}{\sqrt{2}}$	$+3\cdot0$	$+3\cdot0$	$+\tfrac{1}{2}$

We know that

$$\sum\frac{Pkl}{AE}+T\sum\frac{k^2l}{AE}+T'\sum\frac{kk'l}{AE}=-\frac{Tl}{AE}$$

or $\Sigma Pk+T\Sigma k^2+T'\Sigma kk'=-T$ $\left(\text{Dividing by } \dfrac{l}{AE}\right)$

$\therefore\;\;+\dfrac{9}{\sqrt{2}}+3T+\dfrac{1}{2}T'=-T$

or $4T+\dfrac{1}{2}\,T'=-\dfrac{9}{\sqrt{2}}$ ...(i)

We also know that

$$\sum\frac{Pk'l}{AE}+T'\sum\frac{k^2l}{AE}+T\sum\frac{kk'l}{AE}=-\frac{T'l}{AE}$$

or $$\sum Pk' + T' \sum k'^2 + T \sum kk' = -T' \qquad \left(\text{Dividing by } \frac{l}{AE} \right)$$

$$\therefore \quad -\frac{1}{\sqrt{2}} + 3T' + \frac{1}{2} T = -T'$$

$$4T' + \frac{1}{2} T = \frac{1}{\sqrt{2}} \qquad \qquad \dots (ii)$$

Multiplying the equation (i) by 8 and subtracting from equation (ii), we get

$$T = -1 \cdot 62 \text{ t} = 1 \cdot 62 \text{ t (Compn) Ans.}$$

Substituting the value of T in equation (i), we get

$$T' = 0 \cdot 39 \text{ t (Tension) Ans.}$$

18·8. Trussed beams

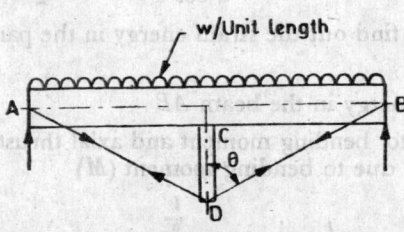

Fig. 18·26

Sometimes a beam is stiffened by a strut (or struts) and tie rods to decrease its deflection. Such a stiffened beam is a statically indeterminate structure and is called a *trussed beam*. Now consider a trussed beam AB with strut CD braced to the beam ends A and B by tie rods AD and BD as shown in Fig. 18·26. Let the beam AB be subjected to a load of w per unit length.

Let
$l =$ Length of beam AB,

$A =$ Cross-sectional area of beam AB,

$E =$ Modulus of elasticity for the beam material,

$I =$ Moment of inertia of the beam section.

$l_1, A_1, E_1, I_1 =$ Corresponding values for the tie rods AD and BD,

$l_2, A_2, E_2, I_2 =$ Corresponding values for the strut CD.

$\theta =$ Inclination of the tie rod with the strut.

As a result of loading on the beam, let P be the thrust in the strut and T be the tension in each tie rod. The values of P and T may be found out by the principle of minimum strain energy. Resolving the forces vertically at D,

$$2 T \cos \theta = P$$

$$\therefore \quad T = \frac{P}{2 \cos \theta} = \frac{P}{2} \sec$$

The reaction at each support due to loading

$$= \frac{wl}{2}$$

The net vertical reaction at each support

$$= \frac{wl}{2} - \text{vertical component of } T$$

$$= \frac{wl}{2} - T \cos \theta = \frac{wl}{2} - \frac{P}{2 \cos \theta} \times \cos \theta$$

$$= \frac{wl - P}{2}$$

The axial thrust exerted by the tie rods in the beam

$$= T \sin \theta = \frac{P}{2 \cos \theta} \times \sin \theta = \frac{P}{2} \tan \theta$$

Now let us find out the strain energy in the parts of the trussed beam.

(i) Strain energy in the beam AB

It is due to bending moment and axial thrust. We know that the strain energy due to bending moment (M)

$$= \int_0^l \frac{M^2 dx}{2EI} = 2 \int_0^{\frac{l}{2}} \frac{M^2 dx}{2EI}$$

$$= \frac{1}{EI} \int_0^{\frac{l}{2}} M^2 dx = \frac{1}{EI} \int_0^{\frac{l}{2}} \left[\left(\frac{wl - P}{2} \right) x - \frac{wx^2}{2} \right]^2 dx$$

$$= \frac{1}{4EI} \int_0^{\frac{l}{2}} \left[(wl - P)^2 x^2 + w^2 x^4 - 2w (wl - P) x^3 \right] dx$$

$$= \frac{1}{4EI} \left[(wl - P)^2 \times \frac{x^3}{3} + \frac{w^2 x^5}{5} - 2w(wl - P) \frac{x^4}{4} \right]_0^{\frac{l}{2}}$$

$$= \frac{l^3}{4EI} \left[\frac{(wl - P)^2}{24} + \frac{w^2 l^2}{160} - \frac{wl (wl - P)}{32} \right] \qquad \ldots(i)$$

and strain energy due to axial thrust $\left(\frac{P}{2} \tan \theta \right)$

$$= \left(\frac{P \tan \theta}{2A} \right)^2 \times \frac{Al}{2E} = \frac{P^2 \tan^2 \theta . l}{8AE} \qquad \ldots(ii)$$

(ii) Strain energy in tie rods AD and BD

We know that the strain energy in two tie rods due to tension (T)

$$=2\left(\frac{T}{A_1}\right)^2\times\frac{A_1l_1}{2E_1}=2\left(\frac{P\sec\theta}{2A_1}\right)^2\times\frac{A_1l_1}{2E_1}$$

$$=\frac{P^2\sec^2\theta.l_1}{4A_1E_1}\qquad\qquad\text{...}(iii)$$

(iii) Strain energy in strut CD

We know that the strain energy in the strut due to thrust (P)

$$=\left(\frac{P}{A_2}\right)^2\times\frac{A_2l_2}{2E_2}=\frac{P^2l_2}{2A_2E_2}\qquad\qquad\text{...}(iv)$$

The total strain energy stored in the trussed beam

$$U=\frac{l^3}{4EI}\left[\frac{(wl-P)^2}{24}+\frac{w^2l^2}{160}-\frac{wl\,(wl-P)}{32}\right]+\frac{P^2\tan^2\theta.l}{8AE}$$
$$+\frac{P^2\sec^2\theta.l_1}{4A_1E_1}+\frac{P^2l_2}{2A_2E_2}$$

The value of P may now be determined by the condition that the strain energy should be a minimum *i.e.*, $\frac{dU}{dP}=0$. Therefore

$$\frac{l^3}{4EI}\left[-\frac{2(wl-P)^2}{24}+\frac{wl}{32}\right]+\frac{2P\tan^2\theta.l}{8AE}+\frac{2P\sec^2\theta.l_1}{4A_1E_1}+\frac{2Pl_2}{2A_2E_2}=0$$

or $\frac{l^3}{4EI}\left[\frac{wl}{32}-\frac{(wl-P)^2}{12}\right]+\frac{P\tan^2\theta.l}{4AE}+\frac{P\sec^2\theta.l}{2A_1E_1}+\frac{Pl_2}{A_2E_2}=0$

Example 18·8. *A trussed beam 5 m long made of 30 cm deep and 20 cm wide timber beam is stiffened with the help of cast iron strut 75 cm long and 50 cm² in area and two tie cast iron rods of 3 cm diameter. The beam carries a uniformly distributed load of 2 t/m. Determine the thrust in strut, tension in the tie rods. Take E for timber, cast iron and steel as 100, 900 and 2000 t/cm² respectively.*

Solution.

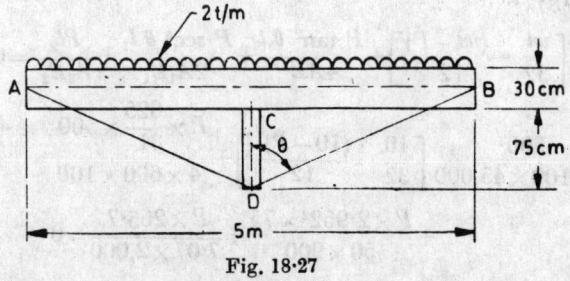

Fig. 18·27

Given. Length of beam,

$$l=5\text{ m}=500\text{ cm}$$

Depth of beam, $d = 30$ cm

Breadth of beam, $b = 20$ cm

$\therefore$ Area of beam, $A = 30 \times 20 = 600$ cm^2

and Moment of Inertia, $I = \dfrac{20 \times 30^3}{12} = 4{,}500$ cm^4

Length of strut. $l_1 = 75$ cm

Area of strut $A_1 = 50$ cm^2

Dia. of rod $= 3$ cm

$\therefore$ Area of tie rod,

$$A_2 = \frac{\pi}{4} \times 3^2 = 7 \cdot 07 \text{ cm}^2$$

and Length of tie rod, $l_2 = \sqrt{250^2 + 90^2} = 265 \cdot 7$ cm

Load on the beam, $w = 2$ t/m

$\therefore$ Total load on the beam

$$= wl = 2 \times 5 = 10 \text{ t}$$

E for beam, $E = 100$ t/cm^2

E for strut, $E_1 = 900$ t/cm^2

E for tie rod, $E_2 = 2{,}000$ t/cm^2

Thrust in the strut

Let $P = $ Thrust in the strut

From the geometry of the figure, we find that

$$\tan \theta = \frac{250}{75 + 15} = \frac{250}{90} = \frac{25}{9}$$

and

$$\sec \theta = \frac{1}{\cos \theta} = \frac{265 \cdot 7}{75 + 15} = 2 \cdot 952$$

Now substituting these values in the condition for minimum strain energy *i.e.*,

$$\frac{l^3}{4EI}\left[\frac{wl}{32} - \frac{(wl - P)}{12} \right] + \frac{P \tan^2 \theta . l}{4AE} + \frac{P \sec^2 \theta . l_1}{2A_1 E_1} + \frac{Pl_2}{A_2 E_2} = 0$$

$$\frac{500^3}{4 \times 100 \times 45{,}000}\left[\frac{10}{32} - \frac{(10 - P)}{12} \right] + \frac{P \times \dfrac{625}{81} \times 500}{4 \times 600 \times 100}$$

$$+ \frac{P \times 2 \cdot 952^2 \times 75}{50 \times 900} + \frac{P \times 265 \cdot 7}{7 \cdot 07 \times 2{,}000} = 0$$

Solving the above equation, we get

$$P = 5 \cdot 33 \text{ t Ans.}$$

Tension in the tie rods

Let $\qquad T=$ Tension in the tie rods

Using the relation,

$$T=\frac{P}{2}\ \sec\ \theta \text{ with usual notations}$$

$$=\frac{5\cdot53}{2}\times2\cdot952=7\cdot87\ \text{t}\quad\textbf{Ans}.$$

Exercise 18-2

1. A pin-jointed frame shown in Fig. 18-28 is made up of bars with 10 cm² cross-sectional area.

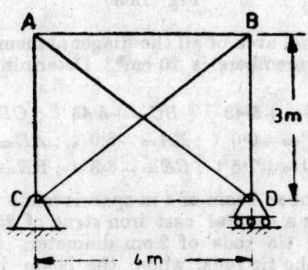

Fig. 18·28

The member *BC* was found to be short by 1 cm and then forced in position. Find the forces in all members of the frame. Take $E=2\cdot0\times10^3$ tonnes/cm².

(Calcutta University)

[**Ans.** AB, $GD=-9\cdot26$ t ; AC, $BD=-6\cdot94$ t ; AD, $BC=+11\cdot57$ t]

2. Fig. 18·29 shows a truss carrying a point load of 12 tonnes at *F*.

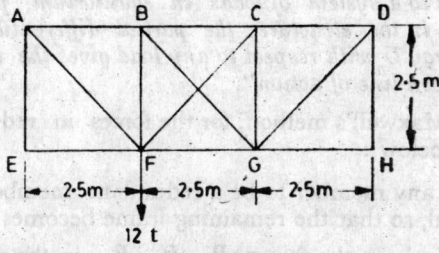

Fig. 18·29

The sectional areas of the top members $=75$ cm², bottom members $=50$ cm², vertical members $=30$ cm² and diagonal members $=25$ cm². Determine the forces in the various members of the truss. *(Nagpur University)*

[**Ans.** $AB=EF=GH=0$; $CD=AE=DH=+4$ t ;

$FG=-5\cdot88$ t ; $BC=+6\cdot12$ t ;

$BF=-1\cdot88$ t ; $CG=+2\cdot12$ t ;

$AF=-8\sqrt2$ t ; $BG=-3$ t ; $DG=-4\sqrt2$ t]

3. A steel frame of span 7 m is carrying a load of 20 T as shown in Fig. 18·30.

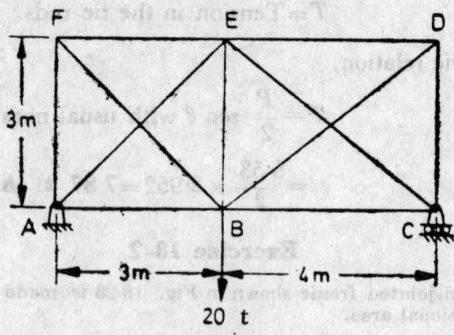

Fig. 18·30

The cross-sectional area of all the diagonal members is 10 cm², whereas that of the other chord members is 20 cm². Determine the forces in all the members of the frame.   (*Rajasthan University*)

[**Ans.** $AB = +5·43$ t ; $BC = +5·43$ t ; $CD = -4·5$ t ; $DE - 6·0$ t ;
$EF = -6·0$ t ; $FA = -6·0$ t ; $AE = -7·7$ t ; $BF = +8·46$ t ;
$BD = +7·5$ t ; $CE = -6·8$ t ; $BE = +9·5$ t]

4. A trussed timber beam of 4 m span is made up of timber 12 cm wide and 25 cm deep. It has a central cast iron strut of 25 cm² sectional area and 62·5 cm long with steel tie rods of 2 cm diameter. Determine the thrust in the strut and pull in the tie rods, when the beam is subjected to a load of 2 t/m.   (*Utkal University*)

[**Ans.** 4·01 t ; 5·81 t]

HIGHLIGHTS

1. A redundant frame is that in which the number of members are more than $(2j-3)$, where j is the number of joints.

2. The Castigliano's first theorem states : "*If an elastic structure is subjected to a system of loads in equilibrium, produce a total strain energy U in the structure, the partial differential coefficient of total strain energy U with respect to any load gives the displacement of the load in its own line of action*".

3. The Maxwell's method, for the forces in redundant frames, is as discussed below :

(*a*) Select any member to be a redundant member and imagine it to be removed, so that the remaining frame becomes a perfect one.

(*b*) Now find out the forces P_1, P_2, P_3...in the members of the perfect frame due to external loads.

(*c*) Now apply a unit pull in the redundant member, which was assumed to be removed and find out the forces k_1, k_2, k_3...in the members of the perfect frame due to unit pull.

(*d*) Now, if T is the actual pull in the redundant member, then actual forces in the members will be

$$(P_1 + k_1 T), (P_2 + k_2 T), (P_3 + k_3 T)...$$

(e) The value of actual pull T in the redundant member,

$$T = - \frac{\sum \frac{Pkl}{A\overline{E}}}{\sum \left(\frac{k^2 l}{A\overline{E}} \right) + \frac{l_0}{A_0 E}}$$

where $\sum \frac{Pkl}{A\overline{E}}$ and $\sum \left(\frac{k^2 l}{A\overline{E}} \right)$ is the sum of these values for all the members of the perfect frame and $\frac{l_0}{A_0 E}$ is the value for redundant member, which is assumed to be removed.

4. If any member of a redundant frame is slightly shorter (δ) than the correct length, then the pull in this member,

$$T = \frac{\delta}{\sum \left(\frac{k^2 l}{A\overline{E}} \right) + \frac{l_0}{A_0 E}}$$

where $\sum \left(\frac{k^2 l}{A\overline{E}} \right)$ is the sum of these values for all the members of the perfect frame and $\frac{l_0}{A_0 E}$ is the value for redundant member.

5. If a frame has two (or more) redundant members, the forces in the various members may be found out by selecting two redundant members and proceeding in the same way as that of a frame, with one redundant member.

Do You Know

1. What do you understand by the term redundant frame ? How will you find out the degree of redundancy of a frame.

2. State and explain the theorem of Castigliano. Show how this theorem is helpful in the analysis of redundant frames ?

3. Explain the Maxwell's method for the determination of forces in a redundant frame.

4. State clearly the procedure of finding out the forces in a redundant frame, when one of its members is (i) slightly shorter and (ii) slightly larger in length than correct length.

5. Explain the procedure for finding out the forces, when a frame has got two redundant members.

6. What is a trussed beam ? How will you find out the tension in the tie rods of a trussed beam ?

19

COLUMNS AND STRUTS

19·1. Introduction

A structural member subjected to an axial compressive force is called a *strut*. As per definition, a strut may be horizontal, inclined or even vertical. But a vertical strut, used in buildings or frames, is called a *column*.

19·2. Failure of a column or strut

It has been observed, that when a column or a strut is subjected to a compressive force, the compressive stress induced,

$$p = \frac{P}{A}$$

where P is the compressive force and A is the area of cross section. A little consideration will show, that if the force or load is gradually increased the column will reach a stage, when it will be subjected to ultimate crushing stress. Beyond this stage the column will fail by crushing. The load corresponding to the **crushing stress is called** *crushing load*.

It has also been experienced, that sometimes, a compression member does not fail entirely by crushing, but also by bending *i.e.*, buckling. This happens in the case of long columns. It has also been observed, that all the short columns fail due to their crushing. But, if a long column is subjected to a compressive load. it is subjected to a compressive stress. If the load is gradually increased, the column will reach a stage, when it will start buckling. The load at which the column just buckles is called *buckling load, critical load* or *crippling load* and the column is said to have developed an elastic instability. A little consideration will show, that for a long column, the value of buckling load will be less than the crushing load. Moreover, the value of buckling load is low for long columns, and relatively high for short columns.

19·3. Euler's column theory

The first rational attempt, to study the stability of long columns, was made by Mr. Euler. He derived an equation. for the buckling load of long columns based on the bending stress. While deriving this equation, the effect of direct stress is neglected. This may be justified with the statement, that the direct stress induced in a long* column is negligible as compared to the bending stress. It may be noted that the Euler's formula cannot be used in the case of short columns, because the direct stress is considerable, and hence cannot be neglected.

19·4. Assumptions in the Euler's column theory

The following simplifying assumptions are made in the Euler's column theory :

1. Initially the column is perfectly straight, and the load applied is truly axial.
2. The cross-section of the column is uniform throughout its length.
3. The column material is perfectly elastic, homogeneous and isotropic ; and thus obey's Hooke's law.
4. The length of column is very large as compared to its cross-sectional dimensions.
5. The shortening of column, due to direct compression (being very small) is neglected.
6. The failure of column occurs due to buckling alone.

19·5. Types of end conditions of columns

In actual practice, there are a number of end conditions, for columns. But, we shall study the Euler's column theory on the following four types of end conditions, which are important from the subject point of view :

1. Both the ends hinged,
2. Both the ends fixed,

---

*As a matter of fact, mere length is not criterion for a column to be called long or short. But it has an important relation with the lateral dimensions of the column.

3. One end is fixed and the other hinged,

4. One end is fixed and the other free.

19·6. Sign conventions

Though there are different signs used for the bending of columns in different books, yet we shall follow the following sign conventions which are commonly used and internationally recognised

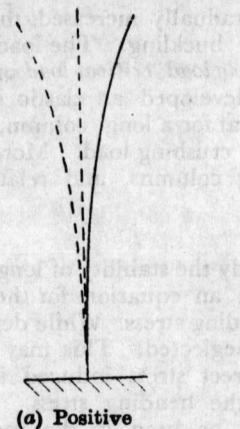 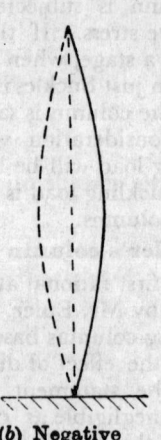

(a) Positive   (b) Negative

Fig. 19·1

1. A moment, which tends to bend the column with *convexity* towards its initial central line, as shown in Fig. 19 1(a), is taken as *positive*.

2. A moment, which tends to bend the column with its *concavity* towards its initial central line, as shown in Fig. 19·1(b), is taken as *negative*.

19·7. Columns with both the ends hinged

Consider a column *AB* of length *l* hinged at both of its ends *A* and *B*, and carrying a critical load at *B*. As a result of loading, let the column deflect into a curved form *AXB* as shown in Fig. 19·2.

Now consider any section *X*, at a distance *x* from *A*,

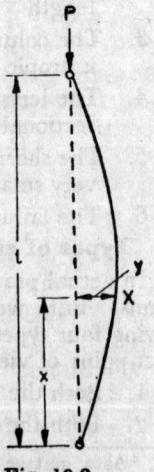

Let   *P* = Critical load on the column,

             y = Deflection of the column at *X*.

The moment due to the critical load *P*,

$$M = -P.y$$

$$\therefore \quad EI \, \frac{d^2y}{dx^2} = -P.y$$

Fig. 19·2

$$\therefore \qquad EI\ \frac{d^2y}{dx^2}+P.y=0$$

or
$$\frac{d^2y}{dx^2}+\frac{P}{EI}\cdot y=0$$

The general solution of the above differential equation is :

$$y=A.\ \cos x\ \sqrt{\frac{P}{EI}}+B\ \sin x\ \sqrt{\frac{P}{EI}}$$

where A and B are the constants of integration. We know that when $x=0$, $y=0$. Therefore $A=0$. Similarly when $x=l$, $y=0$. Therefore

$$0=B\ \sin\ l\ \sqrt{\frac{P}{EI}}$$

A little consideration will show that either B is equal to zero, or $\sin l \sqrt{\dfrac{P}{EI}}$ equal to zero. Now if we consider B to be equal to zero, then it indicates that the column has not bent at all. But if

$$\sin\ l\ \sqrt{\frac{P}{EI}}=0$$

$$\therefore \qquad l\ \sqrt{\frac{P}{EI}}=0=\pi=2\pi=3\pi=\ldots.$$

Now taking the least significant value,

$$l\ \sqrt{\frac{P}{EI}}=\pi$$

or
$$P=\frac{\pi^2 EI}{l^2}$$

19·8. Columns with one end fixed and the other free

Consider a column AB of length l fixed at A and free at B, and carrying a critical load at B. As a result of loading, let the beam deflect into a curved form AXB, such that the free end B deflects through a and occupies a new position B_1 as shown in Fig. 19·3.

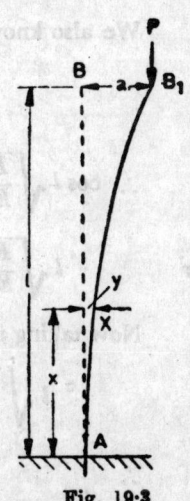

Now consider any section X at a distance x from A.

Let $P=$ Critical load on the column,

      $y=$ Deflection of the column at X,

$\therefore$ Moment due to the critical load P,

$$M=P(a-y)=P.a-Py$$

$$\therefore\ EI\ \frac{d^2y}{dx^2}=P.a-Py$$

Fig. 19·3

$$\therefore \qquad \frac{d^2y}{dx^2} + \frac{P}{EI} \cdot y = \frac{P.a}{EI}$$

The general solution of the above differential equation is,

$$y = A \cos\left(x \sqrt{\frac{P}{EI}} \right) + B \sin\left(x \sqrt{\frac{P}{EI}} \right) + a \qquad \dots(i)$$

where A and B are the constants of integration. We know that when $x=0$, $y=0$ therefore $A=-a$. Now differentiating the above equation,

$$\frac{dy}{dx} = -A \sqrt{\frac{P}{EI}} \sin\left(x \sqrt{\frac{P}{EI}} \right) + B \sqrt{\frac{P}{EI}} \cos\left(x \sqrt{\frac{P}{EI}} \right)$$

We also know that when $x=0$, $\frac{dy}{dx}=0$. Therefore

$$0 = B \sqrt{\frac{P}{EI}}$$

A little consideration will show that either B is equal to zero, or $\sqrt{\frac{P}{EI}}$ is equal to zero. Since the load P is not equal to zero, it is thus obvious, that B is equal to zero. Substituting the values $A=-a$ and $B=0$ in equation (i),

$$y = -a \cos\left(x \sqrt{\frac{P}{EI}} \right) + a$$

$$= a \left[1 - \cos x \sqrt{\frac{P}{EI}} \right]$$

We also know that when $x=l$, $y=a$. Therefore

$$a = a \left[1 - \cos l \sqrt{\frac{P}{EI}} \right]$$

$$\therefore \quad \cos l \sqrt{\frac{P}{EI}} = 0$$

or $\qquad l \sqrt{\frac{P}{EI}} = \frac{\pi}{2} = \frac{3\pi}{2} = \frac{5\pi}{2} \dots$

Now taking the least significant value

$$l \sqrt{\frac{P}{EI}} = \frac{\pi}{2}$$

$$\therefore \qquad P = \frac{\pi^2 EI}{4l^2}$$

19·9. Columns with both the ends fixed

Consider a column AB of length l fixed at both of its ends A and B and carrying a critical load at B. As a result of loading, let the column deflect as shown in Fig. 19·4.

Now consider any section X at a distance x from A.

Let $P =$ Critical load on the column,

$y =$ Deflection of the column at X.

A little consideration will show that since both the ends of the beam AB are fixed, and it is carrying a load, therefore, there will be some fixed ends moments at A and B.

Let $M_0 =$ Fixed end moments at A and B

$\therefore$ Moment due to the critical load P,

$$M = -P.y$$

or $$EI \frac{d^2y}{dx^2} = M_0 - P.y$$

$\therefore$ $$\frac{d^2y}{dx^2} + \frac{P}{EI} \cdot y = \frac{M_0}{EI}$$

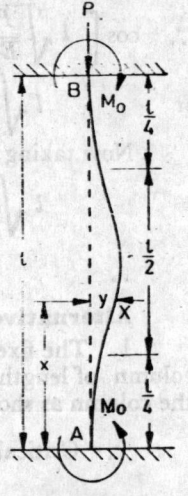

Fig. 19·4

The general solution of the above differential equation is :

$$y = A \cos \left(x \sqrt{\frac{P}{EI}} \right) + B \sin \left(x \sqrt{\frac{P}{EI}} \right) + \frac{M_0}{P} \qquad \ldots(i)$$

where A and B are the constants of integration. We know that when $x=0$, $y=0$. Therefore $A = -\dfrac{M_0}{P}$. Now differentiating the above equation,

$$\frac{dy}{dx} = -A \sqrt{\frac{P}{EI}} \sin \left(x \sqrt{\frac{P}{EI}} \right) + B \sqrt{\frac{P}{EI}} \cos \left(x \sqrt{\frac{P}{EI}} \right)$$

We also know that when $x=0$, $\dfrac{dy}{dx} = 0$. Therefore

$$0 = B \sqrt{\frac{P}{EI}}$$

A little consideration will show that either B is equal to zero or $\sqrt{\dfrac{P}{EI}}$ is equal to zero. Since the load P is not equal to zero, it is thus obvious, that B is equal to zero. Substituting the values $A = \dfrac{M_0}{P}$ and $B=0$ in equation (i),

$$y = -\frac{M_0}{P} \cos \left(x \sqrt{\frac{P}{EI}} \right) + \frac{M_0}{P}$$

$$= \frac{M_0}{P} \left[1 - \cos \left(x \sqrt{\frac{P}{EI}} \right) \right]$$

We also know that when $x=l$, $y=0$. Therefore

$$0=\frac{M_0}{P}\left[\,1-\cos\left(\,l\,\sqrt{\frac{P}{EI}}\,\right)\right]$$

$$\therefore\quad \cos\left[\,l\,\sqrt{\frac{P}{EI}}\,\right]=1$$

or

$$l\,\sqrt{\frac{P}{EI}}=0,\ 2\pi=4\pi=6\pi=\ldots$$

Now taking the least significant value,

$$l\,\sqrt{\frac{P}{EI}}=2\pi$$

$$\therefore\qquad P=\frac{4\pi^2EI}{l^2}$$

Alternative methods

1. The fixed beam AB may be considered as equivalent to a column of length $l/2$ with both ends hinged (*i.e.* middle portion of the column as shown in Fig. 19·4).

$$\therefore\quad \text{Critical load, } P=\frac{\pi^2EI}{\left(\dfrac{l}{2}\right)^2}=\frac{4\pi^2EI}{l^2}$$

2. The fixed beam AB may also be considered as equivalent to a column of length $l/4$ with one end fixed and the other free (*i.e.* lower one-fourth portion of the beam as shown in Fig. 19·4).

$$\therefore\text{Critical load, } P=\frac{\pi^2EI}{4\left(\dfrac{l}{4}\right)^2}=\frac{4\pi^2EI}{l^2}$$

19·10. Columns with one end fixed and the other hinged.

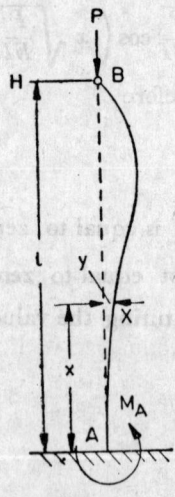

Fig. 19·5

Consider a column AB of length l fixed at A and hinged at B, and carrying a critical load at B. As a result of loading, let the column deflect as shown in Fig. 19·5.

Now consider any section X at a distance x from A.

Let $P=$critical load on the column, and

$y=$Deflection of the beam at X.

A little consideration will show that since the beam AB is fixed at A, and it is carrying a load, therefore there will be some fixed end moment at A. In order to balance the fixing moment at A, there will be a horizontal reaction at B.

Let $M_A=$Fixed end moment at A, and

$H=$Horizontal reaction at B.

$\therefore$ Moment due to critical load P,

$M=-P.y$

or
$$EI \frac{d^2y}{dx^2} = H(l-x) - P.y$$

$$\therefore \quad \frac{d^2y}{dx^2} + \frac{P}{EI} \cdot y = \frac{H(l-x)}{EI}$$

The general solution of the above differential equation is :

$$y = A \cos\left(x\sqrt{\frac{P}{EI}}\right) + B \sin\left(x\sqrt{\frac{P}{EI}}\right) + \frac{H(l-x)}{P} \quad ...(i)$$

where A and B are the constants of integration. We know that when $x=0$, $y=0$. Therefore $A = -\frac{Hl}{P}$. Now differentiating the above equation,

$$\frac{dy}{dx} = -A\sqrt{\frac{P}{EI}} \sin\left(x\sqrt{\frac{P}{EI}}\right) + B\sqrt{\frac{P}{EI}}\cos\left(x\sqrt{\frac{P}{EI}}\right) - \frac{H}{P}$$

We know that when

$$x=0, \frac{dy}{dx}=0. \quad \text{Therefore}$$

$$0 = B\sqrt{\frac{P}{EI}} - \frac{H}{P}$$

$$\therefore \quad B = \frac{H}{P} \times \sqrt{\frac{EI}{P}}$$

We also know that when $x=l$, $y=0$. Therefore substituting these values of x, A and B in equation (i),

$$0 = -\frac{Hl}{P} \cos\left(l\sqrt{\frac{P}{EI}}\right) + \frac{H}{P} \cdot \sqrt{\frac{EI}{P}} \sin\left(l\sqrt{\frac{P}{EI}}\right)$$

$$\frac{H}{P}\sqrt{\frac{EI}{P}} \sin\left(l\sqrt{\frac{P}{EI}}\right) = \frac{Hl}{P} \cos\left(l\sqrt{\frac{P}{EI}}\right)$$

or
$$\tan\left(l\sqrt{\frac{P}{EI}}\right) = \left(l\sqrt{\frac{P}{EI}}\right)$$

A little consideration will show that the value of $\left(l\sqrt{\frac{P}{EI}}\right)$

in radians, has to be such that its tangent is equal to itself. We know that the only angle, the value of whose tangent is equal to itself, is about 4·5 radians.

$$\therefore \quad l\sqrt{\frac{P}{EI}} = 4\cdot5$$

$$l^2 \times \frac{P}{EI} = 20\cdot25$$

or

$$\therefore \quad P = \frac{20\cdot25\ EI}{l^2}$$

$$= \frac{2\pi^2 EI}{l^2}$$

Note. A little consideration will show that $20\cdot25$ is not exactly equal to $2\pi^2$. But it is approximately equal to $2\pi^2$. This has been done to rationalise the value of P *i.e.* crippling load in various cases.

19·11. Equivalent length of a column

In the previous articles, we have derived the relations for the crippling load under various end conditions. Sometimes, all these cases are represented by a general equation,

$$P = \frac{\pi^2 EI}{Cl^2}$$

where C is a constant, representing the end conditions of the column. Thus, we have seen that the value of the constant :

(a) C is 1 for a column with *both ends hinged*.

(b) C is 4 for a column with one *end fixed and the other free*.

(c) C is 1/4 for a column *with both ends fixed*.

(d) C is 1/2 for a column with *one end fixed and the other hinged*.

There is another way of representing the equation, for the crippling load, by an equivalent length, of effective length of a column. The equivalent length of a given column with given end conditions, is the length of an equivalent column of the same material and cross-section with hinged ends, and having the value of the crippling load equal to that of the given column.

The equivalent lengths (L) for the given end conditions are given below.

Table 19–1

S. No.	End conditions	Crippling load	Relation between equivalent length and actual length
1.	Both ends hinged	$\dfrac{\pi^2 EI}{l^2} = \dfrac{\pi^2 EI}{L^2}$	$L = l$
2.	One end fixed and the other free	$\dfrac{\pi^2 EI}{4l^2} = \dfrac{\pi^2 EI}{(2L)^2}$	$L = 2l$
3.	Both ends fixed	$\dfrac{4\pi^2 EI}{l^2} = \dfrac{\pi^2 EI}{\left(\dfrac{L}{2}\right)^2}$	$L = \dfrac{l}{2}$
4.	One end fixed and the other hinged	$\dfrac{2\pi^2 EI}{l^2} = \dfrac{\pi^2 EI}{\left(\dfrac{L}{\sqrt{2}}\right)^2}$	$L = \dfrac{l}{\sqrt{2}}$

Note. The vertical column will have two moments of inertia (*viz.* I_{XX} and I_{YY}). Since the column will tend to buckle in the direction of least moment of inertia, therefore the least value of the two moments of inertia is to be used in the relation.

19·12. Slenderness ratio

We have already discussed in Art. 19·11, that the general equation for the crippling load,

$$P = \frac{\pi^2 EI}{l^2} \qquad \qquad \ldots(i)$$

We know that the buckling of a column under the crippling load will take place about the axis of least resistance. Now substituting $I = Ak^2$ (where A is the area and k is the least radius of gyration of the section) in the above equation,

$$P = \frac{\pi^2 E A k^2}{l^2} = \frac{\pi^2 EA}{\left(\dfrac{l}{k}\right)^2} \qquad \qquad \ldots(ii)$$

where $\dfrac{l}{k}$ is known as slenderness ratio.

Note. It may be noted, that the formulae for crippling load, in the previous articles, have been derived on the assumption that the slenderness ratio l/k is so large, that the failure of the column occurs only due to bending, the effect of direct stress (*i.e.*, P/A) being negligible.

19·13. Limitation of Euler's formula

We have discussed in Art. 19·12 that the general equation for the crippling load,

$$P = \frac{\pi^2 EA}{\left(\dfrac{l}{k}\right)^2}$$

$$\therefore \text{ Crippling stress, } p = \frac{P}{A} = \frac{\pi^2 E}{\left(\dfrac{l}{k}\right)^2}$$

A little consideration will show, that the crippling stress will be high, when the slenderness ratio is small. We know that the crippling stress for a column can not be more than the crushing stress of the column material. It is thus obvious, that the Euler's formula will give the value of crippling stress of the column (equal to the crushing stress of the column material) corresponding to the slenderness ratio. Now consider a mild steel column. We know that the crushing stress for the mild steel is 3,300 kg/cm² and Young's modulus for the mild steel is $2 \cdot 1 \times 10^6$ kg/cm².

Now equating the crippling stress to the crushing stress,

$$\frac{\pi^2 E}{\left(\dfrac{l}{k}\right)^2} = 3,300$$

$$\frac{\pi^2 \times 2 \cdot 1 \times 10^6}{\left(\dfrac{l}{k}\right)^2} = 3,300$$

$$\therefore \qquad \left(\frac{l}{k}\right)^2 = \frac{\pi^2 \times 2 \cdot 1 \times 10^6}{3,300} = 6,282$$

or

$$\frac{l}{k} = 79 \cdot 27 \text{ say } 80$$

Hence, if the slenderness ratio is less than 80, the Euler's formula for a mild steel column is not valid.

Sometimes the columns, whose slenderness ratio is *more* than 80, are known as *long columns* ; and those whose slenderness ratio is *less than* 80 are known as *short columns*. It is thus obvious, that the Euler's formula holds good only for long columns.

Note. In the Euler's formula, for crippling load, we have not taken into account the direct stresses induced in the material due to the load, which increases gradually from zero to its crippling value. As a matter of fact the combined stress, due to direct load and slight bending, reaches its allowable value at a load, lower than that required for buckling ; and therefore this will be the limiting value of the safe load.

Example 19·1. *A steel rod 5 m long and of 4 cm diameter is used as a column, with one end fixed and the other free. Determine the crippling load by Euler's formula. Take E as $2 \cdot 0 \times 10^6$ kg/cm².*

Solution.

Given.   Length, $l = 5$ m $= 500$ cm

Dia. of column $= 4$ cm

$\therefore$ Moment of Inertia,

$$I = \frac{\pi}{64} \times 4^4 = 4\pi \text{ cm}^4$$

Young's modulus, $E = 2 \cdot 0 \times 10^6$ kg/cm²

Let $P = $ Crippling load.

Since the column is fixed at one end, and free at the other, therefore equivalent length of the column,

$$L = 2l = 2 \times 500 = 1,000 \text{ cm}$$

Using the relation,

$$P = \frac{\pi^2 EI}{L^2} \text{ with usual notations.}$$

$$= \frac{\pi^2 \times 2 \cdot 0 \times 10^6 \times 4\pi}{1,000^2} = \textbf{248·1 kg Ans.}$$

Example 19·2. *A hollow alloy tube 5 m long with diameters 4 cm and 2·5 cm respectively was found to extend 6·4 mm under a tensile load of 6 tonnes. Find the buckling load for the tube, when used as a strut with both the ends pinned. Also find the safe load on the tube, taking factor of safety as 4.* (*Vikram University 1976*)

Solution.

Given.   Length, $l = 5$ m $= 500$ cm

Outer diameter, $D = 4$ cm

Inner diameter, $d = 2 \cdot 5$ cm

$\therefore$ Area, $A = \frac{\pi}{4} \ (4^2 - 2 \cdot 5^2) = 7 \cdot 65$ cm² and moment of inertia, $I = \dfrac{\pi}{64} \ (4^4 - 2 \cdot 5^4) = 10 \cdot 65 \ cm^4$

Deflection $\delta l = 6 \cdot 4 \ m = 0 \cdot 64 \ cm$
Tensile load, $= 6 \ T = 6,000 \ kg$

Buckling load

Let $P =$ Buckling load on the tube.

We know that the strain,

$$e = \frac{\delta l}{l} = \frac{0 \cdot 64}{500}$$

and Young's modulus, $E = \dfrac{Load}{Area \times Strain} = -\dfrac{6,000}{7 \cdot 65 \times \dfrac{0 \cdot 64}{500}} \ kg/cm^2$

$$= 6 \cdot 127 \times 10^5 \ kg/cm^2$$

Since the column is pinned at its both ends, therefore equivalent length of the column,

$$L = l = 500 \ cm$$

Using the relation,

$$P = \frac{\pi^2 EI}{L^2} \text{ with usual notations.}$$

$$= \frac{\pi^2 \times 6 \cdot 127 \times 10^5 \times 10 \cdot 65}{500^2} = \textbf{257·6 kg Ans.}$$

Safe load

We know that the safe load on the tube

$$= \frac{Buckling \ load}{Factor \ of \ safety} = \frac{257 \cdot 6}{4} = \textbf{64·4 kg} \quad \textbf{Ans.}$$

Example 19·3. *An I section joist 40 cm × 20 cm × 1 cm and 6 m long is used as a strut with both ends fixed. What is Euler's crippling load for the column ? Take Young's modulus for the joist as* $2 \cdot 0 \times 10^6$ *kg/cm².*

Solution.

Given. Outer width, $b = 20 \ cm$
Outer depth, $d = 40 \ cm$
Inner width, $b_1 = 20 - 1 = 19 \ cm$
Inner depth, $d_1 = 40 - 2 \times 1 = 38 \ cm$
Length, $l = 6 \ m = 600 \ cm$
Young's modulus.

$$E = 2 \cdot 0 \times 10^6 \ kg/cm^2$$

Let $P =$ Euler's crippling load.

We know that the moment of inertia of the joist about $X - X$ axis,

$$I_{XX} = \frac{1}{12} \ (bd^3 - b_1 d_1{}^3)$$

$$= \frac{1}{12} \ (20 \times 40^3 - 19 \times 38^3) \ cm^4$$

$$= 11,452 \cdot 67 \ cm^4$$

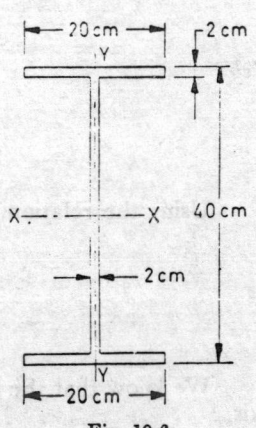

Fig. 19·6

and
$$I_{YY}=2\times\frac{1\times20^3}{12}+\frac{38\times1^3}{12}=1,336\cdot5 \text{ cm}^4$$

Since I_{YY} is less than I_{XX}, therefore the joist will tend to buckle along $y-y$ axis. Thus we shall take the value of I as $I_{YY}=$ 1,336·5 cm⁴.

Since the column is fixed at its both ends, therefore equivalent length of the column,

$$L=\frac{l}{2}=\frac{600}{2}=300 \text{ cm}$$

Now using the relation,

$$P=\frac{\pi^2EI}{L^2} \text{ with usual notations.}$$

$$=\frac{\pi^2\times2\cdot0\times10^6\times1,336\cdot5}{300^2}=2\,93,200 \text{ kg}$$

$$=\mathbf{293\cdot2 \text{ t} \quad Ans.}$$

Example 19·4. *A T-section 15 cm × 12 cm × 2 cm is used as a strut of 4 m long with hinged at its both ends. Calculate the crippling load, if Young's modulus for the material be 2·0 × 10⁶ kg/cm².*

Solution.

Given. Length, $l=4$ m $=400$ cm
Young's modulus,

$$E=2\cdot0\times10^6 \text{ kg/cm}^2$$

Let $P=$ Crippling load.

First of all, let us find out the c.g. of the section. Let $\bar{y}$ be the distance between the c.g. of the section and top of the flange.

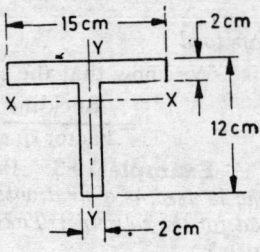

Fig. 19·7

Flange

$$a_1=15\times2=30 \text{ cm}^2$$
$$y_1=\frac{2}{2}=1\cdot0 \text{ cm}$$

Web

$$a_2=10\times2=20 \text{ cm}^2$$
$$y_2=2+\frac{10}{2}=7 \text{ cm}$$

Using the relation,

$$\therefore \qquad \bar{y}=\frac{a_1y_1+a_2y_2}{a_1+a_2} \text{ with usual notations.}$$

$$=\frac{30\times1\cdot0+20\times7}{30+20}=3\cdot4 \text{ cm}$$

We know that the moment of inertia the section about $X-X$ axis,

$$I_{XX}=\left(\frac{15\times2^3}{12}+30\times2\cdot4^2\right)$$

$$+\left(\frac{2\times10^3}{12}+20\times3\cdot6^2\right)\text{cm}^4$$

$$=609\text{ cm}^4$$

and
$$I_{YY}=\frac{2\times15^3}{12}+\frac{10\times2^3}{12}=569\text{ cm}^4$$

Since I_{YY} is less than I_{XX}, therefore the column will tend to buckle along y—y axis. Thus we shall take the value of I as $I_{YY}=569$ cm^4.

Moreover as the column is hinged at its both ends, therefore equivalent length of the column,

$$L=l=400\text{ cm}$$

Now using the relation,

$$P=\frac{\pi^2EI}{L^2}\text{ with usual notations.}$$

$$=\frac{\pi^2\times2\cdot0\times10^6\times569}{400^2}=70{,}200\text{ kg}$$

$$=70\cdot2\text{ t Ans.}$$

Example 19·5. *Determine the ratio of the strengths of a solid steel column to that of a hollow column of the same material and having the same cross-sectional area. The internal diameter of the hollow column is 1/2 of its external diameter. Poth the columns are of the same length and are pinned at their both ends.*

Solution.

Given.   Area of solid column,

$$A_S=A_H$$

Effective length of solid column,

$$L_S=L_H$$

Young's modulus for solid column material,

$$E_S=E_H$$

Let
$P_S=$Crippling load of the solid column,

$P_H=$Crippling load of the hollow column,

$D_1=$Diameter of the solid column, and

$D=$External diameter of the hollow column.

∴   Internal diameter of the hollow column,

$$d=\frac{D}{2}$$

We know that for a solid column,

$$k_S{}^2=\frac{I_S}{A_S}=\frac{\frac{\pi}{64}D_1{}^4}{\frac{\pi}{4}D_1{}^2}=\frac{D_1{}^2}{16}\qquad\ldots(i)$$

Similarly $\qquad k_H{}^2 = \dfrac{I_H}{A_H} = \dfrac{\dfrac{\pi}{64}(D^4 - d^4)}{\dfrac{\pi}{4}(D^2 - d^2)} = \dfrac{D^2 + d^2}{16}$

$$\therefore \qquad k_H{}^2 = \dfrac{D^2 + \left(\dfrac{D}{2}\right)^2}{16} = \dfrac{5D^2}{64} \qquad \qquad ...(ii)$$

We also know that the areas of the two columns are equal. Therefore

$$\dfrac{\pi}{4} D_1{}^2 = \dfrac{\pi}{4}\left[D^2 - \left(\dfrac{D}{2}\right)^2 \right]$$

$$= \dfrac{\pi}{4}\left[D^2 - \dfrac{D^2}{4} \right] = \dfrac{\pi}{4} \times \dfrac{3D^2}{4}$$

$$\therefore \qquad D_1{}^2 = \dfrac{3D^2}{4} \qquad \qquad ...(iii)$$

Now using the relation,

$$P = \dfrac{\pi^2 EI}{L^2} \text{ with usual notations.}$$

$$\therefore \qquad P_S = \dfrac{\pi^2 E_S (A_S k_S{}^2)}{L_S{}^2} \qquad \qquad ...(iv)$$

Similarly, $\qquad P_H = \dfrac{\pi^2 E_H (A_H k_H{}^2)}{L_H{}^2} \qquad \qquad ...(v)$

Dividing equation (v) by (iv),

$$\dfrac{P_H}{P_S} = \dfrac{\dfrac{\pi^2 E_H (A_H k_H)^2}{L_H{}^2}}{\dfrac{\pi^2 E_S (A_S k_S{}^2)}{L_S{}^2}} = \dfrac{k_H{}^2}{k_S{}^2} = \dfrac{\dfrac{5D^2}{64}}{\dfrac{D_1{}^2}{16}}$$

$$= \dfrac{5}{64} \times \dfrac{D^2}{\underset{4 \times 16}{3D^2}} = \dfrac{5}{3} \quad \textbf{Ans.}$$

Exercise 19–1

1. A hollow steel tube of 2·5 cm external diameter and 2·0 cm internal diameter is used as a column 3 m long with both ends hinged. Determine the Euler's crippling load, if modulus of elasticity is $2 \cdot 0 \times 10^6$ kg/cm².

[**Ans.** 248·4 kg]

2. A 4 m long circular bar was found to deflect 2 cm, when it was used as a simply supported beam and subjected to a load of 10 kg at its centre. If this bar is now used as a column, with both ends hinged, determine the crippling load it can carry. (*Gauhati University*)

[**Ans.** 412 kg]

3. A timber column 15 cm × 10 cm in section is 3 m long with both ends hinged. Determine the safe load, the column can carry, with a factor of safety as 10. Take E for timber as $1 \cdot 055 \times 10^2$ t/cm². (*Jabalpur University*)

[**Ans.** 1·44 t]

4. Compare the ratio of strength of a solid steel column to that of a hollow column of internal diameter equal to 3/4 of its external diameter. Both the columns have the same cross-sectional areas, lengths and end conditions.

(Roorkee University)

$$\left[\text{ Ans. } \quad \frac{25}{7} \right]$$

19·14. Empirical formulae for columns

We have already discussed, in the previous articles, that the Euler's formula is valid only for long columns *i.e.* for columns, whose slenderness ratio is greater than a certain value for a particular material. Moreover, it does not take into consideration the direct compressive stress. In order to fill up this lacuna, many more formulae were proposed by different scientists, all over the world. The following empirical formulae, out of those, are important from the subject point of view.

    1. Rankine's formula,

    2. Johnson's formula, and

    3. Indian Standard code.

19·15. Rankine's formula for columns

We have already discussed that the Euler's formula gives correct results only for very long columns. Though this formula is applicable for columns, ranging from very long to short ones, yet it does not give reliable results. Prof. Rankine, after a number of experiments, gave the following empirical formula for columns.

$$\frac{1}{P} = \frac{1}{P_C} + \frac{1}{P_E} \qquad \qquad ...(i)$$

where P = Crippling load by Rankine's formula,

             $P_C = f_C.A$ = Ultimate crushing load for the column,

             $P_E = \dfrac{\pi^2 EI}{L^2}$ = Crippling load, obtained by Euler's

                       formula.

A little consideration will show, that the value of P_C will remain constant irrespect of the fact whether the column is a long one or short one. Moreover, in the case of short columns, the value of P_E will be very high, therefore the value of $\dfrac{1}{P_E}$ will be quite negligible as compared to $\dfrac{1}{P_C}$. It is thus obvious, that the Rankine's formula will give the value of its crippling load *(i.e.* P) approximately equal to the ultimate crushing load (*i.e.* P_C). In case of long columns, the value of P_E will be very small, therefore the value of $\dfrac{1}{P_E}$ will be quite considerable as compared to $\dfrac{1}{P_C}$. It is thus obvious, that the Rankine's formula will give the value of its crippling load (*i.e.* P) approximately equal to the crippling load by Euler's formula (*i.e.* P_E). Thus we see, that the Rankine's formula gives a fairly correct result for all cases of columns, ranging from short to long columns.

From equation (i) we know that

$$\frac{1}{P} = \frac{1}{P_C} + \frac{1}{P_E} = \frac{P_E + P_C}{P_C \cdot P_E}$$

$$\therefore \qquad P = \frac{P_C \cdot P_E}{P_C + P_E} = \frac{P_C}{1 + \dfrac{P_C}{P_E}}$$

Now substituting the values of P_C and P_E in the above equation,

$$P = \frac{f_C \cdot A}{1 + f_C \cdot A \times \dfrac{L^2}{\pi^2 EI}} = \frac{f_C \cdot A}{1 + \dfrac{f_C}{\pi^2 E} \cdot \dfrac{AL^2}{Ak^2}}$$

$$\ldots (\because \ I = Ak^2)$$

$$= \frac{f_C \cdot A}{1 + a\left(\dfrac{L}{k}\right)^2} = \frac{\text{crushing load}}{1 + a\left(\dfrac{L}{k}\right)^2}$$

where
f_C = Crushing stress of the column material,
A = Cross-sectional of the column,
a = Rankine's constant,
L = Equivalent length of the column, and
k = Least radius of gyration.

The following table gives the values of crushing stress and Rankine's constant for various materials.

Table 19-2

S. No.	Material	f_C in kg/cm²	$a = \dfrac{f_C}{\pi^2 E}$
1.	Wrought iron	2.500	$\dfrac{1}{9,000}$
2.	Cast iron	5,500	$\dfrac{1}{1,600}$
3.	Mild steel	3,2 0	$\dfrac{1}{7,500}$
4.	Timber	500	$\dfrac{1}{750}$

Note. The above values are. only for a column with both ends hinged. For other end conditions, the equivalent length should be used.

Example 19·6. *A cast iron hollow column, having 8 cm external diameter and 6 cm internal diameter, is used as a column of 2 m long. Using Rankine's formula, determine the crippling load when both the ends are fixed. Take $f_C = 6,000$ kg/cm².* (A M I E. Winter. 1977)

Solution.

Given. External diameter,
$D = 8$ cm

Internal diameter $d = 6$ cm

$\therefore$ Area, $A = \dfrac{\pi}{4} (8^2 - 6^2) = 7\pi$ cm^2

Moment of inertia, $I = \dfrac{\pi}{64} (8^4 - 6^4) = 43.75\ \pi$ cm^4

and least radius of gyration

$$k = \sqrt{\dfrac{I}{A}} = \sqrt{\dfrac{43.75\ \pi}{7\pi}} = 2.5 \text{ cm}$$

Column length, $l = 2$ m $= 200$ cm

Crushing stress, $f_C = 6{,}000$ kg/cm^2

Let $P =$ crippling load.

From table 17.2, we find that the Rankine's constant,

$$a = \dfrac{1}{1{,}600}$$

Since the column is fixed at its both ends, therefore equivalent length of the column,

$$L = \dfrac{l}{2} = \dfrac{200}{2} = 100 \text{ cm}$$

Using the relation,

$$P = \dfrac{f_C . A}{1 + a\left(\dfrac{L}{k}\right)^2} \text{ with usual notations.}$$

$$= \dfrac{6{,}000 \times 7\pi}{1 + \dfrac{1}{1{,}600}\left(\dfrac{100}{2.5}\right)^2} = \textbf{66,000 kg \ Ans.}$$

Example 19·7. *Find the Euler's crippling load for a hollow cylindrical steel column of 3·8 cm external diameter and ·5 mm thick. Take length of the column as 2·3 m and hinged at its both ends. Take $E = 2·05 \times 10^3$ t/cm^2.*

Also determine the crippling load by Rankine's formula, using constants as 3·35 t/cm^2 ond 1/7,500.

Solution.

Given. Outer diameter,
$$D = 3·8$$

Thickness $= 2·5$ mm $= 0·25$ cm

$\therefore$ Inner diameter, $d = 3·8 - 2 \times 0·25 = 3·3$ cm

Area, $A = \dfrac{\pi}{4} (3·8^2 - 3·3^2) = 2·75$ cm^2

and moment of inertia, $I = \dfrac{\pi}{64} (3·8^4 - 3·3^4) = 4·51$ cm^4

Length of column, $l = 2·3$ m $= 230$ cm

Let $P =$ crippling load.

Crippling load by Euler's formula

Young's modulus, $E = 2.05 \times 10^3$ t/m²

Since the column is hinged at its both ends, therefore effective length of the column,

$$L = l = 230 \text{ cm}$$

Using the relation,

$$P = \frac{\pi^2 EI}{L^2} \text{ with usual notations.}$$

$$= \frac{\pi^2 \times 2.05 \times 10^3 \times 4.51}{230^2} = \mathbf{1.725} \ \mathbf{t} \quad \mathbf{Ans}.$$

Crippling load by Rankine's formula

Yield stress, $f_C = 3.35$ t/cm²

Rankine's constant, $a = \dfrac{1}{7,500}$

We know that the least radius of gyration,

$$k = \sqrt{\frac{I}{A}} = \sqrt{\frac{4.51}{2.75}} = 1.28 \text{ cm}$$

Now using the relation,

$$P = \frac{f_C.A}{1 + a\left(\dfrac{L}{k}\right)^2} \text{ with usual notations.}$$

$$= \frac{3.35 \times 2.75}{1 + \dfrac{1}{7,500}\left(\dfrac{230}{1.28}\right)^2} = \mathbf{1.74} \ \mathbf{t} \quad \mathbf{Ans}.$$

Example 19·8. *Fig. 19·8 shows a built-up column consisting of 150 mm × 100 mm R.S.J. with 120 mm × 12 mm plate riveted to each flange.*

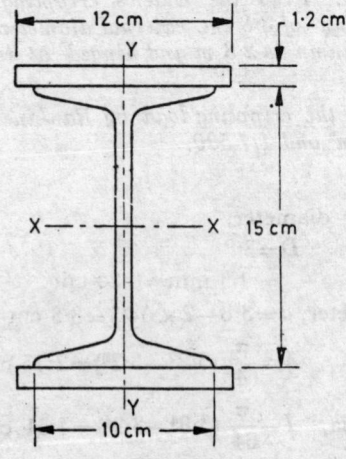

Fig. 19·8

Calculate the safe load, the column can carry, if it is 4 m long having one end fixed and the other hinged with a factor of safety 3·5.

Take the properties of the joist as $A=21.67$ *cm²,* $I_{XX}=839.1$ *cm⁴,* $I_{YY}=94.8$ *cm⁴. Assume Rankine's constant as* 3.15 *t/cm² and* $a=1/7500$.

Solution.

Given.   Length of the column,
$$L=4 \text{ m}=400 \text{ cm}$$
Factor of safety $=3.5$
Yield stress, $f_C=3.15$ t/cm²

Rankine's constant, $a=\dfrac{1}{7,500}$

Let $P=$Crippling load on the column.

From the geometry of the figure, we find that the area of column,
$$A=21.67+2\times12\times1.2=50.47 \text{ cm}^2$$
and moment of inertia of the column section about $X-X$ axis,
$$I_{XX}=839.1+2\left\{\frac{12\times1.2^3}{12}+12\times1.2\times8.1^2\right\} \text{ cm}^4$$
$$=2,732.1 \text{ cm}^4$$
Similarly $I_{YY}=94.8+2\left\{\dfrac{1.2\times12^3}{12}\right\}=440.4 \text{ cm}^4$

Since I_{YY} is less than I_{XX}, therefore the column will tend to buckle along $Y-Y$ axis. Thus we shall take the value of I as $I_{YY}=440.4$ cm⁴ Moreover as the column is fixed at one end, and hinged at the other, therefore equivalent length of the column,
$$L=\frac{l}{\sqrt{2}}=\frac{400}{\sqrt{2}}=282.8 \text{ cm.}$$

We know that the least radius of gyration,
$$k=\sqrt{\frac{I}{A}}=\sqrt{\frac{440.4}{50.47}}=2.95 \text{ cm}$$

Using the relation,
$$P=\frac{f_C.A}{1+a\left(\dfrac{L}{k}\right)^2} \text{ with usual notations}$$
$$=\frac{3.15\times50.47}{1+\dfrac{1}{7,500}\left(\dfrac{282.8}{2.95}\right)^2}=71.5 \text{ t}$$

We know that the safe load on the column
$$=\frac{P}{\text{factor of safety}}$$
$$=\frac{71.5}{3.5}=\mathbf{20.43 \text{ t} \text{ Ans.}}$$

Example 19·9. *A column is made up of two channels ISJC 200 and two 25 cm × 1c m flange plates as shown in Fig. 19·9.*

Determine, by Rankine's formula, the safe load, the column of 6 m length, with both ends fixed, can carry with a factor of safety 4. The

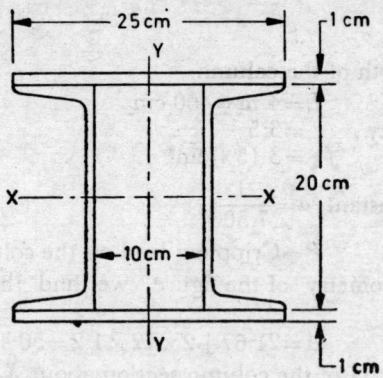

Fig. 19·9

properties of one channel are $A = 17·77$ cm², $I_{XX} = 1,161·2$ cm⁴ and $I_{YY} = 84·2$ cm⁴. Distance of centroid from back of web = 1·97 cm. Take $f_C = 3·2$ t/cm² and Rankine's constant = 1/7,500.

(Panjab University. 1978

Solution.

Given. Length of the column,
$$l = 6 \text{ m} = 600 \text{ cm}$$
Factor of safety $= 4$

Yield stress, $f_C = 3·2$ t/cm²

Rankine's constant, $a = \dfrac{1}{7,500}$

Let $P =$ crippling load on the column.

From the geometry of figure, we find that the area of column
$$A = 2(17·77 + 25 \times 1) = 85·54 \text{ cm}^2$$

and moment of inertia of the column section about $X-X$ axis,

$$I_{XX} = 2 \times 1,161·2 + 2 \left(\frac{25 \times 1^3}{12} + 25 \times 1 \times 10·5^2 \right) \text{ cm}^4$$

$$= 7,839·0 \text{ cm}^4$$

Similarly, $I_{YY} = 2 \left\{ \frac{1 \times 25^3}{12} + 84·2 + 17·77 \times (5 + 1·97)^2 \right\} \text{ cm}^4$

$$= 4,499·0 \text{ cm}^4$$

Since I_{YY} is less than I_{XX}, therefore the column will tend to buckle along $y-y$ axis. Thus we shall take the value of I as $I_{YY} = 4,499$ cm⁴. Moreover as the column is fixed at its both ends, therefore equivalent length of the column,

$$L = \frac{l}{2} = \frac{600}{2} = 300 \text{ cm}$$

We know that the least radius of gyration,

$$k = \sqrt{\frac{I}{A}} = \sqrt{\frac{4,499}{85 \cdot 54}} = 7 \cdot 25 \text{ cm}$$

Using the relation,

$$P = \frac{f_c \cdot A}{1 + a \left(\dfrac{L}{k} \right)^2} \text{---with usual notations.}$$

$$= \frac{3 \cdot 2 \times 85 \cdot 54}{1 + \dfrac{1}{7,500} \left(\dfrac{300}{7 \cdot 25} \right)^2} = 222 \cdot 8 \text{ t}$$

We know that the safe load on the column

$$= \frac{P}{\text{factor of safety}} = \frac{222 \cdot 8}{4} = 55 \cdot 7 \text{ t} \quad \textbf{Ans.}$$

19·16. Johnson's formulae for columns

Prof. Johnson proposed the following two formulae for columns,
(1) Straight line formulae, and
(2) Parabolic formula.

19·17. Johnson's straight line formula for columns

This formula was first proposed by Johnson, which states,

$$P = A \left[f - n \left(\frac{L}{k} \right) \right]$$

where
$P =$ Safe load on the column,
$A =$ Area of the column cross-section,
$f =$ Allowable stress in the column material,
$n = $ A constant, whose value depends upon the column material,
$\dfrac{L}{k} =$ Slenderness ratio.

The values of f and n are given in the following table.

Table 19-3

S. No.	Material	f_c (kg/cm²)	n
1	Mild steel	3,200	0·0053
2	Wrought iron	2,500	0·0053
3	Cast iron	5,500	0·008

A little consideration will show, that for short columns, the safe load $P = f.A$. But for long columns, there is always a possibility of buckling. It is thus obvious, that the safe load will be small, depending upon the slenderness ratio.

Prof. Johnson, while analysing the safe load, used to plot a curve for P/A and L/k. Since he used to get a straight line in this case, so he named this formula as a straight line formula.

19·18. Johnson's parabolic formula for columns

Prof. Johnson, after proposing the straight line formula, found that the results obtained by this formula are very approximate. He then proposed another formula, which states,

$$P = A\left[f - r\left(\frac{L}{k} \right)^2 \right]$$

where
$P =$ Safe load on the column,
$A =$ Area of the column cross-section,
$f =$ Allowable stress in the column material,
$r =$ A constant, whose value depends upon the column material,
$\dfrac{L}{k} =$ Slenderness ratio with equivalent column length.

The values of f and r are given in the following table :

Table 19–4

S. No.	Material	f_c (kg/cm^2)	r
1	Mild Steel	3,200	0·000057
2	Wrought iron	2,500	0·000039
3	Cast iron	5,500	0·00016

Prof. Johnson, while analysing the safe load, used to plot a curve for P/A and L/k. Since he used to get a parabola in this case, so he named this formula as a parabolic formula.

19·19. Indian Standard code for columns

The I.S.I. has also given a code for the safe stress in I.S. 226—1962, which states,

$$p_c = p_c' = \frac{\dfrac{f_y}{m}}{1 + 0·20 \sec \dfrac{L}{k} \sqrt{\dfrac{m p_c'}{4E}}} \quad \dots \left(\text{For } \frac{L}{k} = 0 \text{ to } 160 \right)$$

and
$$p_c = p_c' = \left(1·2 - \frac{L}{800k} \right) \quad \dots \left(\text{For } \frac{L}{k} = 160 \text{ and above} \right)$$

where
$p_c =$ Allowable axial compression stress,
$p_c' =$ A value obtained from the above secant formula,
$f_y =$ The guaranteed minimum yield stress,
$m =$ Factor of safety taken as 1·68.

$$\frac{L}{k} = \text{Slenderness ratio with equivalent column length,}$$

$$E = \text{Modulus of elasticity} = 2 \cdot 047 \times 10^6 \text{ kg/cm}^2.$$

The I S.I. has also given a table in I.S. 800—1962, which gives the values of p_c for mild steel, for slenderness ratio from 0 to 350. The value of f_y i.e. the guaranteed minimum yield stress for mild steel is taken as 2,600 kg/cm². This table is given below :

Table 19·5

$\frac{L}{k}$	p_c (kg/cm²)	$\frac{L}{k}$	p_c (kg/cm²)
0	1,250	140	531
10	1,246	150	474
20	1,239	160	423
30	1,224	170	377
40	1,203	180	336
50	1,172	190	300
60	1,130	200	270
70	1,075	210	243
80	1,007	220	219
90	928	230	199
100	840	240	181
110	753	250	166
120	671	300	109
130	597	350	76

Note. Intermediate values may be obtained by linear interpolation.

Example 19·10. *A hollow cylindrical steel tube of 3·8 cm exter-nal diameter and 2·5 mm thick is used as a column of 2·3 m long with both ends hinged. Determine the safe load by I.S. code.*

Solution.

Given. External diameter,
$$D = 3 \cdot 8 \text{ cm}$$

Thickness $\qquad = 2 \cdot 5 \text{ mm} = 0 \cdot 25 \text{ cm}$

∴ Internal diameter,
$$d = 3 \cdot 8 - 2 \times 0 \cdot 25 = 3 \cdot 3 \text{ cm}$$

Area $A = \dfrac{\pi}{64} (3 \cdot 8^2 - 3 \cdot 3^2) = 2 \cdot 75 \text{ cm}^2$

and moment of inertia, $I = \dfrac{\pi}{64} (3 \cdot 8^4 - 3 \cdot 3^4) = 4 \cdot 51^4 \text{ cm}$

Length of column, $l = 2 \cdot 3 \text{ m} = 230 \text{ cm}$

Let p_c = Allowable compressive stress, and
                             P = Safe load on the column.

We know that the least radius of gyration,

$$k = \sqrt{\dfrac{I}{A}} = \sqrt{\dfrac{4 \cdot 51}{2 \cdot 75}} = 1 \cdot 28 \text{ cm}$$

Since the column is hinged at its both ends, therefore effective length of the column,

$$L = l = 230 \text{ cm}$$

Slenderness ratio $= \dfrac{L}{k} = \dfrac{230}{1 \cdot 28} = 180.$

From table 19·5, we find that the allowable stress for slenderness ratio of 180 is 336 kg/cm².

∴   Safe axial load for the column,

$$P = A \times p_c = 2 \cdot 75 \times 336 = \textbf{924 kg} \textbf{Ans.}$$

19·20   Long columns subjected to eccentric loading

In the previous articles, we have discussed the effect of loading on long columns. We have always referred the cases when the load acts axially on the column (*i.e.* the line of action of the load coincides with the axis of the column). But in actual practice it is not always possible to have an axial load, on the column, and eccentric loading takes place. Here we shall discuss the effect of eccentric loading on the Rankine's and Euler's formula for long columns.

Rankine's formula

Consider on a long column subjected to an eccentric load.

Let P = Load on the column,

                             A = Area of cross-section,

                             e = Eccentricity of the load,

                             Z = Modulus of section,

                             y_c = Distance of the extreme fibre (on compression side) from the axis of the column, and

                             k = Least radius of gyration.

We have already discussed in the previous chapter, that when a column is subjected to an eccentric load, the maximum intensity of compressive stress is given by the relation,

$$p_{max} = \frac{P}{A} + \frac{M}{Z} = \frac{P}{A} + \frac{P.e}{Z}$$

$$= \frac{P}{A} + \frac{P.e.y_c}{Ak^2} \qquad \left(\because \; z = \frac{I}{y_c} = \frac{Ak^2}{y_c} \right)$$

$$= \frac{P}{A}\left(1 + \frac{e.y_c}{k^2}\right)$$

If f_c is the permissible crushing stress for the given material, then the safe load for the given column,

$$f_c = \frac{P}{A}\left(1 + \frac{e.y_c}{k^2}\right)$$

or

$$P = \frac{f_c.A}{\left(1 + \frac{e.y_c}{k^2}\right)}$$

We have already discussed in Art. 20-15, that the safe load by Rankine's formula for long columns and axial load is given by the relation,

$$P = \frac{f_c.A}{1 + a\left(\frac{L}{K}\right)^2}$$

It is thus obvious, that if the effect of buckling is also to be taken into account, the safe axial load with eccentricity

$$P = \frac{f_c.A}{\left(1 + \frac{e.y_c}{k^2}\right)\left[1 + a\left(\frac{L}{K}\right)^2\right]}$$

Euler's formula

Consider a long column AB of length l fixed at A, free at B and carrying an eccentric load as shown in Fig. 19·10. As a result of loading, let the beam deflect into a curved form AXB, such that the free end B deflects through a, and occupies a new position B_1 as shown in Fig. 19·10.

Now consider any section X, at a distance x from A.

Let $P =$ Critical load on the column,

$e =$ Eccentricity of the load,

$y =$ Deflection of the column at X.

Thus the Eccentricity of the load P at X

$$= a + e - y$$

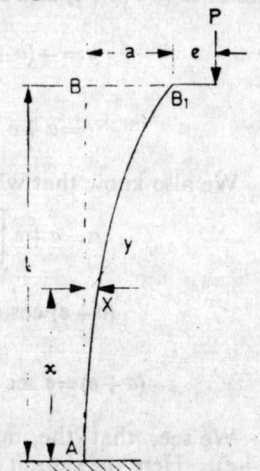

Fig. 19·10

∴ Moment due to the load,

$$M = P(a+e-y) = P(a+e) - P.y$$

$$\therefore \qquad EI\frac{d^2y}{dx^2} = P(a+e) - P.y$$

or

$$\frac{d^2y}{dx^2} + \frac{P.y}{E_I} = \frac{P(a+e)}{E_I}$$

The general solution of the above differential equation is

$$y = A \cos\left(x\sqrt{\frac{P}{EI}}\right) + B \sin\left(x\sqrt{\frac{P}{EI}}\right) + (a+e) \quad \ldots(i)$$

where A and B are the constants of integration. We know that when $x=0$, $y=0$, therefore $A = -(a+e)$. Now differentiating the above equation,

$$\frac{dy}{dx} = -A\sqrt{\frac{P}{EI}} \sin\left(x\sqrt{\frac{P}{EI}}\right)$$
$$+ B\sqrt{\frac{P}{EI}} \cos\left(x\sqrt{\frac{P}{EI}}\right)$$

We also know that when $x=0$, $\frac{dy}{dx}=0$, therefore

$$0 = B\sqrt{\frac{P}{EI}}$$

A little consideration will show, that either B is equal to zero, or $\sqrt{\frac{P}{EI}}$ is equal to zero. Since the load P is not equal to zero, it is thus obvious that B is equal to zero. Substituting the values of $A = -(a+e)$ and $B=0$ in equation (i),

$$y = -(a+e)\cos\left(x\sqrt{\frac{P}{EI}}\right) + (a+e)$$

$$= a+e\left[1 - \cos x\sqrt{\frac{P}{EI}}\right]$$

We also know that when $x=l$, $y=a$. Therefore

$$a = a+e\left[1 - \cos l\sqrt{\frac{P}{EI}}\right]$$

$$\therefore \qquad (a+e)\cos l\sqrt{\frac{P}{EI}} = e$$

or

$$(a+e) = e \sec l\sqrt{\frac{P}{EI}}$$

We see, that the maximum B.M. occurs at B and is equal to $P(a+e)$. Here maximum bending moment,

$$M_{max} = P.e \sec l\sqrt{\frac{P}{EI}}$$

It is thus obvious, that maximum compressive stress will be at A such that

$$p_{max} = \frac{P}{A} + \frac{M}{Z} = \frac{P}{A} + \frac{P.e \sec l \sqrt{\dfrac{P}{EI}}}{Z}$$

Note. It can be easily proved that if both the ends are hinged, the maximum bending moment $= P.e \sec \dfrac{l}{2} \sqrt{\dfrac{P}{EI}}$. Similarly if both the ends are fixed, the maximum bending moment $= P.e \sec \dfrac{l}{4} \sqrt{\dfrac{P}{EI}}$. It is very convenient to remember that the expression for the maximum bending moment for the standard case, of both ends hinged, is $P.e \sec \dfrac{L}{2} \sqrt{\dfrac{P}{EI}}$. Now substitute the equivalent length of the column for the end condition.

Example 19·11. *A hollow circular column of 20 cm external and 16 cm internal diameter is 5 m long and fixed at both of its ends. It subjected to a load of 12 tonnes at an eccentricity of 2 cm from the geometrical axis. Determine the maximum stress induced in the column section. Take E as* $1·2 \times 10^6$ *kg/cm².*

Solution.

Given. External dia, $D = 20$ cm

Internal dia., $d = 16$ cm

Length $l = 5$ m $= 500$ cm

Load, $P = 12$ T $= 12,000$ kg

Eccentricity, $e = 2$ cm

                            $E = 1·2 \times 10^6$ kg/cm²

Let $p_{max} =$ Maximum stress induced in the section.

Since the column is fixed at its both ends, therefore equivalent length of the column,

$$L = \frac{l}{2} = \frac{500}{2} = 250 \text{ cm}$$

We know that the area of column section,

$$A = \frac{\pi}{4} \times (20^2 - 16^2) = 113·2 \text{ cm}^2$$

$$I = \frac{\pi}{64} \times (20^4 - 16^4) = 4,636 \text{ cm}^4$$

$$Z = \frac{I}{10} = \frac{4,636}{10} = 463·6 \text{ cm}^3$$

$$\therefore \quad \frac{L}{2} \sqrt{\frac{P}{EI}} = \frac{250}{2} \sqrt{\frac{12,000}{1·2 \times 10^6 \times 4,636}} = 0·1836 \text{ radian}$$

$$= 10° \, 31'$$

Using the relation,

$$p_{max} = \frac{P}{A} + \frac{P.e \sec \dfrac{L}{2} \sqrt{\dfrac{P}{EI}}}{Z} \quad \text{with usual notations.}$$

$$= \frac{12,000}{113 \cdot 2} + \frac{12,000 \times 2 \cdot 0 \sec 10° \, 31'}{463 \cdot 6} \quad \text{kg/cm}^2$$

$$= 158 \cdot 6 \text{ kg/cm}^2 \quad \textbf{Ans}.$$

Example 19·12. *A cast iron hollow circular column of 20 cm internal diameter and 16 cm internal diameter is 4 m long. It is fixed at its both ends and subjected to an eccentric load of 15 tonnes. Determine the maximum eccentricity, in order that there is no tension anywhere on the section. Take* $E = 9 \cdot 4 \times 10^5$ *kg/cm²*.

Solution.

Given. External dia, $D = 20$ cm
Internal dia, $d = 16$ cm
Length, $l = 4$ m $= 400$ cm
Load, $P = 15$ T $= 15,000$ kg
                       $E = 9 \cdot 4 \times 10^5$ kg/cm²
Let $\epsilon = $ Eccentricity of the load.

Since the column is fixed at its both ends, therefore equivalent length of the column

$$L = \frac{l}{2} = \frac{400}{2} = 200 \text{ cm}$$

We know that area of section,

$$A = \frac{\pi}{4} (20^2 - 16^2) = 113 \cdot 2 \text{ cm}^2$$

$$I = \frac{\pi}{64} (20^4 - 16^4) = 4,636 \text{ cm}^4$$

$$Z = \frac{I}{10} = \frac{4,636}{10} = 463 \cdot 6 \text{ cm}^3$$

$$\therefore \quad \frac{L}{2} \sqrt{\frac{P}{EI}} = \frac{200}{2} \sqrt{\frac{15,000}{9 \cdot 4 \times 10^5 \times 4,636}} = 0 \cdot 1856 \text{ radian}$$

$$= 10° 40'$$

In order to avoid tension anywhere on the section, the direct stress should be equal to the bending stress, *i.e.*

$$\frac{P}{A} = \frac{P.e \sec \dfrac{L}{2} \sqrt{\dfrac{P}{EI}}}{Z}$$

$$\frac{15,000}{113 \cdot 2} = \frac{15,000 \times e \sec 10° 40'}{463 \cdot 6}$$

or $e = 4 \cdot 43$ **cm   Ans**.

Example 19·13. *A column 3 m long hinged at its both ends is made up of two channels ISJC 200 and 225 cm × 1 cm flange plates as shown in Fig. 19·11.*

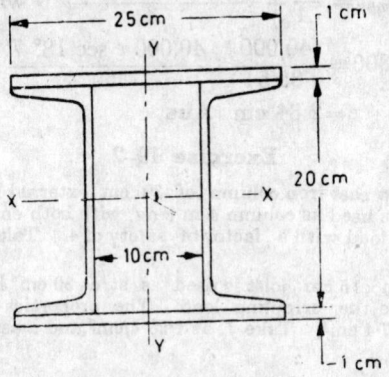

Fig. 19·11

Determine the maximum eccentricity for a load of 40 tonnes from y-y axis, if the maximum permissible stress is 800 kg/cm². Take $E = 2·0 \times 10^6$ kg/cm². The properties of channel section are $A = 17·77$ cm², $I_{YY} = 84·2$ cm⁴. Distance of centroid from back of web = 1·97 cm.

Solution.

Given. Length, $l = 3$ m $= 300$ cm

Load, $P = 40$ T $= 40,000$ kg

Max. stress, $p_{max} = 800$ kg/cm²

               $E = 2·0 \times 10^6$ kg/cm²

Let $e = $ Eccentricity of the load.

Since the column is hinged at its both ends, therefore equivalent length,

$$L = l = 300 \text{ cm}$$

From the geometry of the figure we find that the area of the column,

$$A = 2 (17·77 + 25 \times 1) = 85·54 \text{ cm}^2$$

$$I_{YY} = 2 \left[\frac{1 \times 25^3}{12} + 84·2 + 17·17 (5 + 1·97)^2 \right] \text{ cm}^4$$

$$= 4,499 \text{ cm}^4$$

Section modulus about Y-Y axis,

$$Z = \frac{4,499}{12·5} = 360 \text{ cm}^3$$

$$\therefore \quad \frac{L}{2} \sqrt{\frac{P}{EI}} = \frac{300}{2} \sqrt{\frac{40,000}{2·0 \times 10^6 \times 4,499}} = 0·3163 \text{ radian}$$

$$= 18° \ 7'$$

Using the relation,

$$p_{max} = \frac{P}{A} + \frac{P \cdot e \sec \frac{L}{2} \sqrt{\frac{P}{EI}}}{Z} \quad \text{with usual notations.}$$

$$800 = \frac{40,000}{85 \cdot 54} + \frac{40,000 \, e \sec 18° 7'}{360}$$

∴ $e = 2 \cdot 84$ cm   **Ans**.

Exercise 19-2

1. A hollow cast iron column of 30 cm external diameter and 22 cm internal diameter is used as column 4 m long. with both ends hinged. Determine the Rankine's safe load with a factor of safety of 4.   Take $f_c = 5 \cdot 67$ tonnes/cm² and $a = 1/1,600$.   **[Ans. 214·9 t]**

2. A 20 cm × 15 cm joist is used as strut 30 cm long with both ends hinged. Determine the crippling load. The properties of the joist are $A = 64 \cdot 4$ cm², $I_{min} = 767 \cdot 4$ cm⁴. Take f_c as 1·26 t/cm² and a as 1/9,000.

(Baroda University)
[Ans. 80·5 t]

3. A steel compound column 4 m long is built up of two steel joists of I-section 20 cm × 10 cm × 1 cm joined by two steel plates 30 cm × 1 cm as shown in Fig. 19·12.

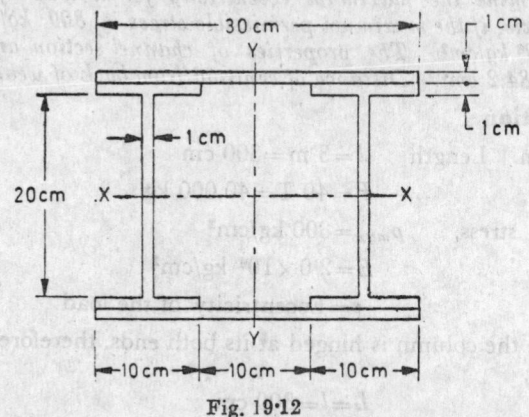

Fig. 19·12

Find the Rankine's crippling load, if both the ends of the column are hinged.   Take $p_c = 3 \cdot 5$ t/cm² and Rankine's constant = 1/7,500.

(Jadavpur University)
[Ans 378·0 t]

HIGHLIGHTS

1. A structural member, subjected to an axial compressive force is called a strut. A vertical strut used in buildings or frames is called a column.

2. A long column fails by buckling; whereas a short column fails by crushing. The load at which a long column starts buckling is called buckling load or crippling load.

3. The Euler's crippling load for a long column,

$$P = \frac{\pi^2 EI}{l^2} \qquad \text{...(when both ends hinged)}$$

$$P = \frac{\pi^2 EI}{4l^2} \qquad \text{...(when one end fixed and the other free)}$$

$$= \frac{4\pi^2 EI}{l^2} \qquad \text{...(when the both ends fixed)}$$

$$= \frac{2\pi^2 EI}{l^2} \qquad \text{...(when one end fixed and the other hinged)}$$

where $\qquad$ E = Young's modulus for the column material

$\qquad\qquad$ I = Least moment of inertia of the column section (*i.e.*, least of I_{XX} and I_{YY}).

$\qquad\qquad$ l = Length of the column.

4. The above four relations may also be used, first by finding out the equivalent length of the column and then using the general relation for columns with both ends hinged.

5. Rankine's crippling load for a column,

$$P = \frac{f_c \cdot A}{1 + a\left(\dfrac{L}{k}\right)^2}$$

where $\qquad$ f_c = crushing stress of the column material,

$\qquad\qquad$ A = Area of the column cross-section,

$\qquad\qquad$ a = Rankine's constant,

$\qquad\qquad$ L = Equivalent length of the column,

$\qquad\qquad$ k = Least radius of gyration.

6. Johnson's relation for safe load on a column,

$$P = A\left[f - n\left(\frac{L}{k}\right) \right] \qquad \text{...(straight line formula)}$$

$$= A\left[F - r\left(\frac{L}{k}\right)^2 \right] \qquad \text{...(Parabolic formula)}$$

where $\qquad$ A = Area of the column cross-section,

$\qquad\qquad$ f = Allowable stress in the column material,

$\qquad\qquad$ n and r = constants, whose values depend upon the column material,

$\qquad\qquad$ $\dfrac{L}{k}$ = Slenderness ratio with equivalent column length.

7. Indian Standard code for columns,

$$p_c = p_c' = \frac{\dfrac{f_y}{m}}{1 + 0.20 \sec \dfrac{L}{K}\sqrt{\dfrac{mp_c'}{4E}}} \qquad \left(\text{For } \frac{l}{k} = 0 \text{ to } 160 \right)$$

$$= p_c \left(1.2 - \frac{L}{800\,k} \right) \qquad \left(\text{For } \frac{l}{k} = 160 \text{ and above} \right)$$

where $\qquad$ p_c = Allowable axial compression stress,

$\qquad\qquad$ p_c' = A value obtained from the above secant formula.

f_y = The guaranteed minimum yield stress,

m = Factor of safety taken as 1·68,

$\dfrac{L}{K}$ = slenderness ratio with equivalent column length,

E = Modulus of elasticity.

8. When a column is subjected to an eccentric load, then maximum stress,

$$p_{max} = \frac{P}{A} + \frac{P.e \sec \dfrac{l}{2}\sqrt{\dfrac{P}{EI}}}{Z}$$

where e = Eccentricity of the load, and

Z = Modulus of column section.

Do You Know ?

1. What do you understand by the terms 'column' and 'strut' ? Distinguish clearly between long columns and short columns.

2. Explain the failure of long columns and short columns.

3. Describe the assumptions in the Euler's column theory.

4. Derive a relation for the Euler's crippling load for a column when (*i*) it has both ends hinged, (*ii*) and both ends fixed.

5. Define the term 'equivalent length'.   Discuss its uses.

6. Explain the term slenderness ratio', and describe with mathematical expression, how it limits the use of Euler's formula for crippling load.

7. Obtain a relation for the Rankine's crippling load for columns.

8. Give the Johnson's straight line and parabolic formula for columns.

9. What is Indian Standard code for columns.   Are you satisfied with the factor of safety of 1·68.   Give explanation to your answer.

10. Explain the effect of eccentric loading on a column.   Derive a relation for the maximum stress in a column eccentrically loaded.

20

PLASTIC THEORY

20·1. Introduction

In the previous chapters, we have discussed the behaviour of structure within the elastic limit. But in the recent years, the engineers have done lot of work to know the behaviour of structures, when stressed beyond the elastic limit called plastic limit. This has led to the development of a new theory popularly known as *plastic theory*. This theory helps us, to a great extent, in the economical design of various types of structures.

749

20·2. Assumptions in plastic theory

The following assumptions are made in the plastic theory :

(1) The material of the beam is perfectly homogeneous (*i.e.*, of the same kind throughout) and isotropic (*i.e.*, of equal elastic properties in all directions).

(2) The beam material is stressed up to the yield value.

(3) The transverse sections (*i.e. AB* or *CD*), which were plane before bending, remain plane after bending.

(4) Each layer of the beam is free to expand or contract, independently of the layer above or below it.

(5) The value of E (*i.e.*, Young's modulus of elasticity) is the same in tension and compression.

20·3. Plastic hinge

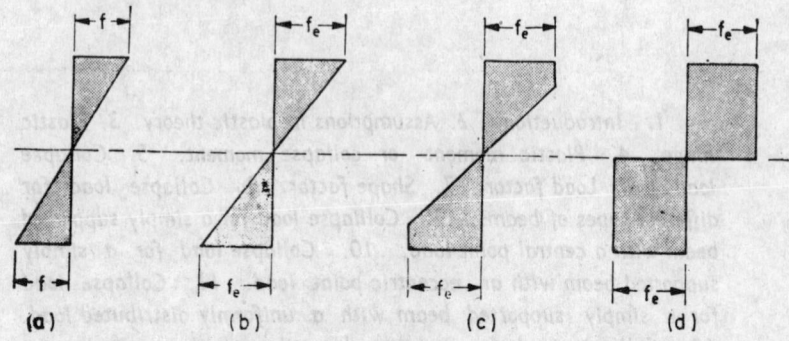

(a)   (b)   (c)   (d)

Fig. 20·1

Consider a beam of symmetrical section, subjected to any moment M. As a result of this moment, the cross-section of the beam will be subjected to some bending stresses as shown in Fig. 20·1 (*a*). We know that if the moment M increases gradually, the stress also increases as discussed below :

1. The bending stress (f) goes on increasing, linearly, until the extreme fibre stress reaches the yield stress (f_e) as shown in Fig. 20·1 (*b*).

2. The increase in moment, cannot produce any increased fibre stress beyond the yield stress. But it causes, the yield stress to spread into the inner fibres as shown in Fig. 20·1 (*c*). At such a stage the beam is said to be in elasto-plastic stage, as part of the beam is in elastic stage and the remaining in plastic stage.

3. The further increase in moment will cause more and more fibres of the section to reach yield stress, until the whole section reaches the yield stress as shown in Fig. 20·1 (*d*).

At this moment, the section has reached its maximum resistance, and thus it cannot be subjected to any more moment or in other words, it cannot carry any further load. A little consideration will show, that any increase in the moment or load will cause the beam to deflect and then to collapse. The beam will behave, as if it is hinged at the plastic section. Such a section is called *plastic hinge*.

20·4. Plastic moment or collapse moment

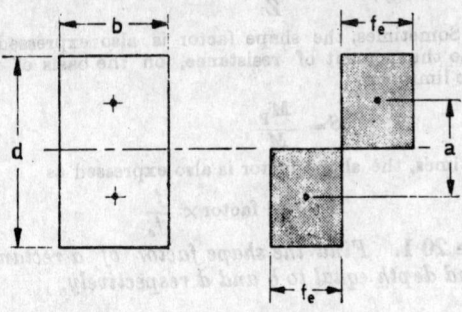

Fig. 20·2

It is the resisting moment of a beam section, which has developed a plastic hinge. Consider a beam, subjected to a moment which has caused a plastic hinge as shown in Fig. 20·2.

Let f_c = Yield stress

a = Lever arm of the couple (*i.e.*, distance between the *c.g.* of the tension and compression areas of the section,

b = Width of the beam,

d = Depth of the beam.

∴ Area, $A = b \times d$

We know that the plastic moment,

$$M_P = \text{Force} \times \text{Area} = f_e \times \frac{A}{2} \times a$$

$$= f_e \times \frac{bd}{2} \times \frac{d}{2} = f_e \times \frac{bd^2}{4}$$

$$= f_e . Z_p$$

where Z_p is the *plastic section modulus or plastic modulus of section and is equal to the moment of the tension and compression areas about the central axis.

20·5. Collapse load

The load, which causes the collapse moment in the section of a beam, is known as *collapse load*.

20·6. Load factor

The ratio of collapse load, to the safe working load, is known as *load factor*.

_____

*It is similar to the relation, in elastic limit

$$M = fZ$$

where M is the moment, f is the stress and Z is the section modulus.

752

20·7. Shape factor

The ratio of plastic modulus of section, to the elastic modulus of section, is known as *shape factor*.

Mathematically, $S = \dfrac{Z_P}{Z}$

Note 1. Sometimes, the shape factor is also expressed as the ratio of plastic moment to the moment of resistance, on the basis of ultimate stress, within the elastic limit, *i.e.*,

$$S = \frac{M_P}{M}$$

2. Sometimes, the shape factor is also expressed as

$$S = \text{load factor} \times \frac{f}{f_e}$$

Example 20·1. *Find the shape factor of a rectangular section having width and depth equal to b and d respectively.*

Solution.

Given. Width $= b$

Depth $= d$

Let $S = $ Shape factor for the section.

We know that the elastic section modulus of a rectangular section,

$$Z = \frac{bd^2}{6}$$

and plastic section modulus,

$$Z_P = \frac{bd^2}{4}$$

Using the relation,

$$S = \frac{Z_P}{Z} \text{ with usual notations.}$$

$$= \frac{\dfrac{bd^2}{4}}{\dfrac{bd^2}{6}} = \frac{6}{4} = 1·5 \text{ Ans.}$$

Example 20·2. *Determine the shape factor for a circular section of 10 cm diameter.*

Solution.

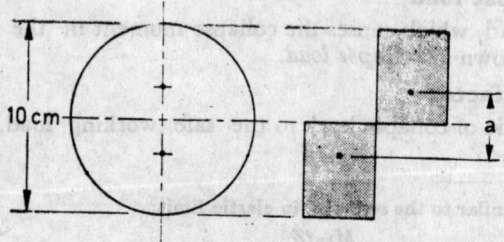

Fig. 20·3

Given. Diameter $d = 10$ cm

Let $S =$ Shape factor for the section,

We know that the elastic section modulus of a circular section,

$$Z = \frac{\pi}{32} \times d^3 = \frac{\pi}{32} \times 10^3 \text{ cm}^3$$

and plastic section modulus,

$$Z_P = \frac{A}{2} \times a = \frac{1}{2} \times \frac{\pi}{4} \times 10^2 \times \frac{2 \times 2 \times 10}{3\pi} \text{ cm}^3$$

$$= \frac{10^3}{6} \text{ cm}^3$$

Using the relation,

$$S = \frac{Z_P}{Z} \text{ with usual notations.}$$

$$= \frac{\dfrac{10^3}{6}}{\dfrac{\pi}{32} \times 10^3} = \frac{32}{6\pi} = 1 \cdot 7 \text{ Ans.}$$

20·8. Collapse load for different types of beams

We have already discussed in Art. 20·5, the collapse load of a beam. In the following pages, we shall discuss the collapse load for the following types of beams :

1. Simply supported beam with a central point load,
2. Simply supported beam with an eccentric point load,
3. Simply supported beam with a uniformly distributed load over the entire span,
4. Propped cantilever with a central point load,
5. Propped cantilever with an eccentric point load,
6. Propped cantilever with a uniformly distributed load,
7. Fixed beam with a central point load,
8. Fixed beam with an eccentric point load,
9. Fixed beam with a uniformly distributed load, over the entire span.

20·9. Collapse load for a simply supported beam with a central point load

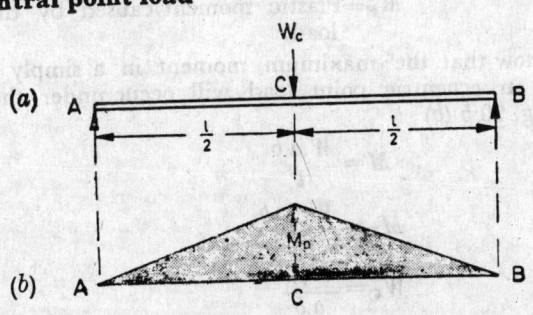

Fig 20·4

Consider a simply supported beam AB carrying a central point load at C as shown in Fig 20·4 (a).

Let
W_C = Collapse load,

l = Span of the beam, and

M_P = Plastic moment caused by the collapse load.

We know that the maximum moment in a simply supported beam, with a central point load will occur at the centre of the span *i.e.* under the load as shown in Fig. 20·4 (b).

$$\therefore \qquad M = \frac{wl}{4}$$

or
$$M_P = \frac{W_C.l}{4}$$

$$W_C = \frac{4M_P}{l}$$

20·10. Collapse load for a simply supported beam with an eccentric point load

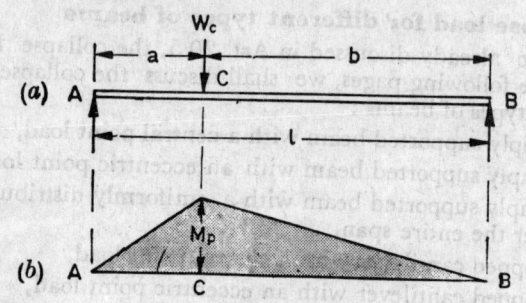

Fig. 20·5

Consider a simply supported beam AB carrying an eccentric point load at distances of a and b from the supports A and B as shown in Fig. 20·5 (a).

Let
W_C = Collapse load,

l = Span of the beam, and

M_P = Plastic moment caused by the collapse load.

We know that the maximum moment in a simply supported beam with an eccentric point load will occur under the load as shown in Fig. 20·5 (b).

$$\therefore \qquad M = \frac{W.a.b}{l}$$

or
$$M_P = \frac{W_C.a.b}{l}$$

$$W_C = \frac{M_P.l}{a.b}$$

20·11. Collapse load for a simply supported beam with a uniformly distributed load

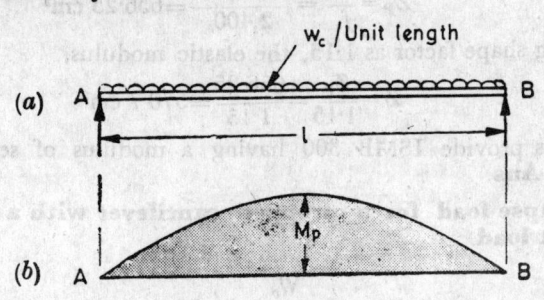

Fig. 20·6

Consider a simply supported beam AB carrying a uniformly distributed load as shown in Fig. 20·6 (a).

Let w_C = Collapse load, per unit length,

                   l = Span of the beam

                   M_P = Plastic moment caused by the collapse load.

We know that the maximum moment, in a simply supported beam with a uniformly distributed load, will occur at the centre of the span as shown in Fig. 20·6 (b).

$$\therefore \qquad M = \frac{wl^2}{8}$$

or $$M_P = \frac{w_C \cdot l^2}{8}$$

or $$w_C = \frac{8 M_P}{l^2}$$

Example 20·3. *A simply supported beam of 6 m span is subjected to a uniformly distributed load of 2 t/m. Design the section on plastic theory. if the yield stress be 2,400 kg/cm² . Take load factor as 1·75.*

Solution.

Given. Span, $l = 6$ m

Load, $w = 2$ t/m

Yield stress, $f_e = 2{,}400$ kg/cm²

Load factor $= 1\cdot75$

$\therefore$ Collapse load, $w_C = 2 \times 1\cdot75 = 3\cdot5$ t/m

We know that the plastic moment caused by the collapse load

$$M_P = \frac{w_C \cdot l^2}{8} = \frac{3\cdot5 \times 6^2}{8} = 15\cdot75 \text{ t-m}$$

$$= 15{,}75{,}000 \text{ kg-cm}$$

∴ Plastic modulus,

$$Z_P = \frac{M_P}{f_e} = \frac{15,75,000}{2,400} = 656 \cdot 25 \text{ cm}^3$$

Taking shape factor as $1 \cdot 15$, the elastic modulus,

$$Z = \frac{Z_P}{1 \cdot 15} = \frac{656 \cdot 25}{1 \cdot 15} = 570 \cdot 7 \text{ cm}^3$$

Let us provide ISMB 300 having a modulus of section as **573·6 cm³ Ans**.

20·12. Collapse load for a propped cantilever with a central point load.

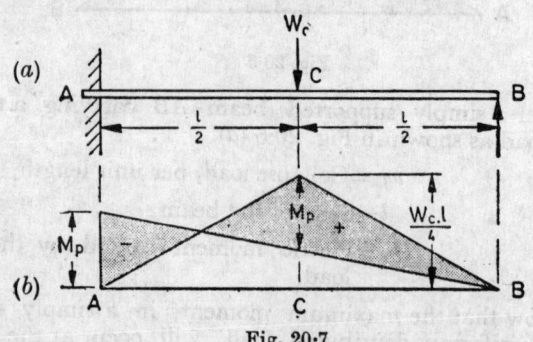

Fig. 20·7

Consider a propped cantilever AB fixed at A, propped at B and carrying a central point load at C as shown in Fig. 20·7(a).

Let W_C = Collapse load.

l = Span of the cantilever,

M_P = Plastic moment caused by the collapse load.

We know that in a propped cantilever with a central point load the maximum negative moment will occur at A, and maximum positive moment will occur at the centre of the span, i.e., under the load as shown in Fig. 20·7 (b).

From the geometry of *moment diagram, we find that

$$\frac{W_C \cdot l}{4} = M_P + \tfrac{1}{2} M_P = \frac{3M_P}{2}$$

$$\therefore \qquad W_C = \frac{6M_P}{l}$$

20·13. Collapse load for a propped cantilever with an eccentric point load

Consider a propped cantilever AB fixed at A, propped at B carrying an eccentric point load at distances of a and b from A and B as shown in Fig. 20·8 (a).

---

*This positive moment at C is due to the load W_C. The total ordinate of moment at C is equal to $\frac{W_C \cdot l}{4}$. This load is also equal to $M_P + \tfrac{1}{2} M_P$.

Let W_C=Collapse load,

l=Span of the cantilever,

M_P=Plastic moment caused by the collapse load.

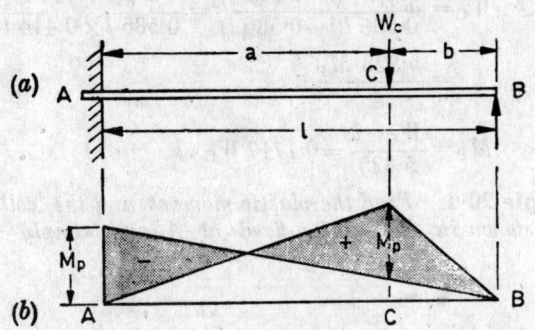

Fig. 20·8

We know that in a propped cantilever, with an eccentric point load, the maximum negative moment will occur at A, and maximum positive moment will occur under the load as shown in Fig. 20·8(b). From the geometry of the moment diagram, we find that

$$\frac{W_C \cdot a \cdot b}{l} = M_P + \frac{M_P \cdot b}{l}$$

$$= M_P \left(1 + \frac{b}{l} \right) = \frac{M_P(l+b)}{l}$$

$$W_C = \frac{M_P(l+b)}{a \cdot b} \qquad \qquad \dots(i)$$

In order to locate the position of the load for minimum collapse load, differentiate the above equation with respect to a and equate the same to zero.

i.e., $$\frac{d}{da}\left[\frac{M_P(l+b)}{a.b} \right] = 0$$

$$\frac{d}{da}\left[\frac{M_P[l+(l-a)]}{a(l-a)} \right] = 0 \qquad \qquad (\because \ a+b=l)$$

$$\frac{d}{da}\left[\frac{M_P(2l-a)}{al-a^2} \right] = 0$$

$$(al-a^2)(-1)-(2l-a)(l-2a)=0$$

$$a^2-al-(2l^2-4al-al+2a^2)=0$$

$$a^2-al-2l^2+4al+al-2a^2=0$$

$$-a^2+4al-2l^2=0$$

or $$a^2-4al+2l^2=0 \qquad \qquad \text{(Multiplying by } -1)$$

$$a^2-4al+4l^2=2l^2 \qquad \qquad \text{(Adding } 2l^2 \text{ on both sides)}$$

$$(a-2l)^2=2l^2$$

$\therefore$ $$a-2l=\pm l \ \sqrt{} \qquad \qquad \text{(Taking minus sign)}$$

or
$$a = 2l - l\sqrt{2}$$
$$= l(2 - \sqrt{2}) = 0.586\ l$$

Substituting this value of a in equation (i),

$$W_C = \frac{M_P[l + (l - 0.586\ l)]}{0.586\ l(l - 0.586\ l)} = \frac{M_P \times 1.414\ l}{0.586\ l \times 0.414\ l}$$

$$= \frac{5.828\ M_P}{l}$$

or
$$M_P = \frac{W_C \cdot l}{5.825} = 0.1717\ W_C \cdot l$$

Example 20·4. *Find the plastic moment and the collapse load for the beam shown in Fig. 20·9, fixed at A and simply supported at B.*

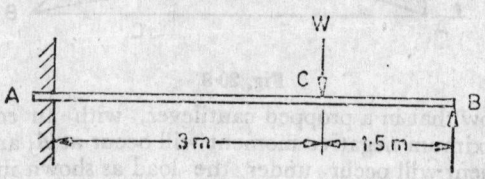

Fig. 20·9

Take working stress as 1·5 t/cm² and yield stress as 2·25 t/cm² and shape factor as 1·15.   (*Roorkee University , 1978*)

Solution

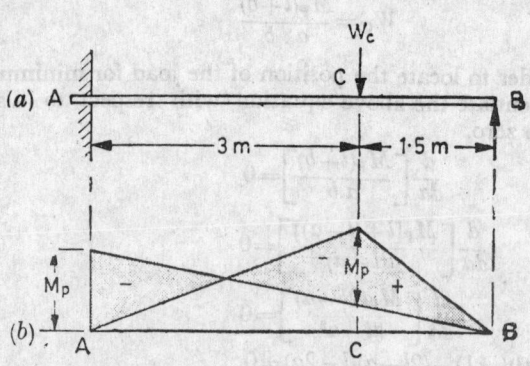

Fig. 20·10

Given. Span, $l = 4.5$ m
Distance AC $a = 3$ m
Distance CB, $b = 1.5$ m
Working stress, $f = 1.5$ t/cm²
Yield stress, $f_s = 2.25$ t/cm²
Shape factor, $S = 1.15$

Collapse load

Let $\qquad W_C$=Collapse load, and

$\qquad\qquad\qquad M_P$=Plastic moment caused by the load.

We know that moment caused by the load W_C at C

$$=\frac{W_C \cdot a \cdot b}{l}=\frac{W_C \cdot 3 \times 1.5}{4.5}=W_C$$

From the geometry of the moment diagram, we find that

$$W_C=M_P+\frac{M_P \times 1.5}{4.5}=\frac{4M_P}{3}$$

$$\therefore \qquad M_P=\frac{3W_C}{4} \qquad\qquad\qquad\qquad\qquad ...(i)$$

We also know that

$$\text{Load factor}=\frac{f_e}{f} \times S$$

$$=\frac{2.25}{1.5} \times 1.15=1.72$$

$\therefore$ Collapse load, $W_C=1.72\ W$ **Ans.**

Plastic moment

Substituting the value of W_C in equation (i),

$$M_P=\frac{3 \times 1.72\ W}{4}=1.29\ W \quad \text{Ans}$$

Example 20·5. *A propped cantilever of rectangular section and of span 6 m is subjected to two point loads of 6 tonnes each at 2 m and 4 m from the fixed end. Design the section, with depth, equal to twice the width with a load factor of 2. Take yield stress for the material as 2,600 kg/cm².* (*Baroda University, 1973*)

Solution.

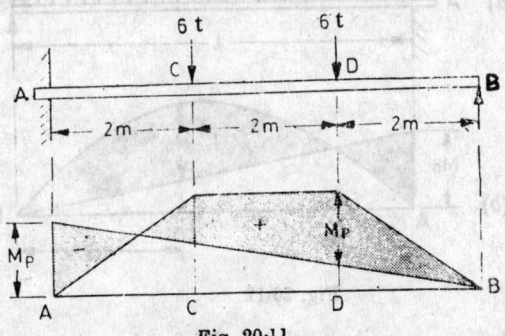

Fig. 20·11

Given. Span, $\qquad l=6$ m

Point loads, $\qquad W=6$ t

Depth of section, $\quad d=2b$ (where b is the width)

Load factor $\qquad\quad =2$

Yield stress, $\qquad f_e=2,600$ kg/cm²$=2.6$ t/cm²

Let $\qquad\qquad M_P$=Plastic moment caused by the load.

Since the load factor is 2, therefore the point loads at C and D will be $2 \times 6 = 12$ tonnes

We know that the moment caused at D
$$= 12 \times 2 = 24 \text{ t-m}$$

From the geometry of the moment diagram, we find that
$$24 = M_P + \frac{M_P}{3} = \frac{4M_P}{3}$$

or
$$M_P = \frac{24 \times 3}{4} = 18.0 \text{ t-m} = 1,800 \text{ t-cm}$$

We also know that
$$M_P = f_P . Z_P$$
$$1,800 = 2.6 \; Z_P$$

or
$$Z_P = \frac{1800}{2.6} = 692.3 \text{ cm}^3 \qquad \qquad \dots(i)$$

Now
$$Z_P = \frac{bd^2}{4} = \frac{b \times (2b)^2}{4} = b^3 \qquad \qquad \dots(ii)$$

Equating (i) and (ii),
$$b^3 = 692.3$$
$$b = 8.845 \text{ cm say } \mathbf{9 \text{ cm}} \quad \textbf{Ans.}$$

and
$$d = 2 \times 9 = \mathbf{18 \text{ cm}} \quad \textbf{Ans.}$$

20·14. Collapse load for a propped cantilever with a uniformly distributed load

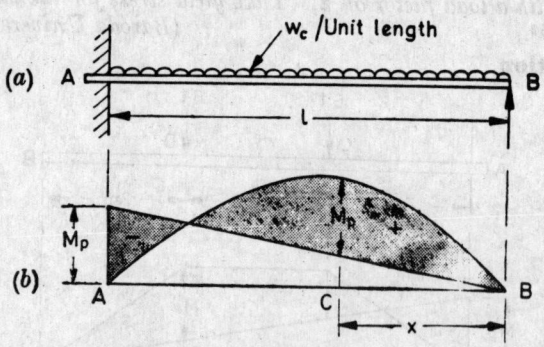

Fig. 20·12

Consider a propped cantilever AB fixed at A, propped at B and carrying a uniformly distributed load as shown in Fig. 20·12 (a).

Let w_C = Collapse load per unit length,

           l = Span of the cantilever.

          M_P = Plastic moment caused by the collapse load.

We know that in a propped cantilever with a uniformly distributed load, the maximum negative moment will occur at A and maximum positive moment will occur at a point C, near the centre of the beam as shown in Fig. 20·11 (b),

Let x = Distance C and B

From the geometry of the moment diagram, we find that

$$\frac{w_C . l . x}{2} - \frac{w_C . x^2}{2} = M_P + \frac{M_P.x}{l} \qquad \text{...}(i)$$

or

$$\frac{w_C . l . x}{2} - \frac{w_C . x^2}{2} - M_P - \frac{M_P . x}{l} = 0$$

In order to find out the maximum value of w_C, differentiate the above equation with respect to x and equate the same to zero.

$$\frac{d}{dx}\left[\frac{w_C . l . x}{2} - \frac{w_C . x^2}{2} - M_P - \frac{M_P . x}{l}\right] = 0$$

$$\frac{w_C . l}{2} - w_C . x - \frac{M_P . x}{l} = 0$$

$$w_C\left(\frac{l}{2} - x\right) = \frac{M_P}{l}$$

or

$$M_P = w_C . l\left(\frac{l}{2} - x\right)$$

Substituting this value of M_P in equation (i)

$$\frac{w_C . l . x}{2} - \frac{w_C . x^2}{2} = w_C . l\left(\frac{l}{2} - x\right) + w_C . l\left(\frac{l}{2} - x\right)\frac{x}{l}$$

$$= \frac{w_C . l \, (l-2x)}{2} + \frac{w_C \, (l-2x)x}{2}$$

or

$$x(l-x) = (l-2x)(l+x)$$

$$lx - x^2 = l^2 + lx - 2lx - 2x^2$$

$$x^2 + 2lx - l^2 = 0$$

or

$$x^2 + 2lx + l^2 = 2l^2$$

$$(x+l)^2 = 2l^2$$

$$x + l = \pm l\sqrt{2}$$

$$\therefore \qquad x = l(\sqrt{2} - 1) = 0·414\, l$$

Substituting the value of x in equation (i),

$$\frac{w_C . l \times 0·414\, l}{2} - \frac{w_C \, (0·414\, l)^2}{2} = M_P + \frac{M_P \times 0·414\, l}{l}$$

$$w_C \, (0·207\, l^2 - 0·085698\, l^2) = M_P + (1+0·414)$$

or

$$w_C = \frac{1·414\, M_P}{0·121302\, l^2} = \frac{11·656\, M_P}{l^2}$$

or

$$M_P = \frac{w_C . l^2}{11·656} = 0·08578\, w_C \, l^2.$$

20·15. Collapse load for a fixed beam with a central point load.

Consider a fixed beam AB carrying a central point load at C as shown in Fig. 20·13 (a).

Let W_C=Collapse load,
                l=Span of the beam,
                M_P=Positive moment caused by the collapse load,

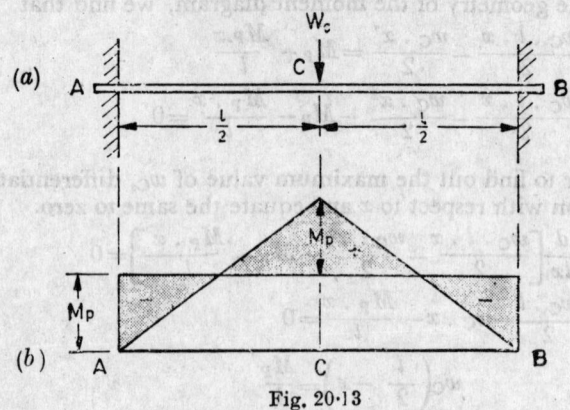

Fig. 20·13

We know that the maximum negative moment, in a fixed beam with a central point, load will occur at A and B; whereas the maximum positive moment will occur under the load as shown in Fig. 20·13 (b). From the geometry of the moment diagram we find that

$$\frac{W_C \cdot l}{4} = M_P + M_P = 2M_P$$

$$\therefore \qquad W_C = \frac{8M_P}{l}$$

20·16. Collapse load for a fixed beam with an eccentric point load

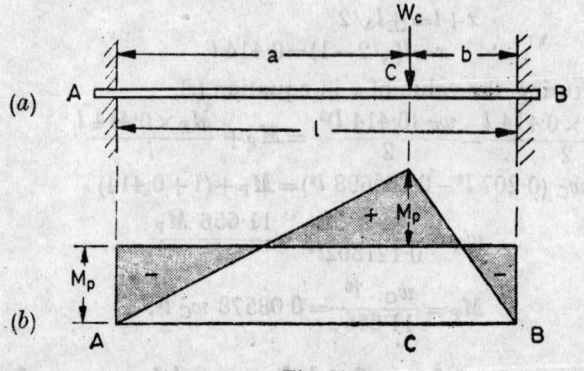

Fig. 20·14

Consider a fixed beam AB carrying an eccentric load at distances of a and b from the supports A and B as shown in Fig. 20·14 (a).

Let $\qquad W_C =$ Collapse load,

$\qquad l =$ Span of the beam,

$\qquad M_P =$ Plastic moment caused by the collapse load.

. We know that the maximum negative moment in a fixed beam, with an eccentric point, load will occur at A and B; whereas the maximum positive moment will occur under the load as shown in Fig. 20·14 (b). From the geometry of the moment diagram, we find that

$$\frac{W_C \cdot a \cdot b}{l} = M_P + M_P = 2M_P$$

or $\qquad\qquad\qquad W_C = \dfrac{2M_P \cdot l}{a \cdot b}$

Example 20·6. *A fixed beam AB carries two-point loads at 1·5 m and 3 m from the support A as shown in Fig. 20·15.*

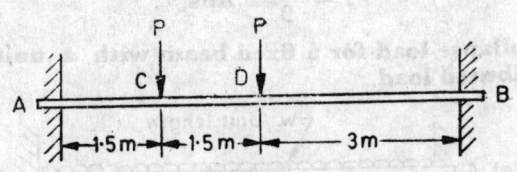

Fig. 20·15

Assuming the beam of constant section, calculate the collapse load for the beam, if the plastic moment is M_P. The point loads are P tonnes each.

(*London University,*)

Solution.

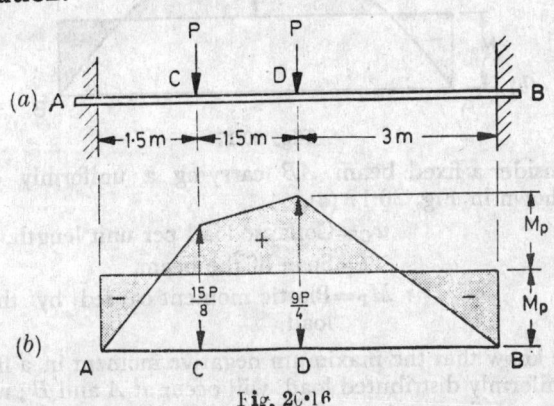

Fig. 20·16

Given. Span, $\qquad\qquad l = 6$ m

Loads at C and $D \qquad = P$

Plastic moment $\qquad = M_P$

We know that in plastic conditions, the negative moment at A and B is M_P.

Now consider tne beam AB as simply supported. Taking moments about A and equating the same.

$$R_B \times 6 = P \times 3 + P \times 1\cdot5 = 4\cdot5\ P$$

$$\therefore \qquad R_B = \frac{4\cdot5\ P}{6} = \frac{3P}{4}$$

and

$$R_A = P + P - \frac{3P}{4} = \frac{5P}{4}$$

$$\therefore \quad \text{Moment at} \quad D = \frac{3P}{4} \times 3 = \frac{9P}{4}$$

and moment at

$$C = \frac{5P}{4} \times 1\cdot5 = \frac{15P}{8}$$

From the geometry of the moment diagram, we find that

$$\frac{9P}{4} = M_P + M_P = 2M_P$$

$$\therefore \qquad\qquad P = \frac{8M_P}{9} \quad \text{Ans.}$$

20·17. Collapse load for a fixed beam with a uniformly distributed load.

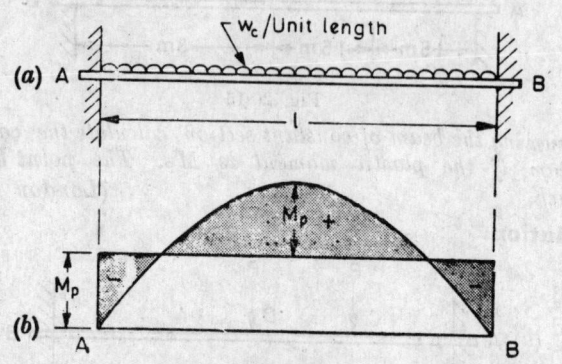

Fig. 20·17

Consider a fixed beam AB carrying a uniformly distributed load as shown in Fig. 20·17 (a).

Let
$$w_C = \text{Collapse load per unit length,}$$
$$l = \text{Span of the beam,}$$
$$M_P = \text{Plastic moment caused by the collapse load.}$$

We know that the maximum negative moment in a fixed beam, with a uniformly distributed load, will occur at A and B; whereas the maximum positive moment will occur at the centre of the beam as shown in Fig. 20·17 (b). From the geometry of the moment diagram, we find that

$$\frac{w_C.l^2}{8} = M_P + M_P = 2M_P$$

$$\therefore \qquad\qquad w_C = \frac{16M_P}{l^2}$$

Example 20·7. *A fixed beam AB of span 6 metres carries a uniformly distributed load of 5 t/m on the right hand 4·5 m as shown in Fig. 20·18.*

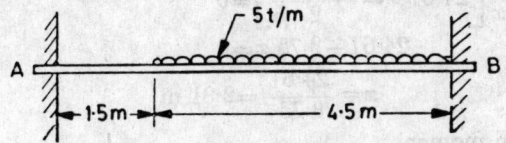

Fig. 20·18

The load factor is 1·75 and shape factor is 1·15, the yield stress is 2·5 tonnes/cm². Calculate the sectional modulus of the beam and locate the positions of the plastic hinges. (*Patna University , 1973*)

Solution.

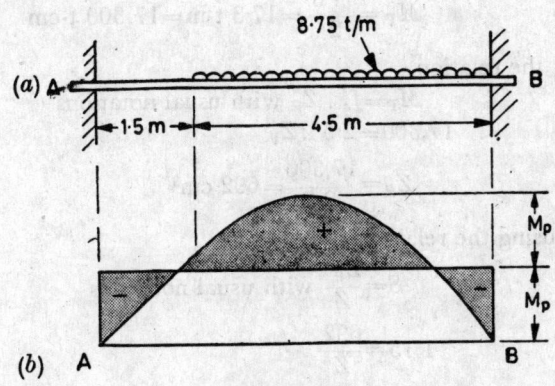

Fig. 20·19

Given. Span, $l = 6$ m
Load, $w = 5$ t/m
Load factor $= 1·75$
∴ Collapse load, $w_C = 5 \times 1·75 = 8·75$ t/m
Shape factor, $S = 1·15$
Yield stress, $f_e = 2·5$ t/m²

Sectional modulus of the beam

Let $Z = $ Sectional modulus of the beam, and
     $Z_P = $ Plastic modulus of the beam.

Now consider the beam *AB* as simply supported. Taking moments about *A* and equating the same,

$$R_B \times 6 = 8·75 \times 4·5 \times 3·75 = 147·66$$

$$\therefore \quad R_B = \frac{147·66}{6} = 24·61 \text{ t}$$

We know that moment at any section, at a distance x from *B*,

$$M_X = R_P \cdot x - \frac{8·75\, x^2}{2} = 24·61 x - \frac{8·75 x^2}{2}$$

For maximum moment, differentiate the above equation with respect to x and equate the same to zero.

$$\frac{d}{dx}\left[24\cdot61\ x-\frac{8\cdot75\ x^2}{2}\right]=0$$

$$24\cdot61-8\cdot75\ x=0$$

$$\therefore\qquad x=\frac{24\cdot61}{8\cdot75}=2\cdot81\ m$$

and maximum moment,

$$M_{max}=24\cdot61\times2\cdot81-\frac{8\cdot75\times2\cdot81^2}{2}=34\cdot6\ t\text{-}m$$

From the geometry of the moment diagram, we find that
$$2M_P=34\cdot6\ t\text{-}m$$

$$\therefore\qquad M_P=\frac{34\cdot6}{2}=17\cdot3\ t\text{-}m=17,300\ t\text{-}cm$$

Using the relation,
$$M_P=f_e\ .\ Z_P\ \text{with usual notations.}$$
$$17,300=2\cdot5\times Z_P$$

$$\therefore\qquad Z_P=\frac{17,300}{2\cdot5}=692\ cm^3$$

Now using the relation,
$$S=\frac{Z_P}{Z}\ \text{with usual notations}$$

$$1\cdot15=\frac{692}{Z}$$

$$\therefore\qquad Z=\frac{692}{1\cdot15}=601\cdot7\ cm^3\quad\textbf{Ans.}$$

Position of plastic hinges

We know that the plastic hinges will take place at the points, where the plastic moment is developed. Therefore plastic hinges will take place at A, B and at a distance of $2\cdot81$ m from B. **Ans.**

Exercise 20-1

1. Show that the shape factor of a rectangular section 10 cm wide and 15 cm deep is 1·5.

2. Determine the shape factor for a hollow circular section having external diameter as D, and internal diameter as d. Take the ratio $d/D=k$.
(Nagpur University)
$$\left[\textbf{Ans.}\quad 1\cdot7\left(\frac{1-k^3}{1-k^4}\right)\right]$$

3. A propped cantilever beam of span 5 m is carrying a point load of 10 T. Determine the plastic moment, when the load is occupying a position to cause collapse of the beam. *(Aligarh University)*
[**Ans.** 8·585 t-m]

4. A fixed beam of 6 m span carries a uniformly distributed load on the left half of the section. If the plastic moment of the section is 10 t-m, find the value of collapse load. *(Bombay University)*
[**Ans.** 23·56 t]

HIGHLIGHTS

1. If a body is strained beyond the elastic limit, a plastic hinge is formed in it as discussed below :

(a) The bending stress goes on increasing linearly, until the extreme fibre stress reaches the yield stress.

(b) The increase in moment causes the yield stress to spread into the inner fibres.

(c) The further increase in the moment causes more and more fibres of the section to yield stress, until the whole section reaches the yield stress. At this stage, the beam behaves, as if the beam is hinged at its plastic section.

2. The resting moment of a beam section, which has developed a plastic hinge, is known as plastic moment.

3. The load, which causes the collapse moment, is known as collapse load. The ratio of collapse load to the safe working load is known as load factor.

4. The ratio of plastic modulus of section to the elastic modulus of section is known as shape factor.

. 5. The collapse load,

$$W_C = \frac{4M_P}{l}$$ (For a simply supported beam with a central point load)

$$= \frac{M_P \cdot l}{ab}$$ (For a simply supported beam with an eccentric point load)

$$= \frac{8 M_P}{l^2}$$ (For a simply supported beam with a uniformly distributed load)

$$= \frac{6 M_P}{l}$$ (For a propped cantilever with a central point load)

$$= \frac{M_P (l+b)}{ab}$$ (For a propped cantilever with an eccentric point load)

$$= \frac{11\cdot656 M_P}{l^2}$$ (For a propped cantilever with a uniformly distributed load)

$$= \frac{8 M_P}{l}$$ (For a fixed beam with a central point load)

$$= \frac{2M_P \cdot l}{ab}$$ (For a fixed beam with an eccentric point load)

$$= \frac{16 M_P}{l^2}$$ (For a fixed beam with a uniformly distributed load)

Do You Know ?

1. Write the assumptions made in the plastic theory.

Define (a) Plastic hinge.

(b) Plastic moment.

(c) Load factor.

(d) Shape factor.

2. Show that the plastic moment of a beam of rectangular section is 50% greater than the bending moment within the elastic limit.

3. A steel beam is fixed at one end, and simply supported at the other. Its fully plastic moment of resistance is M_P and its span is l. A point load W can act on the beam at any point. Calculate the minimum load W, which will cause collapse of the beam.

4. Derive an expression for the minimum value of load w per unit length, which will cause the beam to collapse, which it is fixed at one end, and supported at the other.

3. Obtain a reaction for the collapse load, when it is acting eccentrically on a fixed beam.